Dynamic Systems

Presenting students with a comprehensive and efficient approach to the modeling, simulation, and analysis of dynamic systems, this textbook addresses mechanical, electrical, thermal and fluid systems, feedback control systems, and their combinations. It features a robust introduction to fundamental mathematical prerequisites, suitable for students from a range of backgrounds; clearly established three-key procedures – fundamental principles, basic elements, and ways of analysis – for students to build on in confidence as they explore new topics; over 300 end-of-chapter problems, with solutions available for instructors, to solidify a hands-on understanding; and clear and uncomplicated examples using MATLAB®/Simulink®, and Mathematica®, to introduce students to computational approaches. With a capstone chapter focused on the application of these techniques to real-world engineering problems, this is an ideal resource for a single-semester course in dynamic systems for students in mechanical, aerospace, and civil engineering.

Bingen "Ben" Yang is Professor of Aerospace and Mechanical Engineering at the University of Southern California, where he has taught for more than three decades. Being an active researcher, he has more than 230 publications in the areas of structures, dynamics, vibrations, controls, and mechanics. His current research interest lies in modeling, analysis, control and design of flexible structures, electromechanical systems, and computational methods for problems in engineering applications. A fellow of the American Society of Mechanical Engineers, Dr. Yang received his Ph.D. in mechanical engineering from the University of California at Berkeley.

Inna Abramova is a senior lecturer of Aerospace and Mechanical Engineering at the University of Southern California, where she has taught for seven years. Her research interests lie in the areas of dynamics and controls, with emphasis on collaborative operations of autonomous agents. Dr. Abramova received her Ph.D. in mechanical engineering from the University of Southern California.

Dynamic Systems
Modelling, Simulation, and Analysis

Bingen Yang
University of Southern California

Inna Abramova
University of Southern California

CAMBRIDGE
UNIVERSITY PRESS

University Printing House, Cambridge CB2 8BS, United Kingdom

One Liberty Plaza, 20th Floor, New York, NY 10006, USA

477 Williamstown Road, Port Melbourne, VIC 3207, Australia

314–321, 3rd Floor, Plot 3, Splendor Forum, Jasola District Centre,
New Delhi – 110025, India

103 Penang Road, #05–06/07, Visioncrest Commercial, Singapore 238467

Cambridge University Press is part of the University of Cambridge.

It furthers the University's mission by disseminating knowledge in the pursuit of
education, learning, and research at the highest international levels of excellence.

www.cambridge.org
Information on this title: www.cambridge.org/highereducation/isbn/9781107179790
DOI: 10.1017/9781316841051

© Bingen Yang and Inna Abramova 2022

This publication is in copyright. Subject to statutory exception
and to the provisions of relevant collective licensing agreements,
no reproduction of any part may take place without the written
permission of Cambridge University Press.

First published 2022

Printed in the United Kingdom by TJ Books Limited, Padstow Cornwall

A catalogue record for this publication is available from the British Library.

Library of Congress Cataloging-in-Publication Data
Names: Yang, Bingen, author. | Abramova, Inna, author.
Title: Dynamic systems : modelling, simulation, and analysis / Bingen Yang, University of Southern California,
Inna Abramova, University of Southern California.
Description: Cambridge, United Kingdom ; New York, NY : Cambridge University Press, 2022. | Includes
bibliographical references and index.
Identifiers: LCCN 2021044739 (print) | LCCN 2021044740 (ebook) | ISBN 9781107179790 (hardback) | ISBN
9781316841051 (ebook)
Subjects: LCSH: Dynamics. | BISAC: TECHNOLOGY & ENGINEERING / General
Classification: LCC TA352 .Y36 2022 (print) | LCC TA352 (ebook) | DDC 620.1/04–dc23
LC record available at https://lccn.loc.gov/2021044739
LC ebook record available at https://lccn.loc.gov/2021044740

ISBN 978-1-107-17979-0 Hardback

Cambridge University Press has no responsibility for the persistence or accuracy of
URLs for external or third-party internet websites referred to in this publication
and does not guarantee that any content on such websites is, or will remain,
accurate or appropriate.

Contents

Preface	*page* ix
Acknowledgments	xiii

1 Introduction — 1

1.1 General Concepts	1
1.2 Classification of Dynamic Systems	4
1.3 System Model Representations	6
1.4 Analogous Relations	11
1.5 Software for Computation	13
1.6 Units	15

2 Fundamentals of Mathematics — 16

2.1 Vector Algebra	16
2.2 Matrix Algebra	24
2.3 Complex Numbers	34
2.4 The Laplace Transform	37
2.5 Differential Equations	63
2.6 Solution of Linear Ordinary Differential Equations with Constant Coefficients	70
2.7 Transfer Functions and Block Diagrams of Time-Invariant Systems	95
2.8 Solution of State Equations via Numerical Integration	98
Chapter Summary	102
References	103
Problems	103

3 Mechanical Systems — 114

3.1 Fundamental Principles of Mechanical Systems	115
3.2 Translational Systems	120
3.3 Rotational Systems	144
3.4 Rigid-Body Systems in Plane Motion	169
3.5 Energy Approach	177
3.6 Block Diagrams of Mechanical Systems	181

    3.7 State-Space Representations   185
    3.8 Dynamic Responses via MATLAB   191
    3.9 Dynamic Responses via Mathematica   198
    Chapter Summary   204
    References   204
    Problems   204

4 **Electrical Systems**   222

    4.1 Fundamentals of Electrical Systems   223
    4.2 Concept of Impedance   230
    4.3 Kirchhoff's Laws   242
    4.4 Passive-Circuit Analysis   244
    4.5 State-Space Representations and Block Diagrams   258
    4.6 Passive Filters   271
    4.7 Active-Circuit Analysis   277
    4.8 Dynamic Responses via Mathematica   297
    Chapter Summary   306
    References   306
    Problems   307

5 **Thermal and Fluid Systems**   334

    5.1 Fundamental Principles of Thermal Systems   335
    5.2 Basic Thermal Elements   339
    5.3 Dynamic Modeling of Thermal Systems   346
    5.4 Fundamental Principles of Fluid Systems   353
    5.5 Basic Elements of Liquid-Level Systems   356
    5.6 Dynamic Modeling of Liquid-Level Systems   363
    5.7 Fluid Inertance   368
    5.8 The Bernoulli Equation   370
    5.9 Pneumatic Systems   374
    5.10 Dynamic Response via MATLAB and Simulink   378
    Chapter Summary   384
    References   384
    Problems   384

6 **Combined Systems and System Modeling Techniques**   396

    6.1 Introduction to System-Level Modeling   397
    6.2 System Modeling Techniques   398
    6.3 Fundamentals of Electromechanical Systems   425

6.4	Direct Current Motors	433
6.5	Block Diagrams in the Time Domain	457
6.6	Modeling and Simulation by Simulink	464
6.7	Modeling and Simulation by Mathematica	475
	Chapter Summary	485
	References	486
	Problems	486

7 System Response Analysis — 502

7.1	System Response Analysis in the Time Domain	502
7.2	Stability Analysis	544
7.3	System Response Analysis in the Frequency Domain	558
7.4	Time Response of Linear Time-Varying and Nonlinear Systems	571
	Chapter Summary	577
	References	578
	Problems	578

8 Introduction to Feedback Control Systems — 589

8.1	General Concepts	589
8.2	Advantages of Closed-Loop Control Systems	595
8.3	PID Control Algorithm	601
8.4	Control System Analysis	611
8.5	The Root Locus Method	619
8.6	Analysis and Design by the Root Locus Method	632
8.7	Additional Examples of Control Systems	648
	Chapter Summary	653
	References	654
	Problems	654

9 Application Problems — 666

9.1	Vibration Analysis of a Car Moving on a Bumpy Road	667
9.2	Speed Control of a Coupled Engine–Propeller System	686
9.3	Modeling and Analysis of a Thermomechanical System (a Bimetallic Strip Thermometer)	697
9.4	Modeling and Analysis of an Electro-Thermo-Mechanical System (a Resistive-Heating Element)	702
9.5	Feedback Control of a Liquid-Level System for Water Purification	711
9.6	Sensors, Electroacoustic, and Piezoelectric Devices	723
	References	756

Appendices
A. Units and Conversion Table   758
B. A Brief Introduction to MATLAB and Simulink   761
C. A Brief Introduction to Wolfram Mathematica   771
Index   778

Preface

The subject of dynamic systems deals with modeling and analysis of temporal behaviors of machines, devices, equipment, structures, industrial processes, and beyond. Mathematically, a dynamic system is described by certain time-dependent functions, which are usually governed by a set of differential equations. Physically, a typical dynamic system consists of multiple components or subsystems, which fall in different areas, such as mechanics, thermodynamics, fluid dynamics, electromagnetism, acoustics, and optics. Unlike those single-field courses, a dynamic systems course is focused on interactions among the components of a system, integration of those components from multiple fields in modeling, and overall system-level behaviors in analysis, which is known as system dynamics. Naturally, modeling and analysis of dynamic systems requires a variety of techniques, including transfer function formulations, block diagrams, and state-space representations. Additionally, an in-depth understanding of system dynamics is crucially important to the development of feedback control systems, which nowadays are a must in many applications. Therefore, a dynamic systems course serves as a bridge between single-area studies and system-level investigations, and it is a preparation for research and development on advanced topics in science and technology.

Readership

This book introduces a wide range of dynamic systems encountered in engineering. In particular, mechanical, electrical, thermal and fluid systems, feedback control systems, and their combinations are modeled, simulated, and analyzed. The material is designed for a course at undergraduate level, which is typically found in engineering curricula. The text can also be used by engineers and researchers for self-learning purposes.

The reader is expected to have a background in elementary calculus (differentiation and integration) and some engineering college courses (such as dynamics, the strength of materials, electrical circuits, physics, thermodynamics, and fluid dynamics). A knowledge of differential equations is desirable but not necessary because this topic, along with others, is fully covered in the text.

Special Features

Reflecting the combined 20 plus years of the authors' experiences in teaching dynamic systems, the book has the following special features.

First, three important keys in system modeling (namely, fundamental principles, basic elements, and ways of analysis) are emphasized throughout the text. In the authors' opinion, learning this three-key modeling concept can greatly enhance the reader's skills to develop mathematical models for dynamic systems in various engineering problems.

Second, the relevant mathematical topics, including differential equations, the Laplace transform, matrices, complex numbers, and the numerical solution of state equations, are thoroughly reviewed. Thus, readers of different mathematical backgrounds should be able to master the contents of the book with ease.

Third, computer simulations of dynamic systems using both MATLAB/Simulink and Mathematica are illustrated in examples, and also used to solve a variety of application problems. The coverage of these two powerful software packages, which is not seen in most dynamic systems texts, provides the reader with computational flexibility in investigation of system dynamics.

Last, but not least, a cluster of practical problems of multi-component dynamic systems is presented in the last chapter (Chapter 9). This helps the reader digest the materials that are presented in the previous chapters, and demonstrates the utility of software packages in modeling, simulation, and analysis of complicated dynamic systems in engineering.

Organization

This book comprises nine chapters and three appendices.

Chapter 1 introduces the general concepts and classification of dynamic systems in engineering, and previews commonly used methods and computer software for modeling, simulation, and analysis of dynamic systems. Moreover, analogous relations among mechanical, electrical, thermal, and fluid systems are presented.

Chapter 2 presents the essential mathematical topics that are required for modeling, simulation, and analysis of dynamic systems. These topics include vectors and matrices, differential equations, the Laplace transform, complex numbers, transfer functions, block diagrams, and the solution of state equations by numerical integration.

Chapter 3 covers modeling of mechanical systems, including translational, rotational, and rigid-body-systems. In the derivation of the equations of motion for a system, three keys are emphasized: fundamental principles (such as Newton's laws of motion), basic elements (such as masses, springs, and dampers), and a way of analysis (for example, free-body diagrams). Also, transfer function formulations and state-space representations are introduced to prepare for the analysis and simulation of mechanical systems. In addition, an energy approach with Lagrange's equations is introduced as an alternative method to derive equations of motion for mechanical systems.

Chapter 4 covers modeling of electrical systems, including passive electrical circuits and active electrical circuits with isolation amplifiers and operational amplifiers (op-amps). Just as when dealing with mechanical systems, there are three keys in the modeling of electrical systems: Kirchhoff's current and voltage laws as fundamental principles; resistors, inductors, and capacitors as basic elements; and the loop method and the node method as two ways of analysis. Also, transfer functions and state equations for electrical systems are presented.

Chapter 5 covers modeling of thermal and fluid systems, in which lumped thermal and fluid elements are introduced. The concept of free-body diagrams for mechanical systems is extended to thermal and fluid systems. Furthermore, transfer function formulations, state-space representations, and block diagrams for thermal and fluid systems are presented.

Chapter 6 deals with dynamic systems consisting of multiple components or subsystems that are from different fields. Electromechanical systems are mainly considered as demonstrative examples. To describe the interactions among the components of these systems, transfer functions, s-domain block diagrams, and state equations are used. Furthermore, to deal with time-variant and nonlinear dynamic systems, time-domain block diagrams are introduced, followed by examples of modeling and simulation of combined systems via Simulink and Mathematica.

Chapter 7 presents the methods for the determination of responses of dynamic systems, in both the time domain and the frequency domain, and for stability analysis. With the dynamic models presented in Chapters 3–6, the system response is obtained by either analytical methods or numerical methods.

Chapter 8 gives a brief introduction to feedback control, in which the basic concepts and advantages of closed-loop control systems are discussed, and the proportional–integral–derivative (PID) feedback control algorithm is presented. In addition, the root-locus method, which is a useful tool for the analysis and design of feedback control systems, is presented in some detail.

Chapter 9 assembles some relatively complicated problems from engineering applications, including a vehicle moving on a bumpy road, a coupled engine–propelller system, a bimetallic strip thermometer, a resistive-heating element, a liquid-level system for water purification, sensors, piezoelectric devices, and electroacoustic devices. In solution of these problems, the methods of modeling and analysis that are covered in the previous chapters are used, and the use of the MATLAB/Simulink and Mathematica software packages in simulation is illustrated.

At the end of Chapters 2–8 there is a summary section, reviewing the main contents and learning objectives of the chapter, as well as a comprehensive list of problems for each topic. Solutions for these problems are available at www.cambridge.org/yang-abramova.

Appendix A presents a list of SI units for commonly used physical quantities. Appendices B and C give brief reviews of the basics of MATLAB/Simulink and Mathematica, respectively.

Use of the Book in Teaching

The selection of the topics from the book depends on three factors: the instructor's emphasis, the student's background in mathematics, and the course duration (semester or quarter).

For a semester course at the junior level, the coverage of Chapter 1, part of Chapter 2, Chapters 3 through 7, and part of Chapter 8 is recommended. If time allows, a couple of sections from Chapter 9 can be considered.

Some courses can omit most of Chapter 2 (Fundamentals of Mathematics) if the overall mathematical level of the class is adequate. Other courses may skip Chapter 5 (Thermal and Fluid Systems), in order to address other topics of interest.

With any selected topic, Chapter 2 can always serve as a reference on mathematics.

Instructor's Solutions Manual

A solutions manual for all the listed exercise problems is prepared by the authors.

Website

The publisher maintains a website for this text at www.cambridge.org/yang-abramova. Here, students can get access to the MATLAB/Simulink and Mathematica codes developed for the solutions of the application problems in Chapter 9, a tutorial on usage of MATLAB/Simulink and Mathematica in modeling and simulation, and other resources relevant to the book. Also, at www.cambridge.org/yang-abramova, instructors adopting this text can gain access to the solutions manual.

Acknowledgments

The authors would like to express their deep gratitude to the staff at Cambridge University Press, especially Steven Elliot, Stefanie Seaton, and Charles Howell, and the copy-editor, Zoë Lewin, for their assistance and guidance during various stages of the development of this text.

Many thanks go to the University of Southern California for providing an atmosphere that encourages teaching innovation and curriculum development.

The reviewers of this book provided insightful comments and valuable suggestions, and for that the authors are grateful.

The first author thanks his wife, Haiyan, and daughters, Sonia and Tanya, for their support, patience, and understanding.

The second author thanks her colleagues at the Department of Aerospace and Mechanical Engineering for their invaluable mentorship and support, and her family for their understanding and encouragement all the way.

1 Introduction

Contents

1.1	General Concepts	1
1.2	Classification of Dynamic Systems	4
1.3	System Model Representations	6
1.4	Analogous Relations	11
1.5	Software for Computation	13
1.6	Units	15

In this introductory chapter, the general concepts and classification of dynamic systems in engineering are introduced; commonly used methods and computer software for modeling, simulation, and analysis of dynamic systems are previewed; and the scope of this book is outlined.

1.1 General Concepts

A dynamic system is a combination of components or subsystems, which, with temporal characteristics, interact with each other to perform a specified objective. There exists a variety of dynamic systems in applications, such as machines, devices, appliances, equipment, structures, and industrial processes. Mathematically, a dynamic system is characterized by time-dependent functions or variables, which are governed by a set of differential equations. Physically, the components of a dynamic system may fall in different fields of science and engineering, such as mechanics, thermodynamics, fluid dynamics, vibrations, elasticity, electronics, acoustics, optics, and controls.

As an example, an electric motor is a dynamic system consisting of mechanical components (like rotating shaft, bearing, and housing), electromagnetic components (such as magnets, coils, and electrical interconnects), and components for controlling the motor speed (including speed sensor, control logic board, and driver). These components interact with each other to achieve a desired motor speed. The rotation speed and circuit currents are time-dependent variables of the motor that are governed by differential equations in the fields of dynamics and electromagnetism.

Unlike those single-field courses, a dynamic systems course is focused on the interactions among the components of a dynamic system, the integration of those components from multiple fields in modeling and solution, and the understanding of overall system response in analysis, which is known as system dynamics. Thus, in study of dynamic systems, the knowledge in multiple fields is generally required, and various techniques for modeling and analysis are naturally adopted.

This book introduces a wide range of dynamic systems encountered in engineering applications. In particular, mechanical, electrical, thermal, and fluid systems, feedback control systems, and their combinations are modeled, simulated, and analyzed.

1.1.1 Input and Output

A dynamic system can be graphically illustrated by a diagram. Figure 1.1.1(a) shows a system with one input and one output, which is known as a single-input single-output (SISO) system. Here, the word "input" means a *cause* or *excitation* and the word "output" means an *effect* or *response* due to the input. For example, for an electric motor, the applied voltage is an input, and the motor speed is an output. As another example, for a sailboat, the wind flowing on its sail is an input, and the motion of the boat in the water is the output. The output of a dynamic system is also called a dynamic response or a system response.

A dynamic system can have multiple inputs and multiple outputs, as shown in Figure 1.1.1(b). Such a system is known as a multi-input multi-output (MIMO) system. The specification of inputs and outputs depends on the problem and the investigator's interest. For instance, for an electric motor, the applied voltage and the external load torque, which is exerted by a machine or device that is driven by the motor, are the inputs to the motor; the motor speed and the motor power (a product of speed and torque) can be considered as the outputs. In some applications, like robotics, the motor rotation angle can also be chosen as an output.

Figure 1.1.1 Block diagram representation of dynamic systems: (a) a single-input single-output system; and (b) a multi-input multi-output system

The dynamic behavior of a system can be characterized by its *input–output relations*, which describe the influence of inputs on the system outputs. The input–output relations also provide a means for interconnection of the components of a system in system-level modeling and analysis, and for understanding of the dynamic interactions among these components. Therefore, one major task in the study of dynamic systems is to establish input–output relations.

1.1.2 Modeling, Simulation, and Analysis

In the advent of computer technologies, model-based design has become a must in the development of many products and for the operation of industrial processes. Conventional design often relies on building a prototype system and conducting tests on it, which can be expensive, time-consuming, and in some cases infeasible. Model-based design, on the other hand, does not encounter these issues and, more importantly, it can aid optimal design of complicated dynamic systems, and provide an approach to efficient and reliable operation of industrial processes.

Model-based design engages in the activities of modeling, simulation, and analysis of physical systems. These efforts are explained as follows.

Modeling

Modeling is the effort to develop a mathematical model for a given dynamic system, which is a set of differential equations governing the time-dependent variables and outputs of the system. There are **three important keys in modeling**: *fundamental principles* (like Newton's laws), *basic elements* (such as spring, mass, and damper elements), and *ways of analysis* (for instance, free-body diagrams). By a way of analysis, fundamental principles are applied to basic elements, leading to the governing equations of the system. Throughout this text, these three keys are emphasized in modeling of mechanical, electrical, thermal, and fluid systems, as shown in Chapters 3 to 5.

Simulation

Simulation is the effort to determine the dynamic response of a system, which involves solution of the governing differential equations of the system. For certain simple systems, analytical solutions are available. In general, system response solutions are obtained by numerical methods, such as numerical integration algorithms (Section 2.8). In simulation, the mathematical model of a system is usually converted to one of the following representations: transfer function formulation, state-space representation, and block diagram representation. With these model representations, the dynamic response of a system can be computed through use of software packages, such as MATLAB/Simulink and Mathematica.

Analysis

Analysis is the effort to examine the physical behaviors of a system, to evaluate the performance of the system, and to provide guidance for changing the parameters and configuration of the system in a design process. Analysis is often performed based on system response solutions obtained by simulation. In this text, three main issues in system response analysis are addressed: system response analysis in the time domain, stability analysis, and system response analysis in the frequency domain (see Chapter 7).

1.2 Classification of Dynamic Systems

Dynamic systems can have different types of mathematical models: lumped or distributed; linear or nonlinear; time-invariant or time-variant; and continuous-time or discrete-time. The classification of these models depends on the types of governing equations obtained in system modeling, which is described below.

Distributed Models versus Lumped Models

A distributed model has variables as functions of time and spatial coordinate(s), which are governed by partial differential equations (PDEs). A distributed model is also called an infinite-dimensional system because its response is expressed in terms of an infinite number of coordinates. A lumped model has variables as functions of time only, which are governed by ordinary differential equations (ODEs). A lumped model is also called a finite-dimensional system as its response is expressed by a finite number of coordinates. For example, heat conduction in a continuum is originally described by PDEs. Lumping physical parameters (such as temperature and heat flow rate) leads to element models (thermal capacitance and thermal resistance), which eventually yields a lumped model described by ODEs.

Linear Models versus Nonlinear Models

A linear model has variables that are governed by linear differential equations. A nonlinear model has variables that are governed by nonlinear differential equations. A linear differential equation about a variable is one which only involves a linear combination of the variable and its derivatives. A nonlinear differential equation about a variable is one which involves the products and nonlinear functions of the variable and its derivatives.

For instance, the displacement x of a spring–mass–damper system is modeled by the linear differential equation

$$m\ddot{x} + c\dot{x} + kx = f \tag{1.2.1}$$

where $\dot{x} = dx/dt$, m, c, and k are the mass, damping, and spring parameters of the system, and f is an external force. Equation (1.1.1) represents a linear model. If the system is also subject to dry friction, its governing equation of motion can be written as

$$m\ddot{x} + c\dot{x} + kx + \mu N \, \text{sgn}(\dot{x}) = f \qquad (1.2.2)$$

where μ is a kinetic friction coefficient, N is a normal force, and $\text{sgn}(\dot{x})$ is the sign function. Because $\text{sgn}(\dot{x})$ is a nonlinear function of $\dot{x}$, Eq. (1.1.2) is a nonlinear model.

Time-Invariant Models versus Time-Variant Models

A time-invariant model is one with its governing differential equations having constant coefficients. A time-variant model is one with its governing differential equations containing coefficients that change with time. For example, Eq. (1.2.1), with constant coefficients m, c, and k, describes a time-invariant model. As another example, the differential equation

$$m\ddot{x} + c\dot{x} + (k_0 + \varepsilon \sin \Omega t)x = f \qquad (1.2.3)$$

with the time-varying coefficient $k_0 + \varepsilon \sin \Omega t$, gives a time-variant model.

Continuous-Time Models versus Discrete-Time Models

A continuous-time model is one with its variables as continuous functions time, and it is described by differential equations. A discrete-time model is one with its variables only specified at discrete time points, and it is governed by difference equations. Discrete-time models are usually considered in automatic control and signal processing with digital computers, in which quantization of signals is performed.

It should be noted that for the same physical system, different types of mathematical models can be adopted, depending on the application and interest. For instance, in studying the dynamic behaviors of an airplane, a lumped model may be good enough to describe the flight trajectory and motion of the airplane in space. However, a distributed model is necessary to understand the vibration and stress in the airplane structure, for performance evaluation and failure analysis.

Once a specific model type is selected in design or investigation, the words "model" and "system" are often used without distinction. Thus, for convenience of discussion, terms like lumped system, time-invariant system, time-variant system, and nonlinear system mean the relevant mathematical models.

In this book, lumped and continuous-time systems are considered in modeling, simulation, and analysis. The coverage of distributed systems and discrete-time systems, which requires advanced mathematics in modeling and solution, is beyond the scope of this undergraduate-level text.

1.3 System Model Representations

In simulation and analysis of general dynamic systems, a mathematical model that is obtained from the first laws of nature is usually converted to an equivalent model representation. In this book, four commonly used forms of system model representation are introduced: (i) transfer function formulation, (ii) state-space representation, (iii) block diagram representation in the s-domain, and (iv) block diagram representation in the time domain. The purpose of adopting these forms is three-fold: to have a standard form for numerical solutions, to understand the interactions among the components of a system, and to obtain the information about the internal variables (state variables) of a system. As a matter of fact, software packages MATLAB/Simulink and Mathematica are created based on these model-representation forms.

In the subsequent subsections, the above-mentioned system model representations are illustrated on simple models of mechanical systems, although the basic concepts are applicable to general dynamic systems.

1.3.1 Transfer function Formulation

Consider a spring–mass–damper system shown in Figure 1.3.1, which is a simple mathematical model for many mechanical systems. By Newton's laws and through use of a free-body diagram of the mass (Section 3.2), the equation of the motion governing the displacement x of the system is obtained and it is given in Eq. (1.2.1).

Let the input and output of the system be the external force f and the displacement x of the system, respectively. The transfer function $G(s)$ of a dynamic system is defined as the following input-to-output ratio

$$G(s) = \left. \frac{\mathcal{L}[\text{output}]}{\mathcal{L}[\text{input}]} \right|_{\text{zero initial conditions}} \tag{1.3.1}$$

where $\mathcal{L}$ is the Laplace transform operator, s is the complex Laplace transform parameter, and zero initial conditions mean all the initial values of the output and their time derivatives

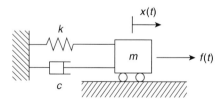

Figure 1.3.1 A spring–mass–damper system

are set to zero. (Refer to Section 2.4 for the Laplace transform.) Perform a Laplace transform of Eq. (1.2.1) with respect to time and use zero initial conditions, to obtain

$$(ms^2 + cs + k)X(s) = F(s) \tag{1.3.2}$$

where $X(s)$ and $F(s)$ are the Laplace transforms of the output x and the input f, respectively. According to the definition (1.3.1), the transfer function of the spring–mass–damper system is

$$G(s) = \frac{X(s)}{F(s)} = \frac{1}{ms^2 + cs + k} \tag{1.3.3}$$

Equation (1.3.3) gives an input–output relation (transfer function formulation):

$$X(s) = G(s)F(s) = \frac{1}{ms^2 + cs + k}F(s) \tag{1.3.4}$$

Thus, in the s-domain (the Laplace transform domain), the system output is the product of its transfer function and the input.

The transfer function formulation (1.3.4) can be used to study the dynamic behavior of the mechanical system, to determine the system response, and to perform system response analysis. With the same concept, a transfer function formulation can be established to a dynamic system of multiple components, by which the dynamic interactions among the components can be investigated and the system-level analysis can be carried out.

The transfer function formulation for mechanical, electrical, thermal, and fluid systems is introduced in Chapters 3 to 5, and its application in multi-component dynamic systems and feedback control systems is covered in Chapters 6 to 8.

It should be pointed out that transfer function formulation is only valid for linear time-invariant systems and it is not applicable to time-variant systems and nonlinear systems. This is because the Laplace transform operator is a linear operator with its utility limited to linear equations with constant coefficients.

1.3.2 State-Space Representation

State-space representation is a modeling technique that is applicable to general dynamic systems, including time-variant systems and nonlinear systems.

A state-space representation consists of two sets of equations in the time domain: state equations and output equations. State equations are a set of coupled first-order differential equations about quantities known as state variables. Output equations are a set of coupled algebraic equations about the state variables. Here, state variables are used to describe the mathematical status of a dynamic system, which is known as the state, and by which the future behavior of the system can be determined with known inputs.

As an example, consider the same spring–mass–damper system as shown in Figure 1.3.1. By selecting two state variables

$$x_1 = x, \quad x_2 = \dot{x} \tag{1.3.5}$$

the differential equation (1.2.1) can be converted to the following two state equations

$$\begin{aligned} \dot{x}_1 &= x_2 \\ \dot{x}_2 &= \frac{1}{m}\left[-kx_1 - cx_2 + f\right] \end{aligned} \tag{1.3.6}$$

Here, the number of state variables is the same as the order of the original differential equation, which is two for the spring–mass–damper system. Refer to Section 3.7 on how to select state variables. The state variables are also known as internal variables, which can completely describe the dynamic behavior of the system, including system outputs and internal forces.

The output equations depend on the selection of system outputs. For instance, if the velocity and spring force are chosen as the outputs, the output equations of the system are of the algebraic form

$$\begin{aligned} y_1 &= x_2 \\ y_2 &= kx_1 \end{aligned} \tag{1.3.7}$$

where y_1 and y_2 are the system outputs. As can be seen from Eq. (1.3.7), the outputs are expressed in terms of the state variables.

To show the versatile utility of state-space representation, consider the spring–mass–damper system subject to dry friction, which is governed by the nonlinear differential equation (1.2.2). With the same state variables as given in Eq. (1.3.5), the state equations are obtained as follows

$$\begin{aligned} \dot{x}_1 &= x_2 \\ \dot{x}_2 &= \frac{1}{m}\left[-kx_1 - cx_2 - \mu N \operatorname{sgn}(x_2) + f\right] \end{aligned} \tag{1.3.8}$$

Equations (1.3.8) are nonlinear state equations. Also, for the time-variant system as described by Eq. (1.2.3), a set of state equations with the state variables defined in Eq. (1.3.5) are as follows

$$\begin{aligned} \dot{x}_1 &= x_2 \\ \dot{x}_2 &= \frac{1}{m}\left[-(k_0 + \varepsilon \sin \Omega t)x_1 - cx_2 + f\right] \end{aligned} \tag{1.3.9}$$

For more details on state-space representation, refer to Sections 2.5 and 6.2, and the examples in Chapters 3 to 5.

State-space representation provides a platform for numerical simulation of the response of dynamic systems. For instance, the state equations (1.3.6), (1.3.8), and (1.3.9) can be solved by standard numerical integration algorithms (see Section 2.8). As a matter of fact, some differential-equation solvers of MATLAB and Mathematica are developed based on state-space representation.

1.3.3 Block Diagram Representation in the *s*-Domain

Block diagrams are widely used in representing dynamic systems in control engineering. A block diagram is an assembly of blocks, each of which describes an input–output relation for a component or subsystem. In the *s*-domain, every block is identified with a transfer function of the component. For instance, for the spring–mass–damper system in Figure 1.3.1, its block is shown in Figure 1.3.2, where the block is a box having input $F(s)$ and output $X(s)$, and inside the box is the transfer function of the system. The input–output relation of the block diagram in general is

$$\text{Output} = \text{Transfer function} \times \text{Input} \quad (1.3.10)$$

By this relation, the block diagram in Figure 1.3.2 indicates

$$X(s) = \frac{1}{ms^2 + cs + k} F(s) \quad (1.3.11)$$

which is equivalent to the transfer function given in Eq. (1.3.3).

To see the interconnection of system components in a diagram, consider a spring–mass–damper system with two masses, shown in Figure 1.3.3, where x_1 and x_2 are the displacements of masses m_1 and m_2, respectively. Following a modeling technique described in Section 3.2, the governing equations of motion of the system are obtained as follows

$$\begin{aligned} m_1\ddot{x}_1 + c\dot{x}_1 + (k + k_1)x_1 - k_1 x_2 &= 0 \\ m_2\ddot{x}_2 + k_1 x_2 - k_1 x_1 &= f \end{aligned} \quad (1.3.12)$$

Taking the Laplace transform of Eq. (1.3.12) with zero initial conditions gives the *s*-domain equations

$$\begin{aligned} (m_1 s^2 + cs + k)X_1(s) &= k_1(X_2(s) - X_1(s)) \\ m_2 s^2 X_2(s) &= F(s) - k_1(X_2(s) - X_1(s)) \end{aligned} \quad (1.3.13)$$

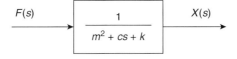

Figure 1.3.2 Block diagram of a spring–mass–damper system

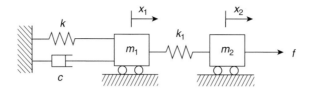

Figure 1.3.3 A mechanical system with two masses

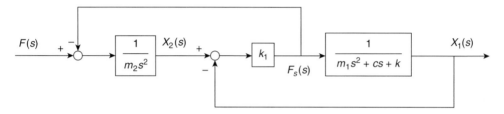

Figure 1.3.4 An *s*-domain block diagram of the two-mass mechanical system in Figure 1.3.3

Let the external force f and the displacement x_1 be the input and output of the system, respectively. By applying the rules given in Section 6.2.3 to Eq. (1.3.13), the block diagram of the two-mass system is constructed in Figure 1.3.4, where the right box (for m_1) and the left box (for m_2) are connected through use of a summation point and a pick-off point; and $F_s(s)$ is the internal force of the spring that couples the two masses by $F_s(s) = k_1 \big(X_2(s) - X_1(s)\big)$. The construction of the block diagram in Figure 1.3.4 is detailed in Section 3.6.

Note that the block diagram representation in the *s*-domain is limited to linear time-invariant systems. This is because the subsystem blocks in such a diagram are created through use of transfer functions.

1.3.4 Block Diagram Representation in the Time Domain

Like state-space representation, block diagram representation in the time domain is applicable to general dynamic systems with time-dependent coefficients and nonlinearities. Time-domain block diagram representation differs from state-space representation in that it provides a graphical way to model multi-component systems. Learning time-domain block diagrams helps one to use the software package MATLAB/Simulink in the modeling and simulation of complicated systems. Indeed, the structure of time-domain block diagrams and that of Simulink models are quite similar (see Sections 6.5 and 6.6).

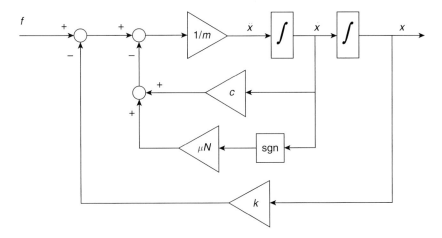

Figure 1.3.5 A time-domain block diagram of the nonlinear system described by Eq. (1.2.2)

As an example, consider the nonlinear differential equation (1.2.2), which models a spring–mass–damper system subject to dry friction. Let f and x be the input and output of the system, respectively. By following the steps in Section 6.5, a time-domain block diagram is constructed in Figure 1.3.5. Refer to Table 6.5.1 for the definitions of the symbols used in the diagram. Based on the time-domain block diagram, a Simulink model for the nonlinear system can be conveniently built and the system response to a given input can be easily computed by running the model.

1.4 Analogous Relations

In this text, four types of dynamic systems are modeled: mechanical systems (Chapter 3), electrical systems (Chapter 4), and thermal and fluid systems (Chapter 5). As mentioned in Section 1.1.2, one of the three important keys in modeling is the usage of basic elements, which are approximate models of components or subsystems and which are the building blocks of a dynamic system. For this, three types of basic lumped-parameter elements are introduced: capacitance, resistance, and inertance. For instance, springs, dampers, and masses in mechanical systems are capacitance, resistance, and inertance elements; capacitors, resistors, and inductors in electric circuits are also capacitance, resistance, and inertance elements.

Flow and Effort

Capacitance, resistance, and inertance elements for mechanical, electrical, thermal, and fluid systems can be similarly treated by equations involving two variables: flow q and effort f. Here, q can be a velocity, a current, heat flow rate, or a mass flow rate while f can be an external force, a voltage, a temperature, or a pressure. Physically, the effort of an element is related to the application of an action to the element, which is an input to the element, and the flow is about the consequence of the action, which is an output of the element.

Flow and effort are also known as a pair of power variables because the power P of an element can be expressed by

$$P = f \times q \tag{1.4.1}$$

For instance, for an electrical element (resistor or capacitor or inductor), its electrical power is

$$P = v \times i$$

where v is the voltage (potential drop) of the element, and i is the current going through it. In this case, current i and voltage v are the effort and flow of the electrical element, respectively. The power $f \times q$ results in storage of potential energy or kinetic energy in the element or in dissipation of energy by the element. As another example, the power of a mechanical element is

$$P = f \times v$$

where f is the force applied to the element, and $v = \dot{x}$ is the velocity of the element. In this case, velocity v and force f are the effort and flow of the mechanical element, respectively.

Element Laws

Denote capacitance, resistance, and inertance of general elements by $\widetilde{C}, \widetilde{R}$ and $\widetilde{I}$, respectively. The governing equations for general elements are stated as follows:

$$\text{For capacitance:} \quad q = \widetilde{C} \frac{df}{dt} \tag{1.4.2}$$

$$\text{For resistance:} \quad f = \widetilde{R} q \tag{1.4.3}$$

$$\text{For inertance:} \quad f = \widetilde{I} \frac{dq}{dt} \tag{1.4.4}$$

where q and f are flow and effort, respectively.

Equations (1.4.2) to (1.4.4) are known as *the element laws* that characterize *analogous relations* among different types of systems by the equivalent coefficients $\widetilde{C}, \widetilde{R}$, and $\widetilde{I}$. In general, a capacitance element stores potential energy; a resistance element dissipates energy; and an inertance element stores kinetic energy.

Consider the spring–mass–damper system in Figure 1.3.1, where the spring (k), mass (m), and damper (c) are capacitance, resistance, and inertance elements. In what follows, we show that the physical parameters of the system are related to the capacitance, resistance, and inertance by

$$\tilde{C} = \frac{1}{k}, \quad \tilde{R} = c, \quad \tilde{I} = m \qquad (1.4.5)$$

The $\tilde{R}$ and $\tilde{I}$ in Eq. (1.4.5) can be easily obtained from the following equations

$$f_d = c\dot{x} = cv \qquad (1.4.6)$$

$$m\ddot{x} = m\dot{v} = f \qquad (1.4.7)$$

where f_d is the internal force (damping force) of the damper, and Eq. (1.4.7) is the momentum equation for the mass according to Newton's second law. The internal force (spring force) f_s of the spring can be written as $f_s = kx$, which after differentiation leads to

$$\frac{df_s}{dt} = k\dot{x} = kv \qquad (1.4.8)$$

It follows that

$$v = \frac{1}{k}\frac{df_d}{dt} \qquad (1.4.9)$$

indicating that $\tilde{C} = 1/k$.

In summary of the previous discussion, the analogous variables and equivalent coefficients of lumped parameter elements of mechanical, electrical, thermal, and fluid systems are listed in Table 1.4.1. In the table, a spring–mass–damper system is used as a representative of mechanical systems and a hydraulic system is used as a representative of fluid systems. Note that thermal systems, by the law of thermodynamics, do not have inertance elements. These analogous relations are helpful for better understanding of the element laws, and they are useful in model development for general dynamic systems.

1.5 Software for Computation

All the numerical results presented in this text are obtained through use of MATLAB, Simulink, and Mathematica. Appendices B and C provide tutorials on these software packages. In Sections 6.6 and 6.7, use of MATLAB, Simulink, and Mathematica is illustrated in examples. Also, several application problems in Chapter 9 are solved by MATLAB, Simulink, and Mathematica.

Note that numerical solution via the above-mentioned software packages adopts transfer function formulation, state-space representation, and block diagrams. Therefore, mastering

Table 1.4.1 Analogous variable and equivalent coefficients

	Mechanical system	Electrical system	Thermal system	Fluid system
Flow	$q = \dfrac{dx}{dt}$ $x =$ displacement	$i = \dfrac{dq_e}{dt}$ $q_e =$ electric charge	Heat flow rate q_h	Mass flow rate q_m
Effort	$f =$ force	$v =$ voltage	Temperature T	Pressure p
Capacitance	Spring: $\tilde{C} = \dfrac{1}{k}$	Capacitance: $\tilde{C} = C$	Thermal capacitance: $\tilde{C} = C = \dfrac{dE}{dt}$	Fluid capacitance: $\tilde{C} = C = \dfrac{dm}{dp}$
Resistance	Damper: $\tilde{R} = c$	Resistance: $\tilde{R} = R$	Thermal resistance: $\tilde{R} = R$	Fluid resistance: $\tilde{R} = R = \dfrac{dp}{dq}$
Inertance	Mass: $\tilde{I} = m$	Inductance: $\tilde{I} = L$		Fluid inertance: $\tilde{I} = I$

these forms of system model representation is essentially important to modeling, simulation, and analysis of dynamic systems.

1.6 Units

In this text, we use the International System of Units (SI), which is also known as the metric system. Refer to Appendix A for the units of commonly used quantities, and a table of unit conversion between the SI system and the US customary system.

2 Fundamentals of Mathematics

Contents

2.1	Vector Algebra	16
2.2	Matrix Algebra	24
2.3	Complex Numbers	34
2.4	The Laplace Transform	37
2.5	Differential Equations	63
2.6	Solution of Linear Ordinary Differential Equations with Constant Coefficients	70
2.7	Transfer Functions and Block Diagrams of Time-Invariant Systems	95
2.8	Solution of State Equations via Numerical Integration	98
	Chapter Summary	102
	References	103
	Problems	103

This chapter provides the reader with a brief refresher course on the mathematical apparatus crucial for modeling of dynamic systems. Sections 2.1 and 2.2 present basic concepts and terminology of vector and matrix algebra. Definitions and basic operations on complex numbers are introduced in Section 2.3. Section 2.4 is devoted to one of the important methods for solving differential equations – the Laplace transform. Sections 2.5 and 2.6 discuss the types of differential equations widely encountered in modeling of common dynamic systems and develop methods for solving these equations. Section 2.7 introduces the mathematical foundation for deriving transfer functions and creating block diagrams of various linear time-invariant dynamic systems. Section 2.8 presents a brief overview of solving differential equations numerically, with MATLAB and Wolfram Mathematica.

2.1 Vector Algebra

Components of a dynamic system can be represented by different physical quantities. Some of these quantities such as, for example, mass or moment of inertia, only have magnitude.

These quantities are called *scalars* and are represented by numbers alone with the appropriate units. The other physical quantities such as force, position, and velocity, for example, are characterized by magnitude and direction; thus, they can be manipulated in a way that takes into account both mentioned parameters. These quantities are represented by *vectors*.

2.1.1 Definition

A vector is a mathematical quantity that has both magnitude and direction. It is customarily depicted as an arrow pointing in the direction of the vector. The length of the arrow corresponds to the magnitude of the vector.

The origin of the vector is called the *tail*, and its end is called the *head* (Figure 2.1.1). Since a vector is unchanged under a translation – a transformation consisting of a constant offset with no rotation or distortion – it is often convenient to consider the vector tail A as located at the origin of the coordinate system.

A vector can be denoted by its tail and head such as $\vec{AB}$, or by a symbol such as $\vec{v}$: $\vec{v} = \vec{AB}$. The arrow above the vector symbol may be omitted, having a vector denoted as just v.

As a formal mathematical object, a vector is commonly defined as an element of a vector space and is specified by its coordinates. In the Cartesian coordinate system, a vector is defined by its three coordinates – x, y, and z – as follows: $\vec{v} = \begin{bmatrix} v_x \\ v_y \\ v_z \end{bmatrix} = \begin{bmatrix} x_B - x_A \\ y_B - y_A \\ z_B - z_A \end{bmatrix}$. The

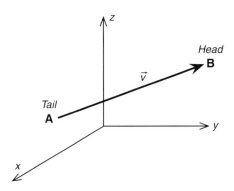

Figure 2.1.1 A vector

magnitude of a vector is called the *length*, or sometimes the *norm*, and is defined as

$$|\vec{v}| = \sqrt{v_x^2 + v_y^2 + v_z^2} \qquad (2.1.1)$$

A vector with unit length is called a *unit vector*, and is denoted with a hat above its symbol such as $\hat{v}$. An arbitrary vector can be converted to a unit vector – normalized – by dividing by its length: $\hat{v} = \frac{\vec{v}}{|\vec{v}|}$. A *zero vector*, denoted as 0, is a vector of arbitrary direction and length zero, having all its coordinate components equal to zero.

Vectors can exist at any point in space. Any two vectors that have the same length and the same direction are considered equal, no matter where in coordinate space they are located.

2.1.2 Algebraic Manipulations: Addition, Subtraction, and Scalar Multiplication

The most important operations on vectors include addition, subtraction, and scalar multiplication. Vector multiplication is not uniquely defined and for any arbitrary pair of vectors includes the dot product and cross product.

Vector addition is performed as shown in Figure 2.1.2(a) – vectors $\vec{p}$ and $\vec{q}$ are placed head to tail, and the resulting vector sum $\vec{v} = \vec{p} + \vec{q}$ is drawn from the free tail to the free head. Vector addition is commutative:

$$\vec{p} + \vec{q} = \vec{q} + \vec{p} \tag{2.1.2}$$

and associative:

$$(\vec{p} + \vec{q}) + \vec{r} = \vec{p} + (\vec{q} + \vec{r}) \tag{2.1.3}$$

For any arbitrary vector $\vec{p}$, a zero vector performs as an identity element for vector addition:

$$\vec{p} + \vec{0} = \vec{0} + \vec{p} = \vec{p} \tag{2.1.4}$$

Also, for any vector $\vec{p}$, there is a unique inverse vector $(-1)\vec{p} \equiv -\vec{p}$ such that $\vec{p} + (-\vec{p}) = \vec{0}$. That means that vectors $\vec{p}$ and $-\vec{p}$ have the same length, $|\vec{p}| = |-\vec{p}|$, but opposite orientations.

Vector subtraction is illustrated in Figure 2.1.2(b). It is an inverse operation to vector addition, and is performed as addition with the orientation of the second vector reversed, such as

$$\vec{v} = \vec{p} - \vec{q} = \vec{p} + (-\vec{q}) \tag{2.1.5}$$

When vectors are defined by their coordinates, vector sums resulting from addition/subtraction are: $\vec{v}_{addition} = \begin{bmatrix} p_x + q_x \\ p_y + q_y \\ p_z + q_z \end{bmatrix}$ and $\vec{v}_{subtraction} = \begin{bmatrix} p_x - q_x \\ p_y - q_y \\ p_z - q_z \end{bmatrix}$

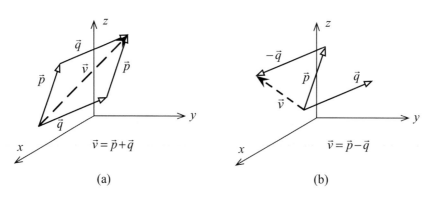

Figure 2.1.2 (a) Addition and (b) subtraction of vectors

Scalar multiplication is the operation of multiplying a vector by some constant a. If $\vec{v} = a * \vec{p}$, then vectors $\vec{v}$ and $\vec{p}$ are of the same direction if $a > 0$ and of opposite direction if $a < 0$, while length of the vector $\vec{v}$ is $|\vec{v}| = a * |\vec{p}|$.

In coordinate notation, the vector $\vec{v}$ is represented as $\vec{v} = \begin{bmatrix} a * p_x \\ a * p_y \\ a * p_z \end{bmatrix}$.

Scalar multiplication is commutative:

$$a * \vec{p} = \vec{p} * a \tag{2.1.6}$$

distributive:

$$(a + b) * \vec{p} = a * \vec{p} + b * \vec{p}$$
$$a * (\vec{p} + \vec{q}) = a * \vec{p} + a * \vec{q} \tag{2.1.7}$$

and associative:

$$b * (a * \vec{p}) = (b * a) * \vec{p} = (a * b) * \vec{p} = a * (b * \vec{p}) \tag{2.1.8}$$

The number 1 acts an identity element for scalar multiplication:

$$1 * \vec{p} = \vec{p}$$

2.1.3 Unit-Vector Decomposition

For a chosen coordinate system with an origin and axes, any vector can be decomposed into component vectors along each coordinate axis, as shown in Figure 2.1.3.

An arbitrary vector $\vec{v}$ in the Cartesian three-dimensional space can be decomposed into the following vector sum: $\vec{v} = \vec{v}_x + \vec{v}_y + \vec{v}_z$ where $\vec{v}_x$, $\vec{v}_y$, and $\vec{v}_z$ are vector components parallel to the x-, y-, and z- coordinate axes, respectively. Let us define unit vectors $\hat{i}, \hat{j}$, and $\hat{k}$ collinear with the x-, y-, and z- coordinate axes, respectively. As mentioned earlier, a unit vector means that its magnitude is one, that is $|\hat{i}| = |\hat{j}| = |\hat{k}| = 1$. As indicated in Figure 2.1.3, the direction of the unit vector $\hat{i}$ points in the direction of the increasing x-coordinate, i.e. it is pointing in the +x direction. Directions of unit vectors $\hat{j}$ and $\hat{k}$ are similarly defined. Having defined the unit vectors, the vector components and the decomposition itself can be expressed as follows:

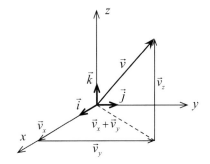

Figure 2.1.3 Unit-vector decomposition

$$\vec{v}_x = v_x \vec{i}$$
$$\vec{v}_y = v_y \vec{j}$$

$$\vec{v}_z = v_z \vec{k}$$
$$\vec{v} = v_x \vec{i} + v_y \vec{j} + v_z \vec{k} \tag{2.1.9}$$

In the expressions above, the terms v_x, v_y, and v_z are called the x-, y-, and z- components of the vectors $\vec{v}_x$, $\vec{v}_y$, and $\vec{v}_z$ respectively. While the absolute value of the term v_x is equal to the magnitude of the vector $\vec{v}_x$, v_x itself can be either positive or negative. Recalling the vector definition in coordinate notation it can be stated that v_x is the x- coordinate of vector $\vec{v}$. The same reasoning applies to the terms v_y and v_z.

In terms of unit vector decomposition, vector addition, subtraction, and scalar multiplication can be expressed as

$$\vec{p} + \vec{q} = (p_x \hat{i} + p_y \hat{j} + p_z \hat{k}) + (q_x \hat{i} + q_y \hat{j} + q_z \hat{k})$$
$$= (p_x + q_x) \hat{i} + (p_y + q_y) \hat{j} + (p_z + q_z) \hat{k} \tag{2.1.10}$$
$$\vec{p} - \vec{q} = (p_x \hat{i} + p_y \hat{j} + p_z \hat{k}) - (q_x \hat{i} + q_y \hat{j} + q_z \hat{k})$$
$$= (p_x - q_x) \hat{i} + (p_y - q_y) \hat{j} + (p_z - q_z) \hat{k} \tag{2.1.11}$$
$$a * \vec{p} = a * (p_x \hat{i} + p_y \hat{j} + p_z \hat{k}) = a * p_x \hat{i} + a * p_y \hat{j} + a * p_z \hat{k} \tag{2.1.12}$$

Example 2.1.1 Algebraic operations on vectors, defined by their coordinates

Given two vectors $\vec{p}$ and $\vec{q}$, defined by their coordinates, compute their sum and difference, using unit-vector decomposition. Also, compute multiplication of vector $\vec{p}$ by the scalar $a = 2$.

Solution For two vectors

$$\vec{p} = \begin{bmatrix} -1.0 \\ 2.1 \\ 4.7 \end{bmatrix} \text{ and } \vec{q} = \begin{bmatrix} 3.2 \\ -1.1 \\ -3.2 \end{bmatrix}$$

the sum is computed by Eq. (2.1.10) as

$$\vec{p} + \vec{q} = (-1.0\hat{i} + 2.1\hat{j} + 4.7\hat{k}) + (3.2\hat{i} - 1.1\hat{j} - 3.2\hat{k})$$
$$= (-1.0 + 3.2)\hat{i} + (2.1 - 1.1)\hat{j} + (4.7 - 3.2)\hat{k}$$
$$= 2.2\hat{i} + 1.0\hat{j} + 1.5\hat{k}$$

while the difference is computed by Eq. (2.1.11) as

$$\vec{p} - \vec{q} = (-1.0\hat{i} + 2.1\hat{j} + 4.7\hat{k}) - (3.2\hat{i} - 1.1\hat{j} - 3.2\hat{k})$$
$$= (-1.0 - 3.2)\hat{i} + (2.1 - (-1.1))\hat{j} + (4.7 - (-3.2))\hat{k}$$
$$= -4.2\hat{i} + 3.2\hat{j} + 7.9\hat{k}$$

Multiplication of vector $\vec{p}$ by the scalar a is performed by Eq. (2.1.12) as

$$a * \vec{p} = 2.0 * (-1.0\hat{i} + 2.1\hat{j} + 4.7\hat{k})$$
$$= 2.0 * (-1.0)\hat{i} + 2.0 * 2.1\hat{j} + 2.0 * 4.7\hat{k}$$
$$= -2.0\hat{i} + 4.2\hat{j} + 9.4\hat{k}$$

2.1.4 Vector Multiplication Operations

Vector multiplication operations such as the *dot product* and *cross product*, as well as the *right-hand rule*, are illustrated in Figure 2.1.4.

The dot-product operation is often called the *scalar product*. It takes two vectors and generates a scalar quantity – a single number. The dot product $\vec{p} \cdot \vec{q}$ of two vectors $\vec{p}$ and $\vec{q}$ is defined as a product of their magnitudes with the cosine of the angle θ between them ($0 \leq \theta \leq \pi$):

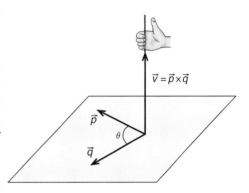

$$\vec{p} \cdot \vec{q} = |\vec{p}| \, |\vec{q}| \cos \theta \qquad (2.1.13)$$

Figure 2.1.4 The right-hand rule

Obviously, depending on the value of the $\cos \theta$ the dot product can be positive (a sharp angle between two vectors), negative (an obtuse angle between two vectors), and zero (the vectors are perpendicular).

The dot product is commutative:

$$\vec{p} \cdot \vec{q} = \vec{q} \cdot \vec{p} \qquad (2.1.14)$$

and distributive:

$$(\vec{p} + \vec{r}) \cdot \vec{q} = \vec{p} \cdot \vec{q} + \vec{r} \cdot \vec{q}$$
$$\vec{q} \cdot (\vec{p} + \vec{r}) = \vec{q} \cdot \vec{p} + \vec{q} \cdot \vec{r} \qquad (2.1.15)$$

With regard to scalar multiplication the following properties hold:

$$(a * \vec{p}) \cdot \vec{q} = a * (\vec{p} \cdot \vec{q})$$
$$\vec{p} \cdot (a * \vec{q}) = a * (\vec{p} \cdot \vec{q}) \qquad (2.1.16)$$

If vectors of the dot-product operation are defined by their unit-vector decompositions, i.e. they are defined in terms of coordinates, the dot product becomes:

$$\vec{p} \cdot \vec{q} = (p_x \hat{i} + p_y \hat{j} + p_z \hat{k}) \cdot (q_x \hat{i} + q_y \hat{j} + q_z \hat{k}) \qquad (2.1.17)$$

To simplify this expression, the dot products of unit vectors need to be derived. Using the dot-product definition, it can be shown that the dot product of the unit vector with itself is unity: $\hat{i} \cdot \hat{i} = |\hat{i}| \, |\hat{i}| \cos(0) = 1$. Similarly, $\hat{j} \cdot \hat{j} = 1$ and $\hat{k} \cdot \hat{k} = 1$. Since two distinct unit vectors (as defined collinear to the coordinate axes) are perpendicular to each other, their dot product is zero: $\hat{i} \cdot \hat{j} = |\hat{i}| \, |\hat{j}| \cos\left(\frac{\pi}{2}\right) = 0$. Similarly, $\hat{i} \cdot \hat{k} = 0$ and $\hat{j} \cdot \hat{k} = 0$. Thus, Eq. (2.1.17) yields:

$$\begin{aligned} \vec{p} \cdot \vec{q} &= (p_x \hat{i} + p_y \hat{j} + p_z \hat{k}) \cdot (q_x \hat{i} + q_y \hat{j} + q_z \hat{k}) \\ &= p_x q_x + p_y q_y + p_z q_z \end{aligned} \qquad (2.1.18)$$

The cross product also takes a pair of vectors, but instead of a scalar it generates a new vector. If some vector $\vec{v}$ represents the cross product of two vectors $\vec{p}$ and $\vec{q}$ with the angle θ between them ($0 \le \theta \le \pi$), i.e. $\vec{v} = \vec{p} \times \vec{q}$, then the direction of this vector $\vec{v}$ is perpendicular to the plane formed by $\vec{p}$ and $\vec{q}$ (as shown in Figure 2.1.4), and its magnitude is $|\vec{v}| = |\vec{p} \times \vec{q}| = |\vec{p}| \, |\vec{q}| \sin\theta$.

The direction of vector $\vec{v} = \vec{p} \times \vec{q}$ is determined using the so-called *right-hand rule*. It reads as follows: if the curl of the right-hand fingers represents a rotation from $\vec{p}$ to $\vec{q}$, then $\vec{v}$ points in the direction of the outstretched right thumb. This is illustrated in Figure 2.1.4.

The cross product is anti-commutative, since changing the order of the vectors in the cross product reverses the direction of the resulting vector by the right-hand rule:

$$\vec{p} \times \vec{q} = -\vec{q} \times \vec{p} \qquad (2.1.19)$$

The cross product is distributive:

$$\begin{aligned} (\vec{p} + \vec{r}) \times \vec{q} &= \vec{p} \times \vec{q} + \vec{r} \times \vec{q} \\ \vec{q} \times (\vec{p} + \vec{r}) &= \vec{q} \times \vec{p} + \vec{q} \times \vec{r} \end{aligned} \qquad (2.1.20)$$

With regard to scalar multiplication the following properties hold:

$$\begin{aligned} (a * \vec{p}) \times \vec{q} &= a * (\vec{p} \times \vec{q}) \\ \vec{p} \times (a * \vec{q}) &= a * (\vec{p} \times \vec{q}) \end{aligned} \qquad (2.1.21)$$

To derive an expression for the cross product of vectors defined by their unit-vector decompositions, cross products of unit-vector pairs need to be evaluated. Using the definition of the cross product, we obtain the following.

- The cross product of a unit vector with itself is zero:

$$\hat{i} \times \hat{i} = |\hat{i}| |\hat{i}| \sin(0) = 0 \quad (2.1.22)$$

Similarly, $\hat{j} \times \hat{j} = 0$ and $\hat{k} \times \hat{k} = 0$.
- The cross product of two distinct unit vectors is computed as follows:

$$\hat{i} \times \hat{j} = \hat{k} \quad (2.1.23)$$

- The magnitude of the cross product $\hat{i} \times \hat{j}$ is unity:

$$\hat{i} \times \hat{j} = |\hat{i}| |\hat{j}| \sin\left(\frac{\pi}{2}\right) = 1 \quad (2.1.24)$$

According to the right-hand rule, the direction of the cross product $\hat{i} \times \hat{j}$ is $+\hat{k}$. Thus, Eq. (2.1.24) holds true.

Similarly, $\hat{j} \times \hat{k} = \hat{i}$ and $\hat{k} \times \hat{i} = \hat{j}$. By the anti-commutative property of the cross product $\hat{j} \times \hat{i} = -\hat{k}$ and $\hat{i} \times \hat{k} = -\hat{j}$.

Considering the above conclusions, we obtain

$$\begin{aligned} \vec{p} \times \vec{q} &= (p_x \hat{i} + p_y \hat{j} + p_z \hat{k}) \times (q_x \hat{i} + q_y \hat{j} + q_z \hat{k}) \\ &= (p_y q_z - p_z q_y) \hat{i} + (p_z q_x - p_x q_z) \hat{j} + (p_x q_y - p_y q_x) \hat{k} \end{aligned} \quad (2.1.25)$$

Example 2.1.2 Computing the dot product and cross product of two vectors

Given two vectors $\vec{p}$ and $\vec{q}$, defined by their coordinates, compute their dot product and cross product, using unit-vector decomposition.

Solution

For two vectors $\vec{p} = \begin{bmatrix} -1.0 \\ 2.1 \\ 4.7 \end{bmatrix}$ and $\vec{q} = \begin{bmatrix} 3.2 \\ -1.1 \\ -3.2 \end{bmatrix}$ the dot product is computed by Eq. (2.1.18) as

$$\begin{aligned} \vec{p} \cdot \vec{q} &= (-1.0\,\hat{i} + 2.1\,\hat{j} + 4.7\,\hat{k}) \cdot (3.2\,\hat{i} - 1.1\,\hat{j} - 3.2\,\hat{k}) \\ &= -1.0 * 3.2 + 2.1 * (-1.1) + 4.7 * (-3.2) \\ &= -20.55 \end{aligned}$$

Their cross product is computed by Eq. (2.1.25) as

$$\begin{aligned} \vec{p} \times \vec{q} &= (-1.0\,\hat{i} + 2.1\,\hat{j} + 4.7\,\hat{k}) \times (3.2\,\hat{i} - 1.1\,\hat{j} - 3.2\,\hat{k}) \\ &= (2.1 * (-3.2) - 4.7 * (-1.1))\,\hat{i} + (4.7 * 3.2 - (-1.0) * (-3.2))\,\hat{j} \\ &\quad + ((-1.0) * (-1.1) - 2.1 * 3.2)\,\hat{k} \\ &= -1.55\,\hat{i} + 11.84\,\hat{j} - 5.62\,\hat{k} \end{aligned}$$

2.2 Matrix Algebra

Matrices are an important element of linear algebra, and are extremely useful in dealing with the systems of equations that are heavily coupled.

2.2.1 Definition

A matrix is a rectangular array of numbers, symbols, or expressions arranged in rows and columns. Matrices are often shown enclosed in square brackets and their elements are denoted by their indices that represent an element *address* in the matrix. This is illustrated in Figure 2.2.1.

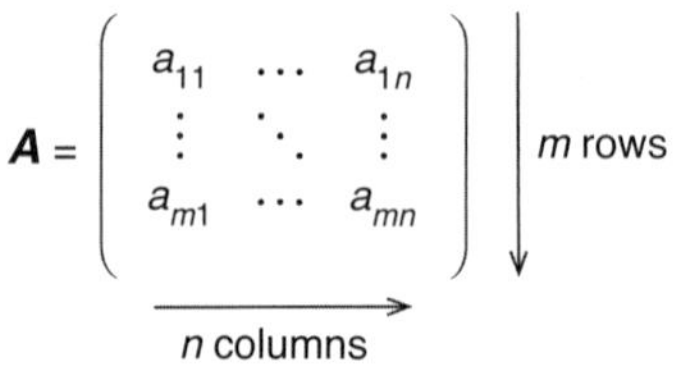

$$A = \begin{pmatrix} a_{11} & \cdots & a_{1n} \\ \vdots & \ddots & \vdots \\ a_{m1} & \cdots & a_{mn} \end{pmatrix} \quad m \text{ rows}$$

n columns

Figure 2.2.1 A matrix

Matrix A, depicted in Figure 2.2.1, has m rows and n columns. It is said to be of size $m \times n$. Obviously, the indices of the last element of the matrix identify the matrix size. When $m = n$, the matrix is square. Otherwise, it is rectangular. An element a_{ij} of matrix A is called the (i, j) entry and it is located at the intersection of the i-th row and the j-th column. In the matrix notation, (i, j) is known as the *address* of the element a_{ij} in matrix A.

A vector is sometimes called an *instance of a matrix*; for example, a matrix of a size $m \times 1$ is a column vector, while a matrix of a size $1 \times n$ is a row vector.

The abbreviated matrix notation represents matrix A of size $m \times n$ as $A = [a_{ij}]_{m \times n}$ or simply as $A_{m \times n}$.

The element the indices of which are the same, i.e. a_{ii}, is called a *diagonal element*. A square matrix, for which the following is true: $a_{ij} = C_i \, \forall \, i = j$ and $a_{ij} = 0 \, \forall \, i \neq j$, where C_i is a number, symbol, or expression, is called a *diagonal matrix*, and is sometimes denoted as $A = Diag[a_{ij}]_{m \times n}$.

A square diagonal matrix for which $a_{ij} = 1 \, \forall \, i = j$ and $a_{ij} = 0 \, \forall \, i \neq j$, is called an *identity matrix*, and is denoted $I_{n \times n}$. In some literature identity matrix may be also called *unit matrix*.

A *zero matrix* or *null matrix* is an $m \times n$ matrix of all zeros.

2.2.2 Single-Matrix Operations

Matrices can undergo several important operations. A single matrix can be transposed, inverted, and multiplied by a scalar. It can also experience the *elementary row and column operations* such as:

(a) interchanging two rows or columns
(b) multiplying any row or column by a nonzero constant
(c) adding a row or column to another
(d) multiplying the *i*-th row by a nonzero constant, adding the result to the *j*-th row, and replacing the *j*-th row with the result of addition

Scalar multiplication with regard to a matrix is the operation of multiplying that matrix by some constant γ. It is done by multiplying each element of the original matrix by the said constant γ:

$$\gamma * \begin{pmatrix} a_{11} & \cdots & a_{1n} \\ \vdots & \ddots & \vdots \\ a_{m1} & \cdots & a_{mn} \end{pmatrix} = \begin{pmatrix} \gamma * a_{11} & \cdots & \gamma * a_{1n} \\ \vdots & \gamma * a_{ij} & \vdots \\ \gamma * a_{m1} & \cdots & \gamma * a_{mn} \end{pmatrix} \qquad (2.2.1)$$

The matrix *transpose* is denoted A^T. It replaces all elements a_{ij} with a_{ji}. Hence, rows become columns, columns become rows, changing the size of the matrix from $m \times n$ to $n \times m$: $[A_{m \times n}]^T = A_{n \times m}$. The matrix is called *symmetric* if $A^T = A$.

The matrix inverse exists only for square matrices. The inverse of matrix $A_{n \times n}$ is a matrix $[A_{n \times n}]^{-1}$ such that $A \cdot A^{-1} = I$, i.e. the product of a square matrix and its inverse is an identity matrix.

Example 2.2.1 Multiplication of a matrix by a scalar, and its transpose

Given the matrix $A = \begin{pmatrix} -1.0 & 2.1 & 4.7 \\ 3.2 & -1.1 & -3.2 \end{pmatrix}$, multiply it by $\gamma = 1.2$ and find the transpose of the resulting matrix.

Solution
Multiplication of the given matrix by a scalar is done by Eq. (2.2.1) as

$$\gamma * A = \begin{pmatrix} 1.2 * (-1.0) & 1.2 * 2.1 & 1.2 * 4.7 \\ 1.2 * 3.2 & 1.2 * (-1.1) & 1.2 * (-3.2) \end{pmatrix} = \begin{pmatrix} -1.2 & 2.52 & 5.64 \\ 3.84 & -1.32 & -3.84 \end{pmatrix}$$

Transposing the resulting matrix yields $(\gamma * A)^T = \begin{pmatrix} -1.2 & 3.84 \\ 2.52 & -1.32 \\ 5.64 & -3.84 \end{pmatrix}$

2.2.3 Matrix Determinants

Computation of the matrix inverse involves a concept of matrix determinants. A determinant is a mathematical object, defined only for square matrices. It is denoted as $\text{Det}(A)$ or $|A|$. For a 2×2 matrix, the determinant is easily computable as

$$\mathrm{Det}\begin{pmatrix} a_{11} & a_{12} \\ a_{21} & a_{22} \end{pmatrix} \equiv \begin{vmatrix} a_{11} & a_{12} \\ a_{21} & a_{22} \end{vmatrix} \equiv a_{11}a_{22} - a_{12}a_{21} \qquad (2.2.2)$$

For larger matrices, computation of the determinant involves expansion by *minors*. M_{ij} denotes a minor of the square matrix $A_{n\times n} = [a_{ij}]_{n\times n}$. This minor is a matrix of the size $(n-1) \times (n-1)$ formed by eliminating row i and column j from the original matrix A. The expansion of the determinant by minors for an arbitrary square matrix $A_{n\times n}$ is as follows:

$$\mathrm{Det}(A) = |A| = \mathrm{Det}\begin{pmatrix} a_{11} & a_{12} & \cdots & a_{1n} \\ a_{21} & a_{22} & \cdots & a_{2n} \\ \vdots & \vdots & \ddots & \vdots \\ a_{n1} & a_{n2} & \cdots & a_{nn} \end{pmatrix}$$

$$= a_{11}\begin{vmatrix} a_{22} & a_{23} & \cdots & a_{2n} \\ a_{32} & a_{33} & \cdots & a_{3n} \\ \vdots & \vdots & \ddots & \vdots \\ a_{n2} & a_{n3} & \cdots & a_{nn} \end{vmatrix} - a_{12}\begin{vmatrix} a_{21} & a_{23} & \cdots & a_{2n} \\ a_{31} & a_{33} & \cdots & a_{3n} \\ \vdots & \vdots & \ddots & \vdots \\ a_{n1} & a_{n3} & \cdots & a_{nn} \end{vmatrix} + \cdots \pm a_{1n}\begin{vmatrix} a_{21} & a_{23} & \cdots & a_{2(n-1)} \\ a_{31} & a_{33} & \cdots & a_{3(n-1)} \\ \vdots & \vdots & \ddots & \vdots \\ a_{n1} & a_{n3} & \cdots & a_{n(n-1)} \end{vmatrix}$$

(2.2.3)

Thus, the determinant of an arbitrary matrix $A_{n\times n}$ is

$$|A_{n\times n}| = \sum_{i=1}^{n} a_{ij}C_{ij} \qquad (2.2.4)$$

where $C_{ij} \equiv (-1)^{i+j}M_{ij}$. C_{ij} is called the *cofactor* of a_{ij}.

The determinant is additive:

$$\left|-A_{n\times n}\right| = (-1)^n \left|A_{n\times n}\right| \qquad (2.2.5)$$

and distributive:

$$\left|A_{n\times n} \cdot B_{n\times n}\right| = \left|A_{n\times n}\right|\left|B_{n\times n}\right| \qquad (2.2.6)$$

With regard to scalar multiplication, the determinant is

$$\left|a * A_{n\times n}\right| = a^n \left|A_{n\times n}\right| \qquad (2.2.7)$$

The determinant of the matrix inverse is

$$\left|(A_{n\times n})^{-1}\right| = \frac{1}{\left|A_{n\times n}\right|} \qquad (2.2.8)$$

The determinant of a transpose equals the determinant of the original matrix:

$$\left|(A_{n\times n})^T\right| = \left|A_{n\times n}\right| \qquad (2.2.9)$$

The determinant is invariant under elementary row and column operations. The other useful properties of determinant are:

(a) switching two rows or columns changes the sign of the determinant
(b) scalars can be factored out from rows and columns
(c) scalar multiplication of a row by a constant multiplies the determinant by that constant
(d) determinant of a matrix with a row or column of zeros equals zero
(e) determinant of a matrix with two equal rows or two equal columns is zero

It can easily be shown that the determinant of a triangular or diagonal matrix equals a product of the elements on its main diagonal. For a 3 × 3 upper triangular matrix, the determinant is computed as follows:

Example 2.2.2 Finding the matrix determinant

Given matrices

$$A = \begin{pmatrix} -1.0 & 2.1 \\ 3.2 & -1.1 \end{pmatrix}$$

and

$$B = \begin{pmatrix} -1.0 & 0.6 & 2.1 \\ 3.2 & 4.7 & -1.1 \\ 2.5 & -1.5 & 1.0 \end{pmatrix}$$

find their determinants.

Solution
The determinant of matrix A can be found by Eq. (2.2.2) as

$$\mathrm{Det}\begin{pmatrix} -1.0 & 2.1 \\ 3.2 & -1.1 \end{pmatrix} = (-1.0)*(-1.1) - 2.1*3.2 = -5.62$$

while to compute the determinant of matrix B we need to use Eq. (2.2.3) that yields

$$\mathrm{Det}\begin{pmatrix} -1.0 & 0.6 & 2.1 \\ 3.2 & 4.7 & -1.1 \\ 2.5 & -1.5 & 1.0 \end{pmatrix}$$

$$= -1.0 * \begin{vmatrix} 4.7 & -1.1 \\ -1.5 & 1.0 \end{vmatrix} - 0.6 * \begin{vmatrix} 3.2 & -1.1 \\ 2.5 & 1.0 \end{vmatrix} + 2.1 * \begin{vmatrix} 3.2 & 4.7 \\ 2.5 & -1.5 \end{vmatrix} = -41.375$$

$$\text{Det}\begin{pmatrix} a_{11} & a_{12} & a_{13} \\ 0 & a_{22} & a_{23} \\ 0 & 0 & a_{33} \end{pmatrix} = a_{11} * \begin{vmatrix} a_{22} & a_{23} \\ 0 & a_{33} \end{vmatrix} - a_{12} * \begin{vmatrix} 0 & a_{23} \\ 0 & a_{33} \end{vmatrix} + a_{13} * \begin{vmatrix} 0 & a_{22} \\ 0 & 0 \end{vmatrix} = a_{11} a_{22} a_{33}$$

For a diagonal matrix, computation of the determinant is even more straightforward:

$$\text{Det}\begin{pmatrix} a_{11} & 0 & 0 \\ 0 & a_{22} & 0 \\ 0 & 0 & a_{33} \end{pmatrix} = a_{11} * \begin{vmatrix} a_{22} & 0 \\ 0 & a_{33} \end{vmatrix} - 0 * \begin{vmatrix} 0 & 0 \\ 0 & a_{33} \end{vmatrix} + 0 * \begin{vmatrix} 0 & a_{22} \\ 0 & 0 \end{vmatrix} = a_{11} a_{22} a_{33}$$

The determinant is very useful for solving linear systems of equations. Cramer's rule derives a solution of a linear system of equations as quotients of determinants. While impractical for the larger systems of equations, it is useful for systems of two and three equations. It also is of interest in differential equations and some engineering applications.

Cramer's rule states the following. Consider a linear system of n equations in the same number of unknowns:

$$\begin{bmatrix} a_{11}x_1 + a_{12}x_2 + \ldots + a_{1n}x_n = b_1 \\ a_{21}x_1 + a_{22}x_2 + \ldots + a_{2n}x_n = b_2 \\ \vdots \\ a_{n1}x_1 + a_{n2}x_2 + \ldots + a_{nn}x_n = b_n \end{bmatrix} \quad (2.2.10)$$

or in matrix form

$$A \cdot \vec{x} = \vec{b}$$

where A is a matrix of coefficients, $A = \begin{pmatrix} a_{11} & \ldots & a_{1n} \\ \vdots & \ddots & \vdots \\ a_{n1} & \ldots & a_{nn} \end{pmatrix}$

$\vec{x}$ is a vector of unknowns, $\vec{x} = [x_1 \, x_2 \ldots x_n]^T$

and $\vec{b}$ is a vector of the results, $\vec{b} = [b_1 \, b_2 \ldots b_n]^T$

Then, if the determinant of matrix A is nonzero:

$$D = \text{Det}(A) \neq 0 \quad (2.2.11)$$

the given linear system of equations has exactly one solution, derived as follows (this expression is typically referred to as Cramer's rule):

$$x_1 = \frac{D_1}{D}, \; x_2 = \frac{D_2}{D}, \ldots x_n = \frac{D_n}{D} \quad (2.2.12)$$

where D_k is obtained from D by replacing the k-th column in D with the vector $\vec{b}$. If the system is homogeneous, i.e. $\vec{b} = 0$, and $D \neq 0$, then this system has only the trivial solution $x_1 = x_2 = \ldots = x_n = 0$. If the homogeneous system has a zero coefficient determinant ($D = 0$), then it has also nontrivial solutions.

Example 2.2.3 Solving a linear system of equations using Cramer's rule

Consider a linear system defined by the matrix equation $A \cdot \vec{x} = \vec{b}$, where matrix $A = \begin{pmatrix} -1.0 & 2.1 \\ 3.2 & -1.1 \end{pmatrix}$, $\vec{x} = [x_1 \; x_2]^T$, and $\vec{b} = [3.2 \; 2.1]^T$.

Use Cramer's rule (Eq. (2.2.11)) to find out whether a unique solution exists.

Solution

$$D = \text{Det}(A) = -1.0 * (-1.1) - 2.1 * 3.2 = -5.62 \neq 0$$

Since the determinant of coefficients matrix is nonzero, the given system has exactly one solution, derived with Cramer's rule:

$$x_1 = \frac{D_1}{D} = \frac{\begin{vmatrix} 3.2 & 2.1 \\ 2.1 & -1.1 \end{vmatrix}}{-5.62} = \frac{-7.93}{-5.62} = 1.41$$

$$x_2 = \frac{D_2}{D} = \frac{\begin{vmatrix} -1.0 & 3.2 \\ 3.2 & 2.1 \end{vmatrix}}{-5.62} = \frac{-12.34}{-5.62} = 2.196$$

This solution can easily be verified by writing the given system in the expanded form:

$$\begin{cases} -1.0x_1 + 2.1x_2 = 3.2 \\ 3.2x_1 - 1.1x_2 = 2.1 \end{cases}$$

deriving the expression for x_1 from the first equation, plugging it into the second equation, deriving the solution for x_2 and subsequently for x_1:

$$x_1 = 2.1x_2 - 3.2$$
$$3.2(2.1x_2 - 3.2) - 1.1x_2 = 2.1 \Rightarrow$$
$$x_2 = \frac{2.1 + 3.2 * 3.2}{3.2 * 2.1 - 1.1} = 2.196 \Rightarrow x_1 = 2.1 * 2.196 - 3.2 = 1.41$$

Proof that a homogeneous system

$$\begin{cases} -1.0x_1 + 2.1x_2 = 0 \\ 3.2x_1 - 1.1x_2 = 0 \end{cases}$$

has only a trivial solution is straightforward:

This homogeneous system, $\begin{pmatrix} 1.0 & 2.1 \\ 2.0 & 4.2 \end{pmatrix} \cdot \begin{pmatrix} x_1 \\ x_2 \end{pmatrix} = \begin{pmatrix} 0 \\ 0 \end{pmatrix}$, has multiple nontrivial solutions in addition to the trivial one:

$$D = \text{Det}(A) = 1.0 * 4.2 - 2.1 * 2.0 = 0$$

As seen from the coefficient matrix, the second equation of the system is obtained from the first one by multiplying it by 2. Hence, this linear system can be described by a single equation

$1.0x_1 + 2.1x_2 = 0$, which has an infinite number of solutions written as $x_1 = -2.1x_2$ with x_2 being an independent variable, or as $x_2 = -\frac{1}{2.1}x_1$ with x_1 being an independent variable.

2.2.4 Matrix Inverse

The application of a concept of determinants to the computation of a matrix inverse allows formulating the following condition of the existence of the inverse. An arbitrary square matrix $A_{n \times n}$ has an inverse if and only if its determinant is nonzero: $|A_{n \times n}| \neq 0$. If this holds true, the matrix A is called *nonsingular* or *invertible*.

For a 2×2 nonsingular matrix

$$A = \begin{bmatrix} a_{11} & a_{12} \\ a_{21} & a_{22} \end{bmatrix}$$

an inverse is easily computed:

$$A^{-1} = \frac{1}{|A|} \begin{pmatrix} a_{22} & -a_{12} \\ -a_{21} & a_{11} \end{pmatrix} = \frac{1}{a_{11}a_{22} - a_{12}a_{21}} \begin{pmatrix} a_{22} & -a_{12} \\ -a_{21} & a_{11} \end{pmatrix} \quad (2.2.13)$$

For a 3×3 nonsingular matrix

$$A = \begin{bmatrix} a_{11} & a_{12} & a_{13} \\ a_{21} & a_{22} & a_{23} \\ a_{31} & a_{32} & a_{33} \end{bmatrix}$$

The computation of an inverse is also straightforward and involves computing determinants of several 2×2 matrices – minors of matrix A
:

$$A^{-1} = \frac{1}{|A|} \begin{pmatrix} \begin{vmatrix} a_{22} & a_{23} \\ a_{32} & a_{33} \end{vmatrix} & \begin{vmatrix} a_{13} & a_{12} \\ a_{33} & a_{32} \end{vmatrix} & \begin{vmatrix} a_{12} & a_{13} \\ a_{22} & a_{23} \end{vmatrix} \\ \begin{vmatrix} a_{23} & a_{21} \\ a_{33} & a_{31} \end{vmatrix} & \begin{vmatrix} a_{11} & a_{13} \\ a_{31} & a_{33} \end{vmatrix} & \begin{vmatrix} a_{13} & a_{11} \\ a_{23} & a_{21} \end{vmatrix} \\ \begin{vmatrix} a_{21} & a_{22} \\ a_{31} & a_{32} \end{vmatrix} & \begin{vmatrix} a_{12} & a_{11} \\ a_{32} & a_{31} \end{vmatrix} & \begin{vmatrix} a_{11} & a_{12} \\ a_{21} & a_{22} \end{vmatrix} \end{pmatrix} \quad (2.2.14)$$

In general, an inverse of an arbitrary nonsingular matrix $A_{n \times n}$ is computed using the *adjoint* of matrix A, which is defined as follows:

$$\mathrm{adj}(A) = \begin{bmatrix} (-1)^{1+1}M_{11} & (-1)^{2+1}M_{21} & \cdots & (-1)^{n+1}M_{n1} \\ (-1)^{1+2}M_{12} & (-1)^{2+2}M_{22} & \cdots & (-1)^{n+2}M_{n2} \\ \vdots & \vdots & \ddots & \vdots \\ (-1)^{1+n}M_{1n} & (-1)^{2+n}M_{2n} & \cdots & (-1)^{n+n}M_{nn} \end{bmatrix} = [c_{ij}]_{n \times n} \qquad (2.2.15)$$

where M_{ij} denotes a minor of the square matrix $A_{n \times n} = [a_{ij}]_{n \times n}$, and every element $c_{ij} = (-1)^{i+j} M_{ij}$ of the adjoint matrix is the *cofactor* of a_{ij}. It shall be noted that c_{ij} occupies the (j, i) position in the adjoint matrix.

In the adjoint-matrix formulation the inverse of matrix $A_{n \times n}$ is

$$A^{-1} = \frac{1}{|A|} \mathrm{adj}(A) \qquad (2.2.16)$$

Inverting an arbitrary nonsingular square matrix $A_{n \times n}$ involves a fair amount of computation and is usually done with software such as MATLAB, Mathematica, or Maple.

Example 2.2.4 Computing the matrix inverse

The inverse of matrix

$$A = \begin{pmatrix} -1.0 & 2.1 \\ 3.2 & -1.1 \end{pmatrix}$$

is computed by Eq. (2.2.13) as

$$A^{-1} = \frac{1}{|A|} \begin{pmatrix} -1.1 & -2.1 \\ -3.2 & -1.0 \end{pmatrix} = \begin{pmatrix} 0.196 & 0.374 \\ 0.569 & 0.178 \end{pmatrix}$$

The inverse of 3×3 matrix

$$A = \begin{pmatrix} -1 & 1 & 2 \\ 3 & -1 & 1 \\ -1 & 3 & 4 \end{pmatrix}$$

is computed by Eq. (2.2.14) as

$$|A| = (-1) * ((-1) * 4 - 1 * 3) - 1 * (3 * 4 - 1 * (-1)) + 2 * (3 * 3 - (-1) * (-1)) = 10$$

$$A^{-1} = \frac{1}{|A|} \begin{pmatrix} \begin{vmatrix} -1 & 1 \\ 3 & 4 \end{vmatrix} & \begin{vmatrix} 2 & 1 \\ 4 & 3 \end{vmatrix} & \begin{vmatrix} 1 & 2 \\ -1 & 1 \end{vmatrix} \\ \begin{vmatrix} 1 & 3 \\ 4 & -1 \end{vmatrix} & \begin{vmatrix} -1 & 2 \\ -1 & 4 \end{vmatrix} & \begin{vmatrix} 2 & -1 \\ 1 & 3 \end{vmatrix} \\ \begin{vmatrix} 3 & -1 \\ -1 & 3 \end{vmatrix} & \begin{vmatrix} 1 & -1 \\ 3 & -1 \end{vmatrix} & \begin{vmatrix} -1 & 1 \\ 3 & -1 \end{vmatrix} \end{pmatrix}$$

$$= \frac{1}{10} \begin{pmatrix} -7 & 2 & 3 \\ -13 & -2 & 7 \\ 8 & 2 & -2 \end{pmatrix} = \begin{pmatrix} -0.7 & 0.2 & 0.3 \\ -1.3 & -0.2 & 0.7 \\ 0.8 & 0.2 & -0.2 \end{pmatrix}$$

2.2.5 Operations on Two Matrices

Two matrices can be added, subtracted, or multiplied together to yield a new matrix. Two matrices are equal if they have the same size and the same elements in the respective locations.

The sum of two matrices $C = A + B$ is defined by adding elements with the same indices: $c_{ij} \equiv a_{ij} + b_{ij}$ over all i, j. For addition to be possible, matrices A and B must be of the same size. Subtraction is a combination of addition and scalar multiplication as follows:

$$A - B = A + (-B) \qquad (2.2.17)$$

Matrix addition is commutative:

$$A + B = B + A \qquad (2.2.18)$$

and associative:

$$(A + B) + C = A + (B + C) \qquad (2.2.19)$$

The transpose of a matrix sum equals the sum of transposes of member matrices:

$$(A + B)^T = A^T + B^T \qquad (2.2.20)$$

Multiplication of two matrices A and B is possible if and only if their dimensions correlate as $(n \times m) \cdot (m \times p) = (n \times p)$, where $n \times m$ are the dimensions of matrix A, $m \times p$ are the dimensions of matrix B, and $n \times p$ are the dimensions of their product $C = A \cdot B$. The product of two matrices $C_{n \times p} = A_{n \times m} \cdot B_{m \times p}$ is defined as

$$c_{ij} = a_{i1}b_{1j} + a_{i2}b_{2j} + a_{i3}b_{3j} + \ldots + a_{im}b_{mj} \qquad (2.2.21)$$

An (i, j) element of the product matrix C is the dot (inner) product of the i-th row of matrix A and the j-th column of matrix B, i.e. a summation of elements of the i-th row of matrix A multiplied one-by-one by their respective elements of the j-th column of matrix B.

The matrix multiplication product can be written explicitly as follows:

$$\begin{pmatrix} a_{11} & a_{12} & \ldots & a_{1m} \\ a_{21} & a_{22} & \ldots & a_{2m} \\ \vdots & \vdots & \ddots & \vdots \\ a_{n1} & a_{n2} & \ldots & a_{nm} \end{pmatrix} \cdot \begin{pmatrix} b_{11} & b_{12} & \ldots & b_{1p} \\ b_{21} & b_{22} & \ldots & b_{2p} \\ \vdots & \vdots & \ddots & \vdots \\ b_{n1} & b_{n2} & \ldots & b_{np} \end{pmatrix} = \begin{pmatrix} c_{11} & c_{12} & \ldots & c_{1p} \\ c_{21} & c_{22} & \ldots & c_{2p} \\ \vdots & \vdots & \ddots & \vdots \\ c_{n1} & c_{n2} & \ldots & c_{np} \end{pmatrix}$$

where

$$c_{11} = a_{11}b_{11} + a_{12}b_{21} + a_{13}b_{31} + \ldots a_{1m}b_{m1}$$
$$c_{12} = a_{11}b_{12} + a_{12}b_{22} + a_{13}b_{32} + \ldots a_{1m}b_{m2}$$
$$\vdots$$
$$c_{1p} = a_{11}b_{1p} + a_{12}b_{2p} + a_{13}b_{3p} + \ldots a_{1m}b_{mp}$$
$$c_{21} = a_{21}b_{11} + a_{22}b_{21} + a_{23}b_{31} + \ldots a_{2m}b_{m1}$$
$$c_{22} = a_{21}b_{12} + a_{22}b_{22} + a_{23}b_{32} + \ldots a_{2m}b_{m2}$$
$$\vdots$$
$$c_{2p} = a_{21}b_{1p} + a_{22}b_{2p} + a_{23}b_{3p} + \ldots a_{2m}b_{mp}$$
$$\vdots$$
$$c_{n1} = a_{n1}b_{11} + a_{n2}b_{21} + a_{n3}b_{31} + \ldots a_{nm}b_{m1}$$
$$c_{n2} = a_{n1}b_{12} + a_{n2}b_{22} + a_{n3}b_{32} + \ldots a_{nm}b_{m2}$$
$$\vdots$$
$$c_{np} = a_{n1}b_{1p} + a_{n2}b_{2p} + a_{n3}b_{3p} + \ldots a_{nm}b_{mp}$$

Matrix multiplication is associative and distributive, but not, in general, commutative:

$$\begin{aligned}(A \cdot B) \cdot C &= A \cdot (B \cdot C) \\ A \cdot (B + C) &= A \cdot B + A \cdot C \\ (A + B) \cdot C &= A \cdot C + B \cdot C\end{aligned} \qquad (2.2.22)$$

Unless A and B are of the same dimension and diagonal, $A \cdot B \neq B \cdot A$.

With regard to the matrix transpose and inverse, the matrix product has the following properties:

$$\begin{aligned}(A \cdot B)^T &= B^T \cdot A^T \\ (A \cdot B)^{-1} &= B^{-1} \cdot A^{-1}\end{aligned} \qquad (2.2.23)$$

The determinant of a matrix product is a product of determinants of its multipliers:

$$|A \cdot B| = |B| \cdot |A| \qquad (2.2.24)$$

Example 2.2.5 Computing the product of two matrices

The product of matrices $A = \begin{bmatrix} -1.0 & 2.1 & 1.0 \\ 3.2 & -1.1 & 2.2 \end{bmatrix}$ and $B = \begin{bmatrix} 2.1 & 1.0 \\ -1.1 & 2.2 \\ 0.5 & -3.0 \end{bmatrix}$ is computed by Eq. (2.2.21) as follows:

$A \cdot B = C$ where every element of the product matrix C is
$c_{11} = (-1.0) * 2.1 + 2.1 * (-1.1) + 1.0 * 0.5 = -3.91$
$c_{12} = (-1.0) * 1.0 + 2.1 * 2.2 + 1.0 * (-3.0) = 0.62$
$c_{21} = 3.2 * 2.1 + (-1.1) * (-1.1) + 2.2 * 0.5 = 9.03$
$c_{22} = 3.2 * 1.0 + (-1.1) * 2.2 + 2.2 * (-3.0) = -5.82$

Thus, the product matrix is $C = \begin{bmatrix} -3.91 & 0.62 \\ 9.03 & -5.82 \end{bmatrix}$

2.3 Complex Numbers

2.3.1 Definition

A complex number z is given as

$$z = x + jy \quad (2.3.1)$$

where x and y are real numbers, and j is the imaginary unit of the form $j = \sqrt{-1}$. The x and y are the real and imaginary parts of z; namely,

$$\mathrm{Re}(z) = x, \ \mathrm{Im}(z) = y \quad (2.3.2)$$

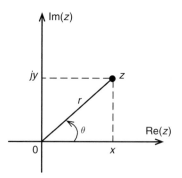

Figure 2.3.1 Definition of a complex number

The complex number can be illustrated in a two-dimensional complex plane, as shown in Figure 2.3.1, where the horizontal and vertical axes are the real and imaginary ones, respectively. Equation (2.3.1) is an expression of complex numbers in Cartesian coordinates. From the figure, z can also be expressed in polar coordinates,

$$z = r(\cos\theta + j\sin\theta) \quad (2.3.3)$$

where

$$x = r\cos\theta, \ y = r\sin\theta \quad (2.3.4)$$

have been used. Furthermore, by Euler's formula,

$$e^{j\theta} = \cos\theta + j\sin\theta \quad (2.3.5)$$

z can be written in the form

$$z = re^{j\theta} \quad (2.3.6)$$

where θ is in radians. The r and θ in the Eq. (2.3.6) are the magnitude (modulus) and phase angle (argument) of a complex number z, which by trigonometry formulas are given by

$$\begin{aligned} |z| &= r = \sqrt{x^2 + y^2} \\ \angle z &= \theta = \tan^{-1}\left(\frac{y}{x}\right) \end{aligned} \quad (2.3.7)$$

Example 2.3.1 Finding the magnitude and phase angle of a complex number

Given a complex number $z = 3 + j4$, its magnitude and phase angle are computed by Eq. (2.3.7) as follows:

$$\begin{aligned} |z| &= \sqrt{3^2 + 4^2} = 5 \\ \angle z &= \tan^{-1}\left(\frac{4}{3}\right) = 53.13° = 0.9273 \ \mathrm{rad} \end{aligned}$$

The complex number can also be written as
$$z = 5\Big(\cos(53.13°) + j\sin(53.13°)\Big) = 5e^{j0.9273}$$

2.3.2 Phase Angles by Quadrants of a Complex Plane

For complex numbers $s_1 = 3 - j4$, $s_2 = -3 + j4$, their phase angles, by Eq. (2.3.7), are given by

$$\theta_1 = \tan^{-1}\left(\frac{-4}{3}\right), \quad \theta_2 = \tan^{-1}\left(\frac{4}{-3}\right)$$

Since $-4/3 = 4/(-3)$, it is a common mistake to conclude that $\theta_1 = \theta_2$. Because the arctangent function is multivalued, these phase angles are not equal: $\theta_1 \neq \theta_2$. Hence, we need to know which quadrants s_1 and s_2 fall in. According to Figure 2.3.2, s_1 is in the second quadrant of the complex plane, while s_2 is in the fourth quadrant.

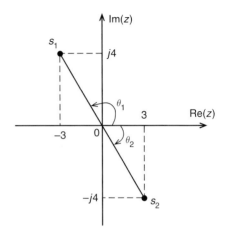

Figure 2.3.2 The phase angle

It follows that
$$\theta_1 = 180° - \tan^{-1}\left(\frac{4}{3}\right) = 126.87°$$
$$\theta_2 = -\tan^{-1}\left(\frac{4}{3}\right) = -53.13°$$

By the same token, one may estimate the following phase angles
$$\angle(-3 - j4) = 180° + \tan^{-1}\left(\frac{4}{3}\right) = 233.13°$$
$$\angle(3) = 0°, \quad \angle(-3) = 180°$$
$$\angle(j4) = 90°, \quad \angle(-j4) = 270°$$

2.3.3 Complex Conjugates

For a complex number $z = x + jy$, its *complex conjugate* is

$$\bar{z} = x - jy \qquad (2.3.8)$$

where the over-bar stands for complex conjugation. The following properties of the complex conjugate can be easily derived:

$$\begin{aligned}
|\bar{z}| &= |z| = \sqrt{x^2 + y^2} \\
\angle \bar{z} &= -\angle z \\
z\bar{z} &= |z|^2 = x^2 + y^2 \\
z + \bar{z} &= 2x = 2\,\mathrm{Re}(z) \\
z - \bar{z} &= 2j\,\mathrm{Im}(z)
\end{aligned} \qquad (2.3.9)$$

2.3.4 Algebraic Manipulations

Complex numbers, just like the real ones, can undergo algebraic manipulations such as addition/subtraction and multiplication/division:

$$\begin{aligned}
(a + jb) \pm (c + jd) &= (a \pm c) + j(b \pm d) \\
(a + jb) * (c + jd) &= (ac - bd) + j(bc + ad) \\
\frac{a + jb}{c + jd} &= \frac{1}{c^2 + d^2}\{(ac + bd) + j(bc - ad)\}
\end{aligned} \qquad (2.3.10)$$

Here in the third line of Eq. (2.3.10), the numerator and denominator of the given fraction were multiplied by the complex conjugate of the denominator, $c - jd$, and the relation $|\bar{z}| = |z| = \sqrt{x^2 + y^2}$ has been used.

Example 2.3.2 Finding real and imaginary parts of a complex number

The complex number $z = \dfrac{3 + j4}{4 - j7}$ can be represented as a sum of its real and imaginary parts:

$$z = \frac{3 + j4}{4 - j7} = \frac{3 + j4}{4 - j7} \cdot \frac{4 + j7}{4 + j7} = \frac{12 - 28 + j(16 + 21)}{4^2 + 7^2} = -\frac{16}{65} + j\frac{37}{65}$$

2.3.5 Trigonometric Manipulations

Since complex numbers can be represented in terms of their magnitudes and phase angles, they can undergo trigonometric manipulations.

For two complex numbers $z_1 = r_1 e^{j\theta_1}$ and $z_2 = r_2 e^{j\theta_2}$,

$$\begin{aligned}
z_1 z_2 &= r e^{j\theta}, \text{ with } r = r_1 r_2,\ \theta = \theta_1 + \theta_2 \\
\frac{z_1}{z_2} &= \rho e^{j\beta}, \text{ with } \rho = \frac{r_1}{r_2},\ \beta = \theta_1 - \theta_2
\end{aligned} \qquad (2.3.11)$$

Example 2.3.3 Multiplication of complex numbers using Euler's formula

Compute the following complex expression

$$z = \frac{3+j4}{4-j7} * \frac{-2-j9}{-10+j6}$$

Solution
Write

$$z = \frac{s_1}{s_3} * \frac{s_2}{s_4} = \rho e^{j\theta}$$

where by Eq. (2.3.6),

$$s_1 = 3+j4 = 5\,e^{j0.9273}, \quad s_2 = -2-j9 = 9.2195\,e^{j4.4937}$$

$$s_3 = 4-j7 = 8.0623\,e^{-j1.0517}, \quad s_4 = -10+j6 = 11.6619\,e^{j2.6012}$$

It follows from Eq. (2.3.11) that

$$\rho = \frac{5}{8.0623} * \frac{9.2195}{11.6619} = 0.490$$

$$\theta = 0.9273 + 4.4937 - (-1.0517) - 2.6012 = 3.399 \text{ rad} = 194.7°$$

Thus,

$$z = 0.490 e^{j3.399} = 0.490\Big(\cos(194.7°) + j\sin(194.7°)\Big) = -0.474 - j\,0.124$$

2.4 The Laplace Transform

The Laplace transform is a mathematical tool that is widely used for the solution of linear ordinary differential equations (sometimes abbreviated to "ODEs") (in particular, linear time-invariant ordinary differential equations), and for modeling and analysis of dynamic systems. The Laplace transform is an integral transform, the utility of which lies in converting linear ordinary differential equations into algebraic equations. It is particularly useful for handling nonhomogeneous differential equations.

Applications of the Laplace transform are numerous. One of the most common is the transfer function formulation, which will be discussed in Chapters 3 to 5. Applications of the Laplace transform to modeling of multi-component dynamic systems, the derivation and analysis of the system response, and the design and analysis of feedback control systems will be discussed in Chapters 6 to 8.

2.4.1 Definition

For a real-valued, piecewise-continuous function $f(t)$ specified for $t \geq 0$, its Laplace transform, denoted by $F(s)$, is defined by

$$F(s) \equiv \mathcal{L}\{f(t)\} = \lim_{T \to \infty} \int_{0^-}^{T} [f(t)] \, e^{-st} dt = \int_{0^-}^{\infty} f(t) \, e^{-st} dt \qquad (2.4.1)$$

where $\mathcal{L}$ is the Laplace transform operator, and s is the complex Laplace transform parameter whose real part is greater than some fixed number σ_0. The operational symbol $\mathcal{L}$ indicates that the expression it prefixes is to be transformed by the Laplace Integral $\int_{0^-}^{\infty} e^{-st} dt$. The variable of integration t is arbitrary, and the transform is a function of the complex variable s. While the original function $f(t)$ operates in the time domain, its transform $F(s)$ operates in the Laplace domain.

It needs to be noted that the initial conditions for the Laplace transform are taken at the time $t = 0^-$ to distinguish them from the initial values taken at the time $t = 0^+$.

When the limit presented by Eq. (2.4.1) is not finite, the Laplace transform for the function $f(t)$ does not exist. Thus, in order for the Laplace transform of a function $f(t)$ to exist this limit must be finite, which implies that the Laplace integral must converge. The sufficient conditions for convergence are:

(a) the function $f(t)$ is piecewise continuous on every finite interval in the range $t > 0$ and
(b) the function $f(t)$ is of exponential order as t approaches infinity, i.e. it satisfies the relation $|f(t)| \leq M e^{\gamma t} \; \forall t \in [0, \infty)$ for some real positive constants M and γ.

Therefore, for the Laplace transform to exist, it is sufficient that

$$\int_{0^-}^{\infty} |f(t)| e^{-\sigma_0 t} dt < \infty$$

The reverse process of finding the time function $f(t)$ from its transform $F(s)$ can be expressed by the inverse Laplace transform:

$$f(t) \equiv \mathcal{L}^{-1}\{F(s)\} = \frac{1}{2\pi j} \int_{\sigma - j\infty}^{\sigma + j\infty} F(s) \, e^{st} ds \qquad (2.4.2)$$

where $\mathcal{L}^{-1}$ is the inverse Laplace transform operator, σ is called an *abscissa of convergence*, $\sigma > \sigma_0$ with σ_0 being a real and positive number, and $j = \sqrt{-1}$. The abscissa of convergence is defined as a real constant and is chosen to be larger than the real parts of all singular points of

$F(s)$. This way the path of integration moves to the right of all singular points and is parallel to the $j\omega$ axis.

The $f(t)$ and $F(s)$ defined in Eqs. (2.4.1) and (2.4.2) form a Laplace transform pair. Listed in Table 2.4.1 are the Laplace transform pairs for some basic functions.

In the table, the unit-impulse function (Dirac delta) is defined by

$$\delta(t) = \begin{cases} \infty, & t = 0 \\ 0, & t \neq 0 \end{cases} \quad (2.4.3)$$

which also satisfies the condition

$$\int_{-b}^{a} \delta(t)dt = 1 \quad (2.4.4)$$

for any positive a and b. The unit step function is defined by

$$u(t) = \begin{cases} 1, & t \geq 0 \\ 0, & t < 0 \end{cases} \quad (2.4.5)$$

In modeling of dynamic systems, the Dirac delta is used to describe impact inputs, while the step function represents constant inputs.

The Laplace transform is unique. If two continuous functions $f_1(t)$ and $f_2(t)$ have Laplace transforms $F_1(s)$ and $F_2(s)$, respectively, and these transforms are identical, $F_1(s) = F_2(s)$, then the functions $f_1(t)$ and $f_2(t)$ are necessarily identical. The unique quality of the Laplace transform is of great practical significance since it allows one to use it for solving differential equations by transforming them into algebraic equations in the Laplace domain, solving for variables in the Laplace domain, and then using the inverse Laplace transform to find the desired time-domain solution.

2.4.2 Poles and Zeros of the Complex Function $F(s)$

In many mathematical problems in science and engineering, the Laplace transformed function $F(s)$ has the following rational form

$$F(s) = \frac{N(s)}{D(s)} = \frac{b_m s^m + b_{m-1} s^{m-1} + \ldots + b_1 s + b_0}{a_n s^n + a_{n-1} s^{n-1} + \ldots + a_1 s + a_0}; \quad m \leq n, \quad a_n \neq 0 \quad (2.4.6)$$

where all the coefficients are real. The denominator $D(s)$ has n roots, p_i, $i = 1, 2, \ldots, n$, which are called the *poles* of $F(s)$. In other words, the poles' p_i values satisfy the algebraic equation

Table 2.4.1 Laplace transform pairs

$f(t)$	$F(s)$
Unit impulse function, $\delta(t)$	1
Unit step function, $u(t)$	$\dfrac{1}{s}$
Unit ramp function, t	$\dfrac{1}{s^2}$
t^n	$\dfrac{n!}{s^{n+1}}$
e^{-at}	$\dfrac{1}{s+a}$
$t^n e^{-at}$	$\dfrac{n!}{(s+a)^{n+1}}$
$\sin \omega t$	$\dfrac{\omega}{s^2+\omega^2}$
$\cos \omega t$	$\dfrac{s}{s^2+\omega^2}$
$t \sin \omega t$	$\dfrac{2\omega s}{(s^2+\omega^2)^2}$
$t \cos \omega t$	$\dfrac{s^2-\omega^2}{(s^2+\omega^2)^2}$
$e^{-at} \sin \omega t$	$\dfrac{\omega}{(s+a)^2+\omega^2}$
$e^{-at} \cos \omega t$	$\dfrac{s+a}{(s+a)^2+\omega^2}$
$\dfrac{1}{b-a}(e^{-at}-e^{-bt})$	$\dfrac{1}{(s+a)(s+b)}$
$\dfrac{1}{b-a}\left[(p-a)e^{-at}-(p-b)e^{-bt}\right]$	$\dfrac{s+p}{(s+a)(s+b)}$
$\dfrac{\omega_n}{\sqrt{1-\zeta^2}}e^{-\zeta\omega_n t}\sin\left(\omega_n\sqrt{1-\zeta^2}\,t\right)$	$\dfrac{\omega_n^2}{s^2+2\zeta\omega_n s+\omega_n^2}$
$e^{-\zeta\omega_n t}\left\{\cos\left(\omega_n\sqrt{1-\zeta^2}\,t\right)-\dfrac{\zeta}{\sqrt{1-\zeta^2}}\sin\left(\omega_n\sqrt{1-\zeta^2}\,t\right)\right\}$	$\dfrac{s}{s^2+2\zeta\omega_n s+\omega_n^2}$
$1-e^{-\zeta\omega_n t}\left\{\cos\left(\omega_n\sqrt{1-\zeta^2}\,t\right)+\dfrac{\zeta}{\sqrt{1-\zeta^2}}\sin\left(\omega_n\sqrt{1-\zeta^2}\,t\right)\right\}$	$\dfrac{\omega_n^2}{s(s^2+2\zeta\omega_n s+\omega_n^2)}$

$$D(s) \equiv a_n s^n + a_{n-1} s^{n-1} + \ldots + a_1 s + a_0 = 0 \tag{2.4.7}$$

The numerator $N(s)$ has m roots, z_i, $i = 1, 2, \ldots, m$, which are called the *zeros* of $F(s)$ and which satisfy the algebraic equation

$$N(s) \equiv b_m s^m + b_{m-1} s^{m-1} + \ldots + b_1 s + b_0 = 0 \tag{2.4.8}$$

It can be shown that

$$\begin{aligned} D(s) &= a_n (s - p_1)(s - p_1) \ldots (s - p_n) \\ N(s) &= b_m (s - z_1)(s - z_1) \ldots (s - z_m) \end{aligned} \tag{2.4.9}$$

With Eq. (2.4.9), the function $F(s)$ can be expressed in a zero-pole-gain form

$$F(s) = k \frac{(s - z_1)(s - z_1) \ldots (s - z_m)}{(s - p_1)(s - p_1) \ldots (s - p_n)}; \quad k = \frac{a_n}{b_m} \tag{2.4.10}$$

Example 2.4.1 Deriving the zero-pole-gain form of a rational function

Consider the rational function

$$F(s) = \frac{N(s)}{D(s)} = \frac{4s + 1}{2s^2 + 5s + 3}.$$

The poles of $F(s)$ are the roots of

$$2s^2 + 5s + 3 = 0 \Rightarrow p_1 = -1.5, \; p_2 = -1$$

and the zero of $F(s)$ is given by

$$4s + 1 = 0 \Rightarrow z_1 = -0.25.$$

Then the function can be written in the zero-pole-gain form

$$F(s) = \frac{N(s)}{D(s)} = 2 \frac{s + 0.25}{(s + 1.5)(s + 1)}.$$

2.4.3 Properties of the Laplace Transform

The Laplace transform is a definite integral; thus, it possesses all the properties of such integrals. The following properties of the Laplace transform are useful in modeling, analysis, and derivation of analytical solutions for dynamic systems in engineering problems.

Factoring

$$\mathcal{L}\{\alpha f(t)\} = \alpha \mathcal{L}\{f(t)\} = \alpha F(s) \qquad (2.4.11)$$

This can be proven using the definition of the Laplace transform

$$\mathcal{L}\{\alpha f(t)\} = \int_{0^-}^{\infty} \alpha f(t) e^{-st} dt = \alpha \int_{0^-}^{\infty} f(t) e^{-st} dt = \alpha \mathcal{L}\{f(t)\} = \alpha F(s)$$

Superposition (Linearity Property)

Because the Laplace transform operator $\mathcal{L}$ is a linear operator, it satisfies the following superposition principle:

$$\mathcal{L}\{\alpha f_1(t) + \beta f_2(t)\} = \alpha \mathcal{L}\{f_1(t)\} + \beta \mathcal{L}\{f_2(t)\} = \alpha F_1(s) + \beta F_2(s) \qquad (2.4.12)$$

The proof is straightforward, using the definition of the Laplace transform and the properties of definite integrals

$$\mathcal{L}\{\alpha f_1(t) + \beta f_2(t)\} = \int_{0^-}^{\infty} [\alpha f_1(t) + \beta f_2(t)] e^{-st} dt$$

$$= \alpha \int_{0^-}^{\infty} f_1(t) e^{-st} dt + \beta \int_{0^-}^{\infty} f_2(t) e^{-st} dt = \alpha \mathcal{L}\{f_1(t)\} + \beta \mathcal{L}\{f_2(t)\} = \alpha F_1(s) + \beta F_2(s)$$

The inverse transform also possesses the linearity property

$$\mathcal{L}^{-1}\{\alpha F_1(s) + \beta F_2(s)\} = \alpha \mathcal{L}^{-1}\{F_1(s)\} + \beta \mathcal{L}^{-1}\{F_2(s)\}$$
$$= \alpha f_1(t) + \beta f_2(t) \qquad (2.4.13)$$

The linearity property is very useful since it allows one to decompose a function into a linear combination of functions whose transforms are known, thus avoiding the necessity of evaluating the Laplace integral.

Example 2.4.2 Evaluating the Laplace transform of a function using the table of Laplace pairs

$$\mathcal{L}\{e^{-4t}(5 \sin 2t + 3t)\} = 5\mathcal{L}\{e^{-4t} \sin 2t\} + 3\mathcal{L}\{te^{-4t}\}$$

$$= \frac{10}{(s+4)^2 + 2^2} + \frac{3}{(s+4)^2}$$

where Table 2.4.1 has been used to evaluate the Laplace transform of the given functions.

Example 2.4.3 Deriving the Laplace transform of the common functions: step, ramp, exponential, and sinusoidal

Deriving the Laplace transform of some common functions.

Step Function

The *step function* is defined as

$$f(t) = \begin{cases} 0 & t < 0 \\ A & t \geq 0 \end{cases}$$

where A is a real positive constant, called the *magnitude of the step function*. If $A = 1$, the function is called the *unit step function*, and it is often denoted as $u(t)$. Hence, any step function can be expressed as

$$f(t) = \begin{cases} 0 & t < 0 \\ A\,u(t) & t \geq 0 \end{cases}$$

Using the definition and the above-described properties, the Laplace transform of a step function is

$$\mathcal{L}\{Au(t)\} = \int_{0^-}^{\infty} Au(t)e^{-st}dt = -\frac{A}{s}e^{-st}\Big|_{0^-}^{\infty} = -\frac{A}{s}(e^{-\infty} - e^0) = \frac{A}{s}$$

Obviously, the Laplace transform of the unit step is as shown in Table 2.4.1

$$\mathcal{L}\{u(t)\} = \int_{0^-}^{\infty} u(t)e^{-st}dt = -\frac{1}{s}e^{-st}\Big|_0^{\infty} = -\frac{1}{s}(e^{-\infty} - e^0) = \frac{1}{s}$$

Ramp Function

The *ramp function* is defined as follows:

$f(t) = \begin{cases} 0 & t < 0 \\ A\,t & t \geq 0 \end{cases}$, where A is a real positive constant. If $A = 1$, the function is called *unit ramp function*.

The Laplace transform of a ramp function is evaluated using the technique of integration by parts:

$$\mathcal{L}\{A\,t\} = \int_{0^-}^{\infty} Ate^{-st}dt = -\frac{A}{s}te^{-st}\Big|_0^{\infty} - \int_{0^-}^{\infty}\left(-\frac{A}{s}e^{-st}\right)dt = \frac{A}{s}\int_{0^-}^{\infty} e^{-st}dt = \frac{A}{s^2}$$

Exponential Function

The *exponential function* is defined as follows:

$f(t) = \begin{cases} 0 & t < 0 \\ A\,e^{-\alpha t} & t \geq 0 \end{cases}$, where A and α are real positive constants.

The Laplace transform of this function is

$$\mathcal{L}\{A\,e^{-at}\} = \int_{0^-}^{\infty} A\,e^{-at}e^{-st}dt = A\int_{0^-}^{\infty} e^{-(a+s)t}dt = \frac{A}{s+a}$$

Sinusoidal Function

The *sinusoidal function* is defined as follows:

$$f(t) = \begin{cases} 0 & t < 0 \\ A\sin\omega t & t \geq 0 \end{cases}$$

where A and ω are real positive constants.

Remembering the Euler's formula, $e^{j\theta} = \cos\theta + j\sin\theta$ and $e^{-j\theta} = \cos\theta - j\sin\theta$, the Laplace transform of the sine and cosine function are derived as follows. From the Euler's formula:

$$\sin\omega t = \frac{1}{2j}(e^{j\omega t} - e^{-j\omega t})$$

$$\cos\omega t = \frac{1}{2}(e^{j\omega t} + e^{-j\omega t})$$

Hence, the Laplace transform of the sine function is

$$\mathcal{L}\{A\sin\omega t\} = \frac{A}{2j}\int_{0^-}^{\infty}(e^{j\omega t} - e^{-j\omega t})e^{-st}dt = \frac{A}{2j}*\frac{1}{s-j\omega} - \frac{A}{2j}*\frac{1}{s+j\omega} = \frac{A\omega}{s^2+\omega^2}$$

and for the cosine function

$$f(t) = \begin{cases} 0 & t < 0 \\ A\cos\omega t & t \geq 0 \end{cases}$$

where A and ω are real positive constants, the Laplace transform is

$$\mathcal{L}\{A\cos\omega t\} = \frac{A}{2}\int_{0^-}^{\infty}(e^{j\omega t} + e^{-j\omega t})e^{-st}dt = \frac{A}{2}*\frac{1}{s-j\omega} + \frac{A}{2}*\frac{1}{s+j\omega} = \frac{As}{s^2+\omega^2}$$

Time Delay (Shifting along the Time Axis)

For a function $f(t)$, define a time-shifted function $f_d(t)$ by

$$f_d(t-T) = \begin{cases} f(t-T), & t \geq T \\ 0, & 0 \leq t < T \end{cases} \qquad (2.4.14)$$

where T is a real positive constant, called *time-delay parameter*.

The time-shifted function is illustrated in Figure 2.4.1.

By definition, the Laplace transform of the translated function $f_d(t-T)$ is computed as

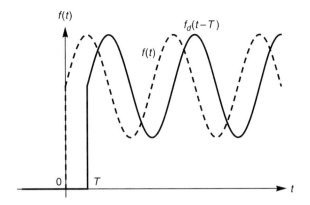

Figure 2.4.1 Time-shifted function

$$\mathcal{L}\{f_d(t-T)\} = \int_{0^-}^{\infty} f(t-T)e^{-st}dt$$

Let the independent variable be defined as $\tau = t - T$. Then the Laplace integral becomes

$$\int_{0^-}^{\infty} f(t-T)e^{-st}dt = \int_{-T}^{\infty} f(\tau)e^{-s(\tau+T)}d\tau$$

Since $f(t-T) = f(\tau) = 0$ for $\tau < 0$, the lower limit of the integration can be changed from $-T$ to 0. Hence, the Laplace integral is evaluated as

$$\int_{-T}^{\infty} f(\tau)e^{-s(\tau+T)}d\tau = \int_{0^-}^{\infty} f(\tau)e^{-s\tau}e^{-sT}d\tau = e^{-sT}\int_{0^-}^{\infty} f(\tau)e^{-s\tau}d\tau$$

$$= e^{-sT}\mathcal{L}\{f(t)\} = e^{-sT}F(s)$$

Therefore, translation of the function $f(t)$ in time domain by $T \geq 0$ corresponds to the multiplication of its Laplace transform $F(s)$ by e^{-sT}

$$\mathcal{L}\{f_d(t-T)\} = e^{-sT}F(s) \tag{2.4.15}$$

Example 2.4.4 Deriving the Laplace transform of a time-shifted function

For a time-shifted step function

$$u(t-T) = \begin{cases} 1, & t \geq T \\ 0, & t < T \end{cases}$$

where $T > 0$, the Laplace transform is

$$\mathcal{L}\{u(t-T)\} = e^{-sT}\mathcal{L}\{u(t)\} = \frac{1}{s}e^{-sT}.$$

Multiplication by the Exponential (Shifting along the *s*-Axis)

Similarly to translation in the time domain, a function may be translated in the Laplace domain. If $f(t)$ has the Laplace transform $F(s)$, then the Laplace transform of a function $e^{-\alpha t}f(t)$ will be

$$\mathcal{L}\{e^{-\alpha t}f(t)\} = \int_{0^-}^{\infty} e^{-\alpha t}f(t)e^{-st}dt = \int_{0^-}^{\infty} f(t)e^{-(\alpha+s)t}dt = F(s+\alpha) \quad (2.4.16)$$

Multiplication by *t*

Let $F(s)$ be the Laplace transform of the function $f(t)$. Differentiating $F(s)$ once with respect to s, we obtain

$$\frac{d}{ds}F(s) = \frac{d}{ds}\int_{0^-}^{\infty} f(t)e^{-st}dt = -\int_{0^-}^{\infty} tf(t)e^{-st}dt = -\mathcal{L}\{tf(t)\}$$

Hence

$$\mathcal{L}\{tf(t)\} = -\frac{d}{ds}F(s) \quad (2.4.17)$$

Extending this approach to the powers of t, the following generic expression is derived

$$\mathcal{L}\{t^n f(t)\} = (-1)^n \frac{d^n}{ds^n} F(s) \quad (2.4.18)$$

Example 2.4.5 Deriving the Laplace transform of various functions using the table of Laplace transform Pairs and Laplace transform properties

Certain functions need to be evaluated to facilitate derivation of their Laplace transform. For example, a function $f(t) = (a+kt)^2$ needs to be converted to a polynomial before its Laplace transform can be derived:

$$\mathcal{L}(f(t)) = \mathcal{L}((a+kt)^2) = \mathcal{L}(a^2 + 2akt + k^2t^2)$$

The linearity property is used:

$$\mathcal{L}(f(t)) = \mathcal{L}(a^2 + 2akt + k^2t^2) = \mathcal{L}(a^2) + \mathcal{L}(2akt) + \mathcal{L}(k^2t^2)$$

Considering that a^2 represents a step function $a^2 u(t)$, apply a factoring property and then use Table 2.4.1 to derive the solution:

$$\mathcal{L}(f(t)) = \mathcal{L}(a^2) + 2ak\mathcal{L}(t) + k^2\mathcal{L}(t^2) = \frac{a^2}{s} + \frac{2ak}{s^2} + \frac{2k^2}{s^3}$$

2.4 The Laplace Transform

Trigonometric identity needs to be used to convert the following function to a form that allows the use of Table 2.4.1 to derive the required Laplace transform:

$$\mathcal{L}(f(t)) = \mathcal{L}(\sin^2(\omega t)) = \mathcal{L}\left(\frac{1}{2}(1 - \cos(2\omega t))\right)$$
$$= \frac{1}{2}\mathcal{L}(u(t)) - \frac{1}{2}\mathcal{L}(\cos(2\omega t)) = \frac{1}{2s} - \frac{s}{2(s^2 + 4\omega^2)}$$

When several properties of the Laplace transform need to be used to derive the transform of a given function, linearity and factoring properties are used first, followed by the other properties as needed.

In some cases, the sequence of applications of the Laplace transform properties is uniquely defined. For example, while proving that

$$\mathcal{L}\left(t\cos(\omega t)\right) = \frac{s^2 - \omega^2}{(s^2 + \omega^2)^2}$$

the transform of a cosine needs to be evaluated first, followed by application of the *multiplication by t* property:

$$\mathcal{L}\left(t\cos(\omega t)\right) = -\frac{d}{ds}\left(\frac{s}{s^2 + \omega^2}\right) = \frac{s^2 - \omega^2}{(s^2 + \omega^2)^2}$$

In other cases, the Laplace transform properties could be applied in a different order. For example, the function $f(t) = 3t\sinh(2t)$ needs first to be represented as a sum of products of exponential functions and t, and then the appropriate Laplace transform properties applied as follows:

$$\mathcal{L}(f(t)) = \mathcal{L}(3t\sinh(2t)) = \mathcal{L}\left(\frac{1}{2}(3te^{2t} - 3te^{-2t})\right)$$

Linearity and factoring properties are applied first:

$$\mathcal{L}(f(t)) = \mathcal{L}\left(\frac{1}{2}(3te^{2t} - 3te^{-2t})\right) = \frac{3}{2}\mathcal{L}(te^{2t}) - \frac{3}{2}\mathcal{L}(te^{-2t})$$

Then two options are available:

- Use the Laplace transform of an exponential function and the *multiplication by t* property

$$\mathcal{L}(f(t)) = \frac{3}{2}\mathcal{L}(te^{2t}) - \frac{3}{2}\mathcal{L}(te^{-2t}) = \frac{3}{2}\left(-\frac{d}{ds}\left(\frac{1}{s-2}\right) + \frac{d}{ds}\left(\frac{1}{s+2}\right)\right)$$
$$= \frac{3}{2}\left(\frac{1}{(s-2)^2} - \frac{1}{(s+2)^2}\right) = \frac{12s}{(s^2 - 4)^2}$$

- Use Laplace transform of a unit ramp function and the *multiplication by an exponential (shifting along the s-axis)* property

$$\mathcal{L}(f(t)) = \frac{3}{2}\mathcal{L}(te^{2t}) - \frac{3}{2}\mathcal{L}(te^{-2t}) = \frac{3}{2}\left(\frac{1}{(s-2)^2} - \frac{1}{(s+2)^2}\right) = \frac{12s}{(s^2-4)^2}$$

Differentiation (Derivative Property)

The Laplace transform of the first derivative of a function $f(t)$ is obtained by using integration by parts to evaluate the Laplace integral:

$$\mathcal{L}\left\{\frac{d}{dt}f(t)\right\} = \int_{0^-}^{\infty}\frac{df}{dt}e^{-st}dt = f(t)e^{-st}\Big|_{0^-}^{\infty} + s\int_{0^-}^{\infty}f(t)e^{-st}dt$$

$$= -f(0^-) + s\mathcal{L}\{f(t)\} = sF(s) - f(0^-) \quad (2.4.19)$$

Similarly, the Laplace transform of the second derivative of a function $f(t)$ is

$$\mathcal{L}\left\{\frac{d^2}{dt^2}f(t)\right\} = s^2 F(s) - sf(0^-) - \frac{d}{dt}f(0^-) \quad (2.4.20)$$

Extending the above procedure to the higher derivatives, the general case expression is obtained:

$$\mathcal{L}\left\{\frac{d^n f(t)}{dt^n}\right\} = s^n F(s) - s^{n-1}f(0^-) - s^{n-2}f^{(1)}(0^-) - \cdots - f^{(n-1)}(0^-) \quad (2.4.21)$$

where $f^{(k)} = \dfrac{d^k f}{dt^k}$.

Equation (2.4.21) can be written in a more compact form

$$\mathcal{L}\left\{\frac{d^n f(t)}{dt^n}\right\} = s^n F(s) - \sum_{k=1}^{n} s^{n-k} g_{k-1}$$

where $g_{k-1} = \dfrac{d^{k-1}f}{dt^{k-1}}\Big|_{t=0^-}$ and, obviously, the sum $\sum_{k=1}^{n} s^{n-k} g_{k-1}$ represents n initial conditions, defined for the n-th order differential equation.

Example 2.4.6 Deriving the Laplace transform of a function derivative

Given derivative $\varphi(t) = \dfrac{d}{dt}\{e^{-4t}(5\sin 2t + 3t)\}$, its Laplace transform is

$$\mathcal{L}\{\varphi(t)\} = s\mathcal{L}\{e^{-4t}(5\sin 2t + 3t)\} - [e^{-4t}(5\sin 2t + 3t)]_{t=0^-} = \frac{10s}{(s+4)^2 + 2^2} + \frac{3s}{(s+4)^2}$$

where the result derived in the Example 2.4.2 has been used.

The utility of the derivative property of the Laplace transform (Eq. (2.4.21)) and its linearity property (Eq. (2.4.12)) lies in the ability to convert a differential equation in the time domain into an algebraic equation in the Laplace domain. This significantly facilitates finding the solution of a given differential equation. Such an approach is illustrated in the Example 2.4.7.

Example 2.4.7 Solving a linear differential equation with a Laplace transform

Consider the following linear second-order differential equation with initial conditions

$$m\ddot{x}(t) + c\dot{x}(t) + kx(t) = f(t)$$
$$x(0^-) = x_0, \; \dot{x}(0^-) = v_0$$

where $\dot{x} = \dfrac{dx}{dt}$, and m, c, and k are constants. The previous equation and initial conditions are often used to describe a spring–mass–damper system, which is to be discussed in Chapter 3. According to Eqs. (2.4.12) and (2.4.21), the Laplace transform of this differential equation yields

$$m\left(s^2 X(s) - sx_0 - v_0\right) + c\left(sX(s) - x_0\right) + kX(s) = F(s)$$

which is reduced to an algebraic equation in terms of $X(s) = \mathcal{L}\{x(t)\}$:

$$\left(ms^2 + cs + k\right)X(s) = F(s) + mx_0 s + mv_0 + cx_0$$

Thus, the s-domain (Laplace transform domain) solution of the given differential equation is

$$X(s) = \frac{F(s) + mx_0 s + mv_0 + cx_0}{ms^2 + cs + k}$$

The inverse Laplace transform of the previous expression gives the time-domain solution of the original differential equation; that is $x(t) = \mathcal{L}^{-1}\{X(s)\}$.

This example clearly demonstrates the procedure of using a Laplace transform to solve linear ordinary differential equations with constant coefficients. This technique shall be investigated further in Section 2.6.3.

Integration (Integral Property)

The Laplace transform of an integral of a function $f(t)$ is derived using integration by parts and defining a function $\tilde{f}(0^-)$ such that

$$\tilde{f}(0^-) = \int f(t)dt \bigg|_{t=0^-}$$

Let $\mathcal{L}\{\int f(t)dt\} = F(s)$. Then the Laplace transform of an indefinite integral of the function $f(t)$ is

$$\mathcal{L}\{\int f(t)dt\} = \int_{0^-}^{\infty} \left(\int f(t)dt\right) e^{-st} dt = \int f(t)dt \frac{e^{-st}}{-s}\bigg|_{0^-}^{\infty} - \int_{0^-}^{\infty} f(t) \frac{e^{-st}}{-s} dt$$

$$= \frac{1}{s} \int f(t)dt\bigg|_{t=0^-} + \frac{1}{s} \int_{0^-}^{\infty} f(t)dt = \frac{1}{s}\tilde{f}(0^-) + \frac{1}{s}F(s) \quad (2.4.22)$$

Since

$$\int_{0^-}^{t} f(t)dt = \int f(t)dt - \tilde{f}(0^-)$$

The application of the result of Eq. (2.4.22) to the definite integral of $f(t)$ derives the integral property as follows:

$$\mathcal{L}\left\{\int_{0^-}^{\infty} f(t)dt\right\} = \mathcal{L}\left\{\int f(t)dt\right\} - \mathcal{L}\{\tilde{f}(0^-)\} \quad (2.4.23)$$

Since $\tilde{f}(0^-)$ is a constant, its Laplace transform is derived using the linearity property and the known Laplace transform of a unit step function

$$\mathcal{L}\{\tilde{f}(0^-)\} = \mathcal{L}\{\tilde{f}(0^-)u(t)\} = \tilde{f}(0^-)\mathcal{L}\{u(t)\} = \frac{1}{s}\tilde{f}(0^-)$$

Applying this result to the Eq. (2.4.23), we obtain the sought integral property

$$\mathcal{L}\left\{\int_{0^-}^{\infty} f(t)dt\right\} = \frac{1}{s}\tilde{f}(0^-) + \frac{1}{s}F(s) - \frac{1}{s}\tilde{f}(0^-) = \frac{1}{s}F(s) \quad (2.4.24)$$

Example 2.4.8 Deriving the Laplace transform of a linear integro-differential expression

Consider the following linear integro-differential expression

$$u(t) = k_p e(t) + k_i \int_{0^-}^{t} e(\tau)d\tau + k_d \frac{de(t)}{dt}$$

where the function $e(t)$ is specified for $t \geq 0$, and k_p, k_i, k_d are constants. Assuming a zero initial value of $e(t)$, namely $e(0) = 0$, the Laplace transform of the above expression gives

$$U(s) = \left(k_p + k_i \frac{1}{s} + k_d s\right) E(s)$$

This *s*-domain expression describes an algorithm widely used in control engineering, called the *PID feedback control law*.

Initial-Value Theorem

The initial-value theorem allows the use of the Laplace transform of a function $f(t)$ to find the function value at a time slightly greater than zero, namely $f(0^+)$. It becomes important in the analysis of dynamic systems under impulse input.

A unit impulse function (Dirac delta), as defined in Eq. (2.4.3), is a mathematical approximation of an infinitely large input of energy into a system, which occurs within an infinitesimally short time.

If the Laplace transform of a given function $f(t)$ and its first derivative with respect to time $\dfrac{df}{dt}$ exist, and $\lim_{s \to 0} sF(s)$ also exists, the initial-value theorem is formulated as

$$f(0^+) = \lim_{t \to 0^+} f(t) = \lim_{s \to \infty} sF(s) \qquad (2.4.25)$$

Example 2.4.9 Finding the system response to a unit impulse using the Laplace transform and the initial-value theorem

A dynamic system is described by a second-order differential equation with initial conditions

$$\ddot{x} = f(t)$$
$$x(0^-) = 5, \; \dot{x}(0^-) = 10$$

Find the system response to a unit impulse, namely values of system parameters $x(t)$ and $\dot{x}(t)$ at $t = 0^+$ when $f(t) = \delta(t)$.

Solution

Since the Dirac delta function is defined for $t = 0$ only, and equals zero for all other values of t, it is logical that the initial conditions are given at a time $t = 0^-$ that immediately precedes $t = 0$, when the impulse input was applied.

Taking the Laplace transform of the given differential equation and considering that $X(s) = \mathcal{L}\{x(t)\}$, we obtain

$$s^2 X(s) - sx(0^-) - \dot{x}(0^-) = s^2 X(s) - 5s - 10 = 1$$
$$X(s) = \frac{5s + 11}{s^2}$$

Using the initial-value theorem, compute

$$x(0^+) = \lim_{s \to \infty} sX(s) = \lim_{s \to \infty} \left(s \frac{5s + 11}{s^2} \right) = \lim_{s \to \infty} 5 + \lim_{s \to \infty} \frac{11}{s} = 5$$

Recalling that

$$x(0^+) = \lim_{s \to \infty} sX(s) = \lim_{s \to \infty}\left(s\mathcal{L}\{x(t)\}\right)$$

derive

$$\dot{x}(0^+) = \lim_{s \to \infty}\left(s\mathcal{L}\{\dot{x}(t)\}\right) = \lim_{s \to \infty}\left(s\left(sX(s) - x(0^-)\right)\right)$$

$$= \lim_{s \to \infty}\left(s\left(s\frac{5s+11}{s^2} - 5\right)\right) = 11$$

Final-Value Theorem

If a function $f(t)$ converges to some constant value after a long time, namely $t \to \infty$, it is said that $f(t)$ attains a *steady state* and that limiting constant is called the *steady-state value*. A counterpart of the initial-value theorem, the final-value theorem allows the Laplace transform of $f(t)$ to be used for finding its steady-state value, $\lim_{t \to \infty} f(t)$.

Let $f(t)$ and $F(s)$ be a Laplace transform pair. If $sF(s)$ is analytic on the imaginary axis and the right half of the s-plane (namely, all the poles of $sF(s)$ have negative real parts), then

$$f_{ss} = \lim_{t \to \infty} f(t) = \lim_{s \to 0} sF(s) \qquad (2.4.26)$$

Here, the poles of a complex function have been defined in Eq. (2.4.7).

Convolution Theorem

If $F_1(s) = \mathcal{L}\{f_1(t)\}$ and $F_2(s) = \mathcal{L}\{f_2(t)\}$, then

$$\mathcal{L}^{-1}\{F_1(s)F_2(s)\} = f_1 * f_2 \equiv \int_{0^-}^{t} f_1(\tau)f_2(t-\tau)d\tau \qquad (2.4.27)$$

where $f_1(t-\tau)$ is time-shifted in relation to $f_1(t)$, and $f_1 * f_2$ is called the *convolution integral* of f_1 and f_2. The convolution integral is commutative, namely

$$f_1 * f_2 \equiv \int_{0^-}^{t} f_1(\tau)f_2(t-\tau)d\tau = f_2 * f_1 \equiv \int_{0^-}^{t} f_2(\tau)f_1(t-\tau)d\tau \qquad (2.4.28)$$

where $f_2(t-\tau)$ is time-shifted in relation to $f_2(t)$.

The utility of the final-value theorem and the convolution theorem will be shown in Chapter 7.

2.4.4 Inverse Laplace Transform via Partial Fraction Expansion

In engineering analysis, the inverse Laplace transform is frequently performed to derive analytical solutions. As mentioned in Example 2.4.7, the inverse Laplace transform of $X(s)$ can yield the solution of the differential equation. Although extensive Laplace-transform tables like Table 2.4.1 are available, they may not be sufficient for finding the inverse Laplace transform of rational functions. Quite often a technique called *partial fraction expansion* (or partial fraction decomposition) is used. Partial fraction expansion has been introduced in integral calculus. This technique is briefly reviewed below.

The utility of partial fraction decomposition lies in the ability of expanding $F(s)$ into a linear combination of components such as $\dfrac{\alpha}{s+p}$ and $\dfrac{\alpha s + \beta}{s^2 + as + b}$, the inverse Laplace transforms of which are available in any table of Laplace transform pairs. From Table 2.4.1

$$\mathcal{L}^{-1}\left\{\frac{\alpha}{s+p}\right\} = \alpha e^{-pt}$$

$$\mathcal{L}^{-1}\left\{\frac{\alpha s + \beta}{s^2 + as + b}\right\} = e^{-\sigma t}(C_1 \cos \omega t + C_2 \sin \omega t)$$

where

$$\sigma = \frac{a}{2}$$

$$\omega = \frac{\sqrt{4b - a^2}}{2}$$

$$C_1 = \alpha \quad \text{and} \quad C_2 = \frac{2\beta - a\alpha}{\sqrt{4b - a^2}}$$

Consider a rational function $F(s)$ of the form

$$F(s) = \frac{N(s)}{D(s)} = \frac{b_m s^m + b_{m-1} s^{m-1} + \ldots + b_1 s + b_0}{s^n + a_{n-1} s^{n-1} + \ldots + a_1 s + a_0}; \quad m < n \qquad (2.4.29)$$

Here, for simplicity in discussion, the unity coefficient for the s^n term and $m < n$ have been assumed. If $n = m$, long division can be applied to obtain

$$F(s) = \gamma + \frac{N_1(s)}{D(s)} \qquad (2.4.30)$$

where γ is a constant, and the order of polynomial $N_1(s)$ is less than n. In this case a partial fraction expansion can be performed on $N_1(s)/D(s)$. Because of this the subsequent discussion will be focused on Eq. (2.4.29), although the results are also applicable to Eq. (2.4.30).

Let the poles of $F(s)$ be p_i, $i = 1, 2, \ldots, n$. These poles can be real and complex. In what follows, the partial fraction expansion and inverse Laplace transform are presented for the two cases of real poles of $F(s)$.

Case 1: All Poles of $F(s)$ are Real and Distinct

Assume that no two poles of $F(s)$ are the same. Then the partial fraction expansion of $F(s)$ is

$$F(s) = \frac{r_1}{s - p_1} + \frac{r_2}{s - p_2} + \ldots + \frac{r_n}{s - p_n} \qquad (2.4.31)$$

where r_i is the residue of $F(s)$ associated with the pole p_i and it is given by

$$r_i = \lim_{s \to p_i} \left((s - p_i) \frac{N(s)}{D(s)} \right) \qquad (2.4.32)$$

Instead of computing the above limit the residues can be determined using the *cover-up method*, which performs cross-multiplication as follows:

$$\begin{aligned} N(s) &= F(s)D(s) \\ &= \left(\frac{r_1}{s - p_1} + \frac{r_2}{s - p_2} + \ldots + \frac{r_n}{s - p_n} \right)(s - p_1)(s - p_2) \ldots (s - p_n) \end{aligned} \qquad (2.4.33)$$

This eventually leads to

$$\begin{aligned} b_m s^m + b_{m-1} s^{m-1} + \ldots + b_1 s + b_0 &= r_1 (s - p_2)(s - p_3) \ldots (s - p_n) \\ &\quad + r_2 (s - p_1)(s - p_3) \ldots (s - p_n) \\ &\quad + \ldots + r_n (s - p_1)(s - p_2) \ldots (s - p_{n-1}) \end{aligned} \qquad (2.4.34)$$

Comparison of the coefficients of the s power terms on the both sides of Eq. (2.4.34) gives coupled algebraic equations in terms of residues, which are then solved for these residues. Upon computing the residues, the inverse Laplace transform of $F(s)$ is found as

$$f(t) = \mathcal{L}^{-1}\{F(s)\} = r_1 e^{p_1 t} + r_2 e^{p_2 t} + \ldots + r_n e^{p_n t} = \sum_{i=1}^{n} r_i e^{p_i t} \qquad (2.4.35)$$

by Eq. (2.4.29) and the fifth entry of Table 2.4.1.

Example 2.4.10 Using partial fraction expansion to derive the inverse Laplace transform of a given rational function

Consider the function $F(s)$ in Example 2.4.1:

$$F(s) = \frac{N(s)}{D(s)} = \frac{4s+1}{2s^2+5s+3} = \frac{2s+0.5}{s^2+2.5s+1.5}$$

As previously found, the poles of $F(s)$ are $p_1 = -1.5$ and $p_2 = -1$. The partial fraction expansion of the function $F(s)$ is then

$$F(s) = \frac{2s+0.5}{(s+1.5)(s+1)} = \frac{r_1}{s+1.5} + \frac{r_2}{s+1}.$$

According to Eq. (2.4.32) residues r_1 and r_2 are computed as

$$r_1 = \lim_{s \to -1.5} \left((s+1.5) \frac{2s+0.5}{(s+1.5)(s+1)} \right) = \frac{-3+0.5}{-0.5} = 5$$

$$r_2 = \lim_{s \to -1} \left((s+1) \frac{2s+0.5}{(s+1.5)(s+1)} \right) = \frac{-2+0.5}{0.5} = -3$$

Subsequently

$$f(t) = \mathcal{L}^{-1} \left\{ \frac{5}{s+1.5} - \frac{3}{s+1} \right\} = 5e^{-1.5t} - 3e^{-t}.$$

On the other hand, the residues can be found by the cover-up method,

$$2s + 0.5 = (s+1.5)(s+1)F(s) = r_1(s+1) + r_2(s+1.5)$$

which, through comparison of s power terms, yields

$$s^1 \text{ terms}: 2 = r_1 + r_2$$
$$s^0 \text{ terms}: 0.5 = r_1 + 1.5 r_2$$

The solution of the above algebraic equations gives the same residues as the previous method: $r_1 = 5$ and $r_2 = -3$, which leads to the same partial fraction decomposition and the same result of the inverse Laplace transform.

Example 2.4.11 Using long division followed by partial fraction expansion to derive the inverse Laplace transform of a given function

Consider the function

$$F(s) = \frac{N(s)}{D(s)} = \frac{3s^2 + 2s + 0.5}{s^2 + 2.5s + 1.5}$$

which has the same denominator as the function in Example 2.4.10, but features an added second-order term $3s^2$ in the numerator. By long division

$$F(s) = 3 - F_1(s), \text{ with } F_1(s) = \frac{5.5s + 4}{s^2 + 2.5s + 1.5}$$

Now perform partial fraction expansion on $F_1(s)$

$$F_1(s) = \frac{5.5s + 4}{(s + 1.5)(s + 1)} = \frac{r_1}{s + 1.5} + \frac{r_2}{s + 1}$$

where

$$r_1 = \lim_{s \to -1.5} \left((s + 1.5) \frac{5.5s + 4}{(s + 1.5)(s + 1)} \right) = 8.5$$

$$r_2 = \lim_{s \to -1} \left((s + 1) \frac{5.5s + 4}{(s + 1.5)(s + 1)} \right) = -3$$

Thus,

$$F(s) = 3 - \frac{8.5}{s + 1.5} + \frac{3}{s + 1}$$

and its inverse Laplace transform is

$$f(t) = 3\delta(t) - 8.5e^{-1.5t} + 3e^{-t}$$

where Table 2.4.1 has been used, and $\delta(t)$ is the Dirac delta function defined in Eqs. (2.4.3) and (2.4.4).

Case 2: Poles of *F(s)* are Real and Repeated

Assume that $F(s)$ has q distinct and k identical poles:

$$D(s) = (s - p_1)(s - p_2) \ldots (s - p_q)(s - p_0)^k \quad (2.4.36)$$

where $k + q = n$, and poles $p_0, p_1, p_2 \ldots p_q$ are distinct. The integer k is the multiplicity of the pole p_0. In this case, the partial fraction expansion of $F(s)$ takes the form

$$F(s) = \frac{r_1}{s - p_1} + \frac{r_2}{s - p_2} + \ldots + \frac{r_q}{s - p_q}$$
$$+ \frac{b_1}{(s - p_0)^k} + \frac{b_2}{(s - p_0)^{k-1}} + \ldots + \frac{b_k}{s - p_0} \quad (2.4.37)$$

where on the right-hand side the first q terms are about the distinct poles $p_1, p_2 \ldots p_q$, and the remaining k terms are related to the k repeated poles (p_0). The $r_1, r_2 \ldots r_q$ are the residues associated with poles $p_1, p_2 \ldots p_q$, while $b_1, b_2 \ldots b_k$ are the residues associated with the pole p_0. The formulae for evaluating the residues are given below:

$$r_i = \lim_{s \to p_i} \left((s - p_i) \frac{N(s)}{D(s)} \right), \quad i = 1, 2, \ldots, q \tag{2.4.38}$$

$$b_i = \frac{1}{(i-1)!} \lim_{s \to p_0} \frac{d^{i-1}\Delta(s)}{ds^{i-1}}, \quad i = 1, \ldots, k \tag{2.4.39}$$

where i is the "running number," and

$$\Delta(s) = F(s)(s - p_0)^k = \frac{N(s)}{(s - p_1)(s - p_2) \ldots (s - p_q)} \tag{2.4.40}$$

with $N(s)$ given in Eq. (2.4.29). Using Table 2.4.1 and Eq. (2.4.32), the inverse Laplace transform of $F(s)$ is found as

$$f(t) = r_1 e^{p_1 t} + r_2 e^{p_2 t} + \ldots + r_q e^{p_q t} +$$
$$+ e^{p_0 t} \left(\frac{b_1}{(k-1)!} t^{k-1} + \frac{b_2}{(k-2)!} t^{k-2} + \ldots + \frac{b_{k-1}}{1!} t + \frac{b_k}{0!} t \right) \tag{2.4.41}$$

Example 2.4.12 Using partial fraction expansion to derive the inverse Laplace transform of a given rational function with repeated poles

The function

$$F(s) = \frac{5(s+2)}{s^2(s+1)(s+3)}$$

has poles $0, 0, -1, -3$, with 0 being a pole of multiplicity 2 ($k = 2$). The partial fraction expansion of $F(s)$ can be written as

$$F(s) = \frac{r_1}{s+1} + \frac{r_2}{s+3} + \frac{b_1}{s^2} + \frac{b_2}{s}$$

The residues associated with -1 and -3 are

$$r_1 = \lim_{s \to -1} \left((s+1) F(s) \right) = \frac{5}{2}$$

$$r_2 = \lim_{s \to -3} \left((s+3) F(s) \right) = \frac{5}{18}$$

For the residues associated with the repeated poles, consider the function
$$\Delta(s) = s^2 F(s) = \frac{5(s+2)}{(s+1)(s+3)}$$

By Eq. (2.4.39), residues associated with the repeated poles are

$$b_1 = \frac{1}{0!} \lim_{s \to 0} \Delta(s) = \frac{10}{3}$$

$$b_2 = \frac{1}{1!} \lim_{s \to 0} \frac{d\Delta(s)}{ds} = \lim_{s \to 0} \left(5 \frac{(s+1)(s+3) - (s+2)(2s+4)}{(s+1)^2(s+3)^2} \right) = -\frac{25}{9}$$

Thus,

$$f(t) = \mathcal{L}^{-1}\left(\frac{r_1}{s+1} + \frac{r_2}{s+3} + \frac{b_1}{s^2} + \frac{b_2}{s} \right) = \frac{5}{2} e^{-t} + \frac{5}{18} e^{-3t} + \frac{10}{3} t - \frac{25}{9}$$

2.4.5 Imaginary and Complex Poles in the Inverse Laplace Transform

In many engineering problems the poles and residues of a rational function $F(s)$ are complex and/or imaginary, which yields complex terms in Eqs. (2.4.35) and (2.4.41). Because $F(s)$ in Eq. (2.4.29) only contains real coefficients (a_i and b_i), its imaginary and complex poles always come in pairs, such as $\pm j\omega$ and $\alpha \pm j\beta$, where $j = \sqrt{-1}$. As a result, the inverse Laplace transform of $F(s)$ is real-valued. In what follows, we derive real expressions for the inverse Laplace transform of $F(s)$ with imaginary or complex poles.

If p is a complex pole and r is the residue of $F(s)$ at this pole, the terms $r\,e^{pt} + \bar{r}\,e^{\bar{p}t}$ will be included in Eq. (2.4.35), where $\bar{p}$ and $\bar{r}$ are the complex conjugates of p and r, respectively. Let $p = \sigma + j\omega$, $r = \alpha + j\beta$, with $j = \sqrt{-1}$. It follows that

$$r\,e^{pt} + \bar{r}\,e^{\bar{p}t} = 2\text{Re}\left((\alpha + j\beta)\,e^{(\sigma + j\omega)t} \right) = 2e^{\sigma t}(\alpha \cos \omega t - \beta \sin \omega t) \tag{2.4.42}$$

If p is an imaginary pole, say, $p = j\omega$, Eq. (2.4.35) will contain the terms $r\,e^{pt} + \bar{r}\,e^{\bar{p}t}$. By writing $r = \alpha + j\beta$,

$$r\,e^{pt} + \bar{r}\,e^{\bar{p}t} = 2\text{Re}\left((\alpha + j\beta)\,e^{j\omega t} \right) = 2(\alpha \cos \omega t - \beta \sin \omega t) \tag{2.4.43}$$

Note that Eq. (2.4.43) is a special case of Eq. (2.4.42) with $\sigma = 0$.

Another way to deal with complex/imaginary poles is to keep quadratic forms in a partial fraction expansion, such as

2.4 The Laplace Transform

$$\frac{as+b}{s^2+cs+d} \text{ with } c^2 < 4d \text{ and } \frac{as+b}{s^2+\omega^2}$$

which have complex and imaginary poles. By proper scaling and through use of Table 2.4.1, the real expression of $\mathcal{L}^{-1}\{F(s)\}$ can be directly obtained without having to deal with complex quantities.

Example 2.4.13 Using partial fraction expansion to derive the inverse Laplace transform of a given rational function with complex poles

We would like to solve the differential equation in Example 2.4.7, with parameters $m=1$, $c=2$, $k=2$, initial conditions $x(0)=1$ and $\dot{x}(0)=0$, and the forcing function being a unit step function, $f(t)=u(t)$. According to the result derived in Example 2.4.7, the s-domain solution of this differential equation is

$$X(s) = \frac{1}{s^2+2s+2}\left\{\frac{1}{s}+s+2\right\} = \frac{s^2+2s+1}{s(s^2+2s+2)}$$

The poles of $X(s)$ are $p_1 = 0$, $p_2 = -1+j$, and $p_3 = \bar{p}_2 = -1-j$. Because the poles are distinct, the inverse Laplace transform of $X(s)$ is of the form presented for Case 1:

$$x(t) = r_1 e^{p_1 t} + r_2 e^{p_2 t} + r_3 e^{p_3 t}$$

where the residues by Eq. (2.4.32) are obtained as

$$r_1 = \frac{1}{2}, \quad r_2 = \frac{1}{4}(1-j), \quad r_3 = \bar{r}_2 = \frac{1}{4}(1+j)$$

According to Eq. (2.4.42),

$$x(t) = \frac{1}{2} + 2\mathrm{Re}\left(\frac{1}{4}(1-j)\,e^{(-1+j)t}\right) = \frac{1}{2} + \frac{1}{2}e^{-t}(\cos t + \sin t).$$

As another method for the inverse Laplace transform, write

$$X(s) = \frac{r}{s} + \frac{as+b}{s^2+2s+2}$$

where r, a, and b are coefficients to be determined. With the method shown in Eq. (2.4.34), we obtain

$$s^2 + 2s + 1 = s(s^2+2s+2)X(s)$$
$$= s(s^2+2s+2)\left\{\frac{r}{s} + \frac{as+b}{s^2+2s+2}\right\}$$
$$= r(s^2+2s+2) + s(as+b)$$

Comparison of the coefficients of the s power terms on the two sides of the previous equation yields

$$s^2 : 1 = r + a$$
$$s^1 : 2 = 2r + b$$
$$s^0 : 1 = 2r$$

The solution of the above coupled algebraic equations gives

$$r = \frac{1}{2}, \ a = \frac{1}{2}, \ b = 1$$

Hence,

$$X(s) = \frac{1}{2s} + \left(\frac{1}{2}\right)\left(\frac{s+2}{s^2+2s+2}\right) = \frac{1}{2s} + \left(\frac{1}{2}\right)\left(\frac{s+2}{(s+1)^2+1}\right)$$
$$= \frac{1}{2s} + \left(\frac{1}{2}\right)\left(\frac{s+1}{(s+1)^2+1}\right) + \left(\frac{1}{2}\right)\left(\frac{1}{(s+1)^2+1}\right)$$

where the expression for $X(s)$ is properly scaled and arranged for utility of Table 2.4.1. It follows from the Laplace transform table that the solution of the differential equation is

$$x(t) = \frac{1}{2} + \frac{1}{2}e^{-t}(\cos t + \sin t)$$

which is the same as obtained previously.

Examples 2.4.7 and 2.4.13 indicate a method for solution of linear differential equations with constant coefficients, which will be further explored later on in Section 2.6.

2.4.6 Piecewise-Continuous Functions in the Laplace Transformation

In many engineering applications, dynamic systems are subject to piecewise-continuous inputs such as, for example, pulses shown in Figure 2.4.2. These pulses physically represent external excitations persisting for the finite time duration.

The Laplace transform of this type of function can be performed through the application of the principle of superposition as described by Eq. (2.4.12) and the formula for time-shifted functions as given in Eq. (2.4.15). In general, a piecewise-continuous function can be decomposed as

$$f(t) = f_0(t)\, u(t) + f_1(t - T_1)\, u(t - T_1) + f_2(t - T_2)\, u(t - T_2) + \ldots \tag{2.4.44}$$

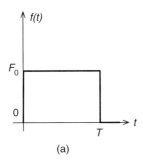
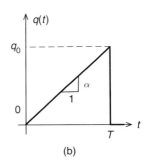

Figure 2.4.2 Piecewise-continuous functions: (a) rectangular pulse $f(t)$, (b) triangular pulse $q(t)$.

where $f_0(t)$, $f_1(t)$, $f_2(t)$, ... are basic functions, as those in Table 2.4.1, $f_1(t - T_1)$, $f_2(t - T_2)$, ... are time-shifted functions, and T_1, T_2, ... are positive time-delay parameters. The Laplace transform of $f(t)$ then has the form

$$F(s) = F_0(s) + F_1(s)e^{-T_1 s} + F_2(s)e^{-T_2 s} + \ldots \tag{2.4.45}$$

where $F_0(s)$, $F_1(s)$, $F_2(s)$, ... are the Laplace transforms of $f_0(t)$, $f_1(t)$, $f_2(t)$, ..., respectively.

Example 2.4.14 Deriving the Laplace transform of a rectangular pulse

Consider the rectangular pulse $f(t)$ in Figure 2.4.2(a), which mathematically is given by

$$f(t) = \begin{cases} F_0, & 0 \leq t \leq T \\ 0, & t > T \end{cases}$$

It is easy to show that $f(t)$ can be expressed in a single expression

$$f(t) = F_0\, u(t) - F_0\, u(t - T)$$

where $u(t)$ is the unit step function defined in Eq. (2.4.5), $u(t - T)$ is the time-shifted unit step function given in Example 2.4.4, and T is the time-delay parameter. Graphically, $f(t)$ is illustrated in Figure 2.4.3(a). The Laplace transform of the rectangular pulse is

$$F(s) = \mathcal{L}\{F_0\, u(t) - F_0\, u(t - T)\} = \frac{F_0}{s}\left(1 - e^{-Ts}\right)$$

where Eq. (2.4.15) has been used.

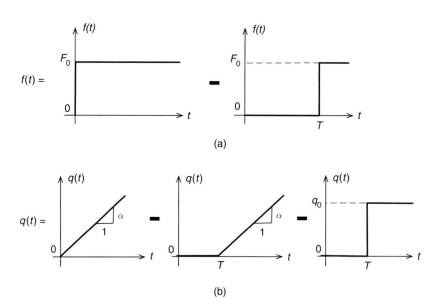

Figure 2.4.3 Graphical derivation of piecewise-continuous functions described in Figure 2.4.2: (a) rectangular pulse, (b) triangular pulse

Example 2.4.15 Deriving the Laplace transform of a triangular pulse

Consider the triangular pulse $q(t)$ in Figure 2.4.2(b), which is defined by

$$q(t) = \begin{cases} \alpha t, & 0 \le t \le T \\ 0, & t > T \end{cases}$$

with the slope $\alpha = q_0/T$. The pulse can be written as

$$q(t) = \alpha\, t\, u(t) - \alpha(t - T)\, u(t - T) - q_0\, u(t - T)$$

which is a linear combination of a ramp function ($\alpha\, t$), a delayed ramp function ($\alpha\, (t - T)$), and a delayed step function ($q_0\, u(t - T)$), as shown in Figure 2.4.3(b). The Laplace transform of this triangular pulse is

$$Q(s) = \frac{\alpha}{s^2}\left(1 - e^{-Ts}\right) - \frac{q_0}{s} e^{-Ts}.$$

2.5 Differential Equations

Differential equations are widely used to describe the behaviors of physical systems that arise in various applications, such as heat and mass transfer, wave propagation, mechanical vibrations, elasticity, combustion, fluid mechanics, gas dynamics, chemical reactions, electrical circuitry, micro-electro-mechanical systems, satellite navigation, robotic systems, and feedback controls. The derivation and solution of differential equations is especially important in modeling and analysis of dynamic systems.

A differential equation is an equation that contains an unknown function and its derivatives. Differential equations can be categorized in several ways, which is briefly discussed in the following subsections.

2.5.1 Linear and Nonlinear Differential Equations

Let $u(x)$ be a function of x, where x is an independent variable. A *linear differential equation* about $u(x)$ is one which only involves a linear combination of $u(x)$ and its derivatives, such as

$$\frac{d^2u}{dx^2} + 2\frac{du}{dx} + 4u = 0$$

$$\frac{du}{dx} + (1 + 0.3e^{-x})u = 5\cos x$$

where the coefficients in linear combination are known functions of x. A *nonlinear differential equation* of $u(x)$ is an equation that contains products and/or nonlinear functions of $u(x)$ and its derivatives, such as

$$\frac{d^2u}{dx^2} + 2u\frac{du}{dx} + 4u = 0$$

$$\frac{d^2u}{dx^2} + x^2\left(\frac{du}{dx}\right)^2 + \frac{xu}{\sqrt{1+u^2}} = 5\cos x$$

In modeling dynamic systems, a general form of linear differential equations is

$$\begin{aligned}a_n(t)\frac{d^n y(t)}{dt^n} + a_{n-1}(t)\frac{d^{n-1}y(t)}{dt^{n-1}} + \ldots + a_1(t)\frac{dy(t)}{dt} + a_0(t)y(t) \\ = b_m(t)\frac{d^m r(t)}{dt^m} + b_{m-1}(t)\frac{d^{m-1}r(t)}{dt^{m-1}} + \ldots + b_1(t)\frac{dr(t)}{dt} + b_0(t)r(t)\end{aligned} \quad (2.5.1)$$

where $y(t)$ is an unknown function, $r(t)$ is a given function, t is an independent temporal variable, and a_k and b_i are known functions of t. Equation (2.5.1) is called an *n-th-order* differential equation because the highest-order derivative of y in it is of order n. The $y(t)$ and $r(t)$ are also called the *output* and *input* of the dynamic system. If $r(t) = 0$, the differential equation is called *homogeneous*, and the corresponding solution $y(t)$ is called a *free response* of the dynamic system. If $r(t) \neq 0$, the differential equation is *nonhomogeneous*, and, provided all initial conditions are zero, the solution $y(t)$ of Eq. (2.5.1) is a *forced response* of the dynamic system. If the a_k values are all constants, the dynamic system is called a time-invariant system. If any of the a_k values are a function of t, the dynamic system is called a time-varying system.

For convenience of analysis, Eq. (2.5.1) is cast as

$$\left(a_n(t)D^n + a_{n-1}(t)D^{n-1} + \ldots + a_1(t)D + a_0(t)\right)y(t)$$
$$= \left(b_m(t)D^m + b_{m-1}(t)D^{m-1} + \ldots + b_1(t)D + b_0(t)\right)r(t) \quad (2.5.2)$$

where D is the differential operator given by

$$D = \frac{d}{dt} \quad (2.5.3)$$

Equation (2.5.2) can be further written in a compact form

$$A(D)y(t) = B(D)r(t) \quad (2.5.4)$$

where $A(D)$ and $B(D)$ are polynomials of the operator D, and they are given by

$$A(D) = \sum_{k=0}^{n} a_k(t)D^k$$

$$B(D) = \sum_{i=0}^{m} b_i(t)D^i \quad (2.5.5)$$

Example 2.5.1 Modeling a spring–mass–damper dynamic system

Application of Newton's second law of motion to the spring–mass–damper system in Figure 2.5.1 results in the second-order linear differential equation, which constitutes a model of the given dynamic system.

This second-order linear differential equation is

$$m\ddot{x}(t) + c\dot{x}(t) + kx(t) = f(t)$$

where $x(t)$ is an unknown displacement, $f(t)$ is a given external force, t is a temporal variable, m, c, and k are the constants that describe mass, damping coefficient, and spring coefficient, respectively, and the over-dot stands for differentiation with respect to t, $\dot{x} = dx/dt$. The derivation of the above equation will be presented in Chapter 3. This differential equation is also shown in Example 2.4.7. The differential equation can be written as

$$A(D)x(t) = B(D)f(t)$$

with

$$A(D) = mD^2 + cD + k, \ B(D) = 1, \ \text{with} \ D = \frac{d}{dt}$$

The spring–mass–damper system is a time-invariant system.

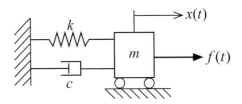

Figure 2.5.1 Spring–mass–damper dynamic system

2.5.2 Systems of Differential Equations

The differential equations in the previous section are single differential equations with one unknown function. A system of differential equations involves multiple unknown functions that are presented in coupled equations.

Example 2.5.2 Modeling of the two-mass dynamic system

Consider two masses (m_1 and m_2) that are coupled by a spring of coefficient k (see Figure 2.5.2). Applying Newton's second law of motion to this system, we obtain the governing differential equations:

$$m_1\ddot{x}_1(t) + k\left(x_1(t) - x_2(t)\right) = f_1(t)$$
$$m_2\ddot{x}_2(t) + k\left(x_2(t) - x_1(t)\right) = f_2(t)$$

where $x_1(t)$ and $x_2(t)$ are the displacements of the masses, and $f_1(t)$ and $f_2(t)$ are external forces applied to the masses. The derivation of these differential equations will be presented in Chapter 3. By comparison, the single differential equation in Example 2.5.1 has one unknown function $x(t)$ while the coupled differential equations for the two-mass system have two unknown functions, $x_1(t)$ and $x_2(t)$.

In many engineering applications, a system of linear differential equations with n unknown functions $x_1(t), x_2(t), \ldots, x_n(t)$ can be written in the following matrix form

$$\{A_m(t)D^m + A_{m-1}(t)D^{m-1} + \ldots + A_1(t)D + A_0(t)\}\boldsymbol{x}(t) = \boldsymbol{f}(t) \quad (2.5.6)$$

where $D = d/dt$; m is the highest order of differentiation; $\boldsymbol{x}(t)$ and $\boldsymbol{f}(t)$ are the vector of unknown functions and the vector of given forcing functions, respectively, that are given by

$$\boldsymbol{x}(t) = \begin{pmatrix} x_1(t) \\ x_2(t) \\ \vdots \\ x_n(t) \end{pmatrix}, \boldsymbol{f}(t) = \begin{pmatrix} f_1(t) \\ f_2(t) \\ \vdots \\ f_n(t) \end{pmatrix} \qquad (2.5.7)$$

and $\boldsymbol{A}_0(t), \boldsymbol{A}_1(t), \ldots, \boldsymbol{A}_m(t)$ are $n \times n$ matrices of known functions. For a set of linear differential equations with constant coefficients, its matrix form is

$$\{\boldsymbol{A}_m D^m + \boldsymbol{A}_{m-1} D^{m-1} + \ldots + \boldsymbol{A}_1 D + \boldsymbol{A}_0\}\boldsymbol{x}(t) = \boldsymbol{f}(t) \qquad (2.5.8)$$

in which all $\boldsymbol{A}_k$ are constant matrices. For instance, the differential equations in Example 2.5.2 are of the matrix form

$$\begin{bmatrix} m_1 & 0 \\ 0 & m_2 \end{bmatrix} \begin{pmatrix} \ddot{x}_1(t) \\ \ddot{x}_2(t) \end{pmatrix} + \begin{bmatrix} k & -k \\ -k & k \end{bmatrix} \begin{pmatrix} x_1(t) \\ x_2(t) \end{pmatrix} = \begin{pmatrix} f_1(t) \\ f_2(t) \end{pmatrix}$$

or

$$\boldsymbol{M}\ddot{\boldsymbol{x}}(t) + \boldsymbol{K}\boldsymbol{x}(t) = \boldsymbol{f}(t)$$

with

$$\boldsymbol{x}(t) = \begin{pmatrix} x_1(t) \\ x_2(t) \end{pmatrix}, \boldsymbol{f}(t) = \begin{pmatrix} f_1(t) \\ f_2(t) \end{pmatrix}, \boldsymbol{M} = \begin{bmatrix} m_1 & 0 \\ 0 & m_2 \end{bmatrix}, \boldsymbol{K} = \begin{bmatrix} k & -k \\ -k & k \end{bmatrix}.$$

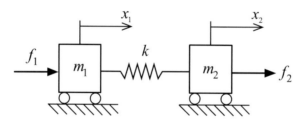

Figure 2.5.2 Two-mass dynamic system

2.5.3 Partial Differential Equations

The differential equations in the previous subsections are called *ordinary differential equations* because the unknown functions in them are functions of one independent variable (like x or t). In engineering problems there are also differential equations with unknowns being functions of two or more independent variables. Two examples are given below.

(i) Equation of motion of an Euler–Bernoulli beam in transverse vibration

$$\rho(x)\frac{\partial^2 w(x,t)}{\partial t^2} + \frac{\partial^2}{\partial x^2}\left(EI(x)\frac{\partial^2 w(x,t)}{\partial x^2}\right) = f(x,t), \quad 0 < x < L$$

where $w(x,t)$, $\rho(x)$, $EI(x)$, and L represent the transverse displacement, linear density (mass per unit length), bending stiffness, and length of the beam, respectively; $f(x,t)$ is an external force; and x and t are spatial and temporal parameters.

(ii) Governing equation for heat conduction in a three-dimensional body (heat equation)

$$c_p \rho \frac{\partial T}{\partial t} = \kappa\left(\frac{\partial^2 T}{\partial x^2} + \frac{\partial^2 T}{\partial y^2} + \frac{\partial^2 T}{\partial z^2}\right) + q(x,y,z,t)$$

where $T = T(x,y,z,t)$ is the temperature distribution; $q = q(x,y,z,t)$ is the internal energy generation; ρ, c_p, and κ are the volume density, specific heat, and thermal conductivity of the body, respectively; x, y, and z are spatial coordinates; and t is a temporal parameter.

The scope of this textbook is about dynamic systems described by ordinary differential equations. Consequently, derivation and solution of partial differential equations will not be covered.

2.5.4 State Equations

State equations are often used in modeling, simulation, and analysis of dynamic systems with multiple inputs and multiple outputs. State equations are a set of coupled first-order ordinary differential equations of the form

$$\begin{cases} \dot{x}_1 = f_1(x_1, x_2, \ldots, x_n; u_1, u_2, \ldots, u_p) \\ \dot{x}_2 = f_2(x_1, x_2, \ldots, x_n; u_1, u_2, \ldots, u_p) \\ \quad \vdots \\ \dot{x}_n = f_n(x_1, x_2, \ldots, x_n; u_1, u_2, \ldots, u_p) \end{cases} \quad (2.5.9)$$

where the unknown functions $x_k = x_k(t)$, $k = 1, 2, \ldots, n$ are state variables; $u_i = u_i(t)$, $i = 1, 2, \ldots p$ are given inputs; and f_1, $f_2, > \ldots, f_n$ are functions of the state variables and inputs. Equation (2.5.9) is a nonlinear form of a set of state equations. Such a set of linear state equations is of the form

$$\begin{cases} \dot{x}_1 = a_{11}x_1 + a_{12}x_2 + \ldots + a_{1n}x_n + b_{11}u_1 + b_{12}u_2 + \ldots + b_{1p}u_p \\ \dot{x}_2 = a_{21}x_1 + a_{22}x_2 + \ldots + a_{2n}x_n + b_{21}u_1 + b_{22}u_2 + \ldots + b_{2p}u_p \\ \quad \vdots \\ \dot{x}_n = a_{n1}x_1 + a_{n2}x_2 + \ldots + a_{nn}x_n + b_{n1}u_1 + b_{n2}u_2 + \ldots + b_{np}u_p \end{cases} \quad (2.5.10)$$

where a_{jk} and b_{ji} are either known functions of t or constants.

It should be noted that, in a state equation, the left-hand side only contains the first derivative of a state variable and the right-hand side does not have any derivative term of x_k or u_i.

Example 2.5.3 Deriving state equations

$$\begin{cases} \dot{x}_1 = 2e^{-x_1}\cos(3x_2) + 4u \\ \dot{x}_2 = \dfrac{2e^{-x_2}\sin(3x_2)}{\sqrt{1+u^2}} \end{cases}$$

is a set of nonlinear state equations, and

$$\begin{cases} \dot{x}_1 = 2x_1 + 5x_2 - 7x_2 + 6u_1 + 3u_1 - 3u_2 \\ \dot{x}_2 = 4x_1 + (1 + 0.2\sin 3t)x_2 + 3x_3 + u_2 \\ \dot{x}_3 = -6x_1 + 10e^{-0.25t}x_3 + 13u_1 - 9u_2 \end{cases}$$

is a set of linear state equations.

State equations are often cast in a matrix form. The nonlinear state equations (2.5.9) can be written as

$$\dot{\boldsymbol{x}}(t) = \boldsymbol{\varphi}(\boldsymbol{x}(t), \boldsymbol{u}(t)) \tag{2.5.11}$$

where $\boldsymbol{x}(t)$ and $\boldsymbol{\phi}$ are given by

$$\boldsymbol{x}(t) = \begin{pmatrix} x_1(t) \\ x_2(t) \\ \vdots \\ x_n(t) \end{pmatrix}, \quad \boldsymbol{u}(t) = \begin{pmatrix} u_1(t) \\ u_2(t) \\ \vdots \\ u_p(t) \end{pmatrix}, \quad \boldsymbol{\varphi}(\boldsymbol{x}(t), \boldsymbol{u}(t)) = \begin{pmatrix} \varphi_1(x_1, x_2, \ldots, x_n; u_1, u_2, \ldots, u_p) \\ \varphi_2(x_1, x_2, \ldots, x_n; u_1, u_2, \ldots, u_p) \\ \vdots \\ \varphi_n(x_1, x_2, \ldots, x_n; u_1, u_2, \ldots, u_p) \end{pmatrix} \tag{2.5.12}$$

with $\varphi_1, \varphi_2, \ldots \varphi_n$ being given nonlinear functions. The $\boldsymbol{x}(t)$ is called the state vector, and $\boldsymbol{u}(t)$ the input vector. Likewise, the linear state equations (2.5.10) can be written as

$$\dot{\boldsymbol{x}}(t) = \boldsymbol{A}\boldsymbol{x}(t) + \boldsymbol{B}\boldsymbol{u}(t) \tag{2.5.13}$$

where $\boldsymbol{x}(t)$ and $\boldsymbol{u}(t)$ are the state and input vectors as given in Eq. (2.5.12), and matrices $\boldsymbol{A}$ and $\boldsymbol{B}$ are as follows:

$$\boldsymbol{A} = \begin{bmatrix} a_{11} & a_{12} & \cdots & a_{1n} \\ a_{21} & a_{22} & \cdots & a_{2n} \\ \vdots & \vdots & \ddots & \vdots \\ a_{n1} & a_{n2} & \cdots & a_{nn} \end{bmatrix}, \quad \boldsymbol{B} = \begin{bmatrix} b_{11} & b_{12} & \cdots & b_{1p} \\ b_{21} & b_{22} & \cdots & b_{2p} \\ \vdots & \vdots & \ddots & \vdots \\ b_{n1} & b_{n2} & \cdots & b_{np} \end{bmatrix} \tag{2.5.14}$$

The linear state equation (2.5.13) is a special case of the nonlinear equation (2.5.11) with $\boldsymbol{\varphi} = \boldsymbol{A}\boldsymbol{x}(t) + \boldsymbol{B}\boldsymbol{u}(t)$.

An n-th-order differential equation can be converted into a set of n state equations through the proper selection of state variables. For instance, consider the governing equation of the spring–mass–damper system in Example 2.5.1:

$$m\ddot{x}(t) + c\dot{x}(t) + kx(t) = f(t)$$

Through selection of state variables as $x_1 = x$ and $x_2 = \dot{x}$, the second-order differential equation is cast into a system of state equations

$$\begin{cases} \dot{x}_1 = x_2 \\ \dot{x}_2 = -\dfrac{k}{m}x_1 - \dfrac{c}{m}x_2 + \dfrac{1}{m}f(t) \end{cases}$$

The deduction of state equations from nonlinear differential equations can be similarly done. For example, let the governing equation for a spring–mass–system subject to nonlinear damping be

$$m\ddot{x}(t) + c\dot{x}(t) + c_1\dot{x}^3(t) + kx(t) = f(t)$$

where the cubic term $c_1\dot{x}^3$ describes the damping nonlinearity. With the state variables defined by $x_1 = x$ and $x_2 = \dot{x}$, the nonlinear state equations are obtained as follows:

$$\begin{cases} \dot{x}_1 = x_2 \\ \dot{x}_2 = -\dfrac{k}{m}x_1 - \dfrac{c}{m}x_2 - \dfrac{c_1}{m}x_2^3 + \dfrac{1}{m}f(t) \end{cases}$$

In general, to convert an n-th-order differential equation into a set of state equations, n state variables must be defined. For the n-th-order linear differential equation

$$a_n(t)\frac{d^n y(t)}{dt^n} + a_{n-1}(t)\frac{d^{n-1} y(t)}{dt^{n-1}} + \ldots + a_1(t)\frac{dy(t)}{dt} + a_0(t)y(t) = r(t) \tag{2.5.15}$$

where $a_k(t)$ are known functions, the selection of state variables by

$$x_1(t) = y(t),\ x_2(t) = \dot{x}_1(t),\ x_3(t) = \dot{x}_3(t),\ \ldots,\ x_n(t) = \dot{x}_{n-1}(t) \tag{2.5.16}$$

eventually yields the matrix state equation

$$\dot{\mathbf{x}}(t) = \mathbf{A}(t)\mathbf{x}(t) + \mathbf{B}r(t) \tag{2.5.17}$$

with matrices $\mathbf{A}$ and $\mathbf{B}$ being

$$\mathbf{A}(t) = \begin{bmatrix} 0 & 1 & 0 & \cdots & 0 \\ 0 & 0 & 1 & \cdots & 0 \\ \vdots & \vdots & \vdots & \ddots & \vdots \\ 0 & 0 & 0 & \cdots & 1 \\ -\dfrac{a_0(t)}{a_n(t)} & -\dfrac{a_1(t)}{a_n(t)} & -\dfrac{a_2(t)}{a_n(t)} & \cdots & -\dfrac{a_{n-1}(t)}{a_n(t)} \end{bmatrix},\ \mathbf{B} = \begin{pmatrix} 0 \\ 0 \\ \vdots \\ 0 \\ \dfrac{1}{a_n(t)} \end{pmatrix} \tag{2.5.18}$$

The derivation of state equations for dynamic systems will be further discussed later on.

2.6 Solution of Linear Ordinary Differential Equations with Constant Coefficients

2.6.1 Free Response and Forced Response

In this section, we present analytical methods for the solution of linear ordinary differential equations with constant coefficients, which, as mentioned previously, are used to model time-invariant dynamic systems. The solution of linear differential equations with nonconstant coefficients and nonlinear equations will be discussed in numerical integration of state equations (see Section 2.8).

The application of the methods described in this section to different types of dynamic systems is further discussed in Chapters 3 to 9.

The differential equation under consideration has the following form

$$A(D)y(t) = B(D)r(t) \tag{2.6.1}$$

and is subject to the initial conditions

$$y(0) = a_{0,0}, \ Dy(0) = a_{0,1}, \ D^2y(0) = a_{0,2}, \ldots, \ D^{n-1}y(0) = a_{0,n-1} \tag{2.6.2}$$

where $D = d/dt$, and $A(D)$ and $B(D)$ are polynomials of the operator D, such as

$$A(D) = \sum_{k=0}^{n} a_k D^k, \ B(D) = \sum_{i=0}^{m} b_i D^i \tag{2.6.3}$$

where a_k and b_i are constant coefficients, and $a_{0,0}, a_{0,1}, \ldots, a_{0,n-1}$ are the prescribed initial values. The given function $r(t)$ is normally referred to as an *input* or a *forcing function*. In general, for a physically viable system, $n \geq m$. Thus, the mathematical problem herein is to solve the differential equation (2.6.1) for the unknown function $y(t)$, for given input $r(t)$ and initial disturbances described in Eq. (2.6.2).

According to the theory of differential equations, the solution of the above-described equation with the initial conditions can be written as

$$y(t) = y_I(t) + y_F(t) \tag{2.6.4}$$

where $y_I(t)$ is the solution of the homogeneous equation

$$A(D)y(t) = 0 \tag{2.6.5}$$

subject to the initial conditions of Eq. (2.6.2), and $y_F(t)$ is the solution of Eq. (2.6.1) subject to the homogeneous initial conditions, such as

$$y(0) = 0, \ Dy(0) = 0, \ D^2y(0) = 0, \ldots, \ D^{n-1}y(0) = 0 \tag{2.6.6}$$

2.6 Solution of Linear Ordinary Differential Equations

In modeling and analysis of dynamic systems, $y_I(t)$ and $y_F(t)$ are called the *free response* and *forced response* respectively. Physically, $y_I(t)$ is the system response caused by the initial disturbances prescribed by Eq. (2.6.2), and $y_F(t)$ is caused by the input $r(t)$. Hence, the complete solution of a differential equation can be viewed as a sum of free response and forced response.

Example 2.6.1 Response of the spring–mass–damper system

Consider the spring–mass–damper system given in Example 2.5.1.

The system response, namely the displacement $x(t)$ of the mass, is governed by the differential equation

$$m\ddot{x}(t) + c\dot{x}(t) + kx(t) = f(t)$$

subject to the initial conditions

$$x(0) = x_0, \quad \dot{x}(0) = v_0$$

where x_0 and v_0 are the initial displacement and the initial velocity of the mass, respectively. The solution of this governing differential equation can be written as

$$x(t) = x_I(t) + x_F(t)$$

in which the free response $x_I(t)$ is the solution of

$$\begin{cases} m\ddot{x}(t) + c\dot{x}(t) + kx(t) = 0 \\ x(0) = x_0, \quad \dot{x}(0) = v_0 \end{cases}$$

and the forced response $x_F(t)$ is the solution of

$$\begin{cases} m\ddot{x}(t) + c\dot{x}(t) + kx(t) = f(t) \\ x(0) = 0, \quad \dot{x}(0) = 0 \end{cases}$$

There exist four widely used analytical methods for the solution of linear differential equations with constant coefficients: (1) direct integration, (2) the method of separation of variables, (3) the Laplace transform, and (4) the method of undetermined coefficients.

2.6.2 Solution by Direct Integration

For a linear differential equation with constant coefficients, where it is possible to isolate the function derivative on the left-hand side

$$D^n y(t) = f(t) \qquad (2.6.7)$$

the solution can be found by integrating both sides of the equation n times

$$y(t) = y(0) + \int_0^t \left(\dot{y}(0) + \ldots \int_0^t \left(y^{(n-1)}(0) + \int_0^t f(t)\,dt \right) dt \ldots \right) dt \qquad (2.6.8)$$

This method is straightforward, but the inherent complexity of the integration operation limits its applicability to simple first- and second-order differential equations.

Example 2.6.2 Solving a first-order differential equation by direct integration

Consider a dynamic system governed by the first-order differential equation with the given initial conditions

$$\begin{cases} m\dot{x}(t) = f(t) \\ x(0) = x_0 \end{cases}$$

Integrate both sides of this equation once to derive the solution

$$\int_{x(0)}^{x(t)} \frac{dx}{dt}\,dt = \frac{1}{m}\int_0^t f(t)\,dt$$

$$x(t) = x_0 + \frac{1}{m}\int_0^t f(t)\,dt$$

If a dynamic system is governed by the second-order differential equation with the given initial conditions

$$\begin{cases} m\ddot{x}(t) = f(t) \\ x(0) = x_0,\ \dot{x}(0) = v_0 \end{cases}$$

integration needs to be done twice as follows:

$$\int_{\dot{x}(0)}^{\dot{x}(t)} \frac{d^2x}{dt^2}\,dt = \frac{1}{m}\int_0^t f(t)\,dt$$

$$\dot{x}(t) = v_0 + \frac{1}{m}\int_0^t f(t)\,dt$$

$$\int_{x(0)}^{x(t)} \frac{dx}{dt}\,dt = \int_0^t \left(v_0 + \frac{1}{m}\int_0^t f(t)\,dt \right) dt$$

$$x(t) = x_0 + v_0 t + \frac{1}{m}\int_0^t \left(\int_0^t f(t)\,dt \right) dt$$

The following numerical example illustrates the direct integration method.

2.6 Solution of Linear Ordinary Differential Equations

Example 2.6.3 Solving a second-order differential equation by direct integration

Consider a dynamic system governed by the second-order differential equation with the given initial conditions

$$\begin{cases} \ddot{x}(t) = 3t^2 \\ x(0) = 1, \; \dot{x}(0) = 2 \end{cases}$$

The solution is found by direct integration as follows:

$$\int_{\dot{x}(0)}^{\dot{x}(t)} \frac{d^2 x}{dt^2} dt = \int_0^t 3t^2 dt$$

$$\dot{x}(t) = 2 + t^3$$

$$\int_{x(0)}^{x(t)} \frac{dx}{dt} dt = \int_0^t \left(2 + t^3\right) dt$$

$$x(t) = 1 + 2t + \frac{1}{4} t^4$$

2.6.3 Solution by the Separation of Variables

This approach can be viewed as an extension of the direct integration method, mostly applicable to first-order differential equations of the following structure:

$$\frac{dx}{dt} = f_1(t) f_2(x) \qquad (2.6.9)$$

For the first-order equation shown above, the solution is found in the following way. First, the equation is reformulated in such a way that its left-hand side is only in terms of x, while the right-hand side is only in terms of t:

$$\frac{dx}{f_2(x)} = f_1(t) dt \qquad (2.6.10)$$

Then, both sides of the resulting equation are integrated:

$$\int_{x(0)}^{x(t)} \frac{dx}{f_2(x)} = \int_0^t f_1(t) dt \qquad (2.6.11)$$

The sought solution $x(t)$ is found by Eq. (2.6.11), provided that integrals on the right-hand side and left-hand side can be evaluated analytically, and $x(t)$ can be derived from the resulting expression.

Example 2.6.4 Solving a first-order differential equation by the separation of variables

Consider a dynamic system governed by the first-order differential equation with the given initial conditions

$$\begin{cases} \dot{x}(t) + x = 3 \\ x(0) = 2 \end{cases}$$

Rearrange this equation into the form, required by the method of separation of variables

$$\frac{dx}{dt} = 3 - x$$

Obviously, $f_1(t) = 1$ and $f_2(x) = 3 - x$. Integrating both sides, we obtain

$$\int_{x(0)}^{x(t)} \frac{dx}{3-x} = \int_0^t 1 \, dt$$

Evaluation of the integral on the right-hand side of the equation is straightforward, while the integral on the left-hand side can be evaluated using integral tables:

$$\int_{x(0)}^{x(t)} \frac{dx}{3-x} = \int_2^{x(t)} \frac{dx}{3-x} = \frac{1}{-1} \ln|3-x| \Big|_2^{x(t)} = -\ln|3-x(t)| = \int_0^t 1 \, dt = t$$

From the resulting expression $x(t)$ can be easily derived:

$$-\ln|3 - x(t)| = t$$
$$3 - x(t) = e^{-t}$$
$$x(t) = 3 - e^{-t}$$

2.6.4 Solution by the Laplace Transform

The Laplace transform is a widely used method in modeling, analysis, and design of control algorithms for linear time-invariant dynamic systems, and in solving the related differential equations with constant coefficients. As mentioned earlier in this chapter (see Section 2.4), the Laplace transform reduces the original problem, described by differential equations, to a set of algebraic equations in the s-domain. The functions of s, which result from solving these algebraic equations, are then transformed into the time domain by using the inverse Laplace transform. To understand how this approach works, consider a simple example:

$$\begin{cases} \dot{x}(t) + 3x(t) = 5 \\ x(0) = 2 \end{cases}$$

Taking the Laplace transform of the above equation gives

$$sX(s) - x(0) + 3X(s) = 5\frac{1}{s} \Rightarrow (s+3)X(s) = 5\frac{1}{s} + 2$$

where the initial condition $x(0) = 2$ has been used. It follows that the s-domain response is

$$X(s) = \frac{2s+5}{s(s+3)}$$

By partial fraction expansion (see Section 2.4.4), the solution (time-domain response) to the differential equation is

$$x(t) = \mathcal{L}^{-1}\left\{\left(\frac{5}{3}\right)\frac{1}{s} + \left(\frac{1}{3}\right)\frac{1}{s+3}\right\} = \left(\frac{1}{3}\right)(5 + e^{-3t})$$

The above example implies three basic steps in the solution of general differential equations via the Laplace transformation, which are outlined below.

Step 1: Laplace Transform of a Differential Equation

Take the Laplace transform of Eq. (2.6.1) and substitute the initial conditions (2.6.2) into the resulting equation. This leads to

$$A(s)Y(s) = B(s)R(s) + I(s) \qquad (2.6.12)$$

where $Y(s)$ and $R(s)$ are the Laplace transforms of $y(t)$ and $r(t)$, respectively; $A(s)$ and $B(s)$ are the polynomials in Eq. (2.6.3) with operator D replaced by the Laplace transform parameter s:

$$A(s) = \sum_{k=0}^{n} a_k s^k, \quad B(s) = \sum_{i=0}^{m} b_i s^i \qquad (2.6.13)$$

and $I(s)$ is an expression consisting of the initial values of $y(t)$, $r(t)$ and its derivatives. In general, $I(s)$ is a polynomial of s and $I(s) = 0$ when all the initial values of $y(t)$, $r(t)$, and their derivatives are set to zero. For instance, consider the Laplace transform of the differential equation

$$\begin{cases} m\ddot{x}(t) + c\dot{x}(t) + kx(t) = f(t) \\ x(0) = x_0, \ \dot{x}(0) = v_0 \end{cases}$$

in Example 2.5.1, which yields Eq. (2.6.12) with

$$A(s) = ms^2 + cs + k, \ B(s) = 1, \ I(s) = mx_0 s + mv_0 + cx_0$$

Note that $I(s) = 0$ if and only if $x_0 = 0$ and $v_0 = 0$.

Step 2: Determination of the s-Domain Response

By Eq. (2.6.12), the s-domain solution is of the form

$$Y(s) = \frac{B(s)R(s) + I(s)}{A(s)} \tag{2.6.14}$$

Step 3: Determination of the Time-Domain Response via the Inverse Laplace Transform

The inverse Laplace transform of Eq. (2.6.14) gives

$$y(t) = y_I(t) + y_F(t) \tag{2.6.15}$$

where

$$y_I(t) = \mathcal{L}^{-1}\left\{\frac{I(s)}{A(s)}\right\}, \ y_F(t) = \mathcal{L}^{-1}\left\{\frac{B(s)}{A(s)}R(s)\right\} \tag{2.6.16}$$

with $y_I(t)$ and $y_F(t)$ being the free response and forced response as described in Section 2.6.1. The inverse Laplace transformation in Eq. (2.6.16) can be performed via partial fraction expansion as described in Section 2.4.4.

Example 2.6.5 Solving a second-order differential equation by the Laplace transform

Consider the differential equation

$$\begin{cases} m\ddot{x}(t) + kx(t) = f_0 \\ x(0) = x_0, \ \dot{x}(0) = v_0 \end{cases}$$

which is a special case of the Example 2.6.1 with $c = 0$ and $f(t)$ being a step (constant) input. Using the results in Example 2.5.1, the s-domain solution is

$$X(s) = \frac{mx_0 s + mv_0}{ms^2 + k} + \frac{f_0}{s(ms^2 + k)} = \frac{x_0 s + v_0}{s^2 + \omega_n^2} + \frac{f_0}{k}\left(\frac{1}{s} - \frac{s}{s^2 + \omega_n^2}\right)$$

where $\omega_n^2 = \dfrac{k}{m}$. With Table 2.4.1, the inverse Laplace transform of the above equation gives the solution of the differential equation:

$$x(t) = \mathcal{L}^{-1}\{X(s)\} = x_0 \cos \omega_n t + \dfrac{v_0}{\omega_n} \sin \omega_n t + \dfrac{f_0}{k}(1 - \cos \omega_n t)$$

According to Section 2.6.1, the free response and forced response of the differential equation are

$$x_I(t) = x_0 \cos \omega_n t + \dfrac{v_0}{\omega_n} \sin \omega_n t$$

$$x_F(t) = \dfrac{f_0}{k}(1 - \cos \omega_n t)$$

Example 2.6.6 Solving a third-order differential equation by the Laplace transform

A third-order differential equation with initial conditions is given as follows:

$$\dddot{x}(t) + 6\ddot{x}(t) + 11\dot{x}(t) + 6x(t) = 6(1 - e^{-4t})$$

$$x(0) = 1, \; \dot{x}(0) = 0, \; \ddot{x}(0) = 0$$

The Laplace transform of the above equation leads to

$$(s^3 + 6s^2 + 11s + 6)X(s) = 6\left(\dfrac{1}{s} - \dfrac{1}{s+4}\right) + s^2 + 6s + 11 = \dfrac{s^4 + 10s^3 + 34s^2 + 44s + 4}{s(s+4)}$$

It can be shown that the algebraic equation $s^3 + 6s^2 + 11s + 6 = 0$ has roots $-1, -2,$ and -3, which indicates that $s^3 + 6s^2 + 11s + 6 = (s+1)(s+2)(s+3)$. From this result it follows that

$$X(s) = \dfrac{s^4 + 10s^3 + 34s^2 + 44s + 24}{s^5 + 10s^4 + 35s^3 + 50s^2 + 24s} = \dfrac{s^4 + 10s^3 + 34s^2 + 44s + 24}{s(s+1)(s+2)(s+3)(s+4)}$$

By partial fraction expansion (Section 2.4.4),

$$X(s) = \dfrac{1}{s} - \dfrac{5}{6(s+1)} + \dfrac{2}{s+2} - \dfrac{3}{2(s+3)} + \dfrac{1}{3(s+4)}$$

Thus, the sought solution (time-domain response) of the given differential equation is

$$x(t) = \mathcal{L}^{-1}\{X(s)\} = 1 - \dfrac{5}{6}e^{-t} + 2e^{-2t} - \dfrac{3}{2}e^{-3t} + \dfrac{1}{3}e^{-4t}$$

The $x(t)$ is plotted against t in Figure 2.6.1.

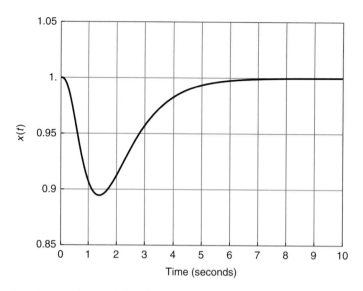

Figure 2.6.1 Time-domain response of the given third-order differential equation

Example 2.6.7 Solving a first-order differential equation with a piecewise-continuous forcing function by the Laplace transform

The Laplace transform is quite useful for solving linear time-invariant differential equations with a piecewise-continuous forcing function.

Consider the system with the rectangular pulse forcing function (see Figure 2.4.2(a)) with the amplitude $F_0 = 3$ and duration $T = 2$:

$$\dot{x}(t) + 4x(t) = f(t) = 3u(t) - 3u(t-2) \text{ for } t > 0, \text{ subject to } x(0) = 1$$

Taking the Laplace transform of this equation, we obtain (the Laplace transform of the rectangular pulse was derived in Example 2.4.14):

$$\mathcal{L}\{\dot{x}(t) + 4x(t)\} = sX(s) - x(0) + 4X(s) = (s+4)X(s) - 1$$

$$= \mathcal{L}\{3u(t) - 3u(t-2)\} = \frac{3}{s}\left(1 - e^{-2s}\right)$$

Then the solution in the Laplace domain is derived from this algebraic equation as

$$X(s) = \frac{3 - 3e^{-2s} + s}{s(s+4)} = \frac{3}{s(s+4)} - \frac{3}{s(s+4)}e^{-2s} + \frac{1}{s+4}$$

To invert this solution to the time domain, use the partial fraction expansion, Table 2.4.1, and the Laplace transform properties

2.6 Solution of Linear Ordinary Differential Equations

$$x(t) = \mathcal{L}^{-1}\{X(s)\} = \mathcal{L}^{-1}\left\{\frac{3}{s(s+4)} - \frac{3}{s(s+4)}e^{-2s} + \frac{1}{s+4}\right\}$$

$$\mathcal{L}^{-1}\left\{\frac{3}{s(s+4)}\right\} = \mathcal{L}^{-1}\left\{\frac{3}{4}\left(\frac{1}{s}\right) - \frac{3}{4}\left(\frac{1}{s+4}\right)\right\} = \frac{3}{4}u(t) - \frac{3}{4}e^{-4t}u(t) = \frac{3}{4}(1-e^{-4t})u(t)$$

$$\mathcal{L}^{-1}\left\{\frac{3}{s(s+4)}e^{-2s}\right\} = \frac{3}{4}\left(1-e^{-4(t-2)}\right)u(t-2)$$

$$\mathcal{L}^{-1}\left\{\frac{1}{s+4}\right\} = e^{-4t}u(t)$$

$$x(t) = \left(e^{-4t} + \frac{3}{4}(1-e^{-4t})\right)u(t) - \frac{3}{4}\left(1-e^{-4(t-2)}\right)u(t-2)$$

which can also be written as

$$x(t) = \begin{cases} e^{-4t} + \dfrac{3}{4}(1-e^{-4t}) & \text{for } 0 \leq t \leq 2 \\ e^{-4t} + \dfrac{3}{4}(1-e^{-4t}) + \dfrac{3}{4}\left(1-e^{-4(t-2)}\right) & \text{for } t > 2 \end{cases}$$

This piecewise-continuous solution is shown in Figure 2.6.2.

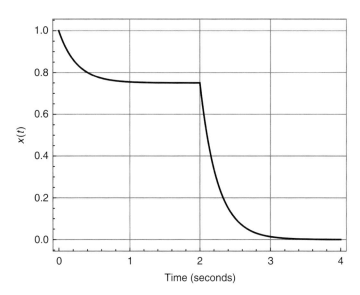

Figure 2.6.2 Piecewise-continuous time-domain response of the given first-order differential equation

2.6.5 The Method of Undetermined Coefficients

Consider the n-th-order differential equation

$$\{a_n D^n + a_{n-1} D^{n-1} + \ldots + a_1 D + a_0\} x(t) = f(t) \quad (2.6.17)$$

subject to the initial conditions

$$x(0) = a_{0,0}, \; Dx(0) = a_{0,1}, \ldots, D^{n-1} x(0) = a_{0,n-1} \quad (2.6.18)$$

with $D = d/dt$. Here, for simplicity of presentation, $f(t) = B(D)r(t)$ has been used. In the context of the method of undetermined coefficients, the solution of Eq. (2.6.17) is written as

$$x(t) = x_h(t) + x_p(t) \quad (2.6.19)$$

where $x_h(t)$ (*homogenous solution*) is a nonzero or nontrivial solution of the complementary homogenous equation

$$\{a_n D^n + a_{n-1} D^{n-1} + \ldots + a_1 D + a_0\} x_h(t) = 0 \quad (2.6.20)$$

and $x_p(t)$ is a *particular solution* of Eq. (2.6.17). Here, neither $x_h(t)$ nor $x_p(t)$ have to satisfy the initial conditions of Eq. (2.6.18).

It should be pointed out that the particular solution $x_p(t)$ is not unique. For instance, consider the first-order differential equation

$$\dot{x}(t) + 2x(t) = 4$$

It is easy to show that both $x_{p1}(t) = 2$ and $x_{p2}(t) = 2 + \alpha e^{-2t}$, with α being an arbitrary number, satisfy the differential equation. Nevertheless, any particular solution that can be used in the format of Eq. (2.6.19) will yield the unique solution of the differential equation (2.6.17) subject to the initial conditions of Eq. (2.6.18).

It needs to be noted that the homogeneous solution $x_h(t)$ and particular solution $x_p(t)$ must be linearly independent, meaning that the expression

$$\alpha_1 x_h(t) + \alpha_2 x_p(t) = 0$$

holds true if and only if $\alpha_1 = \alpha_2 = 0$

Solution of the Complementary Homogenous Equation

To determine the solution of the homogenous equation, write it in an exponential form

$$x_h(t) = A e^{\lambda t} \quad (2.6.21)$$

and substitute it into Eq. (2.6.20). This leads to

2.6 Solution of Linear Ordinary Differential Equations

$$\{a_n\lambda^n + a_{n-1}\lambda^{n-1} + \ldots + a_1\lambda + a_0\}Ae^{\lambda t} = 0$$

which, by the nontrivial solution assumption ($A \neq 0$), yields the *characteristic equation* of the differential equation:

$$a_n\lambda^n + a_{n-1}\lambda^{n-1} + \ldots + a_1\lambda + a_0 = 0 \qquad (2.6.22)$$

Equation (2.6.17), being an n-th-order polynomial of λ, has n roots. Denote the characteristic roots by $\lambda_1, \lambda_2, \ldots, \lambda_n$. The expression for $x_h(t)$ must contain n linearly independent terms. If all the characteristic roots are distinct, the solution of the homogenous equation is

$$x_h(t) = A_1 e^{\lambda_1 t} + A_2 e^{\lambda_2 t} + \ldots + A_n e^{\lambda_n t} \qquad (2.6.23)$$

which has n linearly independent terms. If there exist repeated roots, the expression for $x_h(t)$ will be different in order to have n linearly independent terms. For instance, assume that the characteristic equation (2.6.22) has p identical roots of λ_0 and m distinct roots of $\lambda_1, \lambda_2, \ldots, \lambda_m$, where $\lambda_0 \neq \lambda_k$ for $k = 1, 2, \ldots, m$ and $p + m = n$. Here, the integer p is called root multiplicity. In this case,

$$x_h(t) = A_1 e^{\lambda_1 t} + A_2 e^{\lambda_2 t} + \ldots + A_m e^{\lambda_m t} + \left(B_0 + B_1 t + \ldots + B_{p-1} t^{p-1}\right) e^{\lambda_0 t} \qquad (2.6.24)$$

where the last p terms on the right-hand side of the equation are related to λ_0 of multiplicity p. The coefficients A and B in Eqs. (2.6.23) and (2.6.24) will be determined later on through the use of the initial conditions of Eq. (2.6.18).

The characteristic equation (2.6.22) can be written as

$$A(\lambda) = 0 \qquad (2.6.25)$$

where $A(s)$ is the polynomial given in Eq. (2.6.13), with the s replaced by λ. This means that the roots of the characteristic equation (2.6.22) are the same as the poles of the s-domain solution depicted in Eq. (2.6.14).

Example 2.6.8 Deriving the general homogeneous solution of a linear differential equation with constant coefficients and real distinct roots

Consider the differential equation in the Example 2.6.6, which has the complementary homogeneous equation

$$\dddot{x}(t) + 6\ddot{x}(t) + 11\dot{x}(t) + 6x(t) = 0$$

and the characteristic equation

$$\lambda^3 + 6\lambda^2 + 11\lambda + 6 = 0$$

with roots $\lambda_1 = -1$, $\lambda_2 = -2$, and $\lambda_3 = -3$ The solution of the homogeneous equation then is

$$x_h(t) = A_1 e^{-t} + A_2 e^{-2t} + A_3 e^{-3t}.$$

Example 2.6.9 Deriving the general homogeneous solution of linear differential equation with constant coefficients and real repeated roots

Consider the complementary homogeneous equation

$$\dddot{x}(t) + 5\ddot{x}(t) + 7\dot{x}(t) + 3x(t) = 0$$

which has the characteristic equation

$$\lambda^3 + 5\lambda^2 + 7\lambda + 3 = 0$$

The characteristic roots are -1, -1, and -3, with root -1 having multiplicity 2. By Eq. (2.6.24), the solution of the homogeneous equation is

$$x_h(t) = Ae^{-3t} + (B_0 + B_1 t)e^{-t}$$

Determination of a Particular Solution

In the context of the method of undetermined coefficients, an educated guess of the form of a particular solution $x_p(t)$ is made for the given forcing function $f(t)$ on the right-hand side of Eq. (2.6.17). The guessed $x_p(t)$ contains some coefficients that are to be solved for.

In general, the guessed form of $x_p(t)$ should "mirror" the form of the forcing function as shown in Table 2.6.1 for first-order differential equations.

For the first-order differential equation with an exponential forcing function, the existence of two options for a particular solution is prompted by the necessity of having linearly independent $x_h(t)$ and $x_p(t)$ as explained in detail below.

First-Order Linear Ordinary Differential Equations

A first-order differential equation

$$a\dot{x}(t) + bx(t) = f(t), \quad a \neq 0 \tag{2.6.26}$$

has the corresponding characteristic equation $a\lambda + b = 0$, which has a real root $\lambda_1 = -b/a$. Table 2.6.1 lists polynomial, sinusoidal, and exponential forms of forcing function, and corresponding "guessed" particular solutions. In the table, the coefficients of $f(t)$ are given; the coefficients of $x_p(t)$ are determined by plugging it into Eq. (2.6.26). When $f(t)$ is an exponential function with $\alpha = -b/a$ (the characteristic root), the particular solution is $x_p(t) = Bte^{\alpha t}$. This is because the $e^{\alpha t}$ itself is a solution of the homogeneous equation $a\dot{x}(t) + bx(t) = 0$, which does not satisfy the inhomogeneous equation (2.6.26).

Example 2.6.10 Deriving a particular solution of a first-order linear differential equation with constant coefficients

Consider the differential equation

$$\dot{x}(t) + 3x(t) = 2(1 - e^{-3t})$$

the characteristic equation of which has the root -3. Note that the characteristic root satisfies the condition $\alpha = -b/a$. According to Table 2.6.1 and the principle of superposition, a particular solution of the differential equation is

$$x_p(t) = d_0 + Ate^{-3t}$$

Substituting the guessed form into the differential equation gives

$$Ae^{-3t} - 3Ate^{-3t} + 3(d_0 + Ate^{-3t}) = 2(1 - e^{-3t})$$

which results in

$$3d_0 + Ae^{-3t} = 2 - e^{-3t}$$

Comparison of the coefficients of the relevant terms on the two sides of the equation yields

Constant term : $3d_0 = 2$

Exponential term : $A = -2$

Thus, with $d_0 = 2/3$ and $A = -2$, the particular solution is

$$x_p(t) = \frac{2}{3} - 2te^{-3t}$$

Table 2.6.1 Particular solutions for first-order differential equations

Forcing function $f(t)$	Particular solution $x_p(t)$
(a) Polynomial $f(t) = c_n t^n + c_{n-1} t^{n-1} + \ldots + c_1 t + c_0$	$x_p(t) = d_n t^n + d_{n-1} t^{n-1} + \ldots + d_1 t + d_0$
(b) Sinusoidal $f(t) = A_1 \sin \omega t + A_2 \cos \omega t$	$x_p(t) = B_1 \sin \omega t + B_2 \sin \omega t$
(c) Exponential $f(t) = Ae^{\alpha t}$	Case 1. If $\alpha \neq -b/a$, $x_p(t) = Be^{\alpha t}$ Case 2. If $\alpha = -b/a$, $x_p(t) = Bte^{\alpha t}$ (coefficients a and b are the coefficients of the first-order differential equation $a\dot{x}(t) + bx(t) = f(t)$, $a \neq 0$)

Given a general forcing function

$$f(t) = e^{at}[A_k(t)\cos \omega t + B_m(t)\sin \omega t] \qquad (2.6.27)$$

with $A_k(t)$ and $B_m(t)$ being polynomials of orders k and m, respectively, a particular solution of Eq. (2.6.26) can be written as

$$x_p(t) = t^\mu e^{at}[Q_n(t)\cos \omega t + R_n(t)\sin \omega t] \qquad (2.6.28)$$

where $Q_n(t)$ and $R_n(t)$ are polynomials of order $n = \max(k, m)$ such that

$$\begin{aligned} Q_n(t) &= q_n t^n + q_{n-1} t^{n-1} + \ldots + q_1 t + q_0 \\ R_n(t) &= r_n t^n + r_{n-1} t^{n-1} + \ldots + r_1 t + r_0 \end{aligned} \qquad (2.6.29)$$

and μ is an integer that is given by

$$\mu = \begin{cases} 0, & \text{if } \alpha \neq -b/a \\ 1, & \text{if } \alpha = -b/a \end{cases} \qquad (2.6.30)$$

The coefficients of the polynomials in Eq. (2.6.29) are to be determined. As can be seen, the particular solutions in Table 2.6.1 are special cases of the form given in Eq. (2.6.28).

In the context of the method of undetermined coefficients, the principle of superposition for linear differential equations can be applied. For example, if a forcing function can be written as a linear combination of the functions given in Table 2.6.1 or the form of Eq. (2.6.28), such as

$$f(t) = 2t + 3\cos(4t)$$

a particular solution can be written as

$$x_p(t) = d_1 t + d_0 + A\sin(4t) + B\cos(4t)$$

where the forms in the first and second rows of Table 2.6.1 have been used.

Second-Order Linear Ordinary Differential Equations

The second-order differential equation is of the form

$$a\ddot{x}(t) + b\dot{x}(t) + cx(t) = f(t), a \neq 0 \qquad (2.6.31)$$

The corresponding characteristic equation, by Eq. (2.6.22), is

$$a\lambda^2 + b\lambda + c = 0 \qquad (2.6.32)$$

with its roots given by

$$(\lambda_1, \lambda_2) = -\frac{b}{2a} \pm \frac{\sqrt{b^2 - 4ac}}{2a} \qquad (2.6.33)$$

2.6 Solution of Linear Ordinary Differential Equations

Table 2.6.2 Particular solutions for second-order differential equations

Forcing function $f(t)$	Particular solution $x_p(t)$
(a) Polynomial $f(t) = c_n t^n + c_{n-1} t^{n-1} + \ldots + c_1 t + c_0$	$x_p(t) = d_n t^n + d_{n-1} t^{n-1} + \ldots + d_1 t + d_0$
(b) Sinusoidal $f(t) = A_1 \sin \omega t + A_2 \cos \omega t$	Case 1. If $b \neq 0$ or if $b = 0$ and $\omega^2 \neq c/a$, $x_p(t) = B_1 \sin \omega t + B_2 \cos \omega t$ Case 2. If $b = 0$ and $\omega^2 = c/a$, $x_p(t) = t(B_1 \sin \omega t + B_2 \cos \omega t)$
(c) Exponential $f(t) = A e^{\alpha t}$	Case 1. If $\alpha \neq \lambda_1$ and $\alpha \neq \lambda_2$, $x_p(t) = B e^{\alpha t}$ Case 2. If $\lambda_1 \neq \lambda_2$ and α is equal to one of the roots, $x_p(t) = B t e^{\alpha t}$ Case 3. If $\alpha = \lambda_1 = \lambda_2$, $x_p(t) = B t^2 e^{\alpha t}$

Table 2.6.2 lists polynomial, sinusoidal, and exponential forms of the forcing function, and corresponding "guessed" particular solutions.

As can be seen from Table 2.6.2, the form of a particular solution again "mirrors" the forcing function and is relevant to the coincidence between the characteristic roots (λ_1, λ_2) and certain parameters of a given forcing function. For instance, when $b = 0$, Eq. (2.6.31) becomes

$$a\ddot{x}(t) + cx(t) = f(t), \quad a \neq 0 \qquad (2.6.34)$$

with the characteristic roots

$$(\lambda_1, \lambda_2) = \pm j\sqrt{c/a}, \quad j = \sqrt{-1} \qquad (2.6.35)$$

If $f(t)$ is sinusoidal with $\omega^2 = c/a$, the particular solution by Table 2.6.2 has the form

$$x_p(t) = t(B_1 \sin \omega t + B_2 \cos \omega t) \qquad (2.6.36)$$

indicating an unbounded oscillatory response. This solution is commonly seen in mechanical systems, and is often referred to as *resonant vibration*. One typical example of this phenomenon is seen in the Example 2.6.5 with zero damping, for which the governing differential equation is

$$m\ddot{x}(t) + kx(t) = f(t)$$

and the characteristic roots are

$$(\lambda_1, \lambda_2) = \pm j\omega_n, \text{ with } \omega_n = \sqrt{k/m}$$

where ω_n is called the natural frequency of the spring–mass system. Equation (2.6.36) implies that a resonant response occurs when the frequency ω of the sinusoidal forcing function matches the natural frequency of the system.

Showing that the amplitude of the resonant vibration is unbounded could be done by solving for the coefficients B_1 and B_2, and then taking the limit of the solution with time approaching infinity.

To simplify the derivations, assume zero initial conditions and let the forcing function be sinusoidal. Then,

$$m\ddot{x}(t) + kx(t) = \alpha \sin(\omega t) \tag{2.6.37}$$

where $\omega = \omega_n = \sqrt{k/m}$ and $\alpha \in \mathbb{R}$.

Plugging the derived particular solution (Eq. (2.6.36)) into the governing differential equation, we obtain

$$2B_1 m\omega \cos(\omega t) + B_2(k - m\omega^2)t \cos(\omega t) - 2B_2 m\omega \sin(\omega t) \\ - B_1(k - m\omega^2)t \sin(\omega t) = \alpha \sin(\omega t)$$

which yields the following system of equations for deriving the coefficients B_1 and B_2:

$$\begin{cases} 2B_1 m\omega = 0 \\ -2B_2 m\omega = \alpha \\ B_2(k - m\omega^2)t = 0 \; \forall t > 0 \\ B_1(k - m\omega^2)t = 0 \; \forall t > 0 \end{cases} \Rightarrow \begin{cases} B_1 = 0 \\ B_2 = -\dfrac{\alpha}{2m\omega} = -\dfrac{\alpha}{2\sqrt{km}} \end{cases}$$

Since $\omega = \omega_n = \sqrt{k/m}$, $k - m\omega^2 \equiv 0$ and the last two equalities are always satisfied. Hence, the solution of Eq. (2.6.37) is

$$x(t) = x_p(t) = -\frac{\alpha}{2\sqrt{km}} t \cos(t\sqrt{k/m}) \tag{6.2.38}$$

The upper and lower limits of Eq. (6.2.38) are $+\infty$ and $-\infty$, respectively, i.e. the amplitude of the resonant vibration is unbounded.

Using similar reasoning, the solution for the system with $f(t) = \alpha \cos(\omega t)$ is derived as

$$x(t) = x_p(t) = \frac{\alpha}{2\sqrt{km}} t \sin(t\sqrt{k/m}) \tag{6.2.39}$$

The upper and lower limits of Eq. (6.2.39) are also $+\infty$ and $-\infty$, respectively, showing that this resonant vibration also has an unbounded amplitude.

A numerical example of such system is shown in Example 2.6.11.

2.6 Solution of Linear Ordinary Differential Equations

Example 2.6.11 Deriving a particular solution of a second-order linear differential equation with constant coefficients and complex roots

Consider the differential equation

$$\ddot{x}(t) + 9x(t) = 3 \sin(3t)$$

Its characteristic equation is

$$\lambda^2 + 9 = 0$$

with roots being $\pm j3$, $j = \sqrt{-1}$. According to Table 2.6.2 or Eq. (2.6.36), a particular solution is

$$x_p(t) = t\Big(B_1 \sin(3t) + B_2 \cos(3t)\Big)$$

By using the differentiation formula

$$\frac{d^2}{dt^2}(uv) = \frac{d^2 u}{dt^2} v + 2 \frac{du}{dt}\frac{dv}{dt} + u \frac{d^2 v}{dt^2} \qquad (2.6.40)$$

we obtain

$$\ddot{x}_p(t) = 6\Big(B_1 \cos(3t) - 6B_2 \sin(3t)\Big) - 9t\Big(B_1 \sin(3t) - 6B_2 \cos(3t)\Big)$$
$$= 6B_1 \cos(3t) - 6B_2 \sin(3t) - 9x_p(t) + 9x_p(t) = 3 \sin(3t)$$

Plugging the above expression into the originally given differential equation yields

$$6B_1 \cos(3t) - 6B_2 \sin(3t) - 9x_p(t) + 9x_p(t) = 3 \sin(3t)$$

which is reduced to

$$6B_1 \cos(3t) - 6B_2 \sin(3t) = 3 \sin(3t)$$

Comparison of relevant terms on the two sides of the previous equation gives

$$\text{Sine term}: \quad -6B_2 = 3$$
$$\text{Cosine term}: \quad -6B_1 = 0$$

It follows that $B_1 = 0$ and $B_2 = -\frac{1}{2}$. Thus, the particular solution is

$$x_p(t) = -\frac{1}{2} t \cos(3t).$$

The graph of this solution (Figure 2.6.3) clearly shows the amplitude increasing to $\pm\infty$ with time, demonstrating an unbounded amplitude of resonant vibration.

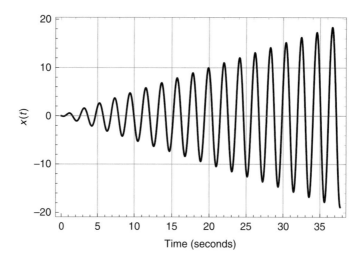

Figure 2.6.3 Resonant vibration of the system in Example 2.6.11

Example 2.6.12 Deriving a particular solution of a second-order linear differential equation with constant coefficients and complex roots, using the principle of superposition

Consider the differential equation

$$\ddot{x}(t) + 2\dot{x}(t) + 14x(t) = 3(1 - e^{-5t})$$

The characteristic equation of the complementary homogeneous equation is

$$\lambda^2 + 2\lambda + 14 = 0$$

with complex roots $-1 \pm j\sqrt{13}$. According to Table 2.6.2 and the principle of superposition,

$$x_p(t) = d_0 + Ae^{-5t}.$$

Substitution of the above form into the differential equation gives

$$25Ae^{-5t} + 2(-5Ae^{-5t}) + 14(d_0 + Ae^{-5t}) = 3(1 - e^{-5t})$$

which is reduced to

$$14d_0 + (25 - 10 + 14)Ae^{-5t} = 3 - 3e^{-5t}.$$

Compare the coefficients of the relevant terms on the two sides of the previous equation to obtain

$$\text{Constant term}: \quad 14d_0 = 3$$
$$\text{Exponential term}: \quad 29A = 3$$

which yields $d_0 = 3/14$ and $A = -3/29$. Hence, the particular solution is

$$x_p(t) = \frac{3}{14} - \frac{3}{29}e^{-5t}.$$

Now, consider a general forcing function

$$f(t) = e^{at}[A_k(t)\cos \omega t + B_m(t)\sin \omega t] \qquad (2.6.41)$$

where $A_k(t)$ and $B_m(t)$ are polynomials of orders k and m, respectively. As can be seen, the three forms of forcing functions in Table 2.6.2 are special cases of Eq. (2.6.41). Define a complex number

$$\gamma = \alpha + j\omega, \quad j = \sqrt{-1} \qquad (2.6.42)$$

where α and ω are the parameters in Eq. (2.6.41) of the forcing function. A particular solution of the differential equation (2.6.31) is

$$x_p(t) = t^\mu e^{\alpha t}[Q_n(t)\cos \omega t + R_n(t)\sin \omega t] \qquad (2.6.43)$$

where $Q_n(t)$ and $R_n(t)$ are polynomials of order $n = \max(k,m)$, as given in Eq. (2.6.29), and μ is an integer that depends on the value of γ as below

Case 1: If $\gamma \neq \lambda_1$ and $\gamma \neq \lambda_2$, then $\mu = 0$.
Case 2: If $\gamma = \lambda_1$ or $\gamma = \lambda_2$ but $\lambda_1 \neq \lambda_2$, then $\mu = 1$.
Case 3: If $\gamma = \lambda_1 = \lambda_2$, then $\mu = 2$.

Example 2.6.13 Deriving a particular solution of a second-order linear differential equation with constant coefficients and real distinct roots, where the parameter of a forcing function coincides with one of the roots

For the differential equation

$$\ddot{x}(t) + 7\dot{x}(t) + 12x(t) = (1 + 2t)e^{-3t}$$

the characteristic equation of the complementary homogeneous equation is

$$\lambda^2 + 7\lambda + 12 = 0$$

with roots -3 and -4. By Eq. (2.6.43), the parameter $\gamma = -3$, which coincides with one of the characteristic roots. By Case 2 of Eq. (2.6.43), a particular solution needs to be used, of the form

$$x_p(t) = te^{-3t}(at+b) = e^{-3t}(at^2 + bt)$$

Substituting the above expression into the differential equation and using Eq. (2.6.40) yields

$$9e^{-3t}(at^2 + bt) - 6e^{-3t}(2at + b) + 2ae^{-3t}$$
$$+ 7\{-3e^{-3t}(at^2 + bt) + e^{-3t}(2at + b)\} + 12e^{-3t}(at^2 + bt) = (1 + 2t)e^{-3t}$$

which is simplified as

$$\{2at + (2a + b)\}e^{-3t} = (1 + 2t)e^{-3t}$$

or

$$2a + b + 2at = 1 + 2t$$

Comparison of the t power terms on the two sides of the previous equation gives

$$t^1 \text{ term}: \quad 2a = 2$$
$$t^0 \text{ term}: \quad 2a + b = 1$$

which leads to $a = 1$ and $b = -1$. So, the particular solution is

$$x_p(t) = te^{-3t}(t - 1)$$

General Differential Equations

Now consider the n-th order differential equation (2.6.17). Given forcing function $f(t)$ of the form of Eq. (2.6.41), define a complex number $\gamma = \alpha + j\omega$ as shown in Eq. (2.6.42). Let the roots of the characteristic equation (2.6.22) be $\lambda_1, \lambda_2, \ldots, \lambda_n$. A particular solution of Eq. (2.6.17) is guessed as belonging to one of the following two cases.

Case 1. The parameter γ does not match any of the roots: $\gamma \neq \lambda_j$ for $j = 1, 2, \ldots, n$.
A particular solution can be written as

$$x_p(t) = e^{\alpha t}[Q_n(t)\cos \omega t + R_n(t)\sin \omega t] \tag{2.6.44}$$

Case 2. The parameter γ is the same as a root of multiplicity m (m identical roots). A particular solution is

$$x_p(t) = t^m e^{\alpha t}[Q_n(t)\cos \omega t + R_n(t)\sin \omega t] \tag{2.6.45}$$

In the above equations $Q_n(t)$ and $R_n(t)$ are polynomials of order $n = \max(k, m)$.

2.6 Solution of Linear Ordinary Differential Equations

Example 2.6.14 Deriving a particular solution of a higher-order linear differential equation with constant coefficients

Consider the following differential equation

$$\dddot{x}(t) + 5\ddot{x}(t) + 7\dot{x}(t) + 3x(t) = 2te^{-t}$$

From Example 2.6.8, the characteristic equation has roots -1, -1, and -3. Because the parameter $\gamma = -1$, a particular solution by Eq. (2.6.45) is

$$x_p(t) = t^2 e^{-t}(at + b)$$

Substituting the above form into the differential equation gives

$$-x_p + e^{-t}(9at^2 + 6bt - 2b - 12at - 4b + 6a)$$
$$+ 5\left\{ x_p + e^{-t}(-6at^2 + 4bt + 6at + 2b) \right\} + 7\left(-x_p + e^{-t}(3at^2 + 2bt) \right) + 3x_p = 2te^{-t}$$

which is reduced to

$$e^{-t}(9at^2 + 6bt - 6at - 2b - 12at - 4b + 6a)$$
$$+ 5e^{-t}(-6at^2 - 4bt + 6at + 2b) + 7e^{-t}(3at^2 + 2bt) = 2te^{-t}$$

This eventually leads to

$$12at + (6a + 4b) = 2t$$

Comparing the coefficients of the t power terms on the two sides of the previous equation gives the coupled algebraic equations

$$12a = 2$$
$$6a + 4b = 0$$

which have the solutions $a = 1/6$ and $b = -1/4$. So, the particular solution of the given differential equation is

$$x_p(t) = \frac{1}{12} e^{-t} t^2 (2t - 3)$$

Complete Solution

Once a particular solution is obtained, the coefficients in Eq. (2.6.23) or Eq. (2.6.24) can be determined through the introduction of the initial conditions of Eq. (2.6.18), For instance, for the solution of Eq. (2.6.23), the coefficients $A_1, A_2, \ldots, A_n$ can be determined by using the coupled algebraic equations below

$$\begin{cases} x(0) = A_1 + A_2 + \ldots + A_n + x_p(0) = a_{0,0} \\ Dx(0) = A_1\lambda_1 + A_2\lambda_2 + \ldots + A_n\lambda_n + Dx_p(0) = a_{0,1} \\ \ldots \ldots \\ D^{n-1}x(0) = A_1\lambda_1^{n-1} + A_2\lambda_2^{n-1} + \ldots + A_n\lambda_n^{n-1} + D^{n-1}x_p(0) = a_{0,n-1} \end{cases} \quad (2.6.46)$$

where $D = d/dt$, and $x_p(t)$ is any particular solution of Eq. (2.6.17). For the coefficients $A_1, A_2, \ldots, A_m$ and $B_0, B_1, \ldots, B_p$, a set of algebraic equations can be similarly set up through the application of the initial conditions of Eq. (2.6.18). The determination of the coefficients in $x_h(t)$ eventually yields the complete solution that satisfies both the differential equation (2.6.17) and the initial conditions of Eq. (2.6.18).

Example 2.6.15 Finding a complete solution of the given linear differential equation with constant coefficients

Consider the differential equation

$$\begin{cases} \ddot{x}(t) + 2\dot{x}(t) + 14x(t) = 3(1 - e^{-5t}) \\ x(0) = 1, \quad \dot{x}(0) = 0 \end{cases}$$

From Example 2.6.11, the characteristic roots are $-1 \pm j\sqrt{13}$. This means that the solution of the complementary homogeneous equation is

$$x_h(t) = e^{-t}\left(B_1 \sin\left(\sqrt{13}\, t\right) + B_2 \cos\left(\sqrt{13}\, t\right)\right)$$

Also from Example 2.6.11, a particular solution is

$$x_p(t) = \frac{3}{14} - \frac{3}{29} e^{-5t}$$

Thus,

$$x(t) = x_h(t) + x_p(t) = e^{-t}\left(B_1 \sin\left(\sqrt{13}\, t\right) + B_2 \cos\left(\sqrt{13}\, t\right)\right) + \frac{3}{14} - \frac{3}{29} e^{-5t}$$

From Eq. (2.6.43)

$$x(0) = 1: \quad B_2 + \frac{3}{14} + \frac{3}{29} = 1$$

$$\dot{x}(0) = 0: \quad \sqrt{13}\, B_1 - B_2 + \frac{15}{29} = 0$$

which have the solutions $B_1 = \frac{1}{\sqrt{13}} \frac{151}{406}$ and $B_2 = \frac{361}{406}$. Therefore, the solution of the given differential equation is

$$x(t) = e^{-t}\left(\frac{1}{\sqrt{13}} \frac{151}{406} \sin(\sqrt{13}\, t) + \frac{361}{406} \cos(\sqrt{13}\, t)\right) + \frac{3}{14} - \frac{3}{29} e^{-5t}.$$

2.6.6 Solution of Differential Equations with Piecewise-Continuous Forcing Functions

In this section, differential equations with continuous piecewise forcing functions, as shown in Section 2.4.6, are considered. We start with an example:

$$\dot{x}(t) + 4x(t) = 3u(t) - 3u(t-2) \text{ for } t > 0, \text{ subject to } x(0) = 1$$

where $u(t)$ is the unit step function, $u(t-2)$ is a delayed unit step function, and the forcing function $3u(t) - 3u(t-2)$ on the right-hand side of the differential equation represents a pulse as described in Example 2.4.14 and shown in Figure 2.4.2(a). By superposition, the solution of the above differential equation with an initial condition can be written as

$$x(t) = x_I(t) + x_{p1}(t) - x_{p2}(t)$$

where $x_I(t)$, $x_{p1}(t)$, $x_{p2}(t)$ are the solutions of the following three problems:

Problem 0 : $\dot{x}_I(t) + 4x_I(t) = 0$, $t \geq 0$; subject to $x_I(0) = 1$

Problem 1 : $\dot{x}_{p1}(t) + 4x_{p1}(t) = 3u(t)$, $t \geq 0$; subject to $x_{p1}(0) = 0$

Problem 2 : $\dot{x}_{p2}(t) + 4x_{p2}(t) = 3u(t-2)$, $t > 2$; subject to $x_{p1}(2) = 0$

The solution of *Problem 0* (free response) is

$$x_I(t) = e^{-4t}$$

The solution of *Problem 1* can be obtained by either the Laplace transform or the method of undermined coefficients, and it is

$$x_{p1}(t) = \frac{3}{4}\left(1 - e^{-4t}\right)$$

Likewise, the solution of *Problem 2* can be derived as

$$x_{p2}(t) = \frac{3}{4}\left(1 - e^{-4(t-2)}\right)u(t-2)$$

It is seen that $x_{p2}(t)$ is a time-shifted function, derived from $x_{p1}(t)$ as

$$x_{p2}(t) = x_{p1}(t-2)u(t-2)$$

Finally, the solution of the differential equation is

$$x(t) = e^{-4t} + \frac{3}{4}\left(1 - e^{-4t}\right) + \frac{3}{4}\left(1 - e^{-4(t-2)}\right)u(t-2)$$

which can also be written as

$$x(t) = \begin{cases} e^{-4t} + \dfrac{3}{4}\left(1 - e^{-4t}\right) & \text{for } 0 \le t \le 2 \\ e^{-4t} + \dfrac{3}{4}\left(1 - e^{-4t}\right) + \dfrac{3}{4}\left(1 - e^{-4(t-2)}\right) & \text{for } t > 2 \end{cases}$$

We now consider the general differential equation

$$A(D)x(t) = f(t) \tag{2.6.47}$$

subject to the initial conditions

$$x(0) = a_{0,0}, \ Dx(0) = a_{0,1}, \ldots, \ D^{n-1}x(0) = a_{0,n-1} \tag{2.6.48}$$

where $A(D) = a_n D^n + a_{n-1} D^{n-1} + \ldots + a_1 D + a_0$, with $D = d/dt$, and the forcing function is

$$\begin{aligned} f(t) &= f_0(t)u(t) + f_1(t - T_1)u(t - T_1) \\ &\quad + f_2(t - T_2)u(t - T_2) + \ldots + f_m(t - T_m)u(t - T_m) \end{aligned} \tag{2.6.49}$$

with $f_j(t - T_j)$ being a time-shifted function and $u(t - T_j)$ being a delayed unit step function. The solution of the differential equation (2.6.47) can be written as

$$\begin{aligned} x(t) &= x_I(t) + x_0(t)u(t) + x_1(t - T_1)u(t - T_1) \\ &\quad + x_2(t - T_2)u(t - T_2) + \ldots + x_m(t - T_m)u(t - T_m) \end{aligned} \tag{2.6.50}$$

where $x_I(t)$ is the free response that is the solution of the homogeneous equation

$$A(D)x(t) = 0 \tag{2.6.51}$$

subject to the initial conditions of Eq (2.6.48), and $x_j(t)$ for $j = 0, 1, 2, \ldots, m$ is the solution of the equation

$$A(D)x(t) = f_j(t), \ t > 0 \tag{2.6.52}$$

subject to zero initial conditions

$$x(0) = 0, \ Dx(0) = 0, \ldots, \ D^{n-1}x(0) = 0 \tag{2.6.53}$$

Note that $x_j(t - T_j)$ in Eq. (2.6.50) is a time-shifted function derived from $x_j(t)$. The solution process thus takes the following steps:

(i) Determine the free response $x_I(t)$.
(ii) Determine the forced responses $x_0(t), x_1(t), \ldots, x_m(t)$ by solving the differential equation (2.6.52) for $j = 0, 1, 2, \ldots, m$. Because no time delays are considered, standard

methods such as the Laplace transform and the method of undetermined coefficients can be applied.

(iii) Time-shift the solutions obtained in Step (ii): $x_1(t-T_1)$, $x_2(t-T_2), \ldots, x_m(t-T_m)$.

(iv) Finally, obtain the complete solution using Eq. (2.6.50).

2.7 Transfer Functions and Block Diagrams of Time-Invariant Systems

Time-invariant dynamic systems are commonly seen in engineering applications. The spring–mass–damper system in Figure 2.5.1 is one example of such a system. A time-invariant system is the one whose physical parameters do not change with time. Time-invariant dynamic systems are governed by differential equations with constant coefficients. In modeling and analysis of time-invariant systems, three types of models are often used: (a) differential equations that are derived based on the second law of Newton (Newton's first law or law of inertia, that is sometimes called "the first law of motion," can be viewed as a special case of the second law with vanishing acceleration or linear momentum); (b) transfer functions that describe the output-to-input ratio in the Laplace transform domain; and (c) block diagrams that facilitate the description of complex dynamic systems with multiple subsystems or components. The derivation of differential equations for a variety of physical systems will be discussed in the subsequent chapters. In this section we introduce basic concepts of transfer functions and block diagrams; the application of these concepts to various dynamic systems are described in Chapters 3 to 9.

2.7.1 Transfer Functions

The transfer function of a time-invariant system is defined as the ratio of the system output to the system input in the Laplace transform domain with zero initial conditions:

$$G(s) = \left. \frac{\mathcal{L}[\text{output}]}{\mathcal{L}[\text{input}]} \right|_{\text{zero initial conditions}} \tag{2.7.1}$$

where $\mathcal{L}$ is the Laplace transform operator, and zero initial conditions mean all the initial values of the input and output and their time derivatives are set to zero.

Assume that a time-invariant system is described by the differential equation (2.6.1), with $r(t)$ and $y(t)$ being the input and output of the system, respectively. The Laplace transform of the differential equation with zero initial conditions gives

$$A(s)Y(s) = B(s)R(s) \tag{2.7.2}$$

where Eq. (2.6.12) with $I(s) = 0$ has been used, and $A(s)$ and $B(s)$ are the polynomials of s given in Eq. (2.6.13). By the definition in Eq. (2.7.1), the transfer function of the system is

$$G(s) = \frac{Y(s)}{R(s)} = \frac{B(s)}{A(s)} = \frac{b_m s^m + b_{m-1} s^{m-1} + \ldots + b_1 s + b_0}{a_n s^n + a_{n-1} s^{n-1} + \ldots + a_1 s + a_0} \tag{2.7.3}$$

As can be seen from Eq. (2.7.3) above, the transfer function does not depend on a particular input; it depends on the coefficients of the differential equation, which are the physical parameters of the system. Furthermore, the transfer function given in Eq. (2.7.3) gives an input-to-output representation as follows:

$$Y(s) = G(s)R(s) \tag{2.7.4}$$

The transfer function given in Eq. (2.7.3) is called an n-th-order transfer function because the denominator $A(s)$ is an n-th-order polynomial of s. In the analysis of dynamic systems, the information on the poles and zeros of the system transfer functions is often acquired. The definition and determination of the poles and zeros of an s-domain function have been discussed in Section 2.4.2, and can be directly applied to transfer functions without further explanation.

Because $A(s)$ and $B(s)$ can be obtained from the operator polynomials $A(D)$ and $B(D)$, with D replaced by s, the transfer function given in Eq. (2.7.3) can be derived as follows:

$$G(s) = \frac{Y(s)}{R(s)} = \frac{B(D)}{A(D)}\bigg|_{D=s} \tag{2.7.5}$$

Take the spring–mass–damper system in Example 2.5.1, which has the governing differential equation

$$(mD^2 + cD + k)x(t) = f(t) \tag{2.7.6}$$

Assume that the system output and system input are the displacement $x(t)$ of the mass and the external force $f(t)$, respectively. By Eq. (2.7.4), the transfer function of the spring–mass–damper system is

$$G(s) = \frac{X(s)}{F(s)} = \frac{1}{mD^2 + cD + k}\bigg|_{D=s} = \frac{1}{ms^2 + cs + k} \tag{2.7.7}$$

The transfer function of a dynamic system is not unique; depending on the definition of system input and system output, the same differential equation can have different transfer functions. For instance, for the previously discussed spring–mass–damper

2.7 Transfer Functions and Block Diagrams of Time-Invariant Systems

system, if the output is defined as the velocity $\dot{x}(t)$ of the mass, the transfer function becomes

$$G_1(s) = \frac{\mathcal{L}[\dot{x}(t)]}{\mathcal{L}[f(t)]} = \frac{sX(s)}{F(s)} = \frac{s}{ms^2 + cs + k} \quad (2.7.8)$$

where Eqs. (2.7.1) and (2.7.7) have been used.

2.7.2 Block Diagrams

Block diagrams are widely used for the representation of dynamic systems in control engineering. For linear time-invariant systems, block diagrams can be viewed as a diagrammatic extension of transfer function models. For example, for the spring–mass–damper system described by the transfer function in Eq. (2.7.7), the block diagram in Figure 2.7.1 relates the input $F(s)$ to the output $X(s)$ by

$$X(s) = \frac{1}{ms^2 + cs + k} F(s)$$

As can be seen from Figure 2.7.1, a block diagram consists of three parts: an arrow pointing to the block representing the input, an arrow departing from the block representing the output, and the block (box) containing the transfer function of the system in it. For general linear time-invariant systems modeled by the transfer function in Eq. (2.7.3), the block diagram is shown in Figure 2.7.2. The algebraic rule of a block diagram is

$$\text{Output} = \text{Transfer function(inside the block)} \times \text{Input} \quad (2.7.9)$$

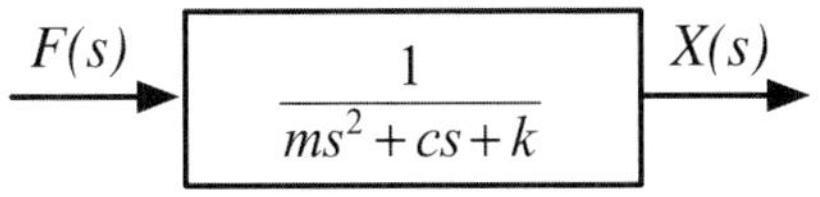

Figure 2.7.1 Block diagram of a spring–mass–damper system

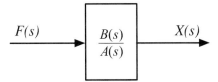

Figure 2.7.2 Block diagram of a generic linear time-invariant system

Example 2.7.1 Deriving governing equations from the given block diagram

For a dynamic system modeled by the block diagram in Figure 2.7.3, derive the governing differential equation.

Solution

According to the rule of Eq. (2.7.9) for a block diagram, the output $Y(s)$ of the system is given by

$$Y(s) = \frac{2s+1}{s^2+3s+6} R(s)$$

or

$$\frac{Y(s)}{R(s)} = \frac{B(s)}{A(s)}, \quad \text{with } B(s) = 2s+1 \text{ and } A(s) = s^2+3s+6$$

This implies that, in the time domain,

$$A(D)y(t) = B(D)r(t), \quad D = d/dt$$

It follows that the governing differential equation of the system is

$$\ddot{y}(t) + 3\dot{y}(t) + 6y(t) = 2\dot{r}(t) + r(t)$$

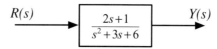

Figure 2.7.3 Block diagram of a dynamic system

Block diagrams can be used to describe the input–output relations for systems with multiple inputs and multiple outputs. Also, block diagrams are the basic element for modeling complex dynamic systems with multiple subsystems. Furthermore, block diagrams go beyond time-invariant systems; the cause-and-effect representation is extended to time-varying systems and nonlinear systems. These modeling issues will be addressed in the subsequent chapters.

2.8 Solution of State Equations via Numerical Integration

Many nonlinear dynamic systems and complex linear systems can be modeled by state equations such as Eq. (2.5.11). Unlike linear time-invariant systems, the analytical methods presented in Section 2.6 are not valid for the solution of nonlinear state equations. In general, exact solutions of nonlinear state equations are difficult and often impossible to obtain.

2.8 Solution of State Equations via Numerical Integration

In this section a numerical integration algorithm is presented for deriving approximate solutions of nonlinear state equations. To this end, write Eq. (2.5.11) as

$$\dot{x}(t) = \Phi\big(t, x(t)\big) \tag{2.8.1}$$

where $\Phi\big(t, x(t)\big) = \phi\big(x(t), u(t)\big)$, with the vector ϕ given in Eq. (2.5.12). The initial condition of the state equation can be written as

$$x(0) = \phi_0 \tag{2.8.2}$$

where ϕ_0 is a given constant vector. The nonlinear equation (2.8.1) can be solved by a fixed-step Runge–Kutta method of order four:

$$x_{k+1} = x_k + \frac{h}{6}(f_1 + 2f_2 + 2f_3 + f_4), \ k = 1, \ 2, \ldots \tag{2.8.3}$$

where $x_k = x(t_k)$ with $t_k = kh$, h is the step size, and

$$\begin{cases} f_1 = \Phi(t_k, x_k) \\ f_2 = \Phi\left(t_k + \dfrac{h}{2}, x_k + \dfrac{h}{2}f_1\right) \\ f_3 = \Phi\left(t_k + \dfrac{h}{2}, x_k + \dfrac{h}{2}f_2\right) \\ f_4 = \Phi(t_k + h, x_k + hf_3) \end{cases} \tag{2.8.4}$$

The above Runge–Kutta algorithm is a method of numerical integration for differential equations. In computation, make sure that the step size h is small enough to guarantee the convergence of the numerical solutions. In advanced numerical integration algorithms, the step size h is adaptively adjusted during the computation, meaning that "if the solution appears to be varying rapidly in a particular region, ... the step size will be reduced to be able to better track the solution" (*Advanced Numerical Differential Equation Solving in Mathematica,* Wolfram Research Institute, 2013). For instance, the MATLAB software has Runge–Kutta solvers like ode45 with automatic step-size adjustment capability, and Wolfram Mathematica uses a NDSolve framework that is also based on explicit and implicit Runge–Kutta methods.

For numerical simulation, the MATLAB commands that implement the algorithm of Eq. (2.8.3) are given as follows:

```
h = tf/(npts - 1); % step size
t1 = linspace(0, tf, npts); % time intervals
x = zeros(n, npts); % n = dimension of the state vector
x(:, 1) = x0; % initial values
for ii = 1:npts-1
```

```
        tt = t1(ii); z = x(:,ii);
        f1 = feval(FunName, tt, z);
        f2 = feval(FunName, tt+h/2, z+h/2*f1);
        f3 = feval(FunName, tt+h/2, x+h/2*f2);
        f4 = feval(FunName, tt+h, x+h*f3);
        x(:,ii+1) = z + h*(f1 + 2*f2 + 2*f3 + f4)/6;
end
```

where feval is a MATLAB function, FunName is a user-provided function for $\Phi\left(t, x(t)\right)$, and npts is the number of points in the region of integration, $0 \leq t \leq tf$ with tf being the total simulation time. In other words, the step size h of the numerical integration can be expressed as

$$h = \frac{tf}{\text{npts} - 1} \qquad (2.8.5)$$

In the context of Wolfram Mathematica, the NDSolve functionality built-in options can be used to specify other than an automatically selected numerical method, absolute and relative errors allowed in the solution, and the maximum number of steps to be taken in attempting to find a solution. The last option is important when dealing with solutions that have a singularity: if an adaptive step size is chosen, the software will try to reduce the step size many times to be able to better track the solution approaching singularity, and will get into an infinite loop.

NDSolve can find a solution for a given nonlinear differential equation directly; prior derivation of the state equations is not necessary. The syntax of a script that finds a solution of the given differential equation and plots it is as follows:

```
NDSolve [{Equation, Initial_Conditions}, x, {t, t_min, t_max}]
Plot [ Evaluate [x[t] /.%], {t, t_min, t_max}]
```

In this script, the first line finds a solution of the given differential equation Equation, which is defined in terms of the function x(t) and subject to initial conditions Initial_Conditions; the solution is found for the independent variable t belonging to the interval $t_{min} \leq t \leq t_{max}$. The second line of the script plots the obtained solution.

The example below illustrates the utility of advanced computational software packages such as MATLAB and Mathematica for finding a numerical solution of a nonlinear differential equation.

Example 2.8.1 Deriving the trajectory of a ball thrown at an elevation with an initial velocity

A ball of the mass of 0.1 kg is thrown at an elevation of 1.8 m with an initial velocity shown in Figure 2.8.1. The aerodynamic drag acting on the rock is $\vec{F}_D = -0.0002\vec{v}|\vec{v}|$ (newtons), where $|\vec{v}| = \sqrt{\dot{x}^2 + \dot{y}^2}$ is the magnitude of the velocity vector. Consider gravity with the gravitational

2.8 Solution of State Equations via Numerical Integration

acceleration $g = 9.81$ m/s. Determine the trajectory of the ball flying in the space for the time period $0 \leq t \leq 5$ s.

The equations of motion of the ball are derived using Newton's second law, such as

$$\begin{cases} m\ddot{x} = -0.0002\dot{x}\sqrt{\dot{x}^2 + \dot{y}^2} \\ m\ddot{y} = -0.0002\dot{y}\sqrt{\dot{x}^2 + \dot{y}^2} - mg \end{cases}$$

which can be cast into the state form of Eq. (2.8.1), with

$$x(t) = \begin{pmatrix} x \\ y \\ \dot{x} \\ \dot{y} \end{pmatrix} = \begin{pmatrix} x_1 \\ x_2 \\ x_3 \\ x_4 \end{pmatrix}, \quad \Phi(t, x(t)) = \begin{pmatrix} x_3 \\ x_4 \\ -0.0002 x_3 \sqrt{x_3^2 + x_4^2}/m \\ -0.0002 x_4 \sqrt{x_3^2 + x_4^2}/m - g \end{pmatrix}$$

In applying the Runge–Kutta integration algorithm (2.8.3), call the MATLAB function `feval(force_ball, tt, z)`, where the function `force_ball` is given as follows.

```
function f = force_ball(tt, z)
m = 0.1; g = 9.81; v = sqrt(vx^2 + vy^2);
f(1) = z(3); f(2) = z(4);
f(3) = -0.0002* x(3)*v/m;
f(4) = -0.0002* x(4)*v/m - g;
```

By using the above-mentioned MATLAB commands the trajectory of the ball flying in the two-dimensional space is obtained in Figure 2.8.2.

To use the Wolfram Mathematica NDSolve functionality, derivation of the state equations is not necessary; the obtained equations of motion can be used directly as follows (automatic selection of numerical method, step size, precision, and maximum number of steps is assumed):

```
(* setting up the numerics *)
m = 0.1; g = 9.81;
vxinit = 35*Cos[47 Degree]; vyinit = 35*Sin[47 Degree];
(* solving the system of differential equations *)
NDSolve
```

$$\left[\{ m * x''[t] + 0.0002 * x'[t] * \sqrt{x'[t]^2 + y'[t]^2} == 0, \right.$$

$$m * y''[t] + 0.0002 * y'[t] * \sqrt{x'[t]^2 + y'[t]^2} + m * g == 0,$$

$$x[0] == 0, y[0] == 1.8, x'[0] == vxinit, y'[0] == vyinit \},$$

$$\left. \{x, y\}, \{t, 0, 5\} \right];$$

```
(* plotting the ball trajectory in x-y coordinate plane *)
ParametricPlot[Evaluate[{x[t],y[t]}/.%],{t,0,5}]
```

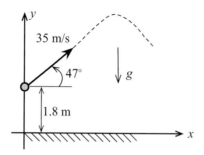

Figure 2.8.1 Problem settings: ball thrown at an elevation with an initial velocity

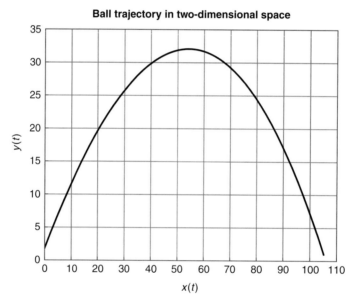

Figure 2.8.2 Derived trajectory of the ball in two-dimensional space

As expected, the solutions derived by both software packages are identical.

CHAPTER SUMMARY

The main objective of this chapter is to provide the reader with a review of mathematical theory and methods instrumental for modeling dynamic systems.

Upon completion of this chapter, you should be able to:

(1) Apply fundamental principles of vector and matrix algebra to equation solving.
(2) Understand the terminology of complex-number operations and be able to perform algebraic and trigonometric manipulations on complex numbers.
(3) Use the Laplace transform and its properties for solving linear ordinary differential equations, commonly encountered in modeling and analysis of dynamic systems.
(4) Use the separation of variables, direct integration, and the method of undetermined coefficients for solving linear ordinary differential equations with constant coefficients, which represent linear time-invariant dynamic systems.
(5) Refresh the knowledge of three different methods for modeling dynamic systems: state-space representation, transfer functions, and block diagrams.
(6) Refresh the knowledge of fundamental principles of solving state equations with numerical integration, using computational software such as MATLAB and Wolfram Mathematica.

REFERENCES

1. S. W. Goode and S. A. Annin, *Differential Equations & Linear Algebra*, 4th ed., Pearson, 2015.
2. G. Strang, *Differential Equations and Linear Algebra*, Wellesley-Cambridge Press, 2014.
3. J. Polking, A. Boggess, and D. Arnold, *Differential Equations: Classic Version*, 2nd ed., Pearson, 2017.
4. T. F. Bogart, Jr., *Laplace Transforms and Control Systems Theory for Technology*, Wiley, 1982.
5. E. Kreyszig, "*Advanced Engineering Mathematics*", 10th ed., Wiley, 2011.
6. K. Ogata, *System Dynamics*, 4th ed., Pearson, 2004 .

PROBLEMS

Section 2.1 Vector Algebra

2.1 Perform the following vector operations:
(a) $\vec{p} = \langle 2.3 \quad 4.1 \rangle$, $\vec{q} = \langle 6.4 \quad -2.5 \rangle$, $\vec{p} + \vec{q}$, $\vec{p} - 2\vec{q}$, $\vec{p} \cdot \vec{q}$, $3\vec{p} \cdot \vec{q}$
(b) $\vec{p} = \langle 7.5 \quad 1.4 \rangle$, $\vec{q} = \langle -3.4 \quad -2.5 \rangle$, $3\vec{p} + \vec{q}$, $\vec{p} - \vec{q}$, $\vec{p} \cdot \vec{q}$, $2.1\vec{p} \cdot (-1.3)\vec{q}$
(c) $\vec{p} = \langle 2.3 \quad 4.1 \quad -3.8 \rangle$, $\vec{q} = \langle 6.4 \quad -2.5 \quad 1.7 \rangle$, $\vec{p} + \vec{q}$, $\vec{p} - 2\vec{q}$, $\vec{p} \cdot \vec{q}$, $3\vec{p} \cdot \vec{q}$
(d) $\vec{p} = \langle 7.5 \quad 1.4 \quad -2.8 \rangle$, $\vec{q} = \langle -3.4 \quad -2.5 \quad 4.9 \rangle$, $\vec{r} = \langle 2.4 \quad -2.1 \quad 3.6 \rangle$, $3\vec{p} + \vec{q}$, $\vec{p} - \vec{q}$, $\vec{p} \cdot \vec{q}$, $2.1\vec{p} \cdot (-1.5)\vec{q} + 3.1\vec{r}$, $\vec{p} \cdot \vec{q} \cdot \vec{r}$, $\vec{r} \cdot \vec{q} \cdot \vec{p}$

2.2 Find the angle between the given vectors:
(a) $\vec{p} = \langle 2.7 \quad 4.3 \rangle$, $\vec{q} = \langle 6.4 \quad -2.5 \rangle$
(b) $1.1\vec{p} = \langle 7.5 \quad 1.4 \rangle$, $1.7\vec{q} = \langle -3.4 \quad 2.5 \rangle$
(c) $\vec{p} = \langle 2.3 \quad 4.1 \quad -3.8 \rangle$, $\vec{q} = \langle 6.4 \quad -2.5 \quad 1.7 \rangle$
(d) $\vec{p} = \langle -3.4 \quad -2.5 \quad 4.9 \rangle$, $2\vec{q} = \langle 2.4 \quad -2.1 \quad 3.6 \rangle$

2.3 Find the cross-product of the given vectors and show the result in the unit-vector decomposition form:
(a) $\vec{p} = \langle 2.3 \quad 4.1 \quad -3.8 \rangle$, $\vec{q} = \langle 6.4 \quad -2.5 \quad 1.7 \rangle$
(b) $\vec{p} = \langle -3.4 \quad -2.5 \quad 4.9 \rangle$, $\vec{q} = \langle 2.4 \quad -2.1 \quad 3.6 \rangle$
(c) $\vec{p} = \langle 7.2 \quad -1.4 \quad -2.9 \rangle$, $\vec{q} = \langle -6.4 \quad 5.3 \quad -8.1 \rangle$
(d) $\vec{p} = \langle -5.3 \quad 3.1 \quad -6.9 \rangle$, $\vec{q} = \langle 3.7 \quad 2.1 \quad -1.8 \rangle$

2.4 Without computing the angle between the vectors, find out whether the given vectors are orthogonal, parallel, or neither:
(a) $\vec{p} = \langle 2.7 \quad 2.0 \rangle$, $\vec{q} = \langle 6.0 \quad -8.1 \rangle$
(b) $\vec{p} = \langle -7.5 \quad 1.5 \rangle$, $\vec{q} = \langle 0.5 \quad 2.5 \rangle$
(c) $\vec{p} = \langle -2.5 \quad 1.5 \rangle$, $\vec{q} = \langle -0.5 \quad 0.3 \rangle$
(d) $\vec{p} = \langle -2.4 \quad 1.5 \rangle$, $\vec{q} = \langle 0.6 \quad -0.3 \rangle$
(e) $\vec{p} = \langle 2.4 \quad 1.0 \quad 0.48 \rangle$, $\vec{q} = \langle 6.0 \quad 2.5 \quad 1.2 \rangle$
(f) $\vec{p} = \langle 1.2 \quad 2.1 \quad -5.0 \rangle$, $\vec{q} = \langle 3.6 \quad 6.2 \quad -15.0 \rangle$
(g) $\vec{p} = \langle -5.0 \quad 1.0 \quad -0.5 \rangle$, $\vec{q} = \langle 4.0 \quad -0.8 \quad 0.4 \rangle$
(h) $\vec{p} = \langle -2.4 \quad 1.5 \quad 0.2 \rangle$, $\vec{q} = \langle 0.6 \quad -0.3 \quad 9.45 \rangle$

Section 2.2 Matrix Algebra

2.5 For the given matrices: $A = \begin{pmatrix} 1.2 & 3.5 \\ 4.8 & -0.2 \end{pmatrix}$, $B = \begin{pmatrix} 3.8 & -5.1 \\ 0.6 & -0.3 \end{pmatrix}$ perform the following operations:
(a) $A+B$
(b) $A^T + B^T$
(c) $2A - 0.1B$
(d) $A \cdot A^T$
(e) $A \cdot B$

2.6 For the given matrices: $A = \begin{pmatrix} 1.2 & 3.5 \\ 4.8 & -0.2 \\ 5.0 & -1.0 \end{pmatrix}$, $B = \begin{pmatrix} 3.8 & -5.1 & 2.0 \\ 0.6 & -0.3 & 4.2 \end{pmatrix}$ perform the following operations. If an operation is impossible, explain why.
(a) $A+B$
(b) $A^T + B$
(c) $2A - 0.1B^T$
(d) $A \cdot A^T$
(e) $B \cdot A$
(f) $A \cdot B$
(g) $B \cdot A^T$

2.7 Find the inverses of the given matrices:

$$A = \begin{pmatrix} 1.2 & 3.5 \\ 0.5 & -1.2 \end{pmatrix}, \quad B = \begin{pmatrix} -5.1 & 2.0 \\ -0.5 & 4.2 \end{pmatrix}, \quad C = \begin{pmatrix} -3.0 & 6.0 \\ -2.1 & 4.2 \end{pmatrix}$$

If the inverse does not exist, explain why.

2.8 Find the inverses of the given matrices:

$$A = \begin{pmatrix} 1.0 & 1.2 & 3.5 \\ 0.7 & 0.5 & -1.2 \\ 1.4 & -2.0 & 2.0 \end{pmatrix}, \quad B = \begin{pmatrix} -5.1 & 1.3 & 2.0 \\ -10.2 & 2.6 & 4.0 \\ -0.5 & 7.1 & 4.2 \end{pmatrix}, \quad C = \begin{pmatrix} 1.0 & -3.0 & 6.0 \\ 0.4 & -1.2 & 2.0 \\ 0.7 & -2.1 & 4.2 \end{pmatrix}$$

If the inverse does not exist, explain why.

2.9 Solve the following systems of linear equations by Cramer's rule:

(a) $\begin{cases} 2x - 5y - 4z = -1 \\ -x + 3y + 4z = 5 \\ 3x - 8y + z = 12 \end{cases}$

(b) $\begin{cases} x - 2y + z = 1 \\ -x + 3y + 4z = 8 \\ 3x - 6y + 3z = 3 \end{cases}$

(c) $\begin{cases} x + 2.1y - 3z = 0 \\ 4.2x - 3y + 5.2z = 6.2 \\ y + z = 0 \end{cases}$

Section 2.3 Complex Numbers

2.10 Find the magnitude and phase for the given complex numbers:
(a) $z = 2 + j7$
(b) $z = -2 + j5$
(c) $z = -4 - j3$
(d) $z = 5 - j9$

2.11 Find real and imaginary parts of the given complex numbers:

(a) $z = \dfrac{2 + j7}{1 + j}$

(b) $z = \dfrac{-2 + j5}{-1 + j3}$

(c) $z = \dfrac{-4 - j3}{2 - j}$

(d) $z = \dfrac{5 - j9}{-3 - 4j}$

2.12 Multiply the given complex numbers using Euler's formula. Find the magnitude and phase of the multiplication result:

(a) $z = \dfrac{2+j7}{1+j} \cdot (-3+j4)$

(b) $z = (-4+j5) \cdot \dfrac{-9+j}{2-j3}$

(c) $z = \dfrac{6-j2}{5+j} \cdot \dfrac{-4-j3}{2-j}$

Section 2.4 The Laplace Transform

2.13 Derive the Laplace transform of the following functions:
(a) $f(t) = 2 + e^{-2t}$
(b) $f(t) = 3t - t^2$
(c) $f(t) = 5te^{4t}$
(d) $f(t) = -6\sin(3t) + 8\cos(5t)$
(e) $f(t) = t\sin(2t)$
(f) $f(t) = (t-3)e^{2(t-3)}u(t-3)$
(g) $f(t) = e^{-2t}(t-4)$
(h) $f(t) = 3\cos(5t)e^{-6t}$
(i) $f(t) = e^{-3t+3}\sin(t-1)u(t-1)$
(j) $f(t) = e^{-3t+5}\sin(t-1)u(t-1)$
(k) $f(t) = 7te^{t+2}\cos(3t)$

2.14 Derive the Laplace transform of the following piecewise-continuous functions:

(a) $f(t) = \begin{cases} 2 & t < 3 \\ -3 & 3 < t \le 5 \\ 0 & t > 5 \end{cases}$

(b) $f(t) = \begin{cases} 0 & t < 3 \\ 2 & 3 < t \le 5 \\ t-4 & t > 5 \end{cases}$

(c) $f(t) = \begin{cases} 1 & t < 2 \\ t-2 & 2 < t \le 6 \\ 4 & t > 6 \end{cases}$

2.15 Derive the Laplace transform of the piecewise-continuous functions defined in the graphs in Figure P2.1:

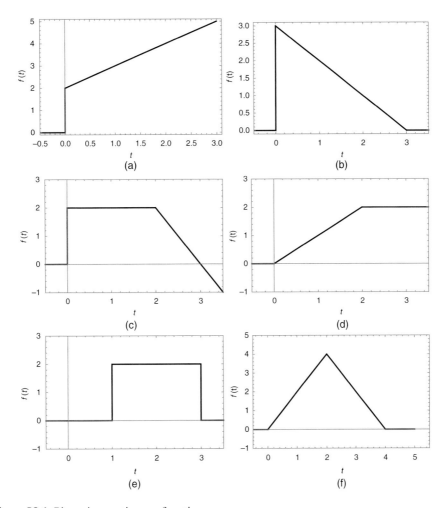

Figure P2.1 Piecewise-continuous functions

2.16 Derive the inverse Laplace transform of the following functions. Use the table of Laplace pairs and/or apply algebraic operations to the given function to represent it as a summation of components, the inverse transform of which is known:

(a) $F(s) = \dfrac{4s}{s^2 + 9}$

(b) $F(s) = \dfrac{5}{s^2 + 4}$

(c) $F(s) = \dfrac{5}{s^2 + 2s + 5}$

(d) $F(s) = \dfrac{s^2 - 9}{(s^2 + 9)^2}$

(e) $F(s) = \dfrac{3}{(s+1)(s+2)}$

(f) $F(s) = \dfrac{3s + 4}{(s+1)(s+2)}$

(g) $F(s) = \dfrac{7}{s(s+5)}$

2.17 Derive the inverse Laplace transform of the following functions. Use the table of Laplace pairs and/or apply algebraic operations to the given function to represent it as a summation of components, the inverse transform of which is known:

(a) $F(s) = \dfrac{8}{s^2 - 9}$

(b) $F(s) = \dfrac{4}{s(s^2 + 6s + 13)}$

(c) $F(s) = \dfrac{2}{s(s+1)^2}$

(d) $F(s) = \dfrac{5s}{(s^2 + 9)^2}$

(e) $F(s) = \dfrac{s^2 + 3s - 1}{(s^2 + 4)^2}$

(f) $F(s) = \dfrac{2s + 3}{s^2(s+5)}$

(g) $F(s) = \dfrac{s + 6}{s^2 + 8s + 20}$

(h) $F(s) = \dfrac{3s + 5}{(s^2 + 4s + 21)(s^2 + 2s + 5)}$

(i) $F(s) = \dfrac{s^2 + 2s + 4}{(s^2 + 4s + 22)(s^2 + 2s + 14)}$

(j) $F(s) = \dfrac{s^2 + 2s + 3}{s^3 + 10s^2 + 31s + 30}$

2.18 Derive the inverse Laplace transform of the following functions using the partial fraction expansion

(a) $F(s) = \dfrac{5}{s(s+2)(s+3)}$

(b) $F(s) = \dfrac{3s + 1}{s(s+1)(s+4)}$

(c) $F(s) = \dfrac{3s+2}{s^2(s+1)(s+4)}$

(d) $F(s) = \dfrac{4s}{s^3(s+2)}$

(e) $F(s) = \dfrac{4s^2+s+7}{s^2(s+2)^2}$

(f) $F(s) = \dfrac{s^2+4s+1}{s^4+6s^3+12s^2+10s+3}$

2.19 Derive the inverse Laplace transform of the following functions using the partial fraction expansion

(a) $F(s) = \dfrac{s^2+2s+2}{s^3+5s^2+17s+13}$

(b) $F(s) = \dfrac{s^3+2s^2+1}{s^4+4s^3+13s^2}$

(c) $F(s) = \dfrac{5s+2}{s^4+8s^3+25s^2+36s+20}$

(d) $F(s) = \dfrac{s^2+3s+7}{s^4+8s^3+26s^2+40s+25}$

(e) $F(s) = \dfrac{2+e^{-2s}}{s^3+11s^2+36s+26}$

Section 2.5 Differential Equations

2.20 Determine whether each of the given differential equations is linear or nonlinear, and provide a rationale for your answer:

(a) $2x\dot{x} + x = e^{-2t}$, $x = x(t)$
(b) $\dot{x} + t^2 x = 0$, $x = x(t)$
(c) $4\ddot{x} + (t-1)\dot{x} + 2x = 0$, $x = x(t)$
(d) $\ddot{x} + 3\dot{x} + \cos(x) = 0$, $x = x(t)$
(e) $e^{4t}\ddot{x} + 3\dot{x} - 12x = 0$, $x = x(t)$

2.21 Write the given equations in D-operator notation:

(a) $5x^{(5)} + 2.1x^{(3)} + \ddot{x} - 4\dot{x} + 12x = f^{(3)} + 7\ddot{f} + 2f$
(b) $x^{(4)} + 0.3x^{(3)} + 1.1\ddot{x} - x = 2\ddot{f} - \dot{f} + f$
(c) $7.2x^{(6)} + 4.3x^{(5)} - 3.1x^{(3)} + 2.8\ddot{x} - 9.2\dot{x} + 12x = 4f^{(4)} + \dddot{f} - 5\dot{f} + f$

2.22 Write the given systems of equations in matrix notation:

(a) $\begin{cases} m_1\ddot{x}_1 + c\dot{x}_1 + (k_1 + k_2)x_1 - k_2 x_2 = f_1(t) \\ m_2\ddot{x}_2 + k_2 x_2 - k_2 x_1 = f_2(t) \end{cases}$

(b) $\begin{cases} m_1\ddot{x}_1 + c\dot{x}_1 + k_1 x_1 - c\dot{x}_2 = f_1(t) \\ m_2\ddot{x}_2 + c\dot{x}_2 - c\dot{x}_1 = f_2(t) \end{cases}$

(c) $\begin{cases} m_1\ddot{x}_1 + (c_1 + c_2)\dot{x}_1 + (k_1 + k_2)x_1 - c_2\dot{x}_2 - k_2 x_2 = f_1(t) \\ m_2\ddot{x}_2 + c_2\dot{x}_2 + k_2 x_2 - c_2\dot{x}_1 - k_2 x_1 = f_2(t) \end{cases}$

Section 2.6 Solution of Linear Differential Equations with Constant Coefficients

2.23 Solve the following first-order linear differential equations by direct integration:
(a) $3\dot{x} = e^{-2t}$, $x(0) = 1$
(b) $2\dot{x} = e^{-t} + 1$, $x(0) = -1$
(c) $4\dot{x} = t^2$, $x(0) = 2$
(d) $\dot{x} = (t + 2)^2$, $x(0) = -4$
(e) $5\dot{x} = \sin(2t)$, $x(0) = 5$
(f) $8\dot{x} = t\cos(t^2)$, $x(0) = 2$

2.24 Solve the following second-order linear differential equations by direct integration or the separation of variables:
(a) $4\ddot{x} = e^{-3t}$, $x(0) = -1$, $\dot{x}(0) = 1$
(b) $2\ddot{x} = e^{-5t} + 3t + t^2$, $x(0) = 1$, $\dot{x}(0) = 2$
(c) $\ddot{x} = 2t^2 - 1$, $x(0) = 2$, $\dot{x}(0) = -1$
(d) $5\ddot{x} = (10t + 3)^2$, $x(0) = -6$, $\dot{x}(0) = 8$

2.25 Solve the following differential equations by the separation of variables:
(a) $2\dot{x} = \dfrac{3}{x}e^{3t}$, $x(0) = 3$
(b) $\dot{x} + 2x = 5$, $x(0) = -1$
(c) $3\dot{x} = x \cdot e^{-2t}$, $x(0) = 2$
(d) $2\dot{x} = \sin(2x)$, $x(0) = \pi/4$
(e) $\dot{x} = x^2 \cos(2t)$, $x(0) = \pi/3$
(f) $\dot{x} = t e^{2x}$, $x(0) = 1$

2.26 Solve the following first-order differential equations by the Laplace transform:
(a) $2\dot{x} + x = 4 + t$, $x(0) = 1$
(b) $\dot{x} + 5x = 8t \sin(3t)$, $x(0) = -1$
(c) $\dot{x} + 6x = 3e^{-t}\cos(2t)$, $x(0) = -2$
(d) $\dot{x} + 3x = 2te^{-t}\sin(3t)$, $x(0) = 37/169$
(e) $2\dot{x} - x = 4t^2 \sin(t/2)$, $x(0) = 4$
(f) $\dot{x} - 2x = 7t^2 e^{-5t} + t\cos(2t)$, $x(0) = 5/49$

2.27 Solve the following differential equations by the Laplace transform:

(a) $5\ddot{x} + 4\dot{x} + 20x = 20$, $x(0) = -1$, $\dot{x}(0) = 1$
(b) $\ddot{x} + 8\dot{x} + 25x = 25t$, $x(0) = 2$, $\dot{x}(0) = -1$
(c) $5\ddot{x} + 24\dot{x} + 180x = 25\,t + 50$, $x(0) = -2$, $\dot{x}(0) = 0$
(d) $\ddot{x} + 4\dot{x} + 16x = 16\,t\,e^{-2t}$, $x(0) = 0$, $\dot{x}(0) = 3$
(e) $\ddot{x} + 3\dot{x} + 9x = 9t + t^2$, $x(0) = 0$, $\dot{x}(0) = 0$
(f) $\ddot{x} + 4\dot{x} + 25x = 25e^{-t}$, $x(0) = -1$, $\dot{x}(0) = 1$
(g) $\ddot{x} + 2\dot{x} + 25x = 5\sin(t)$, $x(0) = 2$, $\dot{x}(0) = 0$
(h) $\ddot{x} + \dot{x} + x = \cos(t)$, $x(0) = 1$, $\dot{x}(0) = 0$

2.28 Solve the following differential equations by the Laplace transform:

(a) $\ddot{x} + 7\dot{x} + 10x = 5$, $x(0) = -1$, $\dot{x}(0) = 1$
(b) $\ddot{x} + 6\dot{x} + 8x = 2t$, $x(0) = 2$, $\dot{x}(0) = -1$
(c) $\ddot{x} + 11\dot{x} + 24x = 12t + 1$, $x(0) = -2$, $\dot{x}(0) = 0$
(d) $\ddot{x} + 4\dot{x} + 4x = 2\sin(t)$, $x(0) = 0$, $\dot{x}(0) = 0$
(e) $\ddot{x} + 16\dot{x} + 64x = 169\cos(t)$, $x(0) = 1$, $\dot{x}(0) = -1$
(f) $\ddot{x} + 4\dot{x} + x = 3e^{-2t}$, $x(0) = 0$, $\dot{x}(0) = 1$
(g) $\ddot{x} + 8\dot{x} + 16x = 16\,t\,e^{-t}$, $x(0) = 1$, $\dot{x}(0) = 2$

2.29 Solve the following differential equations with piecewise-continuous forcing functions and zero initial conditions by the Laplace transform:

(a) $\ddot{x} + 2\dot{x} + 4x = \begin{cases} 1 & t < 2 \\ 0 & t > 2 \end{cases}$, $x(0) = 0$, $\dot{x}(0) = 0$

(b) $\ddot{x} + 4\dot{x} + 3x = \begin{cases} 0 & t < 1 \\ 2 & 1 < t < 3 \\ 0 & t > 3 \end{cases}$, $x(0) = 0$, $\dot{x}(0) = 0$

(c) $\ddot{x} + 8\dot{x} + 16x = \begin{cases} 0 & t < 0 \\ 4t & 0 \le t < 1 \\ 4 & t \ge 1 \end{cases}$, $x(0) = 0$, $\dot{x}(0) = 0$

(d) $\ddot{x} + 9x = \begin{cases} 0 & t < 0 \\ t & 0 \le t < 2 \\ 2 & t \ge 2 \end{cases}$, $x(0) = 0$, $\dot{x}(0) = 0$

(e) $\ddot{x} + 4\dot{x} + 29x = \begin{cases} 0 & t < 0 \\ 1 & 0 \le t < 1 \\ 2 - t & 1 \le t < 3 \\ -1 & t \ge 3 \end{cases}$, $x(0) = 0$, $\dot{x}(0) = 0$

2.30 Solve the following first-order differential equations by the method of undetermined coefficients:

(a) $2\dot{x} + 4x = 0$, $x(0) = 1$

(b) $2\dot{x} + x = \sin(t)$, $x(0) = -1$
(c) $\dot{x} + 3x = 5\cos(2t)$, $x(0) = 2$
(d) $\dot{x} + 2x = e^{-2t}$, $x(0) = 5$
(e) $\dot{x} + 2x = t^2 + 3$, $x(0) = -1$
(f) $\dot{x} + 3x = t^2 e^{-t}$, $x(0) = 1$
(g) $2\dot{x} + 6x = t e^{-3t}$, $x(0) = -2$

2.31 Solve the following homogeneous second-order differential equations by the method of undetermined coefficients:
(a) $\ddot{x} + 5\dot{x} + 6x = 0$, $x(0) = 1$, $\dot{x}(0) = 2$
(b) $\ddot{x} + 8\dot{x} + 16x = 0$, $x(0) = 1$, $\dot{x}(0) = -1$
(c) $\ddot{x} + 4\dot{x} + 13x = 0$, $x(0) = -1$, $\dot{x}(0) = 1$
(d) $\ddot{x} + 6\dot{x} + 5x = 0$, $x(0) = -5$, $\dot{x}(0) = 1$
(e) $\ddot{x} + 6\dot{x} + 9x = 0$, $x(0) = 4$, $\dot{x}(0) = -2$
(f) $\ddot{x} + 2\dot{x} + 17x = 0$, $x(0) = 2$, $\dot{x}(0) = 1$

2.32 Solve the following nonhomogeneous second-order differential equations by the method of undetermined coefficients:
(a) $\ddot{x} + 5\dot{x} + 6x = 3\sin(2t)$, $x(0) = -2$, $\dot{x}(0) = 0$
(b) $\ddot{x} + 5\dot{x} + 6x = 2\cos(3t) + t^2$, $x(0) = 1$, $\dot{x}(0) = -2$
(c) $\ddot{x} + 5\dot{x} + 6x = 4t^2 e^{-2t}$, $x(0) = -3$, $\dot{x}(0) = 2$
(d) $\ddot{x} + 8\dot{x} + 16x = t^2 + 5$, $x(0) = 299/128$, $\dot{x}(0) = 81/16$
(e) $\ddot{x} + 8\dot{x} + 16x = 3\sin(t)$, $x(0) = 61/289$, $\dot{x}(0) = -176/289$
(f) $\ddot{x} + 8\dot{x} + 16x = e^{-4t}(\sin(t) + 2)$, $x(0) = -5$, $\dot{x}(0) = 23$
(g) $\ddot{x} + 6\dot{x} + 5x = 2e^{-5t} + 4te^{-t}$, $x(0) = 27/4$, $\dot{x}(0) = 1$

2.33 Solve the following nonhomogeneous second-order differential equations by the method of undetermined coefficients:
(a) $\ddot{x} + 4\dot{x} + 13x = 45e^{-5t}\cos(3t)$, $x(0) = 5$, $\dot{x}(0) = -1$
(b) $\ddot{x} + 4\dot{x} + 13x = 24e^{-2t}t\sin(3t)$, $x(0) = 9$, $\dot{x}(0) = -3$
(c) $\ddot{x} + 6\dot{x} + 9x = 4e^{-3t} + 27t$, $x(0) = 3$, $\dot{x}(0) = -10$
(d) $\ddot{x} + 6\dot{x} + 9x = 36t^2 e^{-3t}$, $x(0) = -2$, $\dot{x}(0) = 8$
(e) $\ddot{x} + 9x = 2\sin(3t) + 12t\cos(3t)$, $x(0) = 1$, $\dot{x}(0) = -3$
(f) $\ddot{x} + 4x = 17e^{-t}\sin(2t)$, $x(0) = 7$, $\dot{x}(0) = 20$
(g) $\ddot{x} + 2\dot{x} + 17x = 24t^2 e^{-t}\cos(4t)$, $x(0) = 2$, $\dot{x}(0) = 6$

2.34 Solve the equations of Problem 2.29 by the method of undetermined coefficients as described in Section 2.6.6.

Section 2.7 Transfer Functions and Block Diagrams of Time-Invariant Systems

Problems

2.35 Derive transfer functions of the following systems:

(a) $\ddot{x} + 2\dot{x} + 13x = f(t)$
(b) $x^{(3)} + 4\ddot{x} + 6\dot{x} + 2x = f(t)$
(c) $2x^{(3)} + 9\dot{x} + 25x = 4f(t)$
(d) $2x^{(3)} + 7\dot{x} + 3x = 12f(t) - \dot{f}(t)$
(e) $5x^{(4)} - \ddot{x} + 3\dot{x} + 8x = 4\dot{f}(t) - 7f(t)$

2.36 Derive the governing equations for the block diagrams given in Figure P,2,2:

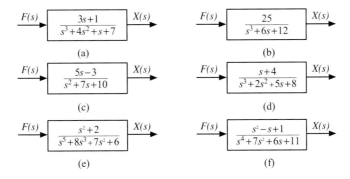

Figure P2.2 Block diagrams

Section 2.8 Solution of State Equations via Numerical Integration

2.37 Solve Problem 2.26 by numerical integration using Mathematica.
2.38 Solve Problem 2.26 by numerical integration using MATLAB.
2.39 Solve Problem 2.30 by numerical integration using Mathematica.
2.40 Solve Problem 2.30 by numerical integration using MATLAB.
2.41 Considering zero initial conditions, solve the system of linear differential equations by numerical integration using Mathematica and plot the variables x_1 and x_2 vs. time

$$\begin{cases} \ddot{x}_1 + 5\dot{x}_1 + 6x_1 - 5\dot{x}_2 = \sin(2t) \\ 4\ddot{x}_2 + 5\dot{x}_2 - 5\dot{x}_1 = 2\cos(3t) \end{cases}$$

2.42 Solve Problem 2.41 by numerical integration using MATLAB and plot the variables x_1 and x_2 vs. time

3 Mechanical Systems

Contents

3.1	Fundamental Principles of Mechanical Systems	115
3.2	Translational Systems	120
3.3	Rotational Systems	144
3.4	Rigid-Body Systems in Plane Motion	169
3.5	Energy Approach	177
3.6	Block Diagrams of Mechanical Systems	181
3.7	State-Space Representations	185
3.8	Dynamic Responses via MATLAB	191
3.9	Dynamic Responses via Mathematica	198
	Chapter Summary	204
	References	204
	Problems	204

Mechanical systems are seen in machines, devices, equipment, and structures in a wide variety of engineering applications. Modeling of a mechanical system involves forces and motion about relevant objects, which can be either solids or fluids. Depending on specific concerns in an application, there are several types of mathematical models for mechanical systems at different levels of complexity, including lumped-parameter models, rigid-body models, and deformable- or flexible-body models. For instance, in studying the dynamic behaviors of an airplane, a lumped-parameter model may be good enough to describe its flight trajectory and a rigid-body model may be sufficient to study its three-dimensional motion in space. However, a flexible-body model is necessary to understand the vibration and stress in the airplane structure in performance evaluation and failure analysis.

Regardless of the level of complexity of the mathematical model to be created, there are three importing keys in modeling of mechanical systems:

- fundamental principles
- basic elements
- ways of analysis

Fundamental principles (such as Newton's laws of motion) are the first laws of nature relating forces to motion. Basic elements (like masses, springs, and dampers) are approximate models of those components that are the building blocks of mechanical systems. A way of analysis (free-body diagram for instance) is an approach by which fundamental principles are applied to basic elements such that a mathematical model is established. The model so obtained is a set of governing equations of motion, which are differential and algebraic equations. The three-key modeling concept will be illustrated throughout this chapter.

In this text, the International System of Units (SI) is used. Refer to Tables A1 and A2 in Appendix A for commonly used quantities for mechanical systems, and Table A5 for unit conversions between the International System and the US customary system.

3.1 Fundamental Principles of Mechanical Systems

Without loss of generality, we shall first consider particles that are objects with negligible dimensions and with masses lumped to points. Such particles are a lumped-parameter model of mass elements of mechanical systems. The fundamental principles of particles shall be extended to rigid and flexible bodies later in the text.

This section provides a quick review of terms, definitions, and formulas in particle dynamics, as a reference for modeling of mechanical systems. Detailed explanations of these contents can be found in standard textbooks on dynamics, such as those listed at the end of the chapter. Readers who are familiar with these contents can skip this section.

3.1.1 Newton's Laws of Motion

Newton's three laws of motion for particles can be stated as follows:

First law: A particle remains at rest or moves with a constant velocity along a straight line if the sum of the forces acting on it is zero.
Second law: The rate of change of the linear momentum of a particle is equal to the sum of the forces acting on it.
Third law: The forces exerted by two particles upon each other, which are called the action and reaction forces, are equal in magnitude, opposite in direction, and collinear.

In the above-mentioned laws of motion, inertia frames of reference are considered. An inertial frame of reference is one in which Newton's first law of motion is valid.

Mathematically, Newton's second law is expressed by

$$\boldsymbol{F} = \frac{d}{dt}(m\boldsymbol{v}) \qquad (3.1.1)$$

where m and v are the mass and velocity of the particle, respectively; F is the resultant or sum of all forces acting on the particle; and mv is the linear momentum of the particle. Newton's second law is also applicable to rigid and flexible bodies.

If the mass m is constant, Eq. (3.1.1) becomes

$$F = ma \tag{3.1.2}$$

where $a = \dfrac{dv}{dt}$ is the acceleration of the particle. Equation (3.1.2) is a commonly used expression of Newton's second law, which indicates that the acceleration of a particle of constant mass is proportional to the resultant force. Also, Newton's first law can be viewed as a special example of the second law with vanishing linear momentum or acceleration

$$F = 0 \tag{3.1.3}$$

Newton's laws can be extended to a system of particles. For a system of N particles, Newton's second law can be written as

$$\sum_{i=1}^{N} F_i^{ext} = \frac{d}{dt}\left\{ \sum_{i=1}^{N} m_i v_i \right\} \tag{3.1.4}$$

where F_i^{ext} is the resultant external force applied to the i-th particle, and m_i and v_i are the mass and velocity of the i-th particle. Here, those action and reaction forces between any two particles are known as *internal forces* and those forces that are not internal forces are called *external forces*. Internal forces do not appear in Eq. (3.1.4) because they cancel each other.

3.1.2 Forces and Moments

According to Newton's second law, the motion of a particle or body is caused by forces. A force is an action of pull or push applied by one body to another. There are various types of forces, including external loads, gravity, friction, and spring forces. A force is characterized by its magnitude, direction of action, and point of application. Thus, forces can be expressed by vectors and the superposition of forces follows the parallelogram rule of vector summation,

$$F = F_1 + F_2 \tag{3.1.5}$$

The moment of a force is a measure of the tendency of the force to rotate a body. The moment M_O of force F about a reference point O is defined by the vector cross-product

$$M_O = r \times F \tag{3.1.6}$$

where r is a position vector that runs from the reference point to the point of application of F; see Figure 3.1.1. The direction of the moment is determined by the right-hand rule, as shown in the figure. Refer to Section 2.1.4 for determination of vector cross-products.

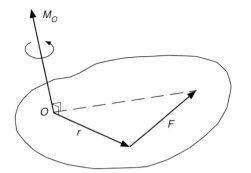

Figure 3.1.1 The moment of a force about point O

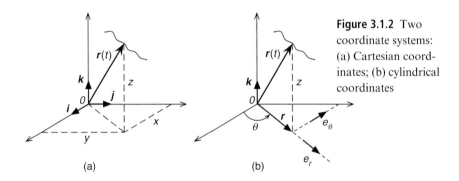

Figure 3.1.2 Two coordinate systems: (a) Cartesian coordinates; (b) cylindrical coordinates

3.1.3 Kinematics of Particles

When a particle moves along a path in a reference frame, its position is described by a position vector $r(t)$, which is a function of time. The velocity $v(t)$ and acceleration $a(t)$ of the particle are the first and second derivatives of $r(t)$

$$v(t) = \frac{dr(t)}{dt} = \dot{r}(t), \quad a(t) = \frac{d^2r(t)}{dt^2} = \frac{dv(t)}{dt} = \ddot{r}(t) \qquad (3.1.7)$$

The motion of a particle can be described in different coordinates. Two commonly used coordinate systems are Cartesian coordinates and cylindrical coordinates; see Figure 3.1.2, where i, j, k and e_r, e_θ, k are unit vectors. Note that the directions of e_r and e_θ change with angle θ.

In Cartesian coordinates, position, velocity, and acceleration vectors are written as

$$\begin{aligned} r(t) &= x\mathbf{i} + y\mathbf{j} + z\mathbf{k} \\ v(t) &= \dot{x}\mathbf{i} + \dot{y}\mathbf{j} + \dot{z}\mathbf{k} \\ a(t) &= \ddot{x}\mathbf{i} + \ddot{y}\mathbf{j} + \ddot{z}\mathbf{k} \end{aligned} \qquad (3.1.8)$$

For plane motion, the expressions of position, velocity, and acceleration vectors can be obtained from Eq. (3.1.8) by neglecting the coordinate z. In the cylindrical coordinates, position, velocity, and acceleration vectors are of the form

$$\begin{aligned} r(t) &= r\mathbf{e}_r + z\mathbf{k} \\ v(t) &= \dot{r}\mathbf{e}_r + r\dot{\theta}\mathbf{e}_\theta + \dot{z}\mathbf{k} \\ a(t) &= (\ddot{r} - r\dot{\theta}^2)\mathbf{e}_r + (r\ddot{\theta} + 2\dot{r}\dot{\theta})\mathbf{e}_\theta + \ddot{z}\mathbf{k} \end{aligned} \qquad (3.1.9)$$

The second and third lines of Eq. (3.1.9) are obtained via differentiation of the first one and through use of the relations

$$\dot{\mathbf{e}}_r = \dot{\theta}\mathbf{e}_\theta, \quad \dot{\mathbf{e}}_\theta = -\dot{\theta}\mathbf{e}_r \qquad (3.1.10)$$

For plane motion, the cylindrical coordinates are reduced to the polar coordinates (r, θ) and position, velocity, and acceleration vectors become

$$\begin{aligned} r(t) &= r\mathbf{e}_r \\ v(t) &= \dot{r}\mathbf{e}_r + r\dot{\theta}\mathbf{e}_\theta \\ a(t) &= (\ddot{r} - r\dot{\theta}^2)\mathbf{e}_r + (r\ddot{\theta} + 2\dot{r}\dot{\theta})\mathbf{e}_\theta \end{aligned} \qquad (3.1.11)$$

3.1.4 Principle of Work and Energy

Consider a particle in motion. Dot-product Eq. (3.1.1) by dr and integrate the resulting equation between positions r_1 and r_2 on a path of the particle, to obtain

$$\int_{r_1}^{r_2} F \cdot dr = \frac{1}{2}\int_{t_1}^{t_2} m \frac{d}{dt}(v \cdot v)\,dt \\ = \frac{1}{2}mv^2 \Big|_{t_1}^{t_2} = \frac{1}{2}mv_2^2 - \frac{1}{2}mv_1^2 \qquad (3.1.12)$$

where $v = |v|$, and v_1 and v_2 are the velocity magnitudes at r_1 and r_2 (at times t_1 and t_2), respectively. Recall that the kinetic energy of the particle is

$$T(t) = \frac{1}{2}m\mathbf{v} \cdot \mathbf{v} = \frac{1}{2}mv^2 \qquad (3.1.13)$$

and that the work done by force $\boldsymbol{F}$ on a path from $\boldsymbol{r}_1$ to $\boldsymbol{r}_2$ is

$$W_{12} = \int_{r_1}^{r_2} \boldsymbol{F} \cdot d\boldsymbol{r} \tag{3.1.14}$$

Thus, Eq. (3.1.12) states the *principle of work and energy*

$$W_{12} = T_2 - T_1 = \Delta T \tag{3.1.15}$$

where $T_1 = T(t_1)$ and $T_2 = T(t_2)$. Equation (3.1.15) shows that the work done by the force applied to a particle is equal to the change of the kinetic energy of the particle.

A force $\boldsymbol{F}$ is *conservative* if its work is independent of any path chosen between two positions $\boldsymbol{r}_1$ and $\boldsymbol{r}_2$. The work of a conservative force can be expressed by

$$W_{12} = -(V_2 - V_1) = -\Delta V \tag{3.1.16}$$

where $V = V(\boldsymbol{r})$, which, being a scalar function of position vector $\boldsymbol{r}$, is called *potential energy*, and $V_1 = V(\boldsymbol{r}_1)$, $V_2 = V(\boldsymbol{r}_2)$. For example, the gravitational force and the spring force are both conservative, with the following potential energy functions:

$$\begin{aligned} \text{Gravity:} \quad & V = mgy \\ \text{Spring force:} \quad & V = \frac{1}{2}kx^2 \end{aligned}$$

where y is a spatial coordinate along an upward axis, and x is the spring elongation. A force $\boldsymbol{F}$ is *non-conservative* if its work is dependent upon a path chosen. No potential energy exists for a non-conservative force. For instance, friction is a non-conservative force.

Let a particle be subject to a conservative force $\boldsymbol{F}^c$ and a non-conservative force $\boldsymbol{F}^{nc}$. The total work done by the forces can be written as

$$W_{12} = W_{12}^{nc} - \Delta V \tag{3.1.17}$$

where W_{12}^{nc} is the work done by the non-conservative force $\boldsymbol{F}^{nc}$. Substitution of Eq. (3.1.17) into Eq. (3.1.15) gives the principle of work and energy

$$W_{12}^{nc} = (T_2 + V_2) - (T_1 + V_1) = \Delta(T + V) \tag{3.1.18}$$

Here, the sum $T + V$ is called the *mechanical energy* of the particle. If only conservative forces are applied to the particle ($\boldsymbol{F}^{nc} = 0$), the mechanical energy of the particle is conserved:

$$T_2 + V_2 = T_1 + V_1 \tag{3.1.19}$$

3.1.5 Principle of Impulse and Momentum

The time integration of Newton's second law, Eq. (3.1.1), leads to the *principle of impulse and momentum*

$$I_{12} = L_2 - L_1 \tag{3.1.20}$$

where $I_{12} = \int_{t_1}^{t_2} F\,dt$ is the impulse of the force applied to the particle, $L(t) = mv$ is the linear momentum of the particle, $L_1 = L(t_1)$ and $L_2 = L(t_2)$. If no force is applied to the particle, which means $I_{12} = 0$, by Eq. (3.1.20), the impulse is conserved:

$$L_2 = L_1 \tag{3.1.21}$$

For a system of N particles $(m_1, m_2, \ldots, m_N)$, the principle of impulse and momentum has the same form of Eq. (3.1.20). However, by Eq. (3.1.4), the impulse and momentum of the system are given as follows:

$$I_{12} = \int_{t_1}^{t_2} \sum_{i=1}^{N} F_i^{ext}\,dt, \quad L(t) = \sum_{i=1}^{N} m_i v_i \tag{3.1.22}$$

3.2 Translational Systems

A translational system is a mechanical system with its components only involved in translational motion. Modeling of translational systems usually results in lumped-parameter models.

3.2.1 Basic Elements

As the first key in modeling of mechanical systems, fundamental principles of particles have been presented in Section 3.1. As the second key in system modeling, three types of basic elements are introduced in this section: mass elements, spring elements, and damping elements.

Lumped Masses

In a lumped-parameter model, lumped masses are used. A lumped mass, which is also called a discrete mass or point mass or simply mass, is a particle of concentrated inertia, as described in Section 3.1. The principles and formulas given in Section 3.1 thus are all applicable to lumped masses. For example, for a mass m in Figure 3.2.1(a), which moves

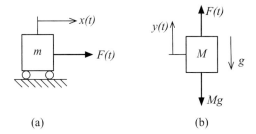

Figure 3.2.1 Lumped masses: (a) horizontal motion; (b) vertical motion

on a smooth horizontal surface and is subject to a force $F(t)$, its displacement $x(t)$ by Newton's second law, Eq. (3.1.2), is governed by the differential equation

$$m\ddot{x} = F$$

In the figure, the two wheels under the moving mass indicate that the supporting horizontal surface is smooth without friction. Similarly, for a mass M in Figure 3.2.1(b), which moves vertically under the influence of gravity and a force $F(t)$, its displacement $y(t)$ is governed by the differential equation

$$M\ddot{y} = F - Mg$$

where g is the gravitational acceleration.

Translational Springs

A translational spring or linear spring is a spring of negligible inertia, with its internal force (spring force) given by Hooke's law

$$\text{Spring force} = k \times \Delta x \qquad (3.2.1)$$

where k is a spring coefficient, in N/m, and Δx is the elongation of the spring. The Δx is also the relative displacement of the two ends of the spring. Figure 3.2.2 shows three cases of springs in deformation: (a) a spring with a fixed end; (b) a spring of two movable ends in tension; and (c) a spring of two movable ends in compression. In case (a), the spring force f_s and the displacement (elongation) x are in the relation

$$f_s = k\Delta x = kx \qquad (3.2.2)$$

where the relative displacement $\Delta x = x - 0 = x$. In case (b), the spring force of the tensioned spring is given by

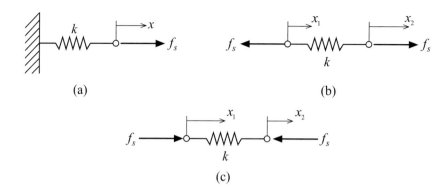

Figure 3.2.2 Springs in deformation: (a) spring with a fixed end; (b) spring of two movable ends in tension; (c) spring of two movable ends in compression

$$f_s = k\,\Delta x = k(x_2 - x_1), \quad \text{for } x_2 > x_1 \tag{3.2.3}$$

In case (c), the spring force of the compressed spring is of the form

$$f_s = k\,\Delta x = k\,(x_1 - x_2), \quad \text{for } x_1 > x_2 \tag{3.2.4}$$

As observed from Figure 3.3.2, in the expression of the relative displacement for a spring that is either tensioned or compressed, the first displacement is the one that has the same direction as the spring force at the point of application. This is because the applied forces and corresponding deformation must be consistent in direction. Hence, $\Delta x = x_2 - x_1$ for a tensioned spring as shown in Figure 3.2.2(b); $\Delta x = x_1 - x_2$ for a compressed spring, as shown in Figure 3.2.2(c). This observation is useful in modeling of mechanical systems.

Note that the forces applied at a massless spring must be balanced at any time. In other words, the forces at the two ends of a spring are always the same in magnitude and opposite in direction, as shown in Figures 3.2.2(b) and 3.2.2(c).

Effective Spring Coefficients of Elastic Bodies

Translational springs in Figure 3.2.2 are lumped-spring models that are deduced from elastic bodies. Indeed, spring elements in engineering applications are made of flexible bodies. Such a flexible body can be described by the theory of the strength of materials, and a spring coefficient so determined is called the effective spring coefficient. In this process, it is often assumed that the inertia of the flexible body is negligible.

As an example, consider a massless cantilever beam subject to a load F at its tip; see Figure 3.2.3(a). By beam theory, the beam tip deflection y is expressed by

$$y = \frac{L^3}{3EI} F$$

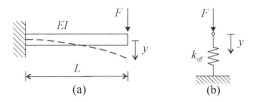

Figure 3.2.3 Effective spring coefficient: (a) massless cantilever beam; (b) beam treated as a spring

where EI and L are the bending stiffness and length of the beam, respectively. The previous equation can be written as

$$F = k_{eff} y$$

with the effective spring coefficient k_{eff} given by

$$k_{eff} = \frac{3EI}{L^3}$$

Thus, the massless cantilever beam can be treated as a spring, as shown in Figure 3.2.3(b).

Table 3.2.1 lists the effective spring coefficients for several elastic continua commonly seen in engineering applications.

Viscous Dampers

A viscous damper or dashpot, which is simply called a damper, is a device generating a force (damping force) that is proportional to the difference Δv in the velocity at the two ends of the device:

$$\text{Damping force} = c \times \Delta v \qquad (3.2.5)$$

where c is a damping coefficient, in N·s/m. The Δv is also called relative velocity. Figure 3.2.4 shows three cases of dampers in deformation: (a) a damper with a fixed end; (b) a damper of two movable ends in tension; and (c) a damper of two movable ends in compression. In case (a), the damping force f_d is related to the velocity at the right end of the damper by

$$f_d = c \Delta v = c \dot{x} \qquad (3.2.6)$$

in which, $\Delta v = \dot{x} - 0 = \dot{x}$. In case (b), the damping force of the tensioned damper is given by

$$f_s = c \Delta v = c (\dot{x}_2 - \dot{x}_1), \text{ for } \dot{x}_2 > \dot{x}_1 \qquad (3.2.7)$$

Table 3.2.1 Effective spring coefficients of certain continua

Cantilever beam

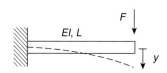

$F = k_{eff}\, y, \ k_{eff} = \dfrac{3EI}{L^3}$

$EI =$ bending stiffness

Simply supported beam

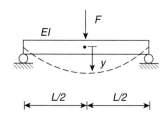

$F = k_{eff}\, y, \ k_{eff} = \dfrac{48EI}{L^3}$

$EI =$ bending stiffness

Fixed-end beam

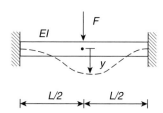

$F = k_{eff}\, y, \ k_{eff} = \dfrac{192EI}{L^3}$

$EI =$ bending stiffness

Rod in longitudinal deformation

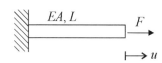

$F = k_{eff}\, u, \ k_{eff} = \dfrac{EA}{L}$

$EA =$ longitudinal rigidity

Taut string

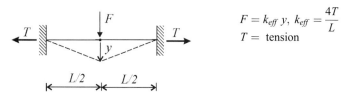

$F = k_{eff}\, y, \ k_{eff} = \dfrac{4T}{L}$

$T =$ tension

Rod in torsion

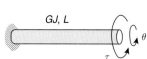

$\tau = k_{eff}\, \theta, \ k_{eff} = \dfrac{GJ}{L}$

$GJ =$ torsional rigidity

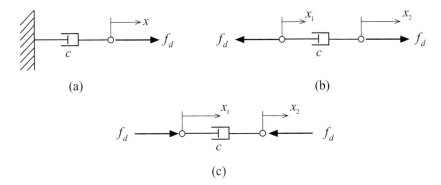

Figure 3.2.4 Dampers in deformation: (a) damper with a fixed end; (b) damper of two movable ends in tension; (c) damper of two movable ends in compression

In case (c), the damping force of the compressed damper is of the form

$$f_d = c\,\Delta v = c\,(\dot{x}_1 - \dot{x}_2), \quad \text{for } \dot{x}_1 > \dot{x}_2 \tag{3.2.8}$$

Similar to springs in deformation, the first velocity in the expression of Δv is the one which has the same direction as the damping force. Thus, $\Delta v = \dot{x}_2 - \dot{x}_1$ for a tensioned damper, as shown in Figure 3.2.4(b); $\Delta v = \dot{x}_1 - \dot{x}_2$ for a compressed damper, as shown in Figure 3.2.4(c). Also, with the massless assumption for dampers, the forces applied at a damper are balanced at any time. Consequently, the damping forces at the two ends of a damper are always the same in magnitude and opposite in direction.

Friction Models

Friction forces are induced by the relative velocity between two moving objects that are in contact during motion. There are two commonly used models of friction: viscous friction (fluid friction) and dry friction (Coulomb damping). In viscous friction, as shown in Figure 3.2.5, the friction force is of the form

$$f_d = B \times \Delta v \tag{3.2.9}$$

where B is a viscous friction coefficient, and Δv is the relative velocity between the objects m_1 and m_2. Note that Eq. (3.2.9) is the same in format as Eq. (3.2.5) for dampers. Figure 3.2.5(a) shows a schematic of viscous friction between two moving objects (m_1 and m_2).

Unlike the damping forces of a damper, the viscous friction forces are the action and reaction forces that are separately applied to the two objects. Figures 3.2.5(b) and 3.2.5(c) show two cases of friction forces in free-body diagrams. In each case, the relative velocity is determined by the relative motion on the two surfaces of the viscous friction layer, which is denoted by symbol ⸺. With the concept of relative velocity presented in Eqs. (3.2.6) to (3.2.8), Δv is the difference between the velocity that is in the same direction as the friction

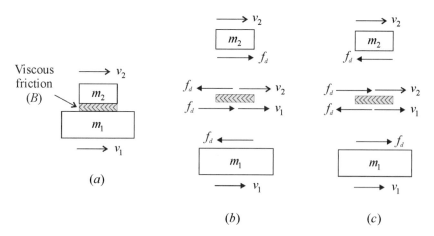

Figure 3.2.5 Viscous friction: (a) schematic; (b) free-body diagrams for $v_1 > v_2$, $\Delta v = v_1 - v_2$; (c)) free-body diagrams for $v_2 > v_1$, $\Delta v = v_2 - v_1$

force f_d and the velocity that is in the opposite direction of the friction force. Here, the velocity and friction force in consideration must be on the same surface of the viscous friction layer. Thus, for $v_1 > v_2$, $\Delta v = v_1 - v_2$, as shown in Figure 3.2.5(b); for $v_2 > v_1$, $\Delta v = v_2 - v_1$, as shown in Figure 3.2.5(c).

Figure 3.2.6 (a) shows a schematic of dry friction between two moving objects (m_1 and m_2). The friction force is of the form

$$f_\mu = \begin{cases} -\mu_D N, & \text{for } \Delta v < 0 \\ \mu_D N, & \text{for } \Delta v > 0 \end{cases} \quad (3.2.10)$$

where μ_D is a coefficient of kinematic friction, N is a normal force, and Δv is the relative velocity between the two objects. For simplicity, Eq. (3.2.10) can written as

$$f_\mu = \mu_D N \operatorname{sgn}(\Delta v) \quad (3.2.11)$$

where $\operatorname{sgn}(a)$ is a sign function, with $\operatorname{sgn}(a) = 1$ for $a > 0$, $\operatorname{sgn}(a) = -1$ for $a < 0$ and $\operatorname{sgn}(a) = 0$ for $a = 0$.

Figures 3.2.6(b) and 3.2.6(c) show two cases of a dry-friction force f_μ and a normal force N in free-body diagrams. In each case, the relative velocity is determined by the relative motion on the two surfaces of the dry-friction region, which is denoted by symbol ⨉⨉⨉⨉⨉. With the concept of relative velocity presented in Eqs. (3.2.6) to (3.2.8), Δv is the difference between the velocity that is in the same direction as the friction force f_μ and the velocity that is in the opposite direction of the friction force. Hence, the velocity and friction force in consideration must be on the same surface of the dry-friction region. Thus, for $v_1 > v_2$, $\Delta v = v_1 - v_2$, as shown in Figure 3.2.6(b); and for $v_2 > v_1$, $\Delta v = v_2 - v_1$, as shown in Figure 3.2.6(c).

3.2 Translational Systems   127

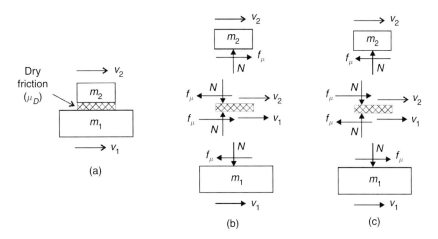

Figure 3.2.6 Dry friction: (a) schematic; (b) free-body diagrams for $v_1 > v_2$, $\Delta v = v_1 - v_2$; (c) free-body diagrams for $v_2 > v_1$, $\Delta v = v_2 - v_1$

As seen from Eq. (3.2.10), dry-friction forces are nonlinear functions of the relative motion between two objects. Different from viscous friction, the magnitude of a dry-friction force f_μ is not related to the relative velocity Δv, but it depends on the normal force N. Also, as depicted in Figures 3.2.6(b) and 3.2.6(c), dry-friction forces are the action and reaction forces that are separately applied to the two objects.

In summary of this subsection, a translational system consists of the above-described lumped masses, springs, dampers, and friction models. Because of this, such a system is often called a spring–mass–damper system. There are other models of spring and damping elements, including nonlinear springs and air damping, some of which shall be introduced later on.

3.2.2 Free-Body Diagrams and Auxiliary Plots

In Sections 3.1 and 3.2.1, the first two keys in system modeling, namely the fundamental principles and basic elements, have been presented. In this section, we consider the last (third) key – ways of analysis. In a way of analysis, the fundamental principles are applied to basic elements, such that the governing equations of a mechanical system in consideration are obtained. There are two typical ways of analysis: free-body diagrams and Lagrange's equations. The former is part of Newtonian mechanics and the latter belongs to Lagrangian mechanics. In this section, free-body diagrams are considered. Lagrange's equations shall be introduced in Section 3.5.

Free-body diagrams are studied in a typical undergraduate dynamics course. A free-body diagram is the diagram of a mass element showing all internal and external forces applied to the element and marking the coordinates or displacements of the element. Here, internal

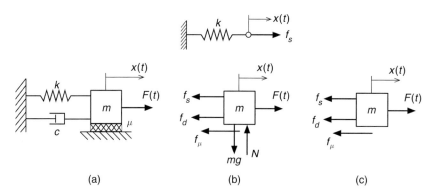

Figure 3.2.7 Free-body diagrams of a spring-mass-damper system (see text for details)

forces refer to those forces applied by spring and damping elements (including friction forces) while external forces, including gravity, are those that are not internal forces. In generating a free-body diagram, no connection between the mass element in consideration and any other element is allowed. In other words, for a system with N particles, N free-body diagrams should be drawn.

As an example, consider a spring–mass–damper system in Figure 3.2.7(a), which is constrained by a spring (k), a damper (c), and dry friction (μ), subject to an external force (F), and under the gravity. The two-dimensional free-body diagram of the mass m is shown Figure 3.2.7(b), where f_s, f_d, and f_μ are the internal forces of the spring, damper, and dry friction, mg is the gravity, and N is the normal force applied by the supporting horizontal surface. If only the motion in the horizontal direction is of interest, the free-body diagram can be simplified to show quantities only in the x direction; see Figure 3.2.7(c).

For convenience in system modeling, plots of spring and damping elements can be drawn, which are called **auxiliary plots**. The plots in Figures 3.2.2 and 3.2.4 are examples of auxiliary plots. Like a free-body diagram, the auxiliary plot of a spring or damping element concerns the forces applied to the element and it does not show the connection between the element and any other elements. Auxiliary plots are usually needed for spring/damping elements with two movable ends.

Auxiliary plots are especially useful in modeling of a system with multiple mass elements. For instance, consider the spring–mass–damper system in Figure 3.2.8(a), where two masses (m_1, m_2) are connected to two springs (k_1, k_2) and one damper (c). Figure 3.2.8(b) shows two free-body diagrams about the masses, and one auxiliary plot about the spring of coefficient k_2, where, by Newton's third law on action and reaction, the directions of the internal forces are properly assigned. Note that the spring is tensioned with the assumed direction of the spring force f_{s2}, indicating that the relative displacement $\Delta x = x_2 - x_1$ by Eq. (3.2.3).

Since the displacements x_1 and x_2 are unknown, the spring is not necessarily tensioned, and it can also be compressed. Figure 3.2.8(c) shows an auxiliary plot of a compressed spring k_2.

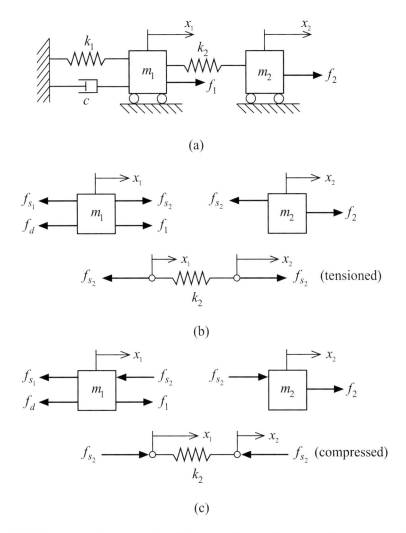

Figure 3.2.8 Free-body diagrams and auxiliary plots for a spring–mass–damper system (see text for details)

In this case, the spring force f_{s2} is in the opposite direction to that in Figure 3.2.7(b) and the relative displacement becomes $\Delta x = x_1 - x_2$ by Eq. (3.2.4). In system modeling, either a tensioned spring or a compressed spring can be assumed, and the governing equations eventually obtained should be the same. This is because the change of the direction of a spring force is matched by the change of sign of the corresponding relative displacement Δx.

In summary, with a system of N particles, N free-body diagrams must be drawn and some auxiliary plots of spring and damping elements may be generated. An auxiliary plot is necessary only if the particular spring or damping element has two movable ends with unknown displacements.

3.2.3 Derivation of Equations of Motion

With the three keys of system modeling (fundamental principles, basic elements, and free-body diagrams) presented in Sections 3.1, 3.2.1, and 3.2.3, the governing equations for a translational system are readily established. This process of system modeling takes the following four steps:

(1) Draw free-body diagrams, one for each mass element, and create auxiliary plots for some spring/damping elements.
(2) Apply Newton's second law to each free-body diagram, yielding the differential equations about the displacements of the mass elements.
(3) Write down the expressions of internal forces (of the spring and damping elements) and use auxiliary plots if necessary.
(4) Obtain the governing equations of motion of the system through substitution of the expressions of internal forces given in Step 3 into the equations obtained in Step 2.

As a convention, the equations of motion so obtained in the above-described process list the terms of unknown displacements and their derivatives on the left-hand sides, and the terms about external forces on the right-hand sides.

In modeling of a mechanical system, a spring or damping element can be assumed either in tension or in compression. The final equations of motion of the system are always the same, provided that the corresponding internal forces and relative motion are treated consistently.

Example 3.2.1

For the spring–mass–damper system in Figure 3.2.8(a), derive the equations of motion.

Solution
The free-body diagrams of the masses and an auxiliary plot of the spring k_2 are drawn in Figure 3.2.8(b), where the spring is assumed to be tensioned. By Newton's second law and with the free-body diagrams, the governing equations of the masses are obtained as follows

$$\text{For } m_1 \quad m_1 \ddot{x}_1 = f_1 + f_{s2} - f_{s1} - f_d \tag{a}$$

$$\text{For } m_2 \quad m_2 \ddot{x}_2 = f_2 - f_{s2} \tag{b}$$

According to Section 3.2.1 and the auxiliary plot in Figure 3.2.8(b), the internal forces are expressed by

$$\begin{aligned} f_1 &= k_1 x_1 \\ f_d &= c \dot{x}_1 \\ f_{s2} &= k_2 (x_2 - x_1) \end{aligned} \tag{c}$$

where Eq. (3.2.3) has been used to obtain the expression of f_{s2}. Substituting Eq. (c) into Eqs. (a) and (b) yields

$$m_1\ddot{x}_1 = f_1 + k_2(x_2 - x_1) - k_1 x - c\dot{x}_1$$
$$m_2\ddot{x}_2 = f_2 - k_2(x_2 - x_1)$$
(d)

Finally, rearranging the previous equations gives the equations of motion of the system
$$m_1\ddot{x}_1 + c\dot{x}_1 + (k_1 + k_2)x_1 - k_2 x_2 = f_1$$
$$m_2\ddot{x}_2 + k_2 x_2 - k_2 x_2 = f_2$$
(e)

If the spring k_2 is assumed to be compressed, the corresponding free-body diagrams and auxiliary plot are shown in Figure 3.2.8(c). The differential equations by Newton's second law become

For m_1 $\quad m_1\ddot{x}_1 = f_1 - f_{s2} - f_{s1} - f_d$ (f)

For m_2 $\quad m_2\ddot{x}_2 = f_2 + f_{s2}$ (g)

in which the signs of the f_{s2}-terms have changed. On the other hand, the force of the spring k_2 by Eq. (3.2.4) becomes $f_{s2} = k_2(x_1 - x_2)$, which differs from the spring force expression in Eq. (c) by a sign. The other internal forces remain the same. This indicates that the same equations (e) of motion can be obtained via substitution and manipulation.

Example 3.2.2

In Figure 3.2.9, a mass is constrained by a spring and a damper that are serially connected. Let the displacement at the point of the spring–damper connection be $y(t)$. Derive the equations of motion for the translational system.

Solution

Figure 3.2.10 shows the free-body diagram of the mass and the auxiliary plots of the damping and spring elements. Note that the spring and damper are in tension. The equations about the three basic elements are as follows:

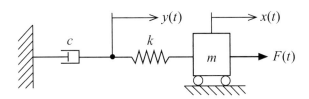

Figure 3.2.9 A mass constrained by serially connected spring and damper

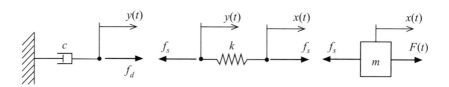

Figure 3.2.10 Free-body diagram and auxiliary plots for the system in Example 3.2.2

$$\text{Mass:} \quad m\ddot{x} = F - f_s$$
$$\text{Damper:} \quad f_d = c\dot{y}$$
$$\text{Spring:} \quad f_s = k(x - y)$$

where Eq. (3.2.3) has been used for the tensioned spring. Because the spring and damper are connected, $f_d = f_s$. It follows that the system is governed by the coupled differential equations

$$m\ddot{x} + kx - ky = F$$
$$c\dot{y} + ky - kx = 0$$

Note that the displacements x and y are independent of each other even though no mass element related to y exists. The coupled differential equations can be solved by the Laplace transform (Section 2.4) or a transfer function formulation (Section 3.2.4).

Example 3.2.3

In Figure 3.2.11, a mass M slides on a slope of angle α, and is constrained by a spring k. There exists dry friction (μ_D) between the slope and mass and gravity is naturally considered. Let x be the displacement of the mass on the slope. Derive the equation of motion of the mass.

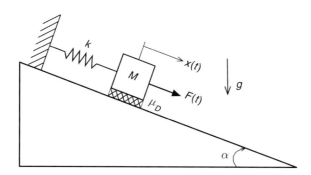

Figure 3.2.11 A mass sliding on a slope

Solution
The free-body diagram of the mass is shown in Figure 3.2.12, where axis y is normal to the slope surface. By Newton's second law, the motion of the mass in the x direction is governed by

$$m\ddot{x} = F + Mg \sin \alpha - f_s - f_\mu$$

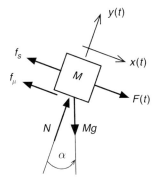

Figure 3.2.12 Free-body diagram of the mass in Example 3.2.3

The spring force is $f_s = kx$. The dry-friction force, by Eq. (3.2.11) with $\Delta v = \dot{x}$, is given by

$$f_\mu = \mu_D N \operatorname{sgn}(\dot{x})$$

Because the mass does not have motion in the y direction, $N = Mg \cos \alpha$. Therefore, the equation of motion of the mass is

$$m\ddot{x} + kx + \mu_D Mg \cos \alpha \operatorname{sgn}(\dot{x}) = F + Mg \sin \alpha$$

which is a nonlinear differential equation.

Example 3.2.4

A translational model of an automotive suspension system is shown in Figure 3.2.13, which is known as quarter-car suspension model because only one wheel is considered. In the figure, spring k_1 models the stiffness of the tire; mass m_1 represents the inertia of the tire and axle, spring k_2 and damper c describe the suspension system, mass m_2 describes the inertia of the supported vehicle components, including the chassis structure, engine, and driver and passengers; and $y_0(t)$

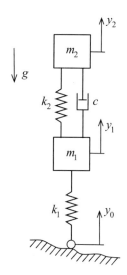

Figure 3.2.13 A quarter-car suspension model

prescribes the surface profile of the road. By considering gravity, derive the equations of motion of the car suspension system.

Solution

The free-body diagrams and auxiliary plots of the system are drawn in Figure 3.2.14, where all the spring and damping elements are in tension. By the free-body diagrams, the differential equations about the displacements $y_1(t)$ and $y_2(t)$ are given by

$$m_1\ddot{y}_1 = f_{s2} + f_d - m_1 g - f_{s1}$$
$$m_2\ddot{y}_2 = -m_2 g - f_{s2} - f_d \tag{a}$$

The internal forces, by the auxiliary plots, are determined as

$$f_{s1} = k_1(y_1 - y_0)$$
$$f_{s2} = k_2(y_2 - y_1)$$
$$f_d = c(\dot{y}_2 - \dot{y}_1) \tag{b}$$

By substituting Eq. (b) into Eq. (a), the governing equations of motion of the vehicle system are obtained as follows:

$$m_1\ddot{y}_1 + c\dot{y}_1 + (k_1 + k_2)y_1 - c\dot{y}_2 - k_2 y_2 = k_1 y_0 - m_1 g$$
$$m_2\ddot{y}_2 + c\dot{y}_2 + k_2 y_2 - c\dot{y}_1 - k_2 y_1 = -m_2 g \tag{c}$$

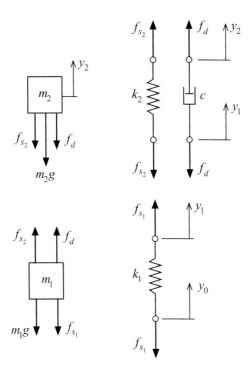

Figure 3.2.14 Free-body diagrams and auxiliary plots of the quarter-car suspension model

where the prescribed road surface profile y_0 serves as a disturbance to the vehicle system.

Example 3.2.5

In Figure 3.2.15, a cart of mass m_1 carries a lumped mass m_2 and it is subject to an external force f. The cart and the mass are connected by a spring (k_2) and there is viscous friction (B) between the two bodies. Develop a mathematical model for the cart–mass system. Also, the cart is constrained by another spring (k_1)

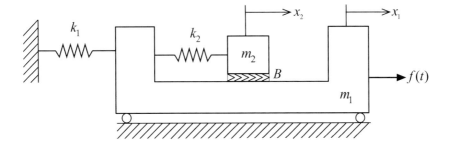

Figure 3.2.15 A cart–mass system in Example 3.2.5

Solution
Figure 3.2.16 shows the free-body diagrams and the auxiliary plots of the system. According to the free-body diagrams, the differential equations of the bodies are

$$m_1\ddot{x}_1 = f - f_{s1} + f_{s2} - f_d$$
$$m_2\ddot{x}_2 = f_d - f_{s2}$$
$$m_1\ddot{x}_1 = f - f_{s1} + f_{s2} - f_d$$
$$m_2\ddot{x}_2 = f_d - f_{s2}$$

By the auxiliary plots, the internal forces are determined as

$$f_{s1} = k_1 x_1, \quad f_{s2} = k_2(x_2 - x_1), \quad f_d = B(\dot{x}_1 - \dot{x}_2)$$

It follows that

$$m_1\ddot{x}_1 = f - k_1 x_1 + k_2(x_2 - x_1) - B(\dot{x}_1 - \dot{x}_2)$$
$$m_2\ddot{x}_2 = B(\dot{x}_1 - \dot{x}_2) - k_2(x_2 - x_1)$$

which, after rearrangement, leads to the mathematical model of the cart–mass system

$$m_1\ddot{x}_1 + B\dot{x}_1 + (k_1 + k_2)x_1 - B\dot{x}_2 - k_2 x_2 = f$$
$$m_2\ddot{x}_2 + B\dot{x}_2 + k_2 x_2 - B\dot{x}_1 - k_2 x_1 = 0$$

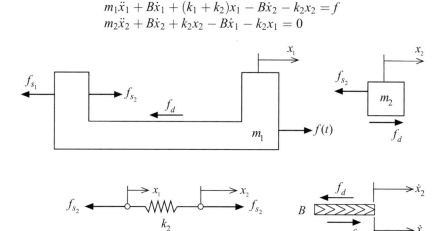

Figure 3.2.16 Free-body diagrams and auxiliary plots for the system in Example 3.2.5

In the previous examples, constant masses are considered. For mechanical systems with time-varying inertia, such as rockets with burning fuel, Eq. (3.1.2) is not directly applicable. A system with time-varying parameters is known as a *time-variant system*. In this case, one may consider the principle of impulse and momentum for a system of particles, as presented in Section 3.1.5. Indeed, the principle of impulse and momentum is the result of the temporal integration of Newton's second law and it serves as a fundamental principle in system modeling. See the next example.

Example 3.2.6

In Figure 3.2.17(a), a rocket is moving upward against gravity. At time t, the rocket has mass $m_R(t)$ and velocity $v(t)$. Let the effective exhaust velocity of burning fuel be $u_e(t)$, which is measured from the rocket frame. Derive the equation of motion for the rocket.

Solution

Assume that a small mass dm is expelled from the rocket during the time interval dt. Thus, at time $t + dt$, we have a two-particle system: the rocket with mass $m_R - dm$ and velocity $v + dv$, and the expelled mass dm with velocity $v - u_e$; see Figure 3.2.17(b). Let f_{drag} be a drag force acting on the rocket body (not shown in Figure 3.2.17). The impulse of the gravity and drag force and the linear momentums at t and $t + dt$, by Eq. (3.1.22), are

$$I_{12} = -(m_R g + f_{drag})\, dt$$
$$L_1 = m_R v, \quad L_2 = (m_R - dm)(v + dv) + dm\,(v - u_e) \tag{a}$$

According to the principle of impulse and momentum, $I_{12} = L_2 - L_1$, we have

$$-(m_R g + f_{drag})dt = (m_R - dm)(v + dv) + dm\,(v - u_e) - m_R v \tag{b}$$

which, after neglecting the higher-order term $dmdv$, is reduced to

$$-(m_R g + f_{drag})dt = m_R\, dv - u_e\, dm \tag{c}$$

Because the mass of the rocket during the interval dt is decreased, $dm_R = -dm$, Eq. (c) is reduced to

$$-(m_R g + f_{drag})dt = m_R\, dv + u_e\, dm_R \tag{d}$$

It follows from Eq. (d) that the equation of motion of the rocket is

$$m_R \frac{dv}{dt} = -u_e \frac{dm_R}{dt} - m_R g - f_{drag} \tag{e}$$

where the term $-u_e \dfrac{dm_R}{dt}$ is called rocket thrust. This equation is known as the Tsiolkovsky rocket equation or the ideal rocket equation.

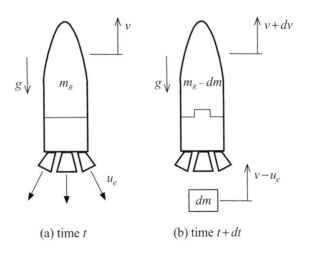

(a) time t (b) time $t + dt$

Figure 3.2.17 The inertia and velocity of a rocket at times (a) t and (b) $t + dt$

Example 3.2.7

In Figure 3.2.18(a), a massless simply supported beam is carrying a spring–mass system at its midpoint. Assume that mass m_1 has the same deflection as the beam at the contact point. This system can be a model of automobile chassis system. With consideration of gravity, develop a mathematical model of the system.

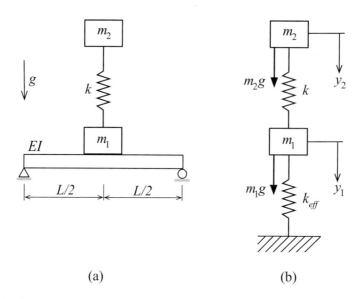

Figure 3.2.18 Simply supporting beam carrying a spring–mass system: (a) schematic; (b) model with the beam treated as a spring

Solution

The massless beam can be treated as a spring of effective coefficient, which by Table 3.2.1 is given by $k_{eff} = 48EI/L^3$. Thus, the combined beam–mass system can be approximated by the one in Figure 3.2.18(b), where y_1 and y_2 are the displacements of the masses. Following the steps in Example 3.2.4 or Example 3.2.5, it can be shown that the governing equations of the combined system are

$$m_1\ddot{y}_1 + (k_{eff} + k)y_1 - ky_2 = m_1 g$$
$$m_2\ddot{y}_2 + ky_2 - ky_1 = m_2 g$$

3.2.4 Transfer function Formulation

As mentioned in Section 1.3, there are four commonly used forms of model representation for modeling, simulation and analysis of dynamic systems: (i) transfer function formulation, (ii) state-space representation, (iii) block diagram representation in the s-domain, and (iv) block diagram representation in the time domain. In this section, a transfer function formulation for mechanical systems is introduced. Block diagram and state-space representations are presented in Sections 3.6 and 3.7. These forms of model representation are convenient for computation with MATLAB and Mathematica, as is shown in Sections 3.8 and 3.9.

A transfer function formulation is a method for modeling and analysis of time-invariant dynamic systems whose parameters do not change with time. As discussed in Sections 1.3 and 2.7, the transfer function of a time-invariant system is defined as the ratio of the system output to the system input in the Laplace transform domain with zero initial conditions:

$$G(s) = \frac{\mathcal{L}[\text{output}]}{\mathcal{L}[\text{input}]}\bigg|_{\text{zero initial conditions}} \quad (3.2.12)$$

where $\mathcal{L}$ is the Laplace transform operator, and zero initial conditions mean all the initial values of the input and output and their time derivatives are set to zero. Thus, a transfer function represents an input–output relation for the system. With a transfer function formulation, the characteristics of a dynamic system, including the system response and stability, can be evaluated, and feedback control laws can be designed and implemented. These topics will be explored in Chapters 7 and 8.

The process of the establishment of transfer functions is called a transfer function formulation. In general, a transfer function can be expressed as

$$G(s) = \frac{B(s)}{A(s)} = \frac{b_m s^m + b_{m-1} s^{m-1} + \cdots + b_1 s + b_0}{a_n s^n + a_{n-1} s^{n-1} + \cdots + a_1 s + a_0} \quad (3.2.13)$$

where $A(s)$ and $B(s)$ are polynomials of the Laplace parameter s. The order of the transfer function is the order of the denominator polynomial $A(s)$. Thus, $G(s)$ in Eq. (3.2.2) is an n-th-order transfer function. A system with an n-th-order transfer function is called an n-th-order system.

For a mechanical system, its input can be an external force or a prescribed displacement, its output can be a displacement or velocity or an internal force. The transfer function for a given physical system is not unique, depending on the specification of input and output. Also, a system may have multiple inputs and multiple outputs. For such a system, multiple transfer functions can be established, with a transfer function for one pair of input and output.

The derivation of a transfer function for a mechanical system generally takes the following steps:

(1) Define a transfer function by selecting an input and an output.
(2) Derive the time-domain governing equations of motion for the system.
(3) Take the Laplace transform of the time-domain equations with zero initial conditions.
(4) With the Laplace transform equations, determine the ratio of a relevant unknown function to a given input function, which may or may not be the transfer function.
(5) By the result obtained in Step 4 and the selected input and output in Step 1, obtain the transfer function of the system.

Of course, Step 5 is not necessary if the ratio obtained in Step 4 is the transfer function.

Example 3.2.8

For the spring–mass–damper system in Figure 3.2.19, derive the transfer function with the external force $f(t)$ as the input and the velocity $\dot{x}(t)$ of the mass as the output.

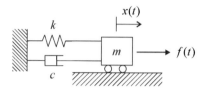

Figure 3.2.19 A spring–mass–damper system

Solution
The transfer function, by Eq. (3.2.12), is of the form

$$G(s) = \frac{\mathcal{L}[\dot{x}]}{\mathcal{L}[f]}\bigg|_{\text{zero initial conditions}} = \frac{sX(s)}{F(s)} \quad \text{(a)}$$

It is easy to show that the equation of motion of the system is

$$m\ddot{x}(t) + c\dot{x}(t) + kx(t) = f(t)$$

Performing Laplace transform of the previous differential equation with zero initial conditions gives

$$(ms^2 + cs + k)X(s) = F(s)$$

from which

$$\frac{X(s)}{F(s)} = \frac{1}{ms^2 + cs + k} \quad \text{(b)}$$

The ratio $X(s)/F(s)$ is not the transfer function as defined in Eq. (a). However, it is useful in the determination of the transfer function as shown below:

$$G(s) = \frac{sX(s)}{F(s)} = \frac{s}{ms^2 + cs + k} \quad \text{(c)}$$

This is a transfer function of order two. Thus, the system is a second-order system.

Example 3.2.9

For the spring–mass–damper system shown in Figure 3.2.9, derive the transfer function with external force F as the input and displacement x as the output. Also, derive another transfer function with F as the input and the spring force as the output.

Solution
With the defined input and output, the first transfer function is $G(s) = X(s)/F(s)$. Taking Laplace transform of the system governing equations in Example 3.2.2, with zero initial conditions, gives the s-domain algebraic equations about $X(s)$ and $Y(s)$

$$(ms^2 + k)X(s) - kY(s) = F(s)$$
$$(cs + k)Y(s) - kX(s) = 0 \qquad (a)$$

Note that $X(s)$ is in the definition of the transfer function. By the second line in Eq. (a)

$$Y(s) = \frac{k}{cs+k}X(s) \qquad (b)$$

Subsitituting Eq. (b) into the first line of Eq. (a) gives

$$X(s) = \frac{1}{(ms^2 + k) - \dfrac{k^2}{cs+k}} F(s)$$

which yields the transfer function of the system as follows:

$$G(s) = \frac{X(s)}{F(s)} = \frac{cs+k}{(ms^2+k)(cs+k) - k^2}$$
$$= \frac{cs+k}{s(mcs^2 + mks + kc)} \qquad (c)$$

Equation (c) represents a third-order system.

If the output is specified as the spring force $f_s = k(x - y)$, with the input being the same, the second transfer function is defined by

$$G_1(s) = \frac{F_s(s)}{F(s)} = \frac{k\big(X(s) - Y(s)\big)}{F(s)} \qquad (d)$$

In this case, both $X(s)$ and $Y(s)$ must be obtained before the transfer function can be determined. By Eqs. (b) and (c),

$$X(s) = \frac{cs+k}{s(mcs^2 + mks + kc)} F(s), \quad Y(s) = \frac{k}{cs+k} X(s) \qquad (e)$$

It follows that the transfer function

$$G_1(s) = k\left(\frac{X(s)}{F(s)} - \frac{Y(s)}{F(s)}\right) = \left(1 - \frac{k}{cs+k}\right)\frac{X(s)}{F(s)}$$
$$= k\frac{cs+k}{s(mcs^2+mks+kc)}\left(1 - \frac{k}{cs+k}\right) = \frac{ck}{mcs^2 + mks + kc} \qquad (f)$$

Unlike $G(s)$, $G_1(s)$ is a second-order transfer function.

Example 3.2.10

For the quarter-car suspension model in Example 3.2.4, derive the transfer function with the prescribed road-surface profile y_0 as the input and the displacement y_2 as the output. Also, identify the order of the system.

Solution

The transfer function is expressed by $Y_2(s)/Y_0(s)$. Taking Laplace transform of Eq. (c) of motion in Example 3.2.4 gives

$$(m_1 s^2 + cs + k_1 + k_2) Y_1(s) - (cs + k_2) Y_2(s) = k_1 Y_0(s)$$
$$(m_2 s^2 + cs + k_2) Y_2(s) - (cs + k_2) Y_1(s) = 0 \tag{a}$$

where the gravitational forces have been eliminated because they are not related to the defined transfer function. (A transfer function can only take one input and one output.) By substitution

$$Y_1(s) = \frac{m_2 s^2 + cs + k_2}{cs + k_2} Y_2(s)$$

$$(m_1 s^2 + cs + k_1 + k_2) \frac{m_2 s^2 + cs + k_2}{cs + k_2} Y_2(s) - (cs + k_2) Y_2(s) = k_1 Y_0(s)$$

which eventually yields the transfer function as follows

$$\frac{Y_2(s)}{Y_0(s)} = k_1 \left\{ (m_1 s^2 + cs + k_1 + k_2) \frac{m_2 s^2 + cs + k_2}{cs + k_2} - (cs + k_2) \right\}^{-1}$$

$$= \frac{k_1 (cs + k_2)}{(m_1 s^2 + k_1 + cs + k_2)(m_2 s^2 + cs + k_2) - (cs + k_2)^2}$$

$$= \frac{k_1 (cs + k_2)}{(m_1 s^2 + k_1)(m_2 s^2 + cs + k_2) + m_2 s^2 (cs + k_2)}$$

Because the transfer function is of order four, the quarter-car suspension model is a fourth-order dynamic system.

Example 3.2.11

For the spring–mass–damper system in Figure 3.2.8(a), determine the transfer functions $X_2(s)/F_1(s)$ and $X_2(s)/F_2(s)$.

Solution
There are two methods for determining the transfer functions:

Method 1. First set $F_2(s) = 0$ and obtain $X_2(s)/F_1(s)$; and then set $F_1(s) = 0$ and obtain $X_2(s)/F_2(s)$. Each time, a pair of input and output is considered.

Method 2. First obtain the complete solutions $X_2(s)$ in terms of $F_1(s)$ and $F_2(s)$, and then determine the transfer functions by properly selecting an input.

For comparison purposes, both the methods are used. To this end, take Laplace transform of Eq. (e) in Example 3.2.1, to obtain

$$(m_1 s^2 + cs + k_1 + k_2)X_1(s) - k_2 X_2(s) = F_1(s)$$
$$(m_2 s^2 + k_2)X_2(s) - k_2 X_1(s) = F_2(s) \qquad \text{(a)}$$

Method 1. For $X_2(s)/F_1(s)$, setting $F_2(s) = 0$ reduces Eq. (a) to

$$(m_1 s^2 + cs + k_1 + k_2)X_1(s) - k_2 X_2(s) = F_1(s)$$
$$(m_2 s^2 + k_2)X_2(s) - k_2 X_1(s) = 0 \qquad \text{(b)}$$

By substitution

$$X_1(s) = \frac{m_2 s^2 + k_2}{k_2} X_2(s)$$

$$(m_1 s^2 + cs + k_1 + k_2)\frac{m_2 s^2 + k_2}{k_2} X_2(s) - k_2 X_2(s) = F_1(s)$$

which leads to the transfer function

$$\frac{X_2(s)}{F_1(s)} = \frac{k_2}{(m_1 s^2 + cs + k_1 + k_2)(m_2 s^2 + k_2) - k_2^2}$$
$$= \frac{k_2}{(m_1 s^2 + cs + k_1)(m_2 s^2 + k_2) + m_2 k_2 s^2} \qquad \text{(c)}$$

Likewise, setting $F_2(s) = 0$ reduces Eq. (a) to

$$(m_1 s^2 + cs + k_1 + k_2)X_1(s) - k_2 X_2(s) = 0$$
$$(m_2 s^2 + k_2)X_2(s) - k_2 X_1(s) = F_2(s) \qquad \text{(d)}$$

By substitution

$$X_1(s) = \frac{k_2}{m_1 s^2 + cs + k_1 + k_2} X_2(s)$$

$$(m_2 s^2 + k_2)X_2(s) - k_2 \frac{k_2}{m_1 s^2 + cs + k_1 + k_2} X_2(s) = F_2(s)$$

which leads to the transfer function

$$\frac{X_2(s)}{F_2(s)} = \frac{m_1 s^2 + cs + k_1 + k_2}{(m_1 s^2 + cs + k_1)(m_2 s^2 + k_2) + m_2 k_2 s^2} \qquad \text{(e)}$$

Method 2. The coupled algebraic equations in Eq. (a) are written in the matrix form

$$\begin{bmatrix} \alpha(s) & -k_2 \\ -k_2 & \beta(s) \end{bmatrix} \begin{pmatrix} X_1(s) \\ X_2(s) \end{pmatrix} = \begin{pmatrix} F_1(s) \\ F_2(s) \end{pmatrix} \qquad \text{(f)}$$

where
$$a(s) = (m_1 s^2 + cs + k_1 + k_2), \quad \beta(s) = (m_2 s^2 + k_2)$$

Application of Cramer's rule to Eq. (f) yields

$$X_2(s) = \frac{1}{\Delta(s)} \begin{vmatrix} a(s) & F_1(s) \\ -k_2 & F_2(s) \end{vmatrix}$$

$$= \frac{1}{\Delta(s)} [k_2 F_1(s) + a(s) F_2(s)] \tag{g}$$

where $\Delta(s)$ is the determinant of the matrix in Eq. (f) and it is given by

$$\begin{aligned}\Delta(s) &= a(s)\beta(s) - k_2^2 \\ &= (m_1 s^2 + cs + k_1)(m_2 s^2 + k_2) + m_2 k_2 s^2 \end{aligned} \tag{h}$$

For $F_2(s) = 0$, Eq. (g) yields $X_2(s)/F_1(s) = \dfrac{k_2}{\Delta(s)}$, which by Eq. (h) gives

$$\frac{X_2(s)}{F_1(s)} = \frac{k_2}{(m_1 s^2 + cs + k_1)(m_2 s^2 + k_2) + m_2 k_2 s^2} \tag{i}$$

For $F_1(s) = 0$, Eq. (g) yields $X_2(s)/F_2(s) = \dfrac{a(s)}{\Delta(s)}$, which gives

$$\frac{X_2(s)}{F_2(s)} = \frac{m_1 s^2 + cs + k_1 + k_2}{(m_1 s^2 + cs + k_1)(m_2 s^2 + k_2) + m_2 k_2 s^2} \tag{j}$$

Obviously, the transfer functions obtained by the two methods are the same.

As mentioned previously, a transfer function formulation is only valid for linear time-invariant systems with linear governing equations having constant coefficients. This is because the Laplace transform operator is linear and because it is not applicable to products of time functions. Hence, no transfer functions can be derived for nonlinear systems (like the system subject to dry friction in Example 3.2.3) and linear time-varying systems (like the rocket in Example 3.2.6). For linear time-varying systems and nonlinear systems, a state-space representation can be used for simulation and analysis; see Section 3.7.

3.3 Rotational Systems

A rotational system is a mechanical system with its components constrained to rotate about an axis. Examples include rotating shafts, spinning disks, slewing arms, pulley systems, and geared systems. In this section, rigid bodies rotating about a fixed axis are considered. General motion of rigid bodies, involving both translation and rotation, is discussed in Section 3.4.

3.3.1 Three Keys in the Derivation of the Equations of Motion

The derivation of the equations of motion for rotational systems, as in modeling of translational systems, have three keys: fundamental principles, basic elements, and a way of analysis (free-body diagrams and auxiliary plots).

Double-Headed Arrow for Rotation

The description of the rotation of a body involves an angle of rotation and torques/moments that cause the rotation. In Figure 3.3.1(a), a rigid body, which is subject to a torque τ and a force F, rotates about a fixed axis O–O with an angle θ of rotation. For convenience in modeling and analysis, the rotation (angular displacement, velocity, and acceleration) and applied forces are represented by double-headed arrows (double-headed vectors); see Figure 3.3.1(b), where $\omega = \dot{\theta}$, $\alpha = \ddot{\theta}$, and M_O is the moment of force F. The direction of a double-headed arrow is determined by the right-hand rule, and the summation of the double-headed arrows is treated as vector summation. Double-headed arrows are useful in dealing with the interactions of multiple rotating bodies. In this text, the conventional notation for rotation and double-headed arrows are in mixed use whenever it is convenient.

Fundamental Principles

Rigid bodies in rotation about a fixed axis are commonly seen in two cases: (i) a three-dimensional rigid body rotating about an axis, as in Figure 3.3.1; and (ii) a slab rotating

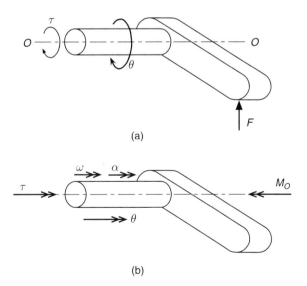

Figure 3.3.1 A rotating rigid body: (a) schematic; (b) double-headed arrows by the right-hand rule

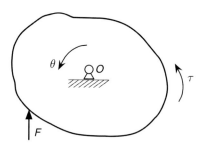

Figure 3.3.2 A slab rotates about hinge O

about a hinge, as shown in Figure 3.3.2, where the rotation axis is perpendicular to the paper, and goes through the hinge point O. A rotating shaft is an example in case (i) and a spinning disk is an example in case (ii). In either case, Newton's second law for particles, Eq. (3.1.2), can be extended as follows

$$I_O \ddot{\theta} = M_O \tag{3.3.1}$$

where M_O is the resultant of torques and moments of forces about the axis or hinge, and I_O is the mass moment of inertia of the body about the axis or hinge. The mass moment of inertia of a body is defined as

$$I_O = \int_V r^2 dm \tag{3.3.2}$$

where r is the distance from the axis or hinge to the infinitesimal mass element dm, and V is the volume of the body.

Parallel-Axis Theorem Assume that a body of mass m has a mass moment of inertia I_G with respect to the axis A that passes through the mass center G of the body. (Refer to Section 3.4.1 for the definition of the mass center.) For the body rotating about axis O that is parallel to axis A at a distance d (see Figure 3.3.3), its mass moment of inertia with respect to axis O can be expressed by

$$I_O = I_G + md^2 \tag{3.3.3}$$

As in Sections 3.1.4 and 3.1.5, the spatial and temporal integration of Eq. (3.3.1) yields the following two principles for a body in rotation.

The Principle of Work and Energy

$$\int_{\theta_1}^{\theta_2} M_O \, d\theta = \frac{1}{2} I_O \omega^2 \bigg|_{\theta_2} - \frac{1}{2} I_O \omega^2 \bigg|_{\theta_1} \tag{3.3.4}$$

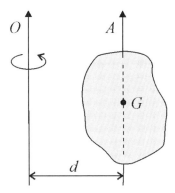

Figure 3.3.3 Parallel axes of a rotating body

where $\omega = \dot{\theta}$, which is the angular velocity of the body. The left-hand side of Eq. (3.3.4) is the work done by the resultant moment during the rotation of the body from angle θ_1 to angle θ_2, and $\frac{1}{2}I_O\omega^2$ is the kinetic energy of the body in rotation.

The Principle of Impulse and Momentum

$$\int_{t_1}^{t_2} M_O \, dt = I_O\omega \Big|_{t_2} - I_O\omega \Big|_{t_1} \qquad (3.3.5)$$

where the left-hand side term is the impulse of the resultant moment in the time interval from t_1 to t_2, and $I_O\omega$ is the angular momentum of the rotating body. These principles can be used to develop mathematical models of rigid bodies in rotation.

Basic Elements

Mass Elements   A rotational system has three types of basic elements: mass elements, spring elements, and damping elements. Mass elements are those bodies to which the rotational version of Newton's second law, Eq. (3.3.1), can be applied. The bodies in Figures 3.3.1 and 3.3.2 are two examples of mass elements. Other examples of mass elements include rotating shafts, spinning disks, slewing robot arms, pulleys, and pendulums, which are discussed in the subsequent sections.

Spring Elements   Torsional springs are spring elements for rotational systems. A torsional spring is a device that produces a resisting torque τ_s (spring torque) given by

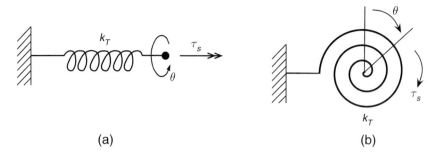

Figure 3.3.4 Torsional springs with one end fixed: (a) spring for three-dimensional bodies; (b) spring for slabs

$$\tau_s = k_T \Delta\theta \qquad (3.3.6)$$

where k_T is a torsional spring coefficient, and $\Delta\theta$ is relative rotation, which is the difference of the angles of twist at the two ends of the spring. Figure 3.3.4 shows two torsional spring models with one fixed end. The spring in Figure 3.3.4(a) is used with three-dimensional rigid bodies like the one in Figure 3.3.1(a). The spring in Figure 3.3.4(b) is for slabs like that in Figure 3.3.2. For either spring, the spring torque is given by

$$\tau_s = k_T \theta \qquad (3.3.7)$$

Torsional springs with two movable ends are usually seen in the rotation of three-dimensional rigid bodies. Unlike translational springs, no tension and compression can be identified for torsional springs. Figure 3.3.5 shows two cases of the application of spring torques. In Figure 3.3.5(a), the double-headed vectors of torques at the two ends of the spring are pointing away from each other and the relative rotation or twist $\Delta\theta = \theta_2 - \theta_1$. In Figure 3.3.5(b), the double-headed vectors of torques at the two ends of the spring are pointing toward each other and the relative rotation $\Delta\theta = \theta_1 - \theta_2$. As observed from the figure, the first angle in the expression of relative rotation $\Delta\theta$ is the one that has the same direction as the torque τ_s at the point of application. This is because the applied torques and corresponding twist must be consistent in direction. Thus, by Eq. (3.3.6),

$$\tau_s = k_T(\theta_2 - \theta_1) \qquad (3.3.8a)$$

for the end torques pointing away from each other as in Figure 3.3.5(a), and by

$$\tau_s = k_T(\theta_1 - \theta_2) \qquad (3.3.8b)$$

for the end torques pointing toward each other as Figure 3.3.5(b).

As shown in Table 3.2.1, a massless rod in torsional deformation can be viewed as a torsional spring with the effective coefficient $k_{eff} = GJ/L$.

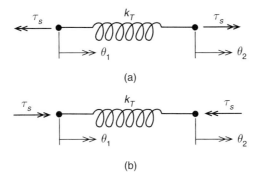

Figure 3.3.5 Relative rotation of a torsional spring with two movable ends: (a) $\Delta\theta = \theta_2 - \theta_1$; (b) $\Delta\theta = \theta_1 - \theta_2$

Damping Elements There are several types of damping element models for rotational systems. A rotational viscous damper is a device with two moving parts that are separated by a thin film of lubricant; see Figure 3.3.6 (a). Due to viscous friction of the lubricant, the damper produces a resisting torque τ_d (damping torque) that is proportional to the relative angular velocity $\Delta\omega$ between the two parts.

$$\tau_d = b\Delta\omega \tag{3.3.9}$$

where b is a torsional damping coefficient. Shown in Figures 3.3.6(b) and (c) are two cases of the relative angular velocity, which depends on the direction of the damping torque. Similar to the relative angular displacement of a torsional spring, the first angular velocity in the expression of $\Delta\omega$ for the rotational viscous damper is the one that has the same direction as the damping torque τ_d at the point of application. It follows that the damping torque is given by

$$\tau_d = b(\dot{\theta}_2 - \dot{\theta}_1) \tag{3.3.10a}$$

for the end torques pointing away from each other as shown in Figure 3.3.6(b), and

$$\tau_d = b(\dot{\theta}_1 - \dot{\theta}_2) \tag{3.3.10b}$$

for the end torques pointing toward each other as shown in Figure 3.3.6(c).

In some applications, one end of a viscous damper is fixed. Figure 3.3.7 shows two examples of such dampers, one for three-dimensional bodies, and the other for slabs. In both the examples, the torque–velocity relation is

$$\tau_d = b\dot{\theta} \tag{3.3.11}$$

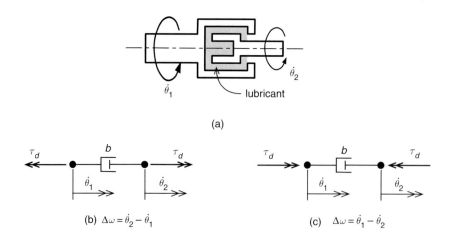

Figure 3.3.6 Rotational viscous damper: (a) schematic; (b) $\Delta\omega = \dot{\theta}_2 - \dot{\theta}_1$; (c) $\Delta\omega = \dot{\theta}_1 - \dot{\theta}_2$

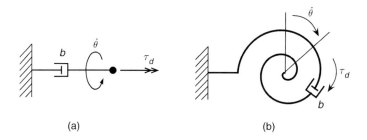

Figure 3.3.7 Rotational viscous damper with one end fixed: (a) damper for three-dimensional bodies; (b) damper for slabs

In rotor machinery, bearings are used to support a rotating shaft and to reduce rotational friction. A bearing can be viewed as a viscous damper in Figure 3.3.7(a). Due to its special configuration and important applications, however, the working principle of bearings deserves description. Without loss of generality, a schematic of a ball bearing is shown in Figure 3.3.8(a), where the outer race of the bearing is stationary, the inner race is attached to a rotating shaft, balls are used to separate the inner and outer races, and lubricant is added between the races to minimize friction and prevent wear.

In the rotation of a bearing-supported shaft, viscous friction forces are induced by the lubricant, ball–shaft contact, and other factors, as demonstrated in Figure 3.3.8(b), where f_d represents viscous friction forces acting on the shaft. These friction forces produce a resisting damping torque τ_d that is described by Eq. (3.3.11). The symbols of bearings adopted in this test are ⌊x⌋ and ⊠, as shown in Figure 3.3.8(c). The damping torque by a bearing is always against the shaft rotation as shown in the free-body diagram in Figure 3.3.8(d).

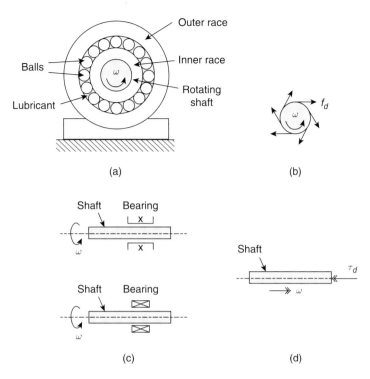

Figure 3.3.8 Ball bearing: (a) schematic; (b) viscous friction forces acting on a rotating shaft; (c) notation of bearing; (d) damping torque applied to the shaft by the bearing

The damping elements presented so far are linear. A nonlinear damping element is considered as follows. For an object moving in the air, a drag force (air resistance) acts in the opposite direction to the motion of the object, with an amplitude proportional to the square of the velocity of the object relative to the air. Thus, the rotation of a body in the air (a propeller for instance), due to an aerodynamic effect, is resisted by a drag torque, which can be described by

$$\tau_d = b_D \, \dot{\theta}^2 \, \mathrm{sgn}(\dot{\theta}) \tag{3.3.12}$$

where the damping coefficient b_D depends on drag coefficient, air mass density, and the shape of the body. The notation of drag torque is shown in Figure 3.3.9.

Free-Body Diagrams and Auxiliary Plots

As a way of analysis (the third key in system modeling), free-body diagrams for mass elements are drawn and an auxiliary plot for a spring or damping element with two movable ends may be sketched. The development of a rotational system model follows the same steps as listed in Section 3.2.3, provided that rotational elements are used.

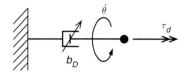

Figure 3.3.9 Drag torque due to air resistance

Example 3.3.1

A rotor (motor–propeller assembly) of a quadcopter is shown in Figure 3.3.10, where the propeller is driven by a torque τ_m of the electric motor, and it is subject to a drag torque due to aerodynamic effect. The motor and the propeller are connected by a flexible shaft, which can be modeled as a torsional spring of coefficient k_T and the motor shaft is supported by a bearing of coefficient b. The mass moment of inertia of the rotating parts of the motor is J_m and the mass moment of inertia of the propeller is J_p. Let the rotation angles of the motor and propeller be θ_m and θ_p, respectively. Derive the equations of motion for the rotor system.

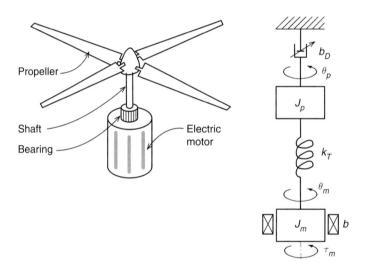

Figure 3.3.10 Simplified model of a motor–propeller assembly

Solution

The free-body diagrams of the motor and propeller and the auxiliary plot of the spring (flexible shaft) are drawn in Figure 3.3.11, where τ_s, τ_b, and τ_D are the spring torque, bearing torque, and drag torque, respectively. The application of Newton's second law, Eq. (3.3.1), to the two mass elements gives

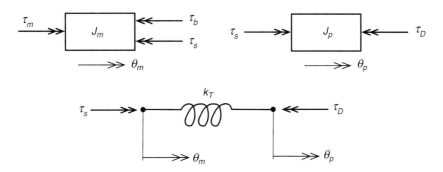

Figure 3.3.11 Free-body diagrams and auxiliary plot of the motor–propeller assembly in Example 3.3.1

$$\text{Motor:} \quad J_m \ddot{\theta}_m = \tau_m - \tau_b - \tau_s \quad (a)$$

$$\text{Propeller:} \quad J_p \ddot{\theta}_p = \tau_s - \tau_D \quad (b)$$

Also, the torques of the spring and damping elements are given by

$$\tau_s = k_T \Delta\theta = k_T(\theta_m - \theta_p)$$
$$\tau_b = b\dot{\theta}_m, \quad \tau_D = b_D \dot{\theta}_p^2 \, \text{sgn}(\dot{\theta}_p) \quad (c)$$

where, according to Figure 3.3.5(b), the relative rotation of the spring is $\Delta\theta = \theta_m - \theta_p$. Substituting Eq. (c) into Eqs. (a) and (b) yields the equations of motion of the motor–propeller assembly as follows

$$J_m \ddot{\theta}_m + b\dot{\theta}_m + k_T \theta_m - k_T \theta_p = \tau_m$$
$$J_p \ddot{\theta}_p + b_D \dot{\theta}_p^2 \, \text{sgn}(\dot{\theta}_p) + k_T \theta_p - k_T \theta_m = 0 \quad (d)$$

3.3.2 Simple Rotational Systems

In this section, six simple rotational systems are modeled: the pulley, simple pendulum, lever, slewing rigid bar, rotating shaft, and spinning rigid disk. These simple systems can be combined with translational systems and rigid bodies to make complex mechanical systems, as shall be shown in the subsequent sections.

Pulley

Pulleys are used to change the direction of movement of a taut cable (rope, belt, or chain), or to transfer power between two elements that are connected by the cable. A pulley is a cylinder that can rotate about its center, and that has a groove for a cable to rest in it. A pulley is shown in Figure 3.3.12, where the center of the pulley is at fixed hinge O and the cable is loaded by forces F_1 and F_2. Assume that the cable is unextendible and that the rope does not slip on the surface of the groove. The governing equation of motion for the pulley is given by

$$I_O \ddot{\theta} = F_1 R - F_2 R \quad (3.3.13)$$

where I_O is the mass moment of inertia of the pulley with respect to the hinge O. If the inertia of the pulley is negligible, $I_O = 0$, and the previous equation yields $F_1 = F_2$. Such a pulley is called an ideal pulley.

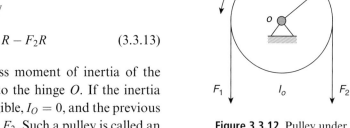

Figure 3.3.12 Pulley under loads

In some applications, the center of a pulley is movable. This renders the motion of the pulley in both translation and rotation. This combination of translation and rotation is addressed in Section 3.4.2.

Simple Pendulum

A simple pendulum is a lumped mass that is suspended by an unstretchable string at a fixed point, as shown in Figure 3.3.13(a). With the gravity as a restoring force, the pendulum sways freely about its equilibrium position ($\theta = 0$). The pendulum can be viewed as a rotational system with the mass moment of inertia $I_O = ml^2$, where l is the string length. The pendulum is subject to a moment $M_O = -mgl \sin \theta$, where the negative sign means that that the movement of gravity is always against the angular motion of the pendulum. By Eq. (3.3.1),

$$ml^2 \ddot{\theta} = -mgl \sin \theta \quad (3.3.14)$$

which leads to the nonlinear equation of motion of the pendulum

$$\ddot{\theta} + \frac{g}{l} \sin \theta = 0 \quad (3.3.15)$$

The pendulum equation can also be obtained by treating the mass as a particle in a circular motion. To this end, consider the free-body diagram in Figure 3.3.13(b). The application of Newton's second law, Eq. (3.1.2), in the tangential direction (perpendicular to the spring tension T) gives

$$ma_\theta = -mg \sin \theta \quad (3.3.16)$$

Because the tangential acceleration is $a_\theta = l\ddot{\theta}$, the same governing equation (3.3.15) for the simple pendulum is obtained.

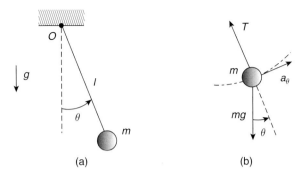

Figure 3.3.13 A simple pendulum: (a) schematic; (b) free-body diagram

If the angle θ of swing is small, $\sin\theta \approx \theta$ and Eq. (3.3.15) is approximated by

$$\ddot{\theta} + \frac{g}{l}\theta = 0 \qquad (3.3.17)$$

This linearized differential equation implies that the oscillation (back and forth swing motion) of the pendulum has a natural period or simply period $T = 2\pi\sqrt{l/g}$, regardless of the inertia or weight of the pendulum. Refer to Section 7.1.3 for the definition of the natural period of a system in free oscillatory motion.

Lever

A lever is a rigid rod that is pivoted at a fixed hinge, and it is used to amplify or transmit forces in machines and equipment. In Figure 3.3.14(a), a lever pivoted at point O is constrained by a spring and is subject to a vertical force f. Assume that the rotation angle θ of the lever is small such that the tip displacements of the lever can be approximated by $y_1 = a\theta$, $y_2 = b\theta$. With the free-body diagram in Figure 3.3.14(b), use of Eq. (3.3.1) gives

$$I_O\ddot{\theta} = fa - f_s b \qquad (3.3.18)$$

Because the spring force is $f_s = ky_2 = kb\theta$, the equation of motion of the lever is

$$I_O\ddot{\theta} + kb^2\theta = fa \qquad (3.3.19)$$

If the inertia of the lever is negligible, the applied force f and spring force are in equilibrium; namely, $kb^2\theta = fa$.

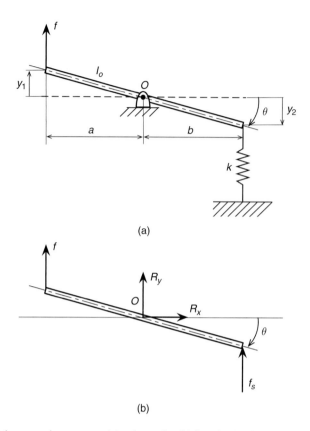

Figure 3.3.14 A lever–spring system: (a) schematic; (b) free-body diagram

Slewing Rigid Bar

In Figure 3.3.15(a), a rigid bar rotates about a pivot point O under a torque τ. For a smooth slewing motion, a bearing is installed at the pivot point. Figure 3.3.15(b) shows the free-body diagram of the system, where τ_d is the damping torque applied by the bearing. It follows from Eq. (3.3.1) that

$$I_O \ddot{\theta} = \tau - \tau_d \tag{3.3.20}$$

which with $\tau_d = b\dot{\theta}$ leads to the governing equation of the slewing bar:

$$I_O \ddot{\theta} + b\dot{\theta} = \tau \tag{3.3.21}$$

The slewing bar is a model of single-link robotic arms.

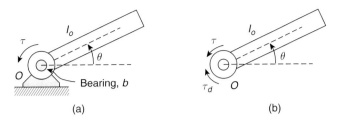

Figure 3.3.15 A slewing rigid bar: (a) schematic; (b) free-body diagram

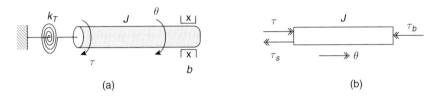

Figure 3.3.16 A rotating shaft: (a) schematic; (b) free-body diagram

Rotating Shaft

Rotating shafts were considered in Example 3.3.1. This type of system is further discussed for the preparation of modeling of geared systems in Section 3.3.4. In Figure 3.3.16(a), a torque τ is applied to a bearing-supported rotating shaft that is constrained by a torsional spring. The free-body diagram of the shaft is shown in Figure 3.3.16(b), where τ_s is the spring torque and τ_b is the bearing torque. By Eq. (3.3.1),

$$J\ddot{\theta} = \tau - \tau_s - \tau_b \tag{3.3.22}$$

which, by $\tau_s = k_T \theta$ and $\tau_b = b\dot{\theta}$, gives the equation of motion of the rotating shaft

$$J\ddot{\theta} + b\dot{\theta} + k_T \theta = \tau \tag{3.3.23}$$

Spinning Rigid Disk

A spinning rigid disk can be viewed as a shortened version of a rotating shaft with or without a bearing. This model is used for a rotating object with its dimension along the rotating axis being relatively small and its mass moment of inertia being relatively large. The governing equation of motion for a spinning disk can be obtained from Eq. (3.3.1).

3.3.3 Combination of Translational and Rotational Elements

In this section, mechanical systems composed of translational elements described in Section 3.2.1 and rotational elements described in Sections 3.3.1 and 3.3.2 are considered. Modeling of this type of systems follows the same steps as described in Section 3.2.3, except that both translational and rotational elements are involved.

Example 3.3.2

A pulley–mass system is shown in Figure 3.3.17(a), where an unextendible cable connects a cylinder and a mass. Consider gravity. Derive a mathematical model of the system.

Solution
The free-body diagrams of the pulley and mass are shown in Figure 3.3.17(b), where T is the tension force of the cable. Because the cable is unextendible, the rotation of the cylinder and the displacement of the mass are related by $y = R\theta$. Note that two free-body diagrams must be drawn here even though the displacements of the cylinder and mass are not independent. By Newton's second law and its rotational version, the equations of the mass elements are

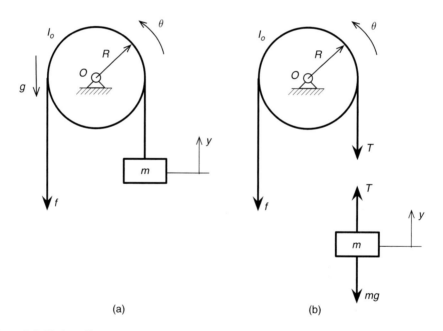

Figure 3.3.17 A pulley–mass system with an unextendible cable: (a) schematic; (b) free-body diagrams

$$I_O\ddot{\theta} = fR - TR$$
$$m\ddot{y} = T - mg$$

It is easy to show that $T = mR\ddot{\theta} + mg$ and that by substitution

$$I_O\ddot{\theta} = fR - (mR\ddot{\theta} + mg)R$$

This yields the governing equation of motion of the pulley–mass system

$$(I_O + mR^2)\ddot{\theta} = fR - mgR$$

Also, by $y = R\theta$, an equivalent equation of motion in terms of y is obtained as follows

$$\left(\frac{I_O}{R^2} + m\right)\ddot{y} = f - mg$$

In Example 3.3.2, the full description of the pulley–mass system motion only requires one displacement parameter (either θ or y). If the cable connecting the cylinder and mass is extendible, two independent displacement parameters are necessary; see Example 3.3.3.

Example 3.3.3

The pulley–mass system in Figure 3.3.18(a) has an extendible cable, which can be modeled as a spring of coefficient k. Derive the equations of motion for the pulley-mass system.

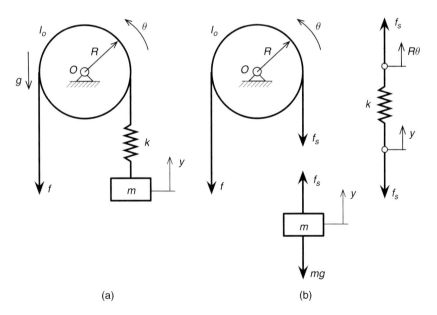

Figure 3.3.18 A pulley–mass system with an extendible cable: (a) schematic; (b) free-body diagrams and auxiliary plot

Solution

Because the cable is extendible, the rotation angle θ and displacement y are independent of each other. In other words, $y \neq R\theta$. Thus, these displacement parameters are all required to fully describe the motion of the pulley–mass system. Figure 3.3.18(b) shows the free-body diagrams of the pulley and mass and the auxiliary plot of the spring (cable). Thus, by Newton's second law,

$$I_0 \ddot{\theta} = fR - f_s R$$
$$m\ddot{y} = f_s - mg$$

With the spring force being $f_s = k(R\theta - y)$, the equations of motion of the system are obtained as follows

$$I_0 \ddot{\theta} + kR^2 \theta - kRy = fR$$
$$m\ddot{y} + ky - kR\theta = -mg$$

Degrees of Freedom

Comparison of Examples 3.3.2 and 3.3.3 shows that the minimum number of governing equations of motion for a mechanical system is the number of independent displacement parameters. This leads to the following definition about the degrees of freedom of mechanical systems.

Definition: The number of degrees of freedom (DOFs) of a mechanical system is the minimum number of independent displacement parameters needed to fully describe the motion of the system. Also, the number of DOFs is equal to the number of independent governing equations of motion.

According to the above definition, the pulley–mass system with an unextendible cable in Example 3.3.2 is a one-DOF system; the pulley–mass system with an extendible cable in Example 3.3.3 is a two-DOF system.

The concept of DOFs is applicable to all mechanical systems.

Example 3.3.4

In Figure 3.3.19(a), a cart of mass M carrying a simple pendulum with mass m and length l moves on a smooth horizontal surface. The cart is constrained by a viscous damper (c) and is subject to an external force f. For the pendulum to work, gravity (g) must be considered. For this two-DOF system, derive the independent governing equations of motion. Also, obtain an expression of the tension of the pendulum cable.

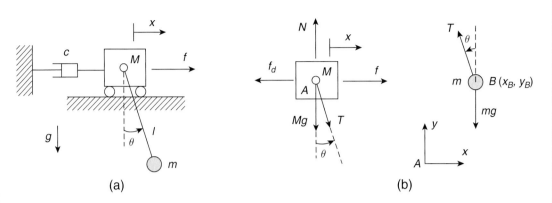

Figure 3.3.19 A cart–pendulum system: (a) schematic; (b) free-body diagrams

Solution
The free-body diagrams of the cart and pendulum are shown in Figure 3.3.19(b), where T is tension of the pendulum cable, N is the normal force by the horizontal surface, A is the point of the cart at which the pendulum is suspended, Axy is the coordinate system with the origin at point A, and x_B and y_B describe the location B of the pendulum mass. By Newton's second law, the displacement x of the cart is governed by

$$M\ddot{x} + c\dot{x} = f + T\sin\theta \tag{a}$$

where $f_d = c\dot{x}$ has been used. Because the cart has no vertical motion, the forces applied to the cart in the y direction are balanced: $N = Mg + T\cos\theta$. Also, the motion of the pendulum mass is governed by

$$ma_{Bx} = -T\sin\theta, \quad ma_{By} = T\cos\theta - mg$$

where a_{Bx} and a_{By} are the acceleration components of mass m in the x and y directions, respectively. The coordinates of the pendulum mass are

$$x_B = x + l\sin\theta, \quad y_B = -l\cos\theta$$

by which

$$a_{Bx} = \ddot{x}_B = \ddot{x} + l(\ddot{\theta}\cos\theta - \dot{\theta}^2\sin\theta)$$
$$a_{By} = \ddot{y}_B = l(\ddot{\theta}\sin\theta + \dot{\theta}^2\cos\theta)$$

Thus, the governing equations of motion for the pendulum become

$$m[\ddot{x} + l(\ddot{\theta}\cos\theta - \dot{\theta}^2\sin\theta)] = -T\sin\theta \tag{b}$$

$$m[l(\ddot{\theta}\sin\theta + \dot{\theta}^2\cos\theta)] = T\cos\theta - mg \tag{c}$$

Of the three equations (a) to (c), only two are independent. This is because the cart–pendulum system has only two independent displacement parameters (x and θ). The derivation of two independent equations can be done by eliminating the tension force T. To this end, adding Eqs. (a) and (b) gives

$$(M+m)\ddot{x} + c\dot{x} + ml(\ddot{\theta}\cos\theta - \dot{\theta}^2\sin\theta) = f \tag{d}$$

Also, multiplying Eq. (b) by $\cos\theta$, multiplying Eq. (c) by $\sin\theta$ and adding the resulting equations, yields

$$m[\ddot{x} + l(\ddot{\theta}\cos\theta - \dot{\theta}^2\sin\theta)]\cos\theta + m[l(\ddot{\theta}\sin\theta + \dot{\theta}^2\cos\theta)]\sin\theta + mg\sin\theta = 0$$

which is further reduced to

$$\ddot{x} + l\ddot{\theta} + g\sin\theta = 0 \tag{e}$$

Equations (d) and (e) are two independent governing equations of motion for the cart–pendulum system. Furthermore, multiplying Eq. (b) by $-\sin\theta$, multiplying Eq. (c) by $\cos\theta$, and adding the resulting equations, gives the expression of the cable tension as follows

$$T = m(l\dot{\theta}^2 - \ddot{x}\sin\theta + g\cos\theta) \tag{f}$$

3.3.4 Geared Systems

Gears are rotating machine parts that are used to change the speed, torque, and direction of rotation, and to transmit power in mechanical systems. Rotating systems consisting of gears are called geared systems. There are several types of gears, including spur, helical, bevel, crown, rack-and-pinion, and worm gears. In this section, spur gears are considered. The approach in modeling and analysis can be applied to other types of gears.

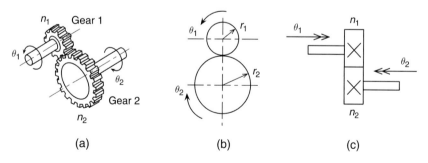

Figure 3.3.20 A pair of spur gears: (a) schematic; (b) side view, (c) notation

Consider a pair of spur gears in Figure 3.3.20(a), where Gears 1 and 2 have n_1 and n_2 teeth, respectively. Gear 1, which is mounted on a shaft with an applied toque τ, is called the input gear or drive gear; Gear 2, which drives a connecting shaft, is called the output gear or driven gear. In modeling, the following *ideal gear assumptions* are made:

(i) gears are massless; and
(ii) gears are in perfect meshing, without friction, backlash, and slipping.

According to these assumptions there is no energy loss in gear engagement.

Gears change speed, torque, and transmitting power through a gear ratio. A side view of the spur gear pair is shown in Figure 3.3.20(b), where r_1 and r_2 are the radii of Gears 1 and 2 respectively. The gear ratio N of the gear pair (from the input gear to the output gear) is defined as

$$N = \frac{n_2}{n_1} = \frac{r_2}{r_1} \qquad (3.3.24)$$

The notation for a gear pair is shown in Figure 3.3.20(c).

There are several formulas about a gear pair in operation, which are presented as follows.

Kinematic Relation

Due to the ideal gear assumptions, the passage of the circumferential arc length of each gear through a teeth engagement point must be the same. This means that

$$r_1 \theta_1 = r_2 \theta_2 \qquad (3.3.25)$$

which can be written as

$$\frac{r_2}{r_1} = \frac{\theta_1}{\theta_2} = N \qquad (3.3.26)$$

Because of Eq. (3.3.25), rotational displacements θ_1 and θ_2 are not independent.

Torque Ratio

A pair of gears under torques are shown in Figure 3.3.21(a), where τ_1 is a torque applied to the drive gear by the connecting shaft; and τ_2 is a resisting torque applied to the driven gear by the connecting shaft. Torques τ_1 and τ_2 are always in the same direction although they are applied to two different bodies. The free-body diagrams of the gears are plotted in Figure 3.3.21(b), where F and N are the tangential and normal forces induced by gear meshing. The equations of motion for the gears by Newton's second law are as follows

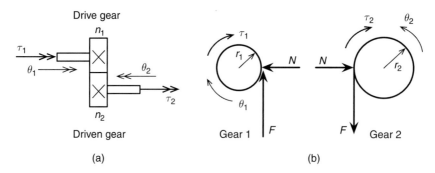

Figure 3.3.21 Gears subject to torques: (a) schematic; (b) free-body diagrams

For Gear 1: $\quad I_1 \ddot{\theta}_1 = \tau_1 - Fr_1$

For Gear 2: $\quad I_2 \ddot{\theta}_2 = Fr_2 - \tau_2$

where I_1 and I_2 are the mass moments of inertia of the gears. Because the gears are massless, $I_1 = 0$ and $I_2 = 0$. This renders zero resultant torque on each gear:

$$\tau_1 = Fr_1, \quad \tau_2 = Fr_2 \tag{3.3.27}$$

It follows that

$$\frac{\tau_2}{\tau_1} = \frac{r_2}{r_1} = N \tag{3.3.28}$$

Thus, for a pair of ideal gears, its torque ratio is the same as the gear ratio N.

Work Done by Torques

By the ideal gear assumptions, the work done by torque τ_1 through rotation θ_1 should be the same as the work done by torque τ_2 through rotation θ_2:

$$\tau_1 \theta_1 = \tau_2 \theta_2 \tag{3.3.29}$$

This makes sense because $\dfrac{\tau_2}{\tau_1} = \dfrac{\theta_1}{\theta_2} = N$, as indicated by Eq. (3.3.28).

With the gear ratio N and the above-mentioned relevant formulas, governing equations of geared systems can be established. In a modeling process, an auxiliary plot for a pair of gears as a whole can be generated, like Figure 3.3.21(a), and Eqs. (3.3.26) and (3.3.28) can be used.

Example 3.3.5

Two bearing-supported rotating shafts are coupled by a pair of gears in Figure 3.3.22, where a torque τ is applied to Shaft 1. This is a system with one DOF. Derive the equation of motion for the geared system in θ_2. Also, obtain an equivalent equation of motion in θ_1.

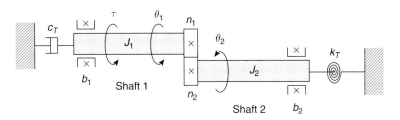

Figure 3.3.22 Schematic of a geared system

Solution

The free-body diagrams of the shafts and the auxiliary plot of the gear pair are shown in Figure 3.3.23, where T_{g1} and T_{g2} are the torques applied to the gears by Shafts 1 and 2, respectively, like τ_1 and τ_2 in Figure 3.3.21(a). The equations of motion of the shafts are obtained as

$$\text{For Shaft 1:} \quad J_1\ddot{\theta}_1 = \tau - T_{b1} - T_d - T_{g1}$$

$$\text{For Shaft 2:} \quad J_2\ddot{\theta}_2 = T_{g2} - T_{b2} - T_s$$

The gear pair, by Eqs. (3.3.26) and (3.3.28), is described by

$$T_{g2} = NT_{g1}, \quad \theta_2 = \frac{1}{N}\theta_1$$

where $N = n_2/n_1$ is the gear ratio, and Eqs. (3.3.19) and (3.3.21) have been used. The internal forces (torques) of the spring and damping elements are

$$\text{For bearings:} \quad T_{b1} = b_1\dot{\theta}_1, \quad T_{b2} = b_2\dot{\theta}_2$$

$$\text{For torsional damper:} \quad T_d = c_T\dot{\theta}_1$$

$$\text{For torsional spring:} \quad T_s = k_T\theta_2$$

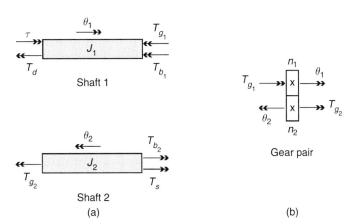

Figure 3.3.23 (a) Free-body diagrams and (b) auxiliary plot of the geared system in Example 3.3.5

It follows that

$$J_1\ddot{\theta}_1 = \tau - b_1\dot{\theta}_1 - c_T\dot{\theta}_1 - T_{g1}$$
$$J_2\ddot{\theta}_2 = T_{g2} - b_2\dot{\theta}_2 - k_T\theta_2$$

The previous equations are not independent. To obtain an independent governing equation in terms of θ_2, rewrite the previous equations by the gear equations, yielding

$$J_1 N\ddot{\theta}_2 = \tau - b_1 N\dot{\theta}_2 - c_T N\dot{\theta}_2 - T_{g1} \tag{a}$$

$$J_2\ddot{\theta}_2 = NT_{g1} - b_2\dot{\theta}_2 - k_T\theta_2 \tag{b}$$

From Eq. (a),

$$T_{g1} = J_1 N\ddot{\theta}_2 - \tau + b_1 N\dot{\theta}_2 + c_T N\dot{\theta}_2 \tag{c}$$

Substituting Eq. (c) into Eq. (b) yields the equation of motion of the geared system in θ_2:

$$(J_1 N^2 + J_2)\ddot{\theta}_2 + (b_1 N^2 + c_T N^2 + b_2)\dot{\theta}_2 + k_T\theta_2 = N\tau \tag{d}$$

Also, by $\theta_2 = \theta_1/N$, Eq. (d) can be reduced to an equivalent equation of motion in θ_1

$$\left(J_1 + \frac{1}{N^2}J_2\right)\ddot{\theta}_1 + \left(b_1 + c_T + \frac{1}{N^2}b_2\right)\dot{\theta}_1 + \left(\frac{1}{N^2}k_T\right)\theta_1 = \tau$$

For simplicity, the previous equation of motion can be written as

$$J_e\ddot{\theta}_1 + b_e\dot{\theta}_1 + k_e\theta_1 = \tau$$

with effective parameters $J_e = J_1 + \frac{1}{N^2}J_2$, $b_e = b_1 + c_T + \frac{1}{N^2}b_2$, and $k_e = \frac{1}{N^2}k_T$.

Example 3.3.6

A motor drives a propeller through a pair of gears, as shown in Figure 3.3.24, where τ_m is the motor torque; J_m and J_p are the mass moments of inertia of the motor–shaft assembly and the

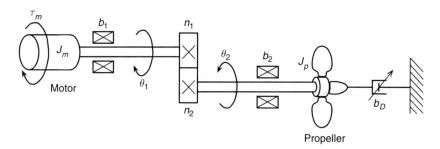

Figure 3.3.24 Schematic of a motor-gear-propeller system

propeller–shaft assembly, respectively; b_1 and b_2 are the bearing coefficients, and θ_1 and θ_2 are the rotational displacements of the motor shaft and propeller shaft, respectively. Assume that the propeller is opposed by a drag torque due to an aerodynamic effect, which is characterized by damping coefficient b_D. Develop a mathematical model for the geared system.

Solution

This example is different from Example 3.3.1 in that the motor and propeller are coupled by a pair of gears, instead of a flexible shaft. Thus, the system in Example 3.3.1 has two DOFs while the geared system in the current example has only one DOF. The free-body diagrams of the mass elements and the auxiliary plot of the gear pair are drawn in Figure 3.3.25, where T_{g1} and T_{g2} are gear torques, and T_D is a drag torque applied to the propeller. With the free-body diagrams in Figure 3.3.25(a), the equations of motion of the system are written as

$$\text{For motor – shaft assembly:} \quad J_m \ddot{\theta}_1 = \tau_m - T_{b1} - T_{g1} \tag{a}$$

$$\text{For propeller – shaft assembly:} \quad J_p \ddot{\theta}_2 = T_{g2} - T_{b2} - T_D \tag{b}$$

The internal forces of the damping elements are given by

$$\text{For bearings:} \quad T_{b1} = b_1 \dot{\theta}_1, \ T_{b2} = b_2 \dot{\theta}_2 \tag{c}$$

$$\text{For aerodynamic effect:} \quad T_D = b_D \dot{\theta}_p^2 \operatorname{sgn}(\dot{\theta}_p) \tag{d}$$

Plugging Eqs. (c) and (d) into Eqs. (a) and (b) yields the governing equations

$$\begin{aligned} J_m \ddot{\theta}_1 + b_1 \dot{\theta}_1 &= \tau_m - T_{g1} \\ J_p \ddot{\theta}_2 + b_2 \dot{\theta}_2 + b_D \dot{\theta}_p^2 \operatorname{sgn}(\dot{\theta}_p) &= T_{g2} \end{aligned} \tag{f}$$

As discussed in Example 3.3.5, the previous two differential equations are not independent. By the auxiliary plot in Figure 3.3.25(b), the gear equations are given by

$$T_{g2} = N T_{g1}, \ \theta_2 = \frac{1}{N} \theta_1, \ \text{with } N = \frac{n_2}{n_1} \tag{g}$$

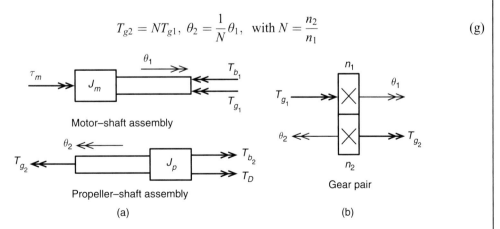

Figure 3.3.25 The geared system in Example 3.3.6: (a) free-body diagrams; (b) auxiliary plot

If θ_2 is chosen as the independent displacement parameter, by substituting Eq. (g) into Eq. (f), the independent governing equation of motion for the geared system is obtained as follows:

$$(J_m N^2 + J_p)\ddot{\theta}_2 + (b_1 N^2 + b_2)\dot{\theta}_2 + b_D \dot{\theta}_p^2 \text{sgn}(\dot{\theta}_2) = N\tau_m \tag{h}$$

3.3.5 Transfer function Formulation

The establishment of transfer function formulations for rotational systems follows the same steps for translational systems, as described in Section 3.2.4.

Example 3.3.7

Consider the motor–propeller assembly in Example 3.3.1. By ignoring drag torque, derive a transfer function for the system, with the motor torque τ_m as the input and the rotation speed $\dot{\theta}_p$ of the propeller as the output. Also, determine the order of the transfer function.

Solution
The governing equations (d) of motion in Example 3.3.1, without the drag torque, are

$$\begin{aligned} J_m \ddot{\theta}_m + b\dot{\theta}_m + k_T \theta_m - k_T \theta_p &= \tau_m \\ J_p \ddot{\theta}_p + k_T \theta_p - k_T \theta_m &= 0 \end{aligned} \tag{a}$$

which are linear differential equations with constant coefficients. Taking Laplace transform of the equations with respect to time and with zero initial values gives

$$\begin{aligned} (J_m s^2 + bs + k_T)\Theta_m(s) - k_T \Theta_p(s) &= T_m(s) \\ (J_p s^2 + k_T)\Theta_p(s) - k_T \Theta_m(s) &= 0 \end{aligned} \tag{b}$$

where $\Theta_m(s)$, $\Theta_p(s)$, and $T_m(s)$ are the Laplace transforms of θ_m, θ_p and τ_m, respectively. Elimination of $\Theta_m(s)$ from Eq. (b) arrives at

$$(J_m s^2 + bs + k_T)\frac{J_p s^2 + k_T}{k_T}\Theta_p(s) - k_T \Theta_p(s) = T_m(s) \tag{c}$$

which leads to

$$s[(J_m s + b)(J_p s^2 + k_T) + J_p k_T s]\Theta_p(s) = k_T T_m(s)$$

By the definition given in Eq. (3.2.12), the transfer function of the motor–propeller system is obtained as follows

$$G(s) = \frac{s\Theta_p(s)}{T_m(s)} = \frac{sk_T}{s[(J_m s + b)(J_p s^2 + k_T) + J_p k_T s]}$$

$$= \frac{k_T}{(J_m s + b)(J_p s^2 + k_T) + J_p k_T s}$$

This is a third-order transfer function.

3.4 Rigid-Body Systems in Plane Motion

The mechanical systems modeled so far are composed of those mass elements which are in either translation or rotation about a fixed axis. Considered in this section are the two-dimensional motion of slab-like rigid bodies, in a combination of translation and rotation. Description of three-dimensional motion of rigid bodies, which is much more involved, is beyond the scope of this text.

3.4.1 Three Keys in System Modeling

As before, modeling of this type of mechanical system emphasizes three keys: fundamental principles, basic elements, and free-body and auxiliary plots.

Mass Center and Relative Motion

Consider a rigid body contained in domain Ω; see Figure 3.4.1. The mass center G of the body (also called the center of gravity) is defined by

$$\boldsymbol{r}_G = \frac{1}{m} \int_\Omega \boldsymbol{r} \, dm \tag{3.4.1}$$

where $m = \int_\Omega dm$ is the mass of the body, and $\boldsymbol{r}$ is the position vector of the differential mass dm. In Cartesian coordinates, the position of the mass center is given by

$$x_G = \frac{1}{m} \int_\Omega x \, dm, \quad y_G = \frac{1}{m} \int_\Omega y \, dm \tag{3.4.2}$$

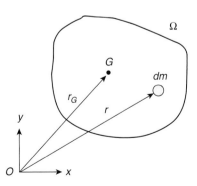

Figure 3.4.1 The mass center of a rigid body

A general plane motion of a rigid body can be decomposed into a translation defined by the motion of its mass center G and a simultaneous rotation about G. As shown in Figure 3.4.2, the velocity v_P of any point P on the rigid body is the sum of the velocity v_G of G and the velocity $v_{P/G}$ relative to G:

$$v_P = v_G + v_{P/G} \tag{3.4.3}$$

The velocity $v_{P/G}$ is a relative velocity vector associated with rotation about the reference point G and it can be expressed by

$$v_{P/G} = \dot{r}_{P/G} = \omega k \times r_{P/G} \tag{3.4.4}$$

where ω is the angular velocity of the body, which is independent of the reference point, and k is a unit vector perpendicular to plane Oxy, which is determined by the right-hand rule. The acceleration of point P is

$$a_P = a_G + a_{P/G} \tag{3.4.5}$$

where a_G is the velocity of G; $a_{P/G}$ is a relative acceleration associated with rotation about the reference point G and it is given by

$$a_{P/G} = \dot{v}_{P/G} = \alpha k \times r_{P/G} - \omega^2 r_{P/G} \tag{3.4.6}$$

with α being the angular acceleration of the body $(\alpha = \dot{\omega})$.

The above description of relative motion is applicable to any reference point selected on a rigid body. For instance, for a reference point A, the velocity and acceleration of point B are determined by Eqs. (3.4.3)–(3.4.6) with G and P replaced by A and B.

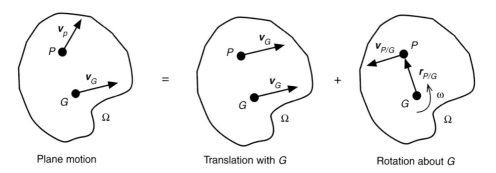

Figure 3.4.2 Plane motion as the sum of translation and rotation

Example 3.4.1

In Figure 3.4.3, a disk is rolling on a horizontal surface with rotation speed ω. There is no slipping at point B so that the velocity at the point of contact is zero. If the center C of the disk is taken as a reference point, the motion of the disk is the sum of the translation with C and rotation about C. For instance, the velocity at B is $v_B = v_C - R\omega$. Because $v_B = 0$, $v_C = R\omega$. Point B with zero velocity is referred to as an instantaneous center of the disk. Also, by choosing B as a reference point, the velocity magnitudes at A, C, and D are determined as follows

$$v_A = |AB|\,\omega = 2R\,\omega, \quad v_C = |CB|\omega = R\,\omega, \quad v_D = |DB|\,\omega$$

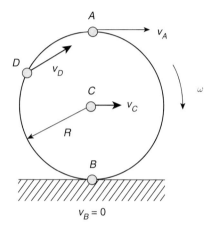

Figure 3.4.3 A disk rolling on a horizontal surface without slipping

As in the modeling of translational and rotational systems, the development of mathematical models for rigid-body systems has the following three keys.

Key 1: Fundamental Principle

The plane motion of a slab-like rigid body, by Newton's second law, is governed by the following three differential equations about the mass center G of the body

$$\begin{aligned}\sum F_x &= ma_{Gx} = m\ddot{x}_G \\ \sum F_y &= ma_{Gy} = m\ddot{y}_G \\ \sum M_G &= I_G\,\alpha = I_G\,\ddot{\theta}\end{aligned} \qquad (3.4.7)$$

where x_G and y_G are the displacement components of G, θ is the angular displacement of the body, m is the mass of the body, I_G is the mass moment of inertia about G, $\sum F_x$ and $\sum F_y$ are the resultants of forces acting on the body, and $\sum M_G$ is the resultant of force moments and torques about G.

Key 2: Basic Elements

Mass Elements A slab-like rigid body of arbitrary shape is a mass element if its inertia (m, I_G) is nonnegligible. For such an element, the equations of motion in Eq. (3.4.7) can be applied.

Lumped Spring and Damping Elements The spring elements and damping elements (translational and torsional springs, translational and torsional dampers, bearings and aerodynamic drag, and flexible bodies with effective spring coefficients) can all be used in the modeling of rigid-body systems. For these elements, the same force–displacement relations, as described in Sections 3.2 and 3.3, are valid.

Key 3: Free-Body Diagrams and Auxiliary Plots

In the modeling of a rigid-body system, a free-body diagram for each mass element is drawn. Also, auxiliary plots for spring and damping elements can be generated if needed.

With the above-mentioned three keys, the system modeling takes the same four steps as described in Section 3.2.3 provided that Eq. (3.4.7) is applied to rigid-body mass elements.

3.4.2 Examples of Rigid-Body Systems

In this section, pulleys or disks with movable rotation centers, wheels in rolling and sliding motion, and a combination of rigid bodies and lumped masses are examined. The equations of motion for these systems shall be derived based on the process described in Section 3.4.1.

Pulleys or Disks with Movable Pins

A pulley with fixed pin (rotation center) was considered in Section 3.3.2. If the pin of a pulley or disk is movable, the motion of the rigid body involves both translation and rotation.

Example 3.4.2

Consider a uniform disk in Figure 3.4.4(a), which is suspended by a spring at its rotation center (also the mass center G). Under forces F_1, F_2 and gravity, the disk experiences translation x and rotation θ, which are independent of each other. The free-body diagram of the disk is shown in Figure 3.4.4(b). The application of Eq. (3.4.7) gives

$$m\ddot{x} + kx = mg + F_1 + F_2$$
$$I_G\ddot{\theta} = F_1 R - F_2 R$$

where $f_s = kx$ has been used. In this example, the disk has no horizontal motion.

If the inertia of the disk is negligible ($m = 0, I_G = 0$), the previous dynamic equations become the following equilibrium equations

$$kx = mg + F_1 + F_2$$
$$F_1 = F_2$$

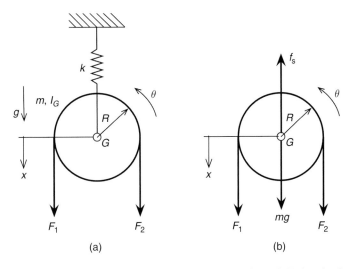

Figure 3.4.4 A disk with movable rotation center: (a) schematic and (b) free-body diagram

Wheel in Rolling and Sliding Motion

A wheel on a flat surface can involve both rolling and sliding motions.

Example 3.4.3

In Figure 3.4.5(a), a uniform wheel moves on a horizontal surface and it is subject to gravity and a horizontal force F. Here, x is the displacement of the mass center G, and θ is the rotational displacement of the wheel. The free-body diagram of the wheel is shown in Figure 3.4.5(b), where f_μ is the friction force between the wheel and surface at the contact point C. The motion of the wheel has the following three possibilities.

(i) Pure Sliding Motion
If there is no friction ($f_\mu = 0$), the wheel is in a translational motion with displacement x, just like a lumped mass. The motion of the wheel is governed by

$$m\ddot{x} = F$$

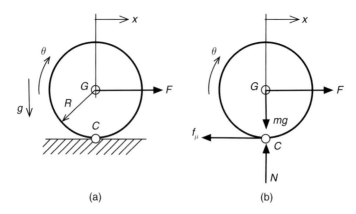

Figure 3.4.5 A disk with movable rotation center: (a) schematic and (b) free-body diagram

(ii) Nonslipping Motion

Friction between the wheel and the surface exits, but the wheel does not slip on the surface. The nonslipping condition is that the velocity of the wheel at the contact point C is zero:

$$v_C = \dot{x} - R\dot{\theta} = 0$$

In this case, the wheel is in both translation and rotation, as described by the following equations

$$m\ddot{x} = F - f_\mu$$
$$I_G\ddot{\theta} = f_\mu R$$

where the friction force $f_\mu \leq \mu_S N$, with μ_S being a coefficient of static friction. Because $x = R\theta$, only one of the displacement parameters (x and θ) is independent. Through elimination of the friction force in the previous equations, an independent equation of motion of the wheel is

$$(I_G + mR^2)\ddot{\theta} = FR$$

or

$$\left(m + \frac{1}{R^2}I_G\right)\ddot{x} = F$$

(iii) Mixed Rolling and Slipping Motion

For wheel slips on a surface that has friction, the governing equations of motion are given by

$$m\ddot{x} = F - \mu_D N \operatorname{sgn}(\dot{x})$$

$$I_G\ddot{\theta} = \mu_D N \operatorname{sgn}(\dot{x})R$$

where Eq. (3.2.11) for friction force has been used. The coefficient μ_D of kinetic friction is usually smaller than the coefficient μ_S of static friction. In this case, with $x \neq R\theta$, x and θ are the independent displacement parameters, which are needed to completely describe the mixed rolling and sliding motion of the wheel.

Combination of Rigid Bodies and Lumped Masses

Models of mechanical systems often consist of both rigid bodies and lumped-mass elements. See Example 3.4.4.

Example 3.4.4

In Figure 3.4.6, a cart carrying a uniform rigid bar moves on a smooth horizontal surface. One end of the bar with mass m and length l is hinged at the point O of the cart with mass M, which is constrained by a viscous damper of coefficient c. Under gravity, the rigid bar is an inverted pendulum. This system has two independent displacement parameters: x and θ. Derive the governing equations of motion of the cart–pendulum system.

Solution

The free-body diagrams of the bar and cart are drawn in Figure 3.4.7, where R_x and R_y are the components of the reaction force at the hinge O, and N is the normal force provided by the surface. By Eq. (3.4.7), the motion of the cart and pendulum is governed by

$$\text{Cart} \quad M\ddot{x} = f - f_d - R_x$$

$$\text{Pendulum} \quad ma_{Gx} = R_x, \quad ma_{Gy} = R_y - mg$$

$$I_G \ddot{\theta} = R_y \frac{l}{2} \sin\theta - R_x \frac{l}{2} \cos\theta$$

where the damping force $f_d = c\dot{x}$, and a_{Gx} and a_{Gy} are the components of the acceleration of the bar at its mass center G, in the x- and y-directions. The coordinates of the mass center can be written as

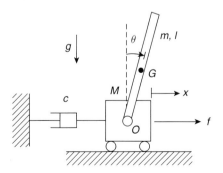

Figure 3.4.6 A cart carrying an inverted pendulum

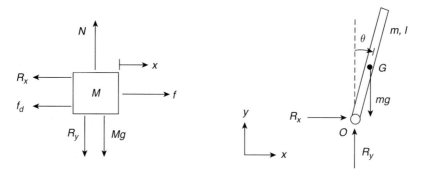

Figure 3.4.7 Free-body diagrams of the cart–pendulum system in Example 3.4.4

$$x_G = x + \frac{l}{2}\sin\theta, \quad y_G = \frac{l}{2}\cos\theta$$

This yields

$$a_{Gx} = \ddot{x}_G = \ddot{x} + \frac{l}{2}\left(\ddot\theta\cos\theta - \dot\theta^2\sin\theta\right)$$
$$\ddot{y}_G = -\frac{l}{2}\left(\ddot\theta\sin\theta + \dot\theta^2\cos\theta\right)$$

It follows that the equations of motion for the system are

$$M\ddot{x} + c\dot{x} = f - R_x \tag{a}$$

$$m\left[\ddot{x} + \frac{l}{2}\left(\ddot\theta\cos\theta - \dot\theta^2\sin\theta\right)\right] = R_x \tag{b}$$

$$-m\frac{l}{2}[\ddot\theta\sin\theta + \dot\theta^2\cos\theta] = R_y - mg \tag{c}$$

$$I_G\ddot\theta = R_y\frac{l}{2}\sin\theta - R_x\frac{l}{2}\cos\theta \tag{d}$$

Because the system has two independent displacement parameters (x, θ), the previous four equations are not independent. By Eq. (c)

$$R_y = mg - m\frac{l}{2}[\ddot\theta\sin\theta + \dot\theta^2\cos\theta] \tag{e}$$

Substituting Eqs. (b) and (e) into Eqs. (a) and (d) gives two independent equations of motion as follows:

$$(M+m)\ddot{x} + c\dot{x} + m\frac{l}{2}\left(\ddot\theta\cos\theta - \dot\theta^2\sin\theta\right) = f$$
$$\frac{ml^2}{3}\ddot\theta + m\ddot{x}\frac{l}{2}\cos\theta - mg\frac{l}{2}\sin\theta = 0 \tag{f}$$

where $I_G = \frac{1}{12}ml^2$ for the uniform bar has been used. The $\frac{1}{3}ml^2$ in the second equation is the mass moment of inertia of the bar about the hinge point O, which can be obtained by the parallel-axis theorem given in Section 3.3.1.

3.5 Energy Approach

In the previous sections, the equations of motion for a mechanical system are derived through use of Newton's laws to the free-body diagrams. This way of system modeling is called the *Newtonian approach*. In this section, an alternative method, which makes use of energy functions and produces a set of differential equations of motion called Lagrange's equations, is introduced. This method is referred to as the *Lagrangian approach* or energy approach. As shown in Section 3.1.4, spatial integration of Newton's second law leads to the principle of work and energy. The Lagrangian approach follows this principle, and hence is completely compatible with the Newtonian approach. The Lagrangian approach in general is more capable of modeling complex mechanical systems with multiple components.

3.5.1 Energy Functions of Mechanical Systems

In this section, kinetic energy, potential energy, and the Rayleigh dissipation function are presented to prepare for the derivation of Lagrange's equations in the next section.

Kinetic Energy

As shown in Section 3.1, the kinetic energy of a particle is

$$T = \frac{1}{2}m\mathbf{v} \cdot \mathbf{v} = \frac{1}{2}mv^2 \qquad (3.5.1)$$

For a rotating body about a fixed axis or pin O, as described in Section 3.3, its kinetic energy is

$$T = \frac{1}{2}I_O\dot{\theta}^2 \qquad (3.5.1)$$

For a rigid body in plane motion, as shown in Figure 3.5.1, its kinetic energy is given by

$$T = \frac{1}{2}mv_G^2 + \frac{1}{2}I_G\dot{\theta}^2 \qquad (3.5.2)$$

where $v_G = |\mathbf{v}_G|$, with $\mathbf{v}_G$ being the velocity of the body at its mass center G.

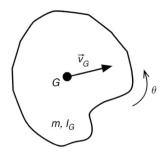

Figure 3.5.1 A rigid body in plane motion

Potential Energy

The potential energy due to gravity and elastic deformation of a spring with a fixed end, as discussed in Section 3.1.4, has the form

$$\text{Gravity:} \quad V = mgy \tag{3.5.3}$$

$$\text{Spring force:} \quad V = \frac{1}{2}kx^2 \tag{3.5.4}$$

where y is a coordinate that is in the opposite direction to the gravitational acceleration. For a spring with two movable ends, as shown in Figure 3.2.2(b, c), its potential energy is given by

$$V = \frac{1}{2}k(x_1 - x_2)^2 \tag{3.5.5}$$

Rayleigh Dissipation Function for Viscous Dampers

For a viscous damper with a fixed end, the Rayleigh dissipation function is

$$R = \frac{1}{2}c\dot{x}^2 \tag{3.5.6}$$

For a viscous damper with two movable ends, as shown in Figure 3.2.4(b, c), the Rayleigh dissipation function is given by

$$R = \frac{1}{2}c(\dot{x}_1 - \dot{x}_2)^2 \tag{3.5.7}$$

Note that a Rayleigh dissipation function is in quadratic form.

Example 3.5.1

For the quarter-car suspension model in Figure 3.2.13, the energy functions are as follows:

$$T = \frac{1}{2}m_1\dot{y}_1^2 + \frac{1}{2}m_2\dot{y}_2^2$$

$$V = \frac{1}{2}k_1(y_1 - y_0)^2 + \frac{1}{2}k_2(y_2 - y_1)^2 + m_1 g y_1 + m_2 g y_2$$

$$R = \frac{1}{2}c(\dot{y}_2 - \dot{y}_1)^2$$

3.5.2 Lagrange's Equations

The establishment of Lagrange's equations requires the usage of a minimum number of displacement parameters or coordinates that are independent and that can fully describe the motion of a mechanical system in consideration. This number is called the number of DOFs of the system, as has been defined in Section 3.3.3.

Let $q_1, q_2, \ldots, q_n$ be a set of n *independent* coordinates for a mechanical system. The system thus has n degrees of freedom (DOFs) and it is thus called an n-DOF system. The energy functions for the system can be written as

$$\begin{aligned} T &= T(q_1, q_2, \ldots, q_n; \dot{q}_1, \dot{q}_2, \ldots, \dot{q}_n) \\ V &= V(q_1, q_2, \ldots, q_n) \\ R &= T(\dot{q}_1, \dot{q}_2, \ldots, \dot{q}_n) \end{aligned} \quad (3.5.8)$$

Note that kinetic energy T can be a function of $q_1, q_2, \ldots, q_n$ due to nonlinearities. With these energy functions, Lagrange's equations read

$$\frac{d}{dt}\left(\frac{\partial T}{\partial \dot{q}_i}\right) - \frac{\partial T}{\partial q_i} + \frac{\partial R}{\partial \dot{q}_i} + \frac{\partial V}{\partial q_i} = Q_i, \quad i = 1, 2, \ldots, n \quad (3.5.9)$$

where Q_i is the resultant force associated with coordinate q_i, and it is called a generalized force. The force Q_i consists of external forces and friction forces, but it must exclude gravitational forces and forces from springs and viscous dampers because they have been considered in the energy functions V and R. For this reason, Q_i is a nonconservative force. In the application of Lagrange's equations, free-body diagrams are not needed.

Generalized forces can be identified from the following incremental work

$$dW_{nc} = \sum_{j=1}^{n} Q_j dq_j \quad (3.5.10)$$

Rigorously speaking, the incremental work should be written as

$$\delta W_{nc} = \sum_{j=1}^{n} Q_j \delta q_j$$

where δ is a variation operator, δW_{nc} is called virtual work, and δq_j is the virtual displacement. Due to the limited mathematical background covered in this undergraduate text, Eq. (3.5.10) is used instead.

Example 3.5.2

Consider again the quarter-car suspension model in Figure 3.2.13, which is described by two independent coordinates, y_1 and y_2. The energy functions of the system were obtained in Example 3.5.1. By Eqs. (3.5.9), the equations of motion for the system are derived as follows:

for y_1

$$\frac{d}{dt}(m_1 \dot{y}_1) - c(\dot{y}_2 - \dot{y}_1) + k_1(y_1 - y_0) - k_2(y_2 - y_1) + m_1 g = 0$$

which yields

$$m_1 \ddot{y}_1 + c \dot{y}_1 + (k_1 + k_2) y_1 - c \dot{y}_2 - k_2 y_2 = -m_1 g + k_1 y_0 \qquad (a)$$

and for y_2

$$\frac{d}{dt}(m_2 \dot{y}_2) + c(\dot{y}_2 - \dot{y}_1) + +k_2(y_2 - y_1) + m_2 g = 0$$

which yields

$$m_2 \ddot{y}_2 + c \dot{y}_2 + k_2 y_2 - c \dot{y}_1 - k_2 y_1 = -m_2 g \qquad (b)$$

Equations (a) and (b) are the same as those obtained by the Newtonian approach in Example 3.2.4.

Example 3.5.3

For the cart–pendulum system in Example 3.4.4, derive the governing equations of motion by the Lagrange approach.

Solution

With Figure 3.4.6, the energy functions of the system are obtained as

$$T = \frac{1}{2} M \dot{x}^2 + \frac{1}{2} m (v_{Gx}^2 + v_{Gy}^2) + \frac{1}{2} \left(\frac{1}{12} m l^2 \right) \dot{\theta}^2$$

$$V = mg \frac{l}{2} \cos \theta, \quad R = \frac{1}{2} c \dot{x}^2$$

where v_{Gx} and v_{Gy} are the components of the velocity of the bar at its mass center G in the x- and y-directions, and $I_G = \frac{1}{12} m l^2$ has been used. Because $x_G = x + \frac{l}{2} \sin \theta$ and $y_G = \frac{l}{2} \cos \theta$,

$$v_{Gx} = \dot{x}_G = \dot{x} + \frac{l}{2}\dot{\theta}\cos\theta$$

$$v_{Gy} = \dot{y}_G = -\frac{l}{2}\dot{\theta}\sin\theta$$

Hence, the kinetic energy is of the form

$$T = \frac{1}{2}M\dot{x}^2 + \frac{1}{2}m\left[\left(\dot{x} + \frac{l}{2}\dot{\theta}\cos\theta\right)^2 + \left(-\frac{l}{2}\dot{\theta}\sin\theta\right)^2\right] + \frac{1}{2}\left(\frac{1}{12}ml^2\right)\dot{\theta}^2$$

$$= \frac{1}{2}(M+m)\dot{x}^2 + \frac{1}{2}\left(\frac{1}{3}ml^2\right)\dot{\theta}^2 + \frac{1}{2}ml\,\dot{x}\dot{\theta}\cos\theta$$

According to Eq. (3.5.10), the incremental work by generalized forces is $dW_{nc} = f dx$, indicating that $Q_x = f$ and $Q_\theta = 0$.

By Eq. (3.5.9), the equations of motion of the cart-pendulum system are obtained as follows: for x,

$$\frac{d}{dt}\left((M+m)\dot{x} + \frac{1}{2}ml\,\dot{\theta}\cos\theta\right) + c\dot{x} = f$$

yielding

$$(M+m)\ddot{x} + c\dot{x} + m\frac{l}{2}\left(\ddot{\theta}\cos\theta - \dot{\theta}^2\sin\theta\right) = f \qquad (a)$$

and for θ,

$$\frac{d}{dt}\left(\frac{1}{3}ml^2\dot{\theta} + \frac{1}{2}ml\,\dot{x}\cos\theta\right) + \frac{1}{2}ml\,\dot{x}\dot{\theta}\sin\theta + -mg\frac{l}{2}\sin\theta = 0$$

yielding

$$\frac{ml^2}{3}\ddot{\theta} + m\ddot{x}\frac{l}{2}\cos\theta - mg\frac{l}{2}\sin\theta = 0 \qquad (b)$$

Note that Eqs. (a) and (b) are the same as Eq. (f) in Example 3.4.4. The latter is obtained after the elimination of the hinge reaction forces N_x and N_y. This example shows that the Lagrangian approach automatically delivers independent equations of motion.

3.6 Block Diagrams of Mechanical Systems

The transfer function formulation described in Sections 3.2 and 3.3 is one form of model representation for modeling and analysis of linear time-invariant mechanical systems. In this section, we introduce another model representation, namely, the block diagram. Block-diagram representation is a useful tool for modeling and analysis of mechanical systems. Block diagrams have been widely used in the system response analysis of dynamic systems

(Chapter 7) and the design of feedback control systems (Chapter 8). There are two types of block diagrams: s-domain block diagrams and time-domain block diagrams. This section concerns s-domain block diagrams.

The concept of s-domain block diagrams was introduced in Section 2.7. There are four basic components of block diagrams: block, signal, summing point, and pick-off point. In addition, for complicated diagrams, a crossing bridge is often used. These components are listed in Table 3.6.1. A block diagram for a dynamic system is constructed through assembly of these basic components. Like in a transfer function formulation, Laplace transform of the governing equations of a system with zero initial conditions is performed in construction of a block diagram. Block diagrams in the s-domain, as suggested by its name, are only valid for linear time-invariant dynamic systems.

Construction of a block diagram for a mechanical system takes the following four steps.

(1) Derive the equations of motion for the system.
(2) Take the Laplace transform of the equations of motion with zero initial conditions.

Table 3.6.1 Basic components of the s-domain block diagram

Component	Symbol	Rule
Block	$R(s) \to \boxed{G(s)} \to Y(s)$	$Y(s) = G(s)R(s)$
Signal	$U(s) \to \quad V(s) \leftarrow$	Variable, input or output
Summing point	$U(s) \xrightarrow{+} \bigcirc \xrightarrow{} W(s)$, $-V(s)$	$W(s) = U(s) - V(s)$
	$X(s) \xrightarrow{+} \bigcirc \xrightarrow{} Z(s)$, $+Y(s)$	$Z(s) = X(s) + V(s)$
Pick-off point	→ $U(s)$; ↓ $V(s)$, → $X(s)$	$V(s) = U(s)$ $X(s) = U(s)$
Crossing bridge (without connection)	$V(s)$ ↑, → $U(s)$	Signals $U(s)$ and $V(s)$ do not interfere with each other

(3) Create parts of the block diagram with the Laplace transform equations and by the basic components in Table 3.6.1.
(4) Assembly the parts obtained in Step 3 and use the preassigned inputs and outputs, to complete the block diagram for the system.

These steps are demonstrated in Example 3.6.1.

Example 3.6.1

In this example, we explain the construction of the block diagram for the mechanical system shown in Figure 1.3.2 in Chapter 1. For this system shown, its equations of motion are given in Eq. (1.3.12), namely,

$$m_1\ddot{x}_1 + c\dot{x}_1 + (k + k_1)x_1 - k_1 x_2 = 0 \quad \text{(a)}$$

$$m_2\ddot{x}_2 + k_1 x_2 - k_1 x_1 = f \quad \text{(b)}$$

Consider a block diagram of the system with external force f as the input and displacement x_1 as the output. By Laplace transform and rearrangement, Eqs. (i) and (ii) are reduced to

$$X_1(s) = \frac{1}{m_1 s^2 + cs + k} k_1 (X_2(s) - X_1(s)) \quad \text{(c)}$$

$$X_2(s) = \frac{1}{m_2 s^2} \left\{ F(s) - k_1 (X_2(s) - X_1(s)) \right\} \quad \text{(d)}$$

With the basic components in Table 3.6.1, Eqs. (c) and (d) are represented by the parts in Figure 3.6.1. Assembly of these parts by using summing point and pick-off point eventually yields the block diagram in Figure 3.6.2, where $F_s(s) = k_1(X_2(s) - X_1(s))$. The block diagram is the same as in Figure 1.3.3.

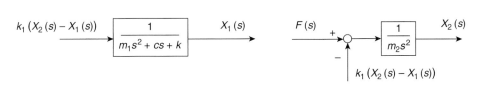

Figure 3.6.1 Parts of the block diagram in Example 3.6.1: (a) representation of Eq. (c); (b) representation of Eq. (d)

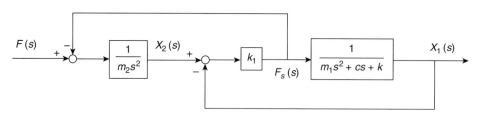

Figure 3.6.2 Block diagram of the mechanical system in Figure 1.3.2

Example 3.6.2

Consider the spring–mass–damper system in Figure 3.2.8(a). Let the system inputs be f_1 and f_2. Let the system outputs be $\dot{x}_1$ and x_2. Draw a block diagram for the system.

Solution

From Example 3.2.1, the governing equations of motion of the system are

$$m_1 \ddot{x}_1 = f_1 + k_2(x_2 - x_1) - k_1 x - c\dot{x}_1 \quad \text{(a)}$$

$$m_2 \ddot{x}_2 = f_2 - k_2(x_2 - x_1) \quad \text{(b)}$$

Performing Laplace transform of Eqs. (a) and (b) with zero initial conditions gives

$$(m_1 s^2 + cs + k_1) X_1(s) = F_1(s) + k_2(X_2(s) - X_1(s))$$
$$m_2 s^2 X_2(s) = F_2(s) - k_2(X_2(s) - X_1(s))$$

or

$$X_1(s) = \frac{1}{m_1 s^2 + cs + k_1}\{F_1(s) + k_2(X_2(s) - X_1(s))\} \quad \text{(c)}$$

$$X_2(s) = \frac{1}{m_2 s^2}\{F_2(s) - k_2(X_2(s) - X_1(s))\} \quad \text{(d)}$$

With the basic components in Table 3.6.1, (c) and (d) are represented by the parts in Figure 3.6.3.

Assembling the parts in Figure 3.6.3 gives the block diagram of the spring–mass–damper system in Figure 3.6.4, where $V_1(s) = \mathcal{L}[\dot{x}_1] = sX_1(s)$ and $F_s(s)$ represents the internal force of the spring (k_2) connecting the two masses.

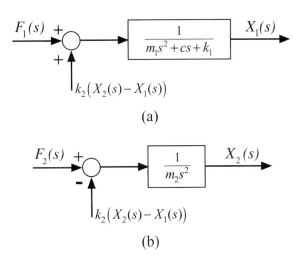

Figure 3.6.3 Parts of the block diagram in Example 3.6.2: (a) representation of Eq. (c); and (b) representation of Eq. (d)

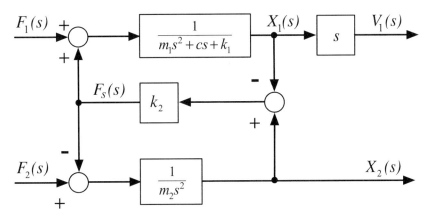

Figure 3.6.4 A block diagram of the mechanical system in Example 3.6.2

3.7 State-Space Representations

Transfer function formulations and s-domain block diagrams are only valid for linear time-invariant systems. On the other hand, linear time-variant systems and nonlinear systems, such as the rocket in Example 3.2.6 and the coupled motor–propeller system in Example 3.3.1, are seen in many engineering applications. For these dynamic systems, modeling,

Table 3.7.1. Comparison of state-space representations and transfer function formulations

	State-space representation	Transfer function formulation
System	Linear or nonlinear Time-invariant or time-variant	Linear and time-invariant
Domain	Time domain	Laplace transform domain
Format	State equations and output equations	Input–output relation via transfer function

simulation, and analysis rely on other forms of model representation. In this section, state-space representations for general dynamic systems, including time-variant and nonlinear mechanical systems, are introduced. Also, as shall be seen in Section 6.5, block diagrams in the time domain are another tool for modeling and simulation of general dynamic systems.

A state-space representation, which is also called a state-space model, consists of two sets of equations in the time domain: state equations and output equations. State equations are a set of coupled first-order differential equations about quantities known as state variables. Output equations are a set of coupled algebraic equations about the state variables. State variables for a dynamic system are independent quantities that can completely describe the future behavior of the system with known inputs and initial disturbances. State equations and output equations can be either linear or nonlinear. A comparison for state-space representations and transfer function formulations is given in Table 3.7.1.

Suppose that a system has n state variables $x_1, x_2, \ldots, x_n$, m inputs $u_1, u_2, \ldots, u_m$ and p outputs $y_1, y_2, \ldots, y_n$, all of which in general are functions of time. If the system is nonlinear, its state equations can be written as

$$\begin{cases} \dot{x}_1 = f_1(x_1, x_2, \ldots, x_n; u_1, u_2, \ldots, u_m) \\ \dot{x}_2 = f_2(x_1, x_2, \ldots, x_n; u_1, u_2, \ldots, u_m) \\ \vdots \\ \dot{x}_n = f_n(x_1, x_2, \ldots, x_n; u_1, u_2, \ldots, u_m) \end{cases} \quad (3.7.1)$$

and its output equations are of the form

$$\begin{cases} y_1 = g_1(x_1, x_2, \ldots, x_n; u_1, u_2, \ldots, u_m) \\ y_2 = g_2(x_1, x_2, \ldots, x_n; u_1, u_2, \ldots, u_m) \\ \vdots \\ y_p = g_p(x_1, x_2, \ldots, x_n; u_1, u_2, \ldots, u_m) \end{cases} \quad (3.7.2)$$

where f_i and g_j are nonlinear functions of the state variables and inputs. Note that in a state equation, the left-hand side only contains the first derivative of a state variable and the right-hand side does not have any derivative term of x_k or u_l.

3.7 State-Space Representations

If the system is linear, which can be either time-invariant or time-variant, its state equations and output equations are given by

$$\begin{cases} \dot{x}_1 = a_{11}x_1 + a_{12}x_2 + \cdots + a_{1n}x_n + b_{11}u_1 + b_{12}u_2 + \cdots + b_{1m}u_m \\ \dot{x}_2 = a_{21}x_1 + a_{22}x_2 + \cdots + a_{2n}x_n + b_{21}u_1 + b_{22}u_2 + \cdots + b_{2m}u_m \\ \quad\vdots \\ \dot{x}_n = a_{n1}x_1 + a_{n2}x_2 + \cdots + a_{nn}x_n + b_{n1}u_1 + b_{n2}u_2 + \cdots + b_{nm}u_m \end{cases} \quad (3.7.3)$$

and

$$\begin{cases} y_1 = c_{11}x_1 + c_{12}x_2 + \cdots + c_{1n}x_n + d_{11}u_1 + d_{12}u_2 + \cdots + d_{1m}u_m \\ y_2 = c_{21}x_1 + a_{22}x_2 + \cdots + c_{2n}x_n + d_{21}u_1 + d_{22}u_2 + \cdots + d_{2m}u_m \\ \quad\vdots \\ y_p = c_{p1}x_1 + c_{p2}x_2 + \cdots + c_{pn}x_n + d_{p1}u_1 + d_{p2}u_2 + \cdots + d_{pm}u_m \end{cases} \quad (3.7.4)$$

where a_{ij}, b_{jk}, c_{kl}, and d_{lm} are either constants or known functions of t. The linear state equations and output equations can be cast in a matrix form as follows:

State equation: $\dot{\boldsymbol{x}}(t) = \boldsymbol{A}\boldsymbol{x}(t) + \boldsymbol{B}\boldsymbol{u}(t)$ (3.7.5)

Output equation: $\boldsymbol{y}(t) = \boldsymbol{C}\boldsymbol{x}(t) + \boldsymbol{D}\boldsymbol{u}(t)$ (3.7.6)

where $\boldsymbol{x}(t)$, $\boldsymbol{u}(t)$, and $\boldsymbol{y}(t)$ are the state, input, and output vectors, respectively, given by

$$\boldsymbol{x}(t) = \begin{pmatrix} x_1(t) \\ x_2(t) \\ \vdots \\ x_n(t) \end{pmatrix}, \quad \boldsymbol{u}(t) = \begin{pmatrix} u_1(t) \\ u_2(t) \\ \vdots \\ u_p(t) \end{pmatrix}, \quad \boldsymbol{y}(t) = \begin{pmatrix} y_1(t) \\ y_2(t) \\ \vdots \\ y_q(t) \end{pmatrix} \quad (3.7.7)$$

and $\boldsymbol{A}$, $\boldsymbol{B}$, $\boldsymbol{C}$, and $\boldsymbol{D}$ are matrices composed of a_{ij}, b_{jk}, c_{kl}, and d_{lm}, respectively.

Assume that a mathematical model for the mechanical system in consideration has been obtained by following the guidelines in Section 3.2.3. Conversion of the model to a state space representation takes the following four steps.

(1) Specify inputs and outputs of the system. The inputs are naturally the disturbances to the system, including external forces and prescribed displacements. The outputs, depending on the application and interest, can be displacements, velocities, accelerations, and internal forces of the spring and damping elements.
(2) Select state variables. For a mechanical system, its displacements and velocities can be conveniently chosen as state variables. The selection of state variables, however, is not unique. Regardless of how state variables are selected, the total number of state variables should be the same and it is the sum of the orders of the governing differential equations of the system model. Refer to Section 2.5.4 for some examples of the selection of state variables.

(3) Differentiate the state variables one by one and perform mathematical manipulations to obtain state equations of the form of Eq. (3.7.1) or Eq. (3.7.3). In this process, the definition of the state variables and the governing differential equations are used.

(4) Derive output equations of the form of Eq. (3.7.2) or Eq. (3.7.4), by the specified outputs and selected state variables.

Example 3.7.1

For the mass on a slope in Figure 3.2.11, establish a state-space representation, with the gravitational force as the input and the velocity of the mass and spring force as the outputs.

Solution
From Example 3.2.3, the equation of motion for the mass is

$$m\ddot{x} + kx + \mu_D Mg \cos \alpha \, \text{sgn}(\dot{x}) = F + Mg \sin \alpha \tag{a}$$

which is a second-order nonlinear differential equation. Accordingly, two state variables are selected as below

$$x_1 = x, \quad x_2 = \dot{x} \tag{b}$$

By differentiating the first state variable, $\dot{x}_1 = \dot{x}$, the first state equation is obtained as

$$\dot{x}_1 = x_2 \tag{c}$$

Differentiate the second state variable, $\dot{x}_2 = \ddot{x}$, and substitute the result into Eq. (a), to obtain

$$m\dot{x}_2 + kx_1 + \mu_D Mg \cos \alpha \, \text{sgn}(x_2) = F + Mg \sin \alpha \tag{d}$$

where Eq. (b) has been used. Thus, the second state equation is

$$\dot{x}_2 = \frac{1}{m}[-kx_1 - \mu_D Mg \cos \alpha \, \text{sgn}(x_2) + F + Mg \sin \alpha] \tag{e}$$

The outputs of the system are

$$y_1 = \dot{x}, \quad y_2 = f_s = kx \tag{f}$$

which, by Eq. (b), yields the output equations as follows

$$y_1 = x_2, \quad y_2 = kx_1 \tag{g}$$

Example 3.7.2

For the cart–mass system in Figure 3.2.15, develop a state-space model, with the external force f as the input and the displacement x_2 of mass m_2 and the force of spring k_2 as the outputs. Also, for this linear system, obtain the state and output equations in matrix form.

Solution

From Example 3.2.5, the coupled equations of motion are

$$m_1\ddot{x}_1 + B\dot{x}_1 + (k_1 + k_2)x_1 - B\dot{x}_2 - k_2 x_2 = f \\ m_2\ddot{x}_2 + B\dot{x}_2 + k_2 x_2 - B\dot{x}_1 - k_2 x_1 = 0 \tag{a}$$

with each being a second-order differential equation. Because the sum of the orders of the equations is $2+2=4$, four state variables are chosen as follows

$$z_1 = x_1, \; z_2 = \dot{x}_1, \; z_3 = x_2, \; z_4 = \dot{x}_2 \tag{b}$$

where the notation z_k is used for the state variables to avoid confusion in symbols. Differentiating z_1 and z_3 immediately gives two state equations $\dot{z}_1 = z_2$, $\dot{z}_3 = z_4$, where the definition of z_2 and z_4 in Eq. (b) has been used. Now, differentiate z_2 and z_4 and substitute the results into the equations (a) of motion, to obtain

$$m_1\dot{z}_2 + Bz_2 + (k_1 + k_2)z_1 - Bz_4 - k_2 z_3 = f \\ m_2\dot{z}_4 + Bz_4 + k_2 z_3 - Bz_2 - k_2 z_1 = 0 \tag{c}$$

It follows that the state equations of the system are

$$\begin{aligned} \dot{z}_1 &= z_2 \\ \dot{z}_2 &= \frac{1}{m_1}\left[-(k_1+k_2)z_1 - Bz_2 + k_2 z_3 + Bz_4 + f\right] \\ \dot{z}_3 &= z_4 \\ \dot{z}_4 &= \frac{1}{m_2}\left[k_2 z_1 + Bz_2 - k_2 z_3 - Bz_4\right] \end{aligned} \tag{d}$$

The outputs are

$$y_1 = x_2, \quad y_2 = f_s = k_2(x_2 - x_1) \tag{e}$$

which gives the output equations as follows:

$$\begin{aligned} y_1 &= z_3 \\ y_2 &= k_2(z_3 - z_1) \end{aligned} \tag{g}$$

The state equations (d) can be cast in matrix form, $\dot{x} = A + Bu$, with

$$x = \begin{pmatrix} z_1 \\ z_2 \\ z_3 \\ z_4 \end{pmatrix}, \quad A = \begin{bmatrix} 0 & 1 & 0 & 0 \\ -\dfrac{k_1+k_2}{m_1} & -\dfrac{B}{m_1} & \dfrac{k_2}{m_1} & \dfrac{B}{m_1} \\ 0 & 0 & 0 & 1 \\ \dfrac{k_2}{m_2} & \dfrac{B}{m_2} & -\dfrac{k_2}{m_2} & -\dfrac{B}{m_2} \end{bmatrix}, \quad B = \begin{bmatrix} 0 \\ \dfrac{1}{m_1} \\ 0 \\ 0 \end{bmatrix}, \quad u = f \qquad \text{(h)}$$

Also, the output equations can be written in the matrix form, $y = Cx + Du$, with

$$y = \begin{pmatrix} y_1 \\ y_2 \end{pmatrix}, \quad C = \begin{bmatrix} 0 & 0 & 1 & 0 \\ -k_2 & 0 & k_2 & 0 \end{bmatrix}, \quad D = \begin{bmatrix} 0 \\ 0 \end{bmatrix} \qquad \text{(i)}$$

Example 3.7.3

Consider the motor–propeller assembly in Figure 3.3.10, which experiences a nonlinear drag torque. Establish a state-space representation, with the motor torque as the input, and the propeller rotation speed and the torque of the spring connecting the motor and propeller as the outputs.

Solution
The equations of motion of the system, from Example 3.3.1, are given by

$$\begin{aligned} J_m \ddot{\theta}_m + b\dot{\theta}_m + k_T \theta_m - k_T \theta_p &= \tau_m \\ J_p \ddot{\theta}_p + b_D \dot{\theta}_p^2 \text{sgn}(\dot{\theta}_p) + k_T \theta_p - k_T \theta_m &= 0 \end{aligned} \qquad \text{(a)}$$

According to these equations, four state variables are selected as follows

$$x_1 = \theta_m, \quad x_2 = \dot{\theta}_m, \quad x_3 = \theta_p, \quad x_4 = \dot{\theta}_p \qquad \text{(b)}$$

Equation (b) automatically gives two state equations: $\dot{x}_1 = x_2$, $\dot{x}_3 = x_4$. The other two state equations are derived by substituting Eq. (b) into Eq. (a), which leads to

$$\begin{aligned} J_m \dot{x}_2 + bx_2 + k_T x_1 - k_T x_3 &= \tau_m \\ J_p \dot{x}_4 + b_D x_4^2 \text{sgn}(x_4) + k_T x_3 - k_T x_1 &= 0 \end{aligned} \qquad \text{(c)}$$

It follows that the four state equations of the motor–propeller system are

$$\dot{x}_1 = x_2$$
$$\dot{x}_2 = \frac{1}{J_m}(-k_T x_1 - b x_2 + k_T x_3 + \tau_m)$$
$$\dot{x}_3 = x_4 \tag{d}$$
$$\dot{x}_4 = \frac{1}{J_p}\left(k_T x_1 - k_T x_3 - b_D x_4^2\, \text{sgn}(x_4)\right)$$

Note that the fourth state equation is nonlinear due to the drag torque. The system outputs are $y_1 = \dot{\theta}_p$ and $y_2 = \tau_s = k_T(\theta_m - \theta_p)$, which by Eq. (b) give the output equations as follows:

$$y_1 = x_4$$
$$y_2 = k_T(x_1 - x_3) \tag{e}$$

3.8 Dynamic Responses via MATLAB

So far, this chapter has been focused on the derivation of mathematical models for mechanical systems and the corresponding transfer function formulations and state-space representations. In this section, the simulation of dynamic responses of mechanical systems by the software MATLAB is presented. The purpose here is twofold: to show the utility of transfer function formulations and state-space representations in the simulation and analysis of dynamic systems, and to investigate the dynamic characteristics of mechanical systems in time-response calculations. One may refer to Appendix B for a quick tutorial on MATLAB.

Example 3.8.1

Consider the spring–mass–damper system:

$$m\ddot{x}(t) + c\dot{x}(t) + kx(t) = f(t)$$

with its parameters given by $m = 20$ kg, $c = 80$ N·s/m, and $k = 750$ N/m. Assume a constant external force (step input), $f(t) = f_0 = 15$ N, and zero initial disturbances, $x(0) = 0$ and $\dot{x}(0) = 0$. Plot the displacement and the resultant of the internal forces (the sum of the spring and damping forces) of the system, for $0 \leq t \leq 3$ s.

Solution

As shown in Example 3.2.8, the s-domain displacement of the system is given by

$$X(s) = \frac{1}{ms^2 + cs + k} F(s) \tag{a}$$

The resultant of the internal forces is $f_R = c\dot{x} + kx$, which in the s-domain is

$$F_R(s) = (cs + k)X(s) \tag{b}$$

It follows from Eqs. (a) and (b) that

$$X(s) = \frac{1}{ms^2 + cs + k}\frac{f_0}{s} \tag{c}$$

$$F_R(s) = \frac{cs + k}{ms^2 + cs + k}\frac{f_0}{s} \tag{d}$$

where $F(s) = f_0/s$ has been used.

With Eqs. (c) and (d), the time histories of $x(t)$ and $f_R(t)$ are computed by the following MATAB commands in a script M-file:

```
m = 20; c = 80; k = 750;
f0 = 15; t_final = 3;
figure(1) % Displacement plot
syst1 = tf([f0],[m c k])
step(syst1, t_final, 'k')
xlabel('t'), ylabel('x (m)'), title(' ')
figure(2) % Resultant force plot
syst2 = tf(f0*[c k],[m c k])
step(syst2, t_final, 'k')
xlabel('t'), ylabel('f_R (N)'), title(' ')
```

This produces Figure 3.8.1. Here, `tf` and `step` are MATLAB functions; see Tables B6 and B8 in Appendix B.

As seen from Figure 3.8.1(a), the displacement x of the mechanical system under a step input is oscillatory at the beginning, and it eventually settles to a constant value, which is known as the steady-state response. This value is the static deflection of the spring at equilibrium; namely, $x_{ss} = \frac{f_0}{k} = \frac{15}{750} = 0.02$ m. Likewise, the resultant force $f_R(t)$ of the spring and damping forces eventually settles at the value of $kx_{ss} = f_0 = 15$ N, as shown in Figure 3.8.1(b).

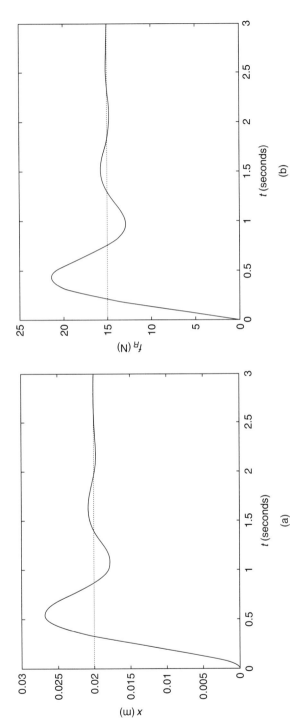

Figure 3.8.1 Dynamic response of the mechanical system in Example 3.8.1: (a) displacement x versus time; (b) resultant force f_R versus time

Example 3.8.2

For the coupled motor–propeller system shown in Figure 3.3.10 of Example 3.3.1, let the parameter values be

$$J_m = 1.2 \text{ kg·m}^2, \; b = 0.5 \text{ N·m·s}, \; k_T = 5000 \text{ N·m}, \; J_p = 1.1 \text{ kg·m}^2, \; b_D = 0.02 \text{ N·m·s}^2$$

Consider a motor torque $\tau_m = 12$ N·m. Assume that the system is initially at rest. By a state-space representation, compute and plot the propeller rotation speed $\dot{\theta}_p$ and the spring torque $\tau_s = k_T(\theta_m - \theta_p)$ of the coupled system, for $0 \le t \le 30$ s. Also, investigate the effects of the torsional spring and air damping on the system response.

Solution

To determine the system response, use the state and output equations obtained in Example 3.7.3 and the fixed-step Runge–Kutta method of order four given in Section 2.8. To this end, the following commands contained in a script M-file:

```
global Jm b_r Kt Jp       % System parameters
global b_D                % Air damping torque
global Tm                 % Motor torque

% Input data
Jm = 1.2; b_r = 0.5; Kt = 5000; Jp = 1.1;
b_D = 0.02;
Tm = 12;
% Compute system response
t_stop = 30;
npts = 10001;
h = t_stop/(npts-1);
z0 = [0; 0; 0; 0]; % initial state
t = linspace(0, t_stop, npts);
z = zeros(4, npts);
z(:,1) = z0;
for i = 1: npts-1
    tt = t(i);
    zz = z(:,i);
    f1 = RKfunc362(tt, zz);
    f2 = RKfunc362(tt+h/2, zz+h/2*f1);
    f3 = RKfunc362(tt+h/2, zz+h/2*f2);
    f4 = RKfunc362(tt+h, zz+h*f3);
    z(:,i+1) = zz +h/6*(f1+2*f2+2*f3+f4);
end

% Plot system response
figure(1)
theta_p = z(4,:);          % Propeller rotation speed
plot(t,theta_p,'k')
grid
xlabel('Time, t (seconds)')
```

```
ylabel('Speed, \theta_p (rad/s)')
figure(2)
Ts = Kt*(z(1,:) - z(3,:));      % Spring torque
plot(t,Ts,'k')
grid
xlabel('Time, t (seconds)')
ylabel('Spring Torque, \tau_s (N-m)')
```

and the following function M-file:

```
function f = RKfunc362(t,z)
global Jm b_r Kt Jp              % System parameters
global b_D                       % Air damping torque
global Tm                        % Motor torque

f = zeros(4,1);
f(1) = z(2);
f(2) = (-Kt*z(1) - b_r*z(2) + Kt*z(3) + Tm)/Jm;
f(3) = z(4);
f(4) = (Kt*z(1) - Kt*z(3) - b_D*z(4)^2*sign(z(4)))/Jp;
```

are created. This plots the propeller speed and the spring torque against time in Figure 3.8.2.

As seen from Figure 3.8.2, the steady-state values of the propeller speed and spring torque are about 15 rad/s and 4.5 N·m, respectively. Due to the elastic coupling between the motor shaft (J_m) and propeller shaft (J_p), high-frequency oscillations appear in the system response; see Figures 3.8.3(a) and 3.8.3(b), which are obtained by zooming in on Figures 3.8.2(a) and 3.8.2(b).

To see the effect of air damping, set $b_D = 0$ and execute the above-mentioned M-files. This gives the propeller speed plot given in Figure 3.8.4, which has a steady-state value of 24 rad/s. Comparison of Figures 3.8.3(a) and 3.8.4 shows that air damping significantly reduces the steady-state speed of the propeller (from 24 rad/s to 15 rad/s). Note that the system without air damping is a linear system. From Example 3.3.7, the transfer function

$$G(s) = \frac{s\Theta_p(s)}{T_m(s)} = \frac{k_T}{(J_m s + b)(J_p s^2 + k_T) + J_p k_T s}$$

can be used to compute the system response. For instance, the MATLAB commands

```
d = [Jm*Jp b_r*Jp (Jm+Jp)*Kt b_r*Kt];
n = Kt*Tm;
sys = tf(n,d);
Tspan = linspace(0, t_stop, npts);
step(sys, Tspan)
```

yield the same speed plot as shown in Figure 3.8.4.

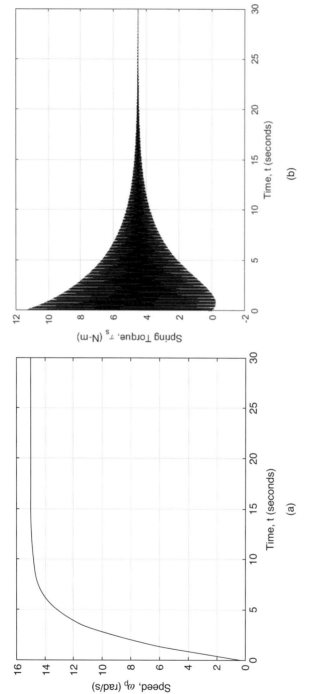

Figure 3.8.2 Dynamic response of the coupled system in Example 3.8.2: (a) the propeller speed, $\omega_p = \dot{\theta}_p$; and (b) the spring torque, $\tau_s = k_T(\theta_m - \theta_p)$.

3.8 Dynamic Responses via MATLAB 197

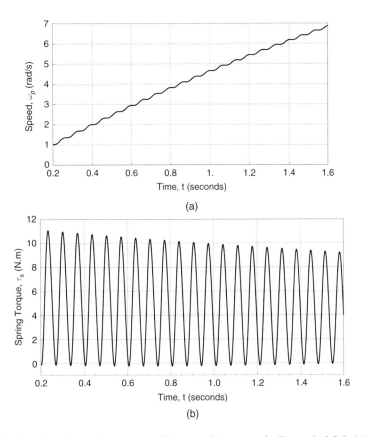

Figure 3.8.3 Zooming in on the response of the coupled system in Example 3.8.2: (a) the propeller speed; and (b) the spring torque

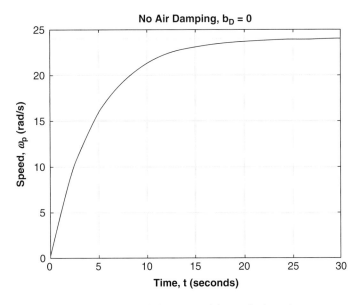

Figure 3.8.4 Propeller speed of the coupled system without air damping

3.9 Dynamic Responses via Mathematica

This section illustrates the use of Wolfram Mathematica for the simulation of dynamic responses of mechanical systems. It presents the implementation of the mathematical models of the dynamic systems presented in Examples 3.8.1 and 3.8.2 in Wolfram Language. One may refer to Appendix C for a tutorial on Wolfram Mathematica.

Example 3.9.1

Repeat Example 3.8.1 by using Mathematica.

Solution

From Example 3.8.1, the s-domain displacement and resultant force are given by

$$X(s) = \frac{1}{ms^2 + cs + k} F(s) \quad \text{(a)}$$

$$F_R(s) = (cs + k)X(s) \quad \text{(b)}$$

Simulation of the system response takes the following two steps:

(i) implement the expressions in Eqs. (a) and (b) by using the built-in function `TransferFunctionModel`; and
(ii) generate a system response by using the built-in function `OutputResponse`.

The code for the previous steps is shown in Figure 3.9.1.

Note the use of a replacement operator instead of a numerical value assignment. This is done in order to be able to print the generated transfer functions in symbolic form and verify them with the hand-derived results.

Since the governing second-order ordinary differential equation is linear with constant coefficients, a closed-form solution is expected to be generated by `OutputResponse`. This solution can be shown by using the `Print` statement, as seen in Figure 3.9.2.

Plotting the generated responses is straightforward with the built-in function `Plot`. Formatting the produced graphics is done by using the available options of the `Plot` function; see Figure 3.9.3. Refer to Wolfram Mathematica documentation for a detailed description of the shown functions and plot-modifying options.

As expected, the computed values correspond to the depiction of responses on the graphs and they are the same as obtained with MATLAB in Example 3.8.1.

3.9 Dynamic Responses via Mathematica

```
ClearAll["Global`*"];
(* create a transfer function X(s)/F(s) *)
tfmx = TransferFunctionModel[{{1/(m*s^2 + c*s + k)}}, s];

(* create a transfer function F_R(s)/F(s) *)
tfmFr = TransferFunctionModel[{{(c*s + k)/(m*s^2 + c*s + k)}}, s];

(* generate the response x(t) *)
solX =
  (OutputResponse[tfmx /. {m → 20, c → 80, k → 750}, 15*UnitStep[t], t] //
     FullSimplify)[[1]];

(* generate the response F_R(t) *)
solFr =
  (OutputResponse[tfmFr /. {m → 20, c → 80, k → 750}, 15*UnitStep[t], t] //
     FullSimplify)[[1]];
```

Figure 3.9.1 Wolfram Language code for creating the required transfer functions and generating the required responses

```
(* print the transfer functions and response expressions *)
Print["Transfer function: X(s)/F(s) = ", tfmx,
  "\nTransfer function: F_R(s)/F(s) = ", tfmFr,
  "\n Response: x(t) = ",
  TraditionalForm[solX /. UnitStep[t] → Style["u(t)", Italic]],
  "\n Response: f_R(t) = ",
  TraditionalForm[solFr /. UnitStep[t] → Style["u(t)", Italic]]]
```
(a)

Transfer function: $\dfrac{X(s)}{F(s)} = \left(\dfrac{1}{k + c\,s + m\,s^2}\right)\mathcal{T}$

Transfer function: $\dfrac{F_R(s)}{F(s)} = \left(\dfrac{k + c\,s}{k + c\,s + m\,s^2}\right)\mathcal{T}$

Response: $x(t) = \dfrac{u(t)\left(67 - e^{-2t}\left(2\sqrt{134}\sin\left(\sqrt{\dfrac{67}{2}}\,t\right) + 67\cos\left(\sqrt{\dfrac{67}{2}}\,t\right)\right)\right)}{3350}$

Response: $f_R(t) = \dfrac{15}{67}u(t)\left(e^{-2t}\left(2\sqrt{134}\sin\left(\sqrt{\dfrac{67}{2}}\,t\right) - 67\cos\left(\sqrt{\dfrac{67}{2}}\,t\right)\right) + 67\right)$

(b)

Figure 3.9.2 Printing implemented transfer functions and responses: (a) code; (b) output

```
In[*]:= (* plotting the response x(t) *)
Plot[solX, {t, 0, 3}, PlotRange → {{0, 3}, {0, 0.03}}, PlotStyle → {Black, Thick},
 Frame → True, GridLines → Automatic, AspectRatio → 1/1.75,
 FrameTicks → {{Automatic, None}, {Automatic, None}},
 FrameLabel → {Style["t (seconds)", 14], Style["x (m)", 14], None, None},
 PlotLabel → Style["Displacement", 14]]
```

(a)

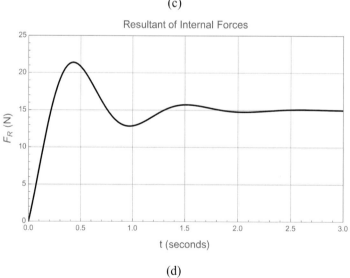

(b)

```
In[*]:= (* plotting the response fR(t) *)
Plot[solFr, {t, 0, 3}, PlotRange → {{0, 3}, {0, 25}}, PlotStyle → {Black, Thick},
 Frame → True, GridLines → Automatic, AspectRatio → 1/1.75,
 FrameTicks → {{Automatic, None}, {Automatic, None}},
 FrameLabel → {Style["t (seconds)", 14], Style["F_R (N)", 14], None, None},
 PlotLabel → Style["Resultant of Internal Forces", 14]]
```

(c)

(d)

Figure 3.9.3 Generating the required plots: (a) code for the plot of displacement vs. time; (b) plot of the displacement vs. time; (c) code for the plot of resultant of internal forces vs. time; (d) plot of resultant of internal forces vs. time

Example 3.9.2

Repeat Example 3.8.2 by Mathematica.

Solution
The coupled motor–propeller system is implemented by using the built-in Mathematica function `NonlinearStateSpaceModel`. To this end, a nonlinear state-space model of the coupled system, which is described by Eqs. (b), (d) and (e) in Example 3.7.3, is used. Implementation of the state-space model in Wolfram Language is shown in Figure 3.9.4.

The specified responses: the propeller rotation speed and the spring torque – are computed by using the `OutputResponse` function, and graphs are generated and formatted to the desired appearance by using `Plot` with its options. The code and its output are shown in Figure 3.9.5.

If desired, we can zoom in on the generated graphs to observe the high-frequency oscillations due to the elastic coupling between the motor shaft and the propeller shaft, as shown in Figure 3.9.6.

To see the effect of air damping, generate a response from the implemented state-space model setting $b_D = 0$. Since the model was created in a symbolic form, with assignment of values to variables that are not expected to change, we can use the replacement syntax to compute the response for a system with no nonlinear air damping. Plotted in Figure 3.9.7 is the propeller speed of the coupled system with and without air damping. Here, the words "linear system" refer to the coupled system without damping, which is seen from Eq. (a) in Example 3.7.3, and the words "nonlinear system" refer to the coupled system with air damping.

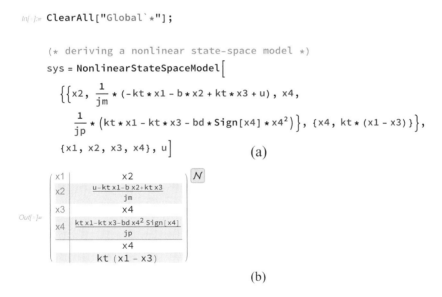

Figure 3.9.4 Implementation of the nonlinear state-space model developed in Example 3.7.3: (a) code; (b) model matrix

```
In[•]:= (* set up the numeric variables *)
jm = 1.2; b = 0.5; kt = 5000; jp = 1.1;

(* generate the response *)
res = OutputResponse[sys /. bd → 0.02, 12*UnitStep[t], {t, 0, 30}];

(* plot θ̇p(t) vs. time and τs(t) vs. time,
   formatting the graphics to achieve desired look *)
plNL = Plot[res[[1]], {t, 0, 30}, PlotRange → {{0, 30}, {0, 16}},
  PlotStyle → {Black, Thick}, Frame → True,
  GridLines → {{0, 5, 10, 15, 20, 25, 30}, {0, 2, 4, 6, 8, 10, 12, 14, 16}},
  AspectRatio → 1/2,
  FrameTicks → {{{0, 2, 4, 6, 8, 10, 12, 14, 16}, None}, {Automatic, None}},
  FrameLabel → {Style["Time, t (seconds)", 14], Style["Speed, ωp (rad/s)", 14],
    None, None}]

Plot[res[[2]], {t, 0, 30}, PlotRange → {{0, 30}, {-2, 12}},
  PlotStyle → {Black, Thick}, Frame → True,
  GridLines → {{0, 5, 10, 15, 20, 25, 30}, {-2, 0, 2, 4, 6, 8, 10, 12}},
  AspectRatio → 1/2,
  FrameTicks → {{{0, 2, 4, 6, 8, 10, 12}, None}, {Automatic, None}},
  FrameLabel → {Style["Time, t (seconds)", 14],
    Style["Spring Torque, τs (N.m)", 14], None, None}]
```

(a)

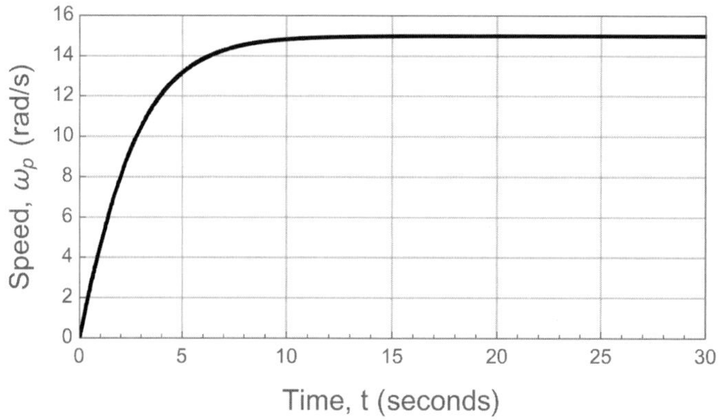

(b)

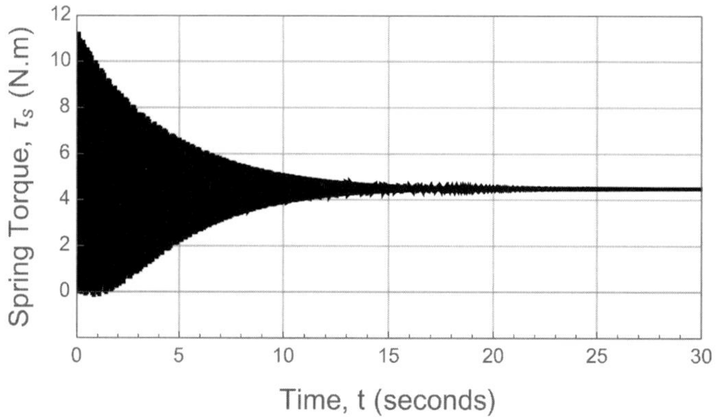

(c)

Figure 3.9.5 Dynamic response of the coupled motor–propeller system: (a) code; (b) the propeller speed; and (c) the spring torque

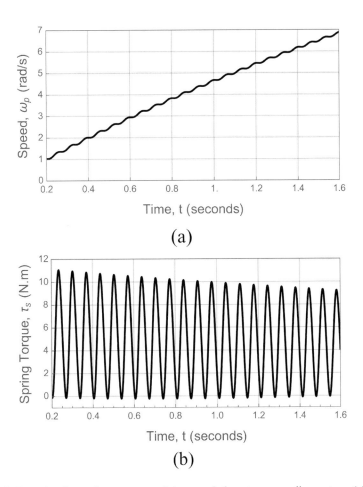

Figure 3.9.6 Zooming in on the response of the coupled motor–propeller system: (a) the propeller speed; (b) the spring torque

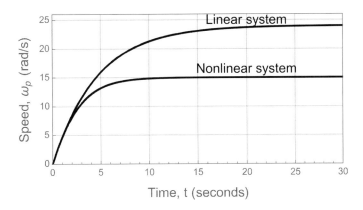

Figure 3.9.7 Comparison of the propeller speed of the coupled system: (a) nonlinear system – the coupled system with air damping; and (b) linear system – the coupled without air damping

CHAPTER SUMMARY

This chapter introduces a systematic method for the derivation of equations of motion for general mechanical systems. In this modeling method, three important keys, namely Newton's laws, basic elements, and free-body diagrams, are emphasized. After a mathematical model is developed, it is converted to two forms of model representation: transfer-function formulation and state-space representation. These model representations are useful for simulation and analysis. In addition, an energy method with Lagrange's equations is introduced for modeling of mechanical systems. Finally, the utility of MATLAB and Mathematica to simulate dynamic responses of mechanical systems is illustrated.

Upon completion of this chapter, you should be able to:

(1) Understand and use the three important keys in modeling of mechanical systems.
(2) Derive equations of motion for spring–mass–damper systems, translational and rotational systems, rigid-body systems in plane motion, and coupled multi-body systems.
(3) Establish transfer function formulations and state-space representations for given equations of motion.
(4) Apply Lagrange's equations to derive the equations of motion of multi-body systems.
(5) Compute the dynamic response of mechanical systems by MATLAB and Mathematica.

REFERENCES

1. F. P Beer, E. R. Johnston, Jr., P. Cornwell, and B. Self, *Vector Mechanics for Engineers: Dynamics*, 11th ed., McGraw-Hill Education, 2015.
2. J. L. Meriam, L. G. Kraige, and J. N. Bolton, *Engineering Mechanics: Dynamics*, 8th ed., Wiley, 2015.

PROBLEMS

Section 3.1 Fundamental Principles of Mechanical Systems

3.1 A truck of mass m moves up on a slope of angle α; see Figure P3.1, where x is the displacement of the truck, f is a pulling force generated by the vehicle engine, and g is the gravitational acceleration. Assume that $f(t) = f_0 e^{-\sigma t} + mg \sin \alpha$, where f_0 is a constant. Let the initial displacement and velocity of the truck be x_0 and v_0, respectively

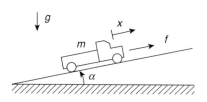

Figure P3.1

(a) Derive the differential equation of motion of the car.
(b) Solve the governing equation of part (a) for the car displacement x.
(c) Consider the parameters of the car: $m = 2000$ kg, $\alpha = 5°$, $f_0 = 1500$ N, $\sigma = 0.3/s$, and $v_0 = 3$ m/s. Plot the velocity of the car for $0 \le t \le 20$ s.

3.2 In Figure P3.2, under the influence of gravity, a ball of mass m slides down on a smooth slope ($H > h$), starting at point A with zero velocity. After leaving point B of the slope, the ball is in a free fall, and it eventually lands on the floor at point C.

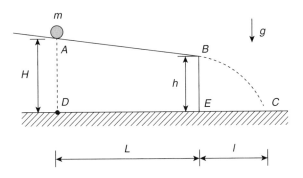

Figure P3.2

(a) Write down the kinetic energy and potential energy of the ball at points A and B. You may take point D as a reference point for potential energy.
(b) For $H = 2$ m, $h = 1.5$ m and $L = 3$ m, determine the horizontal and vertical components of the ball velocity at point C, and compute the distance l between point E and point C.

3.3 In Figure P3.3, a soccer ball of mass m at point O is kicked with an angle θ. The kicking force, which acts at the ball in a very short period of time, can be described by $f(t) = I_0 \delta(t)$, where $\delta(t)$ is the Dirac delta function, and I_0 is the impulse magnitude. Consider the coordinate system Oxy, where the ball is initially at the origin O. Point A where the ball lands on the floor has the coordinates $x = x_A$ and $y = 0$. Ignore aerodynamic effects.

Figure P3.3

(a) Determine the horizontal distance x_A of the ball.

(b) Determine the angle θ such that the ball drops at the farthest point.

(c) Consider the parameter values: $m = 0.4$ kg, $\theta = 43.5°$, $I_0 = 16.8$ kg · m/s. Plot the trajectory of the ball in the xy-plane and determine the distance x_A.

Section 3.2 Translational Systems

3.4 Consider the mechanical system in Figure P3.4.

(a) Draw free-body diagrams for the mass and auxiliary plots for the spring and damping elements.

(b) Derive the governing equations of motion for the system.

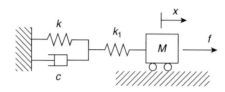

Figure P3.4

3.5 Consider the mechanical system in Figure P3.5, where y is a specified displacement.

(a) Draw free-body diagrams for the mass elements and auxiliary plots for the spring and damping elements.

(b) Derive the governing equations of motion for the system.

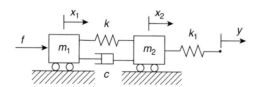

Figure P3.5

3.6 Consider the mechanical system in Figure P3.6, where f and q are external forces.

(a) Draw free-body diagrams for the mass elements and auxiliary plots for the spring and damping elements.

(b) Derive the governing equations of motion for the system.

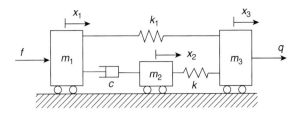

Figure P3.6

3.7 Consider the mechanical system in Figure P3.7, where there exists dry friction (μ_D) between mass m_1 and the slope.

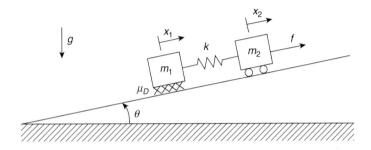

Figure P3.7

(a) Draw free-body diagrams for the mass elements and auxiliary plot for the spring element.
(b) Derive the governing equations of motion for the system.

3.8 Consider the mechanical system in Figure P3.8, where viscous friction (B) exists between m_1 and m_2, and dry friction (μ_D) exists between m_3 and the ground.
(a) Draw free-body diagrams for the mass elements and auxiliary plots for the spring and damping elements.
(b) Derive the governing equations of motion for the system.

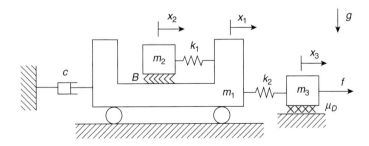

Figure P3.8

3.9 Consider the mechanical system in Figure P3.9, where f and p are external forces; viscous friction (B) exists between m_1 and m_2; and a massless pulley changes the direction of the tension force of an unstretchable rope, which suspends a weight (M).

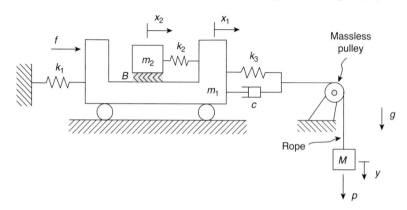

Figure P3.9

(a) Draw free-body diagrams for the mass elements and auxiliary plots for the spring and damping elements.
(b) Derive the governing equations of motion for the system.

3.10 Consider a quarter-car model shown in Figure P3.10, where m_1, m_2, and m_3 represent the masses of the wheel–tire–axle assembly, car body, and engine, respectively; springs k, k_1, and k_2 model the elasticity of the tire, suspension, and engine mounts, respectively; dampers c_1 and c_2 describe the energy dissipation of the suspension and engine mounts, respectively; and y_o is an displacement input describing the condition of road surface.

(a) Draw free-body diagrams for the mass elements and auxiliary plots for the spring and damping elements.
(b) Derive the governing equations of motion for the car.

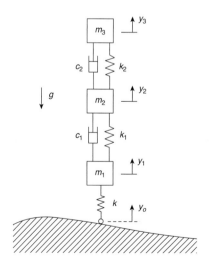

Figure P3.10

3.11 In Figure P3.11, a spring–mass–damper system is connected to a massless rod of length L and longitudinal rigidity EA. By using Table 3.2.1, develop a mathematical model for the system.

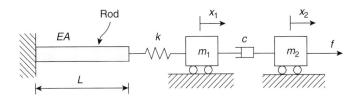

Figure P3.11

3.12 A massless cantilever beam of length L and bending stiffness EI carries a spring–mass–damper subsystem at the tip; see Figure P3.12, where p is an external force. The combined beam–mass system can be used as a simplified model of airplane wings carrying engines or crane arms lifting loads. Consider gravity. Develop a mathematical model of the combined system. Refer to Table 3.2.1 for the effective spring coefficient of the beam.

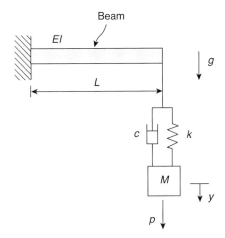

Figure P3.12

3.13 For the system in Figure P3.4, derive the transfer function with force f as the input and velocity $\dot{x}$ as the output. Also, state the order of the transfer function.

3.14 For the system in Figure P3.5, derive the following two transfer functions:

(a) The transfer function $G_1(s)$ with displacement y as the input and velocity $\dot{x}_1$ as the output.

(b) The transfer function $G_2(s)$ with force f as the input and displacement x_2 as the output.

Section 3.3 Rotational Systems

3.15 Power transmission belts are commonly seen in automobiles and machines. Shown in Figure P3.15(a) is a belt–drive system, in which a looped belt of flexible material links two pulleys. Driven by the drive pulley (connected to a motor), the belt brings the driven pulley to rotate. The inertia of the belt, compared to that of the pulleys, is negligible. Therefore, the belt can be viewed as two springs that connect the pulleys; see Figure P3.15(b), where τ is an external torque applied to the drive pulley.

(a) Draw free-body diagrams for the mass elements and auxiliary plots for the spring elements.
(b) Derive the governing equations of motion for the belt–drive system.
(c) Determine the transfer function with τ as the input and $\dot{\theta}_2$ as the output. Also, state the order of the transfer function.

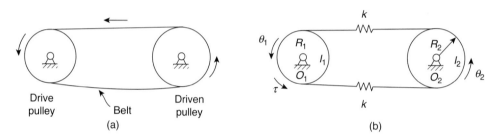

Figure P3.15

3.16 A disk–shaft system is shown in Figure P3.16(a), where a torque τ is applied to the left disk. A model of the system is shown in Figure P3.16(b), where the shafts are modeled as torsional springs (k_1, k_2); the mass moments of inertia of the disks are denoted by I_1 and I_2; and double-headed arrows are used to describe the torque and disk rotations.

(a) Draw free-body diagrams for the mass elements and auxiliary plots for the spring elements.
(b) Derive the governing equations of motion for the disk–shaft system.
(c) Derive the transfer function with τ as the input and the internal torque of spring k_2 as the output.

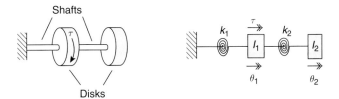

Figure P3.16

3.17 In Figure P3.17, an L-shaped rigid bar is pivoted at point O and constrained by a spring (k) at tip A and a damper (c) at tip B. The mass moment of inertia of the bar with respect to O is I_O. An external (horizontal) force f is applied to the bar at point C. Assume a small rotation, $|\theta| \ll 1$. In this problem, gravity is not considered.

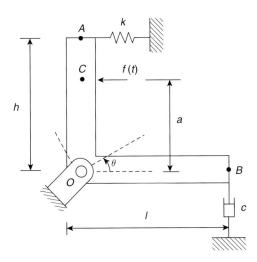

Figure P3.17

(a) Derive the linear differential equation governing the rotation θ of the bar.

(b) Determine the transfer function $\dfrac{\Theta(s)}{F(s)}$.

3.18 In Figure P3.18, a mass (M) is linked to a level by a spring (k_l), which is pinned at point O, is subject to a horizontal force f, and constrained by a viscous damper (c). Denote the mass moment of inertia of the level with respect to point O by I_O. Consider a small rotation of the level.

(a) Derive the linear differential equations of motion for the mass–level system.

(b) Determine the transfer function $\dfrac{X(s)}{F(s)}$. Also, state the order of the dynamic system.

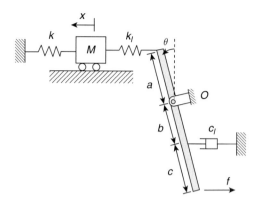

Figure P3.18

3.19 A coupled disk–mass system is shown in Figure P3.19, where p is an external force applied to the mass. In this problem, gravity is not considered.
  (a) Derive the governing equations of motion for the system.
  (b) Derive the transfer function, from force p to rotation θ.

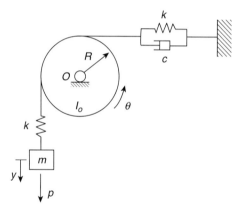

Figure P3.19

3.20 In Figure P3.20, a cart of mass M carrying a single pendulum (m_b, L) moves vertically. Constrained by a spring and damper pair (k, c), the cart is subject to an external force f.
  (a) Draw free-body diagrams for the mass elements.
  (b) Derive the nonlinear equations of motion for the cart–pendulum system.
  (c) With the results from part (b), obtain two independent differential equations of motion about y and θ.
  (d) Determine the expression of the tension force of the pendulum in terms of y and θ.

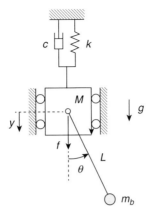

Figure P3.20

3.21 In Figure P3.21, two rotating shafts are coupled by a torsional spring (k). A torque τ is applied to the left shaft and a torsional damper (c) is connected to the right shaft.
   (a) Draw free-body diagrams for the mass elements and auxiliary plots for the spring and damping elements.
   (b) Derive the governing equations of motion for the system.
   (c) Determine the transfer function with the torque as the input and the rotation speed of the right shaft as the output.

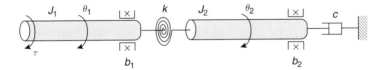

Figure P3.21

3.22 In Figure 3.22(a), a rotating shaft is coupled to a pulley by a torsional spring (k) and the pulley suspends a mass (M) by an unstretchable rope. An external torque τ is applied to the shaft. The side view of the pulley is shown in Figure P3.22(b). Consider gravity.
   (a) Derive the governing equations of motion for the combined shaft–pulley–mass system.
   (b) Determine the transfer function from the torque τ to the displacement y of the mass.

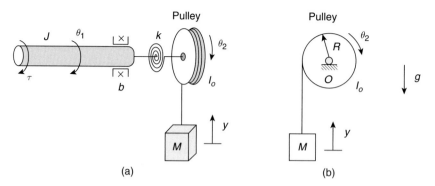

Figure P3.22

3.23 Consider the geared system in Figure P3.23, where double-headed arrows are used to describe torque τ and shaft rotations θ_1 and θ_2.
  (a) Draw free-body diagrams for the mass elements and auxiliary plots for the spring and gear pair.
  (b) Derive two independent equations of motion for the geared system.
  (c) Determine the transfer function from torque τ to rotation speed $\dot{\theta}_2$. Also, state the order of the geared system.

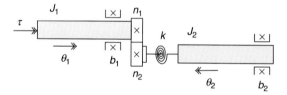

Figure P3.23

3.24 In Figure P3.24(a), two shafts are coupled by a compound gear train. The gear train consists of two pairs of gears, as shown in Figure P3.24(b). The right shaft experiences a drag torque (b_D), which is described by Eq. (3.3.12).
  (a) Draw free-body diagrams for the mass elements and auxiliary plots for the gear train.
  (b) Derive the nonlinear equation of motion for the geared system in terms of $\dot{\theta}_2$.

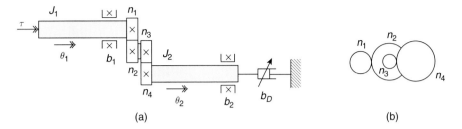

Figure P3.24

Section 3.4 Rigid-Body Systems in Plane Motion

3.25 A rigid body is mounted on two springs at its lower corners; see Figure P3.25, where G is the center of mass of the body, a force f is applied at the upper-right corner A of the body, y_G is the vertical displacement of the mass center, and θ is the rotation angle of the body. Assume that the body moves in the vertical direction, and rotates with small angle, $|\theta| \ll 1$. The springs are undeformed if both y_G and θ are zero. Gravity is not considered. Derive the equations of motion for the elastically supported rigid body, in terms of y_G and θ.

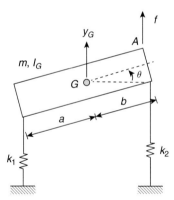

Figure P3.25

3.26 In Figure P3.26, a mass (m) is suspended from a movable pulley (m_p, I_G) that is elastically supported by a spring (k_p). The rope suspending the mass m is unextendible and it is connected to a spring and damper pair (k, c). The spring and damping elements are undeformed when both y_G and θ are zero. Consider gravity. Derive two independent equations of motion for the coupled pulley–mass system about y_G and θ.

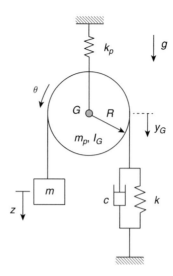

Figure P3.26

3.27 Consider the pulley–mass system in Figure P3.27, where two masses (m_1, m_2) are suspended from a movable pulley (m_p, I_G) that is elastically supported by a spring (k_p). The masses are also constrained by a spring and damper pair (k, c). The spring and damping elements are undeformed when both y_G and θ are zero. Consider gravity. Derive two independent equations of motion for the coupled pulley–mass system about y_G and θ.

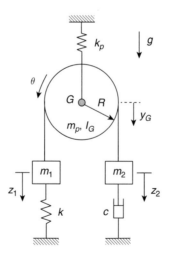

Figure P3.27

3.28 A cylinder rolls on a slope without slipping, as shown in Figure P3.28, where f is an external force, G is the mass center of the cylinder, A is the cylinder–slope contact point, and x is the displacement of the mass center. The spring is undeformed when x is zero. Let m and I_G be the mass and mass moment of inertia with respect to G, respectively. Derive the equation of motion of the system.

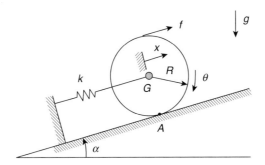

Figure P3.28

3.29 In Figure P3.29, a cart carrying a cylinder is pulled by a force f. The cylinder rolls on the cart without slipping and it is constrained by a spring and damper pair (k, c). Denote the mass and mass moment of inertia with respect to the mass center G of the body by m and I_G, respectively. The spring and damper are undeformed when the rotation angle θ of the cylinder is zero. Derive the equations of motion of the system in terms of x and θ.

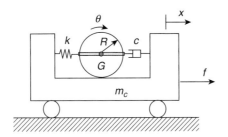

Figure P3.29

3.30 A nonuniform rigid bar is suspended by two strings of equal length l; see Figure P3.30, where a horizontal force f is applied to the bar at point A; the string attachment points A and B and the mass center G of the bar are on a straight line. The distance between points O and O' is the same as the length of the bar; namely, $\overline{OO'} = a + b$. With the geometric configuration, the bar in motion does not have rotation. Denote the mass and mass moment of inertia with respect to the mass center G of the body by m and I_G, respectively. Derive the equation of motion of the bar in θ. Also, determine the tension forces of the strings.

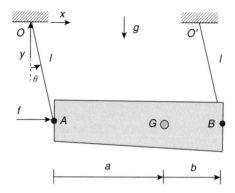

Figure P3.30

3.31 In Figure P3.31, a uniform rigid bar is pinned at point O of a massless cart, which is elastically supported by a spring (k). The cart can only move vertically, and it is subject to an external force f. The spring is undeformed when the cart displacement y is zero. Derive the equations of motion for the system.

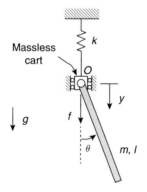

Figure P3.31

3.32 A uniform rigid bar of mass m and length l is suspended by a string of length a; see Figure P3.32, where Oxy is a Cartesian coordinate system, and G is the mass center of the bar. Derive the equations of motion for the string–bar system in terms of β and θ.

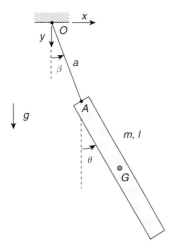

Figure P3.32

Section 3.5 Energy Approach

3.33 Consider the mechanical system in Figure P3.9.

  (a) Write down the kinetic energy, potential energy, and Rayleigh dissipation function of the system.
  (b) Derive the equations of motion by Lagrange's equations.

3.34 Consider the cart–pendulum system in Figure P3.20.
  (a) Write down the kinetic energy, potential energy, and Rayleigh dissipation function of the system.
  (b) Derive the equations of motion by Lagrange's equations.

3.35 Consider the elastically supported rigid body in Figure P3.25, without gravity.
  (a) Write down the kinetic energy, potential energy, and Rayleigh dissipation function of the system.
  (b) Derive the equations of motion by Lagrange's equations.

3.36 Consider the pulley–mass system in Figure P3.27.
  (a) Write down the kinetic energy and potential energy of the system.
  (b) Derive the equations of motion by Lagrange's equations.

3.37 Consider the string-suspended rigid body in Figure P3.30.
  (a) Write down the kinetic energy and potential energy of the system.
  (b) Derive the equations of motion by Lagrange's equations.

Section 3.6 Block Diagrams of Mechanical Systems

3.38 Draw a block diagram for the spring–mass–damper system Figure P3.5, with f and y as the inputs, x_1 and x_2 as the outputs.

3.39 Consider the mechanical system in Figure P3.9. By ignoring gravity, draw a block diagram with p as the input, and y and the spring force of k_2 as the outputs.

3.40 For the belt–drive system in Figure P3.15, draw a block diagram, with the toque τ as the input, and the rotation speed $\dot{\theta}_2$ and the force of the upper spring as the outputs.

3.41 For the disk–mass system in Figure P3.19, draw a block diagram, with the force p as the input, and the rotation speed $\dot{\theta}$ and the displacement y as the outputs.

3.42 For the geared system in Figure P3.23, draw a block diagram, with the torque τ as the input, and the rotation speed $\dot{\theta}_2$ and the torque of the spring k as the outputs.

3.43 For the elastically supported rigid body in Figure P3.25, draw a block diagram, with the force f as the input, and the internal forces of the springs (k_1, k_2) as the outputs

3.44 For the combined cart–cylinder system in Figure P3.29, draw a block diagram, with force f as the input, and the rotation speed $\dot{\theta}$ and the spring force as the outputs.

Section 3.7 State-Space Representations

3.45 For the system in Figure P3.5, develop a state-space model (state equations and output equations), with f and y being the inputs and x_1 and x_2 the outputs. Also, cast the state and output equations in matrix form.

3.46 Consider the system in Figure P3.6. Let forces f and q be inputs. Let the displacements of the masses m_1 and m_2 and the internal force of the damper c be outputs. Obtain a state-space representation (state and output equations) for the system. Also, cast the state and output equations in matrix form.

3.47 For the quarter-car model in Figure P3.10, develop a state-space model, with y_o as the input, the engine displacement (y_3) and the resultant force of the engine mounts as the outputs. For this problem, gravity is ignored. Also, cast the state and output equations in matrix form.

3.48 For the belt–drive system in Figure P3.15, develop a state-space model, with torque τ as the input, and rotation speed $\dot{\theta}_2$ and the internal force of the upper spring as the outputs. Also, cast the state and output equations in matrix form.

3.49 Consider the coupled disk–mass system in Figure P3.19. Ignore gravity. Develop a state-space model, with force p as the input, and the rotation θ of the disk and the internal force of the spring connecting the mass and disk as the outputs. Also, cast the state and output equations in matrix form.

3.50 For the cart–pendulum system in Figure P3.20, develop a state-space model, with force f as the input, and displacement y and rotation θ as the outputs. Note that this is a nonlinear system.

3.51 Consider the combined shaft–pulley–mass system in Figure P3.22. Let the torque τ and gravitational force Mg be the inputs. Let the displacement of the mass and tension force of the rope be the outputs. Develop a state-space model for the system. Also, cast the state and output equations in matrix form.

3.52 Consider the elastically supported rigid body in Figure P3.25. Assume a small rotation of the body. Develop a state-space model for the system, with f as the input and the spring forces as the outputs. Also, cast the state and output equations in matrix form.

Sections 3.8 and 3.9 Simulation of Dynamic Response via Software

3.53 Consider the spring–mass–damper system in Figure P3.5, with the parameter values given by $m_1 = 20$ kg, $m_2 = 30$ kg, $k = 2000$ N/m, $k_1 = 3000$ N/m, $c = 75$ N·s/m. Assume zero initial conditions.
  (a) Assume that $f = 0$ and $y = 0.5$ m (step input). Plot the displacement x_2 versus time by MATLAB.
  (b) Assume that $f = 45\delta(t)$ N·s (impulse input) and $y = 0$, where $\delta(t)$ is the Dirac delta function. Note that the force of the spring k_1 is zero in this case. Plot the displacement x_2 versus time by MATLAB.

  Hint: In parts (a) and (b), you can obtain the corresponding transfer functions first.

3.54 Consider the belt–drive system in Figure P3.15, with the following non-dimensional parameters as follows: $I_1 = 20$, $I_2 = 35$, $k = 600$, $R_1 = 1$ and $R_2 = 1.5$. Assume zero initial conditions. Consider a step input $\tau = 40$. Use MATLAB to plot the following two outputs versus time: (a) the rotation speed $\dot{\theta}_2$ of the driven pulley; and (b) the force of the upper spring. Hint: You can derive the corresponding transfer functions first.

3.55 Consider the elastically supported rigid body in Figure P3.25, with the assigned parameter values: $m = 50$ kg, $I_G = 32$ kg·m², $k_1 = 1200$ N/m, $k_2 = 1600$ N/m, $a = 1.2$ m, $b = 0.8$ m. Assume zero initial conditions. Consider an impulse force $f = 10\delta(t)$N·s. Use MATLAB to plot the time response of the system: $y_G(t)$ and $\theta(t)$, for $0 \le t \le 10$ s.

3.56 Consider the string-suspended rigid bar in Figure P3.30, with the parameters assigned as follows: $m = 40$ kg, $I_G = 18$ kg·m², $l = 0.6$ m, $a = 0.6$ m, $b = 0.4$ m. The initial conditions of the system are given by $\theta(0) = \pi/9$ rad and $\dot{\theta}(0) = 0$.
  (a) Obtain a state-space model for this nonlinear system.
  (b) Let the applied force f at point A be $f = 20\sin(3t)$ N. With the model of part (a), plot the response $\theta(t)$ in degree versus time ($0 \le t \le 10$ s) by the MATLAB function ode45.
  (c) Plot $\theta(t)$ in degree versus time ($0 \le t \le 10$ s) for $f = 20\sin(16t)$ N and compare the result with that of part (b).

3.57 Solve Problem 3.53 by Mathematica.
3.58 Solve Problem 3.54 by Mathematica.
3.59 Solve Problem 3.55 by Mathematica.
3.60 Solve Problem 3.56 by Mathematica.

4 Electrical Systems

Contents

4.1	Fundamentals of Electrical Systems	223
4.2	Concept of Impedance	230
4.3	Kirchhoff's Laws	242
4.4	Passive-Circuit Analysis	244
4.5	State-Space Representations and Block Diagrams	258
4.6	Passive Filters	271
4.7	Active-Circuit Analysis	277
4.8	Dynamic Responses via Mathematica	297
Chapter Summary		306
References		306
Problems		307

Practically every modern engineered dynamic system has electrical components such as motors, sensors, controllers, or, at the very least, power sources. Therefore, an understanding of the physical processes occurring in the typical electrical circuits, and the ability to model behavior of an electrical subsystem are essential for anyone interested in dynamic systems.

Similarly to the treatment of mechanical systems, the three keys in modeling of electrical systems are identified as follows:

- fundamental principles, which are Kirchhoff's current and voltage laws, as presented in Section 4.3;
- basic elements, which include resistors, inductors, and capacitors, as described in Sections 4.1 and 4.2; and
- two ways of analysis, which are the loop method and the node method, as presented in Section 4.4.

By a chosen way of analysis (either the loop method or the node method), Kirchhoff's laws are applied to the basic elements, resulting in a mathematical model for the electrical system in consideration, which is a mixture of a set of differential and algebraic equations. This three-key modeling concept is illustrated throughout this chapter.

A slightly different way of analysis, based on the described three-key modeling concept, is presented in Section 4.5, where alternative models (state-space representations and block diagrams) of the electrical circuits are derived.

Furthermore, in Sections 4.6 and 4.7, Kirchhoff's laws and derived models of basic elements are used to derive mathematical models of passive filters as well as active electrical circuits and glean an insight into their operations. Numerical analysis of electrical circuits behavior is presented in Section 4.8.

4.1 Fundamentals of Electrical Systems

Electrical and electronic components constitute important building blocks of many dynamic systems. The most commonly encountered examples of electrical subsystems include power supplies, controllers, sensors, and motors. Understanding their time-dependent behavior becomes crucial for gaining an ability to model, analyze, design, and control a variety of dynamic systems.

4.1.1 Definitions

An electrical system is often called an *electrical circuit*. A circuit consists of interconnected *circuit elements* such as, for example, resistors, capacitors, inductors, or sources of energy. There exist two main groups of circuit elements: *active* and *passive*.

Passive elements, usually called *loads*, dissipate or temporarily store the energy supplied to the electrical circuit. They are represented by resistors, inductors, and capacitors.

Active elements are the *sources* that supply an electrical circuit with energy. Any type of the energy-generating device such as chemical battery, solar/photovoltaic cells assembly, thermocouple junction, or mechanical generator can serve as an active element in the electrical circuit. Conversion of some kind of energy, for example, chemical for a battery, thermal for a thermocouple, or mechanical for a generator, into electrical charge is the common trait of all sources.

The active elements are commonly modeled as *ideal sources* of either current or voltage. An *ideal current source* always supplies the specified current regardless of the voltage required by the circuit, while an *ideal voltage source* always provides the specified voltage regardless of the current generated in the circuit. A battery is considered an ideal voltage source even though its performance is influenced by the heat generated as current flows through the load circuit. If an impact of the produced heat cannot be ignored, the battery can be modeled as an ideal voltage source connected to an internal resistor.

An electrical circuit generally includes one or more active elements. The energy provided by a source is considered an *input* to the electrical system.

The electrical circuit theory uses *current* and *voltage* as primary variables to describe the behavior of any electrical circuit.

Physically, current is the flow of electrons inside the circuit. Mathematically, it is expressed as the time rate of change of electrical charge through a specific area:

$$i = \frac{dQ}{dt} \qquad (4.1.1)$$

where i is the current, t denotes time, and Q is the electrical charge, or number of electrons passing through the cross-section of a wire. Current is measured in *amperes* (A), and charge is measured in *coulombs* (C). A negative one coulomb is equivalent to the electrical charge in 6.2415×10^{18} electrons. Similarly, a positive one coulomb is equivalent to the electrical charge in 6.2415×10^{18} protons. It can also be defined as the amount of electrical charge transported in one second by a constant current of one ampere.

The charge, expressed in terms of current and time, is

$$Q(t) = \int i\,dt \qquad (4.1.2)$$

The positive direction of the current flow is defined as opposite to the actual movement of electrons inside the circuit. Hence, it can be visualized as a flow of positively charged particles.

Voltage is defined as the work needed to move a unit charge between two points in a circuit:

$$v = \frac{dW}{dQ} \qquad (4.1.3)$$

where v is the voltage, Q is the electrical charge, and W denotes the work. One joule of work performed moving one coulomb of charge along the electrical circuit yields the unit voltage called the *volt* (V).

The voltage between any two points in a circuit can be also defined as the difference between electric potentials of these points as shown in Figure 4.1.1, where e_1 and e_2 denote electric potentials at the entrance and the exit terminals of a circuit element respectively, i is the current flowing across this element, and v is the voltage computed as

$$v = e_1 - e_2 \qquad (4.1.4)$$

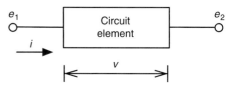

Figure 4.1.1 Electrical circuit terminology

The voltage at any point of an electrical circuit is represented as a difference between electric potentials of this point and some selected reference point. The *ground*, or point of zero electric potential, is commonly chosen as that reference point.

By convention, the element's high-voltage terminal is denoted by the "+" sign, while the low-voltage terminal is denoted by the "−" sign.

The sign of the voltage difference indicates whether the energy was generated or consumed by the circuit element. A negative voltage difference across an element corresponds to energy generation, while a positive difference indicates energy consumption. To illustrate that concept consider the direct current (DC) electrical circuit depicted in Figure 4.1.2.

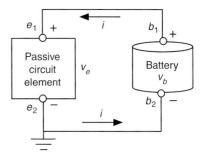

Figure 4.1.2 DC electrical circuit

The electrons traveling across this circuit are repelled by the negative terminal of the battery and attracted by its positive one. Thus, their motion is clockwise, and, by definition, the positive flow of the current becomes counterclockwise (as shown by the arrows). Let v_b and v_e be the respective voltage differences across the battery and across the passive circuit element. By definition, $v_b = b_2 - b_1$ and $v_e = e_1 - e_2$. The current gains energy inside the battery, flowing from the low (b_2) to high (b_1) voltage terminal, so the potential difference across the battery is negative: $v_b < 0$. On the other hand, the current loses energy inside the passive circuit element, which makes $v_e = e_1 - e_2 > 0$, and $e_1 > e_2$ (as shown in Figure 4.1.2). Customarily v_e is referred to as the voltage drop, and v_b – as the voltage gain.

According to the law of conservation of energy, in a closed circuit the energy supplied by the battery must be fully consumed by the circuit elements. Hence, $v_b + v_e = 0$.

While the voltage drop and voltage gain have different signs as shown above, the opinion on which one is considered positive differs from one literature source to another. In this book, we adopt the convention that assigns a positive sign to voltage gain and a negative sign to voltage drop.

In addition to voltage and current, electric circuits are characterized by *power*. The power applied to a passive element, or generated by an active element, is a product of voltage across that element and current flowing through it:

$$P = vi \qquad (4.1.5)$$

This expression can be easily derived using the definition of power as work performed in a unit time:

$$P = \frac{dW}{dt} = \frac{dW}{dt}\frac{dQ}{dQ} = \frac{dW}{dQ}\frac{dQ}{dt} = vi \qquad (4.1.6)$$

The unit of power is the *watt* (W), which corresponds to one joule of work performed in one second.

Accordingly, energy consumed or generated by an element is

$$W = \int_0^t P\,dt \qquad (4.1.7)$$

4.1.2 Basic Elements

A set of symbols, used to indicate active and passive elements on the electric circuit diagrams, is shown in Figure 4.1.3, where the symbols are as follows: (a) current source, (b) voltage source, (c) battery, (d) resistor, (e) inductor, (f) capacitor, (g) ground, and (h) terminals.

Understanding the voltage–current relations of individual circuit elements is paramount to the ability to model and analyze the behavior of any circuit.

Resistors

A *resistor* is a passive circuit element that dissipates energy. A simple light bulb is an example of a resistor, which dissipates electrical energy by converting it into light and heat. The behavior of a common resistor is governed by the empirically derived *Ohm's law*, which states that the electric current flowing through a resistor is directly proportional to the voltage difference across that resistor:

$$i = \frac{v}{R} \qquad (4.1.8)$$

where i is the current, v denotes voltage, and R is the *resistance*.

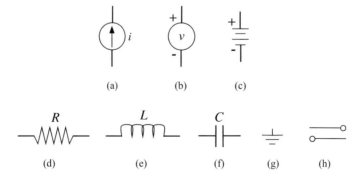

Figure 4.1.3 Basic elements of electrical circuits

Ohm's law specifies the linear relation between voltage and current through a resistor, and is often written in the form:

$$v = iR \qquad (4.1.9)$$

If a resistor obeys Ohm's aw, it is called *linear resistor*. The resistance of a linear resistor is a constant, the value of which depends on the resistor's material, geometry, and temperature. For example, the resistance of a nonideal wire can be expressed as:

$$R = \frac{\rho L}{A} \qquad (4.1.10)$$

where L and A denote length and cross-sectional area of the wire respectively, and ρ is the temperature-dependent resistivity of the wire material.

In this chapter all the resistors are considered linear, and all the connecting wires are assumed to be perfect conductors and their resistance is neglected, i.e. the voltage drop across an ideal wire is zero.

The unit of resistance is the *ohm* (Ω). A resistor with $R = 1\,\Omega$ experiences a voltage drop of 1 V when the current flowing through it equals 1 A.

The power dissipated by a linear resistor is

$$P = vi = \frac{v^2}{R} = i^2 R \qquad (4.1.11)$$

Capacitors

A capacitor is a passive circuit element that stores energy in the form of electrical charge.

The simplest capacitor circuit can be visualized as a pair of parallel plates (a capacitor) connected to a battery as shown in Figure 4.1.4. In this circuit the battery will transport charge from one plate to the other until the voltage difference that results from the charge build-up on the capacitor equals battery voltage v_b.

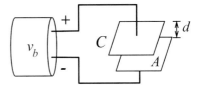

Figure 4.1.4 A capacitor

A capacitor is characterized by its *capacitance* – a measure of how much charge can be stored for a given voltage difference across the element. The mathematical expression of capacitance is

$$C = \frac{Q}{v} \qquad (4.1.12)$$

where Q is the stored charge, v denotes voltage, and C is the *capacitance*. The unit of capacitance is the *farad* (F), which corresponds to a one coulomb of charge stored by the capacitor experiencing a voltage drop of one volt.

For a parallel-plate capacitor, depicted in Figure 4.1.4, the capacitance depends on its geometry and the material such as:

$$C = \frac{\varepsilon A}{d} = \frac{k\,\varepsilon_0 A}{d} \tag{4.1.13}$$

where A is the area of the flat parallel metallic plates of the capacitor, k is the relative permittivity of the dielectric material between the plates ($k \approx 1$ for the air), $\varepsilon_0 = 8.854 \times 10^{-12}$ F/m denotes the free-space permittivity, d is the distance between plates, and $\varepsilon = \varepsilon_0 k$ is the permittivity.

For a capacitor, the relation between a current through it and a voltage difference across it is not linear. According to Eq. (4.1.12), the capacitor voltage is

$$v = \frac{Q}{C} \tag{4.1.14}$$

Substituting the expression for charge, defined by Eq. (4.1.2), we obtain

$$v = \frac{1}{C}\int i\,dt = \frac{Q_0}{C} + \frac{1}{C}\int_0^t i\,dt \tag{4.1.15}$$

where Q_0 is a charge on the capacitor at the time $t = 0$. For simplicity of computations, it is assumed that $Q_0 = 0$. Hence, the relation between capacitor voltage and current is expressed as:

$$\begin{aligned}\text{Integral form} \quad & v = \frac{1}{C}\int_0^t i\,dt \\ \text{Derivative form} \quad & i = C\frac{dv}{dt}\end{aligned} \tag{4.1.16}$$

The energy stored in a capacitor is derived by integrating the expression for power (Eq. (4.1.5)) as stated in Eq. (4.1.7). Zero initial conditions are assumed:

$$W = \int_0^t P\,dt = \int_0^t v i\,dt = \int_0^t v\left(C\frac{dv}{dt}\right)dt = \frac{1}{2}Cv^2(t) \tag{4.1.17}$$

Inductors

An inductor is a passive circuit element that stores energy in the form of a magnetic field.

The simplest inductor circuit can be visualized as a solenoid – a coil of wire wound into a tight helix, which may or may not be wrapped around a ferromagnetic core (inductor), connected to a source, as shown on Figure 4.1.5. An electric current is always accompanied by a magnetic field – *flux* – around the conductor. In the case of a solenoid conductor this flux is concentrated almost entirely inside the solenoid, while the outside field is weak. The longer the solenoid, the more pronounced is this effect of a nearly uniform flux inside the solenoid and negligible one on the outside.

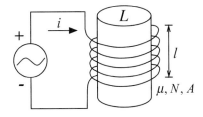

Figure 4.1.5 An inductor

According to Faraday's law (see Section 6.3.1 for more detailed information), the relation between flux and the current flowing through the inductor is

$$\phi = Li \tag{4.1.18}$$

where ϕ is the flux, i denotes current, and L is the inductance. The inductance is a constant that depends on solenoid geometry such as the length l, cross-sectional area of a coil A, and the number of coils N, as well as on the material of the solenoid and its core, if one is present. The mathematical expression for the induction of the solenoid, shown in Figure 4.1.5, is as follows:

$$L = \frac{\mu N^2 A}{l} \tag{4.1.19}$$

where μ denotes the magnetic permeability of the solenoid. The unit of inductance is the henry (H).

The relation between flux and the voltage is expressed in integral form:

$$\phi = \int v dt \tag{4.1.20}$$

Equations (4.1.18) and (4.1.20) can be used to derive the expression for the voltage–current relation of an inductor.

$$\text{Integral form} \quad i = \frac{1}{L} \int_0^t v dt \tag{4.1.21}$$

$$\text{Derivative form} \quad v = L \frac{di}{dt}$$

The energy stored in an inductor is derived using the Eqs. (4.1.5), (4.1.7), and the derivative form of Eq. (4.1.21). Zero initial conditions are assumed:

$$W = \int_0^t P dt = \int_0^t v i dt = \int_0^t i \left(L \frac{di}{dt} \right) dt = \frac{1}{2} L i^2(t) \tag{4.1.22}$$

Active Elements

Active circuit elements include energy sources described in Section 4.1.1, and electronic devices such as integrated circuits, which include amplifiers, transistors, and logic gates. While a detailed discussion on integrated circuits is beyond the scope of this book, amplifiers will be presented later in this chapter (see Section 4.7).

4.2 Concept of Impedance

By consuming energy, passive circuit elements "resist" the flow of current. This opposition to a current is called *impedance*, and is represented by a voltage drop across the element. Impedance is often viewed as a generalization of the resistance concept, and as an analog of a transfer function that takes current as an input and yields voltage as an output, as shown in Figure 4.2.1. Impedance is typically expressed in the Laplace domain, and is derived by taking the Laplace transform of the time domain expression of the element's voltage–current relationship,

Figure 4.2.1 Impedance

4.2.1 Definition

Impedance is defined as the ratio of the voltage drop across the circuit element to the current flowing across that element. Traditionally it is expressed in the Laplace domain as:

$$Z = \frac{\mathcal{L}\{v(t)\}}{\mathcal{L}\{i(t)\}} = \frac{V(s)}{I(s)} \tag{4.2.1}$$

The relations between current and voltage for the described earlier passive circuit elements are used to derive their impedance expressions.

A resistor's impedance is obtained by taking the Laplace transform of Eq. (4.1.9):

$$\mathcal{L}\{v\} = V(s) = \mathcal{L}\{i\,R\} = RI(s) \tag{4.2.2}$$

Then

$$Z_{RESISTOR} = \frac{V(s)}{I(s)} = R \tag{4.2.3}$$

Using the derivative form of Eq. (4.1.16), the capacitor's impedance is obtained:

$$\mathcal{L}\{i\} = I(s) = \mathcal{L}\left\{C\frac{dv}{dt}\right\} = C s\, V(s)$$

$$Z_{CAPACITOR} = \frac{1}{C s} \tag{4.2.4}$$

For an inductor the derivative form of the Eq. (4.1.21) is used:

$$\mathcal{L}\{v\} = V(s) = \mathcal{L}\left\{L\frac{di}{dt}\right\} = L\,s\,I(s)$$

$$Z_{INDUCTOR} = L\,s \qquad (4.2.5)$$

4.2.2 Combination of Impedances

A typical electrical circuit contains interconnected multiple passive and active elements. Knowledge of the elements' arrangement is important for the correct modeling of the circuit since different ways of connecting the same elements yield differently behaving circuits.

Circuit analysis typically pursues two goals, the first one being the derivation of the voltage–current relation for the circuit as an entity, and the second one being the computation of the voltages across and the currents through every component of the circuit. The former task is more easily fulfilled when the circuit in question is transformed into an equivalent

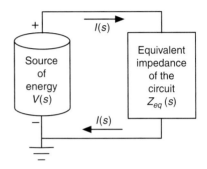

Figure 4.2.2 Equivalent impedance of a circuit

form, as shown in Figure 4.2.2. Then, the computation of currents and voltage drops for every passive circuit element is straightforward with the use of the previously presented voltage–current relations for resistors, capacitors, and inductors.

To find the impedance of a circuit we first need to derive the rules for computing equivalent impedance for two interconnected elements. Obviously, this equivalent impedance depends on the elements' arrangement. The individual impedances can be connected in *series* or in *parallel*, as shown in Figure 4.2.3.

Series Connection

A closed electrical circuit that contains a power source and two loads connected in series, with impedances Z_1 and Z_2 respectively, is presented in Figure 4.2.3(a). According to the law of conservation of charge, the currents flowing through impedances Z_1 and Z_2 are the same: $I_1(s) = I_2(s) = I(s)$. Also, according to the law of conservation of energy the algebraic sum of voltages in a closed circuit equals zero, i.e. all the energy generated by the source must be dissipated or stored by the loads. As illustrated in Section 4.1.1, the voltage gain across a source and the voltage drop across a load have opposite signs. Assigning a positive sign to voltage gain and a negative sign to voltage drop, the law of conservation of energy

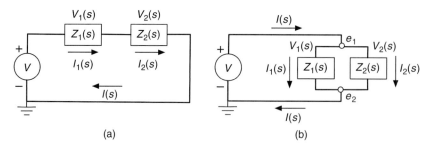

Figure 4.2.3 (a) Series and (b) parallel connections of impedances

for the circuit shown in Figure 4.2.3(a) translates into: $V(s) - V_1(s) - V_2(s) = 0$ or $V_1(s) + V_2(s) = V(s)$.

Then the equivalent impedance of this circuit is derived as follows:

$$Z_{eq}(s) = \frac{V(s)}{I(s)} = \frac{V_1(s) + V_2(s)}{I(s)} = Z_1(s) + Z_2(s) \qquad (4.2.6)$$

Consequently, a series connection is characterized by the same current passing through the connected electrical elements. The equivalent impedance for n circuit elements connected in series is a sum of individual impedances:

$$Z_{eq}(s) = \sum_{i=1}^{n} Z_i(s) \qquad (4.2.7)$$

Parallel Connection

A closed electrical circuit that contains a power source and two loads connected in parallel, with impedances Z_1 and Z_2 respectively, is presented in Figure 4.2.3(b). According to the law of conservation of charge the currents flowing through impedances Z_1 and Z_2 are: $I_1(s) + I_2(s) = I(s)$. The voltage drops across the loads Z_1 and Z_2 are the same: $V_1(s) = e_1 - e_2 = V_2(s) = V(s)$. Then the equivalent impedance is derived as follows:

$$Z_{eq}(s) = \frac{V(s)}{I(s)} = \frac{V(s)}{I_1(s) + I_2(s)}$$
$$\frac{1}{Z_{eq}(s)} = \frac{I_1(s)}{V(s)} + \frac{I_2(s)}{V(s)} = \frac{1}{Z_1(s)} + \frac{1}{Z_2(s)} \qquad (4.2.8)$$

Consequently, parallel connection is characterized by the same voltage drop across the connected electrical elements. The equivalent impedance for n circuit elements connected in parallel is computed as:

$$\frac{1}{Z_{eq}(s)} = \sum_{i=1}^{n} \frac{1}{Z_i(s)} \qquad (4.2.9)$$

Elements connected either in series or in parallel do not have to be of the same type for the computation of equivalent impedance. The following examples illustrate the use of Eqs. (4.2.7) and (4.2.9).

Example 4.2.1 Deriving equivalent impedances for combinations of circuit elements

The connections of several elements and derivations of impedances of these assemblies are shown in Table 4.2.1, where entries (a) and (b) denote series connections, (c) and (d) denote parallel connection, and (e) shows a combination of series and parallel connections.

Table 4.2.1 Equivalent impedances

Circuit	Equivalent Impedance
(a) R_1 — R_2 in series	$Z_{eq} = Z_1 + Z_2 = R_1 + R_2$
(b) R, L, C in series	$Z_{eq} = Z_1 + Z_2 + Z_3 = R + Ls + \dfrac{1}{Cs}$
(c) $R_1 \parallel R_2$	$\dfrac{1}{Z_{eq}(s)} = \dfrac{1}{Z_1(s)} + \dfrac{1}{Z_2(s)} = \dfrac{1}{R_1(s)} + \dfrac{1}{R_2(s)}$ $Z_{eq}(s) = \dfrac{R_1(s) R_2(s)}{R_1(s) + R_2(s)}$
(d) $R \parallel L$	$\dfrac{1}{Z_{eq}(s)} = \dfrac{1}{Z_1(s)} + \dfrac{1}{Z_2(s)} = \dfrac{1}{R(s)} + \dfrac{1}{Ls}$ $Z_{eq}(s) = \dfrac{RLs}{R + Ls}$
(e) $(R \parallel L)$ in series with C	$Z_{eq} = Z_{parallel} + Z_C$ $Z_{parallel} = \dfrac{RLs}{R + Ls}$ $Z_{eq} = \dfrac{RLs}{R + Ls} + \dfrac{1}{Cs} = \dfrac{RLCs^2 + Ls + R}{(R + Ls)Cs}$

Example 4.2.2 Finding the current through a circuit element using equivalent impedances

To illustrate how the use of equivalent impedance simplifies circuit analysis, consider the problem of finding the current through the inductor L and voltage v_1 at terminals A and B for the circuit shown in Figure 4.2.4.

Solution

Derive an expression for equivalent impedance for the elements connected in parallel:

$$\frac{1}{Z_P} = \frac{1}{R} + \frac{1}{1/Cs}$$

$$Z_P = \frac{R}{1 + RCs}$$

According to the law of conservation of energy:

$$V(s) - V_L(s) - V_P(s) = 0$$

where $V(s)$ denotes the voltage supplied by the source, $V_L(s)$ is the voltage drop across the inductor, and $V_P(s)$ represents the voltage drop across the impedance Z_P.

Using the definition of impedance (Eq. (4.2.1)) and the Laplace transform of the derivative form of the inductor's voltage–current relation, we obtain

$$V(s) = I(s)Ls + I(s)Z_P = I(s)\left(Ls + \frac{R}{1 + RCs}\right)$$

The current through the inductor – $I(s)$ – is then derived as

$$I(s) = V(s)\frac{1 + RCs}{RLCs^2 + Ls + R}$$

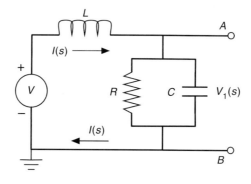

Figure 4.2.4 Circuit for Example 4.2.2

The voltage at terminals A and B equals the voltage drop across the connected in parallel component with impedance Z_P and is

$$V_1(s) = I(s)Z_P = V(s)\frac{R}{RLCs^2 + Ls + R}$$

4.2.3 Divider Rules

The concept of impedance proves very useful for the derivation of two basic rules: the voltage divider and current divider.

The *voltage divider rule* is best demonstrated on the series connection of two impedances. Consider a circuit shown in Figure 4.2.5. The currents through impedances Z_1 and Z_2 respectively are:

$$I_1(s) = I_2(s) = I(s) \qquad (4.2.10)$$

Using the definition of voltage drop as a difference between electric potentials on the element's entrance and exit terminals, we obtain

$$\begin{aligned} I_1 &= \frac{V_1}{Z_1} = \frac{e_A - e_B}{Z_1} \\ I_2 &= \frac{V_2}{Z_2} = \frac{e_B - e_C}{Z_2} \end{aligned} \qquad (4.2.11)$$

The terminals e_1 and C are located on the wire that is connected to the ground; hence, electric potentials of these terminals equal zero. Terminals e_2 and A belong to the same wire; thus, considering the assumption of all wires being ideal, electric potentials of these terminals

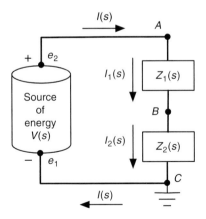

Figure 4.2.5 Voltage divider rule

are the same. Also, assigning a positive sign to voltage gain of the energy source as described in Section 4.2.2, we obtain

$$V(s) = e_2 - e_1 = e_2 \qquad (4.2.12)$$

Then currents through impedances Z_1 and Z_2 respectively are

$$I_1 = \frac{V - e_B}{Z_1} = I_2 = \frac{e_B}{Z_2} = I(s) \qquad (4.2.13)$$

potential of the terminal B is

$$e_B = V \frac{Z_2}{Z_1 + Z_2} \qquad (4.2.14)$$

and the voltage drops across impedances Z_1 and Z_2 are derived as

$$\begin{aligned} V_1 &= V \frac{Z_1}{Z_1 + Z_2} \\ V_2 &= V \frac{Z_2}{Z_1 + Z_2} \end{aligned} \qquad (4.2.15)$$

The voltage divider rule is expressed as:

$$\frac{V_1}{V_2} = \frac{Z_1}{Z_2} \qquad (4.2.16)$$

Hence, the voltage drop across the individual impedance – the member of a series connection – is proportional to the ratio of that impedance and equivalent impedance of the series connection as a whole. For a circuit that contains a single energy source and n impedances connected in series, the voltage drop across the i-th impedance is

$$V_i = V \frac{Z_i}{Z_{eq}} \qquad (4.2.17)$$

where V_i denotes voltage drop across the i-th impedance, V is voltage gain of the source, and $Z_{eq} = \sum_{i=1}^{n} Z_i$ is the equivalent impedance of the circuit.

Note that Eq. (4.2.16) can also be derived using the law of conservation of charge and the definition of impedance. According to the law of conservation of charge, the currents through impedances Z_1 and Z_2 are the same:

$$I_1(s) = I_2(s) = I(s)$$

By the definition of impedance (Eq. (4.2.1)):

$$I_1 = \frac{V_1}{Z_1} = I_2 = \frac{V_2}{Z_2}$$

And thus

$$\frac{V_1}{V_2} = \frac{Z_1}{Z_2}.$$

The *current divider rule* is best demonstrated on the parallel connection of two impedances. Consider a circuit shown in Figure 4.2.6. The voltage drops across the impedances Z_1 and Z_2 are the same:

$$V_1 = V_2 = e_A - e_B \quad (4.2.18)$$

As described earlier in this section, electric potentials at terminals e_1 and B are zero, and electric potentials at terminals e_2 and A are equal:

$$e_2 = e_A = V(s).$$

Thus:

$$V_1 = V_2 = V(s) \quad (4.2.19)$$

Using the definition of impedance and Eq. (4.2.19), we obtain

$$I_1 = \frac{V_1}{Z_1} = \frac{V}{Z_1}$$
$$I_2 = \frac{V_2}{Z_2} = \frac{V}{Z_2} \quad (4.2.20)$$

Therefore, the current divider rule can be expressed as follows:

$$\frac{I_1}{I_2} = \frac{Z_2}{Z_1} \quad (4.2.21)$$

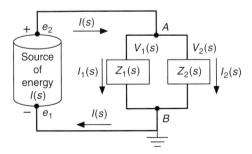

Figure 4.2.6 Current divider rule

According to the law of conservation of charge the currents flowing through impedances Z_1 and Z_2 are: $I_1(s) + I_2(s) = I(s)$. Using this equation together with Eq. (4.2.21), the expressions for currents across the impedances Z_1 and Z_2 respectively, are derived as

$$I_1 = I \frac{Z_2}{Z_1 + Z_2}$$
$$I_2 = I \frac{Z_1}{Z_1 + Z_2}$$
(4.2.22)

where I denotes the current through a circuit supplied by the source of current.

Example 4.2.3 Using voltage and current divider rules to find the current through a specific impedance and the voltage drop across it

To illustrate the use of the voltage and current divider rules in circuit analysis, consider the circuit shown in Figure 4.2.7(a). We need to find the current through the impedance Z_3 and the voltage drop across it.

Impedances Z_2 and Z_3 are connected in parallel; thus, their respective voltages are the same as $V_2 = V_3 = e_A$. According to the current divider rule, their respective currents are:

$$I_2 = I \frac{Z_3}{Z_2 + Z_3}$$

and

$$I_3 = I \frac{Z_2}{Z_2 + Z_3}$$

where I is the current across the circuit. From the circuit schematic it is obvious that $I = I_1$. From the definition of impedance, the expression for the current is easily obtained as $I = \frac{V}{Z}$, where V is the voltage supplied by the energy source, and Z is the equivalent impedance of the circuit.

Let Z_{eq} be the equivalent impedance for the sub-circuit consisting of the impedances Z_2 and Z_3, which are connected in parallel. Considering that the circuit is now transformed as shown in Figure 4.2.7(b), its equivalent impedance is

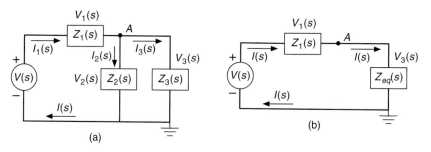

Figure 4.2.7 Simplifying the circuit using equivalent impedances (see text for details)

$$Z = Z_1 + Z_{eq},$$

where

$$Z_{eq} = \frac{Z_2 Z_3}{Z_2 + Z_3}.$$

Hence, the current across the circuit is derived as

$$I = V \frac{Z_2 + Z_3}{Z_1 Z_2 + Z_1 Z_3 + Z_2 Z_3}.$$

Substituting this result into the expression for the sought current through the impedance Z_3 we find that

$$I_3 = V \frac{Z_2}{Z_1 Z_2 + Z_1 Z_3 + Z_2 Z_3}.$$

Applying the voltage divider rule to the transformed circuit, shown in Figure 4.2.7(b), the expression for the sought voltage drop across the impedance Z_3 is derived as

$$V_3 = V_{eq} = V \frac{Z_{eq}}{Z_1 + Z_{eq}} = V \frac{Z_2 Z_3}{Z_1 Z_2 + Z_1 Z_3 + Z_2 Z_3}.$$

Example 4.2.4 Voltage divider rule in a potentiometer

The operation of a popular electrical device – a *potentiometer* – is governed by the voltage divider rule. A potentiometer is a manually adjustable variable resistor with three terminals. Two of the terminals are connected to the ends of the resistor, and the third one is a sliding electrical pick-off or *wiper*. The position of the wiper determines the device's output voltage V_0. Potentiometers are used as linear and angular position sensors in various applications. For example, the volume control knob on the older-model radios is a potentiometer that modifies voltage to the speakers,

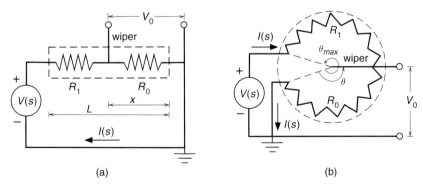

Figure 4.2.8 (a) Linear and (b) rotational potentiometers

thus changing the sound volume. Potentiometers may be linear or rotational, as shown in Figure 4.2.8. The resistance R_0 between the ground and the wiper is a function of the distance x for linear potentiometer (see Figure 4.2.8(a)), or of the rotation angle θ for rotational potentiometer (see Figure 4.2.8(b)).

The voltage output of a potentiometer is computed by applying the voltage divider rule. For the devices shown in Figure 4.2.8, the circuit's resistance equals $(R_0 + R_1)$, and the output voltage is

$$V_0 = V \frac{R_0}{R_0 + R_1}$$

where V is the voltage supplied by the source of energy, and R_0 and R_1 are the partial resistances of the potentiometer as illustrated.

For a linear potentiometer, the resistance R_0 is proportional to the device's length L as follows:

$$R_0 = (R_0 + R_1)\frac{x}{L}$$

Similarly, for a rotational potentiometer:

$$R_0 = (R_0 + R_1)\frac{\theta}{\theta_{max}}.$$

Hence, voltage output is derived as

$$V_0 = V\frac{x}{L} = K_L x \quad \text{for a linear potentiometer}$$

$$V_0 = V\frac{\theta}{\theta_{max}} = K_\theta \theta \quad \text{for a rotational potentiometer}$$

The coefficients K_L and K_θ are called the *potentiometer gain*. They depend on the applied voltage and on the potentiometer geometry, such as the resistor length for a linear potentiometer, and the wiper's maximum rotation angle for a rotational potentiometer.

Given the following numerical values for the resistances at a specific position of the wiper: $R_0 = 500\ \Omega$ and $R_1 = 1\ \text{k}\Omega$, and the input voltage $V = 9$ V, we find the voltage output as a percent of the input:

$$V_0 = V\frac{R_0}{R_0 + R_1} = 9\frac{500}{500 + 1000} = 3\ \text{V}$$

which is approximately 33% of the input voltage.

Example 4.2.5

While application of the voltage divider rule simplifies circuit analysis, it needs to be used with care.

Consider the problem of finding the voltage drop on the resistor R for the circuit shown in Figure 4.2.9. At the first glance, this circuit carries a strong resemblance to the potentiometer shown in Figure 4.2.8(a), and one may be tempted to just use the expression for V_0 derived in the previous example. That would be a mistake since the expression for the voltage output of the linear potentiometer does not take into consideration the influence of the load R. To account for that influence, the equivalent resistance R_{eq} for the connected in parallel resistors R and R_0 needs to be found prior to applying Eq. (4.2.15), where $Z_1 = R_1$ and $Z_2 = R_{eq}$.

The equivalent resistance is found as

$$R_{eq} = \frac{R_0 R}{R_0 + R}$$

and the sought voltage drop is then computed as

$$V_0 = V \frac{R_{eq}}{R_1 + R_{eq}} = V \frac{R_0 R}{R_1 R_0 + R R_0 + R R_1}.$$

This result demonstrates that the addition of a load resistor R to a given potentiometer circuit with fixed wiper position decreases the output voltage by the amount of

$$V_{0\_no\_load} - V_{0\_with\_load} = V \frac{R_1 R_0^2}{(R_1 + R_0)(R_1 R_0 + R R_0 + R R_1)}.$$

The current through the circuit with the load increases due to the decrease of the circuit's equivalent resistance:

$$R_1 + \frac{R_0 R}{R_0 + R} < R_1 + R_0$$

Obviously, the greater the value of R the more pronounced these effects are.

To see the effect of loading the potentiometer wiper use the same values of resistances $R_0 = 500\ \Omega$, $R_1 = 1\ k\Omega$, and the same input voltage $V = 9$ V as in Example 4.2.4, we compute the output voltage after adding the load resistor $R = 3\ k\Omega$:

$$V_0 = V \frac{R_0 R}{R_1 R_0 + R R_0 + R R_1} = 9 \frac{1000 \cdot 3000}{500 \cdot 1000 + 3000 \cdot 1000 + 3000 \cdot 500} = 5.4\ V,$$

which constitutes 60% of the input instead of 33% for the regular potentiometer without the loaded wiper.

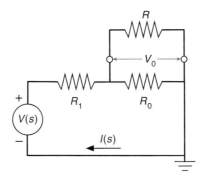

Figure 4.2.9 Finding voltage drop on the resistor R

4.3 Kirchhoff's Laws

Kirchhoff's current law (KCL) and *Kirchhoff's voltage law* (KVL) constitute the fundamental laws of electric circuit modeling. They are derived from the conservation laws of physics. KCL descends from the law of conservation of charge, while KVL descends from the law of conservation of energy.

KCL states that for every node of an electrical circuit an algebraic sum of currents equals zero, i.e. the sum of incoming currents equals the sum of outgoing currents.

The mathematical expression of KCL for a node that has n current elements, from which k currents are incoming (see Figure 4.3.1), is

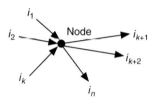

Figure 4.3.1 Kirchhoff's current law

$$\sum_{j=1}^{n} i_j = 0 \qquad (4.3.1)$$

where i_j indicates an individual current.

Assigning a positive sign to an incoming current, and a negative sign to an outgoing current, Eq. (4.3.1) may be re-written as

$$\sum_{j=1}^{k} i_j - \sum_{j=k}^{n} i_j = 0 \quad \text{or} \quad \sum_{j=1}^{k} i_j = \sum_{j=k}^{n} i_j \qquad (4.3.2)$$

KVL states that the algebraic sum of voltages around a closed circuit equals zero, i.e. the sum of the voltage gains equals the sum of the voltage drops. Alternatively, KVL, being an expression of the law of conservation of energy for electrical circuits, is sometimes formulated as: *all the energy produced in a closed circuit must be absorbed by the circuit elements.*

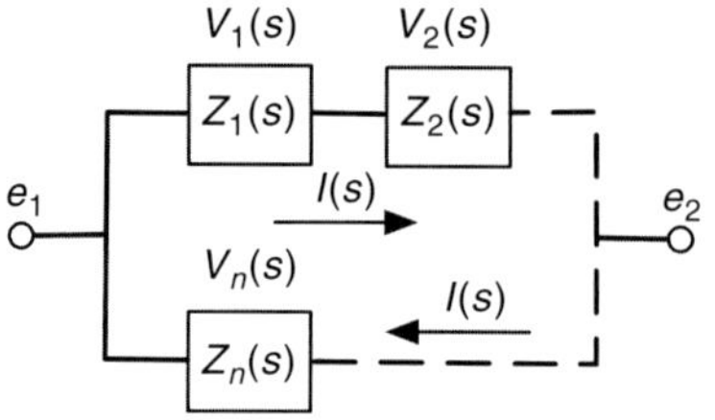

The mathematical expression of KVL for a closed circuit that contains n impedances (see Figure 4.3.2), is

Figure 4.3.2 Kirchhoff's voltage law

$$V_{circuit} = e_1 - e_2 = \sum_{j=1}^{n} V_j = 0 \qquad (4.3.3)$$

where V_j indicates the voltage drop across the j-th impedance.

Example 4.3.1

For a circuit with a single energy source, the application of Kirchhoff's laws is mostly straightforward. Consider the circuit shown in Figure 4.3.3.

Applying KCL to the node C, for which I_1 is the incoming current, and I_2 is the outgoing one, we obtain

$$I_1 - I_2 = 0 \text{ or } I_1 = I_2$$

Applying KCL to the node B, for which I_2 is the incoming current, and I (the circuit current) is the outgoing one, we obtain

$$I_2 - I = 0 \text{ or } I_2 = I$$

Thus, the same current runs through every element of this circuit:

$$I_1 = I_2 = I$$

The direction of this current is clockwise according to the definition of current as flow of the positively charged particles.

In this circuit, a single voltage source generates energy, and two impedances absorb it. In accordance with convention, a voltage gain associated with energy generation is positive, and a voltage drop associated with energy consumption is negative. Then, applying KVL to this circuit, we obtain

$$V - V_1 - V_2 = 0$$

Thus, the system of equations that describes this circuit is

$$\begin{cases} I_1 = I_2 = I \\ V - V_1 - V_2 = 0 \end{cases}$$

It can be easily solved for currents and voltage drops by using definition of impedance (Eq. (4.2.1)).

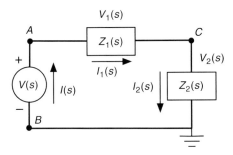

Figure 4.3.3 Circuit for Example 4.3.1

Example 4.3.2

For a circuit with multiple energy sources, determining the direction of a current and the sign of the voltage across a source often presents a challenge. In such a circuit, the energy source may absorb energy instead of generating it, which means that this source experiences a negative voltage drop rather than the expected voltage gain. To illustrate this concept, consider the circuits with two energy sources: a voltage source and a current source, shown in Figure 4.3.4.

The direction of the current is defined by the current source, and is, as shown, counterclockwise.

The voltage source in the circuit in Figure 4.3.4(a), according to its shown polarity, experiences voltage gain: $e_1 < e_2$. So, energy is generated, and v is positive.

Also, since all the connecting wires are considered ideal, $e_4 = e_2 > 0$ and $e_3 = e_1 < 0$. Thus, the current source experiences a voltage drop: $e_4 > e_3$, which means that this element absorbs energy and its voltage v_i is negative.

For the circuit in Figure 4.3.4(b) the situation is reversed. The voltage source consumes energy, experiencing a voltage drop ($e_1 > e_2$), while the current source generates energy, experiencing a voltage gain ($e_4 < e_3$). Therefore, v is negative, and v_i is positive.

Note that Kirchhoff's laws can be expressed either in time or in Laplace domains. The choice of domain depends on the circuit model requirements. Nonetheless, since circuit analysis aims at a finding circuit response, the voltage–current relation expression in the time domain is customarily sought.

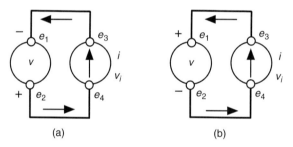

Figure 4.3.4 Circuits with two energy sources (see text for details)

4.4 Passive-Circuit Analysis

The straightforward application of the divider rules and Kirchhoff's laws to a single-loop electrical circuits yields the voltage–current relations for each circuit element with relative ease, as demonstrated earlier in this chapter. For multiple-loop circuits, such a direct approach may fail to derive the correct model. Hence, several methods were developed to simplify the complex multi-loop circuits. While the detailed discussion on the circuit analysis methodology belongs to the field of electrical engineering and is beyond the scope of this book, an introduction to the most widely used approaches is presented.

4.4 Passive-Circuit Analysis

The methods explored further in this chapter include: (a) the loop method, (b) the node method, (c) the superposition theorem, (d) Thevenin's theorem, and (e) Norton's theorem.

4.4.1 Loop Method

The *loop method* of circuit analysis utilizes the concept of *loop currents*, which are the main parameters of the circuit model. A loop current is a current assigned for a specific loop in the circuit. The loop method encompasses the following steps.

First, currents for every loop in the circuit are assigned, using either the marked polarity of the voltage power source, or the current direction indicated by the current power source. For interconnecting elements – elements that simultaneously belong to more than a single loop – *net currents* are used. A net current is an algebraic sum of all the currents passing through the interconnecting element. In evaluation of the element's net current, the direction of the assigned loop current in the examined loop is defined as positive.

After the currents passing through every circuit element are defined, the expressions for the voltages are derived using Ohm's law, the known voltage–current relations, and the divider rules.

The last step of this method includes the application of KVL to every loop in the circuit. The derived KVL equations comprise the sought system model in terms of loop currents as independent variables. The output equation is also formulated in terms of loop currents.

Example 4.4.1 To illustrate operation of the loop method, consider the circuit shown in Figure 4.4.1. This circuit consists of four passive elements: two resistors R_1 and R_2, two capacitors C_1 and C_2, and a single source of energy. The polarity of the voltage source is as shown on the circuit schematic. The system input is the voltage v_{in} supplied by the voltage source, and the output is the capacitor C_2 voltage v_{out}. Zero initial conditions are assumed.

As indicated in the circuit schematic, this circuit consists of two loops, for which the loop currents i_1 and i_2 are defined as shown. The direction of i_1 is clockwise due to the marked polarity of the voltage source. Consequently, the direction of i_2 is also clockwise.

This circuit has a single interconnecting element: capacitor C_1. Its net current is $i_1 - i_2$ for loop 1, and $i_2 - i_1$ for loop 2.

Applying KVL to loop 1 yields

$$v_{in} - v_{R_1} - v_{C_1} = 0$$

and to Loop 2:

$$v_{C_1} + v_{R_2} + v_{C_2} = 0$$

Recalling the voltage–current relations for resistor (Eq. (4.1.9)) and capacitor (Eq. (4.1.16)), the KVL equations become

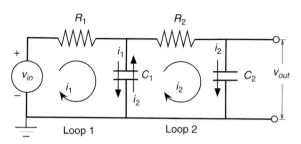

Figure 4.4.1 Modeling a passive circuit using the loop method

$$\text{Loop 1} \quad v_{in}(t) = R_1 i_1 + \frac{1}{C_1} \int_0^t (i_1 - i_2) dt \tag{1}$$

$$\text{Loop 2} \quad R_2 i_2 + \frac{1}{C_1} \int_0^t (i_2 - i_1) dt + \frac{1}{C_2} \int_0^t i_2 dt = 0 \tag{2}$$

The output equation is derived as

$$v_{out}(t) = \frac{1}{C_2} \int_0^t i_2 dt \tag{3}$$

These three equations constitute the required circuit model.

If desired, this model can be transformed into a single differential equation that directly expresses the relation between the input and output voltages. The easiest way to achieve such a transformation is by taking the Laplace transform of Eqs. (1), (2), and (3), performing the necessary algebraic operations, and taking the inverse Laplace transform of the resulting equation. This derivation is as follows (capital letters denote variables in Laplace domain):

$$V_{in} = R_1 I_1 + \frac{1}{C_1 s}(I_1 - I_2) \tag{4}$$

$$R_2 I_2 + \frac{1}{C_1 s}(I_2 - I_1) + \frac{1}{C_2 s} I_2 = 0 \tag{5}$$

$$V_{out} = \frac{1}{C_2 s} I_2 \tag{6}$$

From Eq. (6):

$$I_2 = C_2 s V_{out} \tag{7}$$

Then, substituting this result into Eq. (5), we obtain the expression for I_1:

$$I_1 = C_1 s \left(R_2 + \frac{1}{C_1 s} + \frac{1}{C_2 s} \right) C_2 s V_{out} \tag{8}$$

Finally, substituting Eqs. (7) and (8) into Eq. (4), we derive the sought model:

$$\begin{aligned} V_{in} &= \left(R_1 + \frac{1}{C_1 s} \right) \left(R_2 + \frac{1}{C_1 s} + \frac{1}{C_2 s} \right) C_1 C_2 s^2 V_{out} - \frac{1}{C_1 s} C_2 s V_{out} \\ &= \left(R_1 R_2 C_1 C_2 s^2 + (R_1 C_1 + R_2 C_2 + R_1 C_2) s + 1 \right) V_{out} \end{aligned} \tag{9}$$

Taking the inverse Laplace transform of Eq. (9), the system model is obtained as a differential equation in terms of input and output voltages:

$$v_{in} = R_1 R_2 C_1 C_2 \frac{d^2 v_{out}}{dt^2} + (R_1 C_1 + R_2 C_2 + R_1 C_2) \frac{dv_{out}}{dt} + v_{out} \tag{10}$$

4.4.2 Node Method

The *node method* of circuit analysis operates with the concept of *node voltages*, which are the main parameters of the circuit model. A node is a point in an electric circuit at which connecting wires intersect or branch. The node voltage is an electric potential assigned to a node, and an independent variable in the equations that constitute the circuit model. The node method operates on the circuit as a whole, without subdividing it into loops, and encompasses the following steps.

First, the nodes of interest and their electric potentials – node voltages – are defined. Then, the current for each circuit element is assigned, using the indicated polarity of the voltage source or marked current direction of the current source.

The expressions for the current through each circuit element are written in terms of node voltages, using the known voltage–current relations, Ohm's law, and the divider rules.

After all the expressions for component currents are derived, KCL is applied to each node. These KCL equations comprise the sought system model in terms of node voltages as independent variables. The output equation is also formulated in terms of node voltages.

Example 4.4.2

To illustrate the operation of the node method, consider the circuit used in the Example 4.4.1.

As shown in Figure 4.4.2, three nodes are defined, with assigned node voltages e_0, e_1, and e_2 respectively, and the element currents are as indicated on the circuit schematic.

Remembering that the electric potential of any point on a bottom wire is zero (this is the wire connected to the ground) and using the known voltage–current relations for the resistor and capacitor the element currents in terms of node voltages are derived as:

$$i_{R_1} = \frac{e_0 - e_1}{R_1} = \frac{v_{in} - e_1}{R_1}$$

$$i_{R_2} = \frac{e_1 - e_2}{R_2} = \frac{e_1 - v_{out}}{R_2}$$

$$i_{C_1} = C_1 \frac{d}{dt}(e_1 - 0) = C_1 \frac{de_1}{dt}$$

$$i_{C_2} = C_2 \frac{d}{dt}(e_2 - 0) = C_2 \frac{de_2}{dt} = C_2 \frac{dv_{out}}{dt}$$

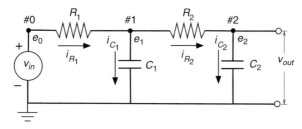

Figure 4.4.2 Modeling a passive circuit using the node method

Applying KCL to nodes 1 and 2 we obtain

$$\text{Node 1 } i_{R_1} - i_{C_1} - i_{R_2} = 0$$
$$\text{Node 2 } i_{R_2} - i_{C_2} = 0$$

Substituting the derived element current expressions into KCL equations, we obtain the required system model:

$$\frac{v_{in} - e_1}{R_1} - C_1 \frac{de_1}{dt} - \frac{e_1 - v_{out}}{R_2} = 0$$
$$\frac{e_1 - v_{out}}{R_2} - C_2 \frac{dv_{out}}{dt} = 0$$

It is a system of two differential equations with two variables: e_1 and v_{out}. Since v_{out} is an output, no separate output equation is needed.

If desired, the intermediate variable e_1 can be eliminated, thus reducing the system to a single differential equation of a single independent variable v_{out}. Expressing e_1 in terms of v_{out}, using the second equation of the above system model yields

$$e_1 = R_2 C_2 \frac{dv_{out}}{dt} + v_{out}$$

Substituting this result into the first equation of the derived system model, we obtain

$$\frac{v_{in}}{R_1} - \left(\frac{1}{R_1} + \frac{1}{R_2}\right)\left(R_2 C_2 \frac{dv_{out}}{dt} + v_{out}\right) - C_1\left(R_2 C_2 \frac{d^2 v_{out}}{dt^2} + \frac{dv_{out}}{dt}\right) + \frac{v_{out}}{R_2} = 0$$

Algebraic transformations result in the following system model:

$$R_1 R_2 C_1 C_2 \frac{d^2 v_{out}}{dt^2} + (R_1 C_1 + R_2 C_2 + R_1 C_2)\frac{dv_{out}}{dt} + v_{out} = v_{in}$$

As expected, this model is identical to that derived in the Example 4.4.1 (see Eq. (10)).

4.4.3 Circuits with Multiple Energy Sources

Straightforward applications of the loop method and the node method to the analysis of a circuit with multiple energy sources may prove difficult. The *Superposition theorem* is one of the approaches that allow efficient dealing with the multiple energy sources within a circuit. *Thevenin's theorem* and *Norton's theorem* constitute the other popular approaches to the simplification of such a circuit. The usability of these three methods is particularly apparent for the equivalent transformation of a circuit that is the combination of resistors and sources.

Superposition Theorem

Superposition theorem states that the current through any passive element of a circuit with multiple energy sources equals the algebraic sum of the currents produced by each individual source independently. To evaluate the current resulting from the i-th source, all the other sources are "turned off." This means replacing the voltage sources by short circuits (zero voltage), and the current sources, by open circuits (zero current). Then, either of the described earlier methods for circuit analysis can be applied.

Example 4.4.3

Consider the circuit shown in Figure 4.4.3(a). It contains two energy sources – a current source $I(s)$ and a voltage source $V(s)$. This problem can be solved either in the time domain or in the Laplace domain. Here, the Laplace domain solution is provided, which is easily inverted into the time domain when required.

We want to evaluate the total current passing through the resistor R_1. According to the superposition theorem, this current (I_1) is a sum of two current components – one due to the current source (I_{I1}), and the other due to the voltage source (I_{V1}):

$$I_1 = I_{V1} + I_{I1}$$

To derive the current I_{V1} resulting from the voltage source, the current source is turned off as shown in Figure 4.4.3(b). Since the right loop of the circuit depicted in Figure 4.4.3(b) is open, there is no current there. Thus, only the left loop consisting of the voltage source and two resistors R_1 and R_3 participates in current generation. The resistors R_1 and R_3 are connected in series, so currents flowing through these elements are the same: $I_{V1} = I_{V3} = I_V$. The current I_{V1} is then derived using Ohm's law and the equivalent resistance of the circuit:

$$I_{V1} = \frac{V}{R_{Veq}} = \frac{V}{R_1 + R_3}$$

To derive the current I_{I1} resulting from the current source, the voltage source is turned off as shown in Figure 4.4.3(c). Here both loops of the circuit have a current.

According to the law of conservation of charge, the current flowing through the resistor R_2 equals the current supplied by the energy source. The resistors R_1 and R_3 are connected in parallel so, using the current divider law (Eq. (4.2.22)), the current across the resistor R_1 is

$$I_{I1} = I \frac{R_3}{R_1 + R_3}$$

Note that the currents I_{I1} and I_{V1} have opposite directions. Assuming the clockwise direction to be positive, the total current across the resistor R_1 is

$$I_1 = I_{V1} + I_{I1} = \frac{V}{R_1 + R_3} - I \frac{R_3}{R_1 + R_3}$$

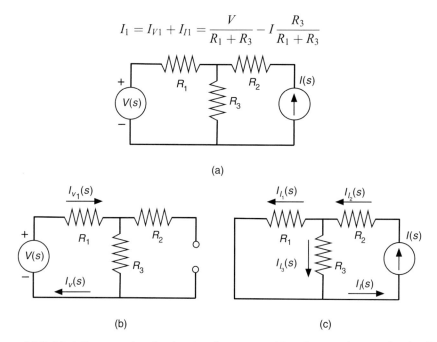

Figure 4.4.3 Modeling a passive circuit using the superposition theorem (see text for details)

Thevenin's and Norton's Theorems

Thevenin's theorem states that any combination of energy sources and resistors with two terminals can be replaced by an equivalent circuit of a single voltage source V_T (*Thevenin voltage*) and a single series resistor R_T (*Thevenin resistance*). The Thevenin voltage is the open-circuit voltage at the terminals. The Thevenin resistance is the resistance of the open circuit measured at the terminals. Its value is obtained by evaluating the equivalent resistance of this circuit, where voltage sources are replaced by short circuits, and current sources are replaced by open circuits.

Norton's theorem states that any combination of energy sources and resistors with two terminals can be replaced by an equivalent circuit of a single current source I_N (*Norton current*) and a single resistor R_N (*Norton resistance*) in parallel with the current source. The Norton resistance is defined in the same way as the Thevenin resistance. The Norton current is the current that would have existed in the wire connecting the terminals of the open circuit. Its value is obtained by Ohm's law, dividing the open-circuit voltage at the terminals by the Norton resistance. If the chosen terminals for the Norton and Thevenin transformations are the same, then $R_N = R_T$ and $I_N = V_T/R_T$.

An illustration of the equivalent transformation of an electrical circuit using Thevenin's and Norton's theorems is presented in the example below.

Example 4.4.4 Consider the two-loop circuit shown in Figure 4.4.4(a). Thevenin's and Norton's equivalents of the right-hand side of this circuit at the terminals A and B are constructed as shown in Figure 4.4.4(b) and Figure 4.4.4(c) respectively.

As stated earlier, $R_T = R_N$. To find its value consider the open circuit shown surrounded with a dashed line in Figure 4.4.4(a). Replacing the voltage source V_1 with a short circuit, it can be easily

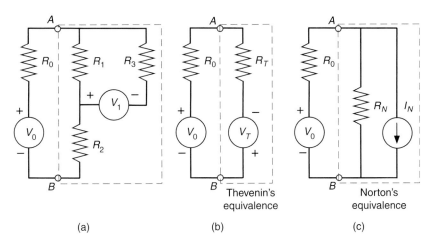

Figure 4.4.4 Modeling a passive circuit using Thevenin's and Norton's theorems (see text for details)

seen that resistors R_1 and R_3 are connected in parallel, and resistor R_2 is in series with their equivalence. Thus:

$$R_T = R_N = R_2 + \frac{R_1 R_3}{R_1 + R_3}$$

The Thevenin voltage is the open-circuit voltage at the terminals A and B. This voltage equals the voltage drop across the resistor R_1. The resistor R_2 has no influence on the Thevenin voltage since there is no current passing through it. To find this voltage consider the loop, consisting of the voltage source V_1 and resistors R_1 and R_3. In this loop, the resistors are connected in series, so the voltage divider rule is used (Eq. (4.2.15)):

$$V_T = V_1 \frac{R_1}{R_1 + R_3}$$

Then, the Norton current is

$$I_N = \frac{V_T}{R_T} = V_1 \frac{R_1}{R_1 R_2 + R_1 R_3 + R_2 R_3}$$

The polarity of the Thevenin's voltage source and the direction of the current provided by the Norton's current source are defined to maintain the direction of current in the transformed equivalent circuit as the same as it was before the transformation.

Example 4.4.5

This example encompasses the application of the circuit analysis methods described in this section to a simple two-loop electrical circuit. The consistency of the derived results is demonstrated.

Consider the two-loop circuit shown in Figure 4.4.5.

We need to derive a circuit model and find the current and the voltage drop for the resistor R_3, if the numerical values of circuit elements are as follows:

$$R_1 = 5\,\Omega,\ R_2 = 3\,\Omega,\ R_3 = 2\,\Omega,$$
$$v_1 = 12\ \text{V},\ v_2 = 8\ \text{V}$$

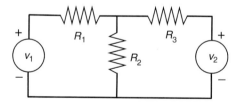

Figure 4.4.5 Circuit for Example 4.4.5

4.4 Passive-Circuit Analysis

Solution by the Loop Method

The loops are defined and the loop currents are assigned in accordance with the indicated polarity of the voltage sources as shown in Figure 4.4.6.

Resistor R_2 is the interconnecting element and its net current is the same for both loops: $i_{R2} = i_1 + i_2$. Recalling the voltage–current relation of a resistor (Eq. (4.1.9)), apply KVL to loop 1 and loop 2:

$$\text{Loop 1} \quad v_1 - i_1 R_1 - (i_1 + i_2) R_2 = 0$$
$$\text{Loop 2} \quad v_2 - i_2 R_3 - (i_1 + i_2) R_2 = 0$$

This system of two equations with two independent variables – i_1 and i_2 – constitutes the system model.

Since the only current that passes through R_3 is i_2, we find the required current and voltage by solving the above system of equations for i_2 and then, using Ohm's law, we derive v_{R_3}:

$$i_{R_3} = i_2 = \frac{-R_2 v_1 + (R_1 + R_2) v_2}{R_1 R_2 + R_1 R_3 + R_2 R_3} = 0.9032 \text{ A}$$

$$v_{R_3} = i_{R_3} R_3 = 1.8064 \text{ V}$$

Considering that system inputs are voltages v_1 and v_2 provided by the voltage sources, and the outputs are the current and the voltage drop for the resistor R_3, the system model can be written in terms of input–output as follows:

$$\text{System model} \qquad i_{R_3} = \frac{-R_2 v_1 + (R_1 + R_2) v_2}{R_1 R_2 + R_1 R_3 + R_2 R_3}$$

$$\text{Output equations} \qquad \{i_{R_3}, \; v_{R_3} = i_{R_3} R_3\}$$

Solution Using Thevenin's Theorem

This solution uses an equivalent transformation of the circuit. We define the terminals A and B as shown in Figure 4.4.7(a), and substitute the left-hand side of the circuit shown encircled in a dashed rectangle with its Thevenin equivalent.

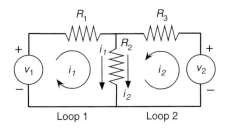

Figure 4.4.6 Solution by the loop method

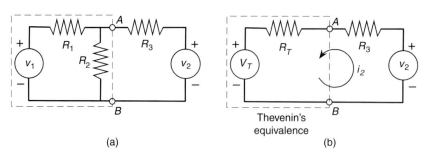

Figure 4.4.7 Solution by Thevenin's theorem (see text for details)

The Thevenin resistance is computed as equivalent resistance for two resistors R_1 and R_2 connected in parallel:

$$R_T = \frac{R_1 R_2}{R_1 + R_2}$$

The Thevenin voltage equals the voltage drop across the resistor R_2 and is derived using the voltage divider rule (Eq. (4.2.15)):

$$V_T = v_1 \frac{R_2}{R_1 + R_2}.$$

The equivalent single-loop circuit is shown in Figure 4.4.7(b). Since the resistors R_T and R_3 are connected in series, the current passing through them is the same and can be derived by direct application of Ohm's law. Considering the direction of this current to be as shown on the circuit schematic, we can state that the voltage source v_2 generates energy, i.e. experiences a voltage gain, while the Thevenin voltage source V_T is absorbing energy, i.e. experiences a voltage drop. Thus, v_2 is positive, and V_T is negative. Then, the system model is

$$v_2 - V_T = i(R_T + R_3)$$

or in extended form:

$$v_2 - v_1 \frac{R_2}{R_1 + R_2} = i\left(\frac{R_1 R_2}{R_1 + R_2} + R_3\right)$$

Then, the required current and voltage for the resistor R_3 are found as:

$$i_{R_3} = i = \frac{v_2 - v_1 \frac{R_2}{R_1 + R_2}}{\frac{R_1 R_2}{R_1 + R_2} + R_3} = \frac{(R_1 + R_2)v_2 - R_2 v_1}{R_1 R_2 + R_1 R_3 + R_2 R_3} = 0.9032 \text{ A}$$

$$v_{R_3} = i_{R_3} R_3 = 1.8064 \text{ V}$$

Considering the specified inputs and outputs of the system, the system model can be defined in terms of input–output as follows:

4.4 Passive-Circuit Analysis

System model
$$i_{R_3} = \frac{(R_1 + R_2)v_2 - R_2 v_1}{R_1 R_2 + R_1 R_3 + R_2 R_3}$$

Output equations
$$\{i_{R_3},\ v_{R_3} = i_{R_3} R_3\}$$

Solution Using Norton's Theorem

This solution also uses an equivalent transformation of the circuit. We define the terminals A and B as shown in Figure 4.4.8(a), and substitute the part of the circuit shown encircled with a dashed line with its Norton equivalent. The equivalent circuit is shown in Figure 4.4.8(b).

Even though the terminals chosen for the Norton transformation are not the same as those chosen for the Thevenin transformation in the previous solution, the Norton resistance is the same as the Thevenin resistance:

$$R_N = \frac{R_1 R_2}{R_1 + R_2}$$

To derive the Norton current, the voltage at the terminals A and B needs to be evaluated first. From the circuit schematic in Figure 4.4.8(a) it is clearly seen that this voltage is a difference between the voltage gain of the source v_2 and the voltage drop across the resistor R_2, which is found using the voltage divider rule:

$$v_{AB} = v_2 - v_{R2} = v_2 - v_1 \frac{R_2}{R_1 + R_2}$$

Hence, Norton current is derived as

$$i_N = \frac{v_{AB}}{R_N} = \frac{v_2 - v_1 \dfrac{R_2}{R_1 + R_2}}{\dfrac{R_1 R_2}{R_1 + R_2}} = \frac{(R_1 + R_2)v_2 - R_2 v_1}{R_1 R_2}$$

The resulting Norton-equivalence circuit has a single current source and two resistors connected in parallel. Then, the current through the resistor R_3 is found using current divider rule:

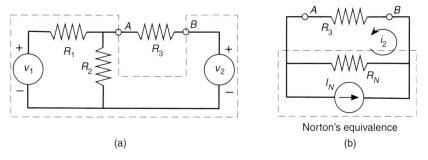

(a)   Norton's equivalence (b)

Figure 4.4.8 Solution by Norton's theorem (see text for details)

$$i_{R3} = i_N \frac{R_N}{R_N + R_3} = \frac{(R_1 + R_2)v_2 - R_2 v_1}{R_1 R_2} \frac{\dfrac{R_1 R_2}{R_1 + R_2}}{\dfrac{R_1 R_2}{R_1 + R_2} + R_3} = \frac{(R_1 + R_2)v_2 - R_2 v_1}{R_1 R_2 + R_1 R_3 + R_2 R_3} = 0.9032 \text{ A}$$

and the voltage drop is computed using Ohm's law:

$$v_{R_3} = i_{R_3} R_{33} = 1.8064 \text{ V}$$

Considering the specified inputs and outputs of the system, the system model can be defined in terms of input–output as follows:

System model
$$i_{R_3} = \frac{(R_1 + R_2)v_2 - R_2 v_1}{R_1 R_2 + R_1 R_3 + R_2 R_3}$$

Output equations
$$\{i_{R_3}, \; v_{R_3} = i_{R_3} R_3\}$$

Solution Using the Superposition Theorem

Since the Superposition theorem deals with currents through circuit elements, the system model will be derived in the form

$$i_{R_3} = i_{13} + i_{23}$$

where i_{R_3} is the current through the resistor R_3, i_{13} is the current through this resistor due to the voltage source v_1, and i_{23} is the current through this resistor due to the voltage source v_2. Note that as discussed earlier in this section, the sum of the component currents is algebraic, which means that the signs of the component currents i_{13} and i_{23} are assigned according to their directions.

The directions of currents i_{13} and i_{23} are defined as shown in Figure 4.4.9.

To find the current i_{13}, the voltage source v_2 is replaced by a short circuit as shown in Figure 4.4.9(a).

Then, the equivalent resistance of this circuit is

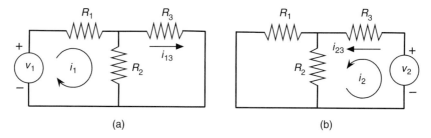

Figure 4.4.9 Solution by the Superposition theorem

$$R_{eq} = R_1 + \frac{R_2 R_3}{R_2 + R_3}$$

and the circuit current is:

$$i_{v_1} = \frac{v_1}{R_1 + \dfrac{R_2 R_3}{R_2 + R_3}} = \frac{v_1 (R_2 + R_3)}{R_1 R_2 + R_1 R_3 + R_2 R_3}$$

Since resistors R_1 and the equivalence $\dfrac{R_2 R_3}{R_2 + R_3}$ are in series, the same current passes through them. Then, i_{13} is found using the current divider rule:

$$i_{13} = i_{v_1} \frac{R_2}{R_2 + R_3} = \frac{v_1 R_2}{R_1 R_2 + R_1 R_3 + R_2 R_3}$$

Similarly, i_{23} is found as the circuit current for the circuit shown in Figure 4.4.9(b):

$$i_{23} = i_{v_2} = \frac{v_2}{R_3 + \dfrac{R_1 R_2}{R_1 + R_2}} = \frac{v_2 (R_1 + R_2)}{R_1 R_2 + R_1 R_3 + R_2 R_3}$$

The currents i_{13} and i_{23} have opposite directions, so they must have different signs in the algebraic sum that represents the current through the resistor R_3. Assume that i_{23} is positive. Then, the required current and voltage for the resistor R_3 are found as:

$$i_{R_3} = i_{23} - i_{13} = \frac{(R_1 + R_2)v_2 - R_2 v_1}{R_1 R_2 + R_1 R_3 + R_2 R_3} = 0.9032 \text{ A}$$

$$v_{R_3} = i_{R_3} R_3 = 1.8064 \text{ V}$$

The computed result for i_{R_3} is positive, which means that the positive current component was chosen correctly. A negative result would have indicated that the original assumption of the direction of current through R_3 is incorrect, and the current actually flows in the opposite direction.

The derived expressions for current and voltage for the resistor R_3 constitute the system model in terms of input and output:

system model $\qquad i_{R_3} = \dfrac{(R_1 + R_2)v_2 - R_2 v_1}{R_1 R_2 + R_1 R_3 + R_2 R_3}$

output equations $\qquad \{i_{R_3}, v_{R_3} = i_{R_3} R_3\}$

As expected, the results derived by all the methods employed in this example are identical.

4.5 State-Space Representations and Block Diagrams

4.5.1 Block Diagrams

As discussed in Section 2.7.2, block diagrams and transfer functions are intimately related. A block diagram is an extremely good tool for describing the input–output relations for a variety of dynamic systems, and electrical circuits in particular. While a circuit schematic depicts the physical connections between the system components, a block diagram deals with variables being transformed and propagated through the system. As such, it is a useful visual medium aimed at better understanding of the relations between the elements of a dynamic system.

The generic block diagram is shown in Figure 4.5.1. Customarily, all the elements of a block diagram are in the Laplace domain.

Figure 4.5.1 Generic block diagram

The presented block diagram may refer to the dynamic system as a whole, in which case it can be viewed as a diagrammatic extension of the system transfer function model. It may also refer to an individual system element. Then the transfer-function block describes the element's dynamics, while the input and output are the intermediate variables of the system.

Consider the simple single-loop electrical circuit shown in Figure 4.5.2. Let the input be the voltage v_{in}, supplied by the voltage source, and the output be the voltage drop v_{out} on the capacitor C.

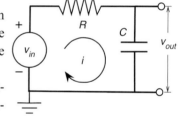

Figure 4.5.2 Simple RC circuit

The construction of a block diagram for this circuit encompasses the following steps: (a) deriving the system model equations in time domain, (b) taking the Laplace transform of the system equations considering zero initial conditions, (c) deriving the component transfer functions and constructing their graphical representation – component block diagrams, as required to comprehensively express the interactions between circuit elements, and (d) assembling the system block diagram from the constructed components.

The equation that constitutes the circuit model is derived by applying KVL:

$$v_{in} - v_R - v_C = v_{in} - iR - \frac{1}{C}\int i\,dt = 0 \qquad (4.5.1)$$

The output equation expresses the voltage–current relation for a capacitor:

$$v_{out} = \frac{1}{C}\int i\,dt \qquad (4.5.2)$$

Taking the Laplace transform of Eqs. (4.5.1) and (4.5.2), we obtain

4.5 State-Space Representations and Block Diagrams

$$V_{in}(s) = I(s)R + \frac{1}{Cs}I(s) = I(s)\frac{RCs+1}{Cs} \qquad (4.5.3)$$

$$V_{out}(s) = \frac{1}{Cs}I(s) \qquad (4.5.4)$$

The sought system block diagram is in the form shown in Figure 4.5.1, where $V_{in}(s)$ is the system input, and $V_{out}(s)$ is the system output. The circuit current $I(s)$ is the intermediate variable. Equation (4.5.3) describes the relation between the variables $V_{in}(s)$ and $I(s)$, where the former is the component input, the latter is the component output, and the transfer function is $\frac{Cs}{RCs+1}$. Similarly, for Eq. (4.5.4) $I(s)$ is the component input, $V_{out}(s)$ is the component output, and the transfer function is $\frac{1}{Cs}$.

The component block diagrams are shown in Figure 4.5.3.

The system block diagram is constructed by combining these component block diagrams; see Figure 4.5.4(a). The intermediate variable $I(s)$ can be removed by the algebraic transformation of this block diagram, shown in Figure 4.5.4(b).

As seen from the final system diagram (Figure 4.5.4(b)), the system transfer function is $\frac{1}{RCs+1}$.

For circuits with multiple energy sources, the construction of a block diagram is more involved. The complexity of a block diagram depends on the circuit inputs and outputs. The procedure is similar to that described in the Example 4.5.1: (a) the equations relating inputs to outputs are derived using the relevant circuit analysis method, (b) the Laplace transform of them is taken, (c) block diagram components are built from individual equations, and finally (d) the components are combined into a finished block diagram.

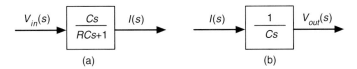

Figure 4.5.3 Component block diagrams for (a) voltage and (b) current

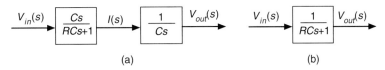

Figure 4.5.4 Construction of a system block diagram (see text for details)

Example 4.5.1

Consider the circuit shown in Figure 4.5.5. It consists of a current source and three passive elements: a resistor R, a capacitor C, and an inductor L. Construct a block diagram of this circuit, considering that the input is represented by the current i_s, and the outputs are voltages v_R, v_C, and v_L across the circuit components.

This problem is approached using the node method. The nodes 1 and 2 are defined as shown in the circuit schematic. Applying KCL to them, we obtain

$$\text{Node 1} \qquad i_S = i_R + i_L \qquad (1)$$
$$\text{Node 2} \qquad i_L = i_C \qquad (2)$$

The next step involves expressing the currents in Eqs. (1) and (2) in terms of appropriate voltages. From Eq. (1):

$$i_R = i_S - i_L \qquad (3)$$

Then the voltage across the resistor is

$$v_R = Ri_R = R(i_S - i_L) \qquad (4)$$

The voltage across the inductor is

$$v_L = v_R - v_C = L\frac{di_L}{dt} \qquad (5)$$

From Eq. (5) the current through the inductor is

$$i_L = \frac{1}{L}\int (v_R - v_C)dt \qquad (6)$$

Substituting this result into Eq. (4), we obtain

$$v_R = R\left(i_S - \frac{1}{L}\int (v_R - v_C)dt\right) \qquad (7)$$

The current through the capacitor is

$$i_C = C\frac{dv_C}{dt} \qquad (8)$$

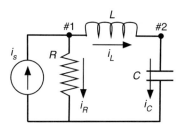

Figure 4.5.5 *RLC* circuit for Example 4.5.1

4.5 State-Space Representations and Block Diagrams

Substituting the results of Eqs. (6) and (8) into Eq. (2), and taking the derivative with respect to time in order to get rid of the integral, we obtain

$$i_C = C\frac{dv_C}{dt} = i_L = \frac{1}{L}\int(v_R - v_C)dt$$

$$LC\frac{d^2 v_C}{dt^2} = v_R - v_C \qquad (9)$$

Hence, the equations needed for construction of a block diagram are (5), (7), and (9). They fully express the relations between the specified input and outputs.

Taking the Laplace transform of these equations, we obtain

$$V_L(s) = V_R(s) - V_C(s) \qquad (10)$$

$$V_R(s) = R\left(I_S(s) - \frac{1}{Ls}\left(V_R(s) - V_C(s)\right)\right) \qquad (11)$$

$$V_R(s) - V_C(s) = LCs^2 V_C(s) \qquad (12)$$

Rearranging the terms in these equations to make the construction of the block diagram easier:

$$V_L(s) = V_R(s) - V_C(s) \qquad (13)$$

$$V_R(s) = \frac{RLs}{Ls + R}\left(I_S(s) + \frac{1}{Ls}V_C(s)\right) \qquad (14)$$

$$V_C(s) = \frac{1}{LCs^2 + 1}V_R(s) \qquad (15)$$

The resulting block diagram is shown in Figure 4.5.6(a). It is one of the possible block diagrams, but hardly the most straightforward one due to the relatively complicated transfer functions of its elements.

A simpler and more intuitive block diagram can be constructed using Eqs. (2), (4), (5), and (8) in the following way:

$$v_L = v_R - v_C = \frac{1}{L}\int i_L dt$$

$$v_R = R(i_S - i_L) \qquad (16)$$

$$i_C = C\frac{dv_C}{dt} = i_L$$

Taking the Laplace transform of Eq. (16), we obtain

$$V_L(s) = V_R(s) - V_C(s) = Ls\, I_L(s)$$

$$V_R(s) = R(I_S(s) - I_L(s)) \qquad (17)$$

$$I_C(s) = CsV_C(s) = I_L(s)$$

The resulting block diagram is shown in Figure 4.5.6(b).

Deriving the transfer function for every output is straightforward. For example, the transfer function for the capacitor is derived from the block diagram in Figure 4.5.6(b) as follows:

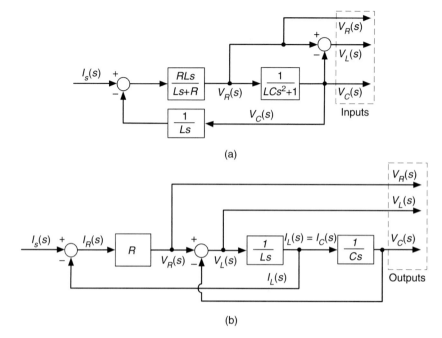

Figure 4.5.6 Block diagrams for the RLC circuit of Example 4.5.1 (see text for details)

$$\left(\left(I_S(s) - I_L(s)\right)R - V_C(s)\right)\frac{1}{Ls}\frac{1}{Cs} = V_C(s)$$

$$\left(I_S(s) - I_L(s)\right)R = \left(LCs^2 + 1\right)V_C(s)$$

Since $I_L(s) = CsV_C(s)$, we obtain

$$RI_S(s) = \left(LCs^2 + RCs + 1\right)V_C(s)$$

$$\frac{V_C(s)}{I_S(s)} = \frac{R}{LCs^2 + RCs + 1} \qquad (18)$$

To demonstrate the correctness of this transfer function, the differential equation for the voltage across the capacitor is derived using Eqs. (7) and (9):

$$LC\frac{d^2v_C}{dt^2} + v_C = R\left(i_S - \frac{1}{L}\int LC\frac{d^2v_C}{dt^2}dt\right)$$

$$LC\frac{d^2v_C}{dt^2} + RC\frac{dv_C}{dt} + v_C = Ri_S \qquad (19)$$

It is clearly seen that Eq. (19) is consistent with the transfer function depicted by Eq. (18).

Example 4.5.2

Consider the circuit shown in Figure 4.5.7. It consists of two voltage source and three passive elements: resistor R, capacitor C, and inductor L.

A block diagram of this circuit can be constructed, considering that the input is represented by the voltages v_1 and v_2, and the outputs are the currents through the circuit components i_R, i_C, and i_L.

The loop method is used to derive the circuit model. The loop currents i_1 and i_2 are assigned as shown in Figure 4.5.7. The currents through the circuit components are expressed in terms of loop currents as follows:

$$i_R = i_1 \tag{1}$$

$$i_L = i_1 + i_2 = i_R + i_C \tag{2}$$

$$i_C = i_2 \tag{3}$$

Applying KVL to the circuit loops, we obtain

$$v_1 = v_R + v_L = i_R R + L\frac{di_L}{dt} \tag{4}$$

$$v_2 = v_C + v_L = \frac{1}{C}\int i_C dt + L\frac{di_L}{dt} \tag{5}$$

Taking the Laplace transform of Eqs. (2), (4), and (5) yields the relations necessary for constructing individual components of the sought block diagram:

$$V_1(s) = I_R(s)R + LsI_L(s) \tag{6}$$

$$V_2(s) = \frac{1}{Cs}I_C(s) + LsI_L(s) \tag{7}$$

$$I_L(s) = I_R(s) + I_C(s) \tag{8}$$

Rearranging Eqs. (6) and (7) to have currents as outputs, the block diagram components are constructed as shown in Figure 4.5.8:

$$I_R(s) = \frac{1}{R}\left(V_1(s) - LsI_L(s)\right) \tag{9}$$

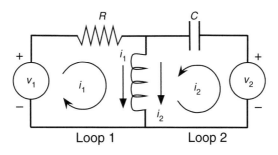

Figure 4.5.7 RLC circuit with multiple energy sources

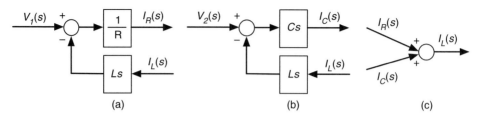

Figure 4.5.8 Component block diagrams for the circuit of Example 4.5.2: (a) corresponds to Eq. (9), (b) to Eq. (10), and (c) corresponds to Eq. (8)

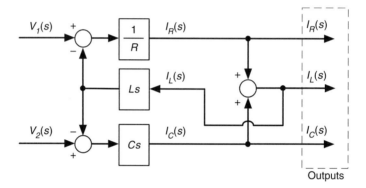

Figure 4.5.9 Block diagram for the circuit of Example 4.5.2

$$I_C(s) = Cs\left(V_2(s) - LsI_L(s)\right) \qquad (10)$$

Combining these components yields the system block diagram as shown in Figure 4.5.9.

For verification purposes, the transfer functions for the current $I_L(s)$ are derived using the block diagram above, and compared with the differential equation where $i_L(t)$ is an independent variable.

From the block diagram, the expression for $I_L(s)$ is derived as follows:

$$I_L(s) = \frac{1}{R}\left(V_1(s) - LsI_L(s)\right) + Cs\left(V_2(s) - LsI_L(s)\right) \qquad (11)$$

$$\left(LRCs^2 + Ls + R\right)I_L(s) = V_1(s) + RCsV_2(s)$$

From Eq. (11), the transfer functions are easily obtainable as follows:

$$\frac{I_L(s)}{V_1(s)} = \frac{1}{LRCs^2 + Ls + R} \qquad (12)$$

$$\frac{I_L(s)}{V_2(s)} = \frac{RCs}{LRCs^2 + Ls + R} \qquad (13)$$

To derive the sought differential equation in terms of $i_L(t)$, we derive the expression for $i_R(t)$ from Eq. (4), and the expression for $i_C(t)$ from Eq. (2):

$$i_R = \frac{1}{R}\left(v_1 - L\frac{di_L}{dt}\right) \quad (14)$$

$$i_C = i_L - i_R = i_L - \frac{1}{R}\left(v_1 - L\frac{di_L}{dt}\right) \quad (15)$$

We substitute the result of Eq. (15) into Eq. (5) and take a derivative with respect to time to get rid of the integral:

$$\frac{dv_2}{dt} = \frac{1}{C}i_C + L\frac{d^2 i_L}{dt^2} = \frac{1}{C}i_L - \frac{1}{RC}\left(v_1 - L\frac{di_L}{dt}\right) + L\frac{d^2 i_L}{dt^2} \quad (16)$$

Then, the system model in terms of the current through the inductor as an independent variable is

$$LRC\frac{d^2 i_L}{dt^2} + L\frac{di_L}{dt} + Ri_L = v_1 + RC\frac{dv_2}{dt} \quad (17)$$

It is obvious that the Eq. (17) is consistent with the transfer functions described with Eqs. (12) and (13).

4.5.2 State-Space Representation

Similarly to the other dynamic systems, an electric circuit can be modeled with differential equations, transfer functions, block diagrams, and state-space representation. Generally, an electric circuit has multiple current and voltage variables as well as more than a single energy source. An incorrectly selected set of independent variables may lead to an unduly complicated model, and, subsequently, difficulties in simulation and analysis of the circuit.

A state-space representation, first introduced in Section 2.5.4, is a way of modeling an electrical circuit simply and efficiently. A linear state-space model can be derived for the circuits discussed in this chapter. These state equations are a set of coupled first-order ordinary differential equations of the form represented by Eq. (2.5.10).

The state variables, capable of comprehensively modeling an electrical circuit, describe the energy stored by the circuit elements. Every energy-storing element is assigned a single state variable.

As described in Section 4.1.2, the only elements that store energy are inductors and capacitors. The energy-storing capability is evidenced by the presence of a derivative in their voltage–current relations. Hence, the number of state variables for a circuit equals the number of inductors and capacitors in that circuit. Resistors do not get a state variable since they don't store any energy.

The form of a linear state equation, where the left-hand side contains the first derivative of a state variable, and the right-hand side does not have any derivatives of state variables and inputs, prompts the choice of the state variables. Examining the derivative form of the voltage–current relation for a capacitor (Eq. (4.1.16)), the assignment of a state variable to obtain the linear state equation is easy – it is a parameter that has the first derivative taken, which is the voltage. Thus, the state variable x and the corresponding state equation for a capacitor are:

$$x = v(t)$$
$$\dot{x} = \frac{1}{C} i(t) \quad (4.5.5)$$

Similarly, for an inductor, the current is selected a state variable (Eq. (4.1.21)):

$$x = i(t)$$
$$\dot{x} = \frac{1}{L} v(t) \quad (4.5.6)$$

Upon the assignment of state variables, any of the presented earlier circuit analysis methods can be used to derive the governing equations. It is customary to present the circuit state-space model, described by the state and output equations, in matrix form.

The following examples illustrate the derivation of a state-space model for electrical circuits.

Example 4.5.3

Derive the state-space model of a circuit shown in Figure 4.5.10, and present it in matrix form. The voltage v_{in}, supplied by the energy source, is the system input, and voltage drop v_{out} across the capacitor is the output.

Solution

This circuit needs two state variables since it has two energy-storing elements – one inductor L and one capacitor C. Therefore, as discussed above, the state variables are: the current i through an inductor, and the voltage v across the capacitor:

$$x_1 = i \quad (1)$$
$$x_2 = v_{out} \quad (2)$$

Before applying KVL to this circuit, we write the voltage–current relations for its components in terms of the above state variables:

$$\text{Resistor} \quad v_R = iR = x_1 R \quad (3)$$

4.5 State-Space Representations and Block Diagrams

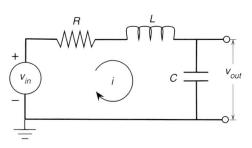

Figure 4.5.10 *RLC* circuit for Example 4.5.3

$$\text{Capacitor} \quad v_C = v_{out} = x_2 = \frac{1}{C}\int i\,dt = \frac{1}{C}\int x_1\,dt \tag{4}$$

$$\frac{dx_2}{dt} = \dot{x}_2 = \frac{1}{C}x_1 \tag{5}$$

$$\text{Inductor} \quad v_L = L\frac{di}{dt} = L\dot{x}_1 \tag{6}$$

Applying KVL to the circuit, we obtain

$$v_{in} = v_R + v_L + v_C = iR + L\frac{di}{dt} + v_{out} \tag{7}$$

In terms of state variables Eq. (7) becomes

$$v_{in} = x_1 R + L\dot{x}_1 + x_2 \tag{8}$$

Then, the state-space model of this circuit is represented as follows:

$$\text{State equations} \quad \begin{cases} \dot{x}_1 = -\frac{R}{L}x_1 - \frac{1}{L}x_2 + \frac{1}{L}v_{in} \\ \dot{x}_2 = \frac{1}{C}x_1 \end{cases}$$

$$\text{Output equation} \quad y = x_2 \tag{9}$$

In matrix form

$$\begin{aligned} \dot{X} &= A \cdot X + B \cdot u \\ y &= C \cdot X \end{aligned} \tag{10}$$

where the components are:

State vector $\quad X = \begin{bmatrix} x_1 \\ x_2 \end{bmatrix}$

Input (scalar) $\quad u = v_{in}$

Matrices $\quad A = \begin{bmatrix} -\dfrac{R}{L} & -\dfrac{1}{L} \\ \dfrac{1}{C} & 0 \end{bmatrix}, \ B = \begin{bmatrix} \dfrac{1}{L} \\ 0 \end{bmatrix}, \ C = \begin{bmatrix} 0 & 1 \end{bmatrix}$

Example 4.5.4

Derive the state-space model of a dual-resistance circuit, discussed in Example 4.4.2, and present it in matrix form. The voltage v_{in}, supplied by the energy source, is the system input, and the voltage drop v_{out} across the capacitor C_2 is the output.

This circuit has two energy-storing components – capacitors C_1 and C_2. Hence, it needs two state variables, defined as:

$$x_1 = v_{C1} = e_1 \\ x_2 = v_{C2} = v_{out} \quad (1)$$

The previously derived component equations are expressed in terms of the state variables defined by Eq. (1):

$$i_{R1} = \frac{v_{in} - e_1}{R_1} = \frac{v_{in} - x_1}{R_1} \quad (2)$$

$$i_{R2} = \frac{e_1 - v_{out}}{R_2} = \frac{x_1 - x_2}{R_2} \quad (3)$$

$$i_{C1} = C_1 \frac{de_1}{dt} = C_1 \frac{dx_1}{dt} = C_1 \dot{x}_1 \quad (4)$$

$$i_{C2} = C_2 \frac{dv_{out}}{dt} = C_2 \frac{dx_2}{dt} = C_2 \dot{x}_2 \quad (5)$$

Re-writing the derived KCL equations in terms of state variables, we obtain

$$\frac{v_{in} - x_1}{R_1} - C_1 \frac{dx_1}{dt} - \frac{x_1 - x_2}{R_2} = 0 \quad (6)$$

$$\frac{x_1 - x_2}{R_2} - C_2 \frac{dx_2}{dt} = 0 \quad (7)$$

Then, the state-space model of this circuit is derived as follows:

State equations
$$\begin{cases} \dot{x}_1 = \frac{1}{C_1}\left(\frac{v_{in} - x_1}{R_1} - \frac{x_1 - x_2}{R_2}\right) \\ \dot{x}_2 = \frac{x_1 - x_2}{R_2 C_2} \end{cases}$$

Output equation $\quad y = x_2 \quad (8)$

In matrix form:

$$\dot{X} = A \cdot X + B \cdot u \\ y = C \cdot X \quad (9)$$

where the components are:

$$\text{State vector} \quad X = \begin{bmatrix} x_1 \\ x_2 \end{bmatrix}$$

$$\text{Input (scalar)} \quad u = v_{in}$$

$$\text{Matrices} \quad A = \begin{bmatrix} -\dfrac{R_1+R_2}{R_1R_2C_1} & \dfrac{1}{R_2C_1} \\ \dfrac{1}{R_2C_2} & -\dfrac{1}{R_2C_2} \end{bmatrix}, \quad B = \begin{bmatrix} \dfrac{1}{R_1C_1} \\ 0 \end{bmatrix},$$

$$C = \begin{bmatrix} 0 & 1 \end{bmatrix}$$

Example 4.5.5

Derive the state-space model of a circuit shown in Figure 4.5.11, and present it in matrix form. The voltage v_{in}, supplied by the energy source, is the system input, and the voltage drop v_{out} across the capacitor C_2 is the output.

This circuit needs three state variables since it has three energy-storing elements – one inductor L and two capacitors C_1 and C_2. Therefore:

$$\begin{aligned} x_1 &= i_L \\ x_2 &= v_{C1} = e_2 \\ x_3 &= v_{C2} = e_3 = v_{out} \end{aligned} \quad (1)$$

Note that e_1 – voltage at the node 1 – is an intermediate variable, which will be eliminated during the derivation of governing equations.

The most efficient way to derive a state-space representation of this circuit is by using the node method. The nodes are defined as shown in the circuit schematic in Figure 4.5.11. The expressions for the current through every circuit element are derived in terms of state variables as follows:

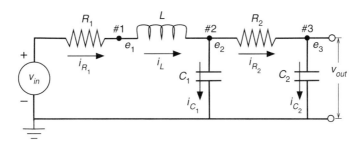

Figure 4.5.11 Circuit for Example 4.5.5

$$i_{R1} = i_L = x_1 = \frac{v_{in} - e_1}{R_1} \tag{2}$$

$$i_{R2} = \frac{e_2 - e_3}{R_2} = \frac{e_2 - v_{out}}{R_2} = \frac{x_2 - x_3}{R_2} \tag{3}$$

$$i_{C1} = C_1 \frac{de_2}{dt} = C_1 \frac{dx_2}{dt} = C_1 \dot{x}_2 \tag{4}$$

$$i_{C2} = C_2 \frac{dv_{out}}{dt} = C_2 \frac{dx_3}{dt} = C_2 \dot{x}_3 \tag{5}$$

From Eq. (2):

$$e_1 = v_{in} - R_1 x_1 \tag{6}$$

Also, from Eq. (6) and the voltage–current relation for an inductor:

$$\frac{di_L}{dt} = \frac{1}{L}(e_1 - e_2) = \dot{x}_1 = \frac{1}{L}(v_{in} - R_1 x_1 - x_2) \tag{7}$$

Apply KCL to the identified nodes:

Node 1	$i_{R1} = i_L$	(8)
Node 2	$i_L = i_{C1} + i_{R2}$	(9)
Node 2	$i_{R2} = i_{C2}$	(10)

Combining the element equations with the node equations and performing the necessary algebraic transformations, the state-space model is derived:

State equations
$$\begin{cases} \dot{x}_1 = -\frac{R_1}{L}x_1 - \frac{1}{L}x_2 + \frac{1}{L}v_{in} \\ \dot{x}_2 = \frac{1}{C_1}x_1 - \frac{1}{R_2 C_1}x_2 + \frac{1}{R_2 C_1}x_3 \\ \dot{x}_3 = \frac{1}{R_2 C_2}x_2 - \frac{1}{R_2 C_2}x_3 \end{cases}$$

Output equation $\quad y = x_3 \tag{11}$

In matrix form:

$$\begin{aligned} \dot{X} &= A \cdot X + B \cdot u \\ y &= C \cdot X \end{aligned} \tag{12}$$

where the components are:

State vector $\quad X = \begin{bmatrix} x_1 \\ x_2 \\ x_3 \end{bmatrix}$

Input (scalar) $\quad u = v_{in}$

Matrices $\quad A = \begin{bmatrix} -\dfrac{R_1}{L} & -\dfrac{1}{L} & 0 \\ \dfrac{1}{C_1} & -\dfrac{1}{R_2 C_1} & \dfrac{1}{R_2 C_1} \\ 0 & \dfrac{1}{R_2 C_2} & -\dfrac{1}{R_2 C_2} \end{bmatrix}, \; B = \begin{bmatrix} \dfrac{1}{L} \\ 0 \\ 0 \end{bmatrix},$

$C = \begin{bmatrix} 0 & 0 & 1 \end{bmatrix}$

4.6 Passive Filters

Filters are important component in the modern electrical and electronic devices. They form an indispensable component of any signal-processing hardware/software assembly, performing a variety of operations such as conditioning signals prior to analog-to-digital conversion, smoothing data sets, blurring of images, reducing high- and low-frequency noise in acoustic applications, and many others. Filters can be represented by passive and active circuits. This section deals with the passive-circuit implementation of common filters, such as low-pass, high-pass, and band-pass filters.

4.6.1 Low-Pass Filter

A *low-pass filter* is a passive circuit, the purpose of which is to pass signals with a frequency lower than the designed *cutoff frequency*, and attenuate signals with higher than cutoff frequencies. Two types of this filter are shown in Figure 4.6.1.

To derive the transfer function for these filters and analyze how they attenuate high-frequency signals, use the impedance method.

The transfer function for an *RL* filter is derived using KVL:

$$V_{in}(s) = V_L + V_R = LsI + RI = I(Ls + R) \qquad (4.6.1)$$

Since the output v_{out} is the voltage drop across the resistor, we find

$$I = \frac{V_{out}}{R}$$

Substituting this expression into Eq. (4.6.1), we derive the transfer function

$$G_{RL}(s) = \frac{V_{out}(s)}{V_{in}(s)} = \frac{R}{Ls + R} \qquad (4.6.2)$$

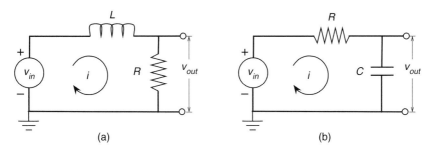

Figure 4.6.1 Passive low-pass filters: (a) series *RL* filter, (b) series *RC* filter

Converting this transfer function to a frequency domain (refer to Chapter 7 for a detailed discussion on frequency response) and finding its magnitude, we obtain

$$|G_{RL}(j\omega)| = \left|\frac{V_{out}(j\omega)}{V_{in}(j\omega)}\right| = \frac{1}{\sqrt{1+L^2R^2\omega^2}} \quad (4.6.3)$$

The cutoff frequency is defined as the frequency at which $V_R = V_L$, and is found as

$$\omega_0 = \frac{1}{RL}$$

Using the same reasoning, derive the transfer function for the RC filter:

$$\begin{aligned} V_{in}(s) &= V_C + V_R = \frac{1}{Cs}I + RI = I\frac{RCs+1}{Cs} \\ V_{out}(s) &= \frac{1}{Cs}I \Rightarrow G_{RC}(s) = \frac{1}{RCs+1} \\ |G_{RC}(j\omega)| &= \left|\frac{V_{out}(j\omega)}{V_{in}(j\omega)}\right| = \frac{1}{\sqrt{1+C^2R^2\omega^2}} \end{aligned} \quad (4.6.4)$$

and its cutoff frequency is

$$\omega_0 = \frac{1}{RC}.$$

To illustrate attenuation of signals with frequency greater than ω_0 generate Bode plots using the `BodePlot` Mathematica function (see Figure 4.6.2).

For both filters, the resistance was the same: $R = 1$ kΩ, and the values of the inductance and capacitance varied as shown in Figure 4.6.2.

As derived in Chapter 7, the amplitude of the frequency response of a dynamic system subject to a periodic input with magnitude A that equals $A|G(j\omega)|$ (see Eqs. (7.3.5) and (7.3.6)). The gain ($|G(j\omega)|$) equals 1 for all input frequencies less than or equal to the cutoff frequency ω_0, which means that these inputs will be passed through the filter without a decrease in amplitude. For signal frequencies greater than ω_0 the amplitude of the response will be decreasing rapidly, approaching zero – this signifies signal attenuation.

The cutoff frequency for both filters increases with the decrease of inductance or capacitance: the lowest ω_0 were registered for the highest inductance $L = 200$ mH and for the highest capacitance $C = 200$ nF.

Mathematica code that generates the Bode plot for RL filter is shown in Figure 4.6.3.

4.6.2 High-Pass Filter

A *high-pass filter* is a passive circuit, the purpose of which is to pass signals with a frequency higher than the designed *cutoff frequency*, and attenuate signals with lower than cutoff frequencies. Two types of this filter are shown in Figure 4.6.4.

4.6 Passive Filters 273

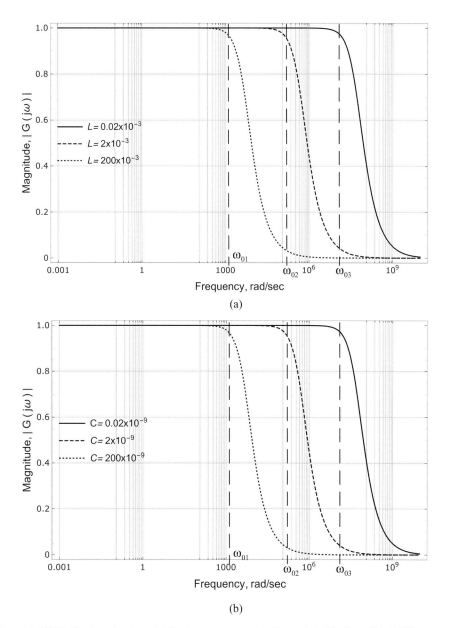

Figure 4.6.2 Bode plots (gain only) for low-pass passive filters: (a) *RL* filter, (b) *RC* filter

Using the same reasoning as in Section 4.6.1, we can derive the transfer functions and generate Bode plots for these filters:

$$|G_{RL}(j\omega)| = \left|\frac{V_{out}(j\omega)}{V_{in}(j\omega)}\right| = \frac{LR\omega}{\sqrt{1+L^2R^2\omega^2}} \qquad (4.6.5)$$

```
tfm = TransferFunctionModel[r/(l*s+r), s];
rule1 = {r → 1000, l → 0.02*10⁻³};
rule2 = {r → 1000, l → 2*10⁻³};
rule3 = {r → 1000, l → 200*10⁻³};
BodePlot[{tfm /. rule1, tfm /. rule2, tfm /. rule3}, {0.001, 10¹⁰}, PlotLayout → "Magnitude",
  PlotStyle → {Black, {Black, Dashed}, {Black, Dotted}}, GridLines → Automatic,
  FrameTicksStyle → Directive["Label", Black, 12],
  FrameLabel → {Style["Frequency, rad/sec", Black, 16],
    Style["Magnitude, | G ( jω ) |", Black, 16]},
  PlotLegends →
   Placed[{Style["L= 0.02x10⁻³", 14], Style["L= 2x10⁻³", 14], Style["L= 200x10⁻³", 14]},
    {Left, Center}], ScalingFunctions → {Automatic, "Absolute"}]
```

Figure 4.6.3 Mathematica code for generating the Bode gain plot for the *RL* filter

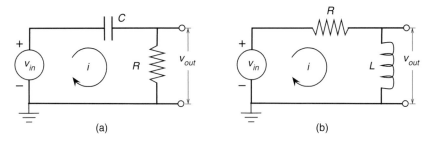

Figure 4.6.4 Passive high-pass filters: (a) series *RC* filter, (b) series *RL* filter

$$|G_{RC}(j\omega)| = \left|\frac{V_{out}(j\omega)}{V_{in}(j\omega)}\right| = \frac{C\omega}{\sqrt{1+C^2R^2\omega^2}} \quad (4.6.6)$$

The Bode plots for these high-pass filters are shown in Figure 4.6.5.

The attenuation of low-frequency signals is clearly seen in these graphs. The same tendency is found: the cutoff frequency ω_0 increases with decreasing inductance or capacitance; and the lowest ω_0 was registered for the highest inductance $L = 200$ mH and for the highest capacitance $C = 200$ nF.

4.6.3 Band-Pass Filter

A *band-pass filter* is a passive circuit, the purpose of which is to pass signals of the frequency being within the specific limits $\omega_{01} \leq \omega \leq \omega_{02}$, where ω_{01} and ω_{02} are the lower and upper cutoff frequencies. It is represented by an *RLC* circuit as shown in Figure 4.6.6.

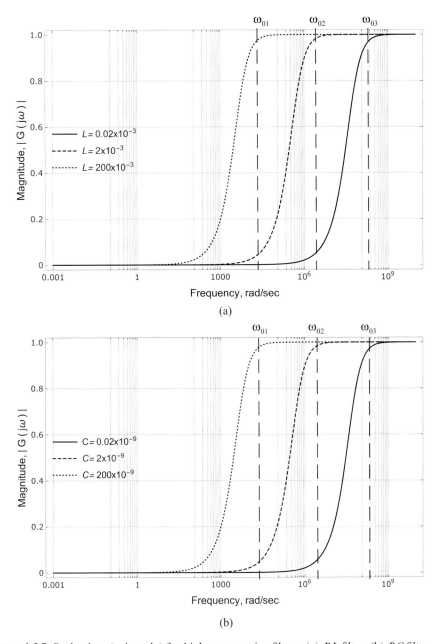

Figure 4.6.5 Bode plots (gain only) for high-pass passive filters: (a) *RL* filter, (b) *RC* filter

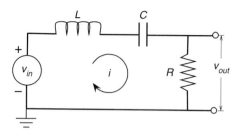

Figure 4.6.6 Band-pass passive filter

The KVL and impedance method is used to derive the transfer function of this circuit:

$$V_{in}(s) = V_C + V_L + V_R = \frac{1}{Cs}I + LsI + RI = I\frac{LCs^2 + RCs + 1}{Cs}$$

$$V_{out}(s) = RI \Rightarrow G_{RLC}(s) = \frac{RCs}{LCs^2 + RCs + 1}$$

$$|G_{RLC}(j\omega)| = \left|\frac{V_{out}(j\omega)}{V_{in}(j\omega)}\right| = \frac{\omega(R/L)}{\sqrt{\left(\frac{1}{LC} - \omega^2\right)^2 + \left(\frac{R}{L}\omega\right)^2}} \qquad (4.6.7)$$

The maximum value of the transfer function is $|G_{RL}(j\omega)| = 1$. The cutoff frequencies are defined as those at which $|G_{RL}(j\omega)| = \frac{1}{\sqrt{2}}$. Solving this equation, we obtain the values of cutoff frequencies:

$$\omega_{01} = -\frac{R}{2L} + \sqrt{\left(\frac{R}{2L}\right)^2 + \frac{1}{LC}}$$

$$\omega_{02} = \frac{R}{2L} + \sqrt{\left(\frac{R}{2L}\right)^2 + \frac{1}{LC}} \qquad (4.6.8)$$

The bandwidth is the difference between the cutoff frequencies:

$$\omega_{01} - \omega_{02} = \frac{R}{L} \qquad (4.6.9)$$

Using the following numeric values for the system parameters: $R = 1\,\text{k}\Omega$, $C = 2\,\text{nF}$, $L = 2\,\mu\text{H}$, the Bode plot shown in Figure 4.6.7 is generated.

This band-pass filter will pass the signals with frequencies between 0.16 MHz and 0.16 GHz, and will attenuate all the others.

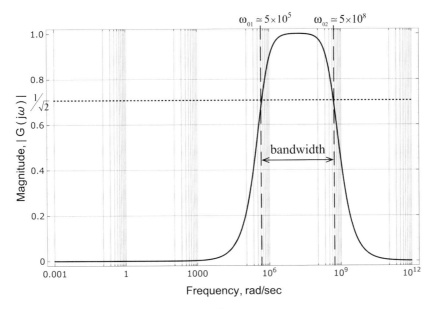

Figure 4.6.7 Bode plot of a band-pass passive filter

4.7 Active-Circuit Analysis

As discussed earlier in this chapter, a generic passive circuit consists of an energy source and passive elements such as resistors, capacitors, and inductors. An *active circuit* also has passive components that constitute a *load*, and an *amplifier* – an integrated circuit that contains transistors, resistors, and capacitors, and is capable of *amplifying* (increasing) an electronic signal without affecting it in any other way.

There exists an extensive classification of amplifiers according to their circuit configurations, type and size of input signal, and frequency of operations. This book is concerned only with the DC small-signal amplifiers, in particular *isolation amplifiers* and *operational amplifiers* (*op-amps*).

4.7.1 Isolation Amplifiers

A voltage-isolation amplifier is an integrated circuit, the purpose of which is to amplify a signal from a low-power source by yielding an output voltage proportional to the input voltage by a predefined factor. Obviously, it requires an external energy source to provide an input voltage. Its design must comply with two important requirements: (a) its operation should not affect the behavior of the energy source circuit, and (b) it should be capable of providing the required voltage independently of the load circuit.

A voltage-isolation amplifier circuit is shown in Figure 4.7.1.

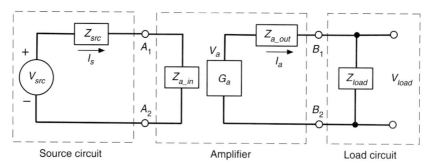

Figure 4.7.1 Voltage isolation amplifier

Here, V_{src} denotes the voltage provided by the energy source, Z_{src} and Z_{load} are the impedances of the source circuit and the load, respectively, V_{load} is the voltage delivered to the load, G_a denotes the voltage gain of the amplifier, and Z_{a_in} and Z_{a_out} are the impedances on the input and output terminals of the amplifier, respectively. All these parameters are in the Laplace domain.

The amplifier input terminals denoted by A_1 and A_2, and the amplifier input voltage, V_a, is the voltage measured at these terminals.

We let I_s be the current through the source circuit impedance Z_{src}. Applying KVL to the source circuit, we obtain

$$V_{src} - I_s Z_{src} - V_a = 0 \qquad (4.7.1)$$

Considering that the impedances Z_{src} and Z_{a_in} are connected in series, I_s is the current drawn by the amplifier circuit. It is obtained using Ohm's law:

$$I_s = \frac{V_a}{Z_{a\_in}} \qquad (4.7.2)$$

Equation (4.7.2) illustrates that, if the amplifier input impedance is very large, I_s will be negligibly small, meaning that the amplifier does not affect the current through the source circuit. Hence, to avoid affecting the behavior of the source circuit, an amplifier must have a large input impedance.

Also, using the voltage divider rule (Eq. (4.2.15)), find that

$$V_a = \frac{Z_{a\_in}}{Z_{src} + Z_{a\_in}} V_{src} \approx V_{src} \qquad (4.7.3)$$

when the amplifier input impedance is sufficiently large, $Z_{a\_in} \gg Z_{src}$.

The amplifier output voltage, measured at the output terminals B_1 and B_2, is found as

$$V_{a\_out} = G_a V_a \approx G_a V_{src} \tag{4.7.4}$$

Then, considering that impedances Z_{a_out} and Z_{load} are connected in series, the application of KVL to the load circuit yields

$$V_{a\_out} - I_a Z_{a\_out} - I_a Z_{load} = 0 \tag{4.7.5}$$

which derives the following expression for the amplifier current I_a:

$$I_a = \frac{V_{a\_out}}{Z_{a\_out} + Z_{load}} = \frac{G_a V_a}{Z_{a\_out} + Z_{load}} \approx \frac{G_a V_{src}}{Z_{a\_out} + Z_{load}} \tag{4.7.6}$$

Thus, the voltage delivered to the load is

$$V_{load} = \frac{Z_{load}}{Z_{a\_out} + Z_{load}} G_a V_{src} \tag{4.7.7}$$

Equation (4.7.7) shows that if the amplifier output impedance, Z_{a_out}, is very small, the voltage delivered to the load becomes independent of the load circuit: $V_{load} \approx G_a V_{src}$.

Hence, a voltage-isolation amplifier must have a large input impedance and a small output impedance in order to avoid affecting the source circuit, while providing the required voltage independently of the load circuit. This voltage is proportional to the voltage, supplied by the energy source, by a factor of G_a – the amplifier gain. Therefore, a voltage-isolation amplifier behaves as a voltage source.

Current-isolation amplifiers provide a current, proportional to the input generated by the energy source. The same essential characteristics apply: the required current is provided regardless of the load circuit, and the amplifier operation does not affect the energy source. Thus, a current-isolation amplifier behaves as a current source. A similar approach can be used to demonstrate that for a current-isolation amplifier:

$$I_{load} \approx G_a I_{src} \tag{4.7.8}$$

Example 4.7.1

Here, we revisit the circuit in the Example 4.4.1, and modify the schematic as shown in Figure 4.7.2, where the amplifier is shown as a simple box with the gain G_a.

In the modified circuit the source and load loops are connected through a voltage-isolation amplifier instead of directly to each other. The amplifier is assumed to comply with the impedance requirements discussed above. Consequently, it does not affect the behavior of the source circuit, while providing the voltage independently of the load circuit. The amplifier then prevents the source circuit (loop 1 in Example 4.4.1) from influencing the load circuit (loop 2 in the Example 4.4.1), thus, creating two separate independent loops.

Consider the source circuit. The voltage across the capacitor C_1 equals the input voltage into the amplifier, V_{a_in}, and is derived using the voltage divider rule (Eq. (4.2.15)):

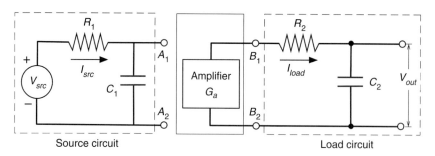

Figure 4.7.2 Circuit for Example 4.7.1

$$V_{C1} = V_{src} \frac{1/C_1 s}{R_1 + 1/C_1 s} = \frac{1}{R_1 C_1 s + 1} V_{src} = V_{a\_in} \quad (1)$$

Note, that the transfer function of this loop is $\frac{1}{R_1 C_1 s + 1}$.

According to Eq. (4.7.4) the amplifier output voltage is

$$V_{a\_out} = G_a V_{a\_in} \quad (2)$$

Consider the load circuit. The output voltage equals the voltage across the capacitor C_2 that is derived using the voltage divider rule (Eq. (4.2.15)):

$$V_{out} = V_{C2} = V_{a\_out} \frac{1/C_2 s}{R_2 + 1/C_2 s} = \frac{1}{R_2 C_2 s + 1} V_{a\_out} \quad (3)$$

Note, that the transfer function of the load loop is $\frac{1}{R_2 C_2 s + 1}$.

Substituting the expressions for the input and output voltages of the amplifier (Eqs. (1) and (2)) into Eq. (3), we obtain the circuit model in Laplace domain:

$$V_{out} = \frac{1}{R_2 C_2 s + 1} V_{a\_out} = \frac{1}{R_2 C_2 s + 1} G_a \frac{1}{R_1 C_1 s + 1} V_{src}$$
$$\left(R_1 R_2 C_1 C_2 s^2 + (R_1 C_1 + R_2 C_2)s + 1\right) V_{out} = G_a V_{src} \quad (4)$$

By taking the inverse Laplace transform of Eq. (4) obtain the system model in the time domain:

$$G_a v_{src} = R_1 R_2 C_1 C_2 \frac{d^2 v_{out}}{dt^2} + (R_1 C_1 + R_2 C_2) \frac{d v_{out}}{dt} + v_{out} \quad (5)$$

Comparing the derived governing equation (Eq. (5)) with that obtained for the unmodified circuit (Eq. (10) of Example 4.4.1), one can clearly see the difference. Even if assuming $G_a = 1$, these models are still not identical. As seen from Eq. (4), the transfer function of the circuit with the isolation amplifier is a product of the transfer functions of two independent circuits – the source loop and the load loop, while for the circuit without the amplifier it is not (see Eq. (9) of

Example 4.4.1). This illustrates the influence two dependent loops exert on each other's voltages and currents, as presented in Example 4.4.1.

Therefore, in any multiple-loop circuit, where the loops are connected end to end, there exists the so-called *loading effect* – the voltage and current in any given loop are affected by the adjacent loops. An isolation amplifier, placed between loops, eliminates this effect, making the loops independent of each other.

4.7.2 Operational Amplifiers

An operational amplifier is capable of almost ideal amplification of DC signals, which makes it one of the major building blocks of analog electronic circuits. In addition to signal amplification, op-amps are used for signal filtering, and for performing mathematical operations such as, for example, addition/ subtraction and integration/ differentiation.

While the concept of an op-amp is not new – the original vacuum-tube-based device was invented at the Bell Laboratories in the early 1940s and became commercially available approximately 10 years later – the op-amps are still widely used. The most common op-amp application is as a component of control systems.

An ideal op-amp is a three-terminal device as shown in Figure 4.7.3. The physical integrated circuit (chip) has a relatively complicated structure and more than three terminals, but this simplified representation is sufficient for analysis and, consequently, for representation on electrical circuit schematics.

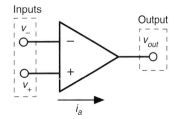

Figure 4.7.3 Ideal inverting op-amp

A typical linear op-amp, as shown in Figure 4.7.3, is a voltage amplifier with a very large gain. It has two high-impedance inputs – an *inverting input* denoted by v_-, and a *non-inverting input* denoted by v_+. Then, the actual voltage input of this op-amp is as follows:

$$v_{in} = v_+ - v_- \qquad (4.7.9)$$

Since only linear op-amps are considered, the output voltage is proportional to the input voltage as follows:

$$v_{out} = K v_{in} = K(v_+ - v_-) \qquad (4.7.10)$$

where K is the op-amp gain. For an ideal device, this gain is assumed to be infinite, while for a physical device it is in the range of 10^5–10^6.

The output terminal, v_{out}, can be viewed as an *amplification factor*, and is often referred to as the *op-amp gain*. The output impedance of an ideal op-amp is assumed to be zero, while the physical devices usually have a small internal resistance in the order of 100–20 Ω.

Due to the very large input impedance, an op-amp draws negligible current:

$$i_a \approx 0 \tag{4.7.11}$$

Therefore, the voltage difference between the input terminals approaches zero:

$$i_a = \frac{v_{out}}{K} = v_+ - v_- \approx 0 \tag{4.7.12}$$

$$v_+ \approx v_-$$

The expression

$$v_+ \approx v_- \tag{4.7.13}$$

is customarily called *the op-amp equation*.

The concept of impedance is instrumental for modeling and analysis of a typical op-amp circuit, as shown in Figure 4.7.4. Since impedance is traditionally expressed in the Laplace domain, in the context of this analysis all the voltages, currents, and electric potentials are also expressed in Laplace domain.

A typical inverting op-amp circuit is presented in Figure 4.7.4, where Z_1 indicates the input impedance, while Z_2 denotes the output or *feedback* impedance; V_{in} is the input voltage, V_{out} is the output voltage, E_{in} signifies the electric potential on the inverting terminal, and E_{out} is the electric potential on the output terminal.

Recalling the voltage–current relation of an impedance (Eq. (4.2.1)):

$$I(s) = \frac{V(s)}{Z} \tag{4.7.14}$$

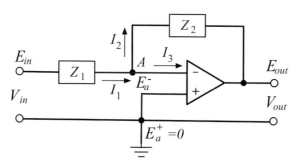

Figure 4.7.4 Typical inverting op-amp circuit

Applying KCL to the node A, we obtain

$$I_1 = I_2 + I_3 \qquad (4.7.15)$$

Since an op-amp draws negligible current, $I_3 = I_a \approx 0$. Then, the currents through the input and output impedances are equal:

$$I_1 = I_2 \qquad (4.7.16)$$

From Eq. (4.6.14), we obtain the expressions for these currents:

$$I_1 = \frac{E_{in} - E_a^-}{Z_1}$$
$$I_2 = \frac{E_a^- - E_{out}}{Z_2} \qquad (4.7.17)$$

and substitute them into Eq. (4.6.16):

$$\frac{E_{in} - E_a^-}{Z_1} = \frac{E_a^- - E_{out}}{Z_2} \qquad (4.7.18)$$

Considering that the non-inverting input terminal of this op-amp is grounded, the input and output voltages are, respectively,

$$V_{in} = E_{in} - E_a^+ = E_{in} - 0 = E_{in}$$
$$V_{out} = E_{out} - E_a^+ = E_{out} - 0 = E_{out} \qquad (4.7.19)$$

From the op-amp equation (Eq. (4.7.13)):

$$E_a^+ \approx E_a^- \approx 0 \qquad (4.7.20)$$

Substituting Eqs. (4.7.19) and (4.7.20) into Eq. (4.7.18), we obtain

$$\frac{V_{in}}{Z_1} = -\frac{V_{out}}{Z_2} \quad \text{or} \quad V_{out} = -\frac{Z_2}{Z_1} V_{in} \qquad (4.7.21)$$

This equation illustrates the concept of the inverting op-amp circuit operation – the amplification and inversion of the input voltage.

Depending on the type of basic elements used as impedances Z_1 and Z_2, an op-amp circuit can perform different functions.

4.7.3 Operational Amplifier Circuits

Multiplier (Amplifier) Circuits

The impedances Z_1 and Z_2 in an op-amp circuit are represented by two resistances R_i and R_f as shown in Figure 4.7.5.

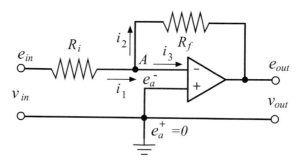

Figure 4.7.5 Multiplier circuit

The multiplier equation can be obtained by using the same reasoning as for the previous derivation. Note that for this derivation all the currents and voltages are expressed in the time domain.

Applying KCL to the node A, we obtain

$$i_1 = i_2 + i_3 \qquad (4.7.22)$$

Since an op-amp draws negligible current, $i_3 = i_a \approx 0$. Therefore:

$$i_1 = i_2 \qquad (4.7.23)$$

Using Eq. (4.1.8) to obtain the expressions for these currents, and substituting them into Eq. (4.7.23):

$$\frac{e_{in} - e_a^-}{R_i} = \frac{e_a^- - e_{out}}{R_f} \qquad (4.7.24)$$

Since the non-inverting input terminal of this op-amp is grounded, the input and output voltages are, respectively,

$$\begin{aligned} v_{in} &= e_{in} - e_a^+ = e_{in} - 0 = e_{in} \\ v_{out} &= e_{out} - e_a^+ = e_{out} - 0 = e_{out} \end{aligned} \qquad (4.7.25)$$

From the op-amp equation (Eq. (4.7.13)):

$$e_a^+ \approx e_a^- \approx 0 \qquad (4.7.26)$$

Substituting Eqs. (4.7.25) and (4.7.26) into Eq. (4.7.24), we obtain

$$\frac{v_{in}}{R_i} = -\frac{v_{out}}{R_f} \quad \text{or} \quad v_{out} = -\frac{R_f}{R_i} v_{in} \qquad (4.7.27)$$

Hence, in a multiplier, the output voltage is proportional to the input voltage by the factor $\dfrac{R_f}{R_i}$, and the sign of the output voltage is inverted.

The multiplier equation (Eq. (4.7.27)) can be alternatively derived using Eqs. (4.7.21) and (4.2.3). Since $Z_1 = R_i$ and $Z_2 = R_f$, substituting into Eq. (4.7.21) yields

$$V_{out} = -\dfrac{R_f}{R_i} V_{in}$$

or in time domain

$$v_{out} = -\dfrac{R_f}{R_i} v_{in}$$

Inverter (Sign-Changer) Circuits

An inverter op-amp circuit is a variation of the multiplier circuit, where the impedances Z_1 and Z_2 are represented by two equal resistances R as shown in Figure 4.7.6. This schematic utilizes the simplified representation of an inverting op-amp – a triangle with the minus sign inside.

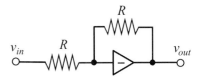

Figure 4.7.6 Inverter circuit

Using Eq. (4.7.27), the inverter equation is easily derived:

$$v_{out} = -\dfrac{R}{R} v_{in} = -v_{in} \qquad (4.7.28)$$

An inverter circuit does not change the magnitude of an input voltage, but changes its sign. It is often used as an inverting buffer in logic circuits.

Integrator Circuits

In an integrator op-amp circuit, the impedance Z_1 is represented by a resistor R, and Z_2 by a capacitor C, as shown in Figure 4.7.7.

The integrator equation can be obtained by using the same reasoning as before. Recalling that $i_1 = i_2$, and using the expressions for the voltage–current relation for resistors and capacitors (Eqs. (4.1.8) and (4.1.16)), we obtain

$$\dfrac{e_{in} - e_a^-}{R} = C \dfrac{d}{dt}(e_a^- - e_{out}) \qquad (4.7.29)$$

Furthermore, recalling that $v_{in} = e_{in}$, $v_{out} = e_{out}$, and $e_a^+ \approx e_a^- \approx 0$:

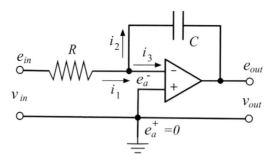

Figure 4.7.7 Integrator circuit

$$\frac{v_{in}}{R} = -C\frac{dv_{out}}{dt} \quad \text{or} \quad v_{out} = -\frac{1}{RC}\int_0^t v_{in}dt \qquad (4.7.30)$$

Alternatively, the integrator equation (Eq. (4.7.30)) is derived using Eqs. (4.7.21), (4.2.3), and (4.2.4). Since $Z_1 = R$ and $Z_2 = \frac{1}{Cs}$, substituting into Eq. (4.7.21) yields:

$$V_{out} = -\frac{1}{RCs}V_{in}$$

or in time domain

$$v_{out} = -\frac{1}{RC}\int_0^t v_{in}dt$$

As illustrated by the integrator equation (Eq. (4.7.30)), this circuit provides the output voltage proportional to the time integral of its input voltage by the factor of $\frac{1}{RC}$, and with the opposite sign. Unlike the multiplier and inverter op-amp circuits, the integrator circuit causes the output voltage to respond to the changes in the input voltage over time rather than to the input magnitude.

Due to the presence of a feedback capacitor, a nonideal integrator experiences the undesirable effect of an output-voltage offset. A capacitor functions as an open circuit until accumulating sufficient charge or *saturating* (see Section 4.1.2). Hence, while the capacitor is charging, the integrator loses its feedback and acts as a regular op-amp, described by Eq. (4.7.10). That means that even a very small voltage input results in a high output due to the high gain of an op-amp. This may push the circuit beyond its operating limits, which is, obviously, unacceptable.

To combat this weakness, the integrator circuit is modified as shown in Figure 4.7.8. In electrical engineering, this design is sometimes called a *lag* or *first-order low-pass filter*.

The addition of a resistor R_f parallel to the capacitor in the feedback line ensures that no feedback is lost while the capacitor is charging, thus minimizing an output-voltage offset.

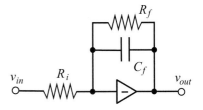

Figure 4.7.8 Lag circuit

The easiest way to derive a governing equation for this op-amp circuit is by using Eq. (4.7.21) and the known expressions for resistor and capacitor impedances as well as that for the parallel connection of impedances (Eqs. (4.2.3), (4.2.4), and (4.2.9)):

$$Z_1 = R_i$$
$$\frac{1}{Z_2} = \frac{1}{R_f} + \frac{1}{1/C_f s} \Rightarrow Z_2 = \frac{R_f}{1 + R_f C_f s} \quad (4.7.31)$$

Then, the relation between the input and output voltages in the Laplace domain is

$$V_{out} = -\frac{Z_2}{Z_1} V_{in} = -\frac{R_f}{R_i(1 + R_f C_f s)} V_{in} \quad (4.7.32)$$

Thus, while the capacitor is charging, the output voltage becomes $V_{out} = -\frac{R_f}{R_i} V_{in}$ instead of the undesirable large-value $V_{out} = KV_{in}$ (see Eq. (4.7.10)) as it was for a simple integrator.

Transforming Eq. (4.7.32) to the time domain, we obtain

$$V_{out} + R_f C_f s V_{out} = -\frac{R_f}{R_i} V_{in}$$
$$v_{out} + R_f C_f \frac{dv_{out}}{dt} = -\frac{R_f}{R_i} v_{in} \quad (4.7.33)$$

Typically, the relation between resistances of the added feedback component R_f and the original resistor R_i is $R_f > 10R_i$.

Another modification of an integrator circuit is called a *bandwidth-limited integrator*, shown in Figure 4.7.9.

The governing equation for this circuit is derived using Eq. (4.7.21) and the known expressions for the resistor and capacitor impedances as well as the one for the series connection of impedances (Eqs. (4.2.3), (4.2.4), and (4.2.7)):

$$Z_1 = R_i$$
$$Z_2 = R_f + \frac{1}{C_f s} = \frac{1 + R_f C_f s}{C_f s} \quad (4.7.34)$$

Then, the relation between input and output voltages in the Laplace domain is

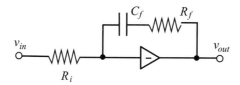

Figure 4.7.9 Bandwidth-limited integrator circuit

$$V_{out} = -\frac{Z_2}{Z_1} V_{in} = -\frac{1 + R_f C_f s}{R_i C_f s} V_{in} \qquad (4.7.35)$$

Transforming Eq. (4.6.35) to the time domain, we obtain

$$R_i C_f \frac{dv_{out}}{dt} = -\left(v_{in} + R_f C_f \frac{dv_{in}}{dt}\right) \qquad (4.7.36)$$

or, integrating once with respect to time and considering zero initial conditions:

$$v_{out} = -\frac{R_f}{R_i} v_{in} - \frac{1}{R_i C_f} \int_0^t v_{in} dt \qquad (4.7.37)$$

Integrator circuits were one of the staples of analog computers. At the present time, they are mostly used in analog-to-digital converters, and various applications that involve wave shaping (for example, a square-wave input yields a triangular-wave output, while a sine input results in a cosine output).

Differentiator Circuits

In a differentiator op-amp circuit, the impedance Z_1 is represented by a capacitor C, and Z_2 by a resistor R, as shown in Figure 4.6.10.

The derivation of its governing equation is straightforward, using the same reasoning as for the circuits discussed above. Since $Z_1 = \frac{1}{Cs}$ and $Z_2 = R$, substitution of these values into Eq. (4.7.21) yields:

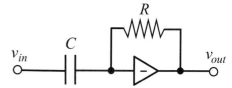

Figure 4.7.10 Differentiator circuit

$$V_{out} = -\frac{Z_2}{Z_1} V_{in} = -RCs V_{in}$$

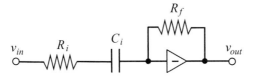

Figure 4.7.11 Bandwidth-limited differentiator circuit

or in time domain

$$v_{out} = -RC\frac{dv_{in}}{dt} \tag{4.7.38}$$

As evidenced by the governing equation (Eq. (4.7.38)), the differentiator circuit reacts to the rate of change of the input voltage. This quality is useful in control applications. When the rate of change of the monitored parameter exceeds the allowed limit, the differentiator circuit yields an output higher than the defined threshold, thus triggering a warning indicator or a specified remedial action.

The sensitivity to the rate of change of the input makes the differentiator circuit very susceptible to high-frequency noise. To combat this inherent weakness, the circuit may be augmented as shown in Figure 4.7.11. This configuration is called a *bandwidth-limited differentiator* or *first-order high-pass filter*.

For this circuit, the impedances Z_1 and Z_2 are evaluated as

$$Z_1 = R_i + \frac{1}{C_i s} = \frac{1 + R_i C_i s}{C_i s}$$

and

$$Z_2 = R_f$$

Substituting these expressions into Eq. (4.7.21), the governing equation in the Laplace domain is

$$V_{out} = -\frac{Z_2}{Z_1} V_{in} = -\frac{R_f C_i s}{1 + R_i C_i s} V_{in} \tag{4.7.39}$$

Transforming Eq. (4.7.31) to the time domain, we obtain

$$v_{out} + R_i C_i \frac{dv_{out}}{dt} = -R_f C_i \frac{dv_{in}}{dt} \tag{4.7.40}$$

Another modification of the differentiator circuit is called the *lead circuit* and is shown in Figure 4.7.12.

For this circuit the impedances Z_1 and Z_2 become

$$\frac{1}{Z_1} = \frac{1}{R_i} + \frac{1}{1/C_i s} \Rightarrow Z_1 = \frac{R_i}{1 + R_i C_i s} \qquad Z_2 = R_f$$

Then, using Eq. (4.7.21), the governing equation is derived as follows:

$$V_{out} = -\frac{Z_2}{Z_1} V_{in} = -\frac{R_f}{R_i}(1 + R_i C_i s) V_{in} \qquad (4.7.41)$$

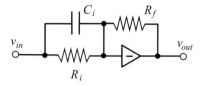

Figure 4.7.12 Lead circuit

Transforming Eq. (4.7.41) to the time domain, we obtain

$$v_{out} = -\frac{R_f}{R_i} v_{in} - R_f C_i \frac{dv_{in}}{dt} \qquad (4.7.42)$$

Integrator–Differentiator Combination Circuits

A *lead-lag* or *lag-lead* op-amp circuit is shown in Figure 4.7.13. In this circuit, the impedances Z_1 and Z_2 are represented by a parallel connection of a resistor and a capacitor.

The impedances are evaluated as follows:

$$\frac{1}{Z_1} = \frac{1}{R_i} + \frac{1}{1/C_i s} \Rightarrow Z_1 = \frac{R_i}{1 + R_i C_i s}$$

$$Z_2 = \frac{1}{R_f} + \frac{1}{1/C_f s} \Rightarrow Z_2 = \frac{R_f}{1 + R_f C_f s}$$

Substituting these expressions into Eq. (4.7.21), we derive the governing equation for this circuit:

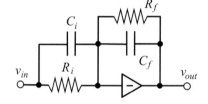

Figure 4.7.13 Lead-lag circuit

$$V_{out} = -\frac{Z_2}{Z_1} V_{in} = -\frac{R_f}{R_i} \frac{1 + R_i C_i s}{1 + R_f C_f s} V_{in} \qquad (4.7.43)$$

The governing equation in the time domain is then

$$\frac{1}{R_f} v_{out} + C_f \frac{dv_{out}}{dt} = -\frac{1}{R_i} v_{in} - C_i \frac{dv_{in}}{dt} \qquad (4.7.44)$$

Another interesting combination of the integrator and differentiator components – the *second-order band-pass filter* – is shown in Figure 4.7.14.

This circuit is very useful in cutting off both high-frequency and low-frequency noise.

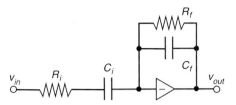

Figure 4.7.14 Second-order band-pass filter

The impedances Z_1 and Z_2 in this circuit are: $Z_1 = R_i + \dfrac{1}{C_i s} \Rightarrow Z_1 = \dfrac{1 + R_i C_i s}{C_i s}$ and $Z_2 = \dfrac{1}{R_f} + \dfrac{1}{1/C_f s} \Rightarrow Z_2 = \dfrac{R_f}{1 + R_f C_f s}$

The governing equation is derived by substituting these expressions into Eq. (4.7.21) as follows:

$$V_{out} = -\dfrac{Z_2}{Z_1} V_{in} = -\dfrac{R_f}{(1 + R_f C_f s)} \dfrac{C_i s}{(1 + R_i C_i s)} V_{in} \quad (4.7.45)$$

In the time domain, the governing equation is a second-order differential equation:

$$\dfrac{1}{R_f} v_{out} + \left(C_f + \dfrac{R_i}{R_f} C_i\right) \dfrac{dv_{out}}{dt} + R_i C_i C_f \dfrac{d^2 v_{out}}{dt^2} = -C_i \dfrac{dv_{in}}{dt} \quad (4.7.46)$$

Adder, Subtractor, and Comparator Circuits

It is possible to add, subtract, and compare signals by using the appropriate op-amp circuits.

The circuit for an *adder* is shown in Figure 4.7.15.

This circuit consists of two component op-amp circuits – a *summing amplifier* and an *inverter*. The inverter was discussed above and its governing equation was derived (Eq. (4.7.28)). To derive the relation between the inputs and the output of the summing amplifier KCL is applied to the node A:

$$i_{R1} + i_{R2} = i_{R3} \quad (4.7.47)$$

Using Ohm's law for a current through a resistor, and remembering that the shown op-amp inverts the sign of the input voltage, Eq. (4.7.47) becomes

$$\dfrac{v_{in1}}{R_1} + \dfrac{v_{in2}}{R_2} = -\dfrac{v_B}{R_3} \quad (4.7.48)$$

where v_B is the voltage at node B, which is an input voltage into an inverter.

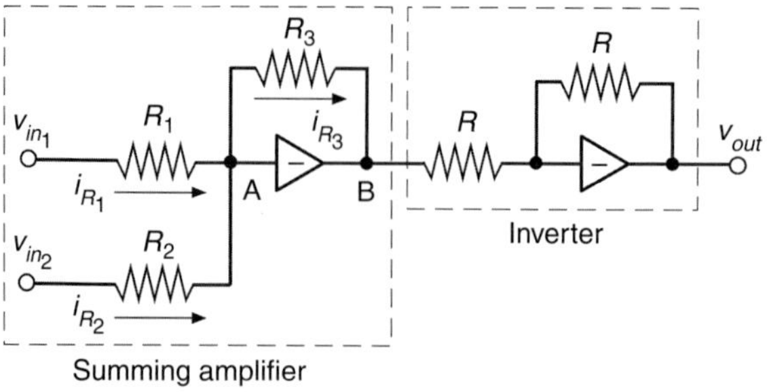

Figure 4.7.15 Adder circuit

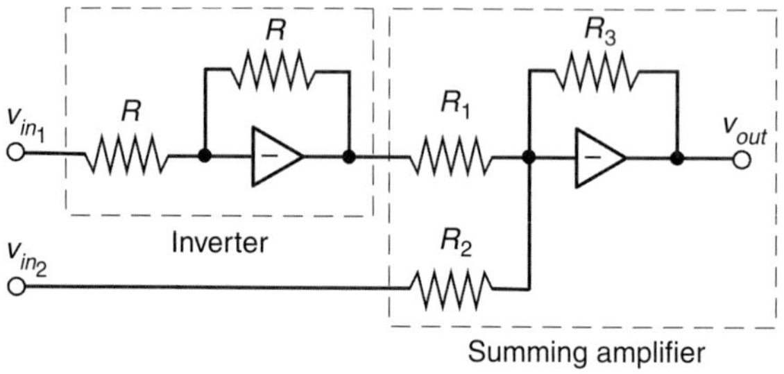

Figure 4.7.16 Subtractor circuit

Since the output of the inverter component is $v_{out} = -v_B$, the governing equation is derived as follows:

$$v_{out} = \frac{R_3}{R_1} v_{in1} + \frac{R_3}{R_2} v_{in2} \qquad (4.7.49)$$

Hence, the output voltage is a weighted sum of input voltages, where weighting factors are defined by the values of the resistances. If all the resistors are the same – $R_1 = R_2 = R_3$ – then the output voltage becomes a straight sum of input voltages, so that $v_{out} = v_{in1} + v_{in2}$.

In the *subtractor* circuit, shown in Figure 4.7.16, one of the input signals is first inverted and then fed into a summing amplifier, while the other input is taken as is. Similarly, the governing equation is derived as follows:

$$-\frac{v_{in1}}{R_1} + \frac{v_{in2}}{R_2} = -\frac{v_{out}}{R_3}$$
$$v_{out} = \frac{R_3}{R_1} v_{in1} - \frac{R_3}{R_2} v_{in2} \qquad (4.7.50)$$

The subtractor with equal resistances – $R_1 = R_2 = R_3$ – is called *comparator*, and its governing equation is then: $v_{out} = v_{in1} - v_{in2}$.

Non-inverting Amplifiers

All the circuits discussed in this section are *inverting voltage amplifiers*, meaning that the output voltage is scaled in relation to the input voltage, and there is a change of sign between the output and the input. In the inverting configuration, the input voltage is applied to the inverting (−) terminal of the op-amp.

The non-inverting op-amp configuration applies the input voltage to the non-inverting (+) terminal of the op-amp, which provides for no sign change between the output and the input, while signal scaling still takes place.

A non-inverting amplifier circuit (*multiplier*) is shown in Figure 4.7.17.

The same procedure as for the inverting op-amp is used to derive the governing equation for this circuit. As before, the grounded wire with zero electric potential is chosen as a reference for the determination of the input and output voltages, v_{in} and v_{out} respectively:

$$v_{in} = e_{in} - 0 = e_{in}$$
$$v_{out} = e_{out} - 0 = e_{out} \quad (4.7.51)$$

Applying KCL to the node A, we obtain

$$i_1 = i_2 + i_3 \quad (4.7.52)$$

Recalling that an op-amp draws negligible current, $i_3 = i_a \approx 0$, the currents through the resistors R_i and R_f are found to be equal:

$$i_1 = i_2 \quad (4.7.53)$$

Ohm's law derives the expressions for these currents:

$$i_1 = \frac{0 - e_a^-}{R_i} = -\frac{e_a^-}{R_i}$$
$$i_2 = \frac{e_a^- - e_{out}}{R_f} = \frac{e_a^- - v_{out}}{R_f} \quad (4.7.54)$$

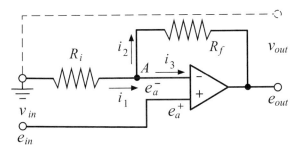

Figure 4.7.17 Non-inverting voltage amplifier

Substituting them into Eq. (4.7.53) yields

$$-\frac{e_a^-}{R_i} = \frac{e_a^- - v_{out}}{R_f} \qquad (4.7.55)$$

Recalling that the electric potentials on the op-amp terminals are the same, $e_a^+ \approx e_a^-$ (Eq. (4.7.13)), and since $e_a^+ = e_{in} = v_{in}$, we obtain

$$-\frac{v_{in}}{R_i} = \frac{v_{in} - v_{out}}{R_f} \qquad (4.7.56)$$

which yields the sought governing equation:

$$v_{out} = \left(1 + \frac{R_f}{R_i}\right) v_{in} \qquad (4.7.57)$$

This equation illustrates the concept of the non-inverting op-amp circuit operation – the amplification of the input voltage by a factor that is always greater than one. There is no sign change between the input and output voltages.

Comparing Eq. (4.7.57) with Eq. (4.7.27) demonstrates that the signal amplification provided by a non-inverting circuit is always greater than that yielded by the similar inverting circuit.

In addition to the discussed multiplier circuit, a non-inverting op-amp can be used in the integrator and differentiator configurations, summing and subtracting circuits, and many others.

Like their inverting counterparts, non-inverting op-amp circuits are used in control systems, analog-to-digital converters, and various analog devices, including analog computers.

Non-inverting Op-Amps as Isolation Amplifiers

With the acquired familiarity with various op-amp circuits, let us revisit the isolation amplifier, presented in Section 4.7.1.

An isolation amplifier with a unity gain, $G_a = 1$, is a non-inverting op-amp with a short-circuited feedback, as shown in Figure 4.7.18. It may also be called *voltage follower* or *unity-gain buffer*.

This circuit is commonly used for isolating circuits from each other in various electronic devices. The derivation of its governing equation is straightforward.

A grounded wire is chosen as a reference for the determination of the input and output voltages,

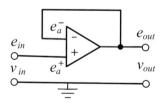

Figure 4.7.18 Non-inverting op-amp as an isolation amplifier

v_{in} and v_{out} respectively, as expressed by Eq. (4.7.51). Since the input voltage is connected directly to the amplifier input terminal,

$$e_{in} = e_a^+ \qquad (4.7.58)$$

Similarly,

$$e_{out} = e_a^- \qquad (4.7.59)$$

Then, recalling that an op-amp draws negligible current, from Eq. (4.7.13) we obtain the governing equation for a voltage follower:

$$v_{out} = v_{in} \qquad (4.7.60)$$

This result can be also derived using Eq. (4.7.57). Since the feedback line in a voltage-follower configuration is just a wire, its resistance equals one, $R_f = 1$. The resistor R_i is replaced by an open circuit, as seen in Figure 4.7.18, hence, $R_i = \infty$. Substituting these values into Eq. (4.7.57), we obtain

$$v_{out} = \left(1 + \frac{R_f}{R_i}\right) v_{in} = \left(1 + \frac{1}{\infty}\right) v_{in} = v_{in}$$

Note that a voltage follower does not provide any signal amplification; its input voltage is passed unchanged to a circuit connected to the output terminals of this amplifier.

Converters

In control systems applications, DC signals are customarily used for the representation of various physical measurements. In such instrumentation circuits, the measuring device acts as a voltage source, and the controller is a load. They are typically connected in series, which keeps the current flowing through the load exactly the same as that leaving the source. Since the parallel connection of these two elements does not allow for equal voltages because of resistive wire losses, a current signal is generally preferable to a voltage signal. Moreover, current-sensing instruments are less susceptible to noise due to their low impedance, which adds to their attractiveness.

A non-inverting amplifier circuit that is a *voltage-to-current converter* or *trans-conductance amplifier* is shown in Figure 4.7.19.

Another integrated circuit – transistor – is added to the output wire of this circuit to increase the range of output currents this op-amp circuit can accommodate. While the detailed derivation of the voltage-to-current converter model is beyond the scope of this chapter, the expression for the output current is as follows:

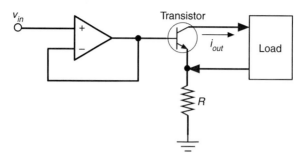

Figure 4.7.19 Voltage-to-current converter circuit

$$i_{out} = \frac{v_{in}}{R}$$

In some cases, an electrical current is produced to represent a physical parameter. For example, a photodiode generates current in response to a flow of electromagnetic radiation. Then, such a device acts as a current source, and it may be desirable to convert this input current to an output voltage. This is done with an inverting op-amp circuit – a *current-to-voltage converter* or *trans-resistance amplifier*. This circuit is shown in Figure 4.7.20.

Its governing equation is

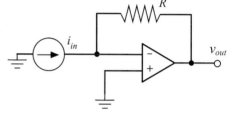

Figure 4.7.20 Current-to-voltage converter

$$v_{out} = -i_{in} R$$

Modifications of this circuit that include a network of resistors are used when a high gain of the converter is required.

4.7.4 Operational Amplifier Circuits with a Load

The analysis of active circuits is often aimed at understanding the evolution of the current and voltage at the load rather than delving into processes that occur within the op-amp component. In such a case, the op-amp circuit is represented by its transfer function (or governing equation), and treated as an isolated voltage source. Then, the approach to passive-circuit analysis, as described earlier in this chapter, applies.

To illustrate this technique, consider the following example.

Example 4.7.2

For the active circuit shown in Figure 4.7.21, find the current passing through the load circuit.

Solution
The op-amp component is a modified differentiator circuit (lead), the governing equation of which is represented by Eq. (4.7.41).

The governing equation for the load circuit is derived as follows:

$$I = \frac{1}{R + Ls} V_{out} \tag{1}$$

From Eq. (4.7.41), the output voltage of the op-amp circuit is

$$V_{out} = -\frac{R_f}{R_i}(1 + R_i C_i s) V_{in} \tag{2}$$

Substituting Eq. (2) into Eq. (1), we obtain

$$I = -\frac{R_f}{R_i}(1 + R_i C_i s)\frac{1}{R + Ls} V_{in} \tag{3}$$

Hence, the transfer function of the circuit as a whole $\dfrac{I}{V_{in}}$ is a product of transfer functions of the circuit components connected in series – the op-amp and the load.

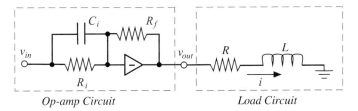

Figure 4.7.21 Op-amp circuit with a load

4.8 Dynamic Responses via Mathematica

In this section, the simulation of dynamic responses of passive and active electrical circuits by Wolfram Mathematica is illustrated in examples. Refer to Appendix C for the information on basic programming in Wolfram Language.

Example 4.8.1

Let us revisit the circuit analyzed in Example 4.4.1 and plot the dynamic response of this system subject to a step input.

Since the governing equations, derived by the loop method, were integral, they need to be transformed into a differential form in order to use the Mathematica family of differential-equation-solver functions. The linear ordinary differential equation input–output model of the circuit in question (see Figure 4.4.1) has been derived as follows (Eq. (10) in the Example 4.4.1):

$$v_{in} = R_1 R_2 C_1 C_2 \frac{d^2 v_{out}}{dt^2} + (R_1 C_1 + R_2 C_2 + R_1 C_2) \frac{d v_{out}}{dt} + v_{out}$$

Given the following numerical values for resistances and capacitances:

$$R_1 = 5 \text{ k}\Omega, \quad R_2 = 2.5 \text{ k}\Omega, \quad C_1 = 5 \text{ μF}, \quad C_2 = 2 \text{ μF}$$

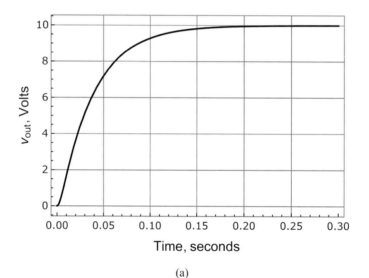

(a)

```
(* setting up numerics and generating the closed-form solution *)
r1 = 5 * 10^3; r2 = 2.5 * 10^3; c1 = 5 * 10^-6; c2 = 2 * 10^-6;
sol = DSolveValue[{r1 * r2 * c1 * c2 * vo''[t] + (r1 * c1 + r2 * c2 + r1 * c2) * vo'[t] + vo[t] == 10,
    vo[0] == vo'[0] == 0}, vo[t], {t, 0, 0.3}]

(* plotting the output voltage *)
Plot[sol, {t, 0, 0.3}, AxesOrigin → {0, 0}, PlotRange → All, PlotStyle → {Black, Thick},
 GridLines → Automatic, GridLinesStyle → Gray, Frame → True, FrameStyle → Gray,
 FrameTicks → {{Automatic, None}, {Automatic, None}},
 FrameTicksStyle → Directive["Label", Black, 12], ImageSize → 400, AspectRatio → 1/1.5,
 FrameLabel → {Style["Time, seconds", Black, 16], Style["Vout, Volts", Black, 16]}]
```

(b)

Figure 4.8.1 Dynamic response of a circuit to the step input: (a) output voltage, (b) Mathematica code

find the system output – voltage across the capacitor C_2 – in response to a constant input voltage $v_{in} = 10$ V.

Since the governing equation is a linear second-order ordinary differential equation and the system is subject to a step input, the closed-form solution is expected. This prompts the use of the Mathematica `DSolve` function or its variant `DSolveValue`. The generated result is shown in Figure 4.8.1(a), and the code is shown in Figure 4.8.1(b).

Plots of the loop currents (see Figure 4.8.2(a)) can be generated using the derived system response. The current in loop 2 is derived from Eq. (3), Example 4.4.1, as $i_2 = C_2 \dot{v}_{out}$. The current in loop 1 is derived from Eq. (2), Example 4.4.1, using the above expression for i_2: $i_1 = C_1 C_2 \ddot{v}_{out} + (C_1 + C_2) \dot{v}_{out}$. These mathematical operations can be done inside the `Plot` function as shown in Figure 4.8.2(b).

As seen in Figure 4.8.2(a), both loop currents are positive, which means that their directions were assumed correctly. Additionally, as the output voltage converges to a constant value equal to the input voltage, the currents in the circuit converge to zero.

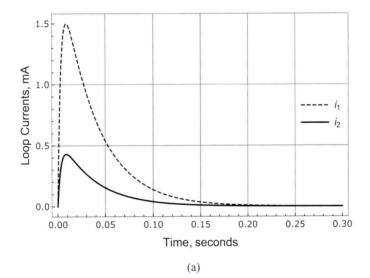

(a)

```
(* plotting currents i1 and i2 *)
Plot[{Evaluate[10^3 * (c1 * c2 * D[sol, {t, 2}] + (c2 + c1) * D[sol, t])],
   Evaluate[10^3 * c2 * D[sol, t]]}, {t, 0, 0.3}, AxesOrigin → {0, 0}, PlotRange → All,
  PlotStyle → {{Black, Dashed}, {Black, Thick}}, GridLines → Automatic,
  GridLinesStyle → Gray, Frame → True, FrameStyle → Gray,
  FrameTicks → {{Automatic, None}, {Automatic, None}},
  FrameTicksStyle → Directive["Label", Black, 12], ImageSize → 400, AspectRatio → 1/1.5,
  FrameLabel → {Style["Time, seconds", Black, 16], Style["Loop Currents, mA", Black, 16]},
  PlotLegends → Placed[{Style["i_1", 14], Style["i_2", 14]}, {Right, Center}]]
```

(b)

Figure 4.8.2 Loop currents in the circuit in response to the step input voltage: (a) loop currents, (b) Mathematica code

Example 4.8.2

Let us revisit the circuit analyzed in Example 4.4.2 and obtain plots of the dynamic responses of this system subject to two types of input: constant (step) and periodic (sinusoidal).

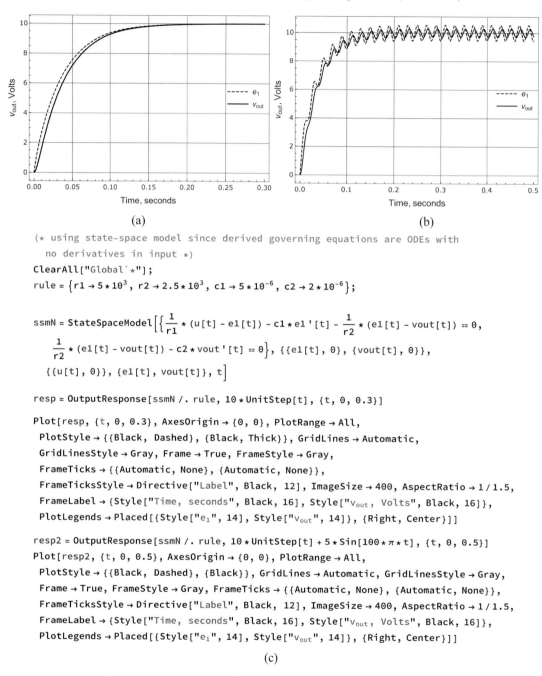

```
(* using state-space model since derived governing equations are ODEs with
   no derivatives in input *)
ClearAll["Global`*"];
rule = {r1 → 5*10^3, r2 → 2.5*10^3, c1 → 5*10^-6, c2 → 2*10^-6};

ssmN = StateSpaceModel[{1/r1 * (u[t] - e1[t]) - c1*e1'[t] - 1/r2 * (e1[t] - vout[t]) == 0,
   1/r2 * (e1[t] - vout[t]) - c2*vout'[t] == 0}, {{e1[t], 0}, {vout[t], 0}},
   {{u[t], 0}}, {e1[t], vout[t]}, t]

resp = OutputResponse[ssmN /. rule, 10*UnitStep[t], {t, 0, 0.3}]

Plot[resp, {t, 0, 0.3}, AxesOrigin → {0, 0}, PlotRange → All,
  PlotStyle → {{Black, Dashed}, {Black, Thick}}, GridLines → Automatic,
  GridLinesStyle → Gray, Frame → True, FrameStyle → Gray,
  FrameTicks → {{Automatic, None}, {Automatic, None}},
  FrameTicksStyle → Directive["Label", Black, 12], ImageSize → 400, AspectRatio → 1/1.5,
  FrameLabel → {Style["Time, seconds", Black, 16], Style["V_out, Volts", Black, 16]},
  PlotLegends → Placed[{Style["e_1", 14], Style["V_out", 14]}, {Right, Center}]]

resp2 = OutputResponse[ssmN /. rule, 10*UnitStep[t] + 5*Sin[100*π*t], {t, 0, 0.5}]
Plot[resp2, {t, 0, 0.5}, AxesOrigin → {0, 0}, PlotRange → All,
  PlotStyle → {{Black, Dashed}, {Black}}, GridLines → Automatic, GridLinesStyle → Gray,
  Frame → True, FrameStyle → Gray, FrameTicks → {{Automatic, None}, {Automatic, None}},
  FrameTicksStyle → Directive["Label", Black, 12], ImageSize → 400, AspectRatio → 1/1.5,
  FrameLabel → {Style["Time, seconds", Black, 16], Style["V_out, Volts", Black, 16]},
  PlotLegends → Placed[{Style["e_1", 14], Style["V_out", 14]}, {Right, Center}]]
```

(c)

Figure 4.8.3 Dynamic response of the circuit to constant and periodic input voltages: (a) response to a step input, (b) response to a sinusoidal input, (c) Mathematica code

4.8 Dynamic Responses via Mathematica

This is the same circuit as in the Example 4.4.1, but the governing equations are derived using the node method. These equations:

$$\begin{cases} \dfrac{v_{in} - e_1}{R_1} - C_1 \dfrac{de_1}{dt} - \dfrac{e_1 - v_{out}}{R_2} = 0 \\ \dfrac{e_1 - v_{out}}{R_2} - C_2 \dfrac{dv_{out}}{dt} = 0 \end{cases}$$

are linear ordinary differential equations with no derivatives in input, so implementation in software with the `StateSpaceModel` and `OutputResponse` functions is possible. Using the same numerical values for resistances and capacitances as in Example 4.8.1, generate the circuit response to two types of input voltage: a step $v_{in} = 10$ V, and one with a periodic component of $v_{in} = 10 + 5\sin(100\pi t)$ V. The sought outputs denoted e_1 and v_{out} are voltage drops across capacitors C_1 and C_2, respectively. They are shown in Figure 4.8.3(a) and 4.8.3(b), and the generating Mathematica code is shown in Figure 4.8.3(c).

As seen in Figure 4.8.3(a) and 4.8.3(b), the voltages across the capacitors converge to the same constant, equal to the magnitude of input voltage for the step input, and converge to an operating trajectory – nondecaying oscillations with constant magnitude for the sinusoidal input. Refer to Chapter 7 for a derivation of the theoretical predictions for the observed results.

Example 4.8.3

Let us revisit the multi-source circuit analyzed in Example 4.5.2.
Given the following values of system parameters:

$$R = 5.4 \text{ k}\Omega, \quad L = 2.3 \text{ mH}, \quad C = 1.8 \text{ }\mu\text{F}$$

and input voltages:

$$v_1 = \begin{cases} 5u(t) & 0 \le t \le 0.05 \\ 0 & t > 0.05 \end{cases}$$

(pulse of the duration of 0.05 s) and $v_2 = 4\sin(60\pi t)$, derive the expression for the desired output – current through the inductor – and obtain a plot of the system response.

Solution

This system is best implemented in software using the transfer matrix. The governing equation for this system in terms of the current through the inductor is derived in Eq. (17) of Example 4.5.2:

$$LRC\frac{d^2 i_L}{dt^2} + L\frac{di_L}{dt} + Ri_L = V_1 + RC\frac{dV_2}{dt}$$

Taking the Laplace transform of this equation, we derive the transfer matrix:

$$G(s) = \begin{bmatrix} \dfrac{1}{LRCs^2 + Ls + R} & \dfrac{RCs}{LRCs^2 + Ls + R} \end{bmatrix}$$

Then, the system response in the Laplace domain is

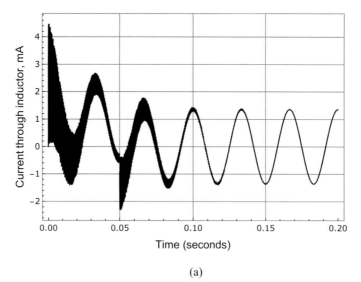

(a)

```
(* transfer matrix of the system *)
tfm = TransferFunctionModel[{{1/(r+l*s+r*l*c*s^2), (r*c*s)/(r+l*s+r*l*c*s^2)}}, s]
(* setting numerical values for system parameters and inputs *)
v1 = {5.0  0 ≤ t ≤ 0.05;
      0    True
v2 = 4*Sin[60*π*t];
rule = {r → 5.4*10^3, l → 2.3*10^-3, c → 1.8*10^-6};
(* generating system response and plotting it *)
resp = OutputResponse[tfm /. rule, {v1, v2}, t];
Plot[Re[10^3 * resp], {t, 0, 0.2}, PlotRange → All, AxesOrigin → {0, 0}, PlotStyle → Black,
  GridLines → Automatic, GridLinesStyle → Gray, Frame → True, FrameStyle → Gray,
  PlotRange → All, FrameTicks → {{Automatic, None}, {Automatic, None}},
  FrameTicksStyle → Directive["Label", Black, 12], ImageSize → 400, AspectRatio → 1/1.5,
  FrameLabel → {Style["Time (seconds)", Black, 16],
    Style["Current through inductor, mA", Black, 16]}]
```

(b)

Figure 4.8.4 Dynamic response of an RLC circuit with two voltage sources: (a) output current, (b) Mathematica code

$$I_L(s) = G(s)R(s) = \begin{bmatrix} \dfrac{1}{LRCs^2 + Ls + R} & \dfrac{RCs}{LRCs^2 + Ls + R} \end{bmatrix} \cdot \begin{bmatrix} 5(1 - e^{-0.05s})/s \\ 4 \cdot 60\pi/(s^2 + 3600\pi^2) \end{bmatrix}$$

$$= \dfrac{5(1 - e^{-0.05s})}{s(LRCs^2 + Ls + R)} + \dfrac{240\pi RCs}{(LRCs^2 + Ls + R)(s^2 + 3600\pi^2)}$$

While this expression can be inverted back to the time domain using partial fractions expansion (refer to Chapter 2 for details), the derivation is very cumbersome and time-consuming. The solution can be generated and plotted using the Mathematica functions `TransferFunctionModel` and `OutputResponse`. The dynamic response of the system is shown in Figure 4.8.4(a) and the generating code is shown in Figure 4.8.4(b).

High-frequency oscillations are observed at the beginning of simulation, where the influences of pulse and periodic input combine. Then, at $t = 0.05$ s, when the action of the pulse input ends, these high-frequency oscillations subside, giving way to a steady nondecaying oscillatory response due to the periodic input. A distinct jump in amplitude of response is noted at $t = 0.05$ s due to end of the pulse input.

Example 4.8.4

This example demonstrates signal attenuation by several active filters, described in Section 4.7.

Consider an active lag circuit (see Figure 4.7.8) that is a first-order low-pass filter, which is expected to attenuate the signals with a frequency higher than the cutoff frequency. Defining numeric values of system parameters as:

$$R_f = 100 \text{ k}\Omega, \quad R_i = 10 \text{ k}\Omega, \quad C_f = 10 \text{ nF}$$

we use Eq. (4.7.32) and the Mathematica functions `TransferFunctionModel` and `BodePlot` to generate a Bode gain plot of this filter (see Figure 4.8.5).

As defined in Section 4.6, at the cutoff frequency ω_0 the magnitude of the transfer function is

$$|G(j\omega)| = \frac{|G(j\omega)|_{\max}}{\sqrt{2}} \cong 7.07$$

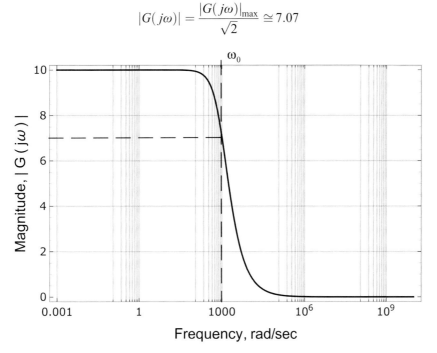

Figure 4.8.5 Bode gain plot of a first-order low-pass active filter (lag circuit)

All the signals with a frequency higher than the computed cutoff frequency (approximately 310 Hz) will be attenuated – this filter is intended to cut off high-frequency noise in various applications. In a loudspeaker, this filter is used on inputs on woofers and subwoofers to block high-frequency sounds that these particular speakers cannot reproduce. Another example of the application of low-pass filters is the tone knob on an electric guitar – here this filter is used to reduce the amount of treble in the generated sound.

The generating Mathematica code is shown in Figure 4.8.6.

Consider a bandwidth-limited differentiator circuit (see Figure 4.7.11) that is a first-order high-pass filter, which is expected to attenuate the signals with a frequency lower than the cutoff frequency. Defining numeric values of system parameters as

$$R_f = 100 \text{ k}\Omega, \quad R_i = 10 \text{ k}\Omega, \quad C_i = 0.02 \text{ }\mu\text{F}$$

we use Eq. (4.7.39) and the Mathematica functions `TransferFunctionModel` and `BodePlot` to generate a Bode gain plot of this filter (see Figure 4.8.7).

All the signals with frequency lower than the computed cutoff frequency (approximately 1.5 kHz) will be attenuated – this filter is intended to cut off low-frequency noise. One of the applications of this filter is an audio crossover – in a loudspeaker it directs high-frequency sounds to a tweeter while attenuating bass (low-frequency) signals that would interfere and potentially damage the speaker.

Consider a second-order band-pass filter (see Figure 4.7.14), which is expected to pass only the signals with a frequency within a predefined bandwidth. Defining numeric values of system parameters as

$$R_f = 10 \text{ k}\Omega, \quad R_i = 1 \text{ k}\Omega, \quad C_i = 20 \text{ }\mu\text{F}, \quad C_f = 15 \text{ nF}$$

we use Eq. (4.7.45) and the Mathematica functions `TransferFunctionModel` and `BodePlot` to generate a Bode gain plot of this filter (see Figure 4.8.8).

```
(* 1st order low-pass filter *)
tfm = TransferFunctionModel[-rf/(ri * (1 + rf * cf * s)), s];
rule1 = {ri -> 10 * 10^3, cf -> 10 * 10^-9, rf -> 100 * 10^3};
BodePlot[tfm /. rule1, {0.001, 10^10}, PlotLayout -> "Magnitude",
  PlotStyle -> Black, GridLines -> Automatic,
  FrameTicksStyle -> Directive["Label", Black, 12],
  FrameLabel -> {Style["Frequency, rad/sec", Black, 16],
    Style["Magnitude, | G ( jω ) |", Black, 16]},
  ScalingFunctions -> {Automatic, "Absolute"}, PlotRange -> Full]
```

Figure 4.8.6 Mathematica code for generating a Bode gain plot for a lag circuit

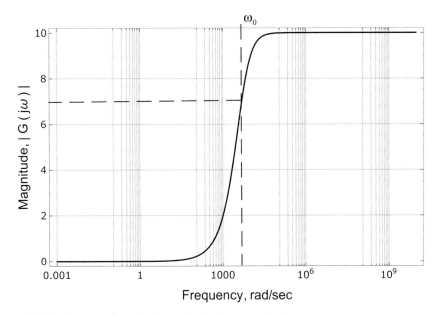

Figure 4.8.7 Bode gain plot of a first-order high-pass active filter

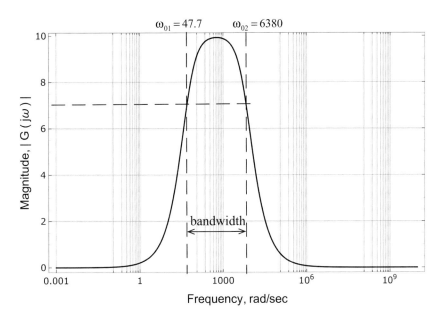

Figure 4.8.8 Bode plot of a second-order band-pass active filter

This band-pass filter is often referred to as *narrow band-pass filter* since its bandwidth, while dependent on the resistances and the capacitances of the passive elements, is typically narrow. One of the main applications for this filter is in communication systems for suppressing all noise and letting the human voice come through without significant distortions.

CHAPTER SUMMARY

The main objective of this chapter is to introduce the basic physics of electrical circuits to help the reader understand their time-dependent behavior and to learn how to model and analyze electrical subsystems, typically encountered in a variety of engineered dynamic systems.

Upon completion of this chapter, you should be able to:

(1) Understand the terminology of electrical circuits; refresh the knowledge of physical laws governing common passive and active elements of electrical circuits, and understand voltage-current relationships of these elements.
(2) Understand the concept of impedance as generalization of the electrical resistance, and use it for modeling of electrical circuits.
(3) Derive the equivalent impedance of a circuit by using the laws of series and parallel connection of impedances.
(4) Derive a model of a circuit by applying voltage and current divider rules and Kirchhoff's current and voltage laws.
(5) Use the loop method, the node method, the superposition theorem, and Norton's and Thevenin's theorems for modeling and analysis of passive electrical circuits.
(6) Derive the state-space representation of electrical circuits and construct their block diagrams.
(7) Model and analyze active electrical circuits, represented by isolation amplifiers and operational amplifiers (op-amps).
(8) Derive transfer functions and governing equation(s) models for common op-amp circuits such as multiplier, inverter, comparator, and low- and high-pass filters.

REFERENCES

1. J. A. Svoboda and R. C. Dorf, *Introduction to Electric Circuits*, Wiley, 9th ed., 2013.
2. V. Hacker and C. Sumereder, *Electrical Engineering: Fundamentals*, De Gruyter Oldenbourg, 2020.
3. K. C. A. Smith and R. E. Alley, *Electrical Circuits: An Introduction*, Cambridge University Press, 1992.

4. C. M. Close, D. K. Frederick, and J. C. Newell, *Modeling and Analysis of Dynamic Systems*, Wiley, 3rd ed., 2002.
5. N. Lobontiu, *System Dynamics for Engineering Students: Concepts and Applications*, Academic Press, 2nd ed., 2017.

PROBLEMS

Section 4.2 Concept of Impedance

4.1 Two passive circuits with the indicated input and output(s) are shown in Figure P4.1.

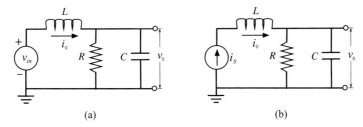

Figure P4.1

Use the impedance method and divider rules to derive the following transfer functions:

(a) $\dfrac{V_o(s)}{V_{in}(s)}$ and $\dfrac{I_o(s)}{V_{in}(s)}$;

(b) $\dfrac{V_o(s)}{I_s(s)}$ and $\dfrac{I_o(s)}{I_s(s)}$.

4.2 Two passive circuits with the indicated input and output(s) are shown in Figure P4.2.

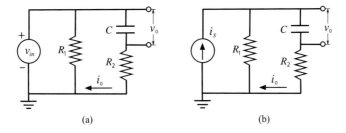

Figure P4.2

Use the impedance method and divider rules to derive the following transfer functions:

(a) $\dfrac{V_0(s)}{V_{in}(s)}$ and $\dfrac{I_0(s)}{V_{in}(s)}$;

(b) $\dfrac{V_0(s)}{I_s(s)}$ and $\dfrac{I_0(s)}{I_s(s)}$.

4.3 Two passive circuits with the indicated input and output(s) are shown in Figure P4.3.

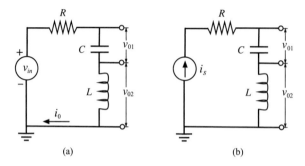

Figure P4.3

Use impedance method and divider rules to derive the following transfer functions:

(a) $\dfrac{V_{01}(s)}{V_{in}(s)}$ and $\dfrac{V_{02}(s)}{V_{in}(s)}$.

(b) $\dfrac{V_{01}(s)}{I_s(s)}$ and $\dfrac{V_{02}(s)}{V_{in}(s)}$.

4.4 A passive circuit with the indicated input and output(s) is shown in Figure P4.4.

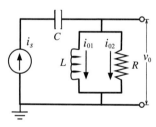

Figure P4.4

Use the impedance method and divider rules to derive the following transfer functions:

$$\dfrac{I_{01}(s)}{I_s(s)}, \quad \dfrac{I_{02}(s)}{I_s(s)} \quad \text{and} \quad \dfrac{V_0(s)}{I_s(s)}.$$

4.5 A passive circuit with the indicated input and output(s) is shown in Figure P4.5.

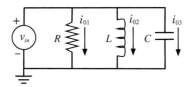

Figure P4.5

Use the impedance method and divider rules to derive the following transfer functions:

$$\frac{I_{01}(s)}{V_{in}(s)}, \quad \frac{I_{02}(s)}{V_{in}(s)}, \quad \text{and} \quad \frac{I_{03}(s)}{V_{in}(s)}$$

4.6 A passive circuit with the indicated input and output(s) is shown in Figure P4.6.

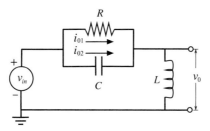

Figure P4.6

Use the impedance method and divider rules to derive the following transfer functions:

$$\frac{I_{01}(s)}{V_{in}(s)}, \quad \frac{I_{02}(s)}{V_{in}(s)}, \quad \text{and} \quad \frac{V_0(s)}{V_{in}(s)}$$

4.7 The passive circuit with the indicated input and output(s) is shown in Figure P4.7.

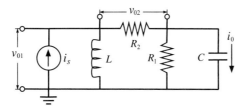

Figure P4.7

Use the impedance method and divider rules to derive the following transfer functions:
$$\frac{V_{01}(s)}{I_s(s)}, \frac{V_{02}(s)}{I_s(s)}, \text{ and } \frac{I_0(s)}{I_s(s)}$$

4.8 A passive circuit with the indicated input and output(s) is shown in Figure P4.8.

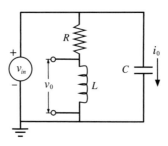

Figure P4.8

Use the impedance method and divider rules to derive the following transfer functions:
$$\frac{V_0(s)}{V_{in}(s)} \text{ and } \frac{I_0(s)}{V_{in}(s)}$$

4.9 Two passive circuits with the indicated input and output(s) are shown in Figure P4.9.

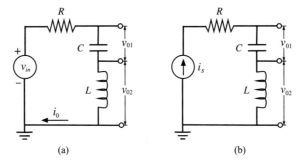

Figure P4.9

Use the impedance method and divider rules to derive the following transfer functions:

(a) $\dfrac{V_{01}(s)}{V_{in}(s)}, \dfrac{V_{02}(s)}{V_{in}(s)}, \text{ and } \dfrac{I_0(s)}{V_{in}(s)};$

(b) $\dfrac{V_{01}(s)}{I_s(s)}, \dfrac{V_{02}(s)}{V_{in}(s)}, \text{ and } \dfrac{I_0(s)}{I_s(s)}.$

4.10 A passive circuit with the indicated input and output(s) is shown in Figure P4.10.

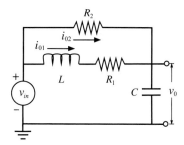

Figure P4.10

Use the impedance method and divider rules to derive the following transfer functions:

$$\frac{I_{01}(s)}{V_{in}(s)}, \quad \frac{I_{02}(s)}{V_{in}(s)}, \quad \text{and} \quad \frac{V_0(s)}{V_{in}(s)}.$$

Section 4.4 Passive-Circuit Analysis

4.11 Use the loop method to derive the governing equations for the circuit shown in Figure P4.8, considering the input (v_{in}) and required outputs (i_0, v_0) indicated on the schematic.

4.12 Use the loop method to derive the governing equations for the circuit shown in Figure P4.10, considering the input (v_{in}) and required outputs (i_{01}, i_{02}, v_0) indicated on the schematic.

4.13 Use the loop method to derive the governing equations for the circuit shown in Figure P4.13, considering the input (v_{in}) and required outputs (i_{01}, i_{02}, v_0) indicated on the schematic

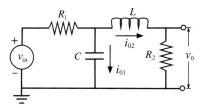

Figure P4.13

4.14 Use the loop method to derive the governing equations for the circuit shown in Figure P4.14, considering the input (v_{in}) and required outputs (i_0, v_0) indicated on the schematic.

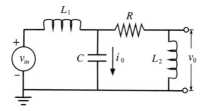

Figure P4.14

4.15 Use the loop method to derive the governing equations for the circuit shown in Figure P4.15, considering the input (v_{in}) and required outputs (i_0, v_{01}, v_{02}) indicated on the schematic.

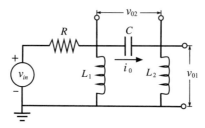

Figure P4.15

4.16 Use the loop method to derive the governing equations for the circuit shown in Figure P4.16, considering the input (v_{in}) and required outputs (i_{01}, i_{02}, v_0) indicated on the schematic.

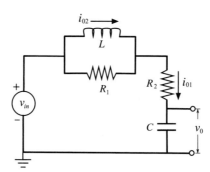

Figure P4.16

4.17 Use the loop method to derive the governing equations for the circuit shown in Figure P4.17, considering the inputs (v_{in1}, v_{in2}) and required outputs (i_0, v_0) indicated on the schematic.

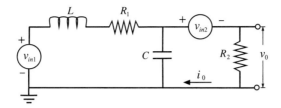

Figure P4.17

4.18 Use the loop method to derive the governing equations for the circuit shown in Figure P4.18, considering the inputs (v_{in1}, v_{in2}) and required outputs (i_0, v_0) indicated on the schematic.

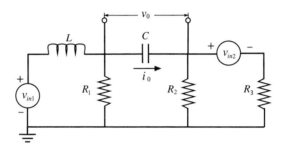

Figure P4.18

4.19 Use the loop method to derive the governing equations for the circuit shown in Figure P4.19, considering the inputs (v_{in1}, v_{in2}) and required outputs ($i_{01}, i_{02}, v_{01}, v_{02}$) indicated on the schematic.

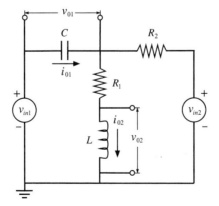

Figure P4.19

4.20 Use the loop method to derive the governing equations for the circuit shown in Figure P4.20, considering the inputs (v_{in1}, v_{in2}) and required outputs (i_{o1}, i_{o2}, v_o) indicated on the schematic.

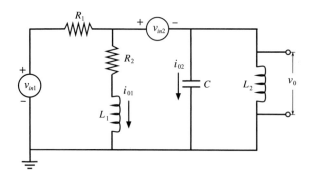

Figure P4.20

4.21 Use the node method to derive the governing equations for the circuits shown in Figure P4.2, considering the input and required outputs indicated on the schematic:
(a) input v_{in}, required outputs i_o, v_o;
(b) input i_s, required outputs i_o, v_o.

4.22 Use the node method to derive the governing equations for the circuits shown in Figure P4.3, considering the input and required outputs indicated on the schematic:
(a) input v_{in}, required outputs i_o, v_{o1}, v_{o2};
(b) input i_s, required outputs v_{o1}, v_{o2}.

4.23 Use the node method to derive the governing equations for the circuit shown in Figure P4.7, considering the input (i_s) and required outputs (i_o, v_{o1}, v_{o2}) indicated on the schematic.

4.24 Use the node method to derive the governing equations for the circuit shown in Figure P4.14, considering the input (v_{in}) and required outputs (i_o, v_o) indicated on the schematic.

4.25 Use the node method to derive the governing equations for the circuit shown in Figure P4.15, considering the input (v_{in}) and required outputs (i_o, v_{o1}, v_{o2}) indicated on the schematic.

4.26 Use the node method to derive the governing equations for the circuit shown in Figure P4.17, considering the inputs (v_{in1}, v_{in2}) and required outputs (i_o, v_o) indicated on the schematic.

4.27 Use the node method to derive the governing equations for the circuit shown in Figure P4.18, considering the inputs (v_{in1}, v_{in2}) and required outputs (i_0, v_0) indicated on the schematic.

4.28 Use the node method to derive the governing equations for the circuit, shown in Figure P4.19, considering the inputs (v_{in1}, v_{in2}) and required outputs (i_{01}, i_{02}, v_{01}, v_{02}) indicated on the schematic.

4.29 Use the node method to derive the governing equations for the circuit shown in Figure P4.20, considering the inputs (v_{in1}, v_{in2}) and required outputs (i_{01}, i_{02}, v_0) indicated on the schematic.

4.30 Use the node method to derive the governing equations for the circuit shown in Figure P4.30, considering the input (i_s) and required outputs (i_{01}, i_{02}, v_0) indicated on the schematic.

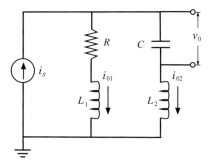

Figure P4.30

For problems 4.31 through 4.35 use Thevenin's theorem to find the current through the load and voltage drop across the load.

Recall the basic procedure for solving a circuit using Thevenin's theorem:

1. Remove the load resistor R_{Load} (open-circuit it, creating two new output terminals; V_T is the voltage drop between these terminals).
2. Find R_T by shorting all voltage sources and by open-circuiting all the current sources, then finding the equivalent resistance of the augmented circuit.
3. Find V_T by the usual circuit analysis methods (node and loop methods).
4. Find the current flowing through the load resistor R_{Load} and the voltage drop across it V_{Load}.

4.31 For the schematic shown in Figure P4.31 use Thevenin's theorem to find the current through the load R_{Load} and the voltage drop across the load R_{Load}.

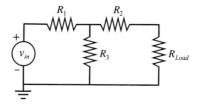

Figure P4.31

4.32 For the schematic shown in Figure P4.32 use Thevenin's theorem to find the current through the load R_{Load} and the voltage drop across the load R_{Load}.

To derive the numerical solution, use the following values for resistances and input voltage:

$$R_1 = 7\,\Omega,\ R_2 = 3\,\Omega,\ R_3 = 2\,\Omega,\ R_4 = 1\,\Omega,\ v_{in} = 15\text{ V},\ R_{Load} = 2.18\,\Omega$$

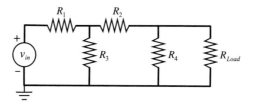

Figure P4.32

4.33 For the schematic shown in Figure P4.33 use Thevenin's theorem to find the current through the load R_{Load} and the voltage drop across the load R_{Load}.

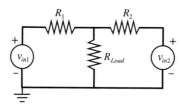

Figure P4.33

4.34 For the schematic shown in Figure P4.34 use Thevenin's theorem to find the current through the load R_{Load} and the voltage drop across the load R_{Load}.

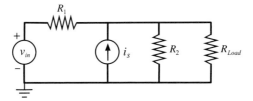

Figure P4.34

4.35 For the schematic shown in Figure P4.35 use Thevenin's theorem to find the current through the load R_{Load} and the voltage drop across the load R_{Load}.

To derive the numerical solution, use the following values for resistances and inputs:

$$R_1 = 8\,\Omega,\ R_2 = 2\,\Omega,\ R_3 = 4\,\Omega,\ i_S = 5\text{ A},\ v_{in} = 10\text{ V},\ R_{Load} = 1.4\,\Omega$$

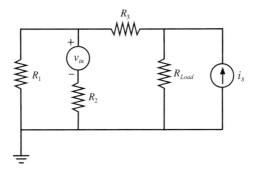

Figure P4.35

For problems 4.36 through 4.40 use Norton's theorem to find the current through the load and voltage drop across the load.

Recall the basic procedure for solving a circuit using Norton's theorem:

1. Remove the load resistor R_{Load} (open-circuit it, creating two new output terminals).
2. Find R_N by shorting all voltage sources and by open-circuiting all the current sources, then finding the equivalent resistance of the augmented circuit.
3. Find I_N by placing a shorting link on the output terminals, created in step 1, and the usual circuit analysis methods (node and loop methods); I_N is the current in the shorting link.
4. Find the current flowing through the load resistor R_{Load} and the voltage drop across it V_{Load}.

4.36 For the schematic shown in Figure P4.31 use Norton's theorem to find the current through the load R_{Load} and the voltage drop across the load R_{Load}.

4.37 For the schematic shown in Figure P4.33 use Norton's theorem to find the current through the load R_{Load} and the voltage drop across the load R_{Load}.

4.38 For the schematic shown in Figure P4.38 use Norton's theorem to find the current through the load R_{Load} and the voltage drop across the load R_{Load}.

To derive the numerical solution, use the following values for resistances and inputs:

$$R_1 = 8\,\Omega,\ R_2 = 12\,\Omega,\ R_3 = 5\,\Omega,\ i_S = 6\,\text{A},\ v_{in} = 7\,\text{V},\ R_{Load} = 23.2\,\Omega$$

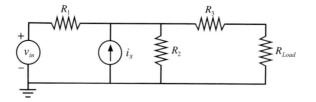

Figure P4.38

4.39 For the schematic shown in Figure P4.39 use Norton's theorem to find the current through the load R_{Load} and the voltage drop across the load R_{Load}.

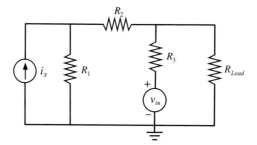

Figure P4.39

4.40 For the schematic shown in Figure P4.40 use Norton's theorem to find the current through the load R_{Load} and the voltage drop across the load R_{Load}.

To derive the numerical solution, use the following values for resistances and inputs:

$$R_1 = 4\,\Omega,\ R_2 = 6\,\Omega,\ R_3 = 10\,\Omega,\ R_4 = 2\,\Omega,\ i_S = 1\,\text{A},\ v_{in} = 45\,\text{V},\ R_{Load} = 3.6\,\Omega$$

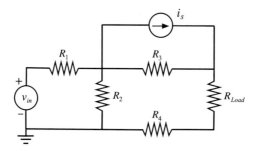

Figure P4.40

Section 4.5 State-Space Representations and Block Diagrams

4.41 Derive the state-space representation for the circuit shown in Figure P4.13, considering the input (v_{in}) and required outputs (i_{01}, i_{02}, v_0) indicated on the schematic.

4.42 Derive the state-space representation for the circuit shown in Figure P4.19, considering the inputs (v_{in1}, v_{in2}) and required outputs (i_{01}, i_{02}, v_{01}, v_{02}) indicated on the schematic.

4.43 Derive the state-space representation for the circuit shown in Figure P4.20, considering the inputs (v_{in1}, v_{in2}) and required outputs (i_{01}, i_{02}, v_0) indicated on the schematic.

4.44 Derive the state-space representation for the circuit shown in Figure P4.44, considering the input (i_s) and required outputs (i_{01}, i_{02}) indicated on the schematic.

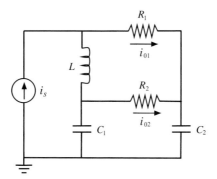

Figure P4.44

4.45 Derive the state-space representation for the circuit shown in Figure P4.45, considering the input (i_s) and required outputs (i_{01}, i_{02}) indicated on the schematic.

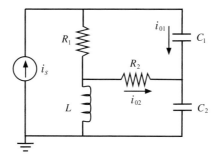

Figure P4.45

4.46 Derive the state-space representation for the circuit shown in Figure P4.46, considering the inputs (v_{in}, i_s) and required outputs (i_{01}, i_{02}, v_o) indicated on the schematic.

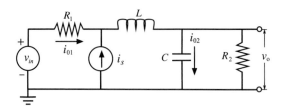

Figure P4.46

4.47 Derive the state-space representation for the circuit shown in Figure P4.47, considering the inputs (v_{in}, i_s) and required output (i_o) indicated on the schematic.

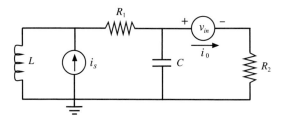

Figure P4.47

4.48 Derive the state-space representation for the circuit, shown in Figure P4.48, considering the inputs (v_{in}, i_s) and required outputs (i_o, v_o) indicated on the schematic.

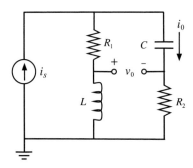

Figure P4.48

4.49 Derive the state-space representation for the circuit shown in Figure P4.49, considering the inputs (v_{in1}, v_{in2}) and required outputs (i_{o1}, i_{o2}) indicated on the schematic.

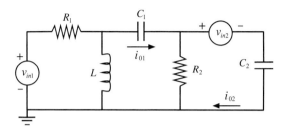

Figure P4.49

4.50 Derive the state-space representation for the circuit shown in Figure P4.50, considering the inputs (v_{in}, i_s) and required outputs (i_{o1}, i_{o2}) indicated on the schematic.

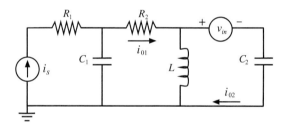

Figure P4.50

4.51 Draw a block diagram for the circuit in Problem 4.11, considering the input (v_{in}) and required outputs (i_o, v_o) indicated on the schematic.

4.52 Draw a block diagram for the circuit in Problem 4.17, considering that the required outputs are the current through the inductor and the voltage drop across the capacitor.

4.53 Draw a block diagram for the circuit in Problem 4.18, considering that the required outputs are the voltage drops across the resistors R_1 and R_2.

4.54 Draw a block diagram for the circuit in Problem 4.28, considering that the required outputs are currents through the capacitor and through the resistor R_2, and the potential of the node on top of the resistor R_1.

4.55 Draw a block diagram for the circuit in Problem 4.45. Use the node method to derive the governing equations, considering that the required output is the current through the capacitor C_1.

Section 4.6 Passive Filters

For every circuit depicted in problems 4.56 through 4.58, find the filter transfer function $G(s) = \dfrac{V_{out}}{V_{in}}$, generate a Bode plot $|G(j\omega)|$ vs. frequency (use either Wolfram Mathematica or MATLAB to obtain the required plots), and indicate which filter type is represented by the given circuit.

Recall the four primary filter types:

1. Low-pass filter – a circuit that offers an easy passage to low-frequency signals ($\omega < \omega_0$) while blocking high-frequency signals ($\omega > \omega_0$); ω_0 is a cutoff frequency in radians per second.
2. High-pass filter – a circuit that blocks low-frequency signals ($\omega < \omega_0$) while offering easy passage to high-frequency signals ($\omega > \omega_0$); ω_0 is a cutoff frequency in radians per second.
3. Band-pass filter – a circuit that passes signals within a specified frequency range $\omega_{01} < \omega < \omega_{02}$ while blocking signals with out-of-range frequencies; ω_{01} and ω_{02} are the cutoff frequencies for the desired bandwidth.
4. Band-stop (notch) filter – a circuit that blocks signals within a specified frequency range $\omega_{01} < \omega < \omega_{02}$ while blocking signals with out-of-range frequencies; ω_{01} and ω_{02} are the cutoff frequencies; frequency of maximum attenuation is called *notch frequency*.

Both low- and high-pass filters may experience a resonance effect at the cutoff frequency, depending on the implementation of the filter.

4.56 The circuits are shown in Figure P4.56.

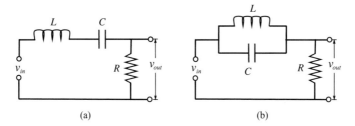

Figure P4.56

The numerical data are as follows:

(a) $L = 5\ \mu\text{H}$, $C = 3\ \text{nF}$, $R = 1\ \text{k}\Omega$;
(b) $L = 2\ \text{mH}$, $C = 0.5\ \text{nF}$, $R = 0.5\ \text{k}\Omega$.

4.57 The circuits are shown in Figure P4.57.

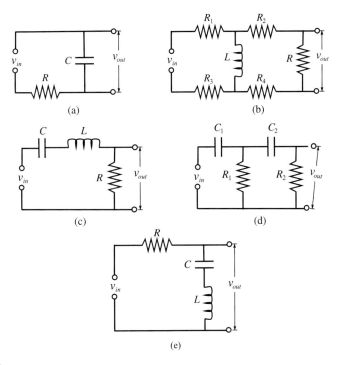

Figure P4.57

The numerical data are as follows:

(a) $C = 3$ nF, $R = 1$ kΩ;
(b) $L = 5$ µH, $R = 1$ kΩ, $R_1 = 30$ Ω, $R_2 = 10$ Ω, $R_3 = 20$ Ω, $R_4 = 15$ Ω;
(c) $L = 7$ µH, $C = 1.5$ nF, $R = 1$ kΩ;
(d) $C_1 = 1.5$ nF , $C_2 = 1.0$ nF, $R_1 = 0.5$ kΩ, $R_2 = 1$ kΩ;
(e) $L = 5$ µH, $C = 3$ nF, $R = 1$ kΩ.

4.58 The circuits are shown in Figure P4.58.
The numerical data are as follows:
(a) $L = 5$ µH, $C = 3$ nF, $R = 20$ Ω;
(b) $C_1 = 1$ nF, $C_2 = 3$ nF, $R_1 = 100$ Ω, $R_2 = 1$ kΩ;
(c) $L_1 = 20$ µH, $L_2 = 3$ µH, $R_1 = 30$ Ω, $R_2 = 1$ kΩ;
(d) $L_1 = 50$ mH, $L_2 = 50$ mH, $C = 0.1$ µF, $R = 1$ kΩ.

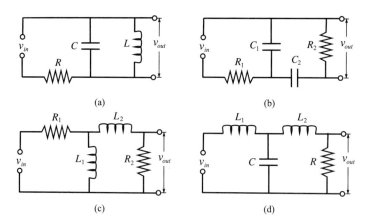

Figure P4.58

Section 4.7 Active-Circuit Analysis

4.59 For every active circuit shown in Figure P4.59, find the transfer function $G(s) = \dfrac{V_{out}}{V_{in}}$ and derive the time-domain relationship between the input (v_{in}) and output (v_{out}) voltages.

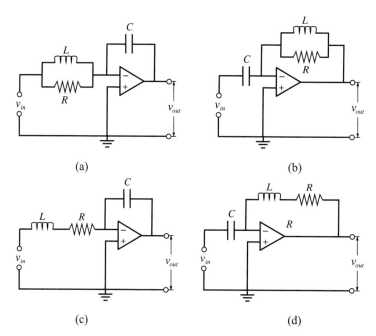

Figure P4.59

4.60 For the differential amplifier circuit shown in Figure P4.60, derive the time-domain relationship between the input (v_{in1}, v_{in2}) and output (v_{out}) voltages. How will this relationship change when $R_1 = R_2$ and $R_3 = R_4$?

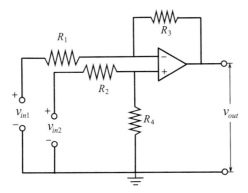

Figure P4.60

4.61 Derive the expression for the current i_0 through the load for the Howland current source – an active circuit, shown in Figure P4.61. Derive the *balanced bridge condition* – a relationship between the resistances R_1, R_2, R_3, and R_4 for i_0 to be independent of the voltage drop v_L across the load.

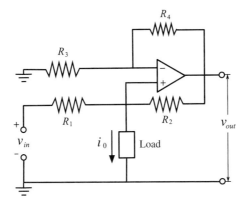

Figure P4.61

4.62 The proportional–integral–derivative (PID) control – one of the commonly used linear controls – can be implemented as an active circuit with inverting op-amps, as shown in Figure 4.62.

Identify each op-amp circuit – the component of the shown control – and derive the PID control law as a time-domain relationship between the input (v_{in}) and output (v_{out}) voltages.

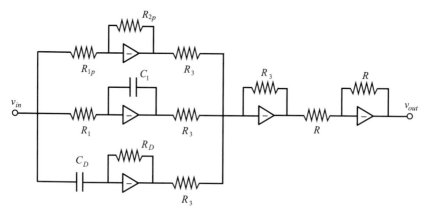

Figure P4.62

4.63 For the active circuit shown in Figure P4.63, derive the time-domain relationship between the input (v_{in1}, v_{in2}, v_{in3}, v_{in4}) and output (v_{out}) voltages. What does this circuit do?

How will the relationship between the circuit input and output voltages change when $R_1 = R_2 = R_3 = R_4 = R_f$?

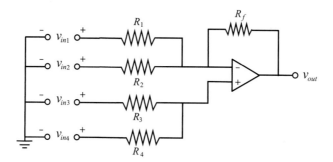

Figure P4.63

4.64 For the active circuits shown in Figure P4.64, derive the time-domain relationships between the input (v_{in1}, v_{in2}, v_{in3}) and output (v_{out}) voltages. To show that both of these circuits are averagers, assume that, for the circuit in Figure P4.64(a), $R_1 = R_2 = R_3 = R$, while for the circuit in Figure P4.64(b), $R_1 = R_2 = R_3 = 3R$.

Hint: consider using Thevenin's theorem for deriving the required expressions.

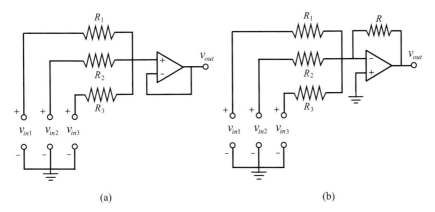

Figure P4.64

4.65 The active circuits shown in Figure P4.65 represent different filters. For each of these circuits, find the filter transfer function $G(s) = \dfrac{V_{out}}{V_{in}}$, generate a Bode plot $|G(j\omega)|$ vs. frequency (use either Wolfram Mathematica or MATLAB to obtain the required plots), and indicate which filter type (low-pass, high-pass, band-pass, or band-stop) is represented by the given circuit.

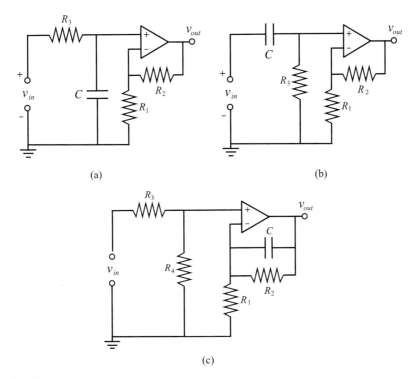

Figure P4.65

The numerical data are as follows:
(a) $C = 100$ µF, $R_1 = 10$ Ω, $R_2 = 9$ Ω, $R_3 = 1$ Ω;
(b) $C = 100$ µF, $R_1 = 10$ Ω, $R_2 = 9$ Ω, $R_3 = 1$ Ω;
(c) $C = 120$ µF, $R_1 = 2$ Ω, $R_2 = 9$ Ω, $R_3 = 10$ Ω, $R_4 = 1$ Ω.

4.66 A non-inverting op-amp can be used as an isolation amplifier in certain implementations of the first-order low-pass and high-pass filters, with and without amplification. For each of the circuits shown in Figure P4.66 derive the filter transfer function $G(s) = \dfrac{V_{out}}{V_{in}}$, generate a Bode plot $|G(j\omega)|$ vs. frequency (use either Wolfram Mathematica or MATLAB to obtain the required plots), and indicate which filter type is represented by the shown circuits.

For both circuits the numerical data are as follows: $C = 3$ nF, $R = 1$ kΩ.

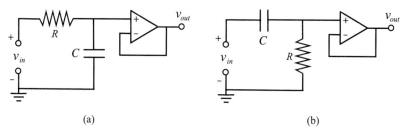

Figure P4.66

4.67 A non-inverting op-amp circuit with dual capacitors is often used to obtain second-order low-pass and high-pass filters, with and without amplification. For each of the circuits shown in Figure P4.67 derive the filter transfer function $G(s) = \dfrac{V_{out}}{V_{in}}$, generate a Bode plot $|G(j\omega)|$ vs. frequency (use either Wolfram Mathematica or MATLAB to obtain the required plots), and indicate which filter type is represented by the shown circuits.

The numerical data are as follows:

(a) $C_1 = C_2 = 2$ µF, $R_1 = 20$ Ω, $R_2 = 5$ Ω, $R_3 = 100$ Ω, $R_4 = 100$ Ω;
(b) $C_1 = 0.5$ µF, $C_2 = 1.5$ µF, $R_1 = 20$ Ω, $R_2 = 1$ Ω, $R_3 = 100$ Ω, $R_4 = 10$ Ω.

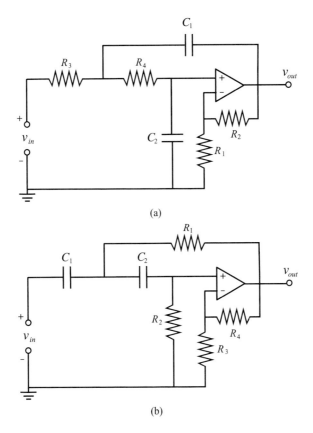

Figure P4.67

4.68 A band-pass filter can be implemented with either inverting or non-inverting op-amps, as shown in Figure P4.68. For each of the circuits shown in Figure P4.68 derive the filter transfer function $G(s) = \dfrac{V_{out}}{V_{in}}$, and generate a Bode plot $|G(j\omega)|$ vs. frequency (use either Wolfram Mathematica or MATLAB to obtain the required plots) to demonstrate the bandwidths passed by these filters. Which one of them passed the wider bandwidth?

Numerical data are as follows:

(a) $C_1 = 1.5\ \mu F$, $C_2 = 1.2\ \mu F$, $R_1 = 20\ \Omega$, $R_2 = 1\ \Omega$, $R_3 = 35\ \Omega$, $R_4 = 2\ \Omega$;
(b) $C_1 = 1.5\ \mu F$, $C_2 = 1.3\ \mu F$, $R_1 = 0.4\ \Omega$, $R_2 = 0.75\ \Omega$.

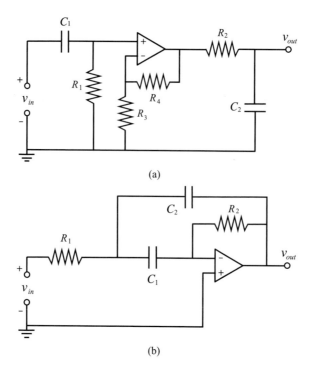

Figure P4.68

4.69 For the active circuit shown in Figure P4.69, derive the time-domain relationship between the input (v_{in1}, v_{in2}) and output (v_{out}) voltages. What does this circuit do?

How will the relationship between the circuit input and output voltages change when $R_1 = R_2 = R_3$?

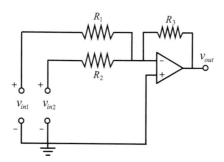

Figure P4.69

4.70 The inverting op-amp active circuit, shown in Figure P4.70, represents a second-order low-pass filter. Derive the filter transfer function $G(s) = \dfrac{V_{out}}{V_{in}}$, and generate a Bode plot $|G(j\omega)|$ vs. frequency (use either Wolfram Mathematica or MATLAB to obtain the required plots), considering the following numerical values:

$$C_1 = 15 \ \mu F, \quad C_2 = 10 \ \mu F, \quad R_1 = 2.5 \ \Omega, \quad R_2 = 0.75 \ \Omega, \quad R_3 = 1 \ \Omega.$$

Will this circuit pass an input signal with a frequency of 0.318 MHz? How about an input signal with a frequency of 0.159 kHz?

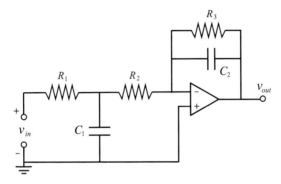

Figure P4.70

4.71 One of the possible implementations of a band-stop filter (notch filter) is shown in Figure P4.71. Derive the filter transfer function $G(s) = \dfrac{V_{out}}{V_{in}}$, and generate a Bode plot $|G(j\omega)|$ vs. frequency (use either Wolfram Mathematica or MATLAB to obtain the required plots), considering the following numerical values:

$$C_1 = 1.5 \ \mu F, \quad C_2 = 100 \ nF, \quad R_1 = 7.5 \ \Omega, \quad R_2 = 0.5 \ \Omega, \quad R_3 = 6 \ \Omega, \quad R_4 = 7.5 \ \Omega, \quad R = 15 \ \Omega.$$

Which frequency (in Hz) will be mostly attenuated by this circuit?

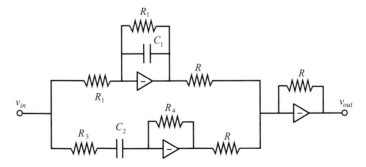

Figure P4.71

Section 4.8 Dynamic Responses via Mathematica

4.72 Consider the passive circuit shown in Figure P4.72, where system input is the voltage v_{in}, and the required outputs are the current i_o through the resistor R_1 and the voltage drop v_o across the inductor L_2.

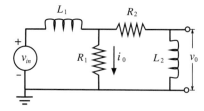

Figure P4.72

Derive the governing equations for this circuit using the loop method and plot the dynamic response of this system to a step input, considering the following numerical values for the system parameters and input voltage:

$L_1 = 2.5$ H, $L_2 = 1.2$ H, $R_1 = 15\ \Omega$, $R_2 = 25\ \Omega$, $v_{in} = 10u(t)$ V, simulation time 2.5 s.

Assume zero initial conditions.

Hint: this system has a closed-form solution for loop currents; hence, the analytical differential equation solver DSolve of Mathematica is applicable.

4.73 Consider the passive circuit described in Problem 4.24. Derive the governing equations for this circuit using the node method. Assuming zero initial conditions, plot the dynamic response of this system to:
(a) a step input $v_{in} = 5u(t)$ V,
(b) a ramp input $v_{in} = 5tu(t)$ V,

considering the following numerical values for the system parameters and input voltage:

$L_1 = 2.5$ H, $L_2 = 1.2$ H, $R = 15\ \Omega$, $C = 0.65$ F, simulation time 115 s.

(c) For the case of the step input, consider that capacitance is no longer constant, but instead is dependent on time as follows: $C = 3(t+1)e^{-t/12}$. How does the introduction of the time-dependent capacitance change the character of system response?
(d) For the case of the step input, consider that capacitance is no longer constant, but instead is dependent on time and the voltage drop on the capacitance as follows: $C = (v_C + \varepsilon)e^{-t/12}$, where $\varepsilon = 10^{-6}$. How does the introduction of this capacitance change the character of the system response?

If the largest current the capacitor can safely handle is 10 A, will the circuits for case (c) and case (d) be safe to operate? Explain your conclusion using the simulation results.

Hint: for this system, the numerical differential equation solver NDSolve of Mathematica is applicable.

4.74 Consider the passive circuit described in Problem 4.44. Derive the state-space model of this system, considering zero initial conditions and the following numerical values of the system parameters: $R_1 = 5\,\Omega$, $R_2 = 1.5\,\Omega$, $L = 2\,H$, $C_1 = 5\,mF$, $C_2 = 2\,mF$.

Plot the dynamic response of this system to the following inputs:
(a) unit step input, simulation duration of 5 s;
(b) sinusoidal input $2\sin(t)$, simulation duration of 25 s;
(c) rectangular pulse input of a duration of 2 sec., described as $i_S(t) = \begin{cases} u(t) & \text{for } t \leq 2 \\ 0 & \text{for } t > 2 \end{cases}$.

Hint: for this system use the Mathematica functions StateSpaceModel and OutputResponse.

4.75 Consider the circuit described in Problem 4.57(a). Plot the dynamic response of this circuit to the sinusoidal input $v_{in}(t) = \sin(10t)$. If the input frequency is increased to 1000 rad/sec, would the periodic output still be expected? What if the input frequency is increased to 10^9 rad/sec?

5 Thermal and Fluid Systems

Contents

5.1	Fundamental Principles of Thermal Systems	335
5.2	Basic Thermal Elements	339
5.3	Dynamic Modeling of Thermal Systems	346
5.4	Fundamental Principles of Fluid Systems	353
5.5	Basic Elements of Liquid-Level Systems	356
5.6	Dynamic Modeling of Liquid-Level Systems	363
5.7	Fluid Inertance	368
5.8	The Bernoulli Equation	370
5.9	Pneumatic Systems	374
5.10	Dynamic Response via MATLAB and Simulink	378
Chapter Summary		384
References		384
Problems		384

Thermal and fluid systems are found in a variety of engineering applications, including machines, devices, automobiles, equipment, buildings, and industrial processes. Thermal and fluid systems are in the areas of thermodynamics, heat transfer, and fluid dynamics, each of which is the subject of a complete text, and for which there is a vast literature. The properties of thermal and fluid systems, such as temperature, pressure, and flow, are spatially distributed in nature, and nonlinearities usually arise in their characterization. Because of this, distributed-parameter models governed by nonlinear partial differential equations are required for an in-depth study of these systems. However, analytical solutions for nonlinear partial differential equations are only limited to a few simple cases. Therefore, approximations have to be made either in modeling or during solution.

Very often, a thermal or fluid system operates with its variables changing in a small neighborhood of a specific operating point. By using incremental variables, a linear or linearized model of the dynamic system can be developed. Also, under certain conditions, the properties of thermal and fluid components can be lumped at a point, resulting in lumped-parameter models. This lumping process is similar to that in the modeling of mechanical systems by using lumped elements, such as point masses, springs, and dampers. Linear lumped-parameter models are governed by linear ordinary differential equations, which can be systematically solved by well-developed analytical and numerical techniques. Indeed, linear lumped-parameter models have been used in many thermal and fluid problems and satisfactory results can be obtained.

In this chapter, we shall cover the basic concepts and important aspects of physics regarding thermal and fluid systems. Due to the limitations of the mathematical tools used in this undergraduate-level textbook, linear lumped-parameter models are mainly covered for certain problems. The chapter is divided into three parts:

- modeling of thermal systems (Sections 5.1 to 5.3)
- modeling of fluid systems (Sections 5.4 to 5.9)
- numerical simulations of system responses by MATLAB and Simulink (Section 5.10)

As in dealing with mechanical systems, three keys are adopted in the modeling of thermal and fluid systems:

- fundamental principles
- basic elements
- ways of analysis

which shall be detailed in the subsequent sections. Here, for the first time, free-body diagrams as a way of analysis for thermal and fluid systems are introduced. The concept of the three-key modeling technique is useful for general thermal and fluid systems, regardless of the level of complexity of the model to be developed.

Once the governing differential equations of a lumped-parameter model are derived, they can be converted to different forms of model representation for analysis and solution, such as transfer function formulations, block diagrams, and state-space representations. Transfer function formulations and state-space representations have been introduced to mechanical systems; see Sections 3.2 and 3.7. For details on block diagrams and the use of Simulink, refer to Sections 6.2 and 6.6. If these sections have not been covered in class or self-learning, the reader may skip the relevant contents in this chapter and revisit them later after mastering these topics.

In this text, the International System of Units (SI) is used. Refer to Tables A1 and A4 in Appendix A for commonly used quantities for thermal and systems, and Table A5 for unit conversion between the SI and the US customary system.

5.1 Fundamental Principles of Thermal Systems

A thermal system is one in which thermal energy is transferred and stored. Examples of thermal systems include heaters, air conditioners, refrigerators, ovens, and cooling systems for automobiles and buildings. Thermal energy is also known as heat or heat energy. Thermal energy transfer takes place due to temperature differences, as heat flows from an object (location) of the higher temperature to an object (location) of the lower temperature. The storage of heat energy in an object depends on the substance and the temperature of the object.

Accordingly, the processes of heat transfer (thermal energy transfer) and heat storage (thermal energy storage) are described by the following two important variables:

T – temperature, in kelvins (K) or degrees in Celsius (°C), with °C = K − 273.15
q_h – heat flow rate, in joules per second (J/s) or watts (W), with 1 W = 1 J/s

Here, temperature is a measure of the potential (heat energy) of an object and heat flow describes the heat energy transfer between two objects or locations. These two variables are analogous to electric potential (e) and current (i) in an electric circuit; see Section 1.4.

5.1.1 Three Ways of Heat Transfer

Heat energy can be transferred in three ways: conduction, convection, and radiation, which are illustrated in Figure 5.1.1.

Conduction

Heat transfer by conduction occurs when two objects (locations) of different temperatures are in physical contact, which leads to diffusion of heat through a substance (either solid or fluid). By Fourier's law of heat conduction, steady-state heat conduction in a one-dimensional slab or a flat plate can be described as follows:

$$q_h = -kA\frac{dT}{dx} \qquad (5.1.1)$$

where q_h is the heat flow rate; k is the thermal conductivity of the material contained in the body, in $W/(m \cdot K)$ or $W/(m\cdot°C)$; A is the cross-sectional area normal to the heat flow direction; and x is a spatial coordinate with its positive direction in that of the heat flow. The negative sign in the previous equation indicates that there is a temperature drop in the direction of heat flow. For a lumped-parameter model of heat conduction through a slab, as shown in Figure 5.1.1(a), Eq. (5.1.1) is approximated by

$$q_h \approx -kA\frac{\Delta T}{\Delta x} = \frac{kA}{L}(T_1 - T_2) \qquad (5.1.2)$$

where L is the thickness of the slab in the direction of the heat flow; $T_1 > T_2$ is assumed to be in line with the direction of q_h.

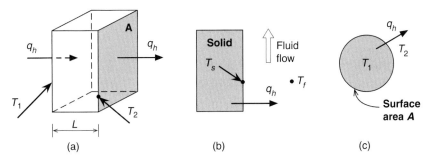

Figure 5.1.1 Heat transfer in three ways: (a) conduction, $T_1 > T_2$; (b) convection, $T_s > T_\infty$; and (c) radiation, $T_1 > T_2$

Convection

Heat transfer by convection involves the flow of thermal energy through the movement of a fluid (liquid or gas). Heat convection occurs within a fluid where warmer areas of the fluid rise to cooler areas of the fluid, as seen in water boiling in a pot. Heat convection also takes place when a moving fluid is in touch with the surface of a solid body whose temperature is different from that of the fluid, as seen in a heat exchanger or a fan blowing cool air through an object. For a fluid flowing over the surface of a solid body, as shown in Figure 5.1.1(b), the heat flow rate is given by Newton's law of cooling,

$$q_h = hA\,(T_s - T_f) \qquad (5.1.3)$$

where h is a convective coefficient, with units of $W/(m^2 \cdot K)$ and $W/(m^2 \cdot {}^\circ C)$; A is the surface area involved in heat transfer; T_s is the temperature of the solid surface; and T_f is the temperature of the fluid at some distance away from the surface.

Radiation

Heat transfer by radiation occurs through electromagnetic waves. Unlike conduction and convection, radiation transfers heat from one body to another without any physical contact in between. In other words, radiation can take place in a vacuum. Examples of radiation include heating by solar radiation and heating by heat lamps. According to the Stefan–Boltzmann law, the heat flow rate by radiation between two separated bodies of temperatures T_1 and T_2, as shown in Figure 5.1.1(c), is given by

$$q_h = \sigma\, F_E F_A\, A(T_1^4 - T_2^4) \qquad (5.1.4)$$

where σ is the Stefan–Boltzmann constant with the value $\sigma = 5.6704 \times 10^{-98}\ kg/(s^3 \cdot K^4)$, F_E is effective emissivity, F_A is a shape factor, and A is the surface area of heat transfer.

As can be seen from the previous discussion, conduction problems and some convection problems can be described by linear equations. Radiation problems in general have the nonlinearity as shown in Eq. (5.1.4), which makes it difficult in modeling and solution. In this chapter, we shall mainly consider linear or linearized problems of heat transfer.

5.1.2 Heat Energy

For an object of mass m and specific heat c, its heat energy E is given by

$$E = mc\,(T - T_0) \tag{5.1.5}$$

where T_0 is an arbitrarily chosen reference temperature. It is assumed that the temperature throughout the object is the same, without spatial distribution. (Refer to Section 5.2.3 for the justification of this uniform temperature assumption.) Note that the absolute value of E is not very meaningful. It is the change in heat energy that affects the dynamics of a thermal system, which is similar to the potential energy of a mass m under gravity, such as $mg(h - h_0)$.

Specific heat, which is also known as specific heat capacity, is the amount of heat per unit mass required to raise the temperature of an object by one unit. In the SI, specific heat has units of J/(kg·K) or J/(kg·°C). The value of the specific heat for an object depends on the thermodynamic process involved. Specific heat is usually measured either at constant volume in an isochoric process, in which c is denoted by c_V, or at constant pressure in an isobaric process, in which c is denoted by c_P. For an incompressible fluid or solid whose volume is unchanged, $c = c_V$. For a constant-pressure process, as seen in the boiling of water to steam or in the freezing of water to ice, $c = c_P$. Either way, Eq. (5.1.5) is valid for the calculation of heat energy.

5.1.3 Conservation of Energy

All thermal systems operate under the law of conservation of energy. For a closed thermal system with a well-defined boundary, the first law of thermodynamics, which is a version of the law of conservation of energy, applies

$$\Delta U = Q - W \tag{5.1.6}$$

where ΔU denotes the change in the energy of the system, Q represents the heat energy supplied to the system, and W is the work done by the system on its surroundings. The change in the energy is expressed by

$$\Delta U = \Delta E + \Delta \Pi_{ME} \tag{5.1.7}$$

where ΔE is the change in the heat energy (also called the internal energy), and $\Delta \Pi_{ME}$ is the change in the mechanical energy of the system (a sum of kinetic energy and potential energy).

Consider a thermal system with negligible change in mechanical energy ($\Delta \Pi_{ME} = 0$). Assume that no work is done by the system on its surroundings ($W = 0$). The heat energy supplied to the system during time interval Δt can be written as

$$\Delta E = [q_{in}(t) - q_{out}(t)]\Delta t \tag{5.1.8}$$

where q_{in} is the rate of heat flow into the system and q_{out} is the rate of heat flow out of the system. Thus, the law of conservation of energy for the system reads

$$\frac{dE}{dt} = q_{in} - q_{out} \tag{5.1.9}$$

or in integral form

$$E = \int_{t_0}^{t} (q_{in} - q_{out})\, dt \tag{5.1.10}$$

where t_0 is an initial time.

5.2 Basic Thermal Elements

A lumped-parameter model of thermal systems has two types of basic elements: thermal capacitance and thermal resistance. As shown in Section 1.4, these thermal elements are analogous to capacitors and resistors in an electric circuit or springs and dampers in a mechanical system. With the law of conservation of energy, the use of thermal capacitance and thermal resistance leads to the establishment of dynamic models for thermal systems (Section 5.3).

5.2.1 Thermal Capacitance

The thermal capacitance of an object is the ability of the material of the object to store heat energy. The thermal capacitance C of an object is defined as the rate of change of the heat energy stored in the object with respect to the object's temperature; that is,

$$C = \frac{dE}{dT} \tag{5.2.1}$$

where E is the stored heat energy as described by Eq. (5.1.5), and T is the temperature of the object. The SI units of thermal capacitance are J/K or J/°C.

If the specific heat c of an object is not a function of temperature, the thermal capacitance of the object by Eqs. (5.1.5) and (5.2.1) is given by

$$C = mc = \rho V c \tag{5.2.2}$$

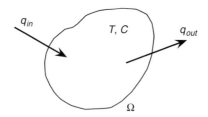

Figure 5.2.1 A body with input and output heat flow rates

where ρ and V are the density and volume of the mass m.

Consider a body Ω of temperature T and thermal capacitance C. Assume a negligible change in the mechanical energy in the body and zero work by the body to its surroundings. By Eq. (5.2.1),

$$\frac{dE}{dt} = \frac{dE}{dT}\frac{dT}{dt} = C\frac{dT}{dt}$$

It follows from the law of conservation of energy as presented in Eq. (5.1.9) that the governing equation of the body is obtained as follows:

$$C\frac{dT}{dt} = q_{in} - q_{out} \qquad (5.2.3)$$

Here, $q_{in} - q_{out}$ is the net heat flow rate into the body, as shown in Figure 5.2.1. So, with given input and output heat flow rates, the time-dependent temperature of the body can be determined by solving the differential equation (5.2.3).

Comparison of Eq. (5.2.3) and Eq. (4.1.16) shows that the temperature T of a thermal capacitance is analogous to the voltage v (potential drop) of an electrical capacitor. Furthermore, because Eq. (5.2.3) is a first-order differential equation, the temperatures of the thermal capacitances can be conveniently selected as state variables in the development of a state-space model of thermal systems, as shown in Section 5.3.

5.2.2 Thermal Resistance

Thermal resistance is about a flow of heat or heat transfer through a body, and it is a property of the material of the body that resists heat transfer. As a result of thermal resistance, the temperature drops as heat flows through a body. This temperature drop in a thermal-resistance element plays a similar role as a potential drop (voltage) in an electrical resistor. In other words, thermal resistance is analogous to electrical resistance.

The thermal resistance in heat transfer is defined as

$$R = \frac{dT}{dq_h} \qquad (5.2.4)$$

which in the SI has units of K/W (or K · s/J) and °C/W (or °C · s/J). As presented in Section 5.1.1, there are three ways of heat transfer: conduction, convection, and radiation. For a linear relationship between the temperature and heat flow rate, as seen in conduction models and some convection models, Newton's law of cooling applies:

$$q_h = \frac{1}{R} \Delta T \tag{5.2.5}$$

where the thermal resistance is a constant for any given temperature, and ΔT is the temperature difference.

If the relationship between the temperature and heat flow rate is nonlinear, as seen in radiation, the corresponding thermal resistance can be determined by a linearized model. To this end, we assume that the temperature varies in a small neighborhood of a specific temperature T_*. By the Taylor series,

$$q_h \approx \frac{1}{R_*} \Delta T \tag{5.2.6}$$

with

$$R_* = \frac{1}{(dq_h/dT)_{T=T_*}} \tag{5.2.7}$$

Thermal resistances in the three forms of heat transfer are obtained in sequel. For heat conduction, as shown in Figure 5.1.1(a), Eqs. (5.1.2) and (5.2.4) give

$$R = \frac{L}{kA} \tag{5.2.8}$$

For heat convection between a fluid and a solid body, as shown in Figure 5.1.1(b), Eqs. (5.1.3) and (5.2.4) yield

$$R = \frac{1}{hA} \tag{5.2.9}$$

For heat transfer by radiation, as shown in Figure 5.1.1(c), the relationship (5.1.4) between the heat flow rate and the temperatures is nonlinear. Assume that the temperature T_2 of the radiated body or position is constant. Consider a small variation of the temperature of the radiating body; that is, $T_1 = T_* + \Delta T$. With the linearized model given by Eq. (5.2.6) and through the use of Eqs. (5.1.4) and (5.2.7), the thermal resistance is obtained as follows:

$$R_* = \frac{1}{4\sigma F_E F_A A T_*^3} \tag{5.2.10}$$

Comparison of Eq. (4.1.8) and Eq. (5.2.5) reveals that the heat flow rate q_h in a thermal system is analogous to the current i in an electrical system. Thus, in developing lumped-parameter models, combinations of several thermal resistances can be replaced by a single equivalent thermal resistance, like combinations of resistors in electrical systems. For two thermal resistances connected in series, as shown in Figure 5.2.2(a), the equations

$$q_h = \frac{1}{R_1}(T_1 - T_c), \quad q_h = \frac{1}{R_c}(T_c - T_2)$$

where T_c is the temperature at the interface between the thermal resistances, can be replaced by

$$q_h = \frac{1}{R_{eq}}(T_1 - T_2) \tag{5.2.11}$$

where the equivalent thermal resistance is given by

$$R_{eq} = R_1 + R_2 \tag{5.2.12}$$

For two thermal resistances in parallel combination, as shown in Figure 5.2.2(b), the equations

$$q_{h1} = \frac{1}{R_1}(T_1 - T_2), \quad q_{h2} = \frac{1}{R_2}(T_1 - T_2), \quad q_h = q_{h1} + q_{h2}$$

can be reduced to Eq. (5.2.11), with the equivalent thermal resistance given by

$$\frac{1}{R_{eq}} = \frac{1}{R_1} + \frac{1}{R_2} \tag{5.2.13}$$

For complicated thermal systems, series combinations and parallel combinations can be further combined, as in the modeling of electrical systems.

The series combination of thermal resistances can be used to model heat conduction through a multilayer composite slab. The parallel combination of thermal resistances can be used to treat heat conduction through a thermal system consisting of multiple parts, as shown in the following example.

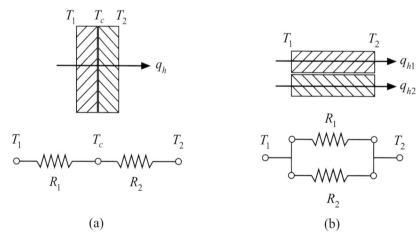

Figure 5.2.2 Combinations of thermal resistances: (a) series combination; (b) parallel combination

Example 5.2.1

Figure 5.2.3 shows a closed and hollow cylindrical vessel, with thickness δ, length L, and outer diameter D. The vessel is made of a material with thermal conductivity k. Let the temperatures inside and outside the vessel be T_i and T_0, respectively. Determine the equivalent thermal resistance of the entire vessel.

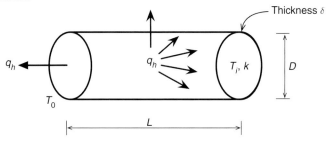

Figure 5.2.3 Heat conduction in a closed cylindrical vessel

Solution

The vessel has three parts: the side of the cylinder and the two circular ends. Denote the thermal resistance of the cylinder side by R_c. Denote the thermal resistance of an end by R_e. The equivalent thermal resistance of the vessel is a parallel combination of three resistances, as shown in Figure 5.2.4. Therefore, the equivalent thermal resistance of the entire vessel is

$$R_{eq} = \left(\frac{1}{R_e} + \frac{1}{R_c} + \frac{1}{R_e}\right)^{-1} = \frac{R_c R_e}{2R_e + R_c} \tag{a}$$

with the resistances R_c and R_e yet to be determined.

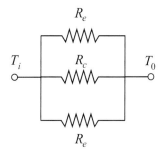

Figure 5.2.4 Parallel combination of thermal resistances for the vessel

There are two cases of wall thickness of the vessel. For a vessel of thin wall ($\delta/D \ll 1$ and $\delta/L \ll 1$), heat conduction through the thickness of the cylinder can be approximated by that through a slab, as shown in Figure 5.1.1(a). Thus,

$$R_c = \frac{\delta}{\pi D L k}, \quad R_e = \frac{4\delta}{\pi D^2 k} \tag{b}$$

where Eq. (5.2.8) has been used. It follows that

$$R_{eq} = \frac{2\beta}{D + 2L}, \quad \text{with } \beta = \frac{\delta}{\pi D k} \tag{c}$$

If the wall of the vessel is thick, Eq. (5.2.8) is not applicable because the wall of the cylinder cannot be approximated as a flat plate. In this case, the radial coordinate version of Fourier's law (5.1.1) can be used, and it is given by

$$q_h = -kA\frac{dT}{dr} = -k(2\pi rL)\frac{dT}{dr}, \quad r_i \leq r \leq r_o \tag{d}$$

where r_i and r_o are the inner and outer radii of the cylinder, and they are given by $r_i = \frac{D}{2} - \delta$ and $r_o = \frac{D}{2}$. Assume that the thermal conductivity k is constant through the thickness of the cylinder wall. Integration of Eq. (d), namely,

$$q_h \int_{r_i}^{r_o} \frac{dr}{r} = -2\pi Lk \int_{T_i}^{T_o} dT$$

gives

$$q_h = \frac{2\pi Lk}{\ln(r_o/r_i)}(T_i - T_o) \tag{e}$$

Thus, the thermal resistance of the cylinder is

$$R_c = \frac{\ln(r_o/r_i)}{2\pi Lk} = \frac{1}{2\pi Lk}\ln\left(\frac{D}{D - 2\delta}\right) \tag{f}$$

The thermal resistance at each end of the vessel is

$$R_e = \frac{h}{\pi r_i^2 k} = \frac{4h}{\pi(D - 2\delta)^2 k} \tag{g}$$

where the inner radius has been used. Substitution of the updated R_c and R_e into Eq. (a) yields the equivalent resistance of the thick-walled vessel.

In this section, two types of lumped-parameter elements have been introduced: thermal capacitances and thermal resistances. A thermal capacitance is governed by an ordinary differential equation like Eq. (5.2.3); a thermal resistance is described by an algebraic equation like Eq. (5.2.5). Thermal capacitance is analogous to electrical capacitance while thermal resistance is analogous to electrical resistance. Because of this, a thermal system can be modeled and simulated through the use of an analogous electrical network.

With the law of thermodynamics, however, there is no heat energy storage mechanism that is directly related to the heat flow rate (though researchers have been trying to establish the concept of thermal inductance in some specific cases of heat transfer). In general, there does not exist a thermal element that is analogous to electrical inductance.

Besides thermal capacitance and resistance, thermal sources (heat energy inputs) are specified to study the dynamic behavior of a thermal system. There are two typical types of thermal resources:

- specified heat flow rate, which is positive when heat energy is added to the system or negative when heat energy is taken away from the system; and
- specified temperature of a body as a heat energy input to the system.

5.2.3 The Biot Criterion for Lumped-Parameter Models

In a lumped-parameter model of thermal systems, the temperature within an object is assumed to be uniform. Physically, heat transfer does occur within the body of the object, rendering the temperature in the body spatially distributed. To justify the utility of the uniform temperature assumption, the object needs to satisfy certain conditions or criteria. In many applications, such criteria can be set up by the Biot number.

For a solid body immersed in a fluid, the Biot number B_i is a nondimensional number that is defined by

$$B_i = \frac{hL_c}{k} \qquad (5.2.14)$$

where h is a convective coefficient as given in Eq. (5.1.3); k is the thermal conductivity of the material contained in the body; and L_c is a characteristic length of the body, which is usually taken to be the ratio of the body volume to the area of heat transfer of the body. For instance, for the slab in Figure 5.1.1(a),

$$L_c = \frac{A \cdot L}{A} = L$$

As another example, for a cylindrical solid of radius R and height H,

$$L_c = \frac{\pi R^2 H}{2\pi R H + 2\pi R^2} = \frac{1}{2}\frac{RH}{H+R}$$

Physically, the Biot number can be viewed as the ratio of the internal conductive thermal resistance to the external conductive thermal resistance. For example, for a solid body submerged in a fluid, the thermal resistance ratio is

$$\frac{(R)_{conduction}}{(R)_{convection}} = \frac{1/kA}{1/hA} = \frac{hL}{k} = B_i \qquad (5.2.15)$$

where Eqs. (5.2.8) and (5.2.9) have been used.

If the Biot number of a body is small enough, say $B_i < 0.1$, the temperature within the body can be assumed to be uniform. This is because the conductive heat transfer within the body experiences much less thermal resistance, compared to the convective heat transfer at the boundary (wall) of the body. As a result, a uniform temperature distribution within the body can be reached through relatively quick heat conduction. If the Biot number is larger than 0.1, a uniform temperature may not be assumed. In this case, a set of coupled lumped-parameter models (with smaller characteristic lengths) or even a distributed-parameter model by partial differential equations is required.

Example 5.2.2
Quenching is a metal heat-treatment process, which involves the rapid cooling of a heated metal in water, oil, or air to improve certain material properties, like hardness and strength. Consider an aluminum cube in quenching. The cube, with a side length of $a = 0.06$ mm, is immersed in an oil tank for which the convective coefficient $h = 350$ W/(m²·K). The thermal conductivity k of aluminum varies from 220 W/(m²·K) to 240 W/(m²·K), in a wide temperature range of 200–700 K. By the Biot criterion, determine if the aluminum cube in quenching can be treated as a lumped-parameter system with a uniform temperature.

Solution
The characteristic length of the cube is $L_c = \dfrac{a^3}{6a^2} = \dfrac{a}{6} = 0.01$ m. Take the median value of the thermal conductivity, $k = 230$ W/(m²·K). By Eq. (5.2.14), the Biot number is

$$B_i = \frac{hL_c}{k} = \frac{350 \times 0.01}{230} = 0.015$$

Because B_i is much less than 0.1, according to the Biot criterion, the aluminum cube can be treated as a lumped-parameter system with a single uniform temperature.

5.3 Dynamic Modeling of Thermal Systems

So far, the first two keys in modeling of thermal systems have been presented: the fundamental principles (Section 5.1) and basic elements (Section 5.2). In this section, the last of the three keys in system modeling is introduced: a way of analysis, which adopts free-body diagrams for thermal capacitances and auxiliary plots for thermal resistances.

Free-body diagrams and auxiliary plots for thermal elements are simular to those for mass, spring, and damping elements in mechanical systems (Section 3.2.2). A free-body diagram is one in which the capacitance element in consideration is isolated from all other thermal elements, but with input and output heat flow rates as interactions with its surroundings. An auxiliary plot is a figure in which the resistance element with a temperature difference undergoes a heat flow rate.

For a given thermal system of multiple thermal capacitances and resistances, the three-key modeling technique takes the following steps:

(1) Draw a free-body diagram for each capacitance element, and an auxiliary plot for each resistance element.
(2) Write a differential equation for each free-body diagram, and an algebraic equation for each auxiliary plot. In this step, the formulas about the basic thermal elements given in Section 5.2 are used.

(3) By substitution, combine the differential and algebraic equations obtained in Step 2, to produce a set of independent differential equations as a mathematical model of the thermal system. The number of independent differential equations is equal to the number of the capacitance elements in the system.

Once the governing equations of a thermal system are established, modeling techniques, such as state-space representations, transfer function formulations, and block diagrams can be applied. In a state-space formulation, it is convenient to select capacitance temperatures as state variables. When constructing a block diagram, follow Example 3.6.1 and use Table 3.6.1. Also, refer to Section 6.2.3 for the basic concepts and rules about block diagrams.

The modeling approach described above is demonstrated in the following examples. For simplicity, the following symbols are used:

▨▨▨▨▨ a wall of perfect insulation material that prevents heat transfer through it

q_{in} ⟶ ⌇ an input heat flow rate to a thermal capacitance

Example 5.3.1

In Figure 5.3.1, a thermal system consists of two capacitances that are separated by two resistances. The system has two thermal inputs: a heat flow rate q_{in} applied to the left capacitance by a heater, and a temperature T_a (say, the ambient temperature) at the right end of the right resistance, where heat is lost to the environment. Derive the differential equations governing the temperatures T_1 and T_2, obtain a state-space representation, and draw a block diagram for the system, with T_1 and the heat flow rate through the right resistance (R_2) as the system outputs.

Solution
In the first step of system modeling, the free-body diagrams of the capacitances and the auxiliary plots of the resistances are drawn in Figure 5.3.2, where q_1 and q_2 are the heat flow rates going through the resistances.

In the second step, the equations of the capacitances and resistance are written down according to the free-body diagrams and auxiliary plots:

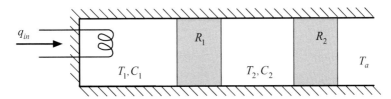

Figure 5.3.1 The thermal system in Example 5.3.1

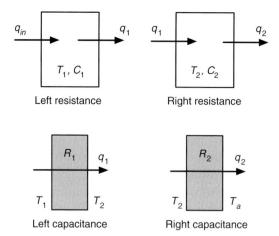

Figure 5.3.2 Free-body diagrams and auxiliary plots of the thermal system in Example 5.3.1

$$\text{Left capacitance:} \quad C_1 \frac{dT_1}{dt} = q_{in} - q_1 \tag{a}$$

$$\text{Right capacitance:} \quad C_2 \frac{dT_2}{dt} = q_1 - q_2 \tag{b}$$

$$\text{Left resistance:} \quad q_1 = \frac{1}{R_1}(T_1 - T_2) \tag{c}$$

$$\text{Right resistance:} \quad q_2 = \frac{1}{R_2}(T_2 - T_a) \tag{d}$$

In the third step, by substituting Eqs. (c) and (d) into Eqs. (a) and (b), two independent differential equations about T_1 and T_2 are obtained as follows:

$$C_1 \frac{dT_1}{dt} = q_{in} - \frac{1}{R_1}(T_1 - T_2) \tag{e1}$$

$$C_2 \frac{dT_2}{dt} = \frac{1}{R_1}(T_1 - T_2) - \frac{1}{R_2}(T_2 - T_a) \tag{e2}$$

For a state-space representation, take the capacitance temperatures as the state variables: $x_1 = T_1$ and $x_2 = T_2$. This, by Eqs. (e1) and (e2), leads to the following state equations

$$\dot{x}_1 = \frac{1}{C_1}\left[-\frac{1}{R_1}x_1 + \frac{1}{R_1}x_2 + q_{in}\right] \tag{f1}$$

$$\dot{x}_2 = \frac{1}{C_2}\left[\frac{1}{R_1}x_1 - \left(\frac{1}{R_1} + \frac{1}{R_2}\right)x_2 + \frac{1}{R_2}T_a\right] \tag{f2}$$

The output equations of the system are

$$y_1 = x_1, \quad y_2 = \frac{1}{R_2}(x_2 - T_a) \tag{g}$$

where Eq. (d) has been used.

Now, for the construction of a block diagram, take Laplace transform of Eqs. (a) to (d) with respect to time, yielding

$$\text{Left capacitance:} \quad C_1 s \hat{T}_1(s) = \hat{q}_{in}(s) - \hat{q}_1(s) \tag{h}$$

$$\text{Right capacitance:} \quad C_2 s \hat{T}_2(s) = \hat{q}_1(s) - \hat{q}_2(s) \tag{i}$$

$$\text{Left resistance:} \quad \hat{q}_1(s) = \frac{1}{R_1}\left(\hat{T}_1(s) - \hat{T}_2(s)\right) \tag{j}$$

$$\text{Right resistance:} \quad \hat{q}_2(s) = \frac{1}{R_2}\left(\hat{T}_2(s) - \hat{T}_a(s)\right) \tag{k}$$

where the over-hat (^) stands for Laplace transformation.

By Eqs. (h) to (k), the parts of the block diagram are plotted in Figure 5.3.3, where the basic components in Table 3.6.1 have been used. Assembly of these parts gives the block diagram of the thermal system in Figure 5.3.4.

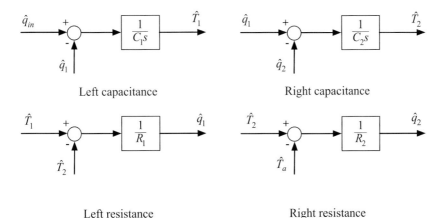

Figure 5.3.3 Parts of the block diagram in Example 5.3.1

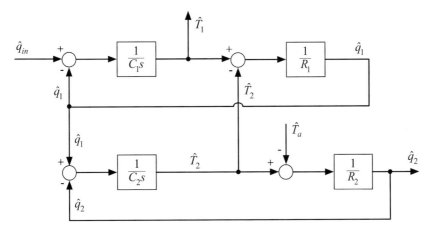

Figure 5.3.4 Block diagram for the thermal system in Example 5.3.1

Example 5.3.2

The thermal system in Figure 5.3.5 consists a capacitance and an assembly of three insulation layers (a series combination of three thermal resistances), where k_i and δ_i are the thermal conductivity and thickness of the i-th layer, respectively. Assume that all the layers have the same cross-sectional area A. A heat flow rate q_{in} is applied to the capacitance. On the right side of the third layer, convective heat transfer takes place, with h being the convective coefficient, T_s the surface temperature, and T_f the temperature of a cold fluid. This thermal system can be viewed as a simplified model of a room or vessel with a double-pane window for insulation and protection. Derive the governing differential equation of the system in terms of the capacitance temperature T and determine the transfer function from q_{in} to T and the transfer function from T_f to T.

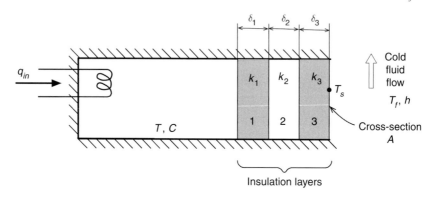

Figure 5.3.5 The thermal system in Example 5.3.2

Solution

The free-body diagram of the capacitance and the auxiliary plot of the resistance assembly are shown in Figure 5.3.6, where q_{out} is the heat flow rate out of the capacitance, R_1, R_2, and R_3 are the thermal resistances of the layers, R_4 is the thermal resistance due to convection heat transfer, and R_{eq} is the combined thermal resistance. The governing equations of the thermal elements by the figure are derived as follows:

$$\text{Capacitance:} \quad C\frac{dT}{dt} = q_{in} - q_{out} \quad (a)$$

$$\text{Combined resistance:} \quad q_{out} = \frac{1}{R_{eq}}(T - T_f) \quad (b)$$

where the equivalent resistance R_{eq}, by Eqs. (5.2.8), (5.2.9), and (5.2.12), is given by

$$R_{eq} = R_1 + R_2 + R_3 + R_4 = \frac{\delta_1}{k_1 A} + \frac{\delta_2}{k_2 A} + \frac{\delta_3}{k_3 A} + \frac{1}{hA} \quad (c)$$

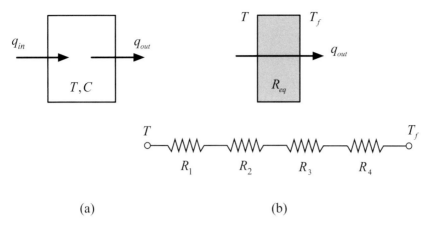

Figure 5.3.6 The thermal system in Example 5.3.2: (a) free-body diagram of the capacitance; (b) auxiliary plot of the resistance

Substituting Eq. (b) into Eq. (a) gives the governing equation of the system

$$R_{eq}C\frac{dT}{dt} + T = R_{eq}q_{in} + T_f \tag{d}$$

Laplace transformation of Eq. (d) leads to

$$(R_{eq}Cs + 1)\hat{T}(s) = R_{eq}\hat{q}_{in}(s) + \hat{T}_f(s) \tag{e}$$

with which the transfer functions of the thermal system are obtained as follows:

$$\left.\frac{\hat{T}(s)}{\hat{q}_{in}(s)}\right|_{\hat{T}_f(s)=0} = \frac{R_{eq}}{R_{eq}Cs + 1}, \quad \left.\frac{\hat{T}(s)}{\hat{T}_f(s)}\right|_{\hat{q}_{in}(s)=0} = \frac{1}{R_{eq}Cs + 1} \tag{f}$$

Example 5.3.3

Consider the quenching process shown in Figure 5.3.7, where a heated metal piece (T_m, C_m) is immersed in a liquid bath (T_b, C_b) of finite size. (Refer to Example 5.2.2 for a brief explanation of quenching.) Heat transfer takes place between the solid and the liquid with thermal resistance R_b and between the liquid and the ambient air of temperature T_a with thermal resistance R_a. Assume that $T_m > T_b$. Derive a dynamic model of the quenching process in terms of the metal and bath temperatures. Also, obtain the state equations of the process.

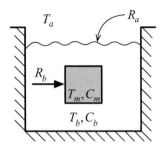

Figure 5.3.7 Quenching in a liquid bath of finite size in Example 5.3.3

Solution

Because the bath is of finite size, its temperature is time-varying. By the free-body diagrams of the capacitances and the auxiliary plots of the resistances shown in Figure 5.3.8, the equations for the elements of the thermal system are obtained as follows:

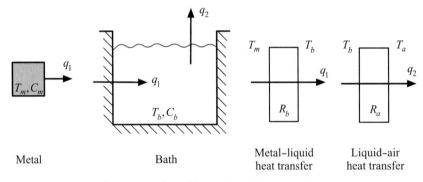

Figure 5.3.8 Free-body diagrams and auxiliary plots of the quenching process

$$\text{Metal:} \quad C_m \frac{dT_m}{dt} = -q_1 \tag{a}$$

$$\text{Bath:} \quad C_b \frac{dT_b}{dt} = q_1 - q_2 \tag{b}$$

$$\text{Solid–liquid heat transfer:} \quad q_1 = \frac{1}{R_b}(T_m - T_b) \tag{c}$$

$$\text{Liquid–air heat transfer:} \quad q_2 = \frac{1}{R_a}(T_b - T_a) \tag{d}$$

By substitution, the dynamic model of the quenching process is described by the following coupled differential equations:

$$\begin{aligned} C_m \frac{dT_m}{dt} &= -\frac{1}{R_b}(T_m - T_b) \\ C_b \frac{dT_b}{dt} &= \frac{1}{R_b}(T_m - T_b) - \frac{1}{R_a}(T_b - T_a) \end{aligned} \tag{e}$$

Now, by selecting two state variables
$$x_1 = T_m, \quad x_2 = T_b \tag{f}$$
and with Eq. (e), the state equations of the process are obtained as follows:
$$\begin{aligned}
\dot{x}_1 &= -\frac{1}{R_b C_m}(x_1 - x_2) \\
\dot{x}_2 &= \frac{1}{R_b C_b}x_1 - \left(\frac{1}{R_b C_b} + \frac{1}{R_a C_b}\right)x_2 + \frac{1}{R_a C_b}T_a
\end{aligned} \tag{g}$$

5.4 Fundamental Principles of Fluid Systems

A fluid system is one in which fluids flow. A fluid may be either a liquid or a gas. Fluid systems can be divided into two categories: (i) *hydraulic systems* in which the working fluids are incompressible; and (ii) *pneumatic systems* in which the working fluids are compressible. An incompressible fluid is a fluid whose density in a range of pressure does not change or has some negligible change. Many liquids, including water, can be viewed as incompressible fluids. On the other hand, gases, including air, are compressible fluids. Hydraulic systems are seen in a variety of applications, including actuators and motors in manufacturing equipment, chemical processes, power steering and shock absorbers in automobiles, and devices in airplanes, rockets, and spaceships. Pneumatic systems are often used in actuators and switches in machines and equipment, air brakes on buses and trucks, and industrial applications such as material handling, drilling, and sawing.

In the study of fluid systems, the following variables are often used:

V – volume, in m^3
ρ – density or mass density, in kg/m^3
q – volume flow rate (volume per unit time), in m^3/s
q_m – mass flow rate, $q_m = \rho q$
h – liquid height, in m
p, P – pressure, in N/m^2 (Pa) or atmospheric pressure (atm), with 1 atm = 101 325 Pa
T – temperature, in kelvins (K) or degrees in Celsius (°C), with °C = K − 273.15

In this section, for dynamic modeling of fluid systems by lumped-parameter models, two types of elements are introduced: fluid capacitance and fluid resistance. Unlike thermal systems, fluid systems also have fluid inertance due to the inertia of a moving fluid. In some applications, like liquid-level systems, fluid inertance can be ignored due to the negligible kinetic energy of a moving fluid. The effects of fluid inertance, however, can be significant in other applications, as seen in piping systems. Fluid inertance is introduced in Section 5.7.

5.4.1 Conservation of Mass

For both hydraulic systems and pneumatic systems, the principle of mass conservation can be stated as follows: *the mass of a body that is closed to all the transfers of matter and energy remains constant over time.* For a body of a fluid with volume V and density ρ, the principle of mass conservation can be expressed by

$$\frac{dM}{dt} = \frac{d}{dt}\int_V \rho \, dv = 0 \qquad (5.4.1)$$

For an incompressible fluid (ρ = constant), the conservation of mass is equivalent to the conservation of volume; that is,

$$\frac{dV}{dt} = \frac{d}{dt}\int_V dv = 0 \qquad (5.4.2)$$

The principle of mass conservation lays out a foundation for the development of models of fluid systems.

5.4.2 Continuity of Fluids

When a fluid moves in a duct or pipe, as shown in Figure 5.4.1(a), Eq. (5.4.1) can be written as

$$q_{m1} = q_{m2} \qquad (5.4.3)$$

where q_{m1} and q_{m2} are the mass flow rates at any two locations 1 and 2. Note that a mass flow rate can be expressed by

$$q_m = \rho q = \rho A v \qquad (5.4.4)$$

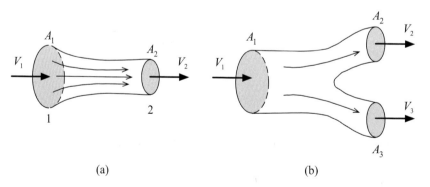

Figure 5.4.1 Fluid moving in a one-dimensional duct: (a) single-body duct; (b) branched duct

where v is the velocity of the fluid, and A is the cross-sectional area with its normal in line with v. We can substitute Eq. (5.4.4) into Eq. (5.4.3), to obtain

$$\rho_1 A_1 v_1 = \rho_2 A_2 v_2 \tag{5.4.5}$$

where ρ_i, A_i and v_i are the density, cross-sectional area, and velocity of the fluid at location i, respectively. If the fluid is incompressible ($\rho_1 = \rho_2$), Eq. (5.4.3) becomes

$$q_1 = A_1 v_1 = A_2 v_2 = q_2 \tag{5.4.6}$$

Equations (5.4.5) and (5.4.6) are known as continuity equations. By the same token, the fluid continuity also holds for a branched duct, as shown in Figure 5.4.1(b), which is

$$\rho_1 A_1 v_1 = \rho_2 A_2 v_2 + \rho_3 A_3 v_3 \tag{5.4.7}$$

or

$$A_1 v_1 = A_2 v_2 + A_3 v_3 \tag{5.4.8}$$

for an incompressible fluid.

These continuity equations are analogous to Kirchhoff's current law for electrical systems (Section 4.3).

5.4.3 Equation of State for Gases

Gases are compressible fluids. The equation of state for an ideal gas, which is also known as the *ideal gas law*, is written as

$$pV = nRT = mR_g T \tag{5.4.9}$$

where p, V, and T are the absolute pressure, volume, and temperature of the gas, respectively; n is the number of moles of the gas; R is the universal gas constant, which is 8.314463 J/(mol·K); m is the mass of the gas contained in V; and R_g is the specific gas constant depending on the particular type of gas. Here, the mole is the unit of the amount of substance and one mole of an element contains 6.023×10^{23} atoms. For dry air, $R_g = 287.06$ J/(kg·K).

Consider a thermodynamic process from State 1 (beginning state) to State 2 (ending state). Based on the ideal gas law, the following processes are commonly used in the development of gas models, with the subscripts 1 and 2 denoting the beginning and ending states:

(1) *Isobaric process* (constant-pressure process) with $p_1 = p_2$
By Eq. (5.4.9),

$$\frac{V_2}{V_1} = \frac{T_2}{T_1} \tag{5.4.10}$$

indicating that addition of heat to the gas can raise the temperature and expand the volume to exert external work.

(2) *Isochoric process* (constant-volume process) with $V_1 = V_2$
By Eq. (5.4.9),
$$\frac{p_2}{p_1} = \frac{T_2}{T_1} \qquad (5.4.11)$$

Because the volume is constant, no external work is done by the gas. Thus, in this process, addition of heat to the gas can only raise the temperature and pressure.

(3) *Isothermal process* (constant-temperature process) with $T_1 = T_2$
By Eq. (5.4.9),
$$\frac{p_2}{p_1} = \frac{V_1}{V_2} \qquad (5.4.12)$$

Because the temperature is constant, the addition of heat to the gas will not increase the internal energy of the gas, but only does external work.

Besides the above three processes, the most general thermodynamic process, which is called the *polytropic process*, obeys the relation

$$p\left(\frac{V}{m}\right)^n = \frac{p}{\rho^n} = \text{constant} \qquad (5.4.13)$$

where ρ and m are the density and mass of the gas, respectively. If m is constant, this general case reduces to an isobaric process for $n = 0$, an isochoric process for $n = \infty$, and an isothermal process for $n = 1$. Furthermore, for $n = \gamma \equiv c_P/c_V$ (heat capacity ratio), Eq. (5.4.13) reduces to an *isentropic process* (revisable adiabatic process), in which no heat is transferred to or from the gas and which can be described by $p_1 V_1^\gamma = p_2 V_2^\gamma$.

The equation of state (5.4.9) and the relation for polytropic process (Eq. (5.4.13)) are useful in the derivation of fluid capacitance for pneumatic systems, as shown in Section 5.9.

5.5 Basic Elements of Liquid-Level Systems

Liquid-level systems are types of hydraulic systems that are assemblies of tanks (containers, reservoirs, vessels), pipes, valves, pumps, and other components. Liquid-level systems have wide applications in industrial processes, water treatment and water supply, and hydraulic actuators. In the subsequent discussion, the working liquid in a liquid-level system is assumed to be incompressible. The dynamic behavior of a liquid-level system can be characterized by the volume flow rate q, liquid height h, and pressure p.

In this section, we introduce two types of lumped-parameter elements: fluid capacitance and fluid resistance. Fluid capacitance corresponds to the gravitational potential energy of the mass of a liquid in a storage tank; fluid resistance is the resistance encountered by a liquid flowing through a pipe, a valve, or an orifice. Besides fluid capacitance and resistance, there is a third type of element – fluid inertance – which is associated with the kinetic energy of the mass of a liquid flowing in a pipe. In modeling liquid-level systems, fluid inertance is usually ignored due to the negligible kinetic energy of the mass of a flowing liquid, compared with the gravitational potential energy of the liquid mass. Fluid inertance is useful in modeling other fluid systems, as shown in Section 5.7.

5.5.1 Storage Tanks

Figure 5.5.1 shows a storage tank (open vessel) of arbitrary shape, which contains a liquid of density ρ and height h. The mass of the liquid in the tank is

$$m = \rho V = \rho \int_0^h A(y)\,dy \tag{5.5.1}$$

where $A(y)$ is the cross-sectional area of the tank at height y, and V is the total volume of the liquid. The absolute pressure of the liquid at the bottom of the tank is

$$p = P_a + \rho g h \tag{5.5.2}$$

where P_a is the atmospheric pressure (1.013×10^5 N/m²), and g is the gravitational acceleration (9.807 m/s²). The gravitational potential energy of the liquid mass is $PE = \rho V g\, h_g$, where h_g is the height of the center of mass of the liquid in the y direction. Because $h_g = \dfrac{1}{V}\int_0^h y A(y)\,dy$, the potential energy is given by

$$PE = \rho g \int_0^h y A(y)\,dy \tag{5.5.3}$$

The kinetic energy of the mass in the tank can be written as $KE = \dfrac{1}{2}\int_V u^2\,dm$, where u is the magnitude of the velocity of dm. In modeling of liquid-level systems, the kinetic energy of the liquid is usually neglected because it is much less than the potential energy given in Eq. (5.5.3).

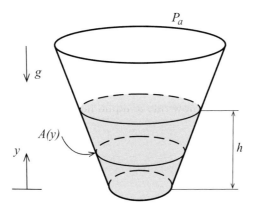

Figure 5.5.1 A storage tank with arbitrary shape

5.5.2 Fluid Capacitance

Fluid capacitance relates the change in stored mass to the change in pressure. The fluid capacitance C of a tank is defined as

$$C = \left.\frac{dm}{dp}\right|_{p=p_r} \quad (5.5.4)$$

where m is the stored mass, and p_r is a reference pressure, which is often taken as the pressure at the bottom of the tank, $p_r = p_a + \rho g h$. The SI unit of C is in kg·m^2/N. By Eqs. (5.5.1) and (5.5.2),

$$dm = \rho A(h) dh, \quad dp = \rho g dh \quad (5.5.5)$$

Thus, the fluid capacitance of the tank is

$$C = \frac{dm}{dp} = \frac{\rho A(h) dh}{\rho g dh} = \frac{A(h)}{g} \quad (5.5.6)$$

Equation (5.5.6) shows that the fluid capacitance of a tank depends on the cross-sectional area. For a tank of constant cross-section, its capacitance C is constant for any liquid height h.

Equation (5.5.6) also indicates that fluid capacitance is independent of the fluid density ρ. Because of this, another commonly used definition of fluid capacitance is defined as follows:

$$C_V = \frac{dV}{dh} \quad (5.5.7)$$

which relates the change in the volume V of the stored mass to the change in the liquid height h. Here, C_V is in m^2. Because $dV = A(y) dh$,

$$C_V = \frac{A(h) dh}{dh} = A(h) \quad (5.5.8)$$

Therefore,

$$C_V = Cg \quad (5.5.9)$$

For clarity, in our discussion, C_V is called the *volume* capacitance and C is called the *mass* capacitance or simply capacitance.

Example 5.5.1

Determine the mass capacitance C and volume capacitance C_V of the tank with the conical frustum shape shown in Figure 5.5.2.

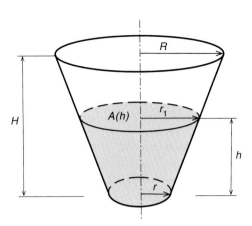

Figure 5.5.2 Conical-frustum-shaped tank in Example 5.5.1

Solution

The cross-sectional area of the tank at $y = h$ (measured from the tank bottom) is $A(h) = \pi r_1^2$, where r_1 is the radius of the liquid surface, and it is in the relation

$$\frac{R-r}{H} = \frac{r_1 - r}{h}$$

This leads to

$$r_1 = \left(\frac{R-r}{H}\right)h + r$$

By Eqs. (5.5.6) and (5.5.8),

$$C = \frac{\pi}{g}\left[\left(\frac{R-r}{H}\right)h + r\right]^2, \quad C_V = \pi\left[\left(\frac{R-r}{H}\right)h + r\right]^2$$

5.5.3 Conservation of Mass

For a storage tank, the principle of mass conservation applies. In Figure 5.5.3, a tank of liquid has volume inflow rate q_{in} and volume outflow rate q_{out}, in m³/s. By the principle of mass conservation, Eq. (5.4.1), the stored mass m in the tank is governed by

$$\frac{dm}{dt} = \rho q_{in} - \rho q_{out} \tag{5.5.10}$$

where ρ is the density of the liquid. Because $\frac{dm}{dt} = \frac{dm}{dp}\frac{dp}{dt}$, Eq. (5.5.10) is written as

$$C\frac{dp}{dt} = \rho q_{in} - \rho q_{out} \tag{5.5.11}$$

where Eq. (5.5.4) has been used. Furthermore, with $dm = \rho A(h)dh$, Eq. (5.5.10) is reduced to

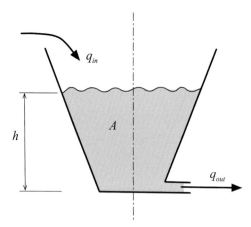

Figure 5.5.3 A storage tank with inflow and outflow

$$C_V \frac{dh}{dt} = q_{in} - q_{out} \qquad (5.5.12)$$

where the volume capacitance $C_V = A(h)$, as given in Eq. (5.5.8). Equation (5.5.12) can be directly obtained through an examination of the change in the volume of stored liquid, which is a special case of Eq. (5.4.2).

5.5.4 Fluid Resistance

As a liquid flows through a pipe or a valve or an orifice, a drop in the pressure of the liquid occurs. This drop in pressure is due to the resistance that is experienced by the liquid. Fluid resistance R can be defined as

$$R = \left. \frac{dp}{dq_m} \right|_{q_m = \bar{q}_m} \qquad (5.5.13)$$

where $\bar{q}_m$ is a reference mass flow rate, and R is in Pa·s/kg or 1/(m·s) in the SI. By the resistance definition and a Taylor series about $\bar{q}_m$, it can be shown that

$$q_m = \frac{1}{R} \Delta p \qquad (5.5.14)$$

where Δp is the change in pressure, and q_m is the mass flow rate through a pipe/valve/orifice. See Figure 5.5.4 for the symbol for fluid resistance, where $\Delta p = p_1 - p_2 > 0$. Fluid resistance is analogous to electrical resistance, which occurs when a current goes through a wire.

5.5 Basic Elements of Liquid-Level Systems

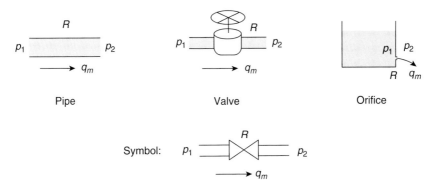

Figure 5.5.4 Symbol of fluid resistance

In determining fluid resistance, the motion of a fluid needs to be considered. There are two types of fluid motion: *laminar flow* and *turbulent flow*. For a laminar flow through a resistance (pipe, valve, or orifice), which is "smooth" with a relatively low Reynolds number, the pressure–mass flow rate relation is given by

$$p = \beta\, q_m \tag{5.5.15}$$

where β is a constant that depends on the geometry and friction of the resistance and the properties of the fluid. According to the definition in Eq. (5.5.13), the resistance for the laminate flow is

$$R = \left.\frac{dp}{dq_m}\right|_{q_m = \bar{q}_m} = \beta \tag{5.5.16}$$

Because Eq. (5.5.15) is linear, the fluid resistance R is the same at any value of $\bar{q}_m$.

For a turbulent flow through a resistance, which is "rough" with a relatively high Reynolds number, the pressure–mass flow rate relation can be described by

$$p = \beta\, q_m^2 \tag{5.5.17}$$

where the constant β depends on the geometry and friction of the resistance and the properties of the fluid. Because Eq. (5.5.17) is nonlinear, the fluid resistance varies as the mass flow rate changes. Consider a reference operating point $(\bar{q}_m, \bar{p})$, with $\bar{p} = \beta \bar{q}_m^2$. In a small neighborhood of the operating point, the fluid resistance, with Eqs. (5.5.13) and (5.5.17), is obtained as

$$R = \left.\frac{dp}{dq_m}\right|_{q_m = \bar{q}_m} = 2\beta\, \bar{q}_m = \frac{2\bar{p}}{\bar{q}_m} \tag{5.5.18}$$

In the literature, the fluid resistance is also defined in terms of the volume flow rate

$$R_V = \left.\frac{dh}{dq}\right|_{q=\bar{q}} \tag{5.5.19}$$

where h is the liquid height, q is the volume flow rate, $\bar{q}$ is a reference flow rate, and R_V in the SI is in s/m². Because $dp = \rho g dh$ and $dq_m = \rho dq$

$$\frac{dp}{dq_m} = \frac{\rho g dh}{\rho dq} = g\frac{dh}{dq} \tag{5.5.20}$$

which yields

$$R = gR_V \tag{5.5.21}$$

With a volume fluid resistance R_V, Eq. (5.5.14) is rewritten as

$$q_m = \rho q = \frac{1}{gR_V}\rho g \Delta h = \rho \frac{1}{R_V}\Delta h \tag{5.5.22}$$

where $\Delta p = (p_a + \rho g h_1) - (p_a + \rho g h_2) = \rho g \Delta h$ has been used. It follows that

$$q = \frac{1}{R_V}\Delta h \tag{5.5.23}$$

Therefore, Eqs. (5.5.14) and (5.5.23) have the same format.

Comparison of Eq. (4.1.8) with Eq. (5.5.14) or Eq. (5.5.23) indicates that the mass flow rate q_m or volume flow rate q in a liquid-level system is analogous to the current i in an electrical system. Thus, combinations of several fluid resistances can be replaced by a single equivalent fluid resistance, such as combinations of resistors in electrical systems. For instance, for two fluid resistances (R_1 and R_2) connected in series, the equivalent fluid resistance is

$$R_{eq} = R_1 + R_2 \tag{5.5.24}$$

and for two resistances in parallel connection, the equivalent fluid resistance is

$$\frac{1}{R_{eq}} = \frac{1}{R_1} + \frac{1}{R_2} \tag{5.5.25}$$

5.5.5 Orifice Resistance Formulas

An orifice can be a hole (opening) in the side of a tank, a nuzzle, or a short pipe segment with a sudden change of cross-sectional area. A flow through an orifice experiences fluid resistance, which is also described by Eq. (5.5.14), with $R = R_o$, where R_o is the orifice resistance. By Torricelli's principle, the orifice resistance can be estimated by

$$R_o = \frac{1}{2\rho C_d^2 A_o^2} \tag{5.5.26}$$

where ρ is the mass density of the fluid, A_o is the orifice area, and C_d is the nondimensional *discharge coefficient*, which is in the range $0 < C_d \leq 1$. The discharge coefficient is the ratio of the actual mass flow rate to the theoretical one, which characterizes a head loss due to friction. The value of C_d for water is around 0.6.

5.5.6 Sources

There are two types of energy sources that are required to operate liquid-level systems:

- flow source – a specified mass flow rate q_m or volume flow rate q, which is positive when liquid is added to a tank or negative when liquid is taken away from a tank
- pressure source – a specified increase in pressure

The flow source and pressure source to a liquid-level system correspond to the current source and voltage source to an electric circuit.

In many liquid-level systems, a pressure source is realized by a pump (say, a centrifugal pump). As the pump is driven by a motor at a constant speed, a pressure difference between the inflow and outflow of the pump is generated. The pressure difference can be adjusted by changing the rotation speed of the motor. Figure 5.5.5 shows the symbol and input–output relationship of a pump, where q_m is the mass flow rate going through the pump, and $\Delta P = P_1 - P_2$ is the pressure difference between the inflow and outflow of the pump.

5.6 Dynamic Modeling of Liquid-Level Systems

As mentioned previously, the modeling of a dynamic system relies on three important keys: fundamental principle(s), basic elements, and a way of analysis. For liquid-level systems, the fundamental principle is the law of conservation of mass, as given in Section 5.4.1 and restated in Eq. (5.5.10) or Eq. (5.5.12); the basic elements, namely fluid capacitance and fluid resistance, were derived in Section 5.5.

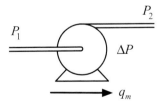

Figure 5.5.5 Symbol and input–output relation of a pump

The last of the three keys in system modeling is a way of analysis, in which free-body diagrams for fluid capacitances and auxiliary plots for fluid resistances are drawn. This is similar to that for thermal systems (Section 5.3). A free-body diagram is one in which the capacitance element is isolated from all other elements, but with input and output flow rates as interactions with its surroundings. An auxiliary plot is one in which the resistance element with a pressure difference undergoes a flow rate.

Therefore, the three-key modeling of a liquid-level system takes the following steps.

(1) Draw a free-body diagram for each fluid capacitance (storage tank), and an auxiliary plot for each fluid resistance (pipe or valve or orifice).
(2) Write a differential equation for each free-body diagram, and an algebraic equation for each auxiliary plot. In this step, the element formulas given in Section 5.5 are used.
(3) By substitution, the differential and algebraic equations obtained in Step 2 are converted to a set of independent differential equations as the dynamic model of the system. The number of independent differential equations is equal to the number of the capacitances in the system.

Once the governing equations of a liquid-level system are derived, state-space representations, transfer function formulations, and block diagrams (refer to Sections 3.2, 3.6, 3.7, and 6.2) can be established. In a state-space representation, it is convenient to select the liquid height h of a storage tank as a state variable.

Example 5.6.1

A tank, with nonconstant cross-sectional area $A(h)$, is connected to a valve at the bottom; see Figure 5.6.1, where q_{in} is a volume inflow rate, q_{out} is the volume outflow rate from the valve, and P_a is the atmospheric pressure. Derive a dynamic model of the liquid-level system, and based on the model, obtain a state equation.

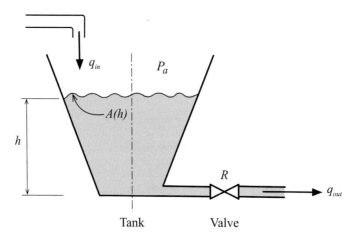

Figure 5.6.1 A tank of nonconstant cross-sectional area

5.6 Dynamic Modeling of Liquid-Level Systems

Solution
The free-body diagram of the tank and the auxiliary plot of the valve are shown in Figure 5.6.2, where q_1 is the volume flow rate from the bottom of the tank to the valve. From the figure, the governing equations of the fluid elements are written as follows:

For the tank,

$$A(h)\frac{dh}{dt} = q_{in} - q_1 \tag{a}$$

where Eqs. (5.5.8) and (5.5.11) have been used.

For the valve:

$$\rho q_{out} = \frac{1}{R}(P_1 - P_a) \tag{b}$$

where P_1 is the pressure at the bottom of the tank given by

$$P_1 = P_a + \rho g h \tag{c}$$

Substituting Eq. (c) into Eq. (b) gives

$$q_{out} = \frac{g}{R}h \tag{d}$$

Equation (d) implies that the outflow rate q_{out} is dependent upon the liquid height h and it cannot be arbitrarily specified. By the continuity equation (5.4.6), $q_{out} = q_1$. It follows that the governing equation of the liquid-level system is

$$A(h)\frac{dh}{dt} = q_{in} - \frac{g}{R}h \tag{e}$$

Now select a state variable as $x = h$. From Eq. (e), a nonlinear state equation is obtained as follows:

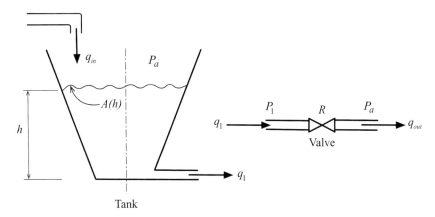

Figure 5.6.2 Free-body diagram and auxiliary plot of the liquid-level system in Example 5.6.1

$$\dot{x} = \frac{1}{A(x)}\left[q_{in} - \frac{g}{R}x\right] \tag{f}$$

Free-body diagrams for fluid capacitances are especially useful in dealing with a system with multiple capacitances (tanks). See the following example.

Example 5.6.2

A liquid-level system of two cylindrical tanks with constant cross-sectional areas is shown Figure 5.6.3, where the tanks are connected by Pipe 1, Tank 1 is connected to a pump (pressure source ΔP) by a valve, Tank 2 has a specified volume inflow rate q_{in}, and P_a is the atmospheric pressure. The fluid resistances in the figure are all related to mass flow rates, as defined by Eq. (5.5.13). Let the density of the fluid be ρ. Derive the governing differential equations of the system in terms of the liquid heights h_1 and h_2. Also, obtain a state-space representation with ΔP and q_{in} as the inputs, and the liquid height of Tank 1 and the volume outflow rate of Tank 2 through Pipe 2 as the outputs.

Solution

The free-body diagrams of the tanks and the auxiliary plots of the pump, valve, and pipes are shown in Figure 5.6.4, where q_v, q_1, and q_{out} are the volume flow rates. By the diagrams and plots, the governing equations of the fluid elements are obtained as follows:

Tank 1:
$$A_1 \frac{dh_1}{dt} = q_v - q_1 \tag{a}$$

Tank 2:
$$A_2 \frac{dh_2}{dt} = q_{in} - q_1 - q_{out} \tag{b}$$

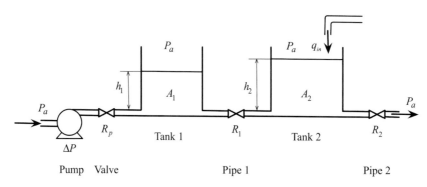

Figure 5.6.3 A two-tank liquid-level system

5.6 Dynamic Modeling of Liquid-Level Systems

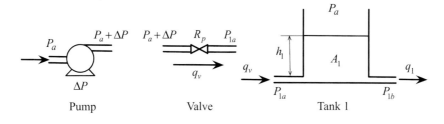

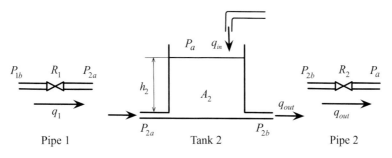

Figure 5.6.4 The free-body diagrams and auxiliary plots for the system in Example 5.6.2

Valve:
$$\rho q_v = \frac{1}{R_p}(P_a + \Delta P - P_{1a}) \tag{c}$$

Pipe 1:
$$\rho q_1 = \frac{1}{R_1}(P_{1b} - P_{2a}) \tag{d}$$

Pipe 2:
$$\rho q_{out} = \frac{1}{R_2}(P_{2a} - P_a) \tag{e}$$

The pressures at the tank bottoms are
$$\begin{aligned} P_{1a} &= P_{1b} = P_a + \rho g h_1 \\ P_{2a} &= P_{2b} = P_a + \rho g h_2 \end{aligned} \tag{f}$$

by which
$$q_v = \frac{\Delta P}{\rho R_p} - \frac{g}{R_p} h_1, \quad q_1 = \frac{g}{R_1}(h_1 - h_2), \quad q_{out} = \frac{g}{R_2} h_2 \tag{g}$$

Substitute Eq. (g) into Eqs. (a) and (b), to obtain the coupled differential equations of the liquid-level system as follows:

$$A_1 \frac{dh_1}{dt} = \frac{\Delta P}{\rho R_p} - g\left(\frac{1}{R_p} + \frac{1}{R_1}\right) h_1 + \frac{g}{R_1} h_2$$

$$A_2 \frac{dh_2}{dt} = q_{in} - \frac{g}{R_1} h_1 + g\left(\frac{1}{R_1} - \frac{1}{R_2}\right) h_2$$

(h)

For a state-space representation, choose two state variables:

$$x_1 = h_1, \quad x_2 = h_2 \tag{i}$$

The state equations are obtained from Eq. (h) as follows:

$$\frac{dx_1}{dt} = -\frac{g}{A_1}\left(\frac{1}{R_p} + \frac{1}{R_1}\right) x_1 + \frac{g}{R_1 A_1} x_2 + \frac{1}{\rho R_p A_1} \Delta P$$

$$\frac{dx_2}{dt} = -\frac{g}{R_1 A_2} x_1 + \frac{g}{A_2}\left(\frac{1}{R_1} - \frac{1}{R_2}\right) x_2 + \frac{1}{A_2} q_{in}$$

(j)

in which ΔP and q_{in} are the inputs. The outputs of the system are

$$y_1 = h_1, \quad y_2 = q_{out} \tag{k}$$

According to Eqs. (g) and (i), the output equations are obtained as follows:

$$y_1 = x_1, \quad y_2 = \frac{g}{R_2} x_2 \tag{l}$$

5.7 Fluid Inertance

In modeling liquid-level systems, the inertia effect of a flowing fluid is neglected because the mass of the fluid moving in pipes is much smaller than the mass of fluid stored in tanks. However, in some other applications, like piping systems, the inertia effect of a moving fluid can be significant and, accordingly, it cannot be ignored. For example, when a plumbing system is suddenly shut off, the large inertia of running water in the pipeline is brought to a quick halt and, as a result, the so-called "water hammer" (a loud bang and then diminishing noises) is generated through shock waves. Because of this, air chambers are usually installed on plumbing systems to reduce to the water hammer effect.

The inertia effect of a moving fluid can be characterized by *fluid inertance*. For instance, the fluid inertance I of an incompressible fluid element moving in a tube of a constant cross-section can be identified from the momentum equation:

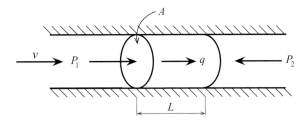

Figure 5.7.1 Fluid inertance of an element moving in a uniform tube

$$I\frac{dq}{dt} = \Delta p \tag{5.7.1}$$

where q is the volume flow rate of the fluid, and Δp is the pressure difference at the two ends of the element. This equation is similar to the equation for an electrical inductor, $L\frac{di}{dt} = \Delta e$, where Δe is the potential difference or the voltage at the two ends of the inductor.

To evaluate fluid inertance, consider an incompressible fluid element of density ρ and length L moving in a uniform tube of cross-sectional area A, as shown in Figure 5.7.1, where v is the velocity of the fluid, q is the volume flow rate of the fluid ($q = Av$), and P_1 and P_2 are the pressures at the two ends of the element. In the theory of fluid mechanics, the element actually is a control volume in space. By Newton's second law, the momentum equation for the element is

$$\frac{d}{dt}(mv) = (P_1 - P_2)A$$

Because $m = \rho AL$ and $v = q/A$, the previous equation is reduced to

$$\frac{\rho L}{A}\frac{dq}{dt} = P_1 - P_2 = \Delta p \tag{5.7.2}$$

It follows from Eqs. (5.7.1) and (5.7.2) that the inertance of the fluid element is

$$I = \frac{\rho L}{A} \tag{5.7.3}$$

With the fluid inertance, the kinetic energy of the element can be expressed as follows:

$$KE = \frac{1}{2}I q^2 \tag{5.7.4}$$

This formula can be easily verified:

$$KE = \frac{1}{2}mv^2 = \frac{1}{2}\rho AL\left(\frac{q}{A}\right)^2 = \frac{1}{2}\frac{\rho L}{A}q^2$$

Note that Eq. (5.7.4) is similar to the energy format of an electrical inductor, $\frac{1}{2}Li^2$.

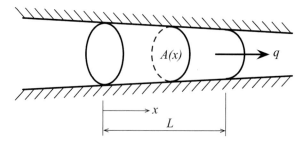

Figure 5.7.2 Fluid inertance of an element moving in a nonuniform uniform tube

The kinetic energy expression given in Eq. (5.7.4) is useful for derivation of the inertance of an incompressible fluid element moving in a nonuniform pipe. Figure 5.7.2 shows a fluid element (control volume: $0 \leq x \leq L$) in a nonuniform tube of cross-sectional area $A(x)$. Recall that the volume flow rate q, due to the continuity equation (5.4.6), is the same at any point of the tube. Therefore, the kinetic energy of the element can be written as

$$KE = \frac{1}{2}\int_0^L \rho A v^2 dx = \frac{1}{2}\int_0^L \rho A \left(\frac{q}{A}\right)^2 dx = \frac{1}{2} I q^2 \qquad (5.7.5)$$

It follows that the fluid inertance I of the element is given by

$$I = \rho \int_0^L \frac{dx}{A(x)} \qquad (5.7.6)$$

The significance of fluid inertance in a hydraulic system must be evaluated relative to the rest of the system and based on the frequency bandwidth of interest. Thus, the decision about whether or not to adopt fluid inertia in system modeling depends on the application. Liquid-level systems and piping systems are two typical examples.

5.8 The Bernoulli Equation

Consider the steady flow of an incompressible fluid, in which the velocity of the fluid at a particular fixed point does not change with time. Steady flow is always laminar flow, but the reverse may not be true because laminar flow can be unsteady. Assume that for a control volume of the fluid, there is no energy exchange with the environment (heat transfer or work) and there is no energy storage or energy dissipation (such as friction). At any point on a streamline in the control volume, the following equation, known as the *Bernoulli equation*, holds:

5.8 The Bernoulli Equation

$$p + \rho g h + \frac{1}{2}\rho v^2 = \text{constant} \tag{5.8.1}$$

where p, ρ, and v are the pressure, density, and velocity of the fluid at the point, respectively; g is the gravitational acceleration; and h is the elevation of the point above a reference plane, with the positive h-direction opposite to the direction of gravity; see Figure 5.8.1. In the equation, p is known static pressure and $\frac{1}{2}\rho v^2$ dynamic pressure.

The Bernoulli equation is a statement of conservation of energy. On the left-hand side of Eq. (5.8.1), the first term is the pressure energy of the fluid per unit volume, the second term is the potential energy per unit volume, and the third term is the kinetic energy per unit volume. The pressure energy per unit volume can be explained by

$$p = \frac{F}{A} = \frac{F \cdot d}{A \cdot d} = \frac{W}{V} = \frac{\text{Energy}}{\text{Volume}}$$

where A is the area of the fluid on which pressure force F is applied, d is a distance in the direction of F, W is the work done by F on d, and V is the volume of the fluid. The potential energy per unit volume and the kinetic energy per unit volume can be similarly explained. Hence, by the Bernoulli equation, the total energy of the fluid, which is the sum of the pressure energy, potential energy, and kinetic energy is conserved.

The Bernoulli equation is usually seen in the following equivalent form

$$p_1 + \rho g h_1 + \frac{1}{2}\rho v_1^2 = p_2 + \rho g h_2 + \frac{1}{2}\rho v_2^2 \tag{5.8.2}$$

where the subscripts 1 and 2 refer to the quantities at Point 1 and Point 2 on a streamline, respectively, as shown in Figure 5.8.1. This form is convenient in the solution of problems in fluid dynamics, as well as in the modeling of fluid systems.

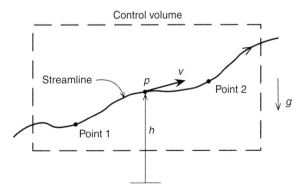

Figure 5.8.1 Pressure, elevation, and velocity of a fluid in steady flow

Example 5.8.1

In Figure 5.8.2, water flows in a pipe of nonconstant cross-section. Assume steady flow and negligible friction in the pipe. The pressure and velocity of water and the cross-sectional area of the pipe at point 1 are known as $P_1 = 2.0 \times 10^5$ Pa, $v_1 = 3$ m/s, $A_1 = 0.4$ m². At point 2, the elevation above Point 1 and the cross-sectional area are $h = 2.8$ m and $A_2 = 0.25$ m², respectively. Determine the pressure and velocity of the water at Point 2.

Solution
The density of water is $\rho = 997$ kg/m³. Take Point 1 as the reference point for elevation: $h_1 = 0$ and $h_2 = h$. By the Bernoulli equation (5.8.2)

$$P_1 + \frac{1}{2}\rho v_1^2 = P_2 + \rho g h + \frac{1}{2}\rho v_2^2 \tag{a}$$

Equation (a) has two unknows: P_2 and v_2. Therefore, one more equation is needed. Because water is an incompressible fluid, the continuity equation (5.4.6) applies

$$A_1 v_1 = A_2 v_2 \tag{b}$$

From Eq. (b), the velocity of the fluid at Point 2 is found as

$$v_2 = \frac{A_1}{A_2}v_1 = \frac{0.4}{0.25} \times 3 = 4.8 \text{ m/s} \tag{c}$$

With the known v_2, the solution of Eq. (a) gives the pressure at point 2 as follows:

$$P_2 = P_1 - \rho g h + \frac{1}{2}\rho(v_1^2 - v_2^2)$$
$$= 2.0 \times 10^5 - 997 \times 9.807 \times 2.8 + \frac{1}{2}997 \times (3^2 - 4.8^2)$$
$$= 1.656 \times 10^5 \text{ Pa}$$

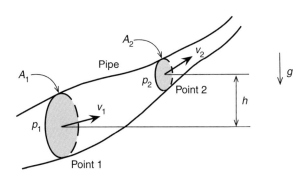

Figure 5.8.2 Steady flow of water in a nonuniform pipe in Example 5.8.1

Modified Bernoulli Equation

Equation (5.8.2) and Example 5.8.1 assume negligible friction in the pipe. In many applications, such as a fluid flowing in a long pipeline, the friction between the fluid and the pipe walls can be significant. Under the circumstances, the Bernoulli equation is often modified. To this end, we define the total head H at a point on a streamline by

$$H = \frac{p}{\rho g} + h + \frac{1}{2g} v^2 \tag{5.8.3}$$

where $\frac{p}{\rho g}$ is called the pressure head, h the elevation head, and $\frac{1}{2g} v^2$ the velocity head. Also, $\frac{p}{\rho g} + h$ is known as the piezometric head or hydraulic head. The H represents the total energy per weight without energy loss, and it is also known as the available head. With H, the Bernoulli equation (5.8.2) can be written in the following equivalent form

$$H_1 = H_2 \tag{5.8.4}$$

Equation (5.8.4) means that the total head of a fluid system without energy loss is constant. For a pipe with energy loss due to friction, a modified Bernoulli equation is written as

$$H_1 = H_2 + h_f \tag{5.8.5}$$

or

$$\frac{p_1}{\rho g} + h_1 + \frac{1}{2g} v_1^2 = \frac{p_2}{\rho g} + h_2 + \frac{1}{2g} v_2^2 + h_f \tag{5.8.6}$$

where h_f is the *head loss due to friction*, characterizing the energy (pressure) loss in the pipe. The h_f is a function of the geometry and internal surface roughness of the pipe and the fluid speed. Depending on applications, the value of h_f can be just a few percent of H in some cases, and it may be between 10% to 20% in other cases. Empirical formulas, such as the Darcy–Weisbach equation, are commonly used to estimate the head loss in a pipeline. Due to limited space, these formulas are not covered here.

Example 5.8.2 Due to friction, the water pipe in Figure 5.8.2 has a head loss $h_f = 0.86$ m from Point 1 to Point 2. Let the pressure and velocity of water at Point 1 be the same as in Example 5.8.1. Determine the pressure and velocity of water at Point 2.

Solution
By the continuity equation (5.4.6), the velocity of water at Point 2 is the same as in Example 5.5.1; that is, $v_2 = 4.8$ m/s. The total head at Point 1 is

$$H_1 = \frac{p_1}{\rho g} + h_1 + \frac{1}{2g}v_1^2$$

$$= \frac{2.0 \times 10^5}{997 \times 9.807} + 0 + \frac{3^2}{2 \times 9.807} = 20.995 \text{ m} \quad \text{(a)}$$

By the modified Bernoulli equation (5.8.5),

$$H_1 = H_2 + h_f = \frac{p_2}{\rho g} + h_2 + \frac{1}{2g}v_2^2 + h_f \quad \text{(b)}$$

which, after algebraic manipulations, gives the pressure at Point 2 as follows:

$$p_2 = \rho g(H_1 - h_f) - \rho g h_2 - \frac{1}{2}\rho v_2^2$$

$$= 997 \times 9.807 \times (20.995 - 0.86) - 997 \times 9.807 \times 2.8 - \frac{1}{2} 997 \times 4.8^2 \quad \text{(c)}$$

$$= 1.576 \times 10^5 \text{ Pa}$$

Comparison of the result with that of Example 5.8.1 shows that the friction in the pipe results in a 4.8% pressure drop (or pressure head loss) at Point 2.

5.9 Pneumatic Systems

The working fluid in a pneumatic system is a gas, which is a compressible fluid. Air is most commonly used in many such systems. Although the properties of pneumatic systems are highly nonlinear and spatially distributed, as in modeling of hydraulic systems, lumped-parameter models can be developed to obtain satisfactory results in many applications. In this section, we follow the same three-key modeling technique as presented in Sections 5.5 and 5.6.

5.9.1 Fundamental Principle

A fundamental principle for pneumatic systems is the principle of mass conservation, which has been presented in Section 5.4.1. There are two things that differentiate a pneumatic system from a hydraulic system. First, the gas in a pneumatic system, due to its compressibility, has changeable density. Second, a pneumatic system exhibits different behaviors in different thermodynamic processes, during which temperature, pressure, volume, and mass are in a certain relationship, such as the ideal gas law.

5.9.2 Basic Elements

Pneumatic systems in general have three types of lumped-parameter elements: pneumatic capacitance, pneumatic resistance, and pneumatic inertance. The kinetic energy of a gas in many applications is negligible, compared to the potential energy due to the pressure.

Because of this, inertance is usually ignored and only capacitance and resistance are considered in system modeling of pneumatic systems.

Similarly to the definition of fluid capacitance in Section 5.5, the pneumatic capacitance C relates the change in a stored gas mass to the change in gas pressure; that is,

$$C = \frac{dm_g}{dp} \tag{5.9.1}$$

where m_g is the stored gas mass; C has units of kg · m²/N or kg/Pa.

Consider a container of constant volume V that stores a gas of density ρ. Because $m_g = \rho V$

$$C = \frac{d(\rho V)}{dp} = V\frac{d\rho}{dp} \tag{5.9.2}$$

For an ideal gas undergoing a polytropic process as described by Eq. (5.4.13),

$$d\left(\frac{p}{\rho^n}\right) = \frac{dp}{\rho^n} - n\frac{pd\rho}{\rho^{n+1}} = 0$$

or

$$\frac{d\rho}{dp} = \frac{\rho}{np}$$

From the ideal gas law of Eq. (5.4.9), $p = \rho R_g T$, the previous equation becomes

$$\frac{d\rho}{dp} = \frac{1}{nR_g T} \tag{5.9.3}$$

It follows from Eqs. (5.9.2) and (5.9.3) that the pneumatic capacitance of the container is

$$C = \frac{V}{nR_g T} \tag{5.9.4}$$

Equation (5.9.4) shows that the capacitance C of a constant-volume container can be different in different thermodynamic processes (n and T) and for different gases (R_g).

Assume that a constant-volume container has a mass inflow rate q_{mi} and a mass outflow rate q_{mo}. According to the principle of mass conservation, Eq. (5.4.1), the mass in the container is governed by the differential equation

$$\frac{dm_g}{dt} = \frac{dm_g}{dp}\frac{dp}{dt} = q_{mi} - q_{mo}$$

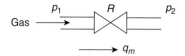

Figure 5.9.1 Pneumatic resistance of a valve or orifice

which, by the definition (5.9.1), gives the governing equation of the container pressure as follows:

$$C\frac{dp}{dt} = q_{mi} - q_{mo} \qquad (5.9.5)$$

where the container pressure p is the dynamic variable. For the gas undergoing a polytropic process, Eq. (5.9.5) can be written as

$$\frac{V}{nR_gT}\frac{dp}{dt} = q_{mi} - q_{mo} \qquad (5.9.6)$$

When a gas flow passes through a valve or an orifice, it meets with resistance. As a result, a pressure drop in the gas occurs, as shown in Figure 5.9.1, where q_m is the mass flow rate, and the pressure drop is $\Delta p = p_1 - p_2$. The pneumatic resistance R in the figure can be expressed by

$$q_m = \frac{1}{R}\Delta p \qquad (5.9.7)$$

and R is in Pa·s/kg. Note that the relation between the mass flow rate and pressure in general is nonlinear. Therefore, the value of R is obtained for a small neighborhood of a reference operating point $(\bar{q}_m, \bar{p})$, which is similar to that of fluid resistance for a turbulent flow (Section 5.5).

Besides capacitance and resistance elements, pneumatic systems require energy sources to operate. There are two types of sources: (i) flow sources, maintaining a specified mass flow rate q_m; and (ii) pressure sources, maintaining a specified pressure or pressure increase. These energy sources can be realized by compressors, ventilators, gas supply tanks, and pressure control valves.

5.9.3 Modeling of Pneumatic Systems

As in the modeling of liquid-level systems, the development of models of pneumatic systems is based on three keys: (i) the fundamental principle (conservation of mass); (ii) basic elements (pneumatic capacitance and resistance); and (iii) a way of analysis (free-body diagrams of capacitances and auxiliary plots of resistances). This modeling technique is demonstrated in the following example.

Example 5.9.1

In Figure 5.9.2, a rigid cylinder of constant volume V is connected to a tank of compressed air by a valve of resistance R. The air-supply tank is a source with a constant pressure P_s. Assume that the gas-filling process in the cylinder is a polytropic process, with $P_s > P$. Derive a model of the pneumatic system in terms of the cylinder pressure P.

Solution

The solution procedure is similar to that for a liquid-level system. From the free-body diagram of the cylinder and the auxiliary plot of the valve shown in Figure 5.9.3, the governing equations of the pneumatic elements are as follows:

$$\text{Cylinder:} \quad \frac{V}{nR_gT}\frac{dP}{dt} = q_m \tag{a}$$

$$\text{Valve:} \quad q_m = \frac{1}{R}(P_s - P) \tag{b}$$

where Eqs. (5.9.6) and (5.9.7) have been used. It follows from Eqs. (a) and (b) that the governing differential equation of the pneumatic system is

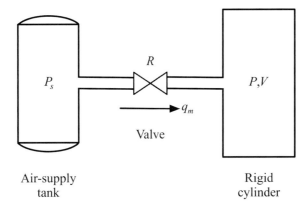

Figure 5.9.2 Pneumatic system for Example 5.9.1

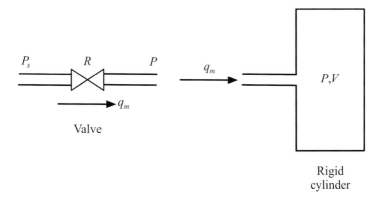

Figure 5.9.3 Free-body diagram and auxiliary plot for Example 5.9.1

$$\frac{RV}{nR_gT}\frac{dP}{dt} + P = P_s \tag{c}$$

For pneumatic systems with multiple capacitances, model development takes steps that are similar to those in the modeling of liquid-level systems (see Section 5.6).

5.10 Dynamic Response via MATLAB and Simulink

In this section, the simulation of dynamic responses of thermal and fluid systems by MATLAB and Simulink is illustrated in examples. Refer to Appendix B for the use of MATLAB and Simulink.

Example 5.10.1

Consider the two-capacitance thermal system in Figure 5.3.1, with parameter values given as: $C_1 = 500$ J/K, $C_2 = 300$ J/K, $R_1 = 12$ K · s/J, $R_2 = 800$ K · s/J. Let the input heat flow rate be $q_{in} = q_0 \, e^{-\sigma t}$, with $q_0 = 326$ J/s and $\sigma = 0.01$/s. Let the outside temperature be $T_a = 250$ K. Assume the initial temperatures as follows: $T_1(0) = T_2(0) = 258$ K. Plot the temperatures $T_1(t)$ and $T_2(t)$ and the heat flow rate $q_{out}(t) = \frac{1}{R_2}(T_2(t) - T_a)$, for $0 \leq t \leq 500$ s.

Solution
From Example 5.3.1, the matrix state equation of the thermal system is

$$\dot{\mathbf{x}} = \mathbf{A}\mathbf{x} + \mathbf{B}\mathbf{u}, \quad \mathbf{x}(0) = \mathbf{x}_0 \tag{a}$$

where

$$\mathbf{x} = \begin{pmatrix} T_1 \\ T_2 \end{pmatrix}, \quad \mathbf{u} = \begin{pmatrix} q_0 \, e^{-\sigma t} \\ T_a \end{pmatrix}, \quad \mathbf{x}_0 = \begin{pmatrix} T_1(0) \\ T_2(0) \end{pmatrix}$$

$$\mathbf{A} = \begin{bmatrix} -\dfrac{1}{C_1 R_1} & \dfrac{1}{C_1 R_1} \\ \dfrac{1}{C_2 R_1} & -\dfrac{1}{C_2}\left(\dfrac{1}{R_1} + \dfrac{1}{R_2}\right) \end{bmatrix}, \quad \mathbf{B} = \begin{bmatrix} \dfrac{1}{C_1} & 0 \\ 0 & \dfrac{1}{C_2 R_2} \end{bmatrix} \tag{b}$$

The state equation (a) can be solved numerically by the fixed-step Runge–Kutta method of order four, as presented in Section 2.8. The following commands, which are contained in a scrip M-file and a function M-file plot the system response (T_1, T_2, q_{out}) in Figure 5.10.1. In the M-file RKfunc5111, A and B are the matrices from Eq. (b).

```
% Input data
C1 = 500; C2 = 300; R1 = 12; R2 = 800;
q0 = 326; sgm = 0.01; Ta = 250;
T10 = 258; T20 = 258;

% Compute system response
B = [1/C1 0; 0 1/C2/R2];
t_stop = 500; npts = 5001; h = t_stop/(npts-1);
z0 = [T10; T20]; % initial state
t = linspace(0, t_stop, npts);
z = zeros(2, npts);
z(:,1) = z0;
for i = 1: npts-1
  tt = t(i);
  zz = z(:,i);
  f1 = RKfunc5111(tt, zz, A, B, q0, sgm, Ta);
  f2 = RKfunc5111(tt+h/2, zz+h/2*f1, A, B, q0, sgm, Ta);
  f3 = RKfunc5111(tt+h/2, zz+h/2*f2, A, B, q0, sgm, Ta);
  f4 = RKfunc5111(tt+h, zz+h*f3, A, B, q0, sgm, Ta);
  z(:,i+1) = zz +h/6*(f1+2*f2+2*f3+f4);
end

% Plot system response
figure(1) % Temperatures
plot(t,z(1,:),'k', t,z(2,:),'k-')
xlabel('Time (s)'), ylabel('Temperature (K)')
grid, legend('T_1', 'T_2')

figure(2) % Heat flow rate
qout = (z(2,:)-Ta)/R2;
plot(t,qout,'k'), grid
xlabel('Time (s)'), ylabel('Heat Flow Rate, q_o_u_t (J/s)')

function f = RKfunc5111(t,z, A, B, q0, sgm, Ta)
u = [q0*exp(-sgm*t); Ta];
f = A*z + B*u;
```

The numerical solutions in this example can also be obtained by Simulink. For this, the basics of block diagrams (Sections 6.2) and Simulink (Section 6.6 and Appendix B) are required. By Eq. (e) of Example 5.3.1, a Simulink model is established in Figure 5.10.2, where the Simulink blocks Clock and Fcn are used to generate the time-dependent input q_{in}. Running a simulation on this model produces the same plots in Figure 5.10.1.

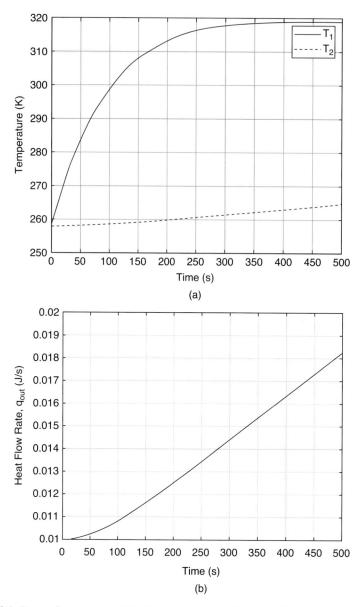

Figure 5.10.1 Dynamic response of the thermal system in Example 5.10.1: (a) the temperatures of the capacitances; and (b) the output heat flow rate of the right capacitance

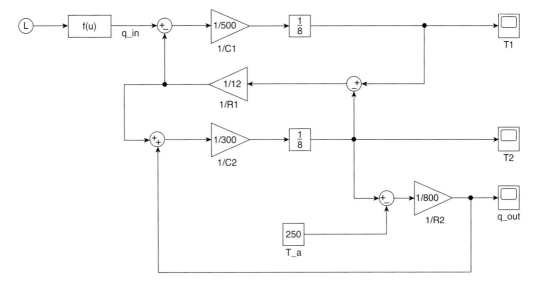

Figure 5.10.2 A Simulink model of the thermal system in Example 5.11.1

Example 5.10.2

Consider the liquid-level system in Figure 5.6.3, with the system parameter values given by: $\rho = 1000 \text{kg/m}^3$, $g = 9.81$ m/s^2, $A_1 = 3.5$ m^2, $A_2 = 4$ m^2, $R_p = 230$ Pa·s/kg, $R_1 = 300$ Pa·s/kg, $R_2 =$ Pa·s/kg. Let the pump pressure difference be $\Delta p = p_0 u(t)$, where $u(t)$ is the unit step function, and $p_0 = 120$ kPa. The volume inflow rate of Tank 2 is $q_{in} = 0.07$ m^3/s. Assume the initial liquid heights $h_1(0) = 1.5$ m and $h_2(0) = 0.3$ m. Plot the liquid heights $h_1(t)$ and $h_2(t)$ of the tanks, for $0 \leq t \leq 1200$ s.

Solution
From Example 5.6.2, the matrix state equation of the liquid-level system is

$$\dot{\mathbf{x}} = \mathbf{A}\mathbf{x} + \mathbf{B}\mathbf{u}, \quad \mathbf{x}(0) = \mathbf{x}_0 \tag{a}$$

where

$$\mathbf{x} = \begin{pmatrix} h_1 \\ h_2 \end{pmatrix}, \quad \mathbf{u} = \begin{pmatrix} \Delta p \\ q_{in} \end{pmatrix}, \quad \mathbf{x}_0 = \begin{pmatrix} h_1(0) \\ h_2(0) \end{pmatrix}$$

and

$$\mathbf{A} = \begin{bmatrix} -\dfrac{g}{A_1}\left(\dfrac{1}{R_p} + \dfrac{1}{R_1}\right) & \dfrac{g}{R_1 A_1} \\ -\dfrac{g}{R_1 A_2} & \dfrac{g}{A_2}\left(\dfrac{1}{R_1} - \dfrac{1}{R_2}\right) \end{bmatrix}, \quad \mathbf{B} = \begin{bmatrix} \dfrac{1}{\rho R_p A_1} & 0 \\ 0 & \dfrac{1}{A_2} \end{bmatrix} \tag{b}$$

By following Example 5.10.1, the use of the fixed-step Runge–Kutta method of order four results in the plots of the liquid heights shown in Figure 5.10.3.

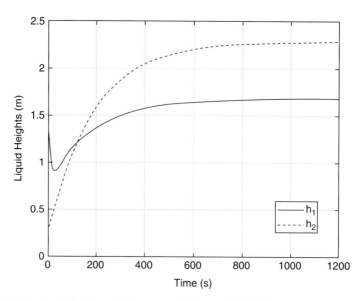

Figure 5.10.3 The liquid heights of the system in Example 5.10.2

Example 5.10.3

Consider the pneumatic system shown in Figure 5.9.2, where a rigid cylinder is connected to a supply tank of dry air by a valve. Assume an isothermal (constant temperature) process. The volume and temperature of the rigid cylinder are $V = 20$ m^3 and $T = 298$ K. The valve resistance is $R = 1200$ Pa·s/kg. Let the pressure of the air-supply tank be $P_s = 265$ kPa. Determine the pressure $P(t)$ of the cylinder, which initially is $P(0) = 101$ kPa.

Solution

From Example 5.9.1, the pressure of the cylinder is governed by

$$\frac{RV}{nR_gT}\frac{dP}{dt} + P = P_s \tag{a}$$

where $n = 1$ for an isothermal process (Section 5.4.3) and $R_g = 287.06$ J/(kg·K) for dry air. Laplace transform of Eq. (a) gives the s-domain cylinder pressure as follows:

$$\hat{P}(s) = \frac{\mu P(0)s + P_s}{s(\mu s + 1)}, \quad \text{with } \mu = \frac{RV}{nR_gT} \tag{b}$$

where $\hat{P}(s)$ is the Laplace transform of $P(t)$. Inverse Laplace transform of $\hat{P}(s)$ gives the solution of Eq. (a).

By the following MATLAB commands:

5.10 Dynamic Response via MATLAB and Simulink

```
% Input data
R = 1200; V = 20; T = 298;
Rg = 287.06; n = 1;
ps = 2.65e5; p0 = 1.01e5;

% Compute and plot response
mu = R*V/(n*Rg*T);
num = [mu*p0 ps];
den = [mu 1 0]/1000;
impulse(num, den);
xlabel('t')
ylabel('Pressure (kPa)')
grid, title('Pressure p(t) of the Rigid Cylinder')
```

the cylinder pressure is plotted in Figure 5.10.4. Here, the MATLAB function impulse from Table B8 in Appendix B has been used for the inverse Laplace transform of $\hat{P}(s)$. Note that the steady-state value of the cylinder pressure is always equal to P_s. This can also be verified by the final-value theorem:

$$P_{ss} = \lim_{t \to \infty} P(t) = \lim_{s \to 0} s\hat{P}(s) = \lim_{s \to 0} \frac{\mu P(0)s + P_s}{\mu s + 1} = P_s$$

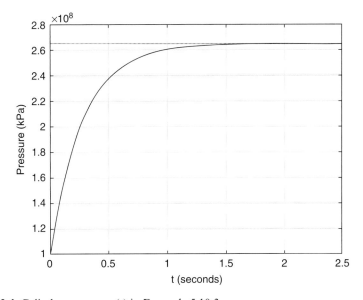

Figure 5.10.4 Cylinder pressure $p(t)$ in Example 5.10.3

CHAPTER SUMMARY

This chapter emphasizes three keys in modeling of thermal and fluid systems: fundamental principles, basic elements, and ways of analysis. For thermal systems, three forms of heat transfer and the law of conservation of energy are reviewed; lumped-parameter thermal elements are presented. For fluid systems, the principle of mass conservation, the continuity equations, the ideal gas law, and the Bernoulli equation are reviewed; lumped-parameter fluid elements are developed for liquid-level systems and pneumatic systems. For both thermal and fluid systems, free-body diagrams, which borrow the concepts from the modeling of mechanical systems, are introduced as a way of analysis. After a mathematical model is devised, it is converted to a transfer function formulation, block diagram and state-space representations, for simulation and analysis. Finally, the utility of MATLAB and Simulink to compute the response of fluid and thermal systems is illustrated.

Upon completion of this chapter, you should be able to:

(1) Understand and use the three important keys in modeling of thermal and fluid systems.
(2) Derive the governing equations of thermal systems, liquid-level systems, and pneumatic systems, with lumped-parameter models.
(3) Establish the transfer function formulation, block diagram, and state-space representation for a given thermal or fluid system.
(4) Compute the dynamic response of modeled thermal and fluid systems by MATLAB and Simulink.

REFERENCES

1. Y. Cengel, M. Boles, and M. Kanoglu, *Thermodynamics: An Engineering Approach*, 9th ed., McGraw-Hill Education, 2018.
2. T. L. Bergman, A. S. Lavine, F. P. Incropera, and D. P. DeWitt, *Introduction to Heat Transfer*, 6th ed., Wiley, 2011.
3. R. C. Hibbeler, *Fluid Mechanics*, 2nd ed., Pearson, 2017.

PROBLEMS

Section 5.1 Fundamental Principles of Thermal Systems

5.1 Consider heat transfer through a single-pane glass window in Figure P5.1, where T_o and T_r are the air temperatures outside and inside the room, respectively; T_1 and T_2 are the temperatures on the outer and inner surfaces of the glass, respectively. The glass layer has thickness l, area A, and thermal conductivity k. Let the heat transfer coefficient for convection between the outside air and the outer glass surface be h_1. Let the heat transfer

coefficient for convection between the inner glass surface and the room air be h_2. Assume that the parameters of the thermal system have the following values: $l = 1$ cm, $A = 1.8$ m^2, $k = 0.85$ W/(m · °C), $h_1 = 12$ W/(m^2 · °C), and $h_2 = 10$ W/(m^2 · °C). Assume that $T_o = 32$ °C and $T_r = 24$ °C.
(a) Determine the heat flow rate q_h through the window.
(b) Determine the temperatures T_1 and T_2.

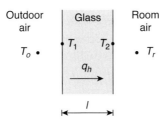

Figure P5.1

5.2 Consider heat transfer through a three-layer wall in Figure P5.2, where T_{a1} and T_{a2} are the air temperatures near the sides of the wall; l_j and k_j are the thickness and thermal conductivity of the j-th layer, for $j = 1, 2, 3$; and h_1 and h_2 are the heat transfer coefficients for convection on the left and right surfaces of the wall, respectively. The layers have the same cross-sectional area A. Assume that $T_{a1} > T_{a2}$.
(a) Determine the heat flow rate q_h through the wall.
(b) Determine the temperature distribution through the wall.

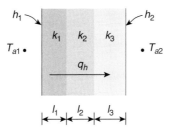

Figure P5.2

Section 5.2 Basic Thermal Elements

5.3 Consider a tank of 1 liter of water, with $\rho = 1000$ kg/m^3 and $c_P = 4.186 \times 10^3$ J/(kg·°C). Here 1 liter = 10^{-3} m^3. A reference temperature is chosen as $T_0 = 0$ °C.
(a) Compute the amount of heat energy stored in the water at temperature $T = 20$ °C.

(b) Compute the thermal capacitance of the water.

(c) How much energy does it take to raise the water temperature from 20 °C to 90 °C?

5.4 For the three-layer wall in Figure P5.2, determine the equivalent thermal resistance R_{eq} in the relation

$$q_h = \frac{1}{R_{eq}} (T_{a1} - T_{a2})$$

5.5 Consider a closed and hollow rectangular container of thickness δ, outer length a, outer width b, and outer height h. The container is thin-walled ($\delta << a, b, c$) and is made of a material with thermal conductivity k. Let the temperatures inside and outside the container be T_i and T_o, respectively. Determine the equivalent thermal resistance of the entire vessel.

5.6 Consider a hollow aluminum cylinder of inner diameter d_i, outer diameter d_o, height H, and thermal conductivity k. The outer surface of the cylinder is exposed to a stream of air with temperature T_o and velocity u_o. On the inner surface a constant temperature T_i is maintained. The convective heat transfer coefficient between the outer surface of the cylinder and the stream of air is approximated by $h = 3.15\, u_o$, in W/(m² · K). Consider the parameter values: $d_i = 0.5$ m, $d_o = 0.75$ m, $H = 1$ m, $k = 240$ W/(m · K).
(a) Compute the characteristic length L_c of the aluminum cylinder.
(b) By the Biot criterion, determine the condition of the velocity u_o of air under which the cylinder can be treated by a lumped-parameter model.

Section 5.3 Dynamic Modeling of Thermal Systems

5.7 A thin-walled container of volume V is fully filled with a liquid, which has mass density ρ and specific heat c. The container wall has a total thermal resistance (convective and conductive) R_c and negligible thermal capacitance. The temperature of the air surrounding the container is T_a. Let the liquid temperature in the container be T_l.
(a) Draw a free-body diagram for the thermal capacitance about the liquid, and an auxiliary plot for the thermal resistance about the container wall.
(b) Derive a mathematical model of the liquid temperature T_l, with T_a as the input.
(c) Let the initial liquid temperature be τ_0. Determine the liquid temperature $T_l(t)$ by solving the differential equation obtained in part (b).

5.8 The thermal system in Figure P5.8 consists of two capacitances that are separated by two resistances. A heat flow rate q_{in} is applied to the left capacitance. On the right side of resistance R_2, convective heat transfer takes place, with h being the convective coefficient and T_f the temperature of a fluid stream.

(a) Draw free-body diagrams for the thermal capacitances, and auxiliary plots for the thermal resistances.
(b) Derive the differential equations governing T_1 and T_2.

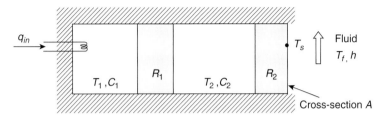

Figure P5.8

5.9 Consider the thermal system in Figure P5.8, with q_{in} and T_f as the inputs.

(a) Derive the transfer function with q_{in} as the input and the heat flow rate through the right resistance (R_2) as the output.
(b) Draw a block diagram for the system, with q_{in} and T_f as the inputs and T_2 as the output.
(c) Obtain a state-space representation for the system, with q_{in} and T_f as the inputs, and the heat flow rate through the left resistance (R_1) and the temperature of the right capacitance (T_2) as the outputs.

5.10 A simplified model of metal heat treatment is shown in Figure P5.10, where a metal part of temperature T_m and thermal capacitance C_m is placed inside an industrial furnace that is heated by a heater at rate q_{in}. In the figure, T_1 and C_1 are the temperature and thermal capacitance of the air inside the furnace; T_2 and C_2 are the temperature and thermal capacitance of the furnace wall; and T_a is the ambient temperature. Assume that heat is transferred by convection only, with h_m and A_m being the convective coefficient and surface area between the metal part and the air inside the furnace; h_1 and A_1, between the air inside the furnace and the inner wall; and h_2 and A_2, between the outer wall and ambient air.

(a) Draw free-body diagrams of the thermal capacitances, and auxiliary plots of the thermal resistances.
(b) Derive the differential equations governing T_m, T_1, and T_2.

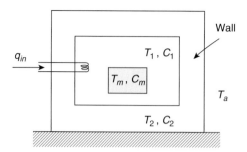

Figure P5.10

5.11 Consider the thermal system in Figure P5.10, with q_{in} and T_a as the inputs.
  (a) Derive the transfer function with q_{in} as the input and T_m as the output.
  (b) Obtain a state-space representation for the system, with T_m and the heat flow rate on the surface of the outer wall as the system outputs.

5.12 Consider the quenching process similar to that in Figure 5.3.7, with the following modification: the volume of the liquid in the bath is large enough so that the bath temperature T_b remains unchanged when the heated metal piece is immersed in the bath. With the modification, the heat transfer between the liquid and the ambient air is ignored, as can be seen from Eq. (b) of Example 5.3.3 with $C_b \to \infty$.
  (a) Derive a mathematical model for the temperature T_m of the metal piece.
  (b) Consider quenching of a steel sphere of radius r. The parameters of the sphere are given as: $r = 0.02$ m, $\rho = 7730$ kg/m^3, $c_P = 420$ J/(kg· °C). The convective coefficient for heat transfer between the sphere and the liquid is $h = 360$ W/(m^2·°C). The temperature of the liquid bath is 60 °C. The initial temperature of the sphere is 820 °C. By the model obtained in part (a), plot T_m versus time and determine the time needed for the sphere temperatures to reach 200 °C.

Section 5.4 Fundamental Principles of Fluid Systems

5.13 A cylinder–piston assembly is connected to a spring–mass system by a rod; see Figure P5.13, where A is the cross-sectional area of the piston, and p_1 and p_2 are the pressures applied to the sides of the piston by two pumps. Under the pressures, the piston slides on the frictionless surface of the cylinder, with displacement x. Let the combined inertia of the piston and rod be M. Assume that the spring is not deformed at $x = 0$.
  (a) Derive a mathematical model for the piston displacement x.
  (b) For $p_1 - p_2 = P_0$, where P_0 is a nonzero constant, determine the displacement $x(t)$ assuming that the piston is initially at rest.

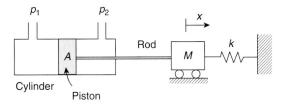

Figure P5.13

5.14 Consider the hydraulic system in Figure P5.14, where an external force f is applied to the left piston of area A_1, and, under the hydraulic pressure, the right piston of area A_2 moves up to lift mass m. The mass is constrained by a spring and damper pair (k, c). Assume that m includes the inertia of the right piston. The inertia effect of the fluid in the hydraulic system is ignored. Assume that the inertia of the left piston is negligible.
(a) Derive a model for the displacement x_2 of the right piston.
(b) Determine the work done by the external force f in terms of the displacement x_2.

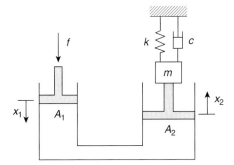

Figure P5.14

Section 5.5 Basic Elements of Liquid-Level Systems

5.15 A spherical tank is shown in Figure P5.15, where q_m is an input mass flow rate, and h is the liquid head. Let the liquid density be ρ. Let the diameter of the tank be D.
(a) Determine the mass capacitance C and volume capacitance C_V for the tank.
(b) Derive a dynamic model of the height h of the tank.

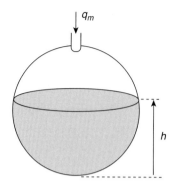

Figure P5.15

5.16 Consider the cylindrical tank in Figure P5.16, where q_m is the input mass flow rate, and h is the liquid head. Let the liquid density be ρ.
  (a) Determine the mass capacitance C and volume capacitance C_V for the tank.
  (b) Derive a dynamic model of the liquid height h of the tank.

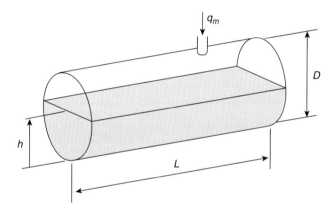

Figure P5.16

5.17 For a long tube with a laminar flow, the fluid resistance can be calculated by the formula

$$R = \frac{32\mu l}{\rho A d_h^2}$$

where l and A are the length and the area of the tube; μ and ρ are the dynamic viscosity (in Pa·s) and density (in kg/m^3) of the fluid, respectively; and d_h is the hydraulic diameter of the tube given by $d_h = 4A/S$, with S being the perimeter of the tube area. Here, μ and ρ are dependent on the temperature.

(a) For a tube with a circular cross-section of diameter D, show that its fluid resistance is

$$R = \frac{128\mu l}{\pi \rho D^4}$$

(b) For a tube with a square cross-section of width w, show that its fluid resistance is

$$R = \frac{32\mu l}{\rho w^4}$$

(c) Consider a circular tube, which is 1 m in length and 5.0 mm in diameter. Determine the fluid resistance of the tube conveying the following two fluids.

Fluid 1: water at 25 °C, with $\mu = 8.90 \times 10^{-4}$ Pa·s and $\rho = 997$ kg/m^3.
Fluid 2: motor oil (SAE 10 W-40) at 60 °C, with $\mu = 37.147 \times 10^{-4}$ Pa·s and $\rho = 838$ kg/m^3.

Section 5.6 Dynamic Modeling of Liquid-Level Systems

5.18 Consider the two-tank liquid-level system in Figure P5.18, where q_{in} is the input volume flow rate (in m^3/s), p_a is the atmospheric pressure, the tanks have constant cross-sectional areas (A_1, A_2), and the fluid flow is laminar in the valves.
  (a) Draw free-body diagrams for the tanks, and auxiliary plots for the valves.
  (b) Derive a model of the system with liquid heights h_1 and h_2 as the dynamic variables.
  (c) Determine the transfer functions $\dfrac{H_2(s)}{Q_{in}(s)}$ and $\dfrac{Q_2(s)}{Q_{in}(s)}$.

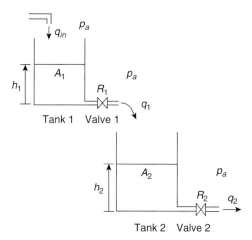

Figure P5.18

5.19 Consider the liquid-level system in Figure P5.19, where a spherical tank of diameter D is connected to an outlet valve (R), q_{in} is the input volume flow rate, and p_a is the atmospheric pressure. Refer to Problem 5.15 for the fluid capacitance of the tank.
(a) Derive a model of the system with liquid height h as the dynamic variable.
(b) Obtain a state-space model of the system, with q_{in} as the input, and q_{out} as the output.

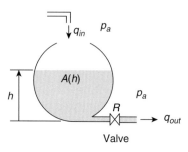

Figure P5.19

5.20 A liquid-level system is shown in Figure P5.20, where two tanks of constant areas (A_1, A_2) are connected by a pump with pressure difference Δp and a valve; Tank 1 has a volume inflow rate q_{in}; Tank 2 has a volume outflow rate q_{out}; and p_a is the atmospheric pressure. The fluid resistances R and R_1 are related to the mass flow rate, as described by Eq. (5.5.14). The density of the fluid is ρ.
(a) Draw free-body diagrams for the tanks, and auxiliary plots for the pump, valve, and pipe.
(b) Derive the governing differential equations of the system with liquid heights h_1 and h_2 as the dynamic variables.
(c) Obtain a state-space model with q_{in} and Δp as the inputs, and h_2 and q_{out} as the outputs.

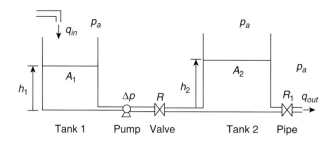

Figure P5.20

5.21 The liquid-level system in Figure P5.21 consists of two tanks of constant areas (A_1, A_2), a branched inflow pipeline with two valves (R_1 and R_2), a pipe (R_3) connecting the tanks, and an outflow pipe (R_4) from tank 2. In the figure, q_{in} and q_{out} are the volume inflow rate and outflow rate, respectively; the fluid resistances are related to the mass flow rate, as described by Eq. (5.5.14); and p_a is the atmospheric pressure. Let the fluid density be ρ.
(a) Draw free-body diagrams for the tanks, and auxiliary plots for the valves and pipes.
(b) Derive the governing differential equations of the system in terms of the liquid heights h_1 and h_2.
(c) Obtain a state-space model with q_{in} as the input, and h_2 and q_{out} as the outputs.

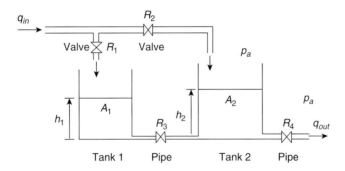

Figure P5.21

5.22 The liquid-level system in Figure P5.22 consists a single tank of constant area A, which has an orifice at height h_o, a pump with pressure difference Δp, and a valve (R); q_{in} is a volume inflow rate, R is related to the mass flow rate, and p_a is the atmospheric pressure. Let the area and discharge coefficient of the orifice be A_o and C_d, respectively. The density of the fluid is ρ. Assume that the liquid height h is always larger than h_o.
(a) Draw a free-body diagram for the tank, and auxiliary plots for the pump, valve, and orifice.
(b) Derive the governing differential equation of the system in terms of h.

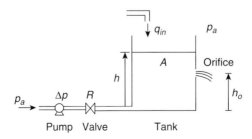

Figure P5.22

Section 5.8 The Bernoulli Equation

5.23 In Figure P5.23, a tank of constant area A has an orifice that is a distance d from the liquid level; p_a is the atmospheric pressure, and q_{in} a volume inflow rate that is properly adjusted to maintain the liquid head H unchanged. At the orifice, there is a head loss h_f due to friction. Given parameter values $H = 1$ m, $d = 0.6$ m, and $h_f = 0.05$ m, determine the speed of the liquid emerging from the orifice.

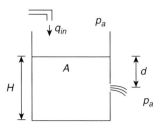

Figure P5.23

Section 5.9 Pneumatic Systems

5.24 A rigid spherical container with a diameter of 0.5 m is filled with dry air. Compute the pneumatic capacitance of the container in an isothermal process, with the air at an ambient temperature of 25 °C.

5.25 Consider the pneumatic system in Figure P5.25, where two rigid cylinders of constant volumes (V_1, V_2) are connected by a valve of resistance R_2, and the left cylinder is connected to an air-supply tank of a constant pressure p_s by a valve of resistance R_1. Let the gas-filling process in the cylinders be an isothermal process, with $p_s > p_1$ and $p_s > p_2$.
(a) Draw free-body diagrams of the cylinders and auxiliary plots of the valves.
(b) Derive the governing differential equations of the cylinder pressures p_1 and p_2.

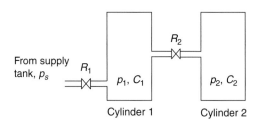

Figure P5.25

Section 5.10 Dynamic Response via MATLAB and Simulink

5.26 Consider the thermal system in Figure P5.8. The right side of resistance R_2 has a cross-sectional area A and a convective coefficient h. The system parameters have the following values: $C_1 = 9500$ J/°C°C/W, $C_2 = 16000$ J/°C, $R_1 = 0.12$ °C/W, $R_2 = 0.082$, $h = 115$ W/(m² · °C) and $A = 1$ m². Let the inputs to the system be $q_{in} = 250$ W and $T_f = 5$ °C. The initial temperatures of the capacitances are $T_1(0) = 15$ °C and $T_2(0) = 10$ °C. Use the MATLAB functions `impulse` and `step` to plot the temperature T_2 of the right capacitance versus time.

5.27 Repeat Problem 5.26 by building a Simulink model and running simulation.

5.28 Consider the liquid-level system in Figure P5.19, with $D = 2$ m and $R = 500$ Pa·s/kg. The liquid density is $\rho = 1000$ kg/m³ and the input volume flow rate $q_{in} = 0.02$ m³/s. The initial liquid height is $h(0) = 0.15$ m. By obtaining a nonlinear state-space model of the system and using the fixed-step Runge–Kutta method of order four (Section 2.8), plot the liquid height $h(t)$ and the output volume flow rate q_{out}, for $0 \leq t \leq 600$ s.

5.29 Consider liquid-level system in Figure P5.21. The system parameters have the following values: $A_1 = 3$ m², $A_2 = 6$ m², $R_1 = 150$ Pa·s/kg, $R_2 = 200$ Pa·s/kg, $R_3 = 500$ Pa·s/kg, $R_4 = 600$ Pa·s/kg. The input volume flow rate is $q_{in} = 0.04$ m³/s. The initial heights of the tanks are $h_1(0) = 0.5$ m and $h_2(0) = 1.2$ m. By Simulink, determine the liquid heights h_1 and h_2.

5.30 Consider the pneumatic system in Figure P5.25. The volumes of the cylinders are $V_1 = 20$ m³ and $V_2 = 40$ m³. The resistances of the valves are $R_1 = 900$ Pa·s/kg, $R_2 = 1200$ Pa·s/kg. Assume an isothermal (constant-temperature) process, with the cylinder temperatures $T_1 = T_2 = 298$ K. The pressure of the air-supply tank is $p_s = 300$ kPa. The initial pressures of the cylinders are $p_1(0) = p_2(0) = 101$ kPa. By Simulink, plot the cylinder pressures p_1 and p_2 versus time.

6 Combined Systems and System Modeling Techniques

Contents

6.1	Introduction to System-Level Modeling	397
6.2	System Modeling Techniques	398
6.3	Fundamentals of Electromechanical Systems	425
6.4	Direct Current Motors	433
6.5	Block Diagrams in the Time Domain	457
6.6	Modeling and Simulation by Simulink	464
6.7	Modeling and Simulation by Wolfram Mathematica	475
Chapter Summary		485
References		486
Problems		486

The main objective of this chapter is to introduce the concept of a dynamic system consisting of multiple subsystems, and to present the reader with the typical approach to modeling and analysis of such a combined system. A detailed discussion on combined systems is conducted using an electromechanical system – direct current (DC) motor – as an example. More instances of combined systems such as sensors and actuators (electromechanical, piezoelectric, piezoresistive), thermomechanical devices, and electroacoustic transducers are presented in Chapter 9.

The first section of this chapter presents terminology and basic concepts of system-level modeling for any combined dynamic system. Section 6.2 offers a detailed discussion on the typical system modeling techniques such as transfer function formulations, state-space representations, and block diagrams in the Laplace domain. Block diagrams in the time domain are discussed further in Section 6.5. Sections 6.3 and 6.4 offer a brief review of the foundations of electromagnetism and describe physical processes occurring in the electromechanical systems using the example of a DC motor. The derived mathematical models of the armature-controlled and field-controlled DC motors are implemented in software:

MATLAB/Simulink (Section 6.6) and Wolfram Mathematica (Section 6.7); system behaviors in response to different inputs – (a) constant input and (b) periodic excitation – are simulated and the generated results are analyzed.

6.1 Introduction to System-Level Modeling

The commonly encountered dynamic systems often comprise multiple subsystems that may belong to different fields. For example, a popular dynamic system such as a DC motor is a combination of two subsystems: electrical and mechanical. Since the majority of engineering systems are controlled, they can be considered as compound systems having at least two different components: a *plant* element, represented by a mechanical, electrical, fluid, or thermal subsystem, or any combination of those; and the *controller* element.

The ability to model and analyze such complex dynamic systems hinges on an understanding of the interactions between subsystems in addition to having a firm grasp on the behavior of the individual system components. A typical approach to system-level modeling and analysis is illustrated in Figure 6.1.1.

The first step, as depicted in Figure 6.1.1, involves deriving a set of governing equations for each subsystem separately, and obtaining mathematical expressions of their *coupling relations* that describe interactions between these subsystems. Obviously, the subsystem's governing equations contain only the variables of the field to which this subsystem belongs. For example, a mechanical subsystem will be described with variables such as displacements,

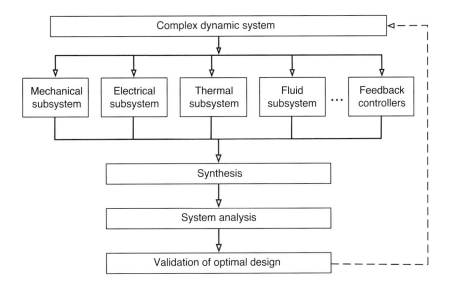

Figure 6.1.1 System-level modeling and analysis flowchart

velocities, and accelerations, while an electrical subsystem is described with currents and voltages. To the contrary, the coupling expressions involve variables of different types.

At the synthesis stage, the subsystems are joined together into a single system model. That is done by the derivation of any of the following three representations: (a) system transfer functions, (b) system state-space representations, or (c) system block diagrams. Since any one of these three representations can be obtained from any other one in a relatively simple and straightforward manner, the derivation of a single representation is usually sufficient for subsequent system analysis.

For the linear time-invariant systems that are the subject of this book, all of the above types of system representation are equally suitable. In general, the choice of the type of system model depends on the nature of the compound system. For example, the applicability of a transfer function formulation is limited by the existence of the Laplace transform of the governing equation. As discussed in detail in Chapter 2, the Laplace transform is best used for linear time-invariant systems that are described by a set of linear ordinary differential equations. Thus, for a system described by a first-order differential equation such as $(4 + \sin(3t))\dot{x} + 5x = F(t)$, the Laplace transform does not exist, making a transfer function formulation unavailable.

Block diagrams and state-space representations are available for both linear and nonlinear systems, whether time-invariant or not. They perform well in modeling multi-input/multi-output systems and are exceptionally well suited for model implementation in modeling and simulation software such as MATLAB/Simulink by MathWorks, or Mathematica/System Modeler by Wolfram Research. They are particularly helpful in modeling nonlinear systems efficiently, and in facilitating their analysis.

The derivation of all three representations from a set of governing equations and the conversion of one type of system model into another is discussed later in this section. The synthesis step concludes system-level modeling.

System analysis includes the implementation of the synthesized system model in the chosen software, followed by simulations. The derived set of system-governing equations is solved to obtain a system response; the system performance in response to the previously defined inputs is studied, and the system stability is analyzed.

The last step in system-level analysis is the validation of the optimal design. On the basis of performance and stability data, obtained during system analysis, the necessary design changes are made and the process is repeated until a satisfactory design is produced.

6.2 System Modeling Techniques

Any dynamic system can be modeled by using at least one of the following four techniques: (a) governing equations by the first laws of nature (input–output differential-equation formulation), (b) state-space representations, (c) transfer function formulations, and (d) block diagram representation. In this section the basic concepts of these modeling techniques are discussed. The derivation of the governing equations for several types of physical systems has been covered in Chapters 3 to 5. The remaining three techniques are presented in Sections 6.3 to 6.5.

The *governing equations* constitute an essential component of a dynamic system model. Typically, they are represented by differential or differential-algebraic equations. The derivation of the governing equations uses the fundamental laws of physics, the choice of which depends on the nature of the dynamic system. For example, for mechanical systems they are Newton's laws, while for electric circuits they are Kirchhoff's current and voltage laws. Detailed discussions on the derivation of the governing equations of mechanical and electrical systems are presented in Chapters 3 and 4, respectively.

The *input–output differential-equation(s)* system model may contain a single equation or a set of coupled equations. If such a model is sufficiently simple and can be easily implemented in software, no other representations are required for system-level analysis. Alas, for the majority of dynamic systems, and, in particular, for compound systems that consist of multiple subsystems of different nature, this is not the case. Hence, this sort of dynamic system needs to be modeled as a *block diagram*, a *transfer function*, or a *state-space representation* to simplify and streamline the system-level analysis.

To demonstrate the process of system-level modeling and analysis consider the simplest compound dynamic system: a controlled mechanical system shown in Example 6.2.1.

Example 6.2.1 The mechanical subsystem shown in Figure 6.2.1 (a) consists of a block of mass m, supported by a spring of stiffness k and a linear damper of the damping coefficient c. The block is subjected to an external force $F(t)$. The motion of the mass is controlled by a differential controller (differentiator, where $D = d/dt$). The sensor measures the displacement $x(t)$ of the mass m. This displacement is processed by a controller and a control force $F_c = \mu \dot{x}$ is then applied to the mass.

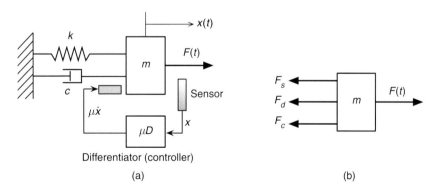

Figure 6.2.1 Controlled dynamic system: (a) description and (b) free-body diagram

A free-body diagram for the mass m is shown in Figure 6.2.1(b). The model of the mechanical subsystem is then

$$m\ddot{x} = F(t) - (F_s + F_d + F_c) \tag{1}$$

where F_s, F_d, and F_c represent the spring, damper, and controller forces respectively.

Recalling the expressions for spring and damper forces – $F_s = kx$ and $F_d = c\dot{x}$ (see Chapter 3) – substitute them into Eq. (1):

$$m\ddot{x} = F(t) - (kx + c\dot{x}) - F_c \tag{2}$$

The controller model is straightforward:

$$F_c = \mu\dot{x} \tag{3}$$

Thus, the system model is described by two equations: Eq. (2) and Eq. (3):

$$\begin{cases} m\ddot{x} = F(t) - (kx + c\dot{x}) - F_c \\ F_c = \mu\dot{x} \end{cases} \tag{4}$$

This model can be easily reduced into a single governing equation:

$$m\ddot{x} + (c + \mu)\dot{x} + kx = F(t) \tag{5}$$

The system model as described by either the set of Eqs. (4) or Eq. (5), is simple, and does not require different representations for system-level analysis. Nonetheless, recalling the fundamentals of state-space representations, transfer function formulations, and block diagram constructions presented in Chapter 2 (Sections 2.5.4 and 2.7), these three types of system model are derived as follows.

State-Space Representation

Select the state variables as $\begin{cases} x_1 = x(t) \\ x_2 = \dot{x}(t) \end{cases}$. Since only the displacement of the mass m is monitored, the output is $y(t) = x(t) = x_1$. Then, the state equations are derived as

$$\begin{cases} \dot{x}_1 = x_2 \\ m\dot{x}_2 + (c + \mu)x_2 + kx_1 = F(t) \end{cases} \Rightarrow \begin{cases} \dot{x}_1 = x_2 \\ \dot{x}_2 = -\dfrac{k}{m}x_1 - \dfrac{c+\mu}{m}x_2 + \dfrac{1}{m}F(t) \end{cases}$$

and the full state representation is

$$\begin{cases} \dot{x}_1 = x_2 \\ \dot{x}_2 = -\dfrac{k}{m}x_1 - \dfrac{c+\mu}{m}x_2 + \dfrac{1}{m}F(t) \end{cases}$$

$$y = x_1$$

Cast in matrix form, it becomes $\begin{cases} \dot{\mathbf{x}}(t) = \mathbf{A}\mathbf{x}(t) + \mathbf{B}\mathbf{u}(t) \\ \mathbf{y}(t) = \mathbf{C}\mathbf{x}(t) + \mathbf{D}\mathbf{u}(t) \end{cases}$, where

$$x(t) = \begin{pmatrix} x_1 \\ x_2 \end{pmatrix}, \quad u(t) = F(t), \quad A = \begin{bmatrix} 0 & 1 \\ -\dfrac{k}{m} & -\dfrac{c+\mu}{m} \end{bmatrix}, \quad B = \begin{pmatrix} 0 \\ \dfrac{1}{m} \end{pmatrix}, \quad C = (1 \ \ 0), \quad D = 0$$

Transfer Function Formulation

To derive the transfer function of this system, the Laplace transform of Eq. (5) is taken, assuming zero initial conditions:

$$ms^2 X(s) + (c+\mu)s X(s) + k X(s) = F(s) \tag{6}$$

From Eq. (6) the system transfer function is derived as follows:

$$G(s) = \frac{X(s)}{F(s)} = \frac{1}{ms^2 + (c+\mu)s + k}$$

Block Diagram

The simplest block diagram is obtained from the system transfer function, and is as shown in Figure 6.2.2.

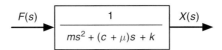

Figure 6.2.2

6.2.1 State-Space Representations

As discussed in Section 2.5.4, state equations are a set of coupled first-order ordinary differential equations of the form described by Eq. (2.5.9). In mathematical literature this representation is often called the *Cauchy form*. As mentioned earlier, this form is well suited for implementation in software to investigate system behaviors in simulation.

The state-space representation of a system consists of two distinct parts as shown in Figure 6.2.3: state equations and output equations.

A state equation is a set of coupled first-order ordinary differential equations, where the left-hand side only contains the first derivative of a state variable, and the right-hand side does not contain any derivative term of a state variable.

Nonlinear state equations are described by Eq. (2.5.9), while linear state equations are described by Eq. (2.5.10). Output equations are in the form of:

$$\begin{cases} y_1 = g_1(x_1, x_2, \ldots, x_n; u_1, u_2, \ldots, u_m) \\ y_2 = g_2(x_1, x_2, \ldots, x_n; u_1, u_2, \ldots, u_m) \\ \quad \vdots \\ y_k = g_k(x_1, x_2, \ldots, x_n; u_1, u_2, \ldots, u_m) \end{cases} \tag{6.2.1}$$

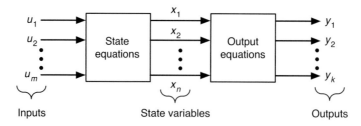

Figure 6.2.3 State-space representation of a dynamic system

where the unknown functions $y_i = y_i(t)$, $i = 1, 2, \ldots, k$ are the specified outputs; $x_i = x_i(t)$, $i = 1, 2, \ldots, n$ are state variables; $u_i = u_i(t)$, $i = 1, 2, \ldots, m$ are given inputs; and $g_1, g_2, \ldots, g_k$ are linear or nonlinear functions of the state variables and inputs.

Hence, the state-space representation of a dynamic system that is described by n state variables, and has m inputs and k outputs is as follows:

$$\text{State equations} \begin{cases} \dot{x}_1 = f_1(x_1, x_2, \ldots, x_n; u_1, u_2, \ldots, u_m) \\ \dot{x}_2 = f_2(x_1, x_2, \ldots, x_n; u_1, u_2, \ldots, u_m) \\ \quad \vdots \\ \dot{x}_n = f_n(x_1, x_2, \ldots, x_n; u_1, u_2, \ldots, u_m) \end{cases} \quad (6.2.2)$$

$$\text{Output equations} \begin{cases} y_1 = g_1(x_1, x_2, \ldots, x_n; u_1, u_2, \ldots, u_m) \\ y_2 = g_2(x_1, x_2, \ldots, x_n; u_1, u_2, \ldots, u_m) \\ \quad \vdots \\ y_k = g_k(x_1, x_2, \ldots, x_n; u_1, u_2, \ldots, u_m) \end{cases}$$

In this formulation functions $f_1, f_2, \ldots, f_n$ and $g_1, g_2, \ldots, g_k$ are of state variables ($x_i = x_i(t)$, $i = 1, 2, \ldots, n$) and inputs ($u_i = u_i(t)$, $i = 1, 2, \ldots, m$); they may be linear or nonlinear.

The scope of this book covers only these dynamic systems that have a linear state-space representation. It is of the form:

$$\text{State equations} \begin{cases} \dot{x}_1 = a_{11}x_1 + a_{12}x_2 + \cdots + a_{1n}x_n + b_{11}u_1 + b_{12}u_2 + \cdots + b_{1m}u_m \\ \dot{x}_2 = a_{21}x_1 + a_{22}x_2 + \cdots + a_{2n}x_n + b_{21}u_1 + b_{22}u_2 + \cdots + b_{2m}u_m \\ \quad \vdots \\ \dot{x}_n = a_{n1}x_1 + a_{n2}x_2 + \cdots + a_{nn}x_n + b_{n1}u_1 + b_{n2}u_2 + \cdots + b_{nm}u_m \end{cases} \quad (6.2.3)$$

$$\text{Output equations} \begin{cases} y_1 = c_{11}x_1 + c_{12}x_2 + \cdots + c_{1n}x_n + d_{11}u_1 + d_{12}u_2 + \cdots + d_{1m}u_m \\ y_2 = c_{21}x_1 + c_{22}x_2 + \cdots + c_{2n}x_n + d_{21}u_1 + d_{22}u_2 + \cdots + d_{2m}u_m \\ \quad \vdots \\ y_k = c_{k1}x_1 + c_{k2}x_2 + \cdots + c_{kn}x_n + d_{k1}u_1 + d_{k2}u_2 + \cdots + d_{km}u_m \end{cases}$$

where a_{ji}, b_{ji}, c_{ji}, and d_{ji} are either known functions of time or constants.

The linear state-space representation of a dynamic system (Eq. (6.2.3)) is often cast in a matrix form:

$$\begin{cases} \dot{x}(t) = Ax(t) + Bu(t) \\ y(t) = Cx(t) + Du(t) \end{cases} \quad (6.2.4)$$

where $x(t)$ and $u(t)$ are the *state* and the *input* column *vectors* respectively, defined as

$$x(t)_{n\times 1} = \begin{pmatrix} x_1(t) \\ x_2(t) \\ \vdots \\ x_n(t) \end{pmatrix}, \quad u(t)_{m\times 1} = \begin{pmatrix} u_1(t) \\ u_2(t) \\ \vdots \\ u_m(t) \end{pmatrix}$$

while matrices A, B, C, and D are as follows:

$$A_{n\times n} = \begin{bmatrix} a_{11} & a_{12} & \cdots & a_{1n} \\ a_{21} & a_{22} & \cdots & a_{2n} \\ \vdots & \vdots & \ddots & \vdots \\ a_{n1} & a_{n2} & \cdots & a_{nn} \end{bmatrix}, \quad B_{n\times m} = \begin{bmatrix} b_{11} & b_{12} & \cdots & b_{1m} \\ b_{21} & b_{22} & \cdots & b_{2m} \\ \vdots & \vdots & \ddots & \vdots \\ b_{n1} & b_{n2} & \cdots & b_{nm} \end{bmatrix}$$

$$C_{k\times n} = \begin{bmatrix} c_{11} & c_{12} & \cdots & c_{1n} \\ c_{21} & c_{22} & \cdots & c_{2n} \\ \vdots & \vdots & \ddots & \vdots \\ c_{k1} & c_{k2} & \cdots & c_{kn} \end{bmatrix}, \quad D_{k\times m} = \begin{bmatrix} d_{11} & d_{12} & \cdots & d_{1m} \\ d_{21} & d_{22} & \cdots & d_{2m} \\ \vdots & \vdots & \ddots & \vdots \\ d_{k1} & d_{k2} & \cdots & d_{km} \end{bmatrix}$$

and column vector $y(t)_{k\times 1}$ is the *output vector*.

A square matrix $A_{n\times n}$ is called a *system matrix*, rectangular matrices $B_{n\times m}$ and $C_{k\times n}$ are the *input matrix* and the *state output matrix*, respectively, and the matrix $D_{k\times m}$ is called either the *control output matrix* or the *direct transmission matrix*.

Obviously, a state-space representation of a system is not unique since it depends on the choice of state variables and outputs.

The general rule of the assignment of state variables to system parameters was discussed in Chapter 2 (see Section 2.5.4). It states that n state variables must be defined to convert an n-th order differential equation into a set of state equations. For an equation

$$a_n(t)\frac{d^n x(t)}{dt^n} + a_{n-1}(t)\frac{d^{n-1} x(t)}{dt^{n-1}} + \ldots + a_1(t)\frac{dx(t)}{dt} + a_0(t)x(t) = u(t) \quad (6.2.5)$$

where $a_k(t)$ are known functions, state variables are selected as follows:

$$x_1(t) = x(t), \; x_2(t) = \frac{dx(t)}{dt}, \; x_3(t) = \frac{d^2 x(t)}{dt^2}, \; \ldots, \; x_n(t) = \frac{d^{n-1} x(t)}{dt^{n-1}} \quad (6.2.6)$$

This selection yields a set of $(n-1)$ state equations:

$$\begin{cases} \dot{x}_1(t) = x_2(t) \\ \dot{x}_2(t) = x_3(t) \\ \dot{x}_3(t) = x_4(t) \\ \vdots \\ \dot{x}_{n-2}(t) = x_{n-1}(t) \\ \dot{x}_{n-1}(t) = x_n(t) \end{cases} \qquad (6.2.7)$$

Rewriting the original differential equation (Eq. (6.2.5)) in terms of the defined state variables:

$$a_n(t)\dot{x}_n(t) + a_{n-1}(t)x_n(t) + a_{n-2}(t)x_{n-1}(t) + \ldots + \\ + a_2(t)x_3(t) + a_1(t)x_2(t) + a_0(t)x_1(t) = u(t) \qquad (6.2.8)$$

We obtain the n-th state equation:

$$\dot{x}_n(t) = -\frac{a_0(t)}{a_n(t)}x_1(t) - \frac{a_1(t)}{a_n(t)}x_2(t) - \frac{a_2(t)}{a_n(t)}x_3(t) - \ldots - \\ - \frac{a_{n-2}(t)}{a_n(t)}x_{n-1}(t) - \frac{a_{n-1}(t)}{a_n(t)}x_n(t) + \frac{1}{a_n(t)}u(t) \qquad (6.2.9)$$

Thus, state equations of such dynamic system are represented by Eqs. (6.2.7) and (6.2.9), while the formulation of the output equations depends on the choice of outputs. For example, if all state variables need to be monitored, the output vector $y(t)$ is the same as the state vector $x(t)$. Then, the familiar state-space representation in matrix form:

$$\begin{cases} \dot{x}(t) = Ax(t) + Bu(t) \\ y(t) = Cx(t) + Du(t) \end{cases}$$

becomes

$$A(t) = \begin{bmatrix} 0 & 1 & 0 & \cdots & 0 \\ 0 & 0 & 1 & \cdots & 0 \\ \vdots & \vdots & \vdots & \ddots & \vdots \\ 0 & 0 & 0 & \cdots & 1 \\ -\frac{a_0(t)}{a_n(t)} & -\frac{a_1(t)}{a_n(t)} & -\frac{a_2(t)}{a_n(t)} & \cdots & -\frac{a_{n-1}(t)}{a_n(t)} \end{bmatrix}, \quad B = \begin{pmatrix} 0 \\ 0 \\ \vdots \\ 0 \\ \frac{1}{a_n(t)} \end{pmatrix}$$

$$C(t) = \begin{bmatrix} 1 & 0 & \cdots & 0 \\ 0 & 1 & 0 & 0 \\ \vdots & \vdots & \ddots & \vdots \\ 0 & 0 & \cdots & 1 \end{bmatrix}, \quad D(t) = 0, \quad x(t) = \begin{pmatrix} x_1(t) \\ x_2(t) \\ \vdots \\ x_n(t) \end{pmatrix} = y(t) \qquad (6.2.10)$$

For a dynamic system described with a set of coupled differential equations, a similar rule for assigning state variable applies. For any independent variable $x_j(t)$ in the set of governing equations, a component state vector is needed. This component state vector is a column vector with the number of rows equal to the highest order of derivative of the considered independent variable $x_j(t)$. The system state vector is a union of all the component vectors. The matrices A, B, C, and D are derived using the same procedure as described above.

Example 6.2.2

Derive a state-space representation for a system described by the following differential equation:

$$\frac{d^3x(t)}{dt^3} + a_1 \frac{d^2x(t)}{dt^2} + a_2 \frac{dx(t)}{dt} + a_3 x(t) = u(t)$$

where $u(t)$ is the system input, the desired output is

$$y(t) = b_1 x(t) + b_2 \frac{dx(t)}{dt}$$

and a_i and b_j are known coefficients.

Solution

Since the differential equation is of third order, three state variables are needed. In accordance with the procedure described above, the state variables are selected as follows:

$$\begin{cases} x_1 = x(t) \\ x_2 = \dot{x}(t) \\ x_3 = \ddot{x}(t) \end{cases}$$

Then, the first two state equations arise from the definition of state variables:

$$\begin{cases} \dot{x}_1 = x_2(t) \\ \dot{x}_2 = x_3(t) \end{cases}$$

and the last one is derived by rewriting the original governing equation in terms of the chosen state variables:

$$\dot{x}_3 + a_1 x_3 + a_2 x_2 + a_3 x_1 = u(t)$$
$$\dot{x}_3 = -a_3 x_1 - a_2 x_2 - a_1 x_3 + u(t)$$

The output equation is

$$y = b_1 x_1 + b_2 x_2$$

Thus, system's state-space representation is

$$\begin{cases} \dot{x}_1 = x_2(t) \\ \dot{x}_2 = x_3(t) \\ \dot{x}_3 = -a_3 x_1 - a_2 x_2 - a_1 x_3 + u(t) \\ y = b_1 x_1 + b_2 x_2 \end{cases}$$

or in matrix form:

$$\begin{cases} \dot{x}(t) = Ax(t) + Bu(t) \\ y(t) = Cx(t) + Du(t) \end{cases}$$

$$A_{3 \times 3} = \begin{bmatrix} 0 & 1 & 0 \\ 0 & 0 & 1 \\ -a_3 & -a_2 & -a_1 \end{bmatrix}, \quad B_{3 \times 1} = \begin{pmatrix} 0 \\ 0 \\ 1 \end{pmatrix} \quad C_{1 \times 3} = (b_1 \quad b_2 \quad 0), \quad D = 0$$

$$x(t) = \begin{pmatrix} x_1 \\ x_2 \\ x_3 \end{pmatrix}, \quad u(t) \text{ and } y(t) \text{ are scalars}$$

Example 6.2.3

Derive the state-space representation for a system described by the following set of coupled differential equations:

$$\begin{cases} a_1 \dfrac{d^2 x(t)}{dt^2} + a_2 \dfrac{dx(t)}{dt} + a_3 x(t) - a_4 \dfrac{dz(t)}{dt} - a_5 z(t) = u_1(t) \\ a_6 \dfrac{dz(t)}{dt} + a_7 z(t) - a_8 \dfrac{dx(t)}{dt} = u_2(t) \end{cases}$$

where $u_1(t)$ and $u_2(t)$ are the system inputs, the desired outputs are

$$\begin{cases} y_1(t) = b_1 x(t) + b_2 \dfrac{dx(t)}{dt} \\ y_2(t) = b_3 z(t) + b_4 u_2(t) \end{cases}$$

and a_i and b_j are known coefficients.

Solution

This dynamic system is described by two independent variables: $x(t)$ and $z(t)$. The highest order of derivative for $x(t)$ is 2, hence two state variables are needed. Similarly, for $z(t)$, a single state variable (first-order derivative) is needed. Then, the system state vector is of the size 3×1, and the state variables are $x_1 = x(t)$, $x_2 = \dot{x}(t)$, $x_3 = z(t)$

From the definition of these state variables, the first state equation is simply obtained:

$$\dot{x}_1 = x_2 \qquad (1)$$

To derive the remaining two state equations, the governing equations are rewritten in terms of state variables:

$$\{a_1 \dot{x}_2 + a_2 x_2 + a_3 x_1 - a_4 \dot{x}_3 - a_5 x_3 = u_1(t) \qquad (2)$$

$$\{a_6\dot{x}_3 + a_7 x_3 - a_8 x_2 = u_2(t) \tag{3}$$

From Eq. (3), we find the expression for $\dot{x}_3$

$$\dot{x}_3 = \frac{a_8}{a_6} x_2 - \frac{a_7}{a_6} x_3 + \frac{1}{a_6} u_2 \tag{4}$$

Plugging this result (Eq. (4)) into Eq. (2) and performing the necessary algebraic operations, the expression for $\dot{x}_3$ is obtained:

$$\dot{x}_2 = -\frac{a_3}{a_1} x_1 + \left(\frac{a_4 a_8}{a_1 a_6} - \frac{a_2}{a_1}\right) x_2 + \left(\frac{a_4 a_7}{a_1 a_6} - \frac{a_5}{a_1}\right) x_3 + \frac{1}{a_1} u_1 + \frac{1}{a_1 a_6} u_2 \tag{5}$$

The output equations in terms of state variables are

$$\begin{aligned} y_1(t) &= b_1 x_1 + b_2 x_2 \\ y_2(t) &= b_3 x_3 + b_4 u_2(t) \end{aligned} \tag{6}$$

Then, the system state-space representation consists of Eqs. (1), (4), (5), and (6) as follows:

State equations
$$\begin{cases} \dot{x}_1 = x_2 \\ \dot{x}_2 = -\dfrac{a_3}{a_1} x_1 + \left(\dfrac{a_4 a_8}{a_1 a_6} - \dfrac{a_2}{a_1}\right) x_2 + \left(\dfrac{a_4 a_7}{a_1 a_6} - \dfrac{a_5}{a_1}\right) x_3 + \dfrac{1}{a_1} u_1 + \dfrac{1}{a_1 a_6} u_2 \\ \dot{x}_3 = \dfrac{a_8}{a_6} x_2 - \dfrac{a_7}{a_6} x_3 + \dfrac{1}{a_6} u_2 \end{cases}$$

Output equations
$$\begin{cases} y_1(t) = b_1 x_1 + b_2 x_2 \\ y_2(t) = b_3 x_3 + b_4 u_2(t) \end{cases}$$

If desired, the derived state-space representation can be cast in the matrix form: $\begin{cases} \dot{x}(t) = Ax(t) + Bu(t) \\ y(t) = Cx(t) + Du(t) \end{cases}$, where the matrices are as follows:

$$A_{3\times 3} = \begin{bmatrix} 0 & 1 & 0 \\ -\dfrac{a_3}{a_1} & \dfrac{a_4 a_8 - a_2 a_6}{a_1 a_6} & \dfrac{a_4 a_7 - a_5 a_6}{a_1 a_6} \\ 0 & \dfrac{a_8}{a_6} & -\dfrac{a_7}{a_6} \end{bmatrix}, \quad B_{3\times 2} = \begin{pmatrix} 0 & 0 \\ \dfrac{1}{a_1} & \dfrac{1}{a_1 a_6} \\ 0 & \dfrac{1}{a_6} \end{pmatrix},$$

$$C_{2\times 3} = \begin{pmatrix} b_1 & b_2 & 0 \\ 0 & 0 & b_3 \end{pmatrix}, \quad D_{2\times 2} = \begin{pmatrix} 0 & 0 \\ 0 & b_4 \end{pmatrix}$$

$$x(t) = \begin{pmatrix} x_1 \\ x_2 \\ x_3 \end{pmatrix}, \quad u(t) = \begin{pmatrix} u_1 \\ u_2 \end{pmatrix}, \quad y(t) = \begin{pmatrix} y_1 \\ y_2 \end{pmatrix}$$

While the above procedure of selecting state variables applies to any linear dynamic system, it is useful to think about the physical meaning of state variables as parameters that describe the system's energy. For example, for mechanical systems discussed in Chapter 3, such state variables are displacement (linear or angular) and velocity (linear or angular). The displacement is used to describe the potential energy of the system, arising due to linear or torsional springs ($V = \frac{1}{2}kx^2$ or $V = \frac{1}{2}k_T\theta^2$). The velocity describes the kinetic energy of the system, arising due to the motion of inertia elements ($T = \frac{1}{2}m\dot{x}^2$ or $T = \frac{1}{2}I_G\dot{\theta}^2$). For electrical elements, discussed in Chapter 4, the state variables are assigned only for the elements that store energy. These state variables are current and voltage, which are both used to describe the electrical energy ($E = \frac{1}{2}Cv^2$ or $E = \frac{1}{2}Li^2$).

For compound dynamic systems that are described by a set of differential equations, state variables are assigned per subsystem, and the resulting state vector will be a union of the subsystems' state vectors. The procedure of deriving a state-space representation of a compound system is similar to that illustrated by Example 6.2.3. The formulation of such state-space representations will be discussed in more detail later in this chapter.

6.2.2 Transfer Function Formulations

As introduced in Chapter 2 (see Section 2.7.1), the transfer function of a time-invariant linear system is defined as the ratio of the system output to system input in the Laplace domain, assuming zero initial conditions (Eq. (2.7.1)).

Consider a dynamic system described by a governing equation

$$a_n \frac{d^n x(t)}{dt^n} + a_{n-1} \frac{d^{n-1} x(t)}{dt^{n-1}} + \cdots + a_1 \frac{dx(t)}{dt} + a_0 x(t) =$$
$$= b_m \frac{d^m u(t)}{dt^m} + b_{m-1} \frac{d^{m-1} u(t)}{dt^{m-1}} + \cdots + b_1 \frac{du(t)}{dt} + b_0 u(t) \tag{6.2.11}$$

where a_i and b_j are known coefficients, $u(t)$ is the system input, and $x(t)$ is the system output. Assuming zero initial conditions, take the Laplace transform of Eq. (6.2.11):

$$\left(a_n s^n + a_{n-1} s^{n-1} + \cdots + a_1 s + a_0\right) X(s) =$$
$$= \left(b_m s^m + b_{m-1} s^{m-1} + \cdots + b_1 s + b_0\right) U(s) \tag{6.2.12}$$

Then the sought system transfer function is

$$G(s) = \frac{X(s)}{U(s)} = \frac{b_m s^m + b_{m-1} s^{m-1} + \cdots + b_1 s + b_0}{a_n s^n + a_{n-1} s^{n-1} + \cdots + a_1 s + a_0} \tag{6.2.13}$$

Recalling the discussion in Section 2.4.2, the poles of this transfer function are defined as the solutions of the equation

$$a_n s^n + a_{n-1} s^{n-1} + \cdots + a_1 s + a_0 = 0 \tag{6.2.14}$$

and its zeros are the solutions of the equation

$$b_m s^m + b_{m-1} s^{m-1} + \cdots + b_1 s + b_0 = 0 \tag{6.2.15}$$

Obviously, the derived transfer function has n poles and m zeros.

Knowledge of the poles and zeros of the system transfer function gives a good insight into system behavior; these parameters are widely used in control design. The order of a transfer function is the number of its poles; hence, the transfer function described by Eq. (6.2.13) is of n-th order.

A transfer function can be expressed in a *zero-pole-gain* form such as

$$G(s) = \frac{X(s)}{U(s)} = k \frac{(s - z_1)(s - z_2) \cdots (s - p_n)}{(s - p_1)(s - p_2) \cdots (s - z_m)} \tag{6.2.16}$$

where p_i and z_j are the transfer function poles and zeros respectively, and $k = \dfrac{b_m}{a_n}$ is the gain. A pole-zero-gain formulation is also extensively used in system analysis and control design.

Since the transfer function form depends on the choice of input and output for the dynamic system, it is not unique.

For any linear time-invariant dynamic system, its transfer function can be obtained from the system state-space representation cast in matrix form:

$$G(s) = C(sI - A)^{-1} B + D \tag{6.2.17}$$

where the matrices A, B, C, and D are defined as shown by Eq. (6.2.4), and I is the identity matrix of the same size as the state matrix A.

Example 6.2.4

Derive transfer functions, find their poles and zeros, and cast the obtained transfer functions into pole-zero-gain form for the dynamic systems described by the following governing equations:

(a) $\dot{x}(t) + ax(t) = u(t)$
(b) $\ddot{x}(t) + a_1 \dot{x}(t) + a_0 x(t) = u(t)$
(c) $\ddot{x}(t) + a_1 x(t) + a_0 \int_0^t x(\tau) d\tau = b_1 \dot{u}(t) + b_0 u(t)$

Solution

For the system input $u(t)$ and output $x(t)$, the transfer functions are obtained by taking the Laplace transform of the governing equations, followed by the algebraic operations as follows:

(a) $\mathcal{L}[\dot{x}(t) + ax(t)] = \mathcal{L}[u(t)]$

$$(s+a)X(s) = U(s)$$
$$G(s) = \frac{X(s)}{U(s)} = \frac{1}{s+a}$$

This is the first-order transfer function ($n = 1$, $m = 0$), with a single pole $p_1 = -a$ and no zeros. It is already in the pole-zero-gain form ($a_n = 1$, $b_m = 1$).

(b) $\mathcal{L}[\ddot{x}(t) + a_1 \dot{x}(t) + a_0 x(t)] = \mathcal{L}[u(t)]$

$$(s^2 + a_1 s + a_0)X(s) = U(s)$$
$$G(s) = \frac{X(s)}{U(s)} = \frac{1}{s^2 + a_1 s + a_0}$$

This is the second-order transfer function ($n = 2$, $m = 0$), with two poles p_1 and p_2 that are the roots of the equation $s^2 + a_1 s + a_0 = 0$ and no zeros. The pole-zero-gain form of transfer function is

$$G(s) = \frac{1}{(s - p_1)(s - p_2)}$$

(c) $\mathcal{L}\left[\ddot{x}(t) + a_1 x(t) + a_0 \int_0^t x(\tau) d\tau\right] = \mathcal{L}[b_1 \dot{u}(t) + b_0 u(t)]$

$$\left(s^2 + a_1 + \frac{a_0}{s}\right) X(s) = (b_1 s + b_0) U(s)$$

$$G(s) = \frac{X(s)}{U(s)} = \frac{b_1 s + b_0}{s^2 + a_1 + \frac{a_0}{s}} = \frac{b_1 s^2 + b_0 s}{s^3 + a_1 s + a_0}$$

This is the third-order transfer function ($n = 3$, $m = 2$), with three poles p_1, p_2, and p_3 that are the roots of the equation $s^3 + a_1 s + a_0 = 0$ and two zeros $z_1 = -b_0/b_1$ and $z_2 = 0$.
Note that any transfer function must be a ratio of two polynomials. Hence, the form

$$G(s) = \frac{b_1 s + b_0}{s^2 + a_1 + a_0/s}$$

is not a proper transfer function since its denominator is not a polynomial. Multiplying the numerator and denominator by s yields the sought transfer function, but with an additional zero ($z_2 = 0$) and of order greater than the order of the highest derivative of $x(t)$.
The pole-zero-gain form of the derived transfer function is

$$G(s) = b_1 \frac{s(s + b_0/b_1)}{(s - p_1)(s - p_2)(s - p_3)}$$

Dynamic systems with multiple inputs and multiple outputs (*MIMO systems*) have a transfer matrix, the elements of which are transfer functions that reflect every input–output relation in the system. The derivation of these transfer functions is done using the same procedure as described above. Every output $Y_i(s)$ is expected to yield an expression of the form

$$Y_i(s) = G_{1i}(s)U_1(s) + G_{2i}(s)U_2(s) + \ldots + G_{(n-1)i}(s)U_{(n-1)}(s) + G_{ni}(s)U_n(s)$$

where $G_{ji}(s)$ denotes the transfer function from the *j*-th input $U_j(s)$ to the *i*-th output $Y_i(s)$. An individual transfer function $G_{ji}(s)$ is derived by setting all the inputs other than $U_j(s)$ to zero, thus obtaining an expression $Y_i(s) = G_{ji}(s)U_j(s)$.

Upon deriving all the transfer functions, the transfer matrix for a system with n inputs and m outputs is constructed. It is of the form

$$G_{m \times n}(s) = \begin{bmatrix} G_{11}(s) & G_{21}(s) & \cdots & G_{n1}(s) \\ G_{12}(s) & G_{22}(s) & \cdots & G_{n2}(s) \\ \vdots & \vdots & \ddots & \vdots \\ G_{1m}(s) & G_{2m}(s) & \cdots & G_{nm}(s) \end{bmatrix}$$

Verifying that multiplication of this matrix by the $n \times 1$ input vector produces the correct expressions for every output $Y_i(s)$ is straightforward.

Even for the low-order MIMO systems, the derivation of the transfer matrix requires a number of algebraic operations and may become quite tedious.

Example 6.2.5

Derive a transfer matrix for the two-input–two-output system described by the following set of governing equations:

$$\begin{cases} \dot{x}(t) + a_1 z(t) = u_1(t) \\ \ddot{z}(t) + a_2 \dot{z}(t) + a_3 x(t) = u_2(t) \end{cases}$$

System inputs are $u_1(t)$ and $u_2(t)$, and the outputs are the independent variables $x(t)$ and $z(t)$.

Solution

Since this dynamic system has two inputs and two outputs, four transfer functions are expected, as shown in Figure 6.2.4.

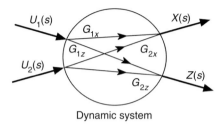

Dynamic system

Figure 6.2.4 Transfer function representation of a two-input–two-output dynamic system

Let the system input vector be

$$U(s) = \begin{pmatrix} U_1(s) \\ U_2(s) \end{pmatrix}$$

and output vector be

$$Y(s) = \begin{pmatrix} X(s) \\ Z(s) \end{pmatrix}$$

Recalling that $Y(s) = G(s)U(s)$, the system transfer matrix has the following form:

$$G(s) = \begin{bmatrix} G_{1X} & G_{2X} \\ G_{1Z} & G_{2Z} \end{bmatrix}$$

where G_{1X}, G_{2X}, G_{1Z}, and G_{2Z} are the individual transfer functions relating every input to every output as shown in Figure 6.2.4.

As discussed earlier, the sought transfer functions are derived taking the Laplace transform of the governing equations assuming zero initial conditions, and then performing the necessary algebraic transformations.

Taking the Laplace transform of the given governing equations, we obtain

$$\{sX(s) + a_1 Z(s) = U_1(s) \tag{1}$$

$$\{(s^2 + a_2 s)Z(s) + a_3 X(s) = U_2(s) \tag{2}$$

To derive the transfer functions for the $X(s)$ output, i.e. G_{1X} and G_{2X}, we need to have a single equation in terms of this output and the inputs. This is done by deriving an expression for $Z(s)$ from Eq. (1):

$$Z(s) = -\frac{s}{a_1} X(s) + \frac{1}{a_1} U_1(s) \tag{3}$$

and substituting Eq. (3) into Eq. (2), obtaining

$$(s^2 + a_2 s)\left(-\frac{s}{a_1} X(s) + \frac{1}{a_1} U_1(s)\right) + a_3 X(s) = U_2(s) \tag{4}$$

Simplifying Eq. (4) yields

$$X(s) = \frac{s^2 + a_2 s}{s^3 + a_2 s^2 - a_1 a_3} U_1(s) - \frac{a_1}{s^3 + a_2 s^2 - a_1 a_3} U_2(s) \qquad (5)$$

Hence, the transfer functions G_{1X} and G_{2X} are found by setting the inputs in Eq. (5) to zero in turn:

$$G_{1X} = \frac{X(s)}{U_1(s)} = \frac{s^2 + a_2 s}{s^3 + a_2 s^2 - a_1 a_3}$$

$$G_{2X} = \frac{X(s)}{U_2(s)} = -\frac{a_1}{s^3 + a_2 s^2 - a_1 a_3} \qquad (6)$$

Similarly, we derive the transfer functions G_{1Z} and G_{2Z}.
From Eq. (2):

$$X(s) = -\frac{s^2 + a_2 s}{a_3} Z(s) + \frac{1}{a_3} U_2(s) \qquad (7)$$

Substituting Eq. (7) into Eq. (1), we derive an expression for $Z(s)$:

$$s\left(-\frac{s^2 + a_2 s}{a_3} Z(s) + \frac{1}{a_3} U_2(s)\right) + a_1 Z(s) = U_1(s)$$

$$Z(s) = -\frac{a_3}{s^3 + a_2 s^2 - a_1 a_3} U_1(s) + \frac{s}{s^3 + a_2 s^2 - a_1 a_3} U_2(s) \qquad (8)$$

Eq. (8) yields the transfer functions G_{1Z} and G_{2Z}:

$$G_{1Z} = \frac{Z(s)}{U_1(s)} = -\frac{a_3}{s^3 + a_2 s^2 - a_1 a_3}$$

$$G_{2Z} = \frac{Z(s)}{U_2(s)} = \frac{s}{s^3 + a_2 s^2 - a_1 a_3} \qquad (9)$$

Having derived the component functions, the system transfer matrix is easily constructed:

$$G(s) = \begin{bmatrix} G_{1X} & G_{2X} \\ G_{1Z} & G_{2Z} \end{bmatrix} = \begin{bmatrix} \dfrac{s^2 + a_2 s}{s^3 + a_2 s^2 - a_1 a_3} & -\dfrac{a_1}{s^3 + a_2 s^2 - a_1 a_3} \\ -\dfrac{a_3}{s^3 + a_2 s^2 - a_1 a_3} & \dfrac{s}{s^3 + a_2 s^2 - a_1 a_3} \end{bmatrix}$$

6.2.3 Block Diagrams

Block diagrams, as introduced in Chapter 2 (see Section 2.7.2) provide a powerful method for describing the input–output relations within a dynamic system. It is particularly useful for MIMO systems as well as compound systems, even though at the system level a compound system may have a single input and a single output. A block diagram representation is a very general cause-and-effect approach to dynamic system modeling and can be applied to many types of systems – linear and nonlinear, time-varying and time-invariant. Additionally, constructing a block diagram for the given dynamic system helps in modeling this system in software such as Simulink by MathWorks, or System Modeler by Wolfram Research.

A block diagram has three major components: the *block*, *summation point*, and *branch* or *takeoff* point, as illustrated in Figure 6.2.5.

In Figure 6.2.5 the arrows represent variables (signals) and the direction of the input–output (cause-and-effect) relation. A block, presented in Figure 6.2.5(a), embodies the transfer function between the depicted input $X(s)$ and output $Y(s)$. Generally, for a dynamic system described by a set of differential equations, the block represents a single equation. Figure 6.2.5(b) shows the summation point. Depending on the signs at the variables entering the summation point (plus or minus at an arrowhead), either addition or subtraction operations may be performed. A summation point may have more than two incoming signals. Figure 6.2.5(c) shows the branch point, from which the signal goes concurrently to the other elements of the block diagram. Note that the branch point does not change the value of a variable that entered it, but only conveys it to the other parts of this block diagram.

It may be desirable to describe a linear time-invariant system that has a state-space representation

$$\begin{cases} \dot{x}(t) = Ax(t) + Bu(t) \\ y(t) = Cx(t) + Du(t) \end{cases}$$

(see Eq. (6.2.4)) by a block diagram. There, the blocks describe input–output relations using matrices A, B, C, D, and an integrator rather than a transfer function (see Figure 6.2.6).

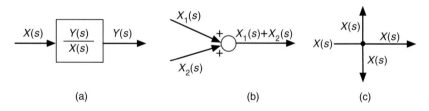

Figure 6.2.5 Basic components of a block diagram: (a) block, (b) summation point, (c) branch point

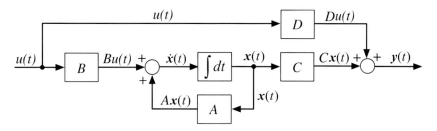

Figure 6.2.6 Block diagram of a linear time-invariant dynamic system

A block diagram represents a powerful visual depiction of a dynamic system. It clearly shows how parameters evolve throughout the system, and how system components interact with each other. Nonetheless, for all but the simplest dynamic systems, the block diagram often becomes a convoluted structure, with a large number of blocks, summation, and branch points, and a tangled web of arrows. A complicated block diagram can be simplified by applying the following rules of equivalent transformation:

(a) a combination of two blocks can be substituted with a single block, the transfer function of which is computed as shown in Table 6.2.1, where G_1 and G_2 denote the transfer functions of connected blocks, while $U(s)$ and $Y(s)$ are the input and output of the block combination, respectively
(b) a branch point can be moved across a block, but never across a summation point; movement of a branch point across any block is accompanied by the addition of an appropriate block element to prevent the loss of the signal

The application of these rules to simplify a block diagram is illustrated in Examples 6.2.7 through 6.2.10.

It needs to be noted that series and parallel connections may include more than two blocks. The equivalent transformation of n blocks in series is still a single block with the transfer function

$$G_{eq} = \prod_n G_i \qquad (6.1.18)$$

where G_i denotes a transfer function of an individual block component.

Similarly, for n blocks in parallel the equivalent block transfer function is

$$G_{eq} = \sum_n G_i \qquad (6.1.19)$$

The derivation of the equivalent transfer functions presented in Table 6.2.1, and expressed with Eqs. (6.1.18) and (6.1.19), is straightforward.

Table 6.2.1 Equivalent transformation of block combinations

Configuration	Equivalent Transformation
1. Series (cascade) connection $U(s) \to \boxed{G_1} \to \boxed{G_2} \to Y(s)$	$U(s) \to \boxed{G_1 G_2} \to Y(s)$
2. Parallel connection $U(s)$ branches to G_1 and G_2, summed to $Y(s)$ (both $+$)	$U(s) \to \boxed{G_1 + G_2} \to Y(s)$
3. Feedback connection (negative feedback) $U(s) \xrightarrow{+} \bigcirc \to \boxed{G_1} \to Y(s)$, feedback through G_2 (with $-$)	$U(s) \to \boxed{\dfrac{G_1}{1 + G_1 G_2}} \to Y(s)$
4. Feedback connection (positive feedback) $U(s) \xrightarrow{+} \bigcirc \to \boxed{G_1} \to Y(s)$, feedback through G_2 (with $+$)	$U(s) \to \boxed{\dfrac{G_1}{1 - G_1 G_2}} \to Y(s)$
5. Feedback connection (negative unity feedback: transfer function of a simple connector line equals 1) $U(s) \xrightarrow{+} \bigcirc \to \boxed{G_1} \to Y(s)$, unity feedback (with $-$)	$U(s) \to \boxed{\dfrac{G_1}{1 + G_1}} \to Y(s)$

Example 6.2.6

Derive the equivalent transfer function of a negative feedback configuration shown in Figure 6.2.7.

Solution

The first step in the derivation includes denoting signals that travel along each shown arrow. Let the outgoing signal of the summation point (in control theory it is often called the *error signal*) be denoted as $E(s) = X(s) - G_2 Y(s)$. Then, the expression for the output signal, shown in Figure 6.2.7, can be rewritten as follows:

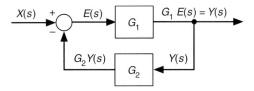

Figure 6.2.7 System for Example 6.2.6

$$Y(s) = G_1 E(s) = G_1(X(s) - G_2 Y(s))$$

Rearranging, we obtain

$$Y(s)(1 + G_1 G_2) = G_1 X(s)$$

Thus, the sought transfer function is

$$G(s) = \frac{Y(s)}{X(s)} = \frac{G_1}{1 + G_1 G_2}$$

Example 6.2.7
Consider a dynamic system, modeled with the block diagram presented in Figure 6.2.8.

Solution
While this block diagram is not exceedingly complex, deriving the system transfer function from it requires an effort. Hence, constructing an equivalent block diagram is necessary. Since none of the block combinations shown in Table 6.2.1 are discernible, the first simplification step encompasses moving the branch point A across the block G_2. This will yield the series connection of the blocks G_1 and G_2, and the parallel connection of the block G_4 with the elements located on the line connecting branch point A and the summation point. Remembering that the signal continuity must be preserved with an equivalent transformation, the signals traveling between the blocks G_2, G_3, and G_4 need to be studied prior to the equivalent transformation.

Let the signal entering the block G_2 be denoted z. Then, the distribution of signals between the blocks G_2, G_3, and G_4 is as shown in Figure 6.2.9. The depicted input and output signals for every block must be preserved in an equivalent transformation. This is achieved by adding a block with the transfer function inverse to that of the block G_2, as illustrated in Figure 6.2.10.

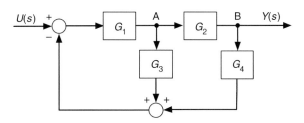

Figure 6.2.8 System for Example 6.2.7

As seen in Figure 6.2.10, the addition of the block with the transfer function $1/G_2$ preserves the input signal of the block $1/G_3$. Hence, it can be deduced that moving a branch point across a block always entails the addition of a *compensating block* with the inverse transfer function right after the moved branch point on the relocated line.

Moving the branch point A across the block G_2 yields two recognizable series connections: one of blocks G_1 and G_2, and the other one of the block G_3 and the newly added block $1/G_2$. The equivalent transformation of these combinations results in the block diagram depicted in Figure 6.2.11.

The equivalent transformation of the shown parallel connection of two blocks G_4 and G_3/G_2 yields the system model in standard negative feedback configuration, illustrated in Figure 6.2.12.

This block diagram is very suitable for implementation in software and subsequent simulation and analysis. Nonetheless, if desired, it can be simplified even further, yielding the system transfer function shown in Figure 6.2.13.

As demonstrated by this example, a dynamic system can be modeled by many equivalent block diagrams. With the given input and output of the dynamic system, the shape of the sought block diagram representation depends on which parameters and component relations need to be studied.

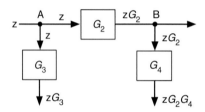

Figure 6.2.9 Distribution of signals

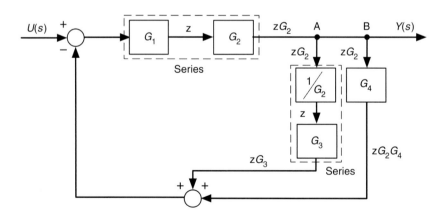

Figure 6.2.10 First equivalent transformation

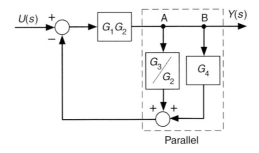

Figure 6.2.11 Second equivalent transformation

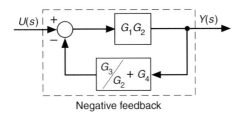

Figure 6.2.12 Third equivalent transformation

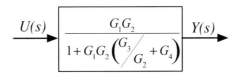

Figure 6.2.13 System transfer function

Example 6.2.8

Consider a typical dynamic system, described by a second-order differential equation with the constant coefficients $\ddot{x}(t) + a\dot{x}(t) + bx(t) = f(t)$. Let the system input be $f(t)$ and the system output be $x(t)$. Consider zero initial conditions: $x(0) = 0$ and $\dot{x}(0) = 0$.

Solution

The system transfer function $G(s) = X(s)/F(s)$ is easily derived by taking the Laplace transform of the governing equation and performing algebraic transformations as follows:

$$s^2 X(s) + asX(s) + bX(s) = F(s)$$
$$X(s)(s^2 + as + b) = F(s)$$
$$\frac{X(s)}{F(s)} = \frac{1}{s^2 + as + b}$$

If we are interested only in the input–output relation at the system level, the block diagram representation is the simplest and most concise one. It utilizes the derived above system transfer function and has only a single block, as shown in Figure 6.2.14.

There are several options of the block diagram representation better suitable for studying the influence of the parameters x and $\dot{x}$ on the system dynamics. Rearranging the original governing equation and taking the Laplace transform to construct these block diagrams, we obtain

$$\ddot{x}(t) = -a\dot{x}(t) - bx(t) + f(t)$$
$$s^2 X(s) = -asX(s) - bX(s) + F(s)$$
$$X(s) = \frac{1}{s^2}\left(-asX(s) - bX(s) + F(s)\right) = \frac{1}{s}\left(\frac{1}{s}\left(-asX(s) - bX(s) + F(s)\right)\right)$$

From these expressions the three-element summation point is easily discernible. Recalling that in the Laplace domain $\dot{x}$ becomes $sX(s)$, the block diagram shown in Figure 6.2.15 is constructed.

If a standard feedback configuration is desired, the inner feedback loop of blocks $1/s$ and a can be simplified as specified in Table 6.2.1. The resulting block diagram is shown in Figure 6.2.16.

Thus, like the state-space representation of a dynamic system, this block diagram model is not unique. The shape of the system's block diagram depends on the system-level inputs and outputs as well as on the elements in which the dynamic behavior needs to be studied.

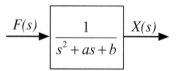

Figure 6.2.14 Second-order system in Example 6.2.8

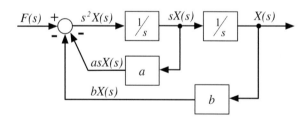

Figure 6.2.15 Expanded block diagram of a second-order system

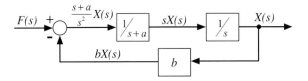

Figure 6.2.16 Second-order system in standard feedback configuration

A block diagram representation is particularly useful for MIMO dynamic systems. As discussed in Section 6.2.2, these systems have a transfer matrix that contains the transfer functions for every input–output relation in the system. While the component transfer functions can be derived directly from the Laplace transform of the governing equations, such derivations involve multiple algebraic operations and are often tedious. Obtaining these transfer functions from a block diagram is much easier. The transfer function from input U_i to output Y_j is obtained by setting all the other inputs to zero and simplifying the block diagram accordingly. The alternative approach includes processing the individual signals to yield the expression for the desired output in terms of all inputs, and then setting all but one inputs to zero.

Example 6.2.9

Consider a two-input–one-output dynamic system represented by the block diagram shown in Figure 6.2.17. Derive the transfer functions

$$G_U(s) = \frac{Y(s)}{U(s)}$$

and

$$G_V(s) = \frac{Y(s)}{V(s)}$$

Let the outgoing signals of the summation points A and B be $E(s)$ and $W(s)$, respectively. According to the block diagram, these signals are

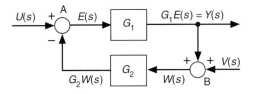

Figure 6.2.17 Two-input–one-output dynamic system for Example 6.2.9

$$E(s) = U(s) - G_2 W(s) = U(s) - G_2\Big(V(s) + Y(s)\Big) \quad (2)$$

Substituting this expression (Eq. (2)) into the equation for $Y(s)$ shown in the block diagram (Figure 6.2.17), we obtain

$$Y(s) = G_1 E(s) = G_1 \Big[U(s) - G_2\big(V(s) + Y(s)\big)\Big] \quad (3)$$

which, after rearranging to separate $Y(s)$ yields

$$Y(s) = \frac{G_1}{1 + G_1 G_2} U(s) - \frac{G_1 G_2}{1 + G_1 G_2} V(s) \quad (4)$$

From Eq. (4) the sought transfer functions are clearly seen:

$$G_U(s) = \frac{G_1}{1 + G_1 G_2}$$

and

$$G_V(s) = -\frac{G_1 G_2}{1 + G_1 G_2}$$

Setting the inputs to zero in turn and simplifying the block diagram accordingly yields this solution with less work.

When $V(s) = 0$, the block diagram becomes a standard negative feedback connection of two blocks G_1 and G_2 (see Figure 6.2.18(a)). Thus, the equivalent transfer function is found as

$$G_U(s) = \frac{G_1}{1 + G_1 G_2}$$

When $U(s) = 0$, the block diagram becomes a positive unity feedback configuration, where the main line contains blocks G_1 and G_2 connected in series. In the original block diagram, the signal originating from the summation point B is transferred to the input of the block G_1 with a negative sign. That sign needs to be preserved in the equivalent transformation. Hence, as seen in Figure 6.2.18(b), the transfer function of the block G_2 assumes negative sign. The equivalent transfer function is then found as

$$G_V(s) = -\frac{G_1 G_2}{1 + G_1 G_2}$$

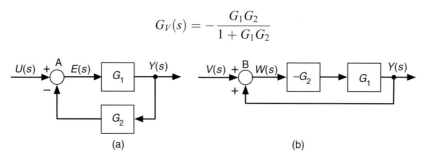

Figure 6.2.18 Derivation of transfer functions: (a) $G_U(s)$ and (b) $G_V(s)$

Example 6.2.10
Revisit the two-input two-output dynamic system described in Example 6.2.5, construct its block diagram, and derive the system transfer matrix using the block diagram.

Solution
The system equations in the Laplace domain were derived as

$$\{sX(s) + a_1 Z(s) = U_1(s) \tag{1}$$

$$\{(s^2 + a_2 s)Z(s) + a_3 X(s) = U_2(s) \tag{2}$$

A block diagram for a system described by a set of equations is constructed by parts, every part corresponding to a single equation. The final step encompasses connecting the individual parts together.

The block diagram for Eq. (1) is shown in Figure 6.2.19(a), and that for Eq. (2) is in Figure 6.2.19(b).

Connecting these partial diagrams, we obtain the system block diagram as illustrated in Figure 6.2.20.

Setting the inputs to zero in turn, we derive the component transfer functions. When $U_1(s) = 0$, the block diagram is in negative feedback configuration, having on the main line a single transfer function $\frac{1}{s^2 + a_2 s}$, and three blocks in series on the feedback line, as shown in Figure 6.2.21(a). Simplifying the diagram by substituting the series connection with a single block

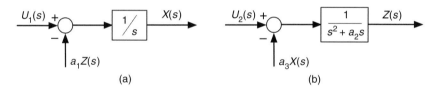

Figure 6.2.19 Component block diagrams for the equations in Example 6.2.10: (a) Eq. (1), (b) Eq. (2)

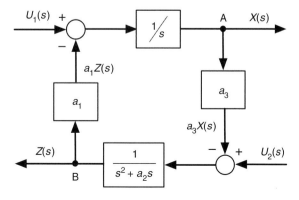

Figure 6.2.20 System block diagrams

$$G_{fb} = \frac{a_1 a_3}{s^3 + a_2 s^2}$$

and moving the branch point B toward the branch point A, we obtain the block diagram shown in Figure 6.2.21(b).

Setting $U_2(s) = 0$ yields the block diagram shown in Figure 6.2.22(a), which is then similarly processed to obtain the block diagram in Figure 6.2.22(b).

Using the equivalent transformations presented in Table 6.2.1, we obtain the sought transfer functions:

$$G_{1X} = \frac{X(s)}{U_1(s)} = \frac{s^2 + a_2 s}{s^3 + a_2 s^2 - a_1 a_3}$$

$$G_{1Z} = \frac{Z(s)}{U_1(s)} = -\frac{a_3}{s^3 + a_2 s^2 - a_1 a_3}$$

$$G_{2X} = \frac{X(s)}{U_2(s)} = -\frac{a_1}{s^3 + a_2 s^2 - a_1 a_3}$$

$$G_{2Z} = \frac{Z(s)}{U_2(s)} = \frac{s}{s^3 + a_2 s^2 - a_1 a_3}$$

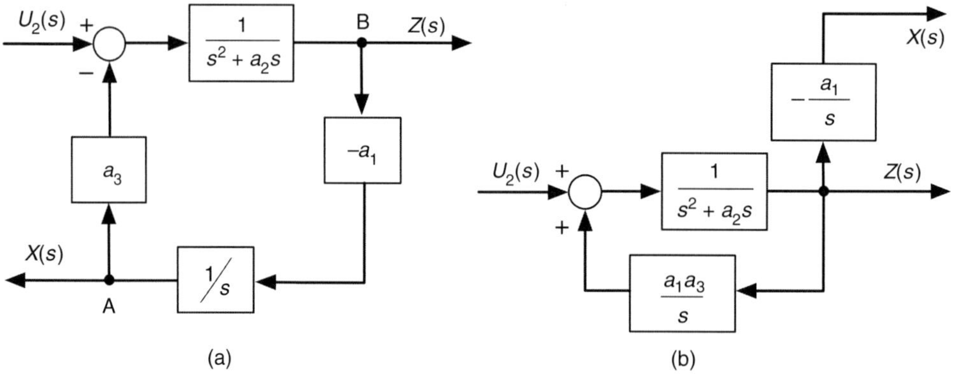

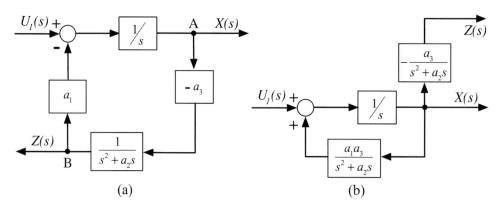

Figure 6.2.21 Equivalent transformations when $U_1(s) = 0$ (see text for details)

Figure 6.2.22 Equivalent transformations when $U_2(s) = 0$ (see text for details)

6.3 Fundamentals of Electromechanical Systems

An electromechanical system, as its name suggests, consists of two coupled subsystems: electrical and mechanical. The structure of such a system is illustrated in Figure 6.3.1.

An electrical subsystem is typically represented by an active or passive circuit, while a mechanical subsystem may contain all three types of basic elements discussed in Chapter 3: inertia, elastic, and damping components (masses, springs, and dampers).

As seen in Figure 6.3.1, energy conversion takes place in any electromechanical system. Either the electrical subsystem generates a force to move the inertial elements of the mechanical subsystem, or the motion of these masses generates current or voltage to feed the electrical subsystem. The former case is exemplified by a *DC motor*, where an applied voltage serves as an input and the parameters describing the rotation of the motor, such as motor torque and angular velocity, are system outputs. The latter case is observed in electroacoustic devices such as a microphone, where the motion of an elastic membrane subject to sound waves generates a voltage for the electrical circuit.

6.3.1 Basic Concepts of Electromagnetism

Typically, coupling between mechanical and electrical subsystems is achieved via electromagnetic effects. Hence, the two fundamental principles of electromagnetism need to be reviewed prior to an exploration of the approaches to modeling of the electromechanical systems. The first principle deals with the force exerted by a magnetic field on the current-carrying wire placed inside this field. The second one concerns the voltage induced in a conductor moving in a magnetic field.

It was found experimentally that when a charged particle moves in a magnetic field, characterized by the flux density $\vec{B}$ (also called *magnetic field strength*), this field exerts a magnetic force $\vec{f}$ on that charge. This force depends on the charge q of the particle, its velocity $\vec{v}$, and flux density of the field as follows:

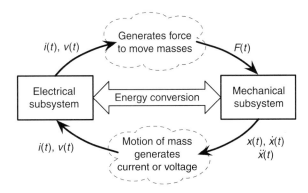

Figure 6.3.1 Typical electromechanical system

$$\vec{f} = q(\vec{v} \times \vec{B}) \tag{6.3.1}$$

Since the electromagnetic force is a vector product of a particle's velocity and field flux density, it is perpendicular to the plane spanned by the vectors $\vec{B}$ and $\vec{v}$, and its direction is found in accordance with the *right-hand rule* (see Section 2.1.4). Note that the right-hand rule applies to a positive charge; thus, for a moving negative charge, the velocity vector needs to be reversed prior to using this rule.

According to the definition of a vector product, the magnitude of this magnetic force is

$$f = qvB \sin \theta \tag{6.3.2}$$

where θ is the angle between the vectors $\vec{B}$ and $\vec{v}$; $0° \leq \theta \leq 180°$.

Obviously, for the magnetic force to be non-zero, the charged particle must cross the magnetic field lines while moving, i.e. vectors $\vec{B}$ and $\vec{v}$ are not parallel ($\theta \neq 0°$ and $\theta \neq 180°$).

Equation (6.3.1) is derived from the *Lorentz force law*:

$$\vec{F} = q\vec{E} + q(\vec{v} \times \vec{B}) \tag{6.3.3}$$

where $\vec{F}$ is the total force exerted by an electromagnetic field of electric strength $\vec{E}$ and magnetic flux density $\vec{B}$ on a charged particle q; $q\vec{E}$ is the electric force.

The magnetic force exerted by a magnetic field on a current-carrying wire is derived from Eq. (6.3.1) by considering the constrained motion of a charge element dq in a wire. The velocity of this charge element is

$$\vec{v} = \frac{d\vec{l}}{dt} \tag{6.3.4}$$

where $d\vec{l}$ is an element of the length of the wire. The direction of this vector is the same as the direction of the electrical current in the wire.

Then, the magnetic force acting on this charge element is

$$d\vec{f} = dq(\vec{v} \times \vec{B}) = dq \left(\frac{d\vec{l}}{dt} \times \vec{B} \right) = \frac{dq}{dt}(d\vec{l} \times \vec{B}) \tag{6.3.5}$$

Recalling the definition of electrical current (Eq. (4.1.1)), the above expression for the magnetic force is rewritten as

$$d\vec{f} = i(d\vec{l} \times \vec{B}) \tag{6.3.6}$$

The entity $i\,d\vec{l}$ is often called the *current element*, which is a vector with the magnitude i and the direction of the electrical current. Integrating the current element for the finite length of the wire ($0 \leq l \leq L$) yields the vector $\vec{Li}$. Obviously, its direction is the direction of the current in the wire, and its magnitude equals Li.

To obtain the magnetic force acting on the wire of length L, Eq. (6.3.6) is integrated for $0 \leq l \leq L$ and becomes

$$\vec{f} = L\vec{i} \times \vec{B} \tag{6.3.7}$$

The direction of the magnetic force is obtained using the right-hand rule as illustrated in Figure 6.3.2, Since the conventional current represents motion of positively charged particles, the application of the right-hand rule is straightforward: when the index finger points in the direction of current $\vec{i}$, and middle finger points in the direction of the magnetic field flux density $\vec{B}$, the thumb will indicate the direction of the exerted magnetic force $\vec{f}$.

The magnitude of the force, exerted by the magnetic field on the current-carrying wire, is

$$f = LiB \sin \theta \tag{6.3.8}$$

where θ is the angle between the vectors $\vec{B}$ and $\vec{i}$; $0° \leq \theta \leq 180°$.

Similarly to the case depicted in Figure 6.3.2, for the systems discussed further in this chapter we assume that the flux density is uniform, the wire is straight, and that the flux density and current vectors are perpendicular. Hence, the expression for the magnetic force becomes

$$f = LBi \tag{6.3.9}$$

Equation (6.3.9) also applies to the circular wires in a radial magnetic field.

In the SI measurement system the magnetic flux density is measured in *webers per square meter* (Wb/m²), while wire length is in *meters* (m) and current is in *amperes* (A). The unit of magnetic flux is also called the *tesla* (T): $T = Wb/m^2 = N/A * m$

A special case of interest represents a current-carrying rectangular wire loop in a uniform magnetic field, as shown in Figure 6.3.3. The axis of the rectangular wire loop $N_1 N_2 N_3 N_4$ is

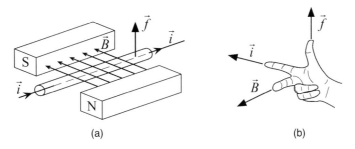

Figure 6.3.2 Right-hand rule: (a) a current-carrying wire in the field of a permanent magnet with the polarity shown; (b) using the right-hand rule to find the direction of the exerted magnetic force

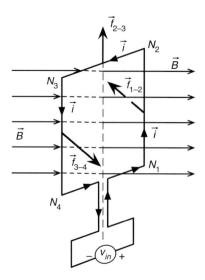

Figure 6.3.3 Current-carrying wire loop in a uniform magnetic field

perpendicular to the flux density vector $\vec{B}$ of the uniform magnetic field. The wire loop carries an electrical current $\vec{i}$ of the direction as shown. Let θ be the angle between $\vec{B}$ and the plane of the loop. Also, let the lengths of the loop rectangle be $|N_1N_2| = |N_3N_4| = L$ and $|N_2N_3| = |N_4N_1| = b$.

This loop experiences a torque, computed in the following way.

The magnetic forces, exerted by the field on the sides of the wire loop are

$$\text{Side } N_2N_3 \quad \vec{f}_{2-3} = b\vec{i}_{2-3} \times \vec{B}$$
$$\text{Side } N_4N_1 \quad \vec{f}_{4-1} = b\vec{i}_{4-1} \times \vec{B} \qquad (6.3.10)$$

The force $\vec{f}_{2-3}$ is shown, while the force $\vec{f}_{4-1}$ is omitted for the clarity of the illustration. These forces have the same magnitude – $|\vec{f}_{2-3}| = |\vec{f}_{4-1}| = biB\sin\theta$ – but opposite directions according to the right-hand rule. Additionally, these force act along the same line – the loop axis. Thus, they cancel each other and do not generate a torque.

For the forces acting on two other sides of the loop the situation is different.

$$\text{Side } N_1N_2 \quad \vec{f}_{1-2} = L\vec{i}_{1-2} \times \vec{B}$$
$$\text{Side } N_3N_4 \quad \vec{f}_{3-4} = L\vec{i}_{3-4} \times \vec{B} \qquad (6.3.11)$$

Since the sides N_1N_2 and N_3N_4 of the loop are parallel to the loop axis, and the axis is perpendicular to the magnetic field flux density, the vectors $\vec{i}_{1-2}$ and $\vec{i}_{3-4}$ are perpendicular to $\vec{B}$. Thus, the forces $\vec{f}_{1-2}$ and $\vec{f}_{3-4}$ have the same magnitude: $|\vec{f}_{1-2}| = |\vec{f}_{3-4}| = LBi$, and opposite directions (as shown in Figure 6.3.3). Since these forces act along different lines, they produce a torque about the loop axis, making the wire loop rotate around its axis. This torque is derived as

$$T = f_{1-2}\frac{b}{2} + f_{3-4}\frac{b}{2} = bLBi \qquad (6.3.12)$$

When a conductor moves in a magnetic field in a certain manner, the field induces a voltage in the conductor. This voltage is sometimes called the *electromotive force (emf)*. The phenomenon of voltage generation (*electromagnetic induction*) is expressed by a fundamental law of electromagnetism – *Faraday's law of induction*. It states that any change in the magnetic environment of a conductor will cause an emf to be induced in that conductor. The induced emf is proportional to the negative rate of change of the magnetic flux.

A change in a magnetic environment can be produced in many different ways, such as, for example, changing the position of a conductor in relation to the magnetic field, or varying the strength of the field. The systems discussed further in this chapter employ conductors moving in the uniform magnetic field ($\vec{B} = const$).

Consider a conductor of length L moving perpendicularly to the magnetic flux lines of a uniform field of a permanent magnet. Vectors of the magnetic field flux density $\vec{B}$ and velocity of the conductor $\vec{v}$ are as indicated in Figure 6.3.4. According to *Lorentz's force law*, a magnetic field exerts a force on any charged particle moving inside that field. Under the influence of this magnetic force (Eq. (6.3.1)) the negatively charged particles inside the conductor move toward the e^- end of the conductor, while the positive charges congregate at the opposite, e^+, end. Movement of the charged particles under the influence of the magnetic force constitutes the electric current $\vec{i}$ inside the conductor. The resulting potential difference between the two ends of the conductor is the induced emf. The direction of the induced current $\vec{i}$ is determined by Lenz's law, which states that the magnetic field, corresponding to the induced current, opposes the change in the magnetic flux that induced the current.

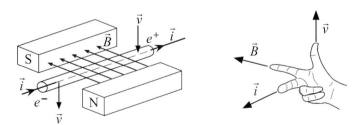

Figure 6.3.4 Direction of the induced current according to Lenz's law

Since the voltage is defined as the work needed to move a unit charge between two points (Eq. (4.1.3)), the induced emf ε_b is the work performed by the magnetic force moving a unit charge along the length of the conductor:

$$\varepsilon_b = \frac{fL}{q} = \frac{qvBL}{q} = vBL \qquad (6.3.13)$$

The direction of the induced current is determined by the right-hand rule as shown in Figure 6.3.4: when index finger points in the direction of the magnetic field flux density $\vec{B}$, and the thumb points in the direction of the conductor's velocity $\vec{v}$, the middle finger will indicate the direction of the induced current $\vec{i}$.

To understand how the electromagnetic laws described by Eqs. (6.3.9) and (6.3.13) are used to express the coupling between mechanical and electrical subsystems of a compound dynamic system, consider the following example. Let the conductor depicted in Figure 6.3.4 have a non-negligible mass m. According to Eq. (4.1.6) the electrical power due to the induced emf ε_b and its corresponding electrical current $\vec{i}$ is

$$P_E = \varepsilon_b i \qquad (6.3.14)$$

Substituting the expression for the induced emf (Eq. (6.3.13)) into Eq. (6.3.14), we obtain

$$P_E = vBLi \qquad (6.3.15)$$

The power applied to the mass m, making it move, is

$$P_M = fv = BLiv \qquad (6.3.16)$$

where f is the magnetic force, described by Eq. (6.3.9).

Assuming that the energy losses due to the conductor resistance and friction are negligible, the power acting on the mass equals the generated electrical power:

$$P_E = P_M \quad \text{or} \quad \varepsilon_b i = fv \qquad (6.3.17)$$

For the known representation of the magnetic force (Eq. (6.3.9)), Eq. (6.3.17) derives the expression for the induced emf (Eq. (6.3.13)).

Also, according to Newton second law, the relation between electrical and mechanical parameters of this dynamic system is

$$ma = f = BLi \qquad (6.3.18)$$

where $a = dv/dt$ is the acceleration of the mass m.

6.3.2 D'Arsonval Meter

The simplest electromechanical system is the *D'Arsonval meter* or *galvanometer*, illustrated in Figure 6.3.5. In the figure, the numbers correspond to the following galvanometer elements:

(1) the moving coil of wire of the size $l \times b$ as shown, and consisting of n loops (often called the *armature*); (2) the assembly of a torsional spring (hairspring) and jewel bearing; (3) the pointer, (4) the scale; (5) the horseshoe magnet with the polarity as indicated; and (6) the soft-iron armature core.

A galvanometer is a current-measuring instrument used in popular analog devices such as ammeters, voltmeters, and ohmmeters. While a detailed discussion on these devices belongs to the field of electrical engineering, understanding of *D'Arsonval movement* or, as it is sometimes called, *permanent-magnet moving-coil movement* is crucial for comprehending DC motor operation.

D'Arsonval movement is based on a stationary permanent magnet and a movable coil of wire that is connected to a source of current to be measured. The coil (armature) is wound around an iron core, forming an inductor that is positioned within a magnetic field, created by the permanent magnet. The iron core of the inductor concentrates the magnetic field, thus strengthening its effect on the current-carrying armature.

Connecting the coil to a voltage source v_{in} closes the coil circuit and produces a current $\vec{i}$ that needs to be measured. The magnetic field of the flux density $\vec{B}$ exerts a torque on the current-carrying coil, as discussed earlier in this section (see Figure 6.3.3). For a single wire loop, the generated torque is expressed by Eq. (6.3.12), while for an armature consisting of n loops it becomes

$$T = nblBi \qquad (6.3.19)$$

where, as shown in Figure 6.3.5, i is the current to be measured, B is the flux density of the magnetic field, b and l are the dimensions of a single loop, and n is the number of loops in the coil.

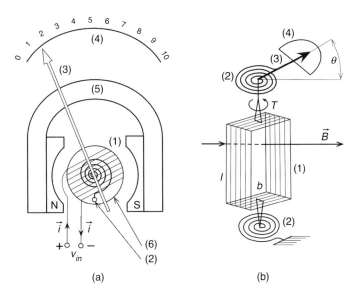

Figure 6.3.5 D'Arsonval meter: (a) construction of the device; (b) moving coil component (see text for further details)

The armature is connected to a pointer, extended out to a scale. Since the angular displacement of the pointer is related to the generated torque, and, in turn, to the amount of current through the armature, the scale is marked to show this applied current. Hairsprings and bearings resist the movement of the coil, and return it to its neutral (initial) position when there is no current. Additionally, when the generated torque is balanced by the restoring forces of the spring-bearing assembly, the pointer stops moving and the reading can be obtained.

Let the stiffness and damping coefficients of the spring-bearing assembly be k_T and c_T, respectively. Let also the inertia of the inductor (core and coil assembly) be I, and the angular displacement of the pointer be θ. Then, the equation of rotational motion of the inductor is

$$I\frac{d^2\theta}{dt^2} = -c_T \frac{d\theta}{dt} - k_T\theta + T \quad (6.3.20)$$

Substituting the expression for the generated torque (Eq. (6.3.19)) into Eq. (6.3.20), we obtain the relation between the measured current and the angular displacement of the pointer:

$$I\frac{d^2\theta}{dt^2} + c_T \frac{d\theta}{dt} + k_T\theta = nblBi \quad (6.3.21)$$

As discussed earlier in this section, when a conductor (the armature) rotates in the magnetic field, an emf is induced, proportional to the conductor's linear velocity and length (see Eq. (6.3.13)). Recalling the relation between linear and angular velocities derived in Chapter 3, the linear velocity of the armature is

$$v = \frac{b}{2}\frac{d\theta}{dt} \quad (6.3.22)$$

The dimension of the armature that is perpendicular to the magnetic flux density vector is $2nl$. Hence, the induced emf is

$$\varepsilon_b = vB(2l) = nblB\frac{d\theta}{dt} \quad (6.3.23)$$

Let the inductance and resistance of the armature solenoid be L and R, respectively. Then, remembering that the induced emf has a polarity that opposes the change in magnetic flux responsible for the induced voltage, the armature circuit can be represented as shown in Figure 6.3.6. Applying Kirchhoff's voltage law to this circuit, we obtain

$$v_{in} - \varepsilon_b - Ri - L\frac{di}{dt} = 0 \quad (6.3.24)$$

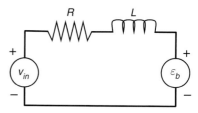

Figure 6.3.6 Electrical subsystem of a galvanometer

Substituting the derived expression for the induced emf (Eq. (6.3.23)), we obtain the relationship between the voltage input and the current to be measured:

$$L\frac{di}{dt} + Ri + nblB\frac{d\theta}{dt} = v_{in} \quad (6.3.25)$$

Hence, the model of a galvanometer consists of two equations – Eqs. (6.3.21) and (6.3.25). When the resistive forces due to the spring–damper assembly and the induced emf become equal to the generated torque, the pointer stops moving. That means that $\frac{di}{dt} = 0$, $\frac{d^2\theta}{dt^2} = 0$, and $\frac{d\theta}{dt} = 0$, which transforms Eqs. (6.3.21) and (6.3.25) into

$$\begin{aligned} Ri &= v_{in} \Rightarrow i = \frac{v_{in}}{R} \\ k_T\theta &= nblBi \end{aligned} \quad (6.3.26)$$

Hence, the pointer displacement can be related to the measured current as

$$\theta = \frac{nblB}{k_T} i \quad (6.3.27)$$

and to the input voltage as

$$\theta = \frac{nblB}{Rk_T} v_{in} \quad (6.3.28)$$

Equations (6.3.27) and (6.3.28) yield the important parameters of a galvanometer: the current sensitivity $\frac{nblB}{k_T}$ – the deflection per unit current, and the voltage sensitivity $\frac{nblB}{Rk_T}$ – the deflection per unit voltage.

The galvanometer is a reliable, sensitive, and accurate measuring device, its accuracy typically being approximately 2–5% of its full-scale deflection. In all three types of analog devices – voltmeters, ammeters, and ohmmeters – the galvanometer element reacts to the amount of current passing through its armature. The principal difference between these meters lies in the manner in which the galvanometer is connected in the device's electrical circuit.

6.4 Direct Current Motors

6.4.1 Fundamentals of DC-Motor Operations

The purpose of any electric motor, whether of DC or alternating current (AC) type, is to convert electric energy into mechanical movement represented by the rotation of the motor's shaft. A DC motor is often called a "mechanical workhorse" since a wide variety of mechanical devices depend on it for the power to move. Easy controllability of the DC motor rotation speed, high torque capabilities, sturdy construction, and relatively low cost make it a ubiquitous component of a wide variety of machines, for example, cranes and winches, cars, and even missile launchers.

The operation of a DC motor is based on D'Arsonval movement, as discussed in the previous section. The construction of a brushed DC motor is illustrated in Figure 6.4.1, in which the numbers denote the following elements: (1) permanent magnets (stator) bonded to the steel enclosure, (2) armature winding, (3) the rotor, (4) the commutator, (5) the brushes, (6) the shaft, (7) the support bearing, and (8) the steel enclosure.

As discussed earlier, for D'Arsonval movement two major components are needed: a stationary magnet and a current-carrying movable inductor, which is a coil of wire wound around an iron core.

The inductor component of a DC motor is an *armature*, which consists of the multiple coils of wire (*armature winding*) attached to a *rotor* – an iron core that is mounted on a *shaft* supported by *bearings* (see Figure 6.4.1). The rotating armature generates the sought mechanical power.

The stationary magnetic field around the inductor (armature) is provided by a *stator* that is either a permanent magnet (see Figure 6.4.1), or an electromagnet. The stator is permanently attached to a steel enclosure.

Recalling the discussion on a single current-carrying wire loop rotating in a uniform magnetic field (see Figure 6.3.3), it can be easily seen that as soon as this coil rotates 180°, the direction of the generated torque will be reversed as illustrated in Figure 6.4.2.

In Figure 6.4.2 $\vec{f}_{1-2}$ and $\vec{f}_{3-4}$ denote the forces exerted by the magnetic field $\vec{B}$ on the $N_1 N_2$ and $N_3 N_4$ sides of the coil, respectively, and T indicates the resulting torque.

In general, this torque reversal is unacceptable; hence, some measures need to be devised to provide continuous rotation. Also, permanent electrical contact between the rotating

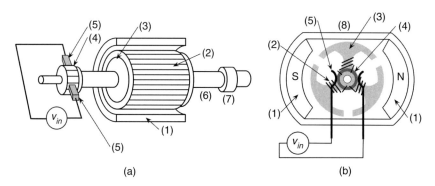

Figure 6.4.1 DC motor: description and operation: (a) cutaway view, (b) cross-section (see text for further details)

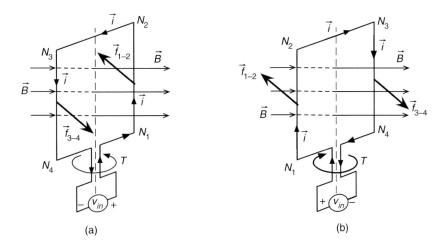

Figure 6.4.2 Current-carrying wire loop rotating in a uniform magnetic field: (a) initial position of a coil and (b) position after the coil has rotated 180°

armature and the motor's power supply must be ensured for the duration of motor operation. For these purposes, *commutator* and *brush* components are devised. A commutator is a segmented sleeve, typically made of copper, mounted on the rotating shaft of a DC motor. The current is supplied to the commutator segments via brushes – two spring-loaded carbon sticks. As the motor's shaft turns, the commutator turns with it and the brushes slide over the commutator segments, supplying electrical energy to the armature via the commutator.

The simplest commutator consists of only two segments. It reverses the direction of the current flowing through the armature every half-rotation (180°), thus providing for a continuous rotation in one direction. While perfectly workable, this type of motor construction is hardly practical due to the inherent problems of a zero-torque position and brush short-circuiting.

The rotation of a single wire loop connected to a two-segment commutator with brushes inside the magnetic field of a permanent magnet with indicated polarity is shown in Figure 6.4.3. In the figure, the label (1) denotes the two-segment commutator and (2) the brushes; b and l are the dimensions of the coil, $\vec{B}$ is the magnetic flux density, $\vec{f}$ is the exerted magnetic force, α is the rotation angle, $\vec{n}$ is the normal to the coil surface, and θ is the angle between the normal and the magnetic flux density.

Recalling the expressions for the exerted magnetic force (Eq. (6.3.11)), the torque is computed as follows:

$$T = 2f\frac{b}{2} = (lBi)b \sin\theta \tag{6.4.1}$$

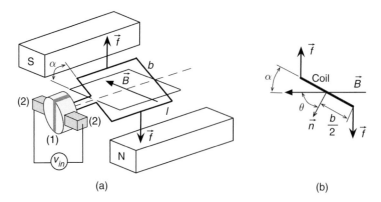

Figure 6.4.3 Rotation of a single wire loop connected to a two-segment commutator: (a) rotating coil; (b) geometry related to the computation of the torque (see text for further details)

Since the area of this coil is $A = bl$, the expression for the generated torque can be rewritten as

$$T = BiA \sin \theta \qquad (6.4.2)$$

From Eq. (6.4.2) it is obvious that the maximum torque is generated when the normal to the coil is perpendicular to the vector of magnetic flux density ($T_{max} = Bi\,A$), and the minimum torque ($T_{min} = 0$) occurs when this normal is parallel to $\vec{B}$. If such a zero-torque position of the motor's armature occurs at powering up, the motor will not be able to start – since no torque is generated, the rotor will stay motionless despite the applied voltage. It is less detrimental when the rotation is already in progress since the rotor's angular momentum will help it pass the dangerous position without stopping.

Commutator and brush positions at maximum and minimum torque generation are illustrated in Figure 6.4.4.

Another weakness of a DC motor with the two-segment commutator is the brush short-circuiting. At a zero-torque position, as shown in Figures 6.4.4(b) and (d), the brushes touch both commutator segments simultaneously, causing a short circuit. Besides wasting energy, these short circuits (one per brush) may result in commutator and brush damage due to overheating and the sparks that result from the short-circuiting, as well as the armature damage.

To avoid these torque ripples and short circuits, modern DC motors use multi-segment commutators if mechanical commutation is desired, or a brushless construction with electronic commutation. Spatial distribution of multiple loops of the armature winding is also helpful in alleviating the mentioned weaknesses.

6.4.2 DC-Motor Modeling

As stated above, there are two ways to manage the performance of a DC motor: either by varying the electrical current through the armature assembly, or by changing the magnetic field flux of the stator. Thus, a DC motor can be either *armature-controlled* or *field-controlled*.

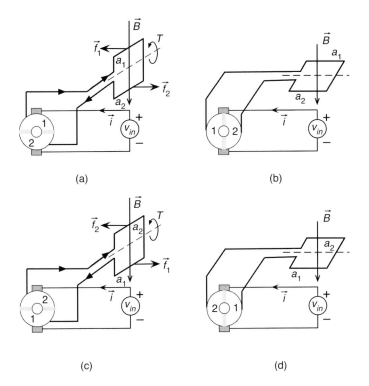

Figure 6.4.4 Rotation of a single wire loop connected to a two-segment commutator – operation: (a) maximum-torque position, $\theta = 90°$; (b) zero-torque position, brush short-circuiting, $\theta = 180°$ (counter-clockwise rotation direction is assumed); (c) maximum-torque position, reversal of current direction, $\theta = 270°$; and (d) zero-torque position, brush short-circuiting, $\theta = 0°$

The electrical subsystem of a typical DC motor, as shown in Figure 6.4.5, is represented by two circuits: an *armature circuit*, responsible for generating an electrical current through the armature assembly, and a *field circuit*, responsible for providing a magnetic field of the stator. Two power supplies are thus required – one for the armature circuit, and one for the field circuit.

For an armature-controlled motor, the field circuit may be substituted with a permanent magnet, leaving this system with a single voltage input. Obviously, for a field-controlled motor, both circuits must be present.

The mechanical subsystem of a DC motor is represented by the inertial element J, which embodies the moment of inertia of the armature about the rotation axis, and damping element c_b that represents the bearings supporting the shaft of the rotating armature (see Figure 6.4.1).

A torque, generated by a DC motor, is proportional to a current in either the armature or field circuit, depending on the employed control method. The relations between the torque and these currents are summarized in Table 6.4.1, where T refers to the produced torque, i_f is

Table 6.4.1 Torque–current relations of a DC motor

Armature-controlled DC motor	Field-controlled DC motor
$i_f = const$	$i_a = const$
$T = Ki_a$	$T = Ki_f$

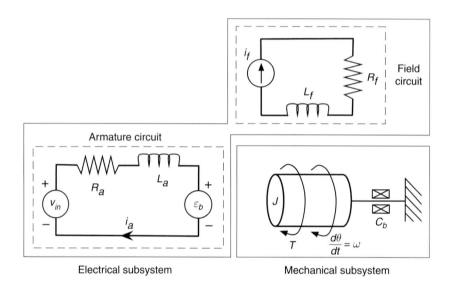

Figure 6.4.5 DC motor: electrical and mechanical subsystems

the current through the field circuit, i_a is the current through the armature circuit, and K denotes the motor *torque constant*.

The expressions for the torque shown in Table 6.4.1 are the coupling relations since they describe the interaction between the electrical and mechanical subsystems of a DC motor. It is clearly seen that these coupling expressions contain the variables of both electrical and mechanical subsystems.

Armature-Controlled DC Motor

Consider an armature-controlled DC motor, where the stator is a permanent magnet, and v_{in} is the input voltage into the armature circuit. In the context of this control scheme, the magnetic flux of the stator field is constant, and the generated torque is regulated by modifying the current through the armature circuit.

Recalling the procedure for modeling the combined systems described in the Section 6.1 we can derive the governing expressions separately for each subsystem.

For the mechanical subsystem, the free-body diagram of which is shown in Figure 6.4.6, the derivation is straightforward.

Let θ be the rotation angle of the armature and T_b the motion-resisting torque generated in the shaft bearings in response to the angular velocity across the bearings. Then, according to the Newton's second law, for a rotational system the equation of motion is

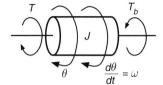

Figure 6.4.6 DC motor: mechanical subsystem

$$J\ddot{\theta} = T - T_b \qquad (6.4.3)$$

Recalling the expression for the torque exerted by a torsional damper:

$$T_b = c_b \dot{\theta} \qquad (6.4.4)$$

the governing equation for the mechanical subsystem is obtained, in terms of angular displacement (rotation angle):

$$J\ddot{\theta} + c_b \dot{\theta} = T \qquad (6.4.5)$$

or in terms of angular velocity $\omega = \dfrac{d\theta}{dt}$:

$$J\dot{\omega} + c_b \omega = T \qquad (6.4.6)$$

To derive the governing equations for the electrical subsystem shown in Figure 6.4.7, recall the discussion about D'Arsonval movement (see Section 6.3.2) and the concept of the induced emf (or *back emf*, as it is often called in the context of a DC motor).

Let v_{in} be the voltage input into the armature circuit, and ε_b be the back emf, or voltage generated in the current-carrying armature in response to the rotation of the armature assembly in the magnetic field of the stator; R_a and L_a denote the resistance and inductance of the circuit elements, respectively, while i_a indicates the current in the armature circuit.

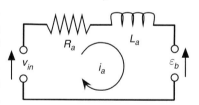

Figure 6.4.7 DC motor: electrical subsystem

As previously discussed, the polarity of the generated back emf is such that it opposes the change in the magnetic flux responsible for the induced voltage. Applying Kirchhoff's voltage law to the armature circuit, we obtain

$$v_{in} - \varepsilon_b - R_a i_a - L_a \frac{di_a}{dt} = 0 \qquad (6.4.7)$$

As derived earlier (see Eq. (6.3.23)), the back emf is

$$\varepsilon_b = nblB\frac{d\theta}{dt} = K_b\frac{d\theta}{dt} = K_b\omega \tag{6.4.8}$$

Equation (6.4.7) is the governing equation for the electrical subsystem, and Eq. (6.4.8) is one of the coupling equations since it expresses the relation between the electrical parameter ε_b and the mechanical parameter $\omega = \frac{d\theta}{dt}$. The second coupling equation is the one that defines the relation between the produced torque T and the current in the armature circuit i_a shown in Table 6.4.1:

$$T = Ki_a \tag{6.4.9}$$

From Eq. (6.3.19) the value of the motor's torque constant is found as

$$K = nblB = K_b \tag{6.4.10}$$

Obviously, the motor's torque constant and its back emf constant are identical. Nonetheless, for further derivations we will use K and K_b so that the influence of each of these constants on the motor performance is more clearly demonstrated.

Thus, the model of an armature-controlled DC motor consists of the following four equations:

$$\begin{cases} J\ddot{\theta} + c_b\dot{\theta} = T \\ v_{in} - \varepsilon_b = R_a i_a + L_a\frac{di_a}{dt} \\ \varepsilon_b = K_b\dot{\theta} \\ T = Ki_a \end{cases} \tag{6.4.11}$$

or in terms of angular velocity:

$$\begin{cases} J\dot{\omega} + c_b\omega = T \\ v_{in} - \varepsilon_b = R_a i_a + L_a\frac{di_a}{dt} \\ \varepsilon_b = K_b\omega \\ T = Ki_a \end{cases} \tag{6.4.12}$$

This system can be further modeled by any of the methods discussed earlier, such as transfer function formulations, block diagrams, and state-space representations.

To derive the transfer function formulation and subsequently construct the block diagram, take the Laplace transform of Eq. (6.4.11), assuming zero initial conditions. Let the voltage $v_{in}(t)$ applied to the armature circuit be the system input, and the motor angular velocity $\omega(t) = \frac{d\theta}{dt}$ be the system output. The governing equations in the Laplace domain are

$$\begin{cases} (Js^2 + c_b s)\Theta(s) = T(s) \\ V_{in}(s) - E_b(s) = (R_a + L_a s)I_a(s) \\ E_b(s) = K_b\, s\Theta(s) \\ T(s) = KI_a(s) \end{cases} \tag{6.4.13}$$

6.4 Direct Current Motors

Table 6.4.2 Armature-controlled DC-motor block diagram components

Governing equation	Block diagram component
$(Js^2 + c_b s)\Theta(s) = T(s)$	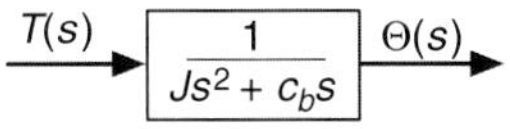 $T(s) \rightarrow \boxed{\dfrac{1}{Js^2 + c_b s}} \rightarrow \Theta(s)$
$V_{in}(s) - E_b(s) = (R_a + L_a s)I_a(s)$	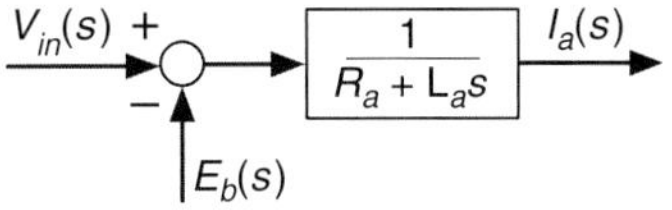 $V_{in}(s) \xrightarrow{+} \bigcirc \rightarrow \boxed{\dfrac{1}{R_a + L_a s}} \rightarrow I_a(s)$, with $E_b(s)$ as negative feedback
$E_b(s) = K_b s\, \Theta(s)$	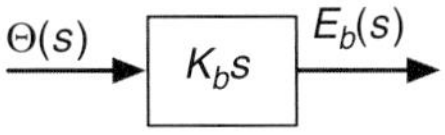 $\Theta(s) \rightarrow \boxed{K_b s} \rightarrow E_b(s)$
$T(s) = K I_a(s)$	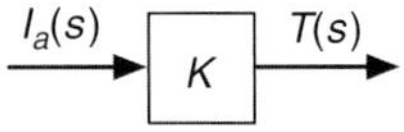 $I_a(s) \rightarrow \boxed{K} \rightarrow T(s)$

$$V_{in}(s) \rightarrow \boxed{G(s) = \frac{K}{(Js + c_b)(R_a + L_a s) + KK_b}} \rightarrow \Omega(s) = s\Theta(s)$$

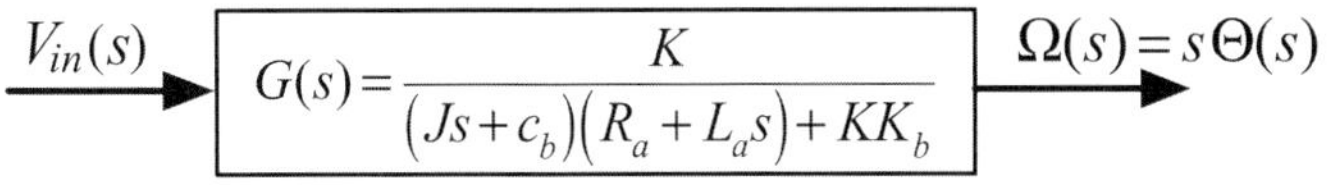

Figure 6.4.8 DC motor: block diagram

The system transfer function then derives as

$$G(s) = \frac{\mathcal{L}[\omega(t)]}{\mathcal{L}[v_{in}(t)]} = \frac{\Omega(s)}{V_{in}(s)} = \frac{s\Theta(s)}{V_{in}(s)} = \frac{K}{(Js + c_b)(R_a + L_a s) + KK_b} \quad (6.4.14)$$

As seen from Eq. (6.4.14), the armature-controlled DC motor is a second-order dynamic system.

Using the derived transfer function (Eq. (6.4.14)), the simplest block diagram for this system is constructed as shown in Figure 6.4.8.

While this block diagram is suitable for implementation in software for simulation and further analysis, its expanded form is more suitable for a deeper understanding of the system behavior and interactions between its subsystems. To construct an expanded block diagram, the components that correspond to each of the four governing equations (Eq. (6.4.13)) are built as shown in Table 6.4.2.

Finally, the expanded block diagram is constructed from the above components as depicted in Figure 6.4.9.

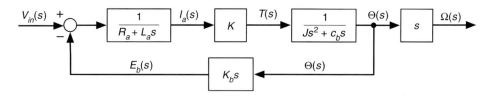

Figure 6.4.9 Armature-controlled DC motor: expanded block diagram

To derive the state-space representation of this dynamic system, consider the governing equations for every individual subsystem since they determine the choice of state variables. Obviously, coupling equations do not introduce any new state variables since they describe the relations between parameters of different subsystems that have already been expressed with the state variables.

The mechanical subsystem is described by Eq. (6.4.5), which is a second-order differential equation in terms of angular displacement (rotation angle) $\theta(t)$. Thus, it needs two state variables: $x_1 = \theta(t)$ and $x_2 = \dot{\theta}(t)$.

The electrical subsystem is described by Eq. (6.4.7), which is a first-order differential equation in terms of the current $i_a(t)$. Hence, it needs a single state variable $x_3 = i_a(t)$. The first two of the selected three state variables

$$\begin{cases} x_1 = \theta(t) \\ x_2 = \dot{\theta}(t) \\ x_3 = i_a(t) \end{cases} \tag{6.4.15}$$

yield the first state equation:

$$\dot{x}_1 = x_2 \tag{6.4.16}$$

Rewriting Eq. (6.4.5) in terms of state variables, we obtain

$$J\dot{x}_2 + c_b x_2 = T \tag{6.4.17}$$

while Eq. (6.4.7) transforms into:

$$v_{in} - \varepsilon_b = R_a x_3 + L_a \dot{x}_3 \tag{6.4.18}$$

Similarly, the coupling relations are expressed. Equation (6.4.9) becomes

$$T = K x_3 \tag{6.4.19}$$

and Eq. (6.4.8) is

$$\varepsilon_b = K_b x_2 \tag{6.4.20}$$

From Eqs. (6.4.17) and (6.4.19) the second state equation is derived as

$$\dot{x}_2 = \frac{1}{J}(-c_b x_2 + T) = -\frac{c_b}{J}x_2 + \frac{K}{J}x_3 \tag{6.4.21}$$

The third state equation is derived from Eqs. (6.4.18) and (6.4.20) as

$$\dot{x}_3 = \frac{1}{L_a}(-\varepsilon_b - R_a x_3 + v_{in}) = -\frac{K_b}{L_a}x_2 - \frac{R_a}{L_a}x_3 + \frac{1}{L_a}v_{in} \tag{6.4.22}$$

The output equation is

$$y = \dot{\theta} = x_2 \tag{6.4.23}$$

Hence, the state-space representation of an armature-controlled DC motor is

$$\text{State equations} \begin{cases} \dot{x}_1 = x_2 \\ \dot{x}_2 = -\dfrac{c_b}{J}x_2 + \dfrac{K}{J}x_3 \\ \dot{x}_3 = -\dfrac{K_b}{L_a}x_2 - \dfrac{R_a}{L_a}x_3 + \dfrac{1}{L_a}v_{in}(t) \end{cases}$$

$$\text{Output equations} \quad y(t) = x_2 \tag{6.4.24}$$

If desired, the derived state-space representation can be cast in the matrix form:

$$\begin{cases} \dot{x}(t) = Ax(t) + Bu(t) \\ y(t) = Cx(t) + Du(t) \end{cases}$$

where matrices are as follows:

$$A_{3\times3} = \begin{bmatrix} 0 & 1 & 0 \\ 0 & -\dfrac{c_b}{J} & \dfrac{K}{J} \\ 0 & -\dfrac{K_b}{L_a} & -\dfrac{R_a}{L_a} \end{bmatrix}, \; B_{3\times1} = \begin{pmatrix} 0 \\ 0 \\ \dfrac{1}{L_a} \end{pmatrix}, \; C_{1\times3} = (0 \; 1 \; 0), \; D = 0$$

$$x(t) = \begin{pmatrix} x_1 \\ x_2 \\ x_3 \end{pmatrix}, \; u(t) = v_{in}(t)$$

Field-Controlled DC Motor

A field-controlled DC motor, shown in Figure 6.4.10, regulates the generated torque by modifying the magnetic field flux of the stator while keeping the current in the armature

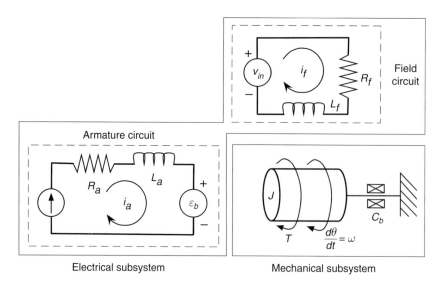

Figure 6.4.10 Field-controlled DC motor: mechanical and electrical subsystems

circuit constant. As depicted, this type of motor has two separate inputs – voltage input v_{in} into the field circuit and current input into the armature circuit.

The mechanical subsystem of a field-controlled DC motor is identical to that of an armature-controlled motor, discussed above. Its governing equation is expressed mathematically by Eqs. (6.4.5) and (6.4.6).

The electrical subsystem consists of two circuits as seen in Figure 6.4.10, where the armature circuit is supplemented by a dedicated control circuit to keep the current i_a constant, counteracting the effect of the back emf ε_b.

Let v_{in} be the voltage input into the field circuit. No back emf is generated since the current-carrying stator assembly does not rotate. R_f and L_f denote the resistance and inductance of the circuit elements, respectively, while i_f indicates the current in the field circuit.

Then, applying Kirchhoff's voltage law to the field circuit, the governing equation for the electrical subsystem is obtained:

$$v_{in} - R_f \, i_f - L_f \frac{di_f}{dt} = 0 \qquad (6.4.25)$$

Typically, the stator's magnetic field flux is a nonlinear function of the current in the field circuit i_f. According to Eq. (6.3.19) the generated torque is

$$T = nblB(i_f)i_a = [nbli_a]B(i_f) \qquad (6.4.26)$$

Since the current in the armature circuit is kept constant, the parameter $[nbli_a]$ is constant, and the torque produced in the rotating armature assembly is also a nonlinear function of the current in the field circuit. For simplicity, an assumption of linear dependence between the torque and current is made, which yields the following coupling relation:

$$T = K i_f \qquad (6.4.27)$$

where K is the motor torque constant.

Hence, there are only three governing equations for a field-controlled DC motor:

$$\begin{cases} J\ddot{\theta} + c_b \dot{\theta} = T \\ v_{in} = R_f\, i_f + L_f \dfrac{di_f}{dt} \\ T = K i_f \end{cases} \qquad (6.4.28)$$

Taking the Laplace transform of Eq. (6.4.28), we obtain

$$\begin{cases} (Js^2 + c_b s)\Theta(s) = T(s) \\ V_{in}(s) = (R_f + L_f s) I_f(s) \\ T(s) = K I_f(s) \end{cases} \qquad (6.4.29)$$

Considering the system input to be the voltage input v_{in} into the field circuit, and the output to be the motor angular velocity $\omega(t) = \dfrac{d\theta}{dt}$, the system transfer function is derived as

$$G(s) = \frac{\mathcal{L}[\omega(t)]}{\mathcal{L}[v_{in}(t)]} = \frac{\Omega(s)}{V_{in}(s)} = \frac{s\Theta(s)}{V_{in}(s)} = \frac{K}{(Js + c_b)(R_f + L_f s)} \qquad (6.4.30)$$

As seen from Eq. (6.4.30), the field-controlled DC motor is also a second-order dynamic system.

The construction of the expanded block diagram for this dynamic system is done using the procedure similar to that described for the armature-controlled motor. The individual components of the sought block diagram are presented in Table 6.4.3, and the complete schematic is shown in Figure 6.4.11.

It is easy to show that the block diagram in Figure 6.4.11 yields the system transfer function expressed by Eq. (6.4.30). Note that unlike an armature-controlled motor a field-controlled one does not have a feedback loop since a back emf is not generated in the field circuit.

The mechanical subsystem of a field-controlled DC motor is a second-order differential equation in terms of angular displacement (rotation angle) $\theta(t)$, while its electrical subsystem is a first-order differential equation in terms of the current $i_f(t)$. Thus, the choice of the three state variables needed is

$$\begin{cases} x_1 = \theta(t) \\ x_2 = \dot{\theta}(t) \\ x_3 = i_f(t) \end{cases} \qquad (6.4.31)$$

Table 6.4.3 Field-controlled DC-motor block diagram components

Governing equation	Block diagram component
$(Js^2 + c_b s)\Theta(s) = T(s)$	$T(s) \rightarrow \boxed{\dfrac{1}{Js^2 + c_b s}} \rightarrow \Theta(s)$
$V_{in}(s) = (R_f + L_f s)I_f(s)$	$V_{in}(s) \rightarrow \boxed{\dfrac{1}{R_f + L_f s}} \rightarrow I_f(s)$
$T(s) = K I_f(s)$	$I_f(s) \rightarrow \boxed{K} \rightarrow T(s)$

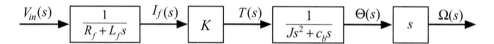

Figure 6.4.11 Field-controlled DC motor: expanded block diagram

Rewriting the governing equations in terms of state variables, we obtain

$$\begin{cases} J\dot{x}_2 + c_b x_2 = T \\ v_{in} = R_f x_3 + L_f \dot{x}_3 \\ T = K x_3 \end{cases} \qquad (6.4.32)$$

Equations (6.4.31) and (6.4.32) together with the output equation (Eq. (6.4.23)) yield the following state-space representation of a field-controlled DC motor:

State equations $\begin{cases} \dot{x}_1 = x_2 \\ \dot{x}_2 = -\dfrac{c_b}{J}x_2 + \dfrac{K}{J}x_3 \\ \dot{x}_3 = -\dfrac{R_f}{L_f}x_3 + \dfrac{1}{L_f}v_{in}(t) \end{cases}$

Output equations $\quad y(t) = x_2 \qquad (6.4.33)$

which may be cast in matrix form

$$\begin{cases} \dot{\boldsymbol{x}}(t) = \boldsymbol{A}\boldsymbol{x}(t) + \boldsymbol{B}\boldsymbol{u}(t) \\ \boldsymbol{y}(t) = \boldsymbol{C}\boldsymbol{x}(t) + \boldsymbol{D}\boldsymbol{u}(t) \end{cases}$$

where matrices are as follows:

$$A_{3\times 3} = \begin{bmatrix} 0 & 1 & 0 \\ 0 & -\dfrac{c_b}{J} & \dfrac{K}{J} \\ 0 & 0 & -\dfrac{R_f}{L_f} \end{bmatrix}, \quad B_{3\times 1} = \begin{pmatrix} 0 \\ 0 \\ \dfrac{1}{L_f} \end{pmatrix}, \quad C_{1\times 3} = (0\ \ 1\ \ 0), \quad D = 0$$

$$x(t) = \begin{pmatrix} x_1 \\ x_2 \\ x_3 \end{pmatrix}, \quad u(t) = v_{in}(t)$$

DC Motor with a Load

In its numerous applications, a DC motor never runs idle. It performs useful work by rotating different mechanical devices such as winches, pumps, or fan blades. Any mechanism that rotates using the torque generated by a DC motor is called *load*. While the interaction between the motor and its load may be quite complex, the most typical interaction is that of a load exerting an additional torque that opposes the motor torque. This simplification applies an assumption that the shaft connecting the motor and the load is rigid, massless, and undamped. In this case, the load and the motor can be lumped together into a single rigid body, the inertia of which is the sum of inertias of the motor and the load. While the motor and the load rotate with the same angular velocity, the load torque acts in the opposite direction to the motor torque.

In some cases, the load torque and motor torque may have the same direction. For example, consider a DC motor used to rotate wheels of a cart. When the cart is moving downhill, the torque generated by gravity makes the wheels descend; thus, the load torque is assisting the motor. If the cart moves uphill, the torque due to gravity is still trying to make cart wheels descend; hence, the load torque resists the motion and opposes the motor.

The presence of a load changes the mechanical subsystem as shown in Figure 6.4.12. As before, θ is the rotation angle of the armature assembly, and T_b is the motion-resisting torque generated in the shaft bearings in response to the angular velocity.

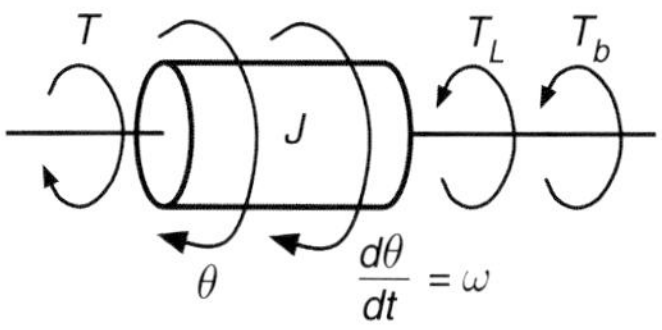

Let T_L denote the load torque. Then, the governing equation for this subsystem becomes

$$J\ddot{\theta} = T - T_b - T_L \qquad (6.4.34)$$

Figure 6.4.12 DC motor with a load: mechanical subsystem

or

$$J\ddot{\theta} + c_b\dot{\theta} = T - T_L \qquad (6.4.35)$$

The governing equation for the electrical subsystem and the coupling relations stay unchanged.

Therefore, the set of equations that describes an armature-controlled motor with a load becomes

$$\begin{cases} J\ddot{\theta} + c_b\dot{\theta} = T - T_L \\ v_{in} - \varepsilon_b = R_a i_a + L_a \dfrac{di_a}{dt} \\ \varepsilon_b = K_b\dot{\theta} \\ T = Ki_a \end{cases} \tag{6.4.36}$$

The load torque depends on the physical characteristics of the load, so in the context of DC-motor performance it is considered an independent input.

The block diagram component that corresponds to the first line of Eq. (6.4.36) is then modified as shown in Figure 6.4.13.

Then, the expanded block diagram of an armature-controlled DC motor with a load is illustrated in Figure 6.4.14.

This block diagram can be reduced as shown in Figure 6.4.15, where two transfer functions:

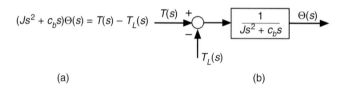

(a)   (b)

Figure 6.4.13 DC motor with a load: (a) equation in the Laplace domain; (b) block diagram component

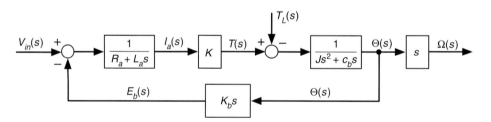

Figure 6.4.14 Armature-controlled DC motor with a load: block diagram

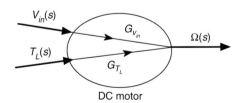

Figure 6.4.15 Armature-controlled DC motor with a load: reduced block diagram

$$G_{V_{in}}(s) = \frac{\Omega(s)}{V_{in}(s)}$$

and

$$G_{T_L}(s) = \frac{\Omega(s)}{T_L(s)}$$

are clearly distinguishable.

From the diagram in Figure 6.4.15, the transfer functions are derived as follows:

$$\left[(V_{in}(s) - K_b\, s\, \Theta(s))\frac{K}{R_a + L_a s} - T_L(s)\right]\frac{s}{Js^2 + c_b s} = \Omega(s)$$

$$V_{in}(s)\frac{K}{(R_a + L_a s)(Js + c_b)} - T_L(s)\frac{1}{Js + c_b} = \left(1 + \frac{KK_b}{R_a + L_a s}\right)\Omega(s)$$

$$G_{V_{in}}(s) = \frac{\Omega(s)}{V_{in}(s)} = \frac{K}{(R_a + L_a s)(Js + c_b) + KK_b}$$

$$G_{T_L}(s) = \frac{\Omega(s)}{T_L(s)} = -\frac{R_a + L_a s}{(R_a + L_a s)(Js + c_b) + KK_b}$$

(6.4.37)

Sometimes the current in the armature circuit is considered to be the second output of this dynamic system (in addition to the angular velocity). Then, the additional two transfer functions are derived as

$$\begin{cases} (V_{in}(s) - K_b s \Theta(s))\dfrac{1}{R_a + L_a s} = I_a(s) \\ (I_a(s)K - T_L(s))\dfrac{s}{Js^2 + c_b s} = \Omega(s) \end{cases}$$

Substituting the value for $\Omega(s) = s\Theta(s)$ of the second equation into the first one, we obtain

$$V_{in}(s) - \frac{I_a(s)K - T_L(s)}{Js + c_b}K_b = (R_a + L_a s)I_a(s)$$

$$V_{in}(s) + \frac{K_b}{Js + c_b}T_L(s) = \frac{(R_a + L_a s)(Js + c_b) + KK_b}{Js + c_b}I_a(s)$$

which derives the sought transfer functions:

$$G_{V_{in}I_a}(s) = \frac{I_a(s)}{V_{in}(s)} = \frac{Js + c_b}{(R_a + L_a s)(Js + c_b) + KK_b}$$

$$G_{T_L I_a}(s) = \frac{I_a(s)}{T_L(s)} = \frac{K_b}{(R_a + L_a s)(Js + c_b) + KK_b}$$

(6.4.38)

The selection of state variables for the case of lumped motor–load inertias is not influenced by the addition of the load. The state-space representation of this dynamic system is obtained as

State equations
$$\begin{cases} \dot{x}_1 = x_2 \\ \dot{x}_2 = -\dfrac{c_b}{J}x_2 + \dfrac{K}{J}x_3 - \dfrac{1}{J}T_L(t) \\ \dot{x}_3 = -\dfrac{K_b}{L_a}x_2 - \dfrac{R_a}{L_a}x_3 + \dfrac{1}{1L_a}v_{in}(t) \end{cases}$$

Output equations $y(t) = x_2$ (6.4.39)

and, in the matrix form:
$$\begin{cases} \dot{x}(t) = Ax(t) + Bu(t) \\ y(t) = Cx(t) + Du(t) \end{cases}$$

where matrices are

$$A_{3 \times 3} = \begin{bmatrix} 0 & 1 & 0 \\ 0 & -\dfrac{c_b}{J} & \dfrac{K}{J} \\ 0 & -\dfrac{K_b}{L_a} & -\dfrac{R_a}{L_a} \end{bmatrix}, \quad B_{3 \times 2} = \begin{pmatrix} 0 & 0 \\ 0 & -\dfrac{1}{J} \\ \dfrac{1}{L_a} & 0 \end{pmatrix}, \quad C_{1 \times 3} = (0 \ 1 \ 0), \quad D = 0$$

$$x(t) = \begin{pmatrix} x_1 = \theta(t) \\ x_2 = \dot{\theta}(t) \\ x_3 = i_a(t) \end{pmatrix}, \quad u(t) = \begin{pmatrix} v_{in}(t) \\ T_L(t) \end{pmatrix}$$

Similarly, for the field-controlled DC motor with a load, the governing equations are

$$\begin{cases} J\ddot{\theta} + c_b\dot{\theta} = T - T_L \\ v_{in} = R_f i_f + L_f \dfrac{di_f}{dt} \\ T = K i_f \end{cases} \quad (6.4.40)$$

The modified block diagram is depicted in Figure 6.4.16.

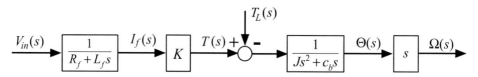

Figure 6.4.16 Field-controlled DC motor with a load: block diagram

The transfer functions are derived as

$$G_{V_{in}}(s) = \frac{\Omega(s)}{V_{in}(s)} = \frac{K}{(R_f + L_f s)(Js + c_b)} \tag{6.4.41}$$

$$G_{T_L}(s) = \frac{\Omega(s)}{T_L(s)} = -\frac{1}{Js + c_b}$$

If the current in the field circuit is considered an additional output, the corresponding transfer function is

$$G_{V_{in}I_f}(s) = \frac{I_f(s)}{V_{in}(s)} = \frac{1}{R_f + L_f s} \tag{6.4.42}$$

Note that due to the absence of a feedback loop (unlike the armature-controlled motor the field-controlled motor does not have a back emf in the field circuit) the load torque has no influence on the current in the field circuit, so $G_{T_L I_f}(s)$ does not exist.

The state-space representation of a field-controlled DC motor with a load is as follows:

State equations
$$\begin{cases} \dot{x}_1 = x_2 \\ \dot{x}_2 = -\frac{c_b}{J}x_2 + \frac{K}{J}x_3 - \frac{1}{J}T_L(t) \\ \dot{x}_3 = -\frac{R_f}{L_f}x_3 + \frac{1}{L_f}v_{in}(t) \end{cases}$$

Output equations $\quad y(t) = x_2 \tag{6.4.43}$

and, in matrix form:

$$\begin{cases} \dot{x}(t) = Ax(t) + Bu(t) \\ y(t) = Cx(t) + Du(t) \end{cases}$$

where the matrices are as follows:

$$A_{3 \times 3} = \begin{bmatrix} 0 & 1 & 0 \\ 0 & -\frac{c_b}{J} & \frac{K}{J} \\ 0 & 0 & -\frac{R_f}{L_f} \end{bmatrix}, \quad B_{3 \times 2} = \begin{pmatrix} 0 & 0 \\ 0 & -\frac{1}{J} \\ \frac{1}{L_f} & 0 \end{pmatrix}, \quad C_{1 \times 3} = (0 \ 1 \ 0), \quad D = 0$$

$$x(t) = \begin{pmatrix} x_1 = \theta(t) \\ x_2 = \dot{\theta}(t) \\ x_3 = i_a(t) \end{pmatrix}, \quad u(t) = \begin{pmatrix} v_{in}(t) \\ T_L(t) \end{pmatrix}$$

Example 6.4.1

Consider an armature-controlled DC motor with a load shown in Figure 6.4.17.

In this assembly, the load is connected to the motor via a flexible damped shaft, with torsional stiffness k_b and torsional damping c_b. The load is supported with the bearings that have a damping coefficient c_L. The inertia of the motor is J, while the inertia of the load is J_L. The stator of the given DC motor is represented by a permanent magnet. R_a and L_a denote the resistance and inductance of the circuit elements respectively, while i_a indicates the current in the armature circuit. The motor torque constant is K, and the back emf constant is K_b. The voltage input into the armature circuit is denoted $v_{in}(t)$.

Considering that the system inputs are $v_{in}(t)$ and the load torque $T_L(t)$, and the system outputs are the angular velocity of the motor ω and the angular velocity of the load ω_L, derive:

(a) the governing equations of this dynamic system
(b) a block diagram
(c) a state-space representation

Solution

As seen in Figure 6.4.17, the mechanical subsystem of this dynamic system is represented with two inertia elements instead of one: the motor and the load cannot be lumped together into a single element because of the flexible damped shaft that connects them. Thus, the mechanical subsystem has two degrees of freedom and needs to be described by two differential equations.

The derivation of the governing equations for the mechanical subsystem starts with drawing a free-body diagram for each inertia element, and indicating all the acting torques. The free-body diagrams are illustrated in Figure 6.4.18. Note, that both the motor and the load rotate in the same direction, while the load torque opposes the motion as discussed earlier.

The inertial element J is subject to three acting torques: $T(t)$ – the generated motor torque; T_k – the restoring torque due to the shaft flexibility that is modeled as a torsional spring; and T_b – the resisting torque due to the shaft torsional damping arising in response to angular velocity. The torque T_k works against the angular displacement $\theta(t)$, while the torque T_b opposes the angular velocity $\omega(t) = \dot{\theta}$.

Assuming that the angular displacement of the motor is greater than that of the load, i.e. $\theta(t) > \theta_L(t)$, the expressions for T_k and T_b are (see Chapter 3)

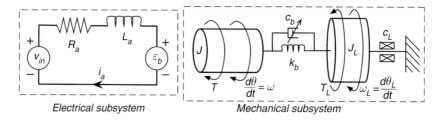

Figure 6.4.17 DC motor with a load assembly for Example 6.4.1

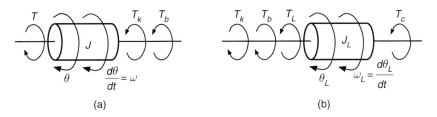

Figure 6.4.18 Free-body diagrams of (a) the motor and (b) the load

$$T_k = k_b(\theta - \theta_L)$$
$$T_b = c_b(\dot\theta - \dot\theta_L)$$

Applying Newton's second law for a rotational system, the equation of motion for the motor is

$$J\ddot\theta = T(t) - T_k - T_b = T(t) - k_b(\theta - \theta_L) - c_b(\dot\theta - \dot\theta_L) \quad (1)$$

Similarly, the governing equation for the load is derived as

$$J_L\ddot\theta_L = -T_L(t) + T_k + T_b - T_c = -T_L(t) + k_b(\theta - \theta_L) + c_b(\dot\theta - \dot\theta_L) - c_L\dot\theta_L \quad (2)$$

Equations (1) and (2) represent the governing equations for the mechanical subsystem.

The governing equation for the electrical subsystem, the mechanism of generating the back emf in response to the rotation of the armature assembly, and the coupling relations are the same as for the armature-controlled DC motor:

$$\begin{aligned} v_{in} - \varepsilon_b &= R_a i_a + L_a \frac{di_a}{dt} \\ \varepsilon_b &= K_b \frac{d\theta}{dt} \\ T &= K i_a \end{aligned} \quad (3)$$

Hence, the governing equations for this dynamic system are

$$\begin{cases} J\ddot\theta + k_b(\theta - \theta_L) + c_b(\dot\theta - \dot\theta_L) = T(t) \\ J_L\ddot\theta_L - k_b(\theta - \theta_L) - c_b(\dot\theta - \dot\theta_L) + c_L\dot\theta_L = -T_L(t) \\ v_{in} - \varepsilon_b = R_a i_a + L_a \frac{di_a}{dt} \\ \varepsilon_b = K_b \frac{d\theta}{dt} \\ T = K i_a \end{cases} \quad (4)$$

To construct a block diagram, take the Laplace transform of Eq. (4) considering zero initial conditions:

$$\begin{cases} (Js^2 + c_b s + k_b)\Theta(s) - (c_b s + k_b)\Theta_L(s) = T(s) \\ (J_L s^2 + (c_b + c_L)s + k_b)\Theta_L(s) - (c_b s + k_b)\Theta(s) = -T_L(s) \\ V_{in}(s) - E_b(s) = (R_a + L_a s)I_a(s) \\ E_b(s) = K_b s \Theta(s) \\ T(s) = K I_a(s) \end{cases} \quad (5)$$

The last three equations in the set of Eq. (5) are easy to transform into block diagram components, as for the armature-controlled DC motor discussed above, and the resulting components are shown in Table 6.4.2.

The construction of the block diagram components that represent the relations between the torques $T(s)$ and $T_L(s)$ and the rotation speeds $\Theta(s)$ and $\Theta_L(s)$ requires a number of algebraic transformations of the first two equations of the set of Eq. (5) to obtain the expressions in the form

$$\Theta(s) = G_1(s)T(s) + G_2(s)T_L(s) \quad (6)$$

and

$$\Theta_L(s) = G_3(s)T(s) + G_4(s)T_L(s) \quad (7)$$

To simplify the derivation, let the following be denoted as

$$Js^2 + c_b s + k_b = g_1$$
$$J_L s^2 + (c_b + c_L)s + k_b = g_2$$
$$c_b s + k_b = g_3$$

Then, the first two equations of Eq. (5) are rewritten as

$$g_1 \Theta(s) - g_3 \Theta_L(s) = T(s) \quad (8)$$
$$g_2 \Theta_L(s) - g_3 \Theta(s) = -T_L(s) \quad (9)$$

From Eq. (9)

$$\Theta_L(s) = -\frac{1}{g_2} T_L(s) + \frac{g_3}{g_2} \Theta(s) \quad (10)$$

Plugging Eq. (10) into Eq. (8), we obtain

$$g_1 \Theta(s) - g_3 \left(-\frac{1}{g_2} T_L(s) + \frac{g_3}{g_2} \Theta(s) \right) = T(s)$$

which yields the desired form of Eq. (6):

$$\Theta(s) = \frac{g_2}{g_1 g_2 - g_3^2} T(s) - \frac{g_3}{g_1 g_2 - g_3^2} T_L(s) \quad (11)$$

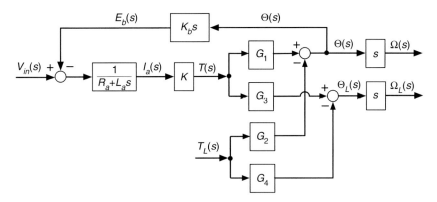

Figure 6.4.19 DC motor with a load assembly: block diagram

In a similar manner, the expression for $\Theta_L(s)$ is derived:

$$\Theta_L(s) = \frac{g_3}{g_1 g_2 - g_3^2} T(s) - \frac{g_1}{g_1 g_2 - g_3^2} T_L(s) \tag{12}$$

The expanded block diagram of this dynamic system is shown in Figure 6.4.19, where

$$G_1(s) = \frac{g_2}{g_1 g_2 - g_3^2}, \quad G_2(s) = \frac{g_3}{g_1 g_2 - g_3^2}, \quad G_3(s) = \frac{g_3}{g_1 g_2 - g_3^2}, \quad G_4(s) = \frac{g_1}{g_1 g_2 - g_3^2}$$

This system has four transfer functions:

$$G_{V_{in}}(s) = \frac{\Omega(s)}{V_{in}(s)}, \quad G_{T_L}(s) = \frac{\Omega(s)}{T_L(s)}, \quad G_{V_{in}L}(s) = \frac{\Omega_L(s)}{V_{in}(s)}, \quad G_{T_L L}(s) = \frac{\Omega_L(s)}{T_L(s)}$$

The derivation of these transfer functions from the constructed block diagram or from the derived governing equations (Eq. (5)) is straightforward, using the approaches discussed in Sections 6.2.2 and 6.2.3.

The state-space representation of this dynamic system is easily obtained from that derived earlier by adding two more state variables $x_4 = \theta_L$ and $x_5 = \dot{\theta}_L$.

State equations
$$\begin{cases} \dot{x}_1 = x_2 \\ \dot{x}_2 = -\frac{k_b}{J} x_1 - \frac{c_b}{J} x_2 + \frac{K}{J} x_3 + \frac{k_b}{J} x_4 + \frac{c_b}{J} x_5 \\ \dot{x}_3 = -\frac{K_b}{L_a} x_2 - \frac{R_a}{L_a} x_3 + \frac{1}{L_a} v_{in}(t) \\ \dot{x}_4 = x_5 \\ \dot{x}_5 = \frac{k_b}{J_L} x_1 + \frac{c_b}{J_L} x_2 - \frac{k_b}{J_L} x_4 - \frac{c_b + c_L}{J_L} x_5 - \frac{1}{J_L} T_L(t) \end{cases}$$

Output equations $\quad y(t) = \begin{pmatrix} x_2 \\ x_5 \end{pmatrix}$

And, in the matrix form:
$$\begin{cases} \dot{x}(t) = Ax(t) + Bu(t) \\ y(t) = Cx(t) + Du(t) \end{cases}$$

where the matrices are

$$A_{5\times 5} = \begin{bmatrix} 0 & 1 & 0 & 0 & 0 \\ -\dfrac{k_b}{J} & -\dfrac{c_b}{J} & \dfrac{K}{J} & \dfrac{k_b}{J} & \dfrac{c_b}{J} \\ 0 & -\dfrac{K_b}{L_a} & -\dfrac{R_a}{L_a} & 0 & 0 \\ 0 & 0 & 0 & 0 & 1 \\ \dfrac{k_b}{J_L} & \dfrac{c_b}{J_L} & 0 & -\dfrac{k_b}{J_L} & -\dfrac{c_b + c_L}{J_L} \end{bmatrix}, \quad B_{5\times 2} = \begin{pmatrix} 0 & 0 \\ 0 & 0 \\ \dfrac{1}{L_a} & 0 \\ 0 & 0 \\ 0 & -\dfrac{1}{J_L} \end{pmatrix},$$

$$C_{2\times 5} = \begin{pmatrix} 0 & 1 & 0 & 0 & 0 \\ 0 & 0 & 0 & 0 & 1 \end{pmatrix}, \quad D = 0$$

$$x(t) = \begin{pmatrix} x_1 = \theta(t) \\ x_2 = \dot{\theta}(t) \\ x_3 = i_a(t) \\ x_4 = \theta_L(t) \\ x_5 = \dot{\theta}_L(t) \end{pmatrix}, \quad u(t) = \begin{pmatrix} V_{in}(t) \\ T_L(t) \end{pmatrix}$$

Simplification of a DC-Motor Model

As can be seen from Eq. (6.4.14), the armature-controlled DC motor is a second-order dynamic system. While the mathematical apparatus for the analysis of such systems is extensively developed, in many cases reduction of this system to a first-order one is desirable.

Such a reduction is reasonable since, typically, the inductance of the armature circuit greatly exceeds the resistance, i.e. $\dfrac{L_a}{R_a} \ll 1$. Then, the expression $(R_a + L_a s)$ in the denominator of the system transfer function (Eq. (6.4.14)) becomes

$$R_a + L_a s = R_a \left(\dfrac{L_a s}{R_a} + 1 \right) \cong R_a$$

since $\dfrac{L_a s}{R_a}$ is very small.

Then, the transfer function reduces as follows:
$$G(s) = \dfrac{\Omega(s)}{V_{in}(s)} = \dfrac{K}{(Js + c_b)(R_a + L_a s) + KK_b} = \dfrac{K}{(Js + c_b)R_a + KK_b}$$

$$= \dfrac{K}{JR_a s + (R_a c_b + KK_b)} = \dfrac{K/(R_a c_b + KK_b)}{\left(JR_a / (R_a c_b + KK_b) \right) s + 1}$$

Setting coefficients

$$K_M = K \big/ (R_a c_b + KK_b)$$

and

$$\tau = JR_a \big/ (R_a c_b + KK_b)$$

presents the transfer function in the familiar form of a first-order system:

$$G(s) = \frac{\Omega(s)}{V_{in}(s)} = \frac{K_M}{\tau s + 1} \qquad (6.4.44)$$

6.5 Block Diagrams in the Time Domain

Transfer function formulations (Section 6.2.2) and block diagrams in the s-domain (Section 6.2.3) are valid for linear time-invariant dynamic systems, for which the governing differential equations are linear with constant coefficients. These techniques are not applicable to linear time-variant systems and nonlinear systems, for which the governing differential equations have time-dependent coefficients and/or nonlinearities. For these systems, state-space representations (Section 6.2.1) are widely used in modeling and simulation.

In this section, another modeling technique, which is called block diagrams in the time domain or time-domain block diagrams, is introduced. This technique is applicable to general dynamic systems, including linear time-invariant systems, linear time-variant systems, and nonlinear systems. One benefit of learning time-domain block diagrams is to prepare for the use of the Simulink software in the modeling and simulation of complicated systems. As will be seen in Section 6.6.1, the structure of time-domain block diagrams and that of Simulink models are quite similar.

Like an s-domain block diagram, a time-domain block diagram is an assembly of basic elements or components. Table 6.5.1 lists some basic components for the construction of time-domain block diagrams. From the table, the signal, summing point, pick-off point, and constant gain follow the same rules as their counterparts in the s-domain; a differential operator D was introduced in Section 2.5.1; and integrator can be symbolically expressed by $\int dt = D^{-1}$. However, three new components (the last three in the table) are not seen in the s-domain block diagrams: the time-varying gain, multiplier, and function block, which are used to describe time-variant properties and nonlinear characteristics. A multiplier can be used to describe the powers of a variable, such as x^2 and x^3. A time varying gain $g(t)$ or nonlinear function $\phi(t)$ can be used to describe the time-variant or nonlinear property of a dynamic system.

Table 6.5.1 Basic components of time-domain block diagram

Component	Symbol	Rule
Signal	$u \rightarrow$ $\leftarrow v$	Variable, input or output
Summing point	$u \xrightarrow{+} \bigcirc \xrightarrow{w}$, v input with $-$	$w = u - v$
Pick-off point	branch: $v \leftarrow$, $\rightarrow u$, $\rightarrow w$	$v = u$, $w = u$
Differentiator	D, overdot	$Du = \dfrac{du}{dt}$, $\dot{x} = \dfrac{dx}{dt}$
Integrator	$y \rightarrow \boxed{\int} \rightarrow z$	$z(t) = \int_{t_0}^{t} y(\tau)\,d\tau$
Constant gain	$x \rightarrow \boxed{k} \rightarrow y$	$y = kx$
Time-varying gain	$x \rightarrow \boxed{g(t)} \rightarrow y$	$y = g(t)x$
Multiplier	$u, v \rightarrow \boxed{\times} \rightarrow w$	$w = u \times v$
	$x_1, x_2, x_3 \rightarrow \boxed{\times} \rightarrow y$	$y = x_1 \times x_2 \times x_3$
Function block	$t \rightarrow \boxed{\phi(t)} \rightarrow y$	$y = \phi(t)$

Consider a dynamic system with its characteristic variable $x(t)$ governed by an n-th-order differential equation. Denote the system input by $r(t)$. A systematic procedure for construction of a time-domain block diagram for the system takes the following five steps.

(1) Rewrite the differential equation in the following form:

$$D^n x = f(x, Dx, \ldots, D^{n-1}x; r) \tag{6.5.1}$$

where f is a function of variable x, its derivatives up to order $n-1$, and input r. For a linear dynamic system, f is a linear function; for a nonlinear system, f is a nonlinear function.

(2) Integrate the left-hand side of Eq. (6.5.1) n times, which gives a part of the block diagram as shown in Figure 6.5.1(a).

(3) By using the summing point and constant gain in Table 6.5.1, obtain a representation of function f, which gives another part of the block diagram, as shown in Figure 6.5.1(b). The representation should take r as an input, and f as an output.

(4) By $D^n x = f$, assemble the two parts obtained in steps 2 and 3 into a block diagram, as shown in Figure 6.5.1(c). During the assembly, the basic components in Table 6.5.1 are used.

(5) Add the system outputs to complete the construction of the block diagram.

The above procedure is illustrated in Examples 6.5.1 through 6.5.4.

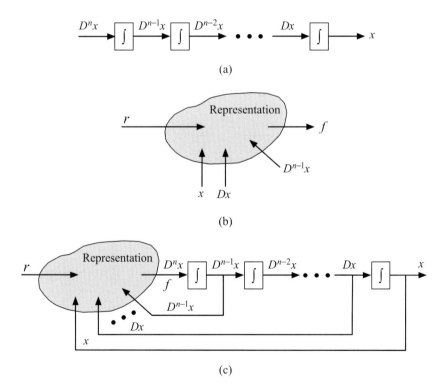

Figure 6.5.1 Construction of a time-domain block diagram: (a) integration of $D^n x$; (b) representation of function f by $x, Dx, \ldots, D^{n-1}x$ and r; (c) a completed block diagram

Example 6.5.1

For a spring–mass–damper system, its displacement $x(t)$ is governed by

$$m\ddot{x} + c\dot{x} + kx = q \tag{a}$$

Let the external force q be the input. Let the displacement x and the damping force ($f_d = c\dot{x}$) be the outputs. Build a time-domain block diagram for the system.

Solution

Apply the five-step procedure. In the first step, Eq. (a) is reduced to

$$D^2x = f \equiv \frac{1}{m}(q - cDx - kx) \tag{b}$$

In the second step, D^2x in Eq. (b) is integrated twice, yielding the part in Figure 6.5.2(a). In the third step, a representation of function f in Eq. (b) is established in Figure 6.5.2(b). In the fourth step, the two parts obtained in the previous steps are assembled according to $D^2x = f$. And in the fifth step, by adding the two designated outputs, the time-domain block diagram for the mechanical system is completed; see Figure 6.5.2(c).

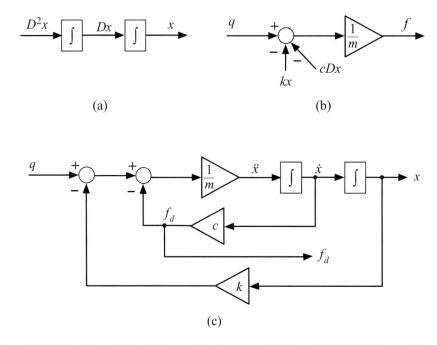

Figure 6.5.2 Time-domain block diagram for the system in Example 6.5.1: (a) integration of D^2x; (b) representation of f; and (c) completion of the block diagram

6.5 Block Diagrams in the Time Domain

The spring–mass–damper system in Example 6.5.1 is a linear time-invariant system, for which an s-domain block diagram can be also obtained (see Section 6.2.3). This, of course, is a trivial case. Now, consider a nonlinear system in the following example, for which the construction of a block diagram in the s-domain is not possible.

Example 6.5.2

The displacement $x(t)$ of a nonlinear mechanical system is governed by

$$m\ddot{x} + c_0 \dot{x} + c_1 \dot{x}^3 + (k_0 + \varepsilon \sin \Omega t) x = q \tag{a}$$

Equation (a) indicates that the system experiences a nonlinear damping force, which is $f_d = c_0 \dot{x} + c_1 \dot{x}^3$, and a time-dependent spring force, which is $f_s = (k_0 + \varepsilon \sin \Omega t) x$. Let the external force q be the input. Let the displacement x and the spring force f_s be the outputs. Build a time-domain block diagram of the system.

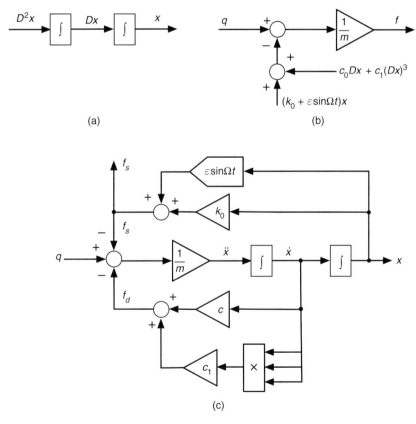

Figure 6.5.3 A time-domain block diagram in Example 6.5.2: (a) integration of $D^2 x$; (b) representation of f, and (c) completion of the block diagram

Solution
Equation (a) is first reduced to

$$D^2 x = f \equiv \frac{1}{m}\left[q - c_0 Dx - c_1(Dx)^3 - (k_0 + \varepsilon \sin\Omega t)x\right] \tag{b}$$

The derivative $D^2 x$ and function f in Eq. (b) are then expressed by the parts in Figures 6.5.3(a) and (b), respectively. Afterwards, these parts are assembled, and the outputs are added, to give a complete time-domain block diagram of the system in Figure 6.5.3(c).

The five-step block diagram construction procedure is applicable to systems with multiple inputs and multiple outputs, which are governed by coupled differential equations. In handling such a system, one just needs to obtain the two parts for each of the differential equations separately in steps 1 to 3, and then assemble all the parts in steps 4 and 5 to complete a block diagram.

Example 6.5.3
A two-input–two-output system is governed by the following coupled differential equations:

$$4\ddot{x} + 5x + 0.1x^3 - y = u \tag{a}$$

$$\dot{y} + 3y - 6xy = v \tag{b}$$

where u and v are the inputs, and x and y are the outputs. Build a block diagram for the system.

Solution
The differential equations in Eqs. (a) and (b) are rewritten as

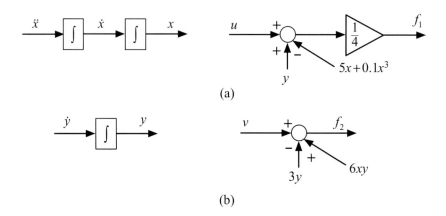

Figure 6.5.4 Parts of a block diagram in Example 6.5.3: (a) for Eq. (c); (b) for Eq. (d)

$$\ddot{x} = f_1 \equiv \frac{1}{4}(u - 5x - 0.1x^3 + y) \tag{c}$$

$$\dot{y} = f_2 \equiv v - 3y + 6xy \tag{d}$$

The two parts of each of the differential equations are drawn in Figure 6.5.4. Assembling all these parts gives the block diagram of the dynamic system in Figure 6.5.5. Finally, the construction of a time-domain block diagram for an armature-controlled DC motor is presented in Example 6.5.4.

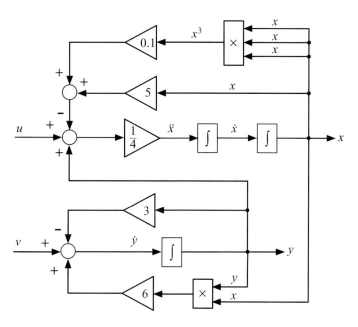

Figure 6.5.5 Time-domain block diagram of the system in Example 6.5.3

Example 6.5.4

For an armature-controlled DC motor, the governing equations (6.4.36) are reduced to

$$D^2\theta = f_1 \equiv \frac{1}{J}(Ki_a - T_L - c_b D\theta) \tag{a}$$

$$Di_a = f_2 \equiv \frac{1}{L_a}(v_{in} - K_b D\theta - R_a i_a) \tag{b}$$

where θ is the rotation angle of the motor; v_{in} and i_a are the voltage input and current of the armature circuit, respectively; T is the motor torque constant; K_b is the back emf constant; and T_L denotes a load torque. As shown in Eq. (6.4.10), $K = K_b$.

Let the inputs of this electromechanical system be v_{in} and T_L. Let the output of the system be $\dot\theta$. By following the five steps listed above, a time-domain block diagram of the motor is obtained in Figure 6.5.6, where $T = Ki_a$ is the motor torque.

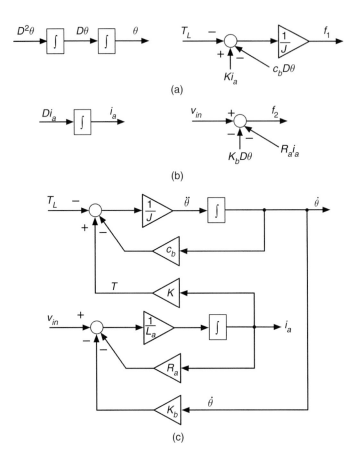

Figure 6.5.6 A time-domain block diagram for the armature-controlled DC motor in Example 6.5.4: (a) two parts from Eq. (a); (b) two parts from Eq. (b); and (c) completion of the block diagram

6.6 Modeling and Simulation by Simulink

In this section, the use of Simulink in the modeling and simulation of dynamic systems is illustrated in examples. Refer to Appendix B.2 on how to use Simulink.

6.6.1 Simulink Models Based on Time-Domain Block Diagrams

A Simulink model can be easily built based on the concept of block diagrams in the time domain, as presented in Section 6.5.

Example 6.6.1

Consider the system in Example 6.5.3. Consider zero initial disturbances. Let the inputs of the system be $u = 1.5\sin(10t)$ and $v = 0.6$. Plot the system outputs x and y for $0 \leq t \leq 20$.

Solution

According to Figure 6.5.5, a Simulink model of the system is constructed, as shown in Figure 6.6.1. Running the model in simulation gives the system response in Figure 6.6.2. Refer to Appendix B.2 for how to obtain simulation results (plots) from the `Scope` block.

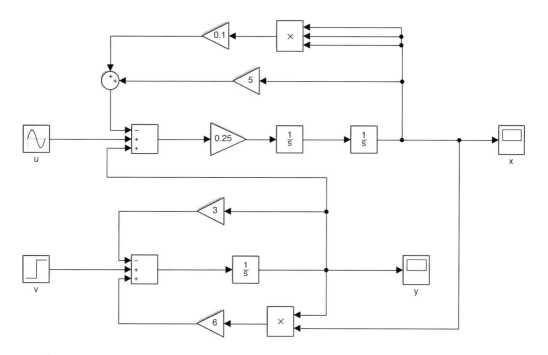

Figure 6.6.1 Simulink model of the system in Example 6.6.1

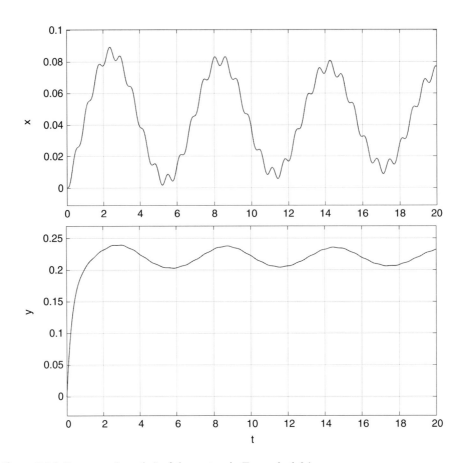

Figure 6.6.2 Response (x and y) of the system in Example 6.6.1

Example 6.6.2

An armature-controlled DC motor, which is governed by Eqs. (a) and (b) in Example 6.5.4, has the following parameter values:

$$L_a = 1.9 \times 10^{-3} \text{ H}, \ R_a = 0.56 \ \Omega, \ K = K_b = 0.06 \text{ N} \cdot \text{m/A}$$
$$J = 9 \times 10^{-5} \text{ kg} \cdot \text{m}^2, \ c_b = 1.2 \times 10^{-4} \text{ N} \cdot \text{m} \cdot \text{s/rad} \quad \text{(a)}$$

Let the applied armature voltage be $v_{in} = 10$ V. Build a Simulink model of the motor, and plot the rotation speed $\omega(t) = \dot{\theta}(t)$ and the armature current $i_a(t)$ of the motor for $0 \leq t \leq 0.1$ s. In simulation, the following two cases are considered:

Case 1: the motor idles with no load torque, $T_L = 0$
Case 2: the motor idles initially and, at $t = 0.04$ s, a constant load torque, $T_L = 0.6$ N·m is applied

Solution
By referencing Figure 6.5.6, a Simulink model of this electromechanical system is built; see Figure 6.6.3. The model has two inputs, which are the load torque T_L and applied voltage v_{in}; and two outputs, which are the rotation speed ω and armature current i_a of the motor. With the Simulink model, the rotation speed and armature current of the motor are plotted against time for the above-mentioned two cases.

In case 1, as shown in Figure 6.6.4, the rotation speed of the idling motor reaches a steady-state value of 163.6 rad/s (1562 rpm). The armature current initially goes up as high as 13.19 A, which is needed to overcome the inertia and damping effects of the motor, and it eventually settles at 0.33 A.

In case 2, the motor behaves in the same way initially. At time $t = 0.04$ s when the load torque T_L is applied, the pattern of the system response starts to change. As shown in

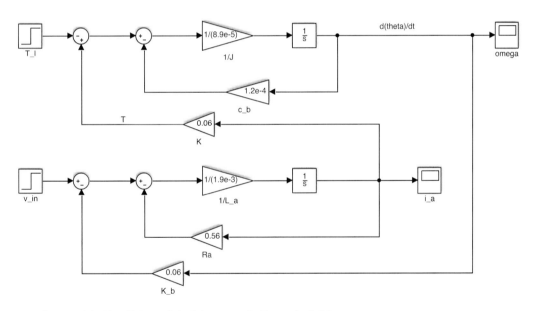

Figure 6.6.3 Simulink model of the motor in Example 6.6.2

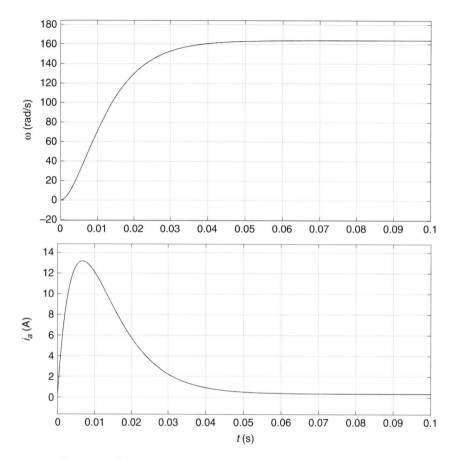

Figure 6.6.4 Response of the motor without a load (case 1)

Figure 6.6.5, the motor speed significantly drops and the armature current quickly climbs up. The steady-state values of the speed and current are 72.0 rad/s (688 rpm) and 10.13 A, respectively, showing the effect of the applied load. As observed, the applied load toque reduces the motor speed by 56%.

In reality, the load torque T_L is usually time varying and the applied voltage v_{in} takes time to reach a designated level, say 10 V. Therefore, to maintain a desired motor speed in operation, a feedback control needs to be implemented. A problem on the feedback control of rotor rotation speed is presented in Section 9.2.

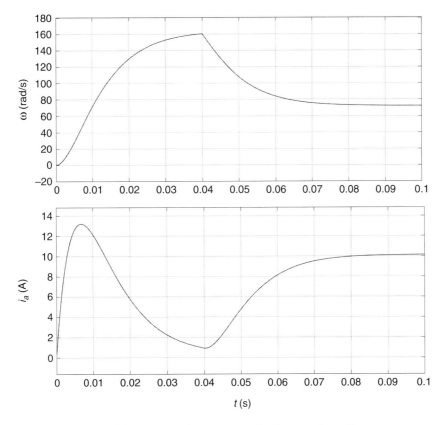

Figure 6.6.5 Response of the motor under a constant load torque (case 2)

6.6.2 Simulink Models with Transfer Functions and State-Space Blocks

Besides building up models from scratch, Simulink can integrate transfer function formulations and state-space representations in modeling and simulation. This is enabled by the `Transfer Fcn` block and the `State-Space` block. See the following two examples.

Example 6.6.3

Consider a time-variant mechanical system with displacement x governed by the following ordinary differential equation:

$$m\ddot{x} + c_0\dot{x} + (k_0 - \varepsilon \sin \Omega t)x = q \tag{a}$$

Assign the system parameter values as follows:

$$m = 2, \quad c_0 = 3, \quad c_1 = 0.1, \quad k_0 = 32, \quad \varepsilon = 12, \quad \text{and } \Omega = 10$$

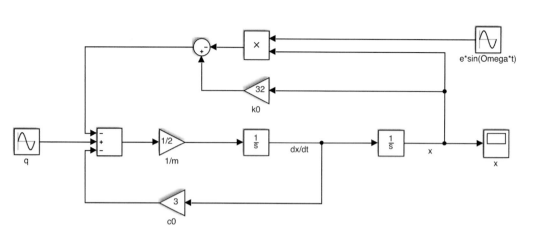

Figure 6.6.6 Simulink model of the system in Example 6.6.3

all of which are in SI units. The external force is given as $q = 2\sin(4t)$. Let q and x be the input and output of the system, respectively. From the time-domain block diagram in Figure 6.5.3(c), a Simulink model of the system is established; see Figure 6.6.6.

On the other hand, Eq. (a) can be written in the symbolic form

$$x = \frac{1}{mD^2 + c_0 D + k_0} f \qquad (b)$$

where

$$D = d/dt$$

and

$$f = q + \varepsilon \sin(\Omega t) x \qquad (c)$$

Equation (b) can be expressed by the `Transfer Fcn` block from the `Continuous Library`, which leads to an equivalent Simulink model of the system in Figure 6.6.7.

Assume zero initial disturbances: $x(0) = \dot{x}(0) = 0$. Simulation with the original model in Figure 6.6.6 and the equivalent model in Figure 6.6.7 gives the same displacement plot as shown in Figure 6.6.8. Refer to Appendix B.2 for how to obtain simulation results (plots) from the `Scope` block.

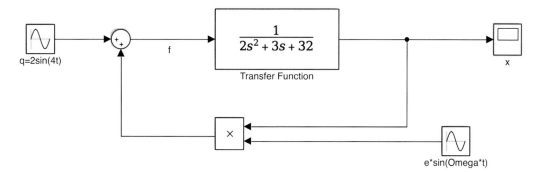

Figure 6.6.7 Equivalent Simulink model with a `Transfer Fcn` block

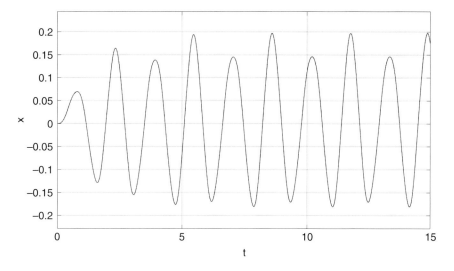

Figure 6.6.8 Displacement of the time-variant system in Example 6.6.3

Example 6.6.4

Consider the armature-controlled DC motor in case 2 of Example 6.6.2. Build a Simulink model for the motor by a `State-Space` block, with the motor rotation speed and the motor torque ($T = Ki_a$) as the system outputs. Also, plot the system outputs for $0 \leq t \leq 0.1$ s.

Solution

According to Eq. (6.4.39), the state equation of the motor is

$$\dot{x} = \begin{bmatrix} 0 & 1 & 0 \\ 0 & -c_b/J & K/J \\ 0 & -K_b/L_a & -R_a/L_a \end{bmatrix} x + \begin{bmatrix} 0 & 0 \\ 0 & -1/J \\ 1/L_a & 0 \end{bmatrix} \begin{pmatrix} v_{in} \\ T_L \end{pmatrix} \quad \text{(a)}$$

and the matrix output equation is

$$y = \begin{bmatrix} 0 & 1 & 0 \\ 0 & 0 & K \end{bmatrix} x \quad \text{(b)}$$

where

$$x = \begin{pmatrix} x_1 \\ x_2 \\ x_3 \end{pmatrix} = \begin{pmatrix} \theta \\ \dot{\theta} \\ i_a \end{pmatrix}, \quad y = \begin{pmatrix} y_1 \\ y_2 \end{pmatrix} = \begin{pmatrix} \dot{\theta} \\ T \end{pmatrix} \quad \text{(c)}$$

Based on the state and output equations, a Simulink model of the motor is generated; see Figure 6.6.9, where a State-Space block with the name DC Motor is connected to a Mux block, which is for inputs v_{in} and T_L, and a Demux block, which is for the outputs $\dot{\theta}$ and T. Here, Mux and Demux are from the Commonly Used Blocks Library and State-Space is from the Continuous Library. Double-click the State-Space block to enter the parameters of the block, which are the elements of the matrices in Eqs. (a) and (b); see Figure 6.6.10, where the parameter values given in Eq. (a) in Example 6.6.2 have been used. Compared with the model in Example 6.6.2, the state-space model here is more efficient in dealing with the multi-input–multi-output system.

Now, run the simulation with $v_{in} = 10$ V, $T_L = 0.6$ N·m applied at $t = 0.04$ s, and zero initial disturbances. The computed motor rotation speed ω and the motor torque T are plotted against time in Figure 6.6.11. The motor speed plot herein is the same as that in Figure 6.6.5. The motor torque reaches a steady-state value of 0.61 N·m, which, by $T = Ki_a$, is in accordance with the current plot in Figure 6.6.5.

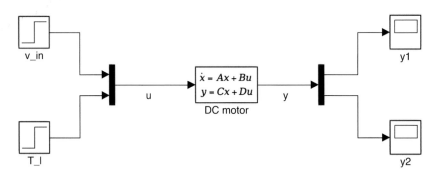

Figure 6.6.9 Simulink model of the DC motor with a State-Space block

6.6 Modeling and Simulation by Simulink

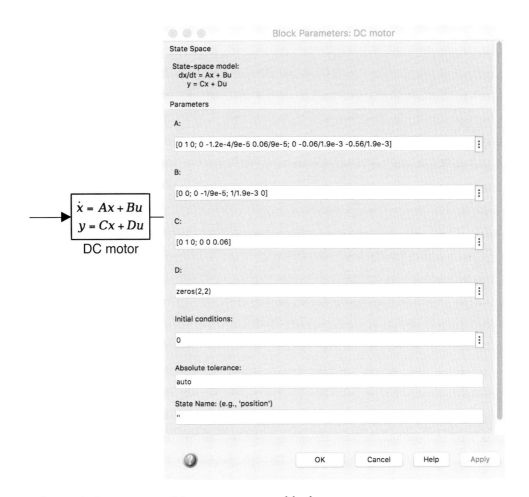

Figure 6.6.10 Parameters of the `State-Space` block

For convenience in data entry and error detection, symbols can be used in specification of the parameters for the `State-Space` block; for example, see Figure 6.6.12. These parameters in the Block Parameters window, however, need to be defined in the MATLAB Command Window before the simulation can be run. (Simulink gets the parameter values from the MATLAB workspace in simulation.) In the current example, this is done by typing:

```
≫ La = 1.9e-3; Ra = 0.56; K = 0.06; Kb = K;
≫ J=9e-5; cb = 1.2e-4;
```

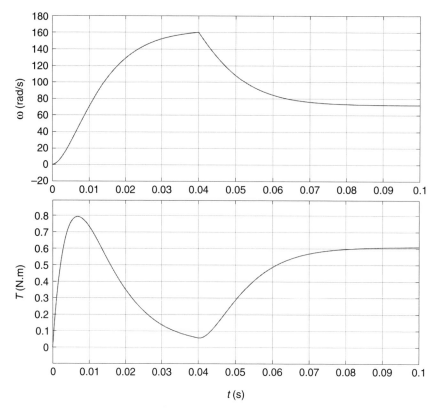

Figure 6.6.11 Response of the motor in case 2 of Example 6.6.2

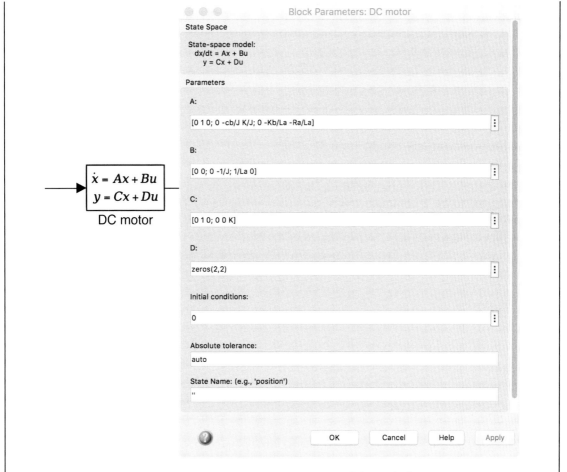

Figure 6.6.12 Parameters of the `State-Space` block with symbols

6.7 Modeling and Simulation by Mathematica

In this section, the use of Wolfram Mathematica in modeling and simulation of dynamic systems is illustrated in the same examples as discussed in Section 6.6. Therefore, this section concentrates on the implementation of the mathematical models of the dynamic systems presented in examples 6.6.1 through 6.6.4 in Wolfram Language. For an analysis of the results, refer to the appropriate examples in Section 6.6. Refer to Appendix C on how to use Mathematica.

Governing equations are key to modeling dynamic systems in Mathematica. The system response can be derived by either directly using differential equation solvers (analytical `DSolve` or numerical `NDSolve`), or by using `StateSpaceModel` function and corresponding `OutputResponse` or `StateResponse`.

Example 6.7.1

Consider the system in Example 6.5.3, with zero initial conditions, and inputs of the system being $u = 1.5\sin(10t)$ and $v = 0.6$:

$$4\ddot{x} + 5x + 0.1x^3 - y = u \tag{a}$$

$$\dot{y} + 3y - 6xy = v \tag{b}$$

Plot the system outputs x and y for $0 \leq t \leq 20$.

Solution

Since this system of differential equations is nonlinear, we do not expect a closed-form solution; hence, we need to use a nonlinear solver, as shown in Figure 6.7.1, obtaining the same plots (see Figure 6.7.2) as previously shown in Figure 6.6.2 for Example 6.6.1.

Note that using a variant NDSolveValue of the Mathematica powerful numerical differential equation solver NDSolve facilitates access to the generated responses $x(t)$ and $y(t)$ and their subsequent visualization with Plot function (see Figure 6.7.2).

```
In[•]:= ClearAll["Global`*"];

In[•]:= (* solve the system:    4x''[t]+5x[t]+0.1*x[t]^3-y[t]== u[t], y'[t]+3y[t]-6*y[t]*x[t]== v[t];
        u[t] and v[t] are inputs, x[t] and y[t] are outputs;
        consider that system starts from equilibrium:  x[0]==x'[0]==y[0]==0;
        and the inputs are:  u[t]==1.5*Sin[10t], v[t]==0.6 *)
        sol = NDSolveValue[{4 x''[t] + 5 x[t] + 0.1*x[t]^3 - y[t] == 1.5*Sin[10 t],
            y'[t] + 3 y[t] - 6*y[t]*x[t] == 0.6, x[0] == x'[0] == y[0] == 0}, {x[t], y[t]}, {t, 0, 20}];

        (* generate the plot of x(t) vs. time, formatting the graphics to achieve desired look *)
        Plot[sol[[1]], {t, 0, 20}, PlotStyle → {Black, Thick}, PlotRange → {{0, 20}, {0, 0.1}},
            Frame → True, GridLines → {{0, 2, 4, 6, 8, 10, 12, 14, 16, 18, 20}, Automatic}, AspectRatio → 1/2,
            FrameTicks → {{All, None}, {{0, 2, 4, 6, 8, 10, 12, 14, 16, 18, 20}, None}},
            FrameLabel → {Style["time", 12], Style["x", 12], None, None}]

        (* generate the plot of y(t) vs. time, formatting the graphics to achieve desired look *)
        Plot[sol[[2]], {t, 0, 20}, PlotStyle → {Black, Thick}, PlotRange → {{0, 20}, {0, 0.25}},
            Frame → True, GridLines → {{0, 2, 4, 6, 8, 10, 12, 14, 16, 18, 20}, Automatic}, AspectRatio → 1/2,
            FrameTicks → {{All, None}, {{0, 2, 4, 6, 8, 10, 12, 14, 16, 18, 20}, None}},
            FrameLabel → {Style["time", 12], Style["y", 12], None, None}]
```

Figure 6.7.1 Code for Example 6.7.1

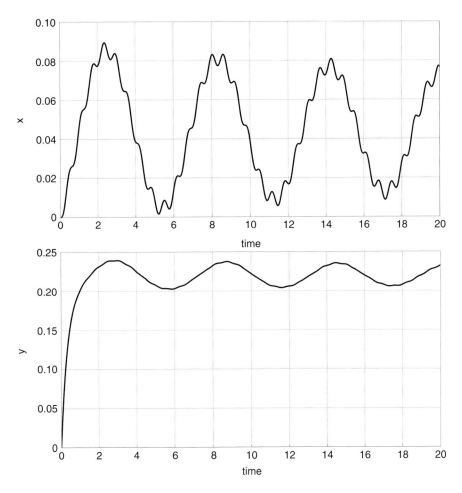

Figure 6.7.2 Response (x and y) of the system in Example 6.7.1

Example 6.7.2

An armature-controlled DC motor is governed by the following differential equations as shown in Example 6.5.4:

$$\begin{cases} J\ddot{\theta} + c_b\dot{\theta} = Ki_a - T_L \\ L_a\dfrac{di_a}{dt} + R_a i_a = v_{in}(t) - K_b\dot{\theta} \end{cases}$$

where θ is the rotation angle of the motor; $v_{in}(t)$ and i_a are the voltage input and current of the armature circuit, respectively; K is the motor torque constant; K_b is the back emf constant; and T_L denotes a load torque. As shown in Eq. (6.4.10), $K = K_b$.

Given the following parameter values:

$$L_a = 1.9 \times 10^{-3} \text{ H}, \ R_a = 0.56 \ \Omega, \ K = K_b = 0.06 \text{ N} \cdot \text{m/A}$$

$$J = 9 \times 10^{-5} \text{ kg} \cdot \text{m}^2, \ c_b = 1.2 \times 10^{-4} \text{ N} \cdot \text{m} \cdot \text{s/rad}$$

and the applied armature voltage $v_{in}(t) = 10u(t)$ V, plot the rotation speed $\omega(t) = \dot{\theta}(t)$ and the armature current $i_a(t)$ of the motor, for $0 \leq t \leq 0.1$ s. In the simulation, the following two cases are considered:

Case 1: the motor idles with no load torque, $T_L = 0$
Case 2: the motor idles initially, and a constant load torque is applied with a delay of 0.04 s, $T_L = 0.6u(t - 0.04)$ N·m

Solution

For this system, a closed-form solution is expected; hence, the analytical solver `DSolve` needs to be used as shown in Figure 6.7.3. Note that the solver generates functions

```
ClearAll["Global`*"];

(* set up the numeric variables *)
la = 1.9*10^-3; ra = 0.56; k = 0.06; kb = 0.06; j = 9*10^-5; cb = 1.2*10^-4; vin = 10;

(* set up the equations in symbolic form *)
eqs = {theta''[t] == 1/j * (k*ia[t] - tl - cb*theta'[t]), ia'[t] == 1/la * (vin - kb*theta'[t] - ra*ia[t]),
    theta[0] == theta'[0] == 0, ia[0] == 0};

(* solve the equations with T_L=0 for θ(t) and i_a(t) *)
idle = DSolveValue[eqs /. tl -> 0, {theta[t], ia[t]}, {t, 0, 0.1}];

(* find the required output by differentiating the derived solution θ(t) *)
toPlot = D[idle[[1]], t];

(* generate the plot of θ̇(t) vs. time, formatting the graphics to achieve desired look *)
Plot[toPlot, {t, 0, 0.1}, PlotRange -> All, PlotStyle -> {Black, Thick}, Frame -> True,
    GridLines -> {Join[{0}, Table[0.01*i, {i, 10}]], {0, 50, 100, 150}}, AspectRatio -> 1/3,
    FrameTicks -> {{{0, 50, 100, 150}, None}, {Join[{0}, Table[0.01*i, {i, 10}]], None}},
    FrameLabel -> {Style["time", 12], Style["ω (rad/s)", 12], None, None}]

(* generate the plot of i_a(t) vs. time, formatting the graphics to achieve desired look *)
Plot[idle[[2]], {t, 0, 0.1}, PlotRange -> All, PlotStyle -> {Black, Thick}, Frame -> True,
    GridLines -> {Join[{0}, Table[0.01*i, {i, 10}]], {0, 5, 10, 15}}, AspectRatio -> 1/3,
    FrameTicks -> {{{0, 5, 10, 15}, None}, {Join[{0}, Table[0.01*i, {i, 10}]], None}},
    FrameLabel -> {Style["time", 12], Style["i_a (A)", 12], None, None}]
```

Figure 6.7.3 Code for Example 6.7.2 (case 1)

$\theta(t)$ and $i_a(t)$ while the rotation speed $\dot{\theta}(t)$ needs to be plotted in addition to the armature current $i_a(t)$. Therefore, the derived solution $\theta(t)$ needs to be differentiated before being used as an input to `Plot`. Also, note that since two cases need to be considered, an efficient code implementation involves setting up equations in a general form, assigning numerical values to the variables and inputs that do not change from case 1 to case 2, but leaving the load torque as a symbolic variable. Its value is assigned with replacement syntax when the solver for case 1 is called.

The generated responses, shown in Figure 6.7.4, are the same as previously shown in Figure 6.6.4 for Example 6.6.2.

To solve case 2, we can use the same Mathematica notebook, immediately calling the solver (see Figure 6.7.5) since the equations and numerical values are already in memory. (Note the replacement syntax and representation of a load torque input with a delayed step function.)

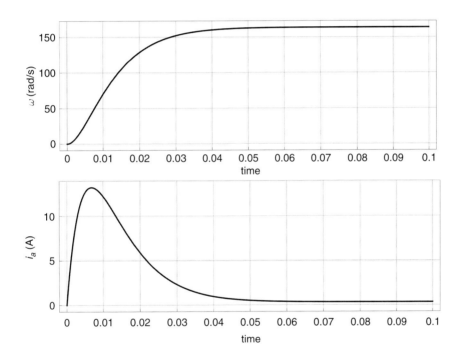

Figure 6.7.4 Response of the motor without a load (case 1)

The plots of $\dot{\theta}(t)$ and $i_a(t)$ are shown in Figure 6.7.6. As expected, the generated responses are the same as previously shown in Figure 6.6.5 for Example 6.6.2.

```
In[•]:= (* solve the equations with T_L=0.6 and is applied with delay of 0.04 seconds,
        for θ(t) and i_a(t) *)
       withTl = DSolveValue[eqs /. tl → 0.6*UnitStep[t - 0.04], {theta[t], ia[t]}, {t, 0, 0.1}];

       (* find the required output by differentiating the derived solution θ(t) *)
       toPlot = D[withTl[[1]], t];

       (* generate the plot of θ̇(t) vs. time, formatting the graphics to achieve desired look *)
       Plot[toPlot, {t, 0, 0.1}, PlotRange → All, PlotStyle → {Black, Thick}, Frame → True,
         GridLines → {Join[{0}, Table[0.01*i, {i, 10}]], {0, 50, 100, 150}}, AspectRatio → 1/3,
         FrameTicks → {{{0, 50, 100, 150}, None}, {Join[{0}, Table[0.01*i, {i, 10}]], None}},
         FrameLabel → {Style["time", 12], Style["ω (rad/s)", 12], None, None}]

       (* generate the plot of i_a(t) vs. time, formatting the graphics to achieve desired look *)
       Plot[withTl[[2]], {t, 0, 0.1}, PlotRange → All, PlotStyle → {Black, Thick}, Frame → True,
         GridLines → {Join[{0}, Table[0.01*i, {i, 10}]], {0, 5, 10, 15}}, AspectRatio → 1/3,
         FrameTicks → {{{0, 5, 10, 15}, None}, {Join[{0}, Table[0.01*i, {i, 10}]], None}},
         FrameLabel → {Style["time", 12], Style["i_a (A)", 12], None, None}]
```

Figure 6.7.5 Code for Example 6.7.2 (case 2)

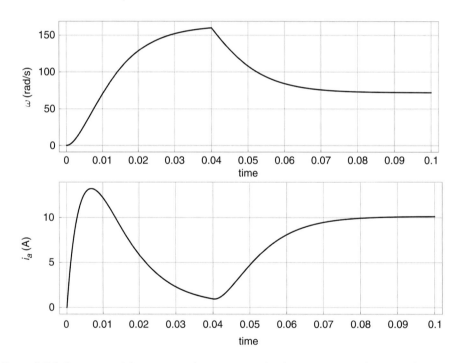

Figure 6.7.6 Response of the motor under a constant load torque (case 2) in Example 6.7.2

Example 6.7.3

Consider a time-variant mechanical system with displacement x governed by the following ordinary differential equation:

$$m\ddot{x} + c_0\dot{x} + (k_0 - \varepsilon \sin \Omega t)x = q \tag{a}$$

and the system parameter values as follows:

$$m = 2, \quad c_0 = 3, \quad c_1 = 0.1, \quad k_0 = 32, \quad \varepsilon = 12, \text{ and } \Omega = 10.$$

The external force is given as $q = 2\sin(4t)$. Plot the displacement of this system for the time period of $0 \leq t \leq 15$.

Solution

This system can be solved using the numerical differential equation solver NDSolve (for a time-variant ordinary differential equation, no closed-form solution is expected) in the same manner as for Example 6.7.2. This code is shown in Figure 6.7.7.

A more efficient implementation uses the Modelica System Model functionality (not available for Mathematica versions older than 12), as shown in Figure 6.7.8. Since this code is in the new notebook, setting up the symbolic equations and assigning numerical values to the variables is necessary.

```
In[ ]:= ClearAll["Global`*"];

(* set up the numeric variables *)
m = 2; c0 = 3; k0 = 32; e = 12; Ω = 10;

(* set up the equations in symbolic form *)
eqs = {m*x''[t] + c0*x'[t] + (k0 - e*Sin[Ω*t])*x[t] == 2*Sin[4 t], x[0] == x'[0] == 0};

(* solve the equations *)
soln = NDSolveValue[eqs, x[t], {t, 0, 16}];

(* generate the plot of x(t) vs. time, formatting the graphics to achieve desired look *)
Plot[soln, {t, 0, 16}, PlotRange → All, PlotStyle → {Black, Thick}, Frame → True,
  GridLines → {{0, 5, 10, 15}, {-0.15, -0.1, -0.05, 0.05, 0.1, 0.15, 0.2}}, AspectRatio → 1/2,
  FrameTicks → {{{-0.2, -0.15, -0.1, -0.05, 0, 0.05, 0.1, 0.15, 0.2}, None}, {Automatic, None}},
  FrameLabel → {Style["time", 12], Style["x", 12], None, None}]
```

Figure 6.7.7 Code for Example 6.7.3: using a numerical differential equation solver

```
In[•]:= (* set up the numeric variables *)
m = 2; c0 = 3; k0 = 32; e = 12; Ω = 10;

(* set up the equations in symbolic form *)
eqs = {m*x''[t] + c0*x'[t] + (k0 - e*Sin[Ω*t])*x[t] == 2*Sin[4 t], x[0] == x'[0] == 0};

(* creating a Modelica system model *)
model = CreateSystemModel[eqs, t];

(* simulating the generated Modelica system model *)
sim = SystemModelSimulate[model, 15];

(* plotting the results of simulation, formatting the graph to achieve desired appearance *)
SystemModelPlot[sim, "x", PlotRange → All, PlotStyle → {Black, Thick}, Frame → True,
 GridLines → {{0, 5, 10, 15}, {-0.15, -0.1, -0.05, 0.05, 0.1, 0.15, 0.2}},
 FrameTicks → {{{-0.2, -0.15, -0.1, -0.05, 0, 0.05, 0.1, 0.15, 0.2}, None}, {Automatic, None}},
 FrameLabel → {Style["time", 12], Style["x", 12], None, None}]
```

Figure 6.7.8 Code for Example 6.7.3: using the Modelica System Model

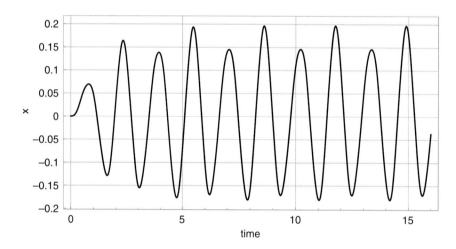

Figure 6.7.9 Displacement of the time-variant system in Example 6.7.3

The same graph is generated by both methods. This plot (see Figure 6.7.9) is identical to the one produced for Example 6.6.3 and shown in Figure 6.6.8.

Example 6.7.4

This example shows a computation of the system response deriving its state-space representation first.

The Mathematica family of functions that generate state-space models is very powerful and can handle a wide range of dynamic systems: linear and nonlinear, time-invariant and time-varying, continuous and discrete. Any of these models is then used as an argument for the response-generating functions: `OutputResponse` (when only a specific output is of interest) and `StateResponse` (when all state variables need to be monitored and analyzed).

Since the armature-controlled DC motor of Example 6.7.2 is a linear dynamic system, it can be modeled using `StateSpaceModel` and the corresponding `OutputResponse` functions.

```
In[·]:= ClearAll["Global`*"];

(* set up the equations in symbolic form *)
eqs = {theta''[t] == 1/j * (k*ia[t] - tl[t] - cb*theta'[t]),
       ia'[t] == 1/la * (vin[t] - kb*theta'[t] - ra*ia[t])};

(* setting up the state-space model directly from the governing equations;
   consider outputs to be θ̇(t) and T=K iₐ(t) *)
ssm = StateSpaceModel[eqs, {{theta[t], 0}, {theta'[t], 0}, {ia[t], 0}}, {{vin[t], 0}, {tl[t], 0}},
      {theta'[t], k*ia[t]}, t];

(* if desired, can make the derived state-space model look exactly as the derived-by-hand one *)
ssm /. {cb → "C_b", k → "K", kb → "K_b", la → "L_a", j → "J", ra → "R_a"}
```

(a)

$$\text{Out[·]} = \left(\begin{array}{ccc|cc} 0 & 1 & 0 & 0 & 0 \\ 0 & -\dfrac{C_b}{J} & \dfrac{K}{J} & 0 & -\dfrac{1}{J} \\ 0 & -\dfrac{K_b}{L_a} & -\dfrac{R_a}{L_a} & \dfrac{1}{L_a} & 0 \\ \hline 0 & 1 & 0 & 0 & 0 \\ 0 & 0 & K & 0 & 0 \end{array}\right)$$

With annotations: A points to top-left block, B points to top-right block, C points to bottom-left block, D points to bottom-right block, S labels the overall matrix.

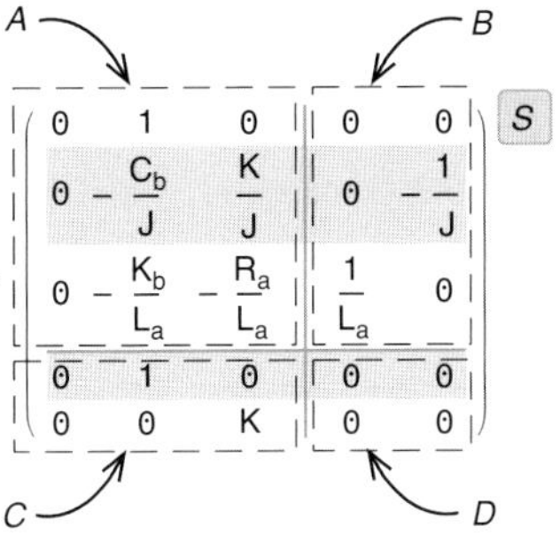

(b)

Figure 6.7.10 Solving Example 6.7.4 with `StateSpaceModel` function: (a) generating the state-space representation from governing equations; (b) visualizing state-space matrices

Nonetheless, if the execution time is crucial, an analytical differential equation solver may yield a better performance.

Unlike with MATLAB/Simulink, the derivation of a state-space model by hand is not necessary – it can be done automatically with the software, as shown in Figure 6.7.10.

The derivation of a state-space model in symbolic form is recommended so that its form can be analyzed, and its correctness verified (see Eqs. (a), (b), and (c) of Example 6.6.4).

To generate and plot the required response, symbolic variables must have assigned values before `OutputResponse` is called. This code is shown in Figure 6.7.11 (it follows the code in Figure 6.7.10 in the same notebook).

The generated plots (see Figure 6.7.12) are identical to those produced for Example 6.6.4 and shown in Figure 6.6.11.

```
In[•]:= (* set up the numeric variables *)
    la = 1.9*10^-3; ra = 0.56; k = 0.06; kb = 0.06; j = 9*10^-5; cb = 1.2*10^-4;

    (* finding θ'(t) for v_in(t)=10u(t) and T_L=0.6u(t-0.04) *)
    response = OutputResponse[ssm, {10 UnitStep[t], 0.6 UnitStep[t - 0.04]}, t];

    (* generate the plot of θ̇(t) vs. time, formatting the graphics to achieve desired look *)
    Plot[response[[1]], {t, 0, 0.1}, PlotRange → All, PlotStyle → {Black, Thick}, Frame → True,
     GridLines → {Join[{0}, Table[0.01*i, {i, 10}]], {0, 50, 100, 150}}, AspectRatio → 1/3,
     FrameTicks → {{{0, 50, 100, 150}, None}, {Join[{0}, Table[0.01*i, {i, 10}]], None}},
     FrameLabel → {Style["time", 12], Style["ω (rad/s)", 12], None, None}]

In[•]:= (* generate the plot of T(t) vs. time, formatting the graphics to achieve desired look *)
    Plot[response[[2]], {t, 0, 0.1}, PlotRange → All, PlotStyle → {Black, Thick}, Frame → True,
     GridLines → {Join[{0}, Table[0.01*i, {i, 10}]], Join[{0}, Table[0.1*i, {i, 8}]]},
     AspectRatio → 1/3,
     FrameTicks → {{Join[{0}, Table[0.1*i, {i, 8}]], None}, {Join[{0}, Table[0.01*i, {i, 10}]], None}},
     FrameLabel → {Style["time", 12], Style["T (N·m)", 12], None, None}]
```

Figure 6.7.11 Code for Example 6.7.4: generating the required outputs from the derived state-space model and plotting them

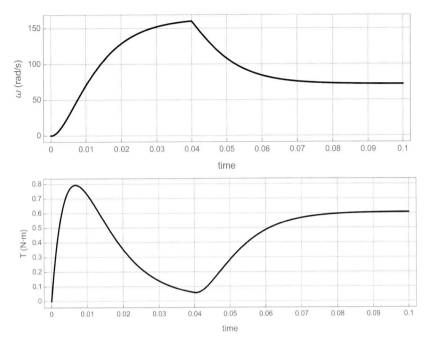

Figure 6.7.12 Response of the motor in case 2 of Example 6.7.2

CHAPTER SUMMARY

This chapter discusses modeling of a dynamic system consisting of multiple subsystems, typically of a different nature. Typical system modeling approaches such as transfer functions, state-space representations, and block diagrams in the s-domain and the time domain are discussed. Foundations of electromechanical systems are presented on the example of a DC motor. The derived mathematical model is implemented in software: MATLAB/Simulink and Wolfram Mathematica; the system behavior in response to several standard inputs is simulated and analyzed.

Upon completion of this chapter the reader should be able to:

(1) Understand the four types of a dynamic system model: governing equations, state-space representations, transfer functions, and block diagrams, and derive any three of them from the given one.
(2) Construct a block diagram of a dynamic system from the given governing equations or state-space model, expanding it to the degree necessary for a good understanding of the input signal propagation through the system and its transformation into system output(s).

(3) Recollect their knowledge of electromagnetism and fundamental physical laws governing D'Arsonval motion and, subsequently, the operation of a DC motor.
(4) Understand the interaction between electrical and mechanical subsystems of an electromechanical system, derive models of armature-controlled and field-controlled DC motors, and of a DC motor with a load.
(5) Implement the derived mathematical models in MATLAB/Simulink and Wolfram Mathematica software, simulate the system behavior in response to standard inputs, and analyze the generated results.

REFERENCES

1. J. L. Meriam, L. G. Kraige, and J. N. Bolton, *Engineering Mechanics: Dynamics*, 8th ed., Wiley, 2015.
2. N. Lobontiu, *System Dynamics for Engineering Students: Concepts and Applications*, 2nd ed., Academic Press, 2017.
3. W. J. Palm, III, *System Dynamics*, 3rd ed. McGraw-Hill, 2014.
4. C. M. Close, D. K. Frederick, and J. C. Newell, *Modeling and Analysis of Dynamic Systems*, Wiley, 3rd ed., 2002.
5. D. G. Luenberger, *Introduction to Dynamic Systems: Theory, Models, and Applications*, Wiley, 1979.
6. R. S. Esfandiari and B. Lu, *Modeling and Analysis of Dynamic Systems*, CRC Press, 2010.
7. R. L. Woods and K. L. Lawrence, *Modeling and Simulation of Dynamic Systems*, Prentice Hall, 1997.
8. K. Ogata, *System Dynamics*, 4th ed., Pearson, 2004.

PROBLEMS

Section 6.2 System Modeling Techniques

6.1 Derive the transfer function for the following dynamic systems, where $f(t)$ is the input and $x(t)$ is the output. Represent these transfer functions in zero-pole-gain form. Indicate the order of each transfer function. Assume that $f(0^-) = \dot{f}(0^-) = \ddot{f}(0^-) = 0$.
(a) $x^{(3)} + 6\ddot{x} + 11\dot{x} + 6x = 3f(t) + \dot{f}(t)$
(b) $x^{(3)} + 13\ddot{x} + 50\dot{x} + 56x = \ddot{f}(t) - f(t)$
(c) $6x^{(4)} + 10x^{(3)} + 15\ddot{x} + 22\dot{x} + 15x = 12f(t)$
(d) $3x^{(3)} + 6\ddot{x} + 7\dot{x} + 5x = 15f(t) + 3\dot{f}(t)$
(e) $6x^{(3)} + 2\ddot{x} + 7\dot{x} + 3x = 9f(t) + 3\dot{f}(t) + 2\ddot{f}(t)$
(f) $x^{(5)} + 7x^{(4)} + 20x^{(3)} + 30\ddot{x} + 24\dot{x} + 8x = 3f(t) - 5\dot{f}(t) + \ddot{f}(t) + f^{(3)}(t)$

6.2 Derive the transfer function for the following dynamic systems, where $f(t)$ is the input and $2x(t) + 3\dot{x}(t)$ is the output. Indicate the order of each transfer function. Assume that $f(0^-) = 0$.
(a) $\ddot{x} + 3\dot{x} + 7x = 5f(t)$
(b) $12x^{(3)} + 5\ddot{x} + 2\dot{x} + 6x = 24\dot{f}(t) - 18f(t)$

(c) $6x^{(4)} + 6x^{(3)} + 9\ddot{x} + 5\dot{x} + 3x = 18f(t) + 3\dot{f}(t)$
(d) $2\ddot{x} + 9\dot{x} + 8x = 8f(t) + 10\dot{f}(t)$

6.3 Derive the following transfer functions:

(a) $\begin{cases} \ddot{x} + 3\dot{x} + 7x - 7y = 5f(t) \\ 2\ddot{y} + 7y - 7x = 0 \end{cases}$
where the input is $f(t)$, the outputs are $x(t)$ and $y(t)$

(b) $\begin{cases} 5\ddot{x} + 2\dot{x} - 2\dot{y} + 11x - 8y = f(t) \\ 3\ddot{y} + 2\dot{y} + 8y - 2\dot{x} - 8x = 0 \end{cases}$
where the input is $f(t)$, the outputs are $x(t) - y(t)$ and $\dot{y}(t)$

(c) $\begin{cases} 2\ddot{x} + \dot{x} - \dot{y} + 4x = f(t) \\ \ddot{y} + 3\dot{y} + y - \dot{x} = g(t) \end{cases}$
where the inputs are $f(t)$ and $g(t)$, the output is $\dot{y}(t)$

(d) $\begin{cases} 2\ddot{x} + 3\dot{x} - \dot{y} + 4x - 3y = f(t) \\ 6\ddot{y} + 2\dot{y} + 5y - \dot{x} = g(t) \end{cases}$
where the inputs are $f(t)$ and $g(t)$, the outputs are $\dot{x}(t)$ and $\dot{y}(t)$

6.4 Derive the state-space representation for the following dynamic systems, where $f(t)$ is the input and $x(t)$ is the output. Also, cast the state and output equations in matrix form.

(a) $x^{(3)} + 6\ddot{x} + 11\dot{x} + 6x = 3f(t)$
(b) $x^{(3)} + 13\ddot{x} + 50\dot{x} + 56x = 15f(t)$
(c) $6x^{(4)} + 10x^{(3)} + 15\ddot{x} + 22\dot{x} + 15x = 12f(t)$
(d) $3x^{(3)} + 6\ddot{x} + 7\dot{x} + 5x = 15f(t) + 3\dot{f}(t)$

6.5 Derive the state-space representation for the following dynamic systems, where $f(t)$ is the input and $5x(t) + \dot{x}(t)$ is the output. Also, cast the state and output equations in matrix form.

(a) $\ddot{x} + 3\dot{x} + 7x = 5f(t)$
(b) $12x^{(3)} + 5\ddot{x} + 2\dot{x} + 6x = 24f(t)$
(c) $6x^{(4)} + 6x^{(3)} + 9\ddot{x} + 5\dot{x} + 3x = 18f(t)$

6.6 Derive the state-space representations for the systems in Problem 6.3; cast the state and output equations in matrix form.

6.7 Derive the state-space representations for the following systems described by their transfer functions:

(a) $\dfrac{Y(s)}{R(s)} = \dfrac{6}{s^2 + 4s + 17}$

(b) $\dfrac{Y(s)}{R(s)} = \dfrac{15}{2s^3 + 5s^2 + 3s + 4}$

(c) $\dfrac{Y(s)}{R(s)} = \dfrac{7}{6s^4 + x^3 + 9s^2 + 2s + 1}$

6.8 Construct a block diagram for the dynamic system represented by the following governing equations:

(a) $\begin{cases} \dot{x} + 3x - 5y = 2f(t) \\ \dot{y} + 3y + 4x = 0 \end{cases}$

where the input is $f(t)$, the output is y

(b) $\begin{cases} \dot{x} + 2x - 6y = 3f(t) \\ \dot{y} + 8y - 4x = -5f(t) \end{cases}$

where the input is $f(t)$, the output is y

(c) $\begin{cases} \dot{x} - 2y = 0 \\ \dot{y} + 7x + 4y - 3w = 0 \\ \dot{w} - 5v = 0 \\ \dot{v} - 6x + 2w = f(t) \end{cases}$

where the input is $f(t)$, the output is x

(d) $2\ddot{x} + 6\dot{x} + 4x = 3f(t) + 5g(t)$

where the inputs are $f(t)$ and $g(t)$, the output is x

(e) where the inputs are $f(t)$ and $g(t)$, the output is y

(f) $\begin{cases} \ddot{x} + 5\dot{x} + 7y = f(t) + 6g(t) \\ \dot{y} + 2x - 3y = 4g(t) \end{cases}$

where the inputs are $f(t)$ and $g(t)$, the output is x

6.9 Derive the transfer function $\dfrac{X(s)}{R(s)}$ for the dynamic systems modeled with the block diagrams shown in Figure P6.9:

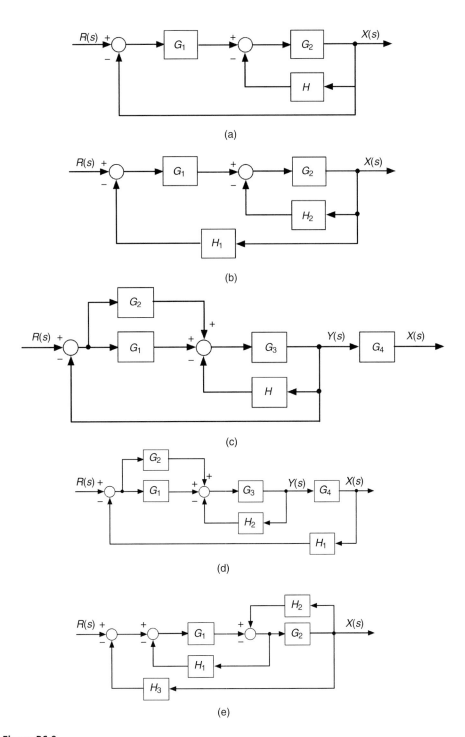

Figure P6.9

6.10 Derive the transfer functions $\dfrac{X(s)}{R(s)}$ and $\dfrac{X(s)}{D(s)}$ for the dynamic systems modeled with the block diagrams shown in Figure P6.10:

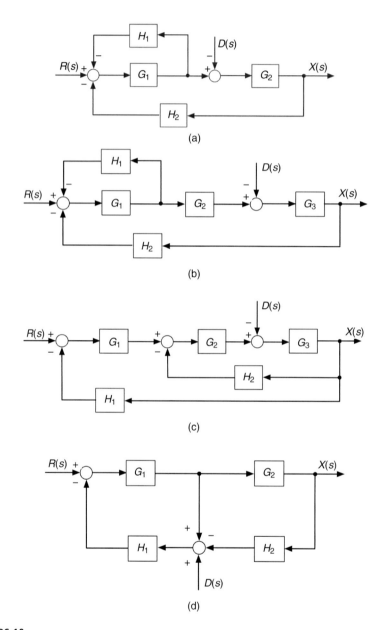

Figure P6.10

Section 6.3 Fundamentals of Electromechanical Systems

6.11 A transducer – an electromechanical system designed to measure translational motion – is shown in Figure P6.11. The permanent magnet produces a uniform magnetic field with magnetic flux density $\vec{B}$. This magnet is in one-dimensional motion, with displacement $x(t)$ as shown and corresponding velocity $\dot{x}(t)$. A conductor is fixed in space in such a way that its portion of the length h is inside the air gap between the north and south poles of the permanent magnet. This conductor is modeled as a series connection of an inductor with inductance L_0 and resistor with resistance R_0. Another resistor with resistance R is connected to this conductor with a lossless wire, forming a closed circuit as shown in Figure P6.11.

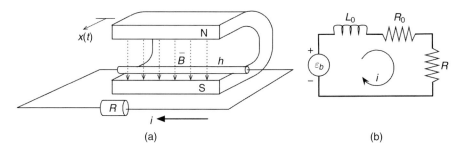

Figure P6.11 Transducer: (a) schematic, (b) electrical subsystem

When the permanent magnet experiences translational motion, the magnetic environment of the conductor changes, hence, an emf is induced in this conductor according to Faraday's law of induction, and current starts flowing in the circuit.

(a) Derive the governing equations for this system.

(b) Derive the transfer function $\dfrac{V_R(s)}{X(s)}$, considering that the input to the system is the displacement of the magnet $x(t)$ and the output is the voltage drop v_R on the resistor R.

6.12 Consider the transducer described in Problem 6.11.

The permanent magnet of mass M is sliding forward along a stationary horizontal surface under the influence of the applied force $f_M(t)$. The direction of motion and the displacement of the magnet $x(t)$ are shown in Figure P6.12. There is an oil film with a viscous friction coefficient c between the magnet and the surface.

(a) Derive the governing equations for this system.

(b) Derive the transfer function $\dfrac{V_R(s)}{F_M(s)}$, considering that the input to the system is the applied force $f_M(t)$ and the output is the voltage drop v_R on the resistor R.

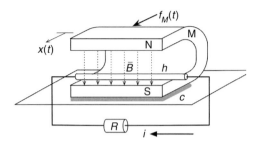

Figure 6.12 Transducer for Problem P6.12

6.13 Consider the system shown in Figure P6.13. A conductor of length h is attached to a block of mass M with a rigid dielectric connector. The block is sliding along a stationary horizontal surface in the direction shown by its displacement $x(t)$. The block is constrained with the linear spring with stiffness coefficient k, and is separated from the surface by an oil film with viscous friction coefficient c.

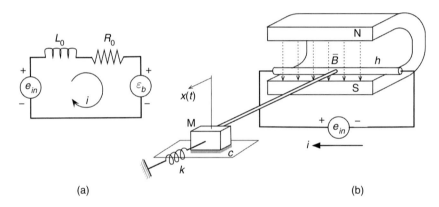

Figure P6.13 Conductor in the magnetic field: (a) electrical subsystem, (b) schematic

The conductor is fixed in space in such a way that its length h is inside the air gap between the north and south poles of the permanent magnet, which produces a uniform magnetic field with magnetic flux density $\vec{B}$. The conductor is modeled as a series connection of an inductor with inductance L_0 and resistor with resistance R_0. The conductor is connected to a voltage source, which provides the input voltage $e_{in}(t)$.
(a) Derive the governing equations for this system.
(b) Derive the state-space representation, considering that the system input is voltage $e_{in}(t)$, and the output is the velocity of the block.

6.14 A setup for a magnetic levitation laboratory experiment (Yeh et al., Sliding Control of Magnetic Bearing Systems, *ASME Journal of Dynamic Systems, Measurement, and Control*, Vol. 123, September 2001) is shown in Figure P6.14.

The current-carrying coil, wound around a metal core, constitutes an electromagnet that exerts an attracting electromagnetic force F_{em} on the metal ball of mass m, making it levitate.

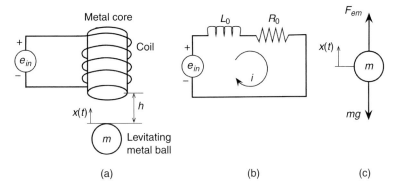

Figure P6.14 Magnetic levitation experiment: (a) schematic, (b) electrical subsystem, (c) mechanical subsystem

The electromagnetic force is modeled as follows:

$$F_{em} = K_F \frac{i^2}{(h-x)^2}$$

where K_F is a constant that depends on the construction of the electromagnet (material of the core and coil, number and geometry of loops in the coil, etc.), $x(t)$ is the position of the levitating ball, measured upward from a fixed at-rest position, and h is a constant air-gap distance from the electromagnet core to the ball.
(a) Derive the governing equations for this system.
(b) Derive the state-space representation, considering that the system input is voltage $e_{in}(t)$, and the output is the displacement of the ball.

6.15 A solenoid actuator, typically used in hydraulic or pneumatic systems, is an electromechanical device that converts electrical energy into a translational force for pushing/pulling a valve.

The schematic of a typical solenoid actuator (Chladny et al., Modeling Automotive Gas-Exchange Solenoid Valve Actuators, *IEEE Transactions on Magnetics*, Vol. 41, March 2005) is shown in Figure P6.15.

The push-type solenoid actuator, shown in Figure P6.15(a), consists of a wire coil, wound around an iron core – a plunger, which is connected to a valve of mass m with a rigid push-pin. Applying voltage $e_{in}(t)$ to the coil makes it into an electromagnet, which exerts a force F_{em} on the plunger, pushing it to the right toward the center of the coil (the displacement of the plunger-valve assembly is $x(t)$ as shown in Figure P6.15(c)). A spring with stiffness coefficient k returns the valve to its original position when the current i no longer flows. A viscous friction force acts on the moving valve; the viscous friction coefficient is c (as shown in Figure P6.15(c)).

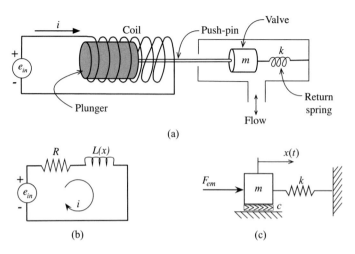

Figure P6.15 Solenoid actuator: (a) schematic, (b) electrical subsystem, (c) mechanical subsystem

The coil-plunger assembly is modeled as a series connection of the resistor R and the inductor. The inductance of the solenoid actuator coil is modeled as a nonlinear function of the plunger displacement: it increases when the plunger moves toward the center of the coil and decreases as the plunger moves out of the coil. This inductance is represented as

$$L(x) = \frac{L_0}{1 - x/d}$$

where L_0 and d are constants that depend on the geometry and material of the coil.

Since the solenoid coil is considered to be an ideal inductor, a linear relationship exists between the current through the inductor i_L and magnetic flux ϕ: $\phi = Li_L$ (see Eq. (4.1.18)). The time derivative of the magnetic flux is equal to the voltage drop across the inductor: $\frac{d\phi}{dt} = v_L$ (see Eq.(4.1.20)).

The electromagnetic force F_{em} is modeled as the first derivative of the energy stored in an inductor with respect to displacement $x(t)$: $F_{em} = \frac{dW}{dx}$, where W is described by Eq. (4.1.22).

Assume that the return spring is at its free length and the electromagnetic force is zero when $x = 0$.

(a) Derive the governing equations for this system.
(b) Derive the state-space representation, considering that the system input is the voltage $e_{in}(t)$, and the output is the displacement of the valve.

6.16 Consider the solenoid actuator described in Problem 6.15.

Experimentation led to development of the following expressions for the solenoid current and the electromagnetic force:

$$i_L(\phi) = a_1\phi^3 + a_2\phi$$
$$F_{em}(\phi) = b_1\phi^6 + b_2\phi^4 + b_3\phi^2$$

where a_1, a_2, b_1, b_2, b_3 are empirical constants (Vaughan, Gamble, The Modeling and Simulation of a Proportional Solenoid Valve, *ASME Journal of Dynamic Systems, Measurement, and Control*, Vol. 118, March 1996).

Assume that the return spring is at its free length and the electromagnetic force is zero when $x = 0$.

(a) Derive the governing equations for this system.
(b) Derive the state-space representation, considering that the system input is the voltage $e_{in}(t)$, and the output is the displacement of the valve.

Section 6.4 Direct Current Motors

6.17 An armature-controlled DC motor is used to rotate the load J_L as shown in Figure P6.17. The load is connected to the motor with a very stiff massless shaft and is supported by bearings with a damping coefficient b_L.

The system input is the voltage $e_{in}(t)$ supplied to the armature circuit, while the output is the rotational velocity of the load ω_L.

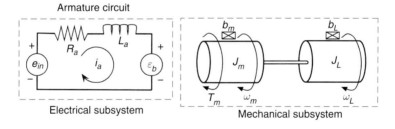

Figure P6.17 Armature-controlled DC motor with a load

(a) Derive the governing equations for this system.
(b) Derive the transfer function $\dfrac{\Omega_L(s)}{E_{in}(s)}$.

6.18 An armature-controlled DC motor is used to rotate the load J_L as shown in Figure P6.18. The load is connected to the motor through a converter K_a and is supported by bearings with a damping coefficient b_L. The converter transforms the rotational velocity ω_m of the motor into torque T_L applied to the rotor in the following manner: $T_L = K_a\omega_m$, where K_a is a known converter constant. The converter and the connector shafts are considered massless.

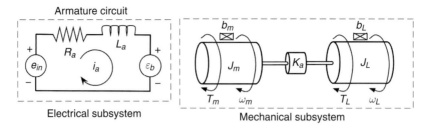

Figure P6.18 Armature-controlled DC motor with a load and converter

The system input is the voltage $e_{in}(t)$ supplied to the armature circuit, while the output is the rotational velocity of the load ω_L.
(a) Derive the governing equations for this system.
(b) Derive the transfer function $\dfrac{\Omega_L(s)}{E_{in}(s)}$.
(c) Construct a modular block diagram and indicate all the relevant signals such as the armature current i_a, back emf ε_b, motor torque T_m, motor velocity ω_m, bearing torques T_{bm} and T_{bL}, load torque T_L, and load velocity ω_L.

6.19 An armature-controlled DC motor is used to rotate the load J_L as shown in Figure P6.19. The load is supported by bearings with a damping coefficient b_L, and is connected to the motor with a flexible shaft, modeled as a torsional spring with a stiffness coefficient k_T.

The system input is the voltage $e_{in}(t)$ supplied to the armature circuit, while output is the rotational velocity of the load ω_L.

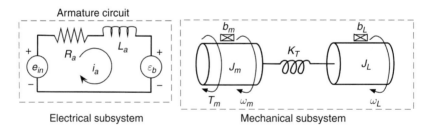

Figure P6.19 Armature-controlled DC motor with a load and a flexible shaft

(a) Derive the governing equations for this system.
(b) Derive the transfer function $\dfrac{\Omega_L(s)}{E_{in}(s)}$.
(c) Derive the state-space representation.
(d) Construct a modular block diagram and indicate all the relevant signals such as the armature current i_a, back emf ε_b, motor torque T_m, motor velocity ω_m, bearing torques T_{bm} and T_{bL}, spring torque T_k, and load velocity ω_L.

6.20 A field-controlled DC motor is used to rotate the load J_L as shown in Figure P6.20. The load is supported by bearings with a damping coefficient b_L, and is connected to the motor through a gear pair with a gear ratio $N = \dfrac{n_2}{n_1}$. The shafts connecting the gear pair with the motor and the load are short, stiff, and considered massless.

The system input is the voltage $e_{in}(t)$ supplied to the field circuit, while output is the rotational velocity of the load ω_L.

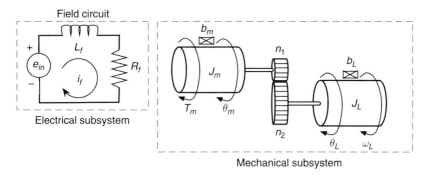

Figure P6.20 Field-controlled DC motor with a load and gear pair

(a) Derive the governing equations for this system.
(b) Derive the transfer function $\dfrac{\Omega_L(s)}{E_{in}(s)}$.
(c) Derive the state-space representation.

6.21 A field-controlled DC motor is used to rotate the load J_L as shown in Figure P6.21. The load is supported by bearings with a damping coefficient b_L, and is connected to the motor with a flexible shaft, modeled as a torsional spring with a stiffness coefficient k_T. The load is subject to an unknown drag torque T_D.

The system inputs are the voltage $e_{in}(t)$ supplied to the field circuit and the drag torque T_D, while output is the rotational velocity of the load ω_L.

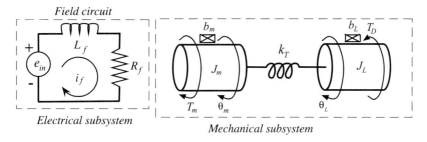

Figure P6.21 Field-controlled DC motor with a load, flexible shaft, and drag torque

(a) Derive the governing equations for this system.
(b) Derive the transfer functions $\dfrac{\Omega_L(s)}{E_{in}(s)}$ and $\dfrac{\Omega_L(s)}{T_D(s)}$.
(c) Derive the state-space representation.
(d) Construct a modular block diagram and indicate all the relevant signals such as the field current i_f, motor torque T_m, motor displacement θ_m, bearing torques T_{bm} and T_{bL}, spring torque T_k, drag torque T_D, load velocity ω_L, and displacement θ_L.

6.22 A field-controlled DC motor is used to rotate the load J_L as shown in Figure P6.22. The load is connected to the motor through a converter K_a and is supported by bearings with a damping coefficient b_L. The converter transforms the rotational velocity ω_m of the motor into the torque T_L applied to the rotor in the following manner: $T_L = K_a \omega_m$, where K_a is a known converter constant. The converter and the connector shafts are considered massless.

The load is subject to an unknown drag torque T_D.

The voltage supplied to the field circuit of the DC motor is driven by a generator. This voltage is proportional to the current in the generator circuit as follows: $e_f = K_g i_g$, where a K_g is a known constant.

The system inputs are the voltage $e_{in}(t)$ supplied to the generator and the drag torque T_D, while output is the rotational velocity of the load ω_L.

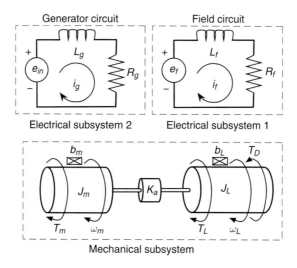

Figure P6.22 Field-controlled DC motor with a load and a generator

(a) Derive the governing equations for this system.
(b) Derive the transfer functions $\dfrac{\Omega_L(s)}{E_{in}(s)}$ and $\dfrac{\Omega_L(s)}{T_D(s)}$.
(c) Derive the state-space representation.
(d) Construct a modular block diagram and indicate all the relevant signals such as the generator current i_g, field circuit input voltage e_f, field current i_f, motor torque T_m, motor velocity ω_m, bearing torques T_{bm} and T_{bL}, drag torque T_D, and load velocity ω_L.

Section 6.5 Block Diagrams in the Time Domain

6.23 Construct time-domain block diagrams for the dynamic systems represented by equations in Problem 6.8.

6.24 Construct a time-domain block diagram for the DC-motor system in Problem 6.18.

6.25 Construct a time-domain block diagram for the DC-motor system in Problem 6.19.

6.26 Construct a time-domain block diagram for the DC-motor system in Problem 6.20.

6.27 Construct a time-domain block diagram for the DC-motor system in Problem 6.21.

Section 6.6 Modeling and Simulation by Simulink

6.28 Consider that the transducer described in Problem 6.11 has the following parameters:

$$L_0 = 0.003 \text{ H}, \ R_0 = 0.8 \ \Omega, \ R = 1.2 \ \Omega, \ h = 0.03 \text{ m}, \ B = 1.25 \text{ T}.$$

The input displacement of the magnet is $x(t) = t$.
Plot the voltage drop on the resistor R (consider a simulation time $0 \le t \le 0.02$ s), find the steady-state value if it exists, and compute the settling time for 0.5% allowed error.

6.29 Consider that the transducer described in Problem 6.12 has the following parameters:

$$L_0 = 0.003 \text{ H}, \ R_0 = 0.8 \ \Omega, \ R = 1.2 \ \Omega, \ h = 0.03 \text{ m}, \ B = 1.25 \text{ T},$$

$$m = 0.25 \text{ kg}, \ c = 7.5 \times 10^{-4} \ \frac{\text{N}}{\text{A} \cdot \text{m}}$$

The external force applied to the magnet is constant: $f_M(t) = 3.1$ N.
Plot the voltage drop on the resistor R (consider a simulation time $0 \le t \le 200$s), find the steady-state value if it exists, and compute the settling time for 1.0% allowed error.

6.30 Consider that the armature-controlled DC motor with a load and converter system described in Problem 6.18 has the following parameters:

$$L_a = 0.004 \text{H}, R_a = 0.75 \ \Omega, \ J_m = 8.0 \times 10^{-3} \text{ kg} \cdot \text{m}^2, \ b_m = 0.005 \ \frac{(\text{N} \cdot \text{m} \cdot \text{s})}{\text{rad}},$$

$$K_m = 0.3 \ \frac{\text{N} \cdot \text{m}}{\text{A}}, \ K_b = 0.2 \ \frac{\text{V} \cdot \text{s}}{\text{rad}}, \ K_a = 0.4 \ \frac{(\text{N} \cdot \text{m} \cdot \text{s})}{\text{rad}},$$

$$J_L = 12.0 \times 10^{-3} \text{ kg} \cdot \text{m}^2, \ b_L = 0.006 \ \frac{\text{N} \cdot \text{m} \cdot \text{s}}{\text{rad}}$$

(a) The voltage applied to the armature circuit of the DC motor is constant: $e_{in}(t) = 0.1$ V. Plot the velocity of the load ω_L (consider a simulation time $0 \le t \le 20$ s), and find the steady-state value if it exists.

(b) The voltage applied to the armature circuit of the DC motor is a positive half-sine periodic function $e_{in}(t) = 0.5|\sin(t)|$ V ($|\sin(t)|$ indicates the absolute value of the sine function). Plot the velocity of the load ω_L (consider simulation time $0 \le t \le 50$ s), and find the largest velocity value after the amplitude of the oscillations settles to a constant sinusoidal pattern.

31 Consider that the field-controlled DC motor with a load and gear pair described in Problem 6.20 has the following parameters:

$$L_f = 0.006 \text{ H}, R_f = 0.56 \text{ Ω}, J_m = 9.1 \times 10^{-4} \text{ kg·m}^2, b_m = 0.0012 \frac{(\text{N·m·s})}{\text{rad}},$$

$$K_m = 0.06 \frac{\text{N·m}}{\text{A}}, N = \frac{n_2}{n_1} = 2.5, J_L = 15.0 \times 10^{-4} \text{ kg·m}^2, b_L = 0.002 \frac{(\text{N·m·s})}{\text{rad}}$$

(a) The voltage applied to the field circuit of the DC motor is constant: $e_{in}(t) = 5.2$ V. Plot the velocity of the load ω_L (consider a simulation time $0 \le t \le 20$ s), and find steady-state value if exists; plot the field current (consider simulation time $0 \le t \le 0.2$ s), and find the steady-state value if it exists.

(b) The voltage applied to the field circuit of the DC motor is a pulse: $e_{in}(t) = 2u(t) - 2u(t-5)$ V. Plot the velocity of the load ω_L (consider a simulation time $0 \le t \le 20$ s), find the maximum velocity and its steady-state value if it exists; plot the field current (consider a simulation time $0 \le t \le 20$ s), find the maximum current and its steady-state value if exists.

(c) The voltage applied to the field circuit of the DC motor is a trapezoidal pulse, shown in Figure P6.31.

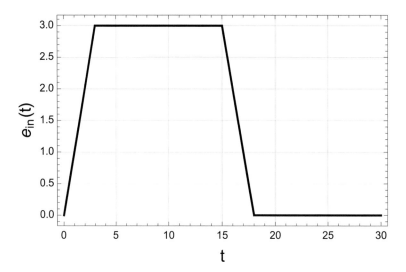

Figure P6.31 Input voltage

Plot the velocity of the load ω_L (consider a simulation time $0 \le t \le 30$ s), find the maximum velocity and its steady-state value if it exists; plot the field current (consider simulation time $0 \le t \le 30$ s), find the maximum current and its steady-state value if it exists.

6.32 Consider that the field-controlled DC motor with a load, flexible shaft, and drag described in Problem 6.21 has the following parameters:

$L_f = 0.006$ H, $R_f = 0.56$ Ω, $J_m = 9.1 \times 10^{-4}$ kg·m², $b_m = 0.0012 \dfrac{\text{N·m·s}}{\text{rad}}$,

$K_m = 0.06 \dfrac{\text{N·m}}{\text{A}}$, $k_T = 0.5 \dfrac{\text{N·m}}{\text{rad}}$, $J_L = 25.0 \times 10^{-4}$ kg·m², $b_L = 0.002 \dfrac{\text{N·m·s}}{\text{rad}}$,

$e_{in}(t) = 5$ V

(a) Plot the velocity of the load ω_L (consider simulation time $0 \leq t \leq 15$ s) for the case when the drag torque $T_D = 0$ and when $T_D = 0.005\, \dot{\theta}_L \text{sgn}(\dot{\theta}_L)$.

$\text{sgn}(\dot{\theta}_L)$ is the signum or sign function, defined as $\text{sgn}(x) = \begin{cases} 1 & \text{when } x > 0 \\ -1 & \text{when } x < 0 \\ 0 & \text{when } x = 0 \end{cases}$.

Do both systems converge to a steady state? Qualitatively describe the influence of the drag torque on the system response.

(b) Plot the velocity of the load ω_L (consider simulation time $0 \leq t \leq 15$ s) for the case when drag torque is applied with 5-s delay: $T_D = 0.005\, \dot{\theta}_L \text{sgn}(\dot{\theta}_L) u(t-5)$. Does the system converge to a steady state? Qualitatively describe influence of the delayed drag torque on system response.

Section 6.7 Modeling and Simulation by Mathematica

6.33 Solve Problem 6.28 by Mathematica.
6.34 Solve Problem 6.29 by Mathematica.
6.35 Solve Problem 6.30 by Mathematica.
6.36 Solve Problem 6.31 by Mathematica.
6.37 Solve Problem 6.32 by Mathematica.

7 System Response Analysis

Contents

7.1	System Response Analysis in the Time Domain	502
7.2	Stability Analysis	544
7.3	System Response Analysis in the Frequency Domain	558
7.4	Time Response of Linear Time-Varying and Nonlinear Systems	571
	Chapter Summary	577
	References	578
	Problems	578

In the previous chapters, mathematical models of various dynamic systems have been developed; several modeling tools, including differential equations by the first laws of nature, transfer function formulations, block diagrams, and state-space representations, have been introduced. With these models and tools, the dynamic response of a system can be determined, and the understanding of its behaviors can be gained through analysis and simulation.

In this chapter, three main issues regarding physical behaviors of dynamic systems are addressed: system response analysis in the time domain, stability analysis, and system response analysis in the frequency domain. In the presentation, analytical methods are used for simple systems; MATLAB and Simulink are used for general linear and nonlinear systems.

7.1 System Response Analysis in the Time Domain

In this section, solutions of the response of linear time-invariant systems in the time domain are determined by Laplace transform. Refer to Section 2.6 for the Laplace transform method.

7.1.1 General Concepts

Consider a general dynamic system governed by an n-th-order linear ordinary differential equation

$$A(D)y(t) = B(D)r(t) \qquad (7.1.1)$$

where $D = d/dt$, $A(D)$ and $B(D)$ are polynomials of operator D

$$A(D) = \sum_{k=0}^{n} a_k D^k, \quad B(D) = \sum_{l=0}^{m} b_l D^l$$

with a_k and b_l being constants, and $r(t)$ and $y(t)$ are the forcing function (system input) and response (system output), respectively. Refer to Section 2.5.1 for details on the operator polynomials. Let the initial conditions of Eq. (7.1.1) be

$$y(0) = a_{0,0}, \quad Dy(0) = a_{0,1}, \ldots, D^{n-1}y(0) = a_{0,n-1} \qquad (7.1.2)$$

where $a_{0,0}, a_{0,1}, \ldots, a_{0,n-1}$ are prescribed values. The basic problem of system response analysis in the time domain is to find the solution of the differential equation (7.1.1) subject to the initial conditions (7.1.2).

Free Response and Forced Response

The solution of the previous differential equation can be obtained by Laplace transforms. To this end, performing Laplace transform of Eq. (7.1.1) yields

$$A(s)Y(s) = I(s) + B(s)R(s) \qquad (7.1.3)$$

where $R(s)$ and $Y(s)$ are the Laplace transforms of the input $r(t)$ and output $y(t)$, respectively, and $I(s)$ is a polynomial of s, which is a linear combination of the initial values $a_{0,0}, a_{0,1}, \ldots, a_{0,n-1}$. It can be shown that $I(s) = 0$ if all the initial values of system response are set to zero. Following Section 2.6, the time response of the system is given by

$$y(t) = y_I(t) + y_F(t) \qquad (7.1.4)$$

with

$$y_I(t) = \mathcal{L}^{-1}\left\{\frac{I(s)}{A(s)}\right\} \qquad (7.1.5a)$$

$$y_F(t) = \mathcal{L}^{-1}\{G(s)R(s)\} \qquad (7.1.5b)$$

where $\mathcal{L}^{-1}$ is the inverse Laplace transform operator, $y_I(t)$ and $y_F(t)$ are the free response and forced response, respectively, and $G(s)$ is a transfer function of the system given by

$$G(s) = \frac{B(s)}{A(s)} = \frac{b_m s^m + b_{m-1} s^{m-1} + \cdots + b_1 s + b_0}{a_n s^n + a_{n-1} s^{n-1} + \cdots + a_1 s + a_0} \qquad (7.1.6)$$

For inverse Laplace transforms, refer to Section 2.4.4.

Example 7.1.1

A spring–mass–damper system is governed by

$$m\ddot{x} + c\dot{x}(t) + kx(t) = f(t)$$
$$x(0) = x_0, \quad \dot{x}(0) = v_0$$

Laplace transform of the previous differential equation yields

$$\underbrace{(ms^2 + cs + k)}_{A(s)} X(s) = \underbrace{F(s)}_{B(s)=1} + \underbrace{m(sx_0 + v_0) + cx_0}_{I(s)}$$

By Eq. (7.1.5a,b), the free response and forced response of the system are given by

$$y_I(t) = \mathcal{L}^{-1}\left\{ \frac{m(sx_0 + v_0) + cx_0}{ms^2 + cs + k} \right\}$$

$$y_F(t) = \mathcal{L}^{-1}\left\{ \frac{1}{ms^2 + cs + k} F(s) \right\}$$

From a physical viewpoint, $y_I(t)$ is the system response excited by initial disturbances and it is described by the differential equation

$$A(D)y_I(t) = 0 \qquad (7.1.7a)$$

subject to initial conditions

$$y_I(0) = a_{0,0}, \quad Dy_I(0) = a_{0,1}, \ldots, \quad D^{n-1}y_I(0) = a_{0,n-1} \qquad (7.1.7b)$$

The $y_F(t)$, on the other hand, is the system response excited by the input $r(t)$ and it is governed by

$$A(D)y_F(t) = B(D)r(t) \qquad (7.1.8a)$$

with zero initial conditions

$$y_F(0) = 0, \quad Dy_F(0) = 0, \ldots, \quad D^{n-1}y_F(0) = 0 \qquad (7.1.8b)$$

Because Eq. (7.1.7a) is a homogeneous differential equation, its solution can be expressed by

$$y_I(t) = A_1 e^{\lambda_1 t} + A_2 e^{\lambda_2 t} + \ldots + A_n e^{\lambda_n t} \qquad (7.1.9)$$

where λ_k are the roots of the characteristic equation

$$A(\lambda) = a_n\lambda^n + a_{n-1}\lambda^{n-1} + \ldots + a_1\lambda + a_0 = 0$$

and A_k are constants that are eventually determined by the initial conditions (7.1.7b). Here, distinct characteristic roots have been assumed. If $A(\lambda)$ has l repeated roots λ_0 (a root of l multiplicity) and m distinct roots $\lambda_1, \lambda_2, \ldots, \lambda_m$, with $l+m=n$ and $\lambda_0 \neq \lambda_k$ for $j=1,2,\ldots,m$, the free response is of the form

$$y_I(t) = A_1 e^{\lambda_1 t} + A_2 e^{\lambda_2 t} + \ldots + A_m e^{\lambda_m t} + \left(B_1 t^{l-1} + B_2 t^{l-2} + \ldots + B_l\right) e^{\lambda_0 t} \qquad (7.1.10)$$

Refer to Section 2.6 for more on the complementary homogeneous equation.

The solution of Eq. (7.1.8a), through the application the convolution theorem (Section 2.4.3) to Eq. (7.1.5b), can be expressed by the convolution integral

$$y_F(t) = \int_0^t g(t-\tau)r(\tau)d\tau \qquad (7.1.11)$$

Here $g(t)$ is the impulse response function of the system, which is the solution of

$$A(D)g(t) = B(D)\delta(t)$$
$$g(0) = 0, \quad Dg(0) = 0, \ldots, \quad D^{n-1}g(0) = 0$$

with $\delta(t)$ being the Dirac delta function. Physically, $g(t)$ is the response of the system under an impulse of unit amplitude, which is often used to describe collision, impact, and shock. It is easy to show that the impulse response function is related to the system transfer function by

$$G(s) = \mathcal{L}\{g(t)\} \qquad (7.1.12)$$

where $\mathcal{L}$ is the Laplace transform operator.

Thus, the free and force responses of a dynamic system can be obtained either by direct inverse Laplace transform, as shown in Eqs. (7.1.5a) and (7.1.5b), or by the homogeneous solution of Eq. (7.1.9) and the convolution integral of Eq. (7.1.11). These analytical methods shall be used alternatively for convenience in analysis.

Transient Response and Steady-State Response

The response (output) $y(t)$ of a dynamic system can be viewed as a sum of a transient response $y_{tr}(t)$ and a steady-state response y_{ss}. The *transient response* $y_{tr}(t)$ describes the change and transition of the system with respect to time and it is a function of time t. If a system is stable and is under a bounded input, its transient response disappears after a long enough time. (The stability of a dynamic system is examined in Section 7.2.) The *steady-state response* y_{ss} is the remaining part of the system response after all transients have died out and it is the time limit

$$y_{ss} = \lim_{t \to \infty} y(t) \tag{7.1.13}$$

The y_{ss} is also known as a final value of the output. A transient response can be excited by initial disturbances or by the application of an input. A steady-state response, however, can only be excited by an input as it physically represents a balance between input energy and output energy. In other words, a steady-state response is contained in the forced response $y_F(t)$.

It should be pointed out that a steady-state response may not exist if the system is unstable or if the input is unbounded. Figure 7.1.1 shows three cases of system responses: (a) response y settles to a steady-state value y_{ss} after the transients die out; (b) response y has no steady state because the limit $\lim_{t \to \infty} y(t)$ does not exist; and (c) response y, being unbounded, has no steady state.

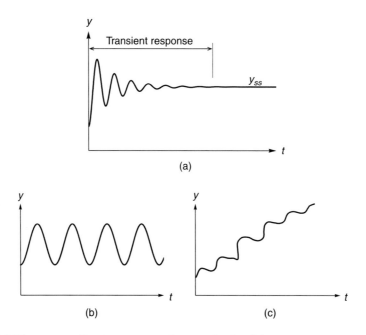

Figure 7.1.1 Three cases of system response (see text for details)

Example 7.1.2
Consider a system with response y governed by

$$\dot{y} + \sigma y = r_0, \quad \text{with } y(0) = y_0$$

where $\sigma > 0$ and r_0 is a constant. The solution of the differential equation by any method given in Chapter 2 yields

$$y(t) = \left(y_0 - \frac{r_0}{\sigma}\right)e^{-\sigma t} + \frac{r_0}{\sigma}$$

On the right-hand side of the previous equation, the first term is the transient response, which vanishes as t goes to infinity; the second term is the steady-state response:

$$y_{ss} = \lim_{t \to \infty} y(t) = \frac{r_0}{\sigma}.$$

Final-Value Theorem for Steady-State Response

The steady-state response of a dynamic system can be determined through direct use of system response in the s-domain. According to the previous discussion, a steady-state response can only be contained in the forced response $y_F(t)$, which in the s-domain is

$$Y(s) = G(s)R(s) \qquad (7.1.14)$$

where Eq. (7.1.3) with $I(s) = 0$ and Eq. (7.1.6) have been used. If all the poles of $sY(s)$ have negative real parts, by the final-value theorem described in Section 2.4.3, the steady-state response of the system exists, and its final value is given by

$$y_{ss} = \lim_{t \to \infty} y(t) = \lim_{s \to 0} sY(s) \qquad (7.1.15)$$
$$= \lim_{s \to 0} sG(s)R(s)$$

When using Eq. (7.1.15), it is extremely important to check *the condition of negative real parts for all the poles of $sY(s)$* of the final-value theorem.

Equation (7.1.15) delivers y_{ss} without the need to perform inverse Laplace transform and to take the time limit, which is convenient in transfer function formulation.

Example 7.1.3

Consider the same system in Example 7.1.2. The s-domain response of the system is $Y(s) = \dfrac{r_0}{s(s+\sigma)}$, which is obtained via the Laplace transform of the differential equation for the system with $y_0 = 0$. Thus, $sY(s) = \dfrac{r_0}{s+\sigma}$ has a pole $p_1 = -\sigma < 0$, satisfying the condition of the final value theorem. It follows that the steady-state response of the system is

$$y_{ss} = \lim_{s \to 0} \frac{r_0}{s+\sigma} = \frac{r_0}{\sigma}$$

which is the same as the result obtained in Example 7.1.2.

System Order and Input Type

For convenience in analysis and for quantification of system response, system order and input type are specified. The order of a dynamic system modeled by the transfer function in Eq. (7.1.6) is the order n of the transfer function, which is also the order of its denominator $A(s)$ or the order of the relevant differential equation. Thus, $G(s)$ in Eq. (7.1.6) is an n-th-order transfer function.

Some of commonly used forcing functions are listed as follows:

Impulse function: $r(t) = I_0 \delta(t)$
Step function: $r(t) = r_0$
Ramp function: $r(t) = \alpha t$
Sinusoidal function: $r(t) = A \sin \omega t + B \cos \omega t$

where $\delta(t)$ is the Dirac delta function, which is described in Section 2.4.1. Sinusoidal forcing functions are mostly considered in frequency response analysis; see Section 7.3.

7.1.2 Time Responses of First-Order Systems

First-order systems are seen in various applications, including mass–damper systems, rotating shafts, resistor–capacitor (RC) and inductor–resistor (LR) circuits, liquid-level systems, and thermal systems. A typical first-order system is governed by the differential equation

$$a\dot{y}(t) + by(t) = f(t) \tag{7.1.16}$$

where $f(t)$ and $y(t)$ are the input and output, respectively. For example, the current of an LR circuit is governed by

$$L \frac{di(t)}{dt} + Ri(t) = v_i(t)$$

where $v_i(t)$ is an input voltage. If $b = 0$, Eq. (7.1.16) can be solved via direct integration:

$$y(t) = y(0) + \frac{1}{a}\int_0^t f(\tau)d\tau$$

Thus, in the subsequent discussion, we will consider $b \neq 0$, which is seen in many applications. For convenience in analysis, Eq. (7.1.16) is converted to a standard form

$$T\dot{y}(t) + y(t) = r(t) \qquad (7.1.17)$$

where $T = \frac{a}{b}$ and $r(t) = \frac{1}{b}f(t)$. The parameter T is known as the time constant of the system. The transfer function of the system is

$$G(s) = \frac{Y(s)}{R(s)} = \frac{1}{Ts+1} \qquad (7.1.18)$$

Free Response

The free response of a first-order system is governed by

$$T\dot{y}(t) + y(t) = 0, \quad y(0) = y_0 \qquad (7.1.19)$$

By either the Laplace transform shown in Eq. (7.1.5a) or the formula in Eq. (7.1.9), the solution of Eq. (7.1.19) is obtained as

$$y(t) = y_0\, e^{-\frac{t}{T}} \qquad (7.1.20)$$

Thus, the system free response decays exponentially. The value of the time constant T determines the rate of decay of $y(t)$. The smaller the T, the faster the free response $y(t)$ decays, as shown in Figure 7.1.2.

Forced Response

The forced response of the first-order system is described by

$$T\dot{y}(t) + y(t) = r(t), \quad y(0) = 0 \qquad (7.1.21)$$

By Eqs. (7.1.5b) and (7.1.18), the forced response is given by

$$y(t) = \mathcal{L}^{-1}\left\{\frac{1}{Ts+1}R(s)\right\} \qquad (7.1.22)$$

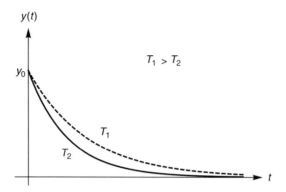

Figure 7.1.2 Free-response curves of two first-order systems with different time constants

Now consider the forced response of a standard first-order system subject to the inputs specified in Section 7.1.1.

Impulse Response
For a system under an impulse input $r(t) = I_0\delta(t)$, its response is

$$y(t) = \mathcal{L}^{-1}\left\{\frac{1}{Ts+1}I_0\right\} = I_0\,g(t), \text{ with } g(t) = \frac{1}{T}e^{-\frac{t}{T}} \tag{7.1.23}$$

where $g(t)$ is the impulse response of the system. The final value y_{ss} of the response is zero. The impulse response curves of two systems with different time constants ($T_1 = 2$ and $T_2 = 1$) are plotted in Figure 7.1.3 with MATLAB. As seen from the figure, a smaller time constant leads to the response reaching zero more quickly.

It is seen from Figures 7.1.2 and 7.1.3 that the impulse response and the free response are similar in pattern. This can be explained as follows. The impulse response is governed by

$$T\dot{y}(t) + y(t) = I_0\delta(t), \text{ for } t > 0-$$
$$\text{with } y(0-) = 0$$

Integration of the previous equation in the domain $0- \leq t \leq 0+$

$$\int_{0-}^{0+}\left(T\dot{y}(t) + y(t)\right)dt = \int_{0-}^{0+} I_0\delta(t)\,dt = I_0$$

yields

$$Ty(0+) = Ty(0-) + I_0 = I_0$$

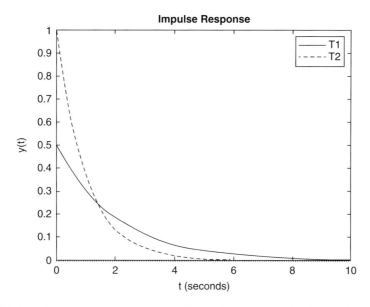

Figure 7.1.3 Impulse response curves of two first-order systems of different time constants: $T_1 > T_2$, generated by MATLAB

where, by the continuity of $y(t)$, $\int_{0-}^{0+} y(t)dt = 0$. Thus, the original impulse response problem is converted to the following free-response problem

$$T\dot{y}(t) + y(t) = 0, \text{ for } t > 0+$$
$$\text{with } y(0+) = I_0/T$$

This explanation can be extended to general linear time-invariant systems.

Step Response

For a system under a step input $r(t) = r_0$, its response is

$$y(t) = \mathcal{L}^{-1}\left\{\frac{1}{Ts+1}\frac{r_0}{s}\right\} = r_0\left(1 - e^{-\frac{t}{T}}\right) \quad (7.1.24)$$

The final value of the system response (the steady-state response) is $y_{ss} = r_0$. Figure 7.1.4 shows the step-response curves of two systems with different time constants. As can be seen from the figure, a smaller time constant results in the response reaching the final value more quickly.

Ramp Response

For a system under a ramp forcing function, $r(t) = \alpha t$, its response is

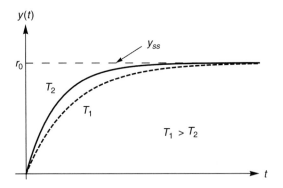

Figure 7.1.4 Step-response curves of two first-order systems with different time constants

$$y(t) = \mathcal{L}^{-1}\left\{\frac{1}{Ts+1}\frac{\alpha}{s^2}\right\} = \alpha\left\{t - T\left(1 - e^{-\frac{t}{T}}\right)\right\} \tag{7.1.25}$$

As time goes by, the forced response becomes unbounded, $y(t) \to \alpha(t - T)$. Equation (7.1.25) also implies that the response lags the input by a time T.

Now define an error function as the difference between the input and response

$$e(t) \equiv r(t) - y(t) = \alpha T\left(1 - e^{-\frac{t}{T}}\right) \tag{7.1.26}$$

According to Eq. (7.1.26), the steady-state error exists:

$$e_{ss} = \lim_{t \to \infty} e(t) = \alpha T \tag{7.1.27}$$

See Figure 7.1.5 for the input and response curves, time lag, and steady-state error.

Response to General Inputs

The impulse response function of a system by Eq. (7.1.12) is

$$g(t) = \mathcal{L}^{-1}\left\{\frac{1}{Ts+1}\right\} = \frac{1}{T}e^{-\frac{t}{T}} \tag{7.1.28}$$

which is also shown in Eq. (7.1.23). The forced response of the system subject to an arbitrary input can be obtained by the convolution integral of Eq. (7.1.11), and it is given by

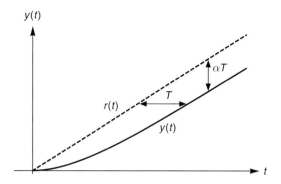

Figure 7.1.5 Ramp response and steady-state error of a first-order system

$$y(t) = \int_0^t g(t-\tau)r(\tau)d\tau = \int_0^t \frac{1}{T} e^{-\frac{t-\tau}{T}} r(\tau)d\tau \qquad (7.1.29)$$

Example 7.1.4

Under an input $r(t) = r_0(1 - e^{-\sigma t})$, with $\sigma > 0$ and $\sigma \neq \frac{1}{T}$, the system response by Eq. (7.1.29) is determined as

$$y(t) = \int_0^t \frac{1}{T} e^{-\frac{t-\tau}{T}} r_0 (1 - e^{-\sigma\tau}) d\tau$$

$$= \frac{r_0}{T} e^{-\frac{t}{T}} \left[T e^{\frac{\tau}{T}} - \frac{T}{1-\sigma T} e^{\frac{1}{T}(1-\sigma T)\tau} \right]_{\tau=0}^{\tau=t}$$

$$= r_0 + r_0 \left(\frac{\sigma T}{1-\sigma T} e^{-\frac{t}{T}} - \frac{1}{1-\sigma T} e^{-\sigma t} \right)$$

On the right-hand side of the previous equation, the first term is the steady-state response, $y_{ss} = r_0$; the second term is the transient response, which vanishes as time goes by.

Estimation of the Time Constant

As shown in Figures 7.1.2 to 7.1.5, the time constant T of a first-order system characterizes its free and forced responses. Hence, modeling and analysis of a first-order system requires the knowledge of its time constant. The time constant of a first-order system can be estimated from its response curves. For instance, consider the step response shown in Figure 7.1.6. At $t = T$, the system response by Eq. (7.1.24) is $y(T) = r_0(1 - e^{-1}) = 0.632 y_{ss}$. In other words, T is the time at which the step response is 63.2% of the steady-state value. Also, the slope of the step response at the initial time is $\dot{y}(0) = r_0 \frac{1}{T}$, indicating that $T = \frac{y_{ss}}{\dot{y}(0)}$. Additionally, the steady-state error shown in Figure 7.1.5 can be used to estimate the time constant. Indeed, Eq. (7.1.27) leads to $T = e_{ss}/\alpha$.

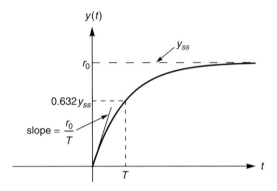

Figure 7.1.6 Step response at $t = T$ and slope at $t = 0$

Example 7.1.5
The output of a first-order system subject to a step input has a final value $y_{ss} = 2.5$. At $t = 2$, the system response is 1.0. Determine the time constant of the system.

Solution
By Eq. (7.1.24),

$$y(2) = 2.5 \times \left(1 - e^{-\frac{2}{T}}\right) = 1$$

where $y_{ss} = r_0$ has been used. Thus,

$$e^{-\frac{2}{T}} = 1 - \frac{1}{2.5} = 0.6$$

It follows that the time constant of the system is

$$T = \frac{-2}{\ln(0.6)} = 3.915$$

7.1.3 Time Responses of Second-Order Systems

Second-order systems are seen in many applications, including spring–mass–damper systems, rotating shafts with torsional springs and dampers, *RCL* circuits, direct current (DC) motors, liquid-level systems, thermal systems, and first-order systems with feedback control. A typical second-order system is governed by the differential equation

$$a\ddot{y} + b\dot{y}(t) + cy(t) = f(t) \tag{7.1.30}$$

with $a > 0$, $b \geq 0$, $c > 0$, where $f(t)$ and $y(t)$ are the system input and output, respectively. For convenience in analysis, Eq. (7.1.30) is converted to a standard form

$$\ddot{y} + 2\zeta\omega_n\dot{y}(t) + \omega_n^2 y(t) = \omega_n^2 r(t) \tag{7.1.31}$$

where ω_n and ζ are the natural frequency and damping ratio of the system, respectively, and they are called *model parameters*. The model parameters are related to the physical parameters a, b, c by

$$\omega_n = \sqrt{\frac{c}{a}}, \quad \zeta = \frac{1}{2}\frac{b}{\sqrt{ac}} \tag{7.1.32}$$

and the scaled input $r(t) = \frac{1}{c}f(t)$. In addition, the following two parameters are often used in analysis:

$$\text{Natural frequency} \quad f_n = \frac{\omega_n}{2\pi}, \quad \text{in Hz}$$

$$\text{Natural period} \quad T = \frac{1}{f_n} = \frac{2\pi}{\omega_n}, \quad \text{in seconds}$$

The transfer function of the system, from Eq. (7.1.31), is

$$G(s) = \frac{Y(s)}{R(s)} = \frac{\omega_n^2}{s^2 + 2\zeta\omega_n s + \omega_n^2} \tag{7.1.33}$$

The poles of the transfer function are the roots of the characteristic equation

$$s^2 + 2\zeta\omega_n s + \omega_n^2 = 0 \tag{7.1.34}$$

and they are

$$\begin{matrix} s_1 \\ s_2 \end{matrix} = -\zeta\omega_n \pm \sqrt{\zeta^2 - 1}\,\omega_n \tag{7.1.35}$$

Example 7.1.6

The equation of motion of a spring–mass–damper system is

$$m\ddot{x} + c\dot{x}(t) + kx(t) = f(t)$$

which can be converted to Eq. (7.1.31), with

$$\omega_n = \sqrt{\frac{k}{m}}, \quad \xi = \frac{1}{2}\frac{c}{\sqrt{mk}}, \quad r(t) = \frac{1}{k}f(t).$$

The transfer function of the system in terms of the physical parameters is

$$\frac{X(s)}{F(s)} = \frac{1}{ms^2 + cs + k}$$

which can be converted to the standard form of Eq. (7.1.33) if the scaled input $r(t)$ is used. The transfer function poles are the roots of the characteristic equation $ms^2 + cs + k = 0$, which are obtained as

$$\begin{matrix} s_1 \\ s_2 \end{matrix} = \frac{1}{2m}\left(-c \pm \sqrt{c^2 - 4mk}\right)$$

A more general form of second-order systems is

$$a\ddot{y} + b\dot{y}(t) + cy(t) = \alpha \dot{f}(t) + \beta f(t) \tag{7.1.36}$$

The transfer function for such a system is

$$\frac{Y(s)}{R(s)} = \frac{(\alpha s + \beta)\,\omega_n^2}{s^2 + 2\xi\omega_n s + \omega_n^2} \tag{7.1.37}$$

if the scaled input $r(t) = f(t)/c$ is used. It is observed that the transfer function has a zero $s = -\beta/\alpha$, which induces numerator dynamics. Also, a comparison of the transfer functions in Eqs. (7.1.33) and (7.1.37) shows that the two systems have the same characteristic equation (7.1.34). This indicates that the input $f(t)$ does not alter the transfer function poles.

Without loss of generality, the standard form of Eq. (7.1.33) will now be used to examine the system response. The effect of the zero on the system response, as shown in the general form of Eq. (7.1.36), shall be discussed in Section 7.1.3.

Four Damping States

According to Eq. (7.1.35), there are four cases or states of damping in the system.

(a) No damping ($\xi = 0$)
The poles of the system are of the imaginary form

$$\begin{matrix} s_1 \\ s_2 \end{matrix} = \pm j\omega_n, \quad j = \sqrt{-1} \qquad (7.1.38)$$

In this case, the system is undamped, and it is called an undamped system.

(b) Underdamping ($0 < \xi < 1$)
The poles of the system are a complex conjugate pair with negative real part:

$$\begin{matrix} s_1 \\ s_2 \end{matrix} = -\sigma \pm j\omega_d, \quad j = \sqrt{-1} \qquad (7.1.39)$$

with $\sigma = \xi\omega_n$ and $\omega_d = \sqrt{1-\xi^2}\ \omega_n$. Here σ and ω_d are the decay factor and damped frequency, respectively. In this case, the system is underdamped, and it is called an underdamped system.

(c) Critical damping ($\xi = 1$)
The poles of the system are identical and they are real and negative as follows:

$$\begin{matrix} s_1 \\ s_2 \end{matrix} = -\omega_n \qquad (7.1.40)$$

In this case, the system is critically damped, and it is called a critically damped system. Critical damping is the border between underdamping and overdamping cases.

(d) Overdamping ($\xi > 1$)
The poles of the system are two distinct real and negative numbers, as given by (7.1.35), and they can be written as

$$\begin{matrix} s_1 \\ s_2 \end{matrix} = -\sigma \pm \beta \qquad (7.1.41)$$

with $\sigma = \xi\omega_n$ and $\beta = \sqrt{\xi^2 - 1}\ \omega_n$. In this case, the system is overdamped, and it is called an overdamped system.

The above four damping cases are applicable to a second-order system of the general form of Eq. (7.1.37) because it has the same characteristic equation (Eq. (7.1.34)).

The four damping cases of second-order systems show that the value of the damping ratio ζ determines the pole locations in the complex plane; see Figure 7.1.7, where crosses ($\times$) indicate the pole locations. As shown in the figure, the poles of the system in cases (b) to (d) all have negative real parts. As shall be seen subsequently, different pole locations give different patterns of system response.

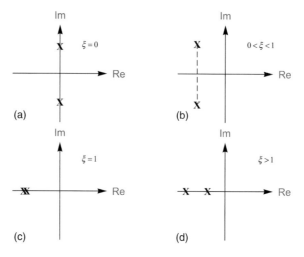

Figure 7.1.7 Damping ratio and pole locations: cross $\times$ = pole location

Example 7.1.7
For a spring–mass–damper system, $m\ddot{x} + c\dot{x}(t) + kx(t) = f(t)$, derive the conditions on the system parameters (m, c, k) such that the system is underdamped.

Solution

From Example 7.1.6, the damping ratio is given by $\zeta = \dfrac{1}{2}\dfrac{c}{\sqrt{mk}}$. For underdamping, $0 < \zeta < 1$, which requires that $0 < \dfrac{1}{2}\dfrac{c}{\sqrt{mk}} < 1$. Thus, the conditions of underdamping are

$$c > 0 \text{ and } c^2 < 4mk$$

Another way to solve this problem is to consider the locations of the transfer function poles, without having to derive the damping ratio in terms of the system parameters. From Example 7.1.6, the characteristic equation of the system is $ms^2 + cs + k = 0$, from which the poles are

$$\begin{matrix} s_1 \\ s_2 \end{matrix} = \dfrac{1}{2m}\left(-c \pm \sqrt{c^2 - 4mk}\right)$$

For the system to be underdamped, its poles s_1 and s_2 must be a complex-conjugate pair with a negative real part (Figure 7.1.7b). This leads to the conditions of $c > 0$ and $c^2 < 4mk$.

Example 7.1.8
Given a system

$$\frac{X(s)}{F(s)} = \frac{8}{s^2 + 3s + 8}$$

compute its natural frequency and damping ratio, and comment on the damping characteristics of the system.

Solution
Comparing the coefficients of the characteristic equation for the system

$$s^2 + 3s + 8 = 0$$

with those of Eq. (7.1.34), yields

$$2\xi\omega_n = 3, \quad \omega_n^2 = 8$$

Solution of the previous algebraic equations gives

$$\omega_n = 2\sqrt{2} = 2.8284, \quad \xi = \frac{3\sqrt{2}}{8} = 0.5303$$

Because $0 < \xi < 1$, the system is underdamped.

Free Response

The free response of a second-order system is governed by

$$\ddot{y} + 2\xi\omega_n \dot{y}(t) + \omega_n^2 y(t) = 0 \tag{7.1.42}$$

with the initial conditions $y(0) = y_0$, $\dot{y}(0) = v_0$. Substituting $y = Ae^{\lambda t}$ into the homogeneous equation yields the characteristic equation

$$\lambda^2 + 2\xi\omega_n \lambda + \omega_n^2 = 0 \tag{7.1.43}$$

Through comparison of Eq. (7.1.34) and Eq. (7.1.43), it is concluded that the poles s_1, s_2 of the transfer functions are the same as the characteristic roots λ_1, λ_2 for the homogeneous solution. Thus, the free-response solution can be expressed as

$$y(t) = A_1 e^{s_1 t} + A_2 e^{s_2 t} \tag{7.1.44}$$

with A_1 and A_2 to be determined by the initial conditions.

The free response of a system is determined in the following four damping cases.

(a) No damping ($\xi = 0$)
The homogeneous solution can be written as

$$y(t) = A_1 e^{j\omega_n t} + A_2 e^{-j\omega_n t}$$
$$= B_1 \cos \omega_n t + B_2 \sin \omega_n t$$

where Euler's formula $e^{j\theta} = \cos\theta + j\sin\theta$ has been used. The constants B_1 and B_2 can be directly determined by the initial conditions, without the need to know A_1 and A_2. It follows that the free response is given by

$$y(t) = y_0 \cos \omega_n t + \frac{v_0}{\omega_n} \sin \omega_n t \tag{7.1.45}$$

(b) Underdamping ($0 < \xi < 1$)
The homogeneous solution is given by

$$y(t) = A_1 e^{(-\sigma+j\omega_d)t} + A_2 e^{(-\sigma-j\omega_d)t}$$
$$= e^{-\sigma t}(B_1 \cos \omega_d t + B_2 \sin \omega_d t)$$

with $\sigma = \xi\omega_n$ and $\omega_d = \sqrt{1-\xi^2}\,\omega_n$. From the initial conditions, the free response is obtained as

$$y(t) = e^{-\sigma t}\left(y_0 \cos \omega_d t + \frac{v_0 + \sigma y_0}{\omega_d} \sin \omega_d t\right) \tag{7.1.46}$$

(c) Critical damping ($\xi = 1$)
Because the system has repeated poles as shown in Eq. (7.1.40), Eq. (7.1.44) is not usable. From differential-equation theory, the solution of Eq. (7.1.42) in this case is of the form

$$y(t) = e^{-\omega_n t}(B_1 t + B_0)$$

Thus, with the initial conditions, the free response is determined as follows:

$$y(t) = e^{-\omega_n t}\left(y_0 + (v_0 + \omega_n y_0)t\right) \tag{7.1.47}$$

(d) Overdamping ($\xi > 1$)
The homogeneous solution is given by

$$y(t) = A_1 e^{(-\sigma+\beta)t} + A_2 e^{(-\sigma-\beta)t}$$
$$= e^{-\sigma t}(B_1 \cosh \beta t + B_2 \sinh \beta t)$$

with $\sigma = \xi\omega_n$ and $\beta = \sqrt{\xi^2-1}\,\omega_n$. From the initial conditions, the free response is obtained as

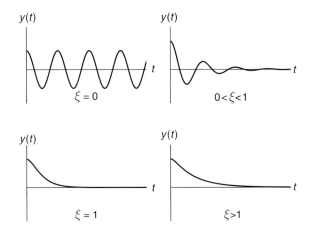

Figure 7.1.8 The influence of the damping ratio ξ on the free response of a second-order system

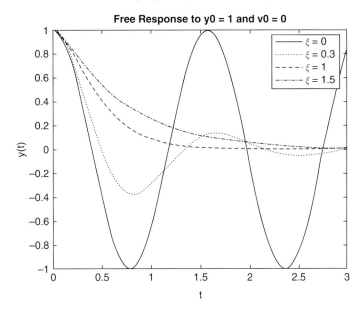

Figure 7.1.9 Free responses with $\omega_n = 4$ and $\xi = 0$, 0.3, 1.0, and 1.5, subject to $y_0 = 1$, $v_0 = 0$, plotted with MATLAB

$$y(t) = e^{-\sigma t}\left(y_0 \cosh \beta t + \frac{v_0 + \sigma y_0}{\beta} \sinh \beta t\right) \qquad (7.1.48)$$

The free-response curves of a second-order system in the four damping cases are shown in Figure 7.1.8. As can be seen from the figure, the damping ratio of a system has significant influence on the pattern of the free response: from oscillation to decayed oscillation to exponential decay, as ξ increases.

To see the effects of damping in further detail, consider four systems with the same natural frequency $\omega_n = 4$ and different damping ratios $\xi = 0$, 0.3, 1.0, and 1.5. Assume the same initial conditions for all the systems: $y(0) = 1$ and $\dot{y}(0) = 0$. The free-response curves are computed from Eqs. (7.1.45) to (7.1.48), and they are plotted in Figure 7.1.9. As seen from the figure, in the first half cycle, the undamped system ($\xi = 0$) takes a shorter time ($t = \pi/(2\omega_n) = 0.393\text{s}$) to reach the equilibrium position ($y = 0$) than the underdamped system ($\xi = 0.3$). The critically damped system ($\xi = 1.0$) and the overdamped system ($\xi = 1.5$), on the other hand, never reach the equilibrium position because their responses are nonoscillatory.

Forced Response

The forced response of a second-order system is governed by the differential equation

$$\ddot{y} + 2\xi\omega_n \dot{y}(t) + \omega_n^2 y(t) = r(t) \tag{7.1.49}$$

with the zero initial conditions $y(0) = 0$, $\dot{y}(0) = 0$. The solution of Eq. (7.1.49) by the transfer function formulation (7.1.5b), is

$$y(t) = \mathcal{L}^{-1}\left\{ \frac{\omega_n^2}{s^2 + 2\xi\omega_n s + \omega_n^2} R(s) \right\} \tag{7.1.50}$$

where the transfer function given in Eq. (7.1.33) has been used. Refer to Section 2.4 for inverse Laplace transforms.

We now examine the impulse and step responses of second-order systems.

Response to an Impulse Input

For a system under an impulse input $r(t) = I_0 \delta(t)$, its forced response is

$$y(t) = I_0 \, g(t) \tag{7.1.51}$$

where $g(t)$ is the impulse response function given by

$$g(t) = \mathcal{L}^{-1}\{G(s)\}$$
$$= \mathcal{L}^{-1}\left\{ \frac{\omega_n^2}{s^2 + 2\xi\omega_n s + \omega_n^2} \right\} \tag{7.1.52}$$

The impulse response function is obtained in the following four damping cases.

(a) Undamped system ($\xi = 0$)

$$g(t) = \omega_n \sin \omega_n t \tag{7.1.53a}$$

(b) Underdamped system $(0 < \xi < 1)$

$$g(t) = \frac{\omega_n^2}{\omega_d} e^{-\sigma t} \sin \omega_d t = \frac{\omega_n}{\sqrt{1-\xi^2}} e^{-\sigma t} \sin \omega_d t \qquad (7.1.53b)$$

(c) Critically damped system $(\xi = 1)$

$$g(t) = \omega_n^2 \, t \, e^{-\omega_n t} \qquad (7.1.53c)$$

(d) Overdamped system $(\xi > 1)$

$$g(t) = \frac{\omega_n^2}{\beta} e^{-\sigma t} \sinh \beta t = \frac{\omega_n}{\sqrt{\xi^2 - 1}} e^{-\sigma t} \sinh \beta t \qquad (7.1.53d)$$

In the above formulas, $\sigma = \xi \omega_n$, $\omega_d = \sqrt{1-\xi^2}\, \omega_n$ and $\beta = \sqrt{\xi^2 - 1}\, \omega_n$.

In Case (b) to iv ($\xi > 0$), the steady-state response of the system is zero. The steady-state response in Case (a) ($\xi = 0$) does not exist.

The impulse response function $g(t)$ in Eqs. (7.1.53a–d) can be obtained by the inverse Laplace transform of the system transfer function $G(s)$. In Cases (a), (b), and (d), $G(s)$ has two distinct poles p_1 and p_2, and it can be expressed by partial fraction expansion:

$$G(s) = \frac{K_1}{s - p_1} + \frac{K_2}{s - p_2}$$

In Case (c), $G(s)$ has two identical poles, $p_1 = p_2 = p$, and it can be written as

$$G(s) = \frac{b_1}{s - p} + \frac{b_2}{(s - p)^2}$$

Here, the coefficients K_1, K_2, b_1, and b_2 are the residues of $G(s)$. Thus, determination of the transfer function residues gives the impulse response function as follows:

$$g(t) = \begin{cases} K_1 e^{p_1 t} + K_2 e^{p_2 t} & \text{for Cases (a), (b), and (d)} \\ e^{pt}(b_1 + b_2 t) & \text{for Case (c)} \end{cases}$$

The curves of the impulse response function $g(t)$ are plotted in Figure 7.1.10 for the four damping cases: $\omega_n = 4$ and $\xi = 0, 0.3, 1.0, 1.5$.

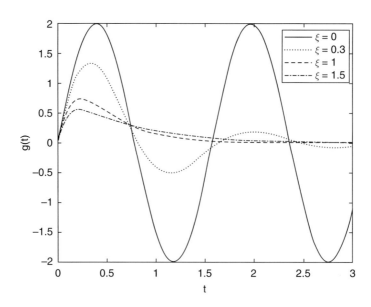

Figure 7.1.10 Impulse response function $g(t)$ with $\omega_n = 4$ and $\xi = 0, 0.3, 1.0$ and 1.5, plotted with MATLAB

Example 7.1.9

Given a system $\dfrac{X(s)}{F(s)} = \dfrac{8}{s^2 + 3s + 8}$, determine its forced response subject to an impulse input $r(t) = 5\delta(t)$.

Solution
From Example 7.1.8, the damping ratio of the system is $\xi = 0.5303$, indicating that the system is underdamped. Also, by the characteristic equation

$$s^2 + 3s + 8 = 0$$

the poles of the transfer function are

$$\begin{matrix} p_1 \\ p_2 \end{matrix} = \frac{1}{2}(-3 \pm j\sqrt{23}) = -1.5 \pm j2.3979$$

which gives $\sigma = 1.5$, $\omega_d = 2.3979$. It follows from Eqs. (7.1.51) and (7.1.53b) that the impulse response of the system is given by

$$y(t) = 5 \times \frac{1}{2.3979} e^{-1.5t} \sin 2.3979t$$
$$= 2.0851\, e^{-1.5t} \sin 2.3979t$$

Response to a Step Input

For a system subject to a step input $r(t) = r_0$, its forced response is given by

$$y(t) = \mathcal{L}^{-1}\left\{G(s)\frac{r_0}{s}\right\} = \mathcal{L}^{-1}\left\{\frac{\omega_n^2}{s^2 + 2\zeta\omega_n s + \omega_n^2}\frac{r_0}{s}\right\} \quad (7.1.54)$$

The inverse Laplace transform in Eq. (7.1.26) can be carried out through the use of partial fraction expansion of the system transfer function (Section 2.4):

$$\frac{\omega_n^2}{s^2 + 2\zeta\omega_n s + \omega_n^2}\frac{r_0}{s} = \frac{K_0}{s} + \frac{K_1}{s - p_1} + \frac{K_2}{s - p_2}$$

where p_1 and p_2 are the poles of $G(s)$, as given in Eq. (7.1.35). If the poles are repeated (as in the case of critical damping), $p_1 = p_2 = p$ and,

$$\frac{\omega_n^2}{s^2 + 2\zeta\omega_n s + \omega_n^2}\frac{r_0}{s} = \frac{K_0}{s} + \frac{b_1}{s - p} + \frac{b_2}{(s - p)^2}$$

With the above expressions of partial fraction expansion, the step response of a second-order system is obtained in the following four damping cases.

(a) Undamped system ($\zeta = 0$)

$$y(t) = r_0 (1 - \cos\omega_n t) \quad (7.1.55a)$$

(b) Underdamped system ($0 < \zeta < 1$)

$$x(t) = r_0\left\{1 - e^{-\sigma t}\left(\cos\omega_d t + \frac{\sigma}{\omega_d}\sin\omega_d t\right)\right\} \quad (7.1.55b)$$

(c) Critically damped system ($\zeta = 1$)

$$y(t) = r_0\left\{1 - e^{-\omega_n t}(1 + \omega_n t)\right\} \quad (7.1.55c)$$

(d) Overdamped system ($\zeta > 1$)

$$y(t) = r_0\left\{1 - e^{-\sigma t}\left(\cosh\beta t + \frac{\sigma}{\beta}\sinh\beta t\right)\right\} \quad (7.1.55d)$$

In the above formulas, $\sigma = \zeta\omega_n$, $\omega_d = \sqrt{1 - \zeta^2}\,\omega_n$ and $\beta = \sqrt{\zeta^2 - 1}\,\omega_n$. In Cases (b) to (d) ($\zeta > 0$), the system response has a final value $y_{ss} = r_0$. The steady-state response in Case (a) ($\zeta = 0$) does not exist.

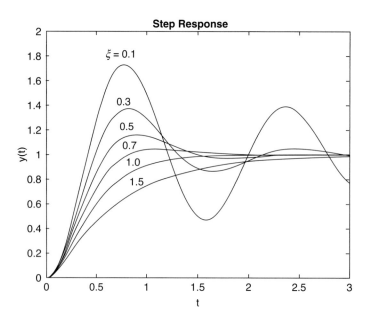

Figure 7.1.11 Step responses with $\omega_n = 4$ and $\xi = 0.1,\ 0.3,\ 0.5,\ 0.7,\ 1.0,\ 1.5$, subject to unit step input $r_0 = 1$

Figure 7.1.11 shows the curves of the step responses of second-order systems with $\omega_n = 4$ and $\xi = 0.1,\ 0.3,\ 0.5,\ 0.7,\ 1.0,\ 1.5$, subject to the unit step input $r_0 = 1$. The effect of damping on the system response is observed. With underdamping ($\xi = 0.1,\ 0.3,\ 0.5,\ 0.7$), the system response is oscillatory; with critical damping ($\xi = 1.0$) and over damping ($\xi = 1.5$), the system is nonoscillatory. All the response curves approach the same steady-state value, $y_{ss} = 1$.

General Second-Order Systems

For a second-order system in the general form of Eq. (7.1.37), its free response is the same as a system in the standard form because the two systems have the same governing equation (7.1.42). The forced response of a general second-order system in the s-domain can be written as

$$Y(s) = \frac{(\alpha s + \beta)\,\omega_n^2}{s^2 + 2\xi\omega_n s + \omega_n^2} R(s)$$

$$= \frac{\omega_n^2}{s^2 + 2\xi\omega_n s + \omega_n^2} \{\alpha R_1(s) + \beta R_2(s)\}$$

(7.1.56)

with $R_1(s) = sR(s)$ and $R_2(s) = R(s)$. This indicates that the forced response of a general second-order system can be viewed as the response of the standard system subject to an input that is the linear combination of $R_1(s)$ and $R_2(s)$. For instance, if a general system is subject to a step input $r(t) = r_0$, $r_1(t) = r_0\delta(t)$ and $r_2(t) = r_0$. Thus, the step response $y(t)$ of a general system is a linear combination of the impulse response $z_{imp}(t)$ and the step response $z_{stp}(t)$ of the standard system:

$$y(t) = \alpha\, z_{imp}(t) + \beta\, z_{stp}(t) \tag{7.1.57}$$

where

$$z_{imp}(t) = \mathcal{L}^{-1}\left\{\frac{\omega_n^2}{s^2 + 2\zeta\omega_n s + \omega_n^2}\, r_0\right\}$$

$$z_{stp}(t) = \mathcal{L}^{-1}\left\{\frac{\omega_n^2}{s^2 + 2\zeta\omega_n s + \omega_n^2}\, \frac{r_0}{s}\right\}$$

Also, by the final-value theorem in Section 7.1.1, the step response has a final value $y_{ss} = \beta r_0$.

Example 7.1.10

For a system $\dfrac{X(s)}{F(s)} = \dfrac{4s+1}{s^2+3s+8}$ subject to a step input $r(t) = 3$, obtain a mathematical expression of the forced response and plot the response.

Solution
Using Eq. (7.1.56)

$$Y(s) = \frac{4s+1}{s^2+3s+8}\frac{3}{s} = \frac{8}{s^2+3s+8}\frac{3}{2} + \frac{8}{s^2+3s+8}\frac{3}{8s} \tag{a}$$

From Examples 7.1.8 and 7.1.9, the system is underdamped, with

$$\omega_n = 2\sqrt{2} = 2.8284, \quad \zeta = \frac{3\sqrt{2}}{8} = 0.5303$$

$$\sigma = \zeta\omega_n = 1.5, \quad \omega_d = \sqrt{1-\zeta^2}\,\omega_n = 2.3979$$

Thus, application of Eqs. (7.1.53b) and (7.1.55b) to Eq. (a) yields the step response of the system as follows:

$$y(t) = \frac{3\omega_n^2}{2\omega_d}e^{-\sigma t}\sin\omega_d t + \frac{3}{8}\left\{1 - e^{-\sigma t}\left(\cos\omega_d t + \frac{\sigma}{\omega_d}\sin\omega_d t\right)\right\}$$

$$= 5.0043 e^{-1.5t}\sin 2.3979 t + \frac{3}{8}\{1 - e^{-1.5t}(\cos 2.3979 t + 0.6255\sin 2.3979 t)\} \tag{b}$$

By Eq. (b), the step response is plotted in Figure 7.1.12. As can be seen from the figure, the step response approaches a steady-state value $y_{ss} = 3/8 = 0.375$.

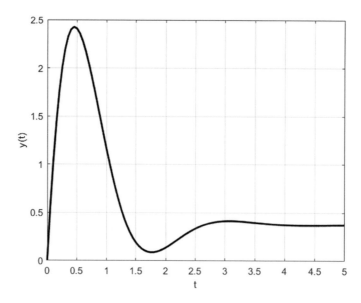

Figure 7.1.12 The step response of the system in Example 7.1.10

Note that the step response in Figure 7.1.12 can be easily generated by the MATLAB function step, with the transfer function formulation given in Eq. (a). This is done by the following MATLAB commands

```
sys = tf(3*[0 4 1], [1 3 8])
[y, t] = step(sys, 5);
plot(t, y, 'k', 'LineWidth', 2);
grid; xlabel('t'); ylabel('y(t)')
```

7.1.4 Specification of Step Responses of Underdamped Second-Order Systems

Step responses of underdamped second-order systems are commonly considered in dynamic analysis and feedback control in engineering applications. For an underdamped system $(0 < \xi < 1)$, its step response by Eq. (7.1.55b) is

$$y(t) = r_0 \left\{ 1 - e^{-\sigma t} \left(\cos \omega_d t + \frac{\sigma}{\omega_d} \sin \omega_d t \right) \right\}$$

$$= r_0 \left\{ 1 - \frac{1}{\sqrt{1 - \xi^2}} e^{-\xi \omega_n t} \cos \left(\sqrt{1 - \xi^2} \, \omega_n t - \phi \right) \right\} \qquad (7.1.58)$$

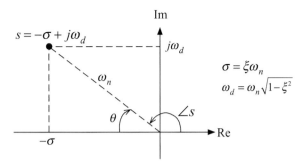

Figure 7.1.13 Pole plot of an underdamped system in the complex plane

where $\phi = \tan^{-1}(\xi/\sqrt{1-\xi^2})$. Step-response curves for some values of ξ are plotted in Figure 7.1.11.

Graphic Interpretation of Model Parameters

To better understand the influence of the model parameters ω_n and ξ on the step response, the pole $s = -\sigma + j\omega_d$ of an underdamped system in the complex plane is shown in Figure 7.1.13, where from Eq. (7.1.39) $\sigma = \xi\omega_n$ and $\omega_d = \sqrt{1-\xi^2}\,\omega_n$. From the figure, it is easy to show that the magnitude and phase angle of the pole are related to the model parameters by

$$|s| = \sqrt{\sigma^2 + \omega_d^2} = \omega_n$$
$$\angle s = \tan^{-1}\left(\frac{\omega_d}{-\sigma}\right) = \pi - \theta \tag{7.1.59}$$

where $\theta = \tan^{-1}(\omega_d/\sigma)$, in rad. It follows that

$$\cos\theta = \frac{\sigma}{|s|} = \xi \tag{7.1.60}$$

Two observations can be made on the above results: (i) the further away the pole s is from the origin, the larger the natural frequency ω_n; and (ii) the smaller the angle θ, the larger the damping ratio ξ. Also, it is seen from Eq. (7.1.60) that overdamped and undamped cases occur when θ is zero and $\pi/2$, respectively.

Performance Specification Parameters

To quantify the performance of an undamped second-order system, its step response described by Eq. (7.1.58) is specified in Figure 7.1.14, where t_r, t_p, M_p, and t_s are the rise

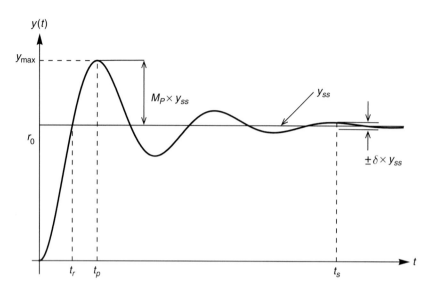

Figure 7.1.14 Specification of a step response

time, peak time, maximum overshoot, and settling time, respectively. These *performance specification parameters* are defined as follows:

(a) *Rise time* t_r, the time required for the output $y(t)$ to rise from 0% to 100% of its final value, $y_{ss} = r_0$. There are other definitions of the rise time, such as the time required for the output to go from 10% to 90% of y_{ss}. In this text, the 100% rise-time definition is used.
(b) *Peak time* t_p, the time required for the output $y(t)$ to reach the first and maximum peak; that is, $y(t_p) = y_{max}$.
(c) *Maximum overshoot* M_p, a non-dimensional number specifying the amount of the output $y(t)$ over its final value y_{ss} at the peak time t_p; namely,

$$M_p = \frac{y(t_p) - y_{ss}}{y_{ss}} \tag{7.1.61}$$

In other words, $y(t_p)/y_{ss} = 1 + M_p$. In applications, the maximum overshoot is often specified in percentage as follows:

$$\%OS = M_p \times 100\% \tag{7.1.62}$$

Here %OS is called the percent maximum overshoot.
(d) *Settling time* t_s, the time required for the output oscillations to stay with the specified percentage ($\pm\delta \times 100\%$) of the final value y_{ss}. The commonly used values of $\delta \times 100\%$ are 1%, 2%, and 5%. Here $\pm\delta$ specifies an error band or allowable tolerance.

The rise time, peak time, and settling time of a system tell how fast the system is in response to a step input. The maximum overshoot is a measure of the oscillatory character of the system response, and it is also an indicator of the relative stability of the system. These performance specification parameters are useful in the design of second-order systems.

We now evaluate the performance specification parameters in terms of ω_n and ξ.

Rise Time

According to the rise-time definition and from Eq. (7.1.58), we have

$$y(t_r) = r_0 \left\{ 1 - e^{-\sigma t} \left(\cos \omega_d t_r + \frac{\sigma}{\omega_d} \sin \omega_d t_r \right) \right\} = r_0$$

which implies that

$$\cos \omega_d t_r + \frac{\sigma}{\omega_d} \sin \omega_d t_r = 0$$

The solution of the previous trigonometry equation gives

$$\begin{aligned} t_r &= \frac{1}{\omega_d} \left(\pi - \tan^{-1}(\omega_d/\sigma) \right) \\ &= \frac{1}{\sqrt{1-\xi^2}\,\omega_n} \left(\pi - \tan^{-1}(\sqrt{1-\xi^2}/\xi) \right) \end{aligned} \quad (7.1.63)$$

Equation (7.1.63) indicates that for a given value of damping ratio ξ, the rise time is inversely proportional to the natural frequency ω_n. Note that $\pi - \tan^{-1}(\omega_d/\sigma)$ is the phase angle of the pole $s = -\sigma + j\omega_d$ and, from Eq. (7.1.59), the rise time can also be written as

$$t_r = \frac{\pi - \theta}{\omega_d} \quad (7.1.64)$$

where the angle θ is given by Eq. (7.1.60) and it is in radians.

Maximum Overshoot

The maximum overshoot occurs at the peak time, which satisfies $dy(t_p)/dt = 0$. From Eq. (7.1.58),

$$\begin{aligned} \frac{dy(t)}{dt} &= r_0 \left\{ \sigma e^{-\sigma t} \left(\cos \omega_d t + \frac{\sigma}{\omega_d} \sin \omega_d t \right) - e^{-\sigma t} (-\omega_d \sin \omega_d t + \sigma \cos \omega_d t) \right\} \\ &= r_0 \frac{1}{\omega_d} (\sigma^2 + \omega_d^2) e^{-\sigma t} \sin \omega_d t = 0 \end{aligned}$$

This implies that $\sin \omega_d t_p = 0$, which has solutions $\omega_d t_p = n\pi$, $n = 1, 2, \ldots$ Because of the exponentially decaying factor $e^{-\sigma t}$ in Eq. (7.1.58), $y(t_p) = y_{\max}$ occurs in the first cycle of oscillation, $n = 1$ (see Figure 7.1.14). Thus,

$$t_p = \frac{\pi}{\omega_d} = \frac{\pi}{\sqrt{1-\xi^2}\,\omega_n} \tag{7.1.65}$$

It follows from Eq. (7.1.58) that

$$\frac{y(t_p)}{r_0} = 1 + M_p = 1 - e^{-\sigma t_p}\left(\cos\omega_d t_p + \frac{\sigma}{\omega_d}\sin\omega_d t_p\right) = 1 + e^{-\sigma t_p}$$

in which Eq. (7.1.65) has been used. Consequently,

$$M_p = e^{-\sigma t_p} = e^{-\pi\xi/\sqrt{1-\xi^2}} \tag{7.1.66}$$

Equation (7.1.66) shows that the maximum overshoot is only dependent on the damping ratio, and that an increase in ξ decreases M_p.

If the maximum overshoot of a step response is known, either experimentally measured or prescribed in design, the damping ratio ξ, according to Eq. (7.1.66), is related to M_p by

$$\xi^2 = \frac{(\ln M_p)^2}{\pi^2 + (\ln M_p)^2} \tag{7.1.67}$$

Settling Time

The oscillations of the step response about its final value ($y_{ss} = r_0$) are described by

$$\frac{y(t) - y_{ss}}{y_{ss}} = -\frac{1}{\sqrt{1-\xi^2}} e^{-\xi\omega_n t}\cos\left(\sqrt{1-\xi^2}\,\omega_n t - \phi\right)$$

where Eq. (7.1.58) has been used. Because $|\cos\theta|\leq 1$ for any θ, the curves $\pm\dfrac{1}{\sqrt{1-\xi^2}}e^{-\xi\omega_n t}$ define an envelope for the step response; see Figure 7.1.15, which also shows that the oscillations after the settling time t_s are confined in the error band

$$\left|\frac{y(t)-y_{ss}}{y_{ss}}\right| \leq \frac{1}{\sqrt{1-\xi^2}}e^{-\xi\omega_n t} \leq \delta, \quad \text{for } t \geq t_s$$

Accordingly, the settling time with the $\delta \times 100\%$ criterion can be estimated by

$$\frac{1}{\sqrt{1-\xi^2}}e^{-\xi\omega_n t_s} = \delta \tag{7.1.68}$$

which gives

$$t_s = -\frac{1}{\xi\omega_n}\ln\left(\delta\sqrt{1-\xi^2}\right) = -\frac{1}{\xi\omega_n}\left\{\ln\delta + \ln\sqrt{1-\xi^2}\right\} \tag{7.1.69}$$

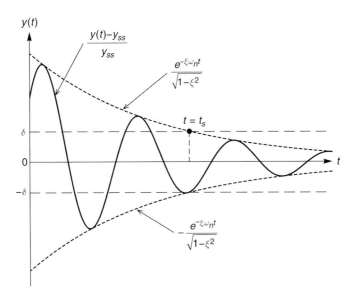

Figure 7.1.15 Envelope of the step response and the settling time

In practice, $|\ln \delta|$ is much larger than $|\ln \sqrt{1-\xi^2}|$. Thus, the following simplified formula for the settling time is often used

$$t_s = -\frac{\ln \delta}{\xi \omega_n} \tag{7.1.70}$$

which is obtained from Eq. (7.1.69) by neglecting $\ln \sqrt{1-\xi^2}$. Now, we calculate three numbers: $\ln(0.05) = -2.9957$, $\ln(0.02) = -3.9120$, and $\ln(0.01) = -4.6051$. Rounding these numbers and substituting the results into Eq. (7.1.70), yields three commonly used formulas for estimation of the settling time:

$$\begin{aligned} 5\% \text{ criterion } (\delta = 0.05): \quad & t_s = \frac{3}{\xi \omega_n} \\ 2\% \text{ criterion } (\delta = 0.02): \quad & t_s = \frac{4}{\xi \omega_n} \\ 1\% \text{ criterion } (\delta = 0.01): \quad & t_s = \frac{4.6}{\xi \omega_n} \end{aligned} \tag{7.1.71}$$

Other criteria with different values of δ, of course, can be derived from Eq. (7.1.70). For instance, for $\delta = 0.5\%$, $\ln(0.005) = -5.2983$ and $t_s = \frac{5.3}{\xi \omega_n}$.

Equations (7.1.70) and (7.1.71) are good for a relatively small damping ratio. If the damping ratio is relatively large, say $0.7 \leq \xi \leq 0.9$, Eq. (7.1.69) may be used to obtain a modified formula:

$$t_s = -\frac{1}{\xi\omega_n}\left\{\ln\delta + \ln\sqrt{1-0.8^2}\right\} = -\frac{1}{\xi\omega_n}\{\ln\delta - 0.5108\}$$

Here, 0.8 is the midpoint of the range $0.7 \leq \xi \leq 0.9$. In this case, the settling time with a 2% criterion becomes $t_s = \dfrac{4.51}{\xi\omega_n}$.

Example 7.1.11

For an underdamped system $\dfrac{X(s)}{F(s)} = \dfrac{8}{s^2 + 3s + 8}$, compute the rise time, maximum overshoot, and settling time with a 2% criterion of its step response.

Solution

The natural frequency and damping ratio of the system are

$$\omega_n = 2\sqrt{2} = 2.8284, \quad \xi = \frac{3\sqrt{2}}{8} = 0.5303$$

which were obtained in Example 7.1.8. Thus,

$$\sigma = \xi\omega_n = 1.5, \quad \omega_d = \sqrt{1-\xi^2}\,\omega_n = 2.3979$$

By Eqs. (7.1.63), (7.1.66), and (7.1.71), the rise time, maximum overshoot, and settling time are obtained as follows:

$$t_r = \frac{1}{\omega_d}\left(\pi - \tan^{-1}(\omega_d/\sigma)\right)$$
$$= \frac{1}{2.3979}\left(\pi - \tan^{-1}(2.3979/1.5)\right) = 0.888 \text{ s}$$

$$M_p = e^{-\pi\xi/\sqrt{1-\xi^2}}$$
$$= \exp\left(-\pi \times 0.5303/\sqrt{1-0.5303^2}\right) = 14.0\%$$

$$t_s = \frac{4}{\xi\omega_n} = \frac{4}{1.5} = 2.667 \text{ s}$$

Example 7.1.12

For the step response of an underdamped spring–mass–damper system, $m\ddot{x} + c\dot{x} + kx = f$, derive its rise time, maximum overshoot, and settling time (1% criterion) in terms of the system parameters (m, c, k).

Solution
According to Example 7.1.7, the natural frequency and damping ratio of the system are given by

$$\omega_n = \sqrt{\frac{k}{m}}, \quad \xi = \frac{1}{2}c/\sqrt{mk}$$

Also, the roots of the characteristic equation $ms^2 + cs + k = 0$ are

$$\begin{matrix} s_1 \\ s_2 \end{matrix} = -\sigma \pm j\omega_d = \frac{1}{2m}\left(-c \pm j\sqrt{4mk - c^2}\right)$$

with $j = \sqrt{-1}$. This implies that

$$\sigma = \frac{c}{2m}, \quad \omega_d = \frac{\sqrt{4mk - c^2}}{2m}$$

where the conditions $2\sqrt{mk} > c > 0$ for underdamping have been applied. It follows that

$$t_r = \frac{1}{\omega_d}\left(\pi - \tan^{-1}(\omega_d/\sigma)\right) = \frac{\{\pi - \tan^{-1}(\sqrt{4mk/c^2 - 1})\}}{\sqrt{k/m - (c/m)^2/4}}$$

$$M_p = e^{-\pi\sigma/\omega_d} = \exp\left\{\frac{-\pi}{\sqrt{4mk/c^2 - 1}}\right\}$$

$$t_s = \frac{4.6}{\sigma} = \frac{9.2m}{c}$$

Three Sets of Parameters in Design

In modeling and analysis of underdamped second-order systems, there are three sets of parameters: (i) *physical parameters*, such as m, c, and k; (ii) *model parameters* ω_n and ξ; and (iii) *performance specification parameters*, such as t_r, M_p, and t_s. As shown in Example 7.1.12, the physical parameters and performance specification parameters are linked by the model parameters:

$$\left\{\begin{array}{c}\text{Physical}\\\text{parameters}\end{array}\right\} \Leftrightarrow \left\{\begin{array}{c}\text{Model}\\\text{parameters: } \omega_n, \xi\end{array}\right\} \Leftrightarrow \left\{\begin{array}{c}\text{Performance specification}\\\text{parameters : } t_r, M_p, t_s\end{array}\right\}$$

In design of an underdamped second-order system based on its step response, certain performance specifications are first prescribed, such as

$$t_r \leq t_r^o; \quad m_L \leq M_p \leq m_H; \quad t_s \leq t_s^o \tag{7.1.72}$$

where t_r^o, m_L, m_H, t_s^o are specified values. The performance specification parameters are then related to the physical parameters through the model parameters, converting the performance specifications (7.1.72) into a set of conditions on the physical parameters. With these conditions, the values and/or ranges of the physical parameters are determined so that the system response satisfies the performance specifications.

Example 7.1.13

For a system $\dfrac{Y(s)}{R(s)} = \dfrac{b}{s^2 + (2+a)s + b}$, with $b > 0$, determine the parameters a and b, such that its step response meets the following two performance specifications: (i) $M_p \leq 10\%$; and (ii) $t_s = 3$ s with a 2% criterion. Plot the response of the system with the selected a and b, subject to a unit step input $r(t) = 1$.

Solution
The characteristic equation of the system is $s^2 + (2+a)s + b = 0$, which by comparison with Eq. (7.1.34) gives $2\xi\omega_n = (2+a)$, $\omega_n^2 = b$, or

$$\xi\omega_n = \frac{1}{2}(2+a), \quad \xi^2 = \frac{1}{4b}(2+a)^2$$

For the condition on the settling time,

$$t_s = \frac{4}{\xi\omega_n} = \frac{4}{\frac{1}{2}(2+a)} = 3$$

which gives $a = 2/3$. For the condition on the maximum overshoot

$$M_p = e^{-\pi\xi/\sqrt{1-\xi^2}} \leq 0.1$$

By Eq. (7.1.67), the above condition can be written as

$$\xi^2 \geq \frac{(\ln 0.1)^2}{\pi^2 + (\ln 0.1)^2} = 0.3494$$

Thus,

$$\xi^2 = \frac{1}{4b}(2+a)^2 = \frac{1}{4b}(2+2/3)^2 \geq 0.3494$$

where $a = 2/3$ has been used. This leads to $b \leq 5.088$. Hence, the selection of the parameters:

$$a = \frac{2}{3}, \quad 0 < b \leq 5.088$$

meets the performance specifications. For a simulation, select $a = 2/3$ and $b = 4$. The transfer function then becomes $\frac{Y(s)}{R(s)} = \frac{4}{s^2 + (8/3)s + 4}$. The step response is plotted by MATLAB in Figure 7.1.16, showing a maximum overshoot of 6% and a settling time of 3 s.

7.1.5 Time Responses of Second-Order Systems to a Sinusoidal Input

So far, we have considered time responses subject to impulse, step, and ramp inputs. In this section, time responses of second-order systems to a sinusoidal input are examined. A general treatment of dynamic systems under sinusoidal excitations is presented in Section 7.3.

Undamped Systems under a Sinusoidal Input

The forced response of an undamped system ($\xi = 0$) under a sinusoidal input is governed by

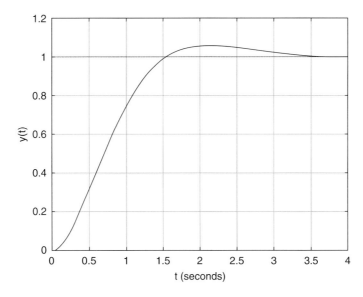

Figure 7.1.16 Step response of the system in Example 7.1.11

$$\ddot{y} + \omega_n^2 y(t) = \omega_n^2 r_0 \sin \omega t$$
$$y(0) = 0, \quad \dot{y}(0) = 0 \tag{7.1.73}$$

Here, without a loss of generality, the input $r(t) = r_0 \sin \omega t$ is used. The solution of Eq. (7.1.73), according to Section 2.6, is of the form

$$y(t) = A \cos \omega_n t + B \sin \omega_n t + y_p(t) \tag{7.1.74}$$

where y_p is a particular solution, and A and B are constants that are determined by the zero initial conditions. If the excitation frequency is not the same as the natural frequency of the system, $\omega \neq \omega_n$, $y_p(t) = \dfrac{\omega_n^2 r_0}{\omega_n^2 - \omega^2} \sin \omega t$, and the response is determined as

$$y(t) = \frac{\omega_n^2 r_0}{\omega_n^2 - \omega^2} \left(\sin \omega t - \frac{\omega}{\omega_n} \sin \omega_n t \right) \tag{7.1.75}$$

If the excitation frequency is identical to the natural frequency of the system, $\omega = \omega_n$, an unbounded oscillatory response called *resonant response* occurs. This is seen from Table 2.6.2, from which, $y_p(t) = -\dfrac{1}{2}\omega_n r_0 \, t \cos \omega_n t$. Thus, with Eq. (7.1.74) and the zero initial conditions, the response of the system is obtained as

$$y(t) = \frac{1}{2} \omega_n r_0 (\sin \omega_n t - t \cos \omega_n t) \tag{7.1.76}$$

As can be seen, the term $t \cos \omega_n t$ in Eq. (7.1.76) renders the response unbounded and oscillatory. Resonant vibrations of mechanical systems are commonly seen in engineering applications.

For comparison, the response of a system with $\omega_n = 4$ is plotted versus time in Figure 7.1.17 in two sinusoidal input cases: (a) $\omega = 3$; and (b) $\omega = 4$. Here, a unity input amplitude ($r_0 = 1$) is assumed and Eqs. (7.1.75) and (7.1.76) are used in computation.

Damped Systems under a Sinusoidal Input

The forced response of a damped system ($\xi > 0$) under a sinusoidal input is described by

$$\ddot{y} + 2\xi\omega_n \dot{y}(t) + \omega_n^2 y(t) = \omega_n^2 r_0 \sin \omega t$$
$$y(0) = 0, \quad \dot{y}(0) = 0 \tag{7.1.77}$$

From Section 2.6, it can be shown that the forced response is of the form

$$y(t) = r_0 H \sin(\omega t - \theta) + e^{-\sigma t} \phi(t) \tag{7.1.78}$$

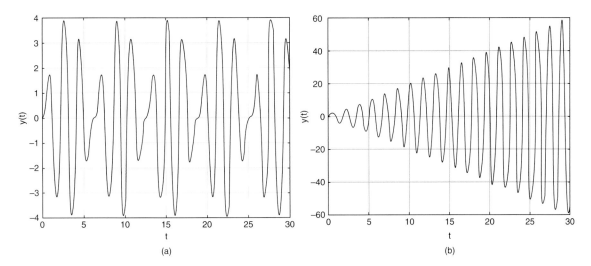

Figure 7.1.17 Response of an undamped system ($\omega_n = 4$) to a sinusoidal excitation ($r_0 = 1$): (a) $\omega = 3$; (b) $\omega = 4$, plotted with MATLAB

where

$$H = \frac{\omega_n^2}{\sqrt{(\omega_n^2 - \omega^2)^2 + 4\zeta^2 \omega_n^2 \omega^2}}, \quad \theta = \tan^{-1}\left(\frac{2\zeta\omega_n\omega}{\omega_n^2 - \omega^2}\right) \quad (7.1.79)$$

and $\phi(t)$ is a known function such that $e^{-\sigma t}\phi(t) \to 0$ as t goes to infinity. Note that with a nonzero ζ, H is finite at any frequency ω. Therefore, the system response is always bounded.

It is easy to see that a steady-state response in the sense of $\lim_{t \to \infty} y(t)$ does not exist. However, another type of steady-state response can be defined by the remaining part of Eq. (7.1.78) as $t \to \infty$, which is given by

$$y_{ss}(t) = r_0 H \sin(\omega t - \theta) \quad (7.1.80)$$

This indicates that after a long enough time, the response of a damped system becomes sinusoidal with the same frequency as the input. This type of steady-state response is studied further in Section 7.3.

7.1.6 Time Responses of Higher-Order Systems

We first examine third-order systems, and then consider general high-order systems. The free response of a high-order system can be obtained by the methods mentioned in Section 7.1.1. In this section, we focus on forced responses, especially step responses and steady-state responses.

Third-Order Systems

For a third-order system, its transfer function is of the general form

$$\frac{Y(s)}{R(s)} = G(s) = \frac{b_2 s^2 + b_1 s + b_0}{s^3 + a_2 s^2 + a_1 s + a_0} \tag{7.1.81}$$

By partial fraction expansion (Section 2.4.4), the transfer function falls into the following two cases.

Case 1. $G(s)$ with Three Real Roots
In this case, the transfer function is in one of the following three forms

$$G(s) = \frac{K_1}{s + \sigma_1} + \frac{K_2}{s + \sigma_2} + \frac{K_1}{s + \sigma_3} \tag{7.1.82a}$$

$$G(s) = \frac{B_1}{(s + \sigma_1)^2} + \frac{B_2}{s + \sigma_1} + \frac{K}{s + \sigma_2} \tag{7.1.82b}$$

$$G(s) = \frac{B_1}{(s + \sigma_1)^3} + \frac{B_2}{(s + \sigma_1)^2} + \frac{B_3}{s + \sigma_1} \tag{7.1.82c}$$

where $-\sigma_1$, $-\sigma_2$, and $-\sigma_3$ are the transfer function poles. Without a loss of generality, assume that $\sigma_k > 0$. Note that Eq. (7.1.82a) is about the system with distinct poles, and that Eqs. (7.1.82b) and (7.1.82c) are about repeated poles. For a system of distinct poles subject to a step input $r(t) = r_0$, the system response in the s-domain, from Eq. (7.1.82a), is

$$Y(s) = \frac{\beta_1}{T_1 s + 1} \frac{r_0}{s} + \frac{\beta_2}{T_2 s + \sigma_2} \frac{r_0}{s} + \frac{\beta_3}{T_3 s + 1} \frac{r_0}{s} \tag{7.1.83}$$

where $T_j = 1/\sigma_j$ and $\beta_j = K_j/\sigma_j$, $j = 1, 2, 3$. It follows that the step response of the third-order system is

$$y(t) = r_0 \beta_1 \left(1 - e^{-\frac{t}{T_1}}\right) + r_0 \beta_2 \left(1 - e^{-\frac{t}{T_2}}\right) + r_0 \beta_3 \left(1 - e^{-\frac{t}{T_3}}\right) \tag{7.1.84}$$

The steady-state response of the system is then given by

$$y_{ss} = \lim_{t \to \infty} y(t) = r_0 \left(\frac{K_1}{\sigma_1} + \frac{K_2}{\sigma_2} + \frac{K_3}{\sigma_3}\right) \tag{7.1.85}$$

Systems with repeated poles can be treated similarly. The step response of the system with repeated poles can be similarly obtained.

Case 2. $G(s)$ with a Real Root and a Pair of Complex/Imaginary Poles

The system transfer function can be written as

$$G(s) = \frac{K}{s+a} + \frac{bs+c}{s^2 + 2\xi\omega_n s + \omega_n^2} \tag{7.1.86}$$

where $0 \leq \xi < 1$. Without loss of generality, we assume that $a > 0$. The poles of the system are distinct: $-a$, $-\sigma \pm j\omega_d$, with $\sigma = \xi\omega_n$, $\omega_d = \sqrt{1-\xi^2}\,\omega_n$, and $j = \sqrt{-1}$. Under a step input $r(t) = r_0$, the system response in the s-domain is written as

$$Y(s) = \frac{\beta}{Ts+1}\frac{r_0}{s} + \frac{\omega_n^2(Bs+C)}{s^2 + 2\xi\omega_n s + \omega_n^2}\frac{r_0}{s} \tag{7.1.87}$$

with $T = 1/a$, $\beta = K/a$, $B = b/\omega_n^2$, and $C = c/\omega_n^2$. Thus, by Eqs. (7.1.24) and (7.1.57), the step response of the system is

$$
\begin{aligned}
y(t) = r_0\beta\left(1 - e^{-\frac{t}{T}}\right) &+ r_0 B \frac{\omega_n}{\sqrt{1-\xi^2}} e^{-\sigma t} \sin\omega_d t \\
&+ r_0 C \left\{1 - e^{-\sigma t}\left(\cos\omega_d t + \frac{\sigma}{\omega_d}\sin\omega_d t\right)\right\}
\end{aligned}
\tag{7.1.88}
$$

The steady-state response of the system is $y_{ss} = \lim_{t\to\infty} y(t) = r_0(\beta + C)$.

As shown in Eqs. (7.1.84) and (7.1.88), the step response of a third-order system is a sum of the responses of first-order and/or second-order systems.

General High-Order Systems

For an n-th-order system, its transfer function given in Eq. (7.1.6) can be written as

$$G(s) = \sum_{i=1}^{n_1}\frac{k_i}{s+a_i} + \sum_{i=1}^{n_2}\frac{b_i s + c_i}{s^2 + 2\xi_i\omega_i s + \omega_i^2}, \quad n = n_1 + n_2 \tag{7.1.89}$$

where partial fraction expansion has been applied. This implies that the system is a combination (parallel connection) of n_1 first-order subsystems and n_2 second-order subsystems; see Figure 7.1.18. Due to the complexity of such a system, analytical expressions of performance specification parameters, such as the rise time, maximum overshoot, and settling time for a second-order system, are extremely difficult to obtain, if not impossible. Under the circumstances, the time response of a high-order system is usually determined numerically by software packages, such as MATLAB and Mathematica.

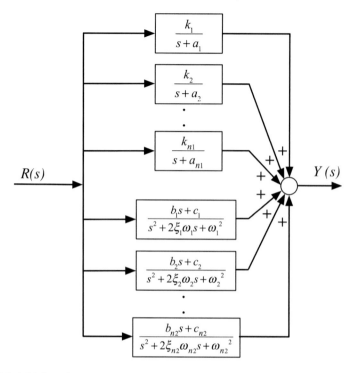

Figure 7.1.18 A high-order system as a combination of first- and second-order subsystems

Example 7.1.14

For a fourth-order system $\dfrac{Y(s)}{R(s)} = \dfrac{3s+5}{s^4 + 2s^3 + 8s^2 + 4s + 5}$, its response to a unit step input is plotted in Figure 7.1.19 by MATLAB, with the following commands:

```
>> num = [3 5]; den = [1 2 8 4 5];
>> step(num, den, 'k')
```

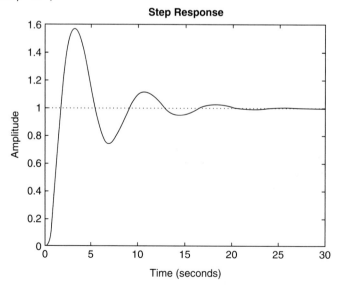

Figure 7.1.19 Step response of the system in Example 7.1.14

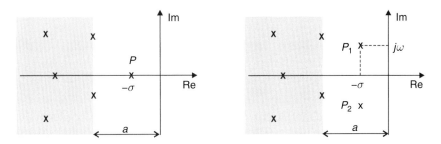

Figure 7.1.20 Dominant poles: (a) one real dominant pole, $p = -\sigma$; (b) two complex dominant poles, $p_{1,2} = -\sigma \pm j\omega$.

Dominant Poles and Model Reduction

Assume that all the poles of the transfer function $G(s)$ given in Eq. (7.1.89) have negative parts. (The system is stable; see Section 7.2.2.) The pole that is closest to the imaginary axis is called the *dominant pole* of the system. There are two cases of dominant poles: (i) one real pole $p = -\sigma$, and (ii) two complex conjugate poles $p_{1,2} = -\sigma \pm j\omega$; see Figure 7.1.20, where $\sigma > 0$, and a is the shortest distance between all other poles and the imaginary axis.

As a rule of thumb, if $a \geq 10\sigma$, the n-th-order system described by Eq. (7.1.89) can be approximated as a first-order system

$$G(s) \approx \frac{K}{s-p} = \frac{K}{s+\sigma} \qquad (7.1.90a)$$

or as a second-order system

$$G(s) \approx \frac{K_1}{s-p_1} + \frac{K_2}{s-p_2} = \frac{As+B}{(s+\sigma)^2 + \omega^2} \qquad (7.1.90b)$$

where K, K_1, and K_2 are the transfer function residues determined by partial fraction expansion.

Even if the condition of $a \geq 10\sigma$ is not met, the concept of dominant poles is still useful in design of feedback-control systems; see Section 8.6, where the root locus method is presented.

Example 7.1.15

Consider a third-order system with the transfer function

$$G(s) = \frac{4}{s^2 + 1.2s + 4} \cdot \frac{1}{s/a + 1}, \quad a > 1 \qquad (a)$$

which has poles $-0.6 \pm j1.9079$ and $-a$. The dominant poles of the system are $-0.6 \pm j1.9079$, with $\sigma = 0.6$. For $a \geq 10\sigma = 6$, the system can be approximated as a second-order system

$$G(s) \approx \frac{4}{s^2 + 1.2s + 4} \qquad \text{(b)}$$

Note that the third-order transfer function in Eq. (a) is reduced to the second-order transfer function in Eq. (b) when $a \to \infty$. This means that the pole $-a$ has an insignificant effect on the system response if it is located far away from the imaginary axis.

To validate the approximate model, consider the response of the third-order system subject to a step input $r(t) = 1$, with $a = 0.6, 1.2, 6, 12$, and ∞. Here, $a = \infty$ represents the second-order system given in Eq. (b). The step response of the system with different values of a is plotted in Figure 7.1.21. As seen from the figure, the step response curves with $a = 6$ and 12 are close to the solid curve, which is for the second-order system ($a = \infty$). On the other hand, with $a = 0.6$ or 1.2, the step response curve has much deviation from the solid curve. Therefore, the approximate model given in Eq. (b) gives acceptable response results for $a \geq 6$.

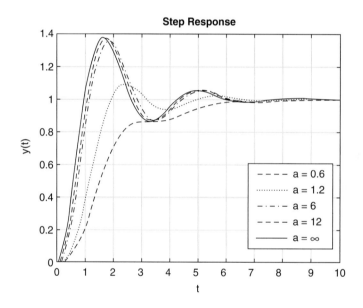

Figure 7.1.21 Step response of the third-order system in Example 7.1.15: $a = 0.6, 1.2, 6, 12,$ and ∞

7.2 Stability Analysis

In analysis and design of dynamic systems, especially feedback control systems, stability is essentially important. The stability of a system is related to the boundedness of its response to initial and external disturbances. An unstable system generally has an unbounded response.

Such an unbounded response makes it difficult to achieve desired system performance, can cause malfunction and failure of the system, and in a worst-case scenario even leads to catastrophic consequences, including property damage and loss of human life. Therefore, in design and operation of dynamic systems, assurance of stability is a must.

7.2.1 Stability Definitions

In this text, we shall only consider stability analysis of linear time-invariant systems. According to Section 7.1, the total response of a linear time-invariant system is the sum of its free response and forced response; namely,

$$y(t) = y_I(t) + y_F(t)$$

Stability can be defined by either the free response or the forced response.

First, consider stability definitions by forced responses. A system is *stable* if its response is bounded in magnitude to *every* bounded input. A system is *unstable* if there is a bounded input that yields an unbounded output. These are stability definitions in the sense of bounded input–bounded output (BIBO).

For a bounded input, say $|r(t)| \leq M < \infty$, Eq. (7.1.11) gives

$$|y_F(t)| \leq \int_0^t |g(t-\tau)||r(\tau)|d\tau \leq M \int_0^t |g(t-\tau)|d\tau$$

where $g(t)$ is the impulse response function of the system. This implies that the system is BIBO stable if $\int_0^\infty |g(t)|d\tau$ is finite. As an example, consider an underdamped second-order system with the impulse response

$$g(t) = \frac{\omega_n}{\sqrt{1-\xi^2}} e^{-\sigma t} \sin \omega_d t$$

which is given in Eq. (7.1.52b). Because

$$\int_0^\infty |g(t)|d\tau \leq \frac{\omega_n}{\sqrt{1-\xi^2}} \int_0^\infty e^{-\sigma t} d\tau = \frac{1}{1-\xi^2} < \infty, \quad \text{for } 0 < \xi < 1$$

the system is BIBO stable.

Next, consider stability definitions with free responses. A linear time-invariant system can have the following three states of stability:

(i) A system is *asymptotically stable* if its free response approaches zero as time goes to infinity.

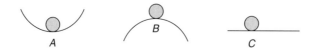

Figure 7.2.1 Illustration of stability states on a ball

(ii) A system is *unstable* if its free response grows unbounded as time goes to infinity.
(iii) A system is *marginally stable* if its free response does not vanish and remains bounded as time goes to infinity.

It can be shown that an asymptotically stable system is also a BIBO stable system. Thus, an asymptotically stable system is simply called a stable system.

It should be pointed out that the above stability definitions about a linear time-invariant system can also be given by its impulse response. In other words, equivalent stability conditions can be stated with "free response" replaced by "impulse response."

Physically, initial disturbances, such as the initial displacement and velocity of a mechanical system, describe a finite energy input at the initial time. A bounded input, on the other hand, can input infinite energy to a system in the entire time region, $0 < t < \infty$. Examples include a step input and a sinusoidal input of properly tuned frequency. So, a bounded free response does not necessarily guarantee a bounded forced response. This implies that a marginally stable system may be BIBO unstable. For instance, an undamped spring–mass system has bounded free response, but it experiences unbounded resonant vibration under a sinusoidal input with the frequency identical to the natural frequency of the system, as shown in Eq. (7.1.75).

The above three stability states with respect to free response can be illustrated in Figure 7.2.1, where a ball, under gravity, sits at different positions of equilibrium: point A, which is the bottom of a valley; point B, which is the peak of a mountain; and point C on a horizontal surface. Assume that the ball experiences friction and damping in motion. If the ball is slightly disturbed at point A, it moves back and forth in the valley, and eventually returns to the original equilibrium position. Point A and the response are said to be stable. If the ball is disturbed at point B, it falls off the peak, and never comes back to the original equilibrium point. Point B and the response are said to be unstable. If the ball is disturbed at point C, it travels a finite distance, and eventually settles at a new equilibrium position on the horizontal plane. Point C and the response are said to be marginally stable.

7.2.2 Stability Conditions in Terms of Pole Locations

In the subsequent stability analysis, the three stability definitions based on free response are used. For an n-th-order system, its characteristic equation is

$$A(s) = a_n s^n + a_{n-1} s^{n-1} + \ldots + a_1 s + a_0 = 0 \qquad (7.2.1)$$

with the roots being the poles of the transfer function given in Eq. (7.1.6). According to Eqs. (7.1.9) and (7.1.10), the system free response can be written as

$$y_I(t) = k_1 e^{p_1 t} + k_2 e^{p_2 t} + \ldots + k_n e^{p_n t} \tag{7.2.2}$$

if the system has a distinct pole $p_1, p_2, \ldots, p_n$, or

$$y_I(t) = \left(b_1 t^{l-1} + b_2 t^{l-2} + \ldots + b_{l-1} t + b_l\right) e^{p_0 t} + k_1 e^{p_1 t} + k_2 e^{p_2 t} + \ldots + k_m e^{p_m t} \tag{7.2.3}$$

if the system has l identical poles p_0 and m distinct poles $p_1, p_2, \ldots, p_l$, with $l + m = n$ and $p_0 \neq p_k$. Here, the constants k_i and b_j are determined by the initial conditions, as given by Eq. (7.1.2). A system with more repeated roots can be treated similarly.

According to Eqs. (7.2.2) and (7.2.3), the free response of a system vanishes as t approaches infinity if all the poles of the system have negative real parts. This means that the poles of a stable system must lie in the left half-plane, excluding the imaginary axis. For the free response to be bounded, all the poles must have nonpositive real parts and there are no repeated poles on the imagery axis ($j\omega$-axis), including the origin of the complex plane. Furthermore, the free response is unbounded if at least one pole has a positive real part or if there are repeated poles on the imaginary axis. The correlation between system stability and pole locations in the complex plane are illustrated in Figure 7.2.2, where the crosses ($\times$) indicate a pole location.

Based on the above analysis and discussion, the following three stability conditions in terms of pole locations are established.

Condition 1 for stability. The necessary and sufficient condition for a system to be stable is that all the poles of the system must have negative real parts. In other words, a system is

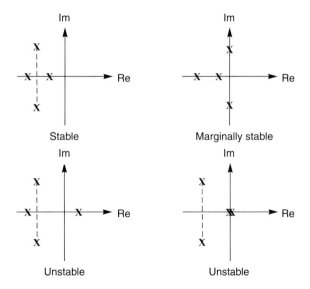

Figure 7.2.2 Stability and poles: ×– pole location in the complex plane

stable if and only if all the poles of the system are in the left half-plane, excluding the imaginary axis.

Condition 2 for instability. A system is unstable if one of the following two conditions is met:

(i) at least one pole is in the right half-plane, with positive real part; or
(ii) there are repeated poles on the imaginary axis, including the origin of the complex plane.

Condition 3 for marginal stability. A system is marginally stable if some simple (distinct) poles are on the imaginary axis (including the origin, $s = 0$) and if all other poles are in the open left half-plane.

Example 7.2.1

Consider the characteristic equations of the following four systems

System 1: $(s+1)(s^2 + 2s + 5) = 0$, with poles -1, $-1 \pm 2j$
System 2: $(s+1)(s^2 - 2s + 5) = 0$, with poles -1, $1 \pm 2j$
System 3: $(s+1)(s^2 + 4) = 0$, with poles -1, $\pm 2j$
System 4: $s^2(s+1) = 0$, with poles $0, 0, -1$

From the stability conditions, System 1 is stable, Systems 2 and 4 are unstable, and System 3 is marginally stable.

Example 7.2.2

For the four mechanical systems in Figure 7.2.3, the transfer functions are as follows:

System 1: $G_1(s) = \dfrac{X_1(s)}{F(s)} = \dfrac{1}{ms^2 + cs + k}$, with poles $\dfrac{1}{2m}\left(-c \pm \sqrt{c^2 - 4mk}\right)$

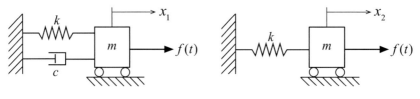

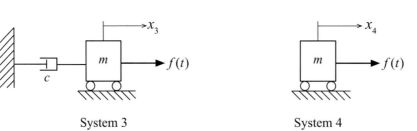

Figure 7.2.3 Four mechanical systems in Example 7.2.2

System 2: $G_2(s) = \dfrac{X_2(s)}{F(s)} = \dfrac{1}{ms^2 + k}$, with poles $\pm j\sqrt{k/m}$

System 3: $G_3(s) = \dfrac{X_3(s)}{F(s)} = \dfrac{1}{ms^2 + cs}$, with poles $0, \ -c/m$

System 4: $G_4(s) = \dfrac{X_4(s)}{F(s)} = \dfrac{1}{ms^2}$, with poles $0, \ 0$

From the stability conditions, System 1 is stable, Systems 2 and 3 are marginally stable, and System 4 is unstable.

To show the stability states of these systems, consider $m = 1$, $c = 2$, and $k = 16$. Also, consider an impulsive force $f(t) = \delta(t)$. Here, the nondimensional system parameters are used. The impulse response, which can also be used to define stability, is plotted for the four systems by MATLAB in Figure 7.2.4.

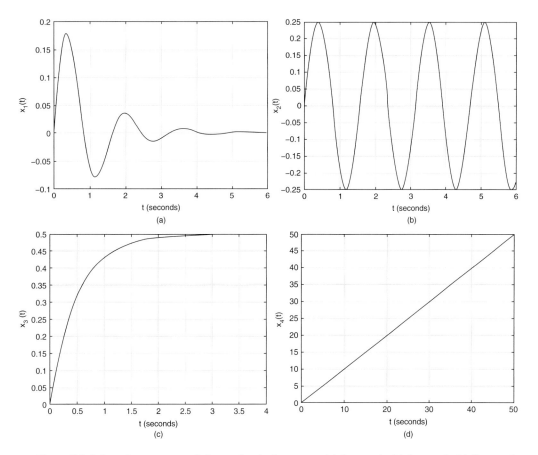

Figure 7.2.4 Impulse response of the mechanical systems: (a) System 1; (b) System 2; (c) System 3; and (d) System 4

Estimation of the Boundedness of Forced Response

The three stability conditions can be used to check the boundedness of a forced response subject to a specific input. As matter of fact, Eqs. (7.2.2) and (7.2.3) are the results of partial fraction expansion of $I(s)/A(s)$, where $I(s)$ is a polynomial of s given in Eq. (7.1.5a), with the order of less than n. Let the Laplace transform of the input $r(t)$ be $R(s) = B_R(s)/A_R(s)$, where $A_R(s)$ and $B_R(s)$ are polynomials of s. The s-domain forced response by Eq. (7.1.5b) is

$$Y_F(s) = G(s)R(s) = \frac{N(s)}{D(s)}$$

with $D(s) = A(s)A_R(s)$ and $N(s) = B(s)B_R(s)$. If the order of $D(s)$ is larger than that of $N(s)$, Eqs. (7.2.2) and (7.2.3) are directly applicable to $N(s)/D(s)$ and, as such, the stability conditions can be used to check the boundedness of the forced response $y_F(t) = \mathcal{L}^{-1}\{N(s)/D(s)\}$.

Example 7.2.3

Let System 2 in Example 7.2.2 be subject to a sinusoidal input $f(t) = f_0 \sin \omega_n t$, with $\omega_n = \sqrt{k/m}$. The s-domain system response is

$$X_2(s) = G_2(s)F(s) = \frac{1}{m\left(s^2 + \omega_n^2\right)} \frac{f_0 \, \omega_n}{\left(s^2 + \omega_n^2\right)}.$$

which has repeated poles on the imaginary axis: $j\omega_n, j\omega_n, -j\omega_n, -j\omega_n$, $j = \sqrt{-1}$. From Condition 2, the time response $x_2(t)$ is unbounded.

Also, let System 3 in Example 7.2.2 be subject to a step input $f(t) = f_0$. The s-domain response is

$$X_3(s) = G_2(s)F(s) = \frac{1}{s(ms+c)}\frac{f_0}{s}$$

which has repeated poles on the imaginary axis: 0, 0. From Condition 2, the time response $x_3(t)$ is unbounded.

Because the sinusoidal and step inputs are bounded, Systems 2 and 3, which are marginally stable, are unstable in the sense of BIBO.

7.2.3 The Routh–Hurwitz Stability Criterion

The stability conditions in the previous section require the knowledge of system poles to determine the stability status. In this section we study a method that does not require solving the characteristic equation (7.2.1) for the system poles. This method, namely, the Routh–Hurwitz stability criterion, has two special features: first, it does not require the knowledge of system poles to gain stability information; second, it can deal with a characteristic equation with coefficients consisting of unknown parameters, and it can obtain the ranges of these parameters for stability. Therefore, the Routh–Hurwitz stability criterion is a useful method in analysis and design of dynamic systems, especially feedback-control systems.

Let the characteristic equation of a system be written as

$$A(s) = s^n + a_{n-1}s^{n-1} + \ldots a_1 s + a_0 = 0 \tag{7.2.4}$$

where, for convenience of analysis, the polynomial $A(s)$ from Eq. (7.2.1) has been scaled such that $a_n = 1$. With a data table called a Routh array, which is constructed by using the coefficients of $A(s)$, the Routh–Hurwitz criterion tells how many poles are in the left half-plane, in the right half-plane, and on the imaginary axis.

We first present a necessary condition for stability, then show how to construct a Routh array, and finally introduce the Routh–Hurwitz stability criterion, which is a necessary and sufficient condition for stability.

Necessary Condition for Stability

A necessary condition for all the roots of the characteristic equation (7.2.4) to have negative real parts is that all the coefficients of $A(s)$ are positive; that is,

$$a_i > 0, \quad i = 0, 1, 2, \ldots, n-1 \tag{7.2.5}$$

This means that for a stable system, all the coefficients of its characteristic equation, Eq. (7.2.4), must be positive, and none of the coefficients vanish.

The proof of the necessary condition is straightforward. Let a stable system have l real roots $(-\sigma_i, i = 1, 2, \ldots, l)$ and $2m$ complex roots $(-\alpha_k \pm j\beta_k, k = 1, 2, \ldots, m, j = \sqrt{-1})$, with $l + 2m = n$. Because the system is stable, $\sigma_i > 0$ for all i and $\alpha_k > 0$ for all k. The characteristic equation becomes

$$A(s) = \prod_{i=1}^{l}(s + \sigma_i) \prod_{k=1}^{m}\left((s + \alpha_k)^2 + \beta_k^2\right) = 0$$

Because σ_i, α_k and β_k^2 are positive numbers for all i and k, a comparison of the previous equation with Eq. (7.2.4) reveals that all the coefficients of $A(s)$ are positive.

The necessary condition can be used as means of initial screening for instability. In addition, one can make use of the following two observations.

(1) If some s-power terms of $A(s)$ disappear and if the remaining terms all have positive coefficients, the system is either marginally stable or unstable.
(2) If at least one coefficient of $A(s)$ is negative, the system is unstable.

However, even if the condition (7.2.5) is met, the stability of the system still cannot be concluded. Further effort, for example analysis by the Routh–Hurwitz criterion, is required to detect the system stability.

Example 7.2.4
Consider the characteristic equations of the following three systems

System 1: $s^3 + 2s^2 + 5s + 6 = 0$
System 2: $s^3 + 2s^2 + 6 = 0$
System 3: $s^4 + 3s^3 - 2s^2 + 6s + 20 = 0$

System 1 satisfies the necessary condition, and as such its stability needs further investigation. System 2, with a missing s^1-term, is either marginally stable or unstable. System 3, with the coefficient of the s^2-term being negative, is unstable. For validation purposes, the poles of these systems are computed by the MATLAB function roots and the corresponding stability states are concluded as follows:

System 1: $-0.2836 \pm 2.0266i, -1.4329$, stable
System 2: $0.3888 \pm 1.4174i, -2.7777$, unstable
System 3: $1.0832 \pm 1.5365i, -1.5761, -3.5904$, unstable

Routh Array

The application of the Routh–Hurwitz criterion takes two steps: (i) to generate a Routh array; and (ii) to examine the array to detect the system stability. We now show how to generate such an array.

The Routh array for the characteristic equation (7.2.4) is a table of $n + 1$ rows, as shown below:

s^n	1	a_{n-2}	a_{n-4}	a_{n-6}	...
s^{n-1}	a_{n-1}	a_{n-3}	a_{n-5}	a_{n-7}	...
s^{n-2}	b_1	b_2	b_3	b_4	...
s^{n-3}	c_1	c_2	c_3	c_4	...
⋮	⋮	⋮	⋮		
s^2	d_1	d_2			
s^1	e_1				
s^0	f_1				

The rows are labeled by the powers of s, from s^n to s^0. The first row (the s^n row) starts with the coefficient 1 of the s^n term (the highest s-power term) of $A(s)$, and lists every other coefficient of the polynomial. The second row (the s^{n-1} row) starts with the coefficient a_{n-1} of the s^{n-1} term of $A(s)$, and lists every other coefficient of the polynomial. The remaining rows (from the s^{n-2} row to the s^0 row) are formed as follows:

$$b_1 = \frac{-1}{a_{n-1}}\begin{vmatrix} 1 & a_{n-2} \\ a_{n-1} & a_{n-3} \end{vmatrix}, \quad b_2 = \frac{-1}{a_{n-1}}\begin{vmatrix} 1 & a_{n-4} \\ a_{n-1} & a_{n-5} \end{vmatrix}, \quad b_3 = \frac{-1}{a_{n-1}}\begin{vmatrix} 1 & a_{n-6} \\ a_{n-1} & a_{n-7} \end{vmatrix}, \ldots$$

$$c_1 = \frac{-1}{b_1}\begin{vmatrix} a_{n-1} & a_{n-3} \\ b_1 & b_2 \end{vmatrix}, \quad c_2 = \frac{-1}{b_1}\begin{vmatrix} a_{n-1} & a_{n-5} \\ b_1 & b_3 \end{vmatrix}, \quad c_3 = \frac{-1}{b_1}\begin{vmatrix} a_n & a_{n-7} \\ b_1 & b_4 \end{vmatrix}, \ldots$$

......

$$f_1 = \frac{-1}{e_1}\begin{vmatrix} d_1 & d_2 \\ e_1 & 0 \end{vmatrix} = d_2$$

As can be seen, the calculation for each element involves a two-by-two determinant and a common factor.

The formation of the rows beneath the s^{n-1} row follows several rules, as stated below:

Rule 1. The current row is generated by using the elements in the previous two rows, and the creation of new elements stops when all the elements in the previous two rows have been exhausted.

Rule 2. The elements in a row share the same factor, which is -1 divided by the first element in the previous row. For instance, in the s^{n-3} row, the elements $c_1, c_2, c_3, \ldots$ have a common factor $-1/b_1$.

Rule 3. In the two-by-two determinant for an element, the left-hand column is always the first column of the previous two rows, and the right-hand column is a column selected from the previous two rows, from the second column to the last one. For example, in the s^{n-3} row, the left-hand column for each determinant is $\begin{pmatrix} a_{n-1} \\ b_1 \end{pmatrix}$, and the right-hand columns for elements $c_1, c_2, c_3, \ldots$ are $\begin{pmatrix} a_{n-3} \\ b_2 \end{pmatrix}, \begin{pmatrix} a_{n-5} \\ b_3 \end{pmatrix}, \begin{pmatrix} a_{n-7} \\ b_4 \end{pmatrix}, \ldots$, respectively.

Rule 4. A zero is added in the determinant for the last element in a row if the previous two rows have unequal numbers of elements. See the calculation of f_1 for instance.

Rule 5. For convenience in calculations, any row can be multiplied or divided by a positive number.

According to Rule 3, the number of elements in a row decreases as the s-power label index decreases. In particular, the s^2 row always has two elements, and the s^1 and s^0 rows each only have one element. Because of this, the single element f_1 of the last row is equal to d_2, which is the second element in the s^2 row.

Example 7.2.5

Given the characteristic equation $A(s) = s^4 + 20s^3 + 8s^2 + 200s + 14 = 0$, the corresponding Routh array is created as follows:

$$\begin{array}{c|ccc} s^4 & 1 & 8 & 14 \\ s^3 & \cancel{20}\ 1 & \cancel{200}\ 10 & \\ s^2 & \frac{-1}{1}\begin{vmatrix} 1 & 8 \\ 1 & 10 \end{vmatrix} = -2 & \frac{-1}{1}\begin{vmatrix} 1 & 14 \\ 1 & 0 \end{vmatrix} = 14 & \\ s^1 & \frac{-1}{-2}\begin{vmatrix} 1 & 10 \\ -2 & 14 \end{vmatrix} = 17 & & \\ s^0 & 14 & & \end{array}$$

In construction of the previous array, Rule 5 is applied to the s^3 row, Rule 4 is applied to the s^2 row, and $f_1 = d_2$ is directly used for the s^0 row without calculation.

Routh–Hurwitz Criterion

Once the Routh array for a system is generated, the stability of the system can be examined by the Routh–Hurwitz criterion as follows:

The number of the roots of $A(s)$ with positive real parts is equal to the number of sign changes in the first column (the far-left column) of the Routh array.

For instance, in Example 7.2.5, the first column of the Routh array has two sign changes: one from 1 of the s^3 row to -2 of the s^2 row, and another from -2 of the s^2 row to 17 of the s^1 row. Therefore, the system is unstable with two poles in the right half-plane. To validate this, using the MATLAB function `roots` gives the roots of the characteristic equation as: -0.0702, -20.0954, and $0.0828 \pm 3.15i$, showing two poles with positive real parts.

According to the Routh–Hurwitz criterion, a system is stable if there is no sign change in the first column of the Routh array. Therefore, *a necessary and sufficient condition for a system to be stable* is that all the elements in the first column of the Routh array are positive:

$$a_{n-1} > 0, \ b_1 > 0, \ c_1 > 0, \ \ldots, \ d_1 > 0, \ e_1 > 0, \ f_1 > 0 \tag{7.2.6}$$

There are some cases in which the first element of a row is zero or all the elements in a row are zeros. Should this happen, the system in consideration is not stable. Under the circumstances, the construction of such a Routh array can proceed with some methods, as shown in the references at the end of this chapter.

In summary, the application of the Routh–Hurwitz criterion in stability analysis takes the following three steps:

(i) Use the necessary condition (7.2.5) to screen for instability.
  (ii) Construct a Routh array.
  (iii) Apply the Routh–Hurwitz criterion to conclude the stability status.

Furthermore, in design of a stable system, the necessary and sufficient condition (7.2.6) must be satisfied.

Example 7.2.6
Consider a system with the characteristic equation

$$s^3 + (a-6)s^2 + 5s + b = 0$$

Determine the ranges of parameters a and b for the system to be stable.

Solution
First, check the necessary condition (7.2.5), which indicates that for the system to be stable, the parameters should satisfy the conditions $a > 6$ and $b > 0$. Second, construct the Routh array as follows:

$$\begin{array}{c|cc} s^3 & 1 & 5 \\ s^2 & a-6 & b \\ s^1 & \Delta & \\ s^0 & b & \end{array}$$

with

$$\Delta = \frac{-1}{a-6}\begin{vmatrix} 1 & 5 \\ a-6 & b \end{vmatrix} = \frac{5a - b - 30}{a - 6}$$

According to the Routh–Hurwitz criterion (7.2.6), the system is stable if and only if

$$a - 6 > 0, \ \Delta > 0, \ b > 0 \qquad \text{(a)}$$

which can be further written as

$$a > 6, \ 0 < b < 5a - 30 \qquad \text{(b)}$$

As can be seen, the conditions (a) include the necessary conditions.

Stability Conditions of Systems of Orders Two to Four

The Routh–Hurwitz criterion is applied to second-, third-, and fourth-order systems, and the stability conditions are listed in Table 7.2.1. The derivation of the stability conditions is given below.

Table 7.2.1. Stability conditions for systems of order two to four

	Characteristic equation	Stability conditions
Second-order system	$s^2 + a_1 s + a_0 = 0$	$a_1 > 0, \; a_0 > 0$
Third-order system	$s^3 + a_2 s^2 + a_1 s + a_0 = 0$	$a_0 > 0, \; a_1 > \dfrac{a_0}{a_2} > 0, \; a_2 > 0$
Fourth-order system	$s^4 + a_3 s^3 + a_2 s^2 + a_1 s + a_0 = 0$	$a_0 > 0, \; a_1 > \dfrac{a_0 a_3}{a_2 - a_1/a_3} > 0$ $a_2 > \dfrac{a_1}{a_3} > 0, \; a_3 > 0$

For a second-order system, the Routh array is

$$\begin{array}{c|cc} s^2 & 1 & a_0 \\ s^1 & a_1 & \\ s^0 & a_0 & \end{array}$$

From the Routh–Hurwitz criterion, the system is stable if and only if $a_1 > 0$ and $a_0 > 0$. Note that for a second-order system the necessary and sufficient conditions are the same.

For a third-order system, the Routh array is

$$\begin{array}{c|cc} s^3 & 1 & a_1 \\ s^2 & a_2 & a_0 \\ s^1 & \Delta & \\ s^0 & a_0 & \end{array}$$

with

$$\Delta = \frac{-1}{a_2} \begin{vmatrix} 1 & a_1 \\ a_2 & a_0 \end{vmatrix} = \frac{a_1 a_2 - a_0}{a_2}$$

From the Routh–Hurwitz criterion, the stability conditions for the system are

$$a_2 > 0, \; \Delta > 0, \; a_0 > 0$$

which, after algebraic manipulations, give

$$a_0 > 0, \; a_2 > 0, \; a_1 > a_0/a_2$$

Here, the condition $a_1 > 0$ is implied by $a_1 > a_0/a_2$ because $a_0 > 0$ and $a_2 > 0$.

For a fourth-order system, the Routh array is

$$\begin{array}{c|ccc} s^4 & 1 & a_2 & a_0 \\ s^3 & a_3 & a_1 & \\ s^2 & b_1 & b_2 & \\ s^1 & c_1 & & \\ s^0 & b_2 & & \end{array}$$

with
$$b_1 = \frac{a_2 a_3 - a_1}{a_3}, \quad b_2 = a_0, \quad c_1 = \frac{a_1 b_1 - a_3 b_2}{b_1}$$

From the Routh–Hurwitz criterion, the stability conditions for the system are
$$a_3 > 0, \; b_1 > 0, \; c_1 > 0, \; b_2 > 0$$
which, after algebraic manipulations, lead to
$$a_0 > 0, \; a_1 > \frac{a_0 a_3}{a_2 - a_1/a_3}, \; a_2 > a_1/a_3, \; a_3 > 0$$
The conditions $a_1 > 0$ and $a_2 > 0$ are imbedded in the above conditions.

Example 7.2.7
In Figure 7.2.5, a second-order plant is under proportional and integral (PI) control (see Section 8.3), where k_P and k_I are gain constants. Determine the ranges of the gain constants for the stability of the closed-loop system.

Solution
The characteristic equation of the closed-loop system is
$$1 + \left(k_P + k_I \frac{1}{s}\right) \frac{1}{s(s+5)} = 0$$
or
$$s^3 + 5s^2 + k_P s + k_I = 0$$
This is a third-order system. From Table 7.2.1, the stability conditions of the closed-loop system are given by
$$k_P > \frac{1}{5} k_I, \quad k_I > 0$$

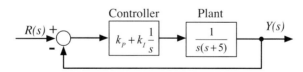

Figure 7.2.5 PI control of a second-order plant

Besides the Routh–Hurwitz stability criterion, the root locus method can be applied for stability analysis. This method, which is a graphic technique, is especially useful in the design of feedback control systems; see Sections 8.5 and 8.6.

7.3 System Response Analysis in the Frequency Domain

Dynamic systems subject to sinusoidal or harmonic excitations are seen in a variety of engineering applications, including rotating machinery with unbalanced mass and electric circuits with alternating current (AC). The term *frequency response* arises in the description of the steady-state behavior of a dynamic system to a harmonic input in a range of input frequencies. For a stable linear time-invariant system, its steady-state response to a sinusoidal input is also sinusoidal, with the same frequency, as illustrated in Figure 7.3.1, where the output differs from the input in both amplitude y_0 and phase ϕ, which are functions of the input frequency ω. As shown in Section 7.3.1, the frequency-dependent amplitude and phase can be determined from the system transfer function $G(s)$ with the complex parameter s replaced by $j\omega$, $j = \sqrt{-1}$. The $G(j\omega)$ is called the frequency response or frequency response function of the system.

Note that the steady-state response mentioned herein is a sinusoidal function of time, and it is not the same as $y_{ss} = \lim_{t \to \infty} y(t)$, which is a constant as defined in Section 7.1.1. These two kinds of steady-state responses have different implications.

Although frequency response was initially used to determine the steady-state response of stable systems, the concept of frequency response is applicable to general systems, including unstable systems, of which a steady-state response may not exist at all. Indeed, a set of frequency-domain tools have been developed for system modeling, stability analysis, and feedback control design. Furthermore, frequency-domain analysis has been extended to linear time-varying systems and even nonlinear systems.

Methods based on frequency responses have two advantages in practice. First, frequency response designs can deliver good results for system models with uncertainties. Second, frequency responses can be conveniently integrated with experimental data in the modeling and feedback control of complex dynamic systems. Because of these advantages, frequency-domain methods are widely used in industry.

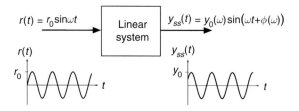

Figure 7.3.1 Steady-state response of a linear time-invariant system

7.3 System Response Analysis in the Frequency Domain

7.3.1 General Concepts

Consider a stable system subject to a sine input $r(t) = r_0 \sin \omega t$. The s-domain response of the system, by partial fraction expansion, can be expressed as

$$Y(s) = G(s) \frac{r_0 \omega}{s^2 + \omega^2} = \frac{b_1}{s - j\omega} + \frac{b_2}{s + j\omega} + \sum_{i=1}^{n} \frac{k_i}{s - p_i} \quad (7.3.1)$$

where $G(s)$ is the system transfer function as given in Eq. (7.1.6), and p_i are the poles of the transfer function. Here, without loss of generality, distinct poles have been assumed. Also note that b_1 and b_2 are a pair of complex conjugates, $b_2 = \bar{b}_1$. The time response of the system then is obtained by inverse Laplace transform as follows:

$$y(t) = b_1 e^{j\omega t} + b_2 e^{-j\omega t} + \sum_{i=1}^{n} k_i e^{p_i t} = 2\text{Re}\left(b_1 e^{j\omega t}\right) + \sum_{i=1}^{n} k_i e^{p_i t} \quad (7.3.2)$$

The residue b_1 is computed by

$$b_1 = \lim_{s \to j\omega} G(s) \frac{r_0 \omega}{s + j\omega} = -\frac{j}{2} r_0 G(j\omega) \quad (7.3.3)$$

Write

$$G(j\omega) = A\, e^{j\phi} \quad (7.3.4)$$

with

$$\begin{aligned} \text{Magnitude:} \quad & A = |G(j\omega)| = \sqrt{[\text{Re}(G(j\omega))]^2 + [\text{Im}(G(j\omega))]^2} \\ \text{Phase:} \quad & \phi = \angle G(j\omega) = \tan^{-1}\left[\frac{\text{Im}(G(j\omega))}{\text{Re}(G(j\omega))}\right] \end{aligned} \quad (7.3.5)$$

which are functions of the excitation frequency ω. Because the system is stable, the $e^{p_i t}$ related terms in Eq. (7.3.2) vanish as time t approaches infinity. Consequently, the remaining part of $y(t)$, which is called the steady-state response (output) and denoted by $y_{ss}(t)$, is given by

$$y_{ss}(t) = 2\,\text{Re}\left(b_1 e^{j\omega t}\right) = 2\,\text{Re}\left(-\frac{j}{2} r_0 A e^{j\phi} e^{j\omega t}\right) = r_0 A\,\text{Re}\left(-j e^{j(\omega t + \phi)}\right)$$

It follows that the steady-state response of the system under the sinusoidal input is

$$y_{ss}(t) = r_0 A \sin(\omega t + \phi) \quad (7.3.6)$$

where the magnitude A and phase ϕ are given by Eq. (7.3.5).

Similarly, for a more general sinusoidal input $r(t) = r_0 \sin(\omega t + \theta_0)$, it can be shown that the steady-state response of the system is given by

$$y_{ss}(t) = r_0 A \sin(\omega t + \theta_0 + \phi) \tag{7.3.7}$$

Equivalently, for a sinusoidal input $r(t) = r_0 \cos(\omega t + \theta_0)$, it can be shown that the steady-state response of the system is

$$y_{ss}(t) = r_0 A \cos(\omega t + \theta_0 + \phi) \tag{7.3.8}$$

Thus, for a system with the transfer function $G(s)$, its steady-state response, which is given by either Eq. (7.3.7) or Eq. (7.3.8), is fully determined by its *frequency response* (also called the frequency response function)

$$G(j\omega) = G(s)|_{s=j\omega} \tag{7.3.9}$$

Although the frequency response was originally derived from the steady-state response of stable systems, the definition (7.3.9) has been extended to general systems, including unstable systems, for other purposes. For an unstable system, of which a steady-state response to a sinusoidal input generally does not exist, its frequency response is still of the form of $G(j\omega)$. In general, frequency responses have the following three utilities: (a) the determination of steady-state response for stable systems; (b) stability analysis; and (c) feedback control design. In this chapter, we focus on the first utility, namely, the determination of steady-state response. For the second and third utilities, refer to standard control texts, some of which are listed at the end of the chapter.

Example 7.3.1
For the system

$$G(s) = \frac{2s - 1}{s^2 + 2s + 5}$$

which is stable according to Table 7.2.1, plot the magnitude and phase of the system frequency response against the frequency parameter ω, and determine the steady-state response of the system subject to a cosine input $r(t) = 3 \cos(4t)$.

Solution
The frequency response of the system, by the definition given in Eq. (7.3.9), is

$$G(j\omega) = \frac{2j\omega - 1}{5 - \omega^2 + 2j\omega}$$

for which the magnitude and phase are

$$A = \frac{\sqrt{4\omega^2 + 1}}{\sqrt{(5 - \omega^2)^2 + 4\omega^2}} \tag{a}$$

$$\phi = \angle(2j\omega - 1) - \angle(5 - \omega^2 + 2j\omega) = \tan^{-1}\left(\frac{2\omega}{-1}\right) - \tan\left(\frac{2\omega}{5 - \omega^2}\right)$$

7.3 System Response Analysis in the Frequency Domain

The magnitude and phase versus the frequency parameter are plotted in Figure 7.3.2. For the given sinusoidal input, $r_0 = 3$ and $\omega = 4$, which by Eq. (a) yields

$$A = \frac{\sqrt{4 \times 4^2 + 1}}{\sqrt{(5 - 4^2)^2 + 4 \times 4^2}} = \sqrt{\frac{13}{37}} = 0.5927$$

$$\phi = \tan^{-1}\left(\frac{2 \times 4}{-1}\right) - \tan\left(\frac{2 \times 4}{5 - 4^2}\right)$$
$$= \pi - \tan^{-1}(8) - \{\pi - \tan^{-1}(8/11)\} = -0.818 \text{ rad}$$

(b)

According to Eq. (7.3.8), the steady-state response of the system is

$$y_{ss}(t) = 1.778 \cos(4t - 0.818) \tag{c}$$

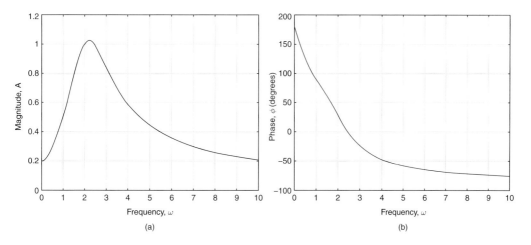

Figure 7.3.2 Frequency response plots of the system in Example 7.3.1: (a) magnitude; (b) phase

7.3.2 Frequency Response Plots

In analysis and design of dynamic systems, frequency response is usually displayed in plots, with frequency ω as an independent variable. There are three types of commonly used frequency response plots:

- plots of M and ϕ against ω from Eq. (7.3.5), as shown in Figure 7.3.2
- Bode diagrams
- Nyquist plots

As mentioned previously, these frequency response plots are applicable to both stable and unstable systems. In this section, Bode and Nyquist plots are introduced. The utility of these plots in control system design is covered in standard control textbooks.

Bode Diagrams

The Bode diagram (or Bode plot) of a frequency response function $G(j\omega)$ consists of the logarithmic plots of its magnitude and phase, which are defined as follows:

$$\begin{aligned} M &= 20 \log_{10} |G(j\omega)| \quad \text{dB} \\ \phi &= \angle G(j\omega) \end{aligned} \tag{7.3.10}$$

Here, the unit for M is decibel (dB). In a Bode diagram, the horizontal axis is about the frequency parameter ω in terms of the logarithm to the base 10, namely $\log_{10}\omega$. On the horizontal axis, an interval between two frequencies with a ratio of 10 is called a *decade*.

Bode diagrams can be easily generated by MATLAB. For instance, the Bode diagram for the system

$$G(s) = \frac{2s - 1}{s^2 + 2s + 5}$$

which is considered in Example 7.3.1, is plotted in Figure 7.3.3 by the MATLAB commands:

```
>> num = [2 -1]; den = [1 2 5];
>> bode(num, den)
>> grid
```

Here, bode is a MATLAB function for plotting Bode diagrams; see Table B9 in Appendix B.

Bode diagrams have the following *additive feature*. For a frequency response as a product of r factors

$$G(j\omega) = G_1(j\omega) \, G_2(j\omega) \, \ldots \, G_r(j\omega) \tag{7.3.11}$$

its Bode diagram can be written as

$$\begin{aligned} M &= M_1 + M_2 + \ldots + M_r \\ \phi &= \phi_1 + \phi_2 + \ldots + \phi_r \end{aligned} \tag{7.3.12}$$

where M_k and ϕ_k are the magnitude and phase of the k-th factor, namely,

$$\begin{aligned} M_k &= 20 \log_{10} |G_k(j\omega)| \quad \text{dB} \\ \phi_k &= \angle G_k(j\omega) \end{aligned} \tag{7.3.13}$$

The additive feature makes the use of Bode diagrams convenient in design of feedback control systems.

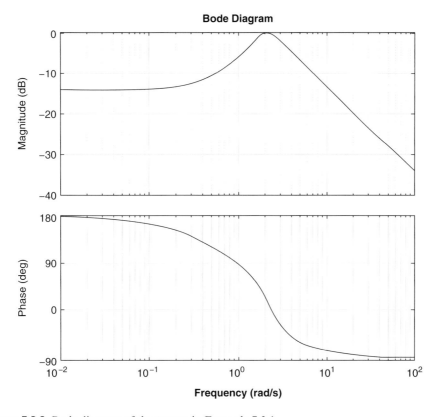

Figure 7.3.3 Bode diagram of the system in Example 7.3.1

Nyquist Plots

The Nyquist plot of a frequency response is a plot in the complex plane that is obtained from

$$G(j\omega) = R(\omega) + jX(\omega), \quad \text{for } -\infty < \omega < \infty \quad (7.3.14)$$

where $R(\omega)$ and $X(\omega)$ are the real and imaginary parts of the frequency response $G(j\omega)$, for any given value of ω. Nyquist plots are widely used in stability analysis and the design of feedback control systems.

Nyquist plots can also be generated by MATLAB. For instance, for the transfer function considered in Example 7.3.1, its Nyquist plot is plotted in Figure 7.3.4 by the commands:

```
>> syst = tf([2 -1], [1 2 5])
>> nyquist(syst)
```

Here, `nyquist` is a MATLAB function for plotting Nyquist plots; see Table B9 in Appendix B.

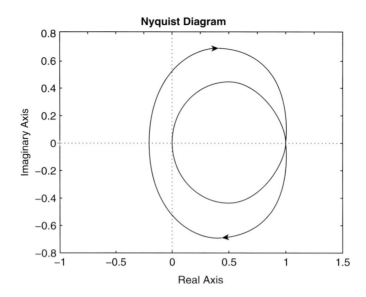

Figure 7.3.4 Nyquist plot of the system in Example 7.3.1

7.3.3 Frequency Response of First-Order Systems

For a standard first-order system, as described by Eq. (7.1.18), its frequency response is

$$G(j\omega) = \frac{1}{Ts+1}\bigg|_{s=j\omega} = \frac{1}{1+j\omega T} \quad (7.3.15)$$

with the magnitude and phase given by

$$A = |G(j\omega)| = \frac{1}{\sqrt{1+\omega^2 T^2}}$$
$$\phi = \angle G(j\omega) = -\tan^{-1}(\omega T) \quad (7.3.16)$$

The magnitude of the Bode diagram is

$$M = 20\log_{10}\left(\frac{1}{\sqrt{1+\omega^2 T^2}}\right) = -10\log_{10}(1+\omega^2 T^2) \quad (7.3.17)$$

The phase of the Bode diagram is the same as that given in Eq. (7.3.16).

The Bode diagram of the system is shown in Figure 7.3.5, where the MATLAB function bode has been used. As can be seen from the figure and Eq. (7.3.17), the magnitude curve has two asymptotes:

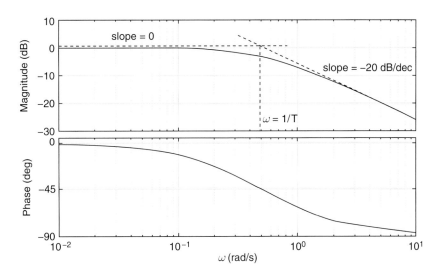

Figure 7.3.5 Bode plot of the first-order system $1/(Ts+1)$

(i) as ω approaches zero, $M \to 0$, with slope $= 0$
(ii) as ω approaches infinity, $M \to -20\log_{10}(\omega T)$, with slope $= -20$ dB/decade

It can be shown that these two asymptotes intersect at $\omega = 1/T$, which is called the *break frequency* or *corner frequency*.

Because the first-order system is stable with the pole in the open left half-plane, its steady-state response to a sinusoidal input $r(t) = r_0 \sin \omega t$ exists. By Eq. (7.3.6) the steady-state response is

$$y_{ss}(t) = \frac{r_0}{\sqrt{1+\omega^2 T^2}} \sin(\omega t - \psi) \qquad (7.3.18)$$

with $\psi = \tan^{-1}(\omega T)$.

For a first-order system that is not in the standard form (7.1.18), its frequency response can be similarly treated, as shown in Example 7.3.2.

Example 7.3.2

A phase-lag circuit is shown in Figure 7.3.6. It can be shown that the transfer function of the circuit, from the input voltage v_{in} to the output voltage v_o, is

$$\frac{V_o(s)}{V_{in}(s)} = \frac{R_2 Cs + 1}{(R_1 + R_2)Cs + 1}$$

Let the system parameters have the following values: $R_1 = 120\ \Omega$, $R_2 = 180\ \Omega$, $C = 0.014$ F. The frequency response of the system then becomes

$$G(j\omega) = \left.\frac{V_o(s)}{V_{in}(s)}\right|_{s=j\omega} = \frac{1 + 2.52j\omega}{1 + 4.2j\omega}.$$

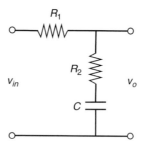

Figure 7.3.6 A phase-lag circuit in Example 7.3.2

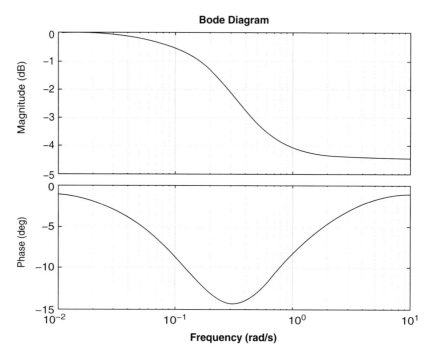

Figure 7.3.7 Bode diagram of the phase-lag circuit in Example 7.3.2

The Bode diagram of the system frequency response is shown in Figure 7.3.7.

Furthermore, for the circuit subject to a sinusoidal input $v_{in}(t) = 10 \cos(3t)$, the magnitude and phase of the system frequency response are found from the Bode diagram as follows:

$$A = G(j3) = 0.6033, \quad \phi = \angle G(j3) = -0.0523 \text{ rad}$$

It follows from Eq. (7.3.8) that the steady-state output voltage of the circuit is given by

$$v_{o,\,ss}(t) = 6.033 \cos(3t - 0.0523)$$

7.3.4 Frequency Response of Second-Order Systems

For a second-order system of the standard form (7.1.33), its frequency response is

$$G(j\omega) = \frac{\omega_n^2}{\omega_n^2 - \omega^2 + 2\xi\omega_n j\omega} \quad (7.3.19)$$

By introducing the frequency ratio

$$r = \frac{\omega}{\omega_n} \quad (7.3.20)$$

the magnitude and phase of the frequency response are given by

$$A = |G(j\omega)| = \frac{1}{\sqrt{(1-r^2)^2 + 4\xi^2 r^2}}$$
$$\phi = \angle G(j\omega) = -\tan^{-1}\left(\frac{2\xi r}{1-r^2}\right) \quad (7.3.21)$$

With the magnitude and phase, the steady-state response of the system subject to a sinusoidal input $r(t) = r_0 \sin\omega t$ is determined by

$$y_{ss}(t) = \frac{r_0}{\sqrt{(1-r^2)^2 + 4\xi^2 r^2}} \sin(\omega t + \phi) \quad (7.3.22)$$

Resonance

In Figure 7.3.8, the magnitude A and phase ϕ given in Eq. (7.3.21) are plotted against the frequency ratio ω/ω_n for different values of damping ratio ξ. The damping ratio has a significant effect on the peak of the magnitude curve: the larger the damping ratio, the lower the peak. The peak magnitude can be determined by $dA/dr = 0$, which by Eq. (7.3.21) gives $r = 1 - 2\xi^2$. It follows that at the frequency

$$\omega_r = \omega_n\sqrt{1 - 2\xi^2} \quad (7.3.23)$$

the magnitude A peaks with the value

$$A_r = \frac{1}{2\xi\sqrt{1-\xi^2}} \quad (7.3.24)$$

The parameter ω_r is known as the *resonant frequency* because at this frequency the steady-state amplitude of the system is significantly large. Note that Eq. (7.3.23) only makes sense when the damping ratio falls in the range of $0 \leq \xi \leq \sqrt{2}/2$. This indicates that the magnitude

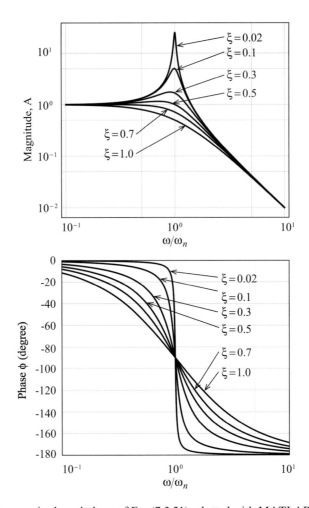

Figure 7.3.8 The magnitude and phase of Eq. (7.3.21), plotted with MATLAB

A has no peak if the damping ratio is larger than $\sqrt{2}/2$, as shown in Figure 7.3.8 (for $\xi = 1$). Also, by Eq. (7.3.24), the resonant magnitude A_r approaches infinity as the damping ratio ξ vanishes. This means that small damping can cause a large amplitude, which may be undesirable for some systems, and desirable for others, depending on the application.

Bode Diagrams

The magnitude of the Bode diagram of the system is

7.3 System Response Analysis in the Frequency Domain

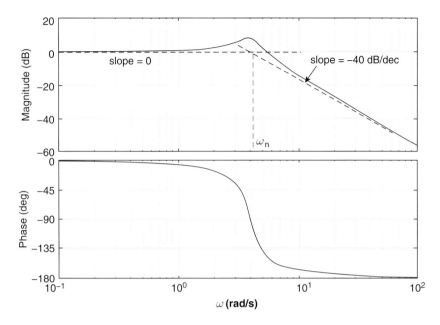

Figure 7.3.9 Bode plot of the second-order system $G(j\omega) = \omega_n^2/(s^2 + 2\xi\omega_n s + \omega_n^2)$, plotted with MATLAB

$$M = 20\log_{10} A = -20\log_{10}\sqrt{(1-r^2)^2 + 4\xi^2 r^2}, \quad r = \omega/\omega_n \quad (7.3.25)$$

and the phase angle ϕ is the same as that given in Eq. (7.3.21). As shown in Figure 7.3.9, the magnitude curve has two asymptotes:

(i) as ω approaches zero, $M \to 0$, with slope = 0
(ii) as ω approaches infinity, $M \to -40\log_{10}(\omega/\omega_n)$, with slope = −40 dB/decade

Also, it can be shown that the two asymptotes intersect at $\omega = \omega_n$, which is called *break frequency* or *corner frequency*.

Example 7.3.3

In Figure 7.3.10, a spring–mass–damper system is subject to a base (foundation) excitation $y(t)$. The equation of motion of the system is

$$m\ddot{x}(t) + c\dot{x}(t) + kx(t) = c\dot{y}(t) + ky(t) \quad (a)$$

The transfer function from the base excitation to the displacement of the mass then is

$$G(s) = \frac{X(s)}{Y(s)} = \frac{cs + k}{ms^2 + cs + k} \quad (b)$$

For a harmonic base excitation $y(t) = y_0 \sin \omega t$, the steady-state response of the system is

$$x_{ss}(t) = X \sin(\omega t + \phi) \qquad (c)$$

where the magnitude and phase are given by

$$X = y_0 \frac{\sqrt{k^2 + (c\omega)^2}}{\sqrt{(k - \omega^2 m)^2 + c^2 \omega^2}} \qquad (d)$$

$$\phi = \tan^{-1}\left(\frac{c\omega}{k}\right) - \tan^{-1}\left(\frac{c\omega}{k - m\omega^2}\right)$$

In the analysis and design of systems subject to base excitation, in such vehicle suspension systems the following two transmissibility parameters are often considered.

Displacement Transmissibility The displacement transmissibility T_d is a ratio of the magnitude of the displacement of the mass to that of the base excitation. In other words,

$$T_d = \frac{X}{y_0} = \frac{\sqrt{k^2 + (c\omega)^2}}{\sqrt{(k - \omega^2 m)^2 + c^2 \omega^2}} = \frac{\sqrt{1 + 4\zeta^2 r^2}}{\sqrt{(1 - r^2)^2 + 4\zeta^2 r^2}} \qquad (e)$$

with the frequency ratio $r = \omega/\omega_n$.

Force Transmissibility The total force transmitted to the vibrating mass that is caused by the base excitation is

$$F_{trans}(t) = -k\left(x(t) - y(t)\right) - c\left(\dot{x}(t) - \dot{y}(t)\right) = -m\omega^2 X \sin(\omega t + \phi) \qquad (f)$$

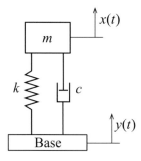

Figure 7.3.10 A system subject to a harmonic base excitation

The force transmissibility T_f is a ratio of the magnitude of the transmitted force to ky_0

$$T_f = \frac{F_T}{ky_0} = \frac{m\omega^2 X}{ky_0} = \frac{r^2\sqrt{1+4\xi^2 r^2}}{\sqrt{(1-r^2)^2 + 4\xi^2 r^2}}, \quad r = \frac{\omega}{\omega_n} \quad (g)$$

Figure 7.3.11 shows the displacement and force transmissibility parameters versus the frequency ratio r for damping ratio $\xi = 0.05, 0.1, 0.2, 0.5, 0.7$, and 1.0, which are plotted by using Eqs. (e) and (g). In applications, the values of the system parameters (m, c, k) are selected such that these transmissibility parameters are in acceptable ranges.

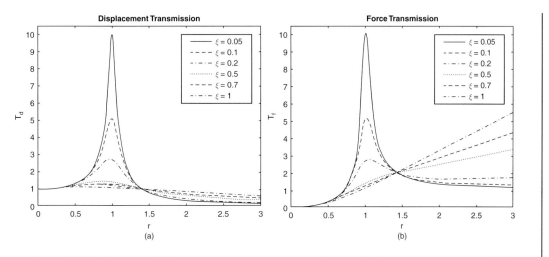

Figure 7.3.11 Transmissibility parameters due to base excitation: (a) displacement transmissibility; (b) force transmissibility

7.4 Time Response of Linear Time-Varying and Nonlinear Systems

So far in this chapter, we have only considered linear time-invariant systems. In engineering applications, however, many systems are time-varying and/or nonlinear. For example, a spring–mass–damper system with a spring of a time-varying coefficient can be described by

$$m\ddot{x} + c\dot{x} + \left(k + p(t)\right)x = f \quad (7.4.1)$$

where the spring force is given by $f_s = \left(k + p(t)\right)x$, with $p(t)$ being a known function of time. As another example, a spring–mass–damper system with a nonlinear spring can be modeled by

$$m\ddot{x} + c\dot{x} + k_0 x + k_1 x^3 = f \qquad (7.4.2)$$

where the nonlinear spring force is given by $f_s = k_0 x + k_1 x^3$, with k_0 and k_1 being constants. Additionally, the rocket equation in Example 3.2.6 is another time-varying system and the rotor system in Example 3.3.1 is another nonlinear system. For these systems, the solution methods for system responses that are presented in the previous sections are invalid.

Analytical solutions for general linear time-varying systems and nonlinear systems are difficult to obtain. Numerical methods are usually applied to determine the time response for these systems. One approach to time response involves two steps: first, formulating equivalent state equations, and then solving the state equations by numerical integration. Another approach is to use software packages, such as MATLAB and Simulink. In this section, both the approaches are demonstrated. The reader may review Appendix B and Sections 6.5 and 6.6 for the utility of MATLAB and Simulink.

7.4.1 Solution of State Equations via Numerical Integration

The governing differential equations of a dynamic system can be converted to an equivalent matrix state equation of the form

$$\dot{z}(t) = \Phi(t, z(t)); \quad z(0) = z_0 \qquad (7.4.3)$$

where the state vector $z(t) = (z_1 \ z_2 \ \cdots \ z_n)^T$, with n being the number of state variables. Refer to Sections 3.7, 4.5, and 6.2 on how to derive state equations for dynamic systems. For the linear time-varying system governed by Eq. (7.4.1), an equivalent state equation in vector form is

$$\dot{z}(t) = \begin{pmatrix} z_2(t) \\ \dfrac{1}{m}\{f(t) - cz_2(t) - (k + p(t))z_1(t)\} \end{pmatrix} \qquad (7.4.4)$$

For the nonlinear system described by Eq. (7.4.2), an equivalent state equation is

$$\dot{z}(t) = \begin{pmatrix} z_2(t) \\ \dfrac{1}{m}\{f(t) - cz_2(t) - k_0 z_1(t) - k_1 z_1^3(t)\} \end{pmatrix} \qquad (7.4.5)$$

In Eqs. (7.4.4) and (7.4.5), the state vector is $z(t) = (z_1(t) \ z_2(t))^T = (x(t) \ \dot{x}(t))^T$.

The state equation (7.4.3) can be solved via numerical integration algorithms. A commonly used algorithm is the fixed-step Runge–Kutta method of order four, which is described as follows:

$$z_{k+1} = z_k + \frac{h}{6}(f_1 + 2f_2 + 2f_3 + f_4) \qquad (7.4.6)$$

See Section 2.8 for details of the numerical integration method.

7.4.2 Solution by MATLAB and Simulink

Equation (7.4.3) can be solved using the MATLAB function ode45, which is a Runge–Kutta solver with an automatic step-size adjustment capability. Also, the software Simulink can be used to build up a model according to Eq. (7.4.3), for simulation of system response.

Example 7.4.1
Consider the time-varying system described by Eq. (7.4.1), with the following parameters

$$m = 1, \ c = 1.2, \ k = 16, \ p(t) = 0.3 \sin(3.8t)$$

Here, for convenience, nondimensional values of the system parameters have been assigned. Determine the time response of the system subject to a step input $f(t) = 16$ and zero initial displacement and velocity. Compare the result with the step response of the corresponding time-invariant system, which is governed by Eq. (7.4.1) with $p(t) = 0$.

Solution
From Eq. (7.4.4), the equivalent state equation is

$$\dot{z} = \begin{pmatrix} z_2 \\ 16 - 1.2 z_2 - (16 + 0.3 \sin(3.9t)) z_1 \end{pmatrix}, \text{ with } z(0) = \begin{pmatrix} 0 \\ 0 \end{pmatrix}$$

The displacement $x(t)$ of the time-varying system, which is the first element of the state vector $z(t)$, is computed by the fixed-step Runge–Kutta method and plotted in Figure 7.4.1. For comparison, the step response $y(t)$ of the corresponding time-invariant system

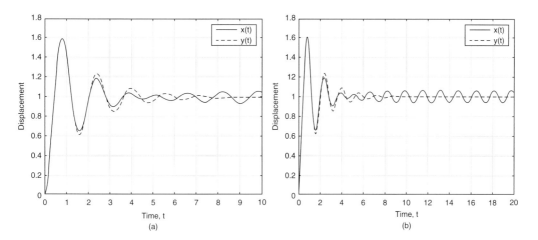

Figure 7.4.1 Figure 7.4.1 Step responses: solid line – the time-varying system; dashed line – the time-invariant system for (a) $0 \le t \le 10$ and (b) $0 \le t \le 20$

$$y(t) = 1 - e^{-\sigma t}\left(\cos\omega_d t + \frac{\sigma}{\omega_d}\sin\omega_d t\right)$$

with $\sigma = 0.6$ and $\omega_d = 3.9547$ is also plotted in the figure. Here, $y(t)$ is obtained by using Eq. (7.1.57). As seen from Figure 7.4.1(a), the displacement of the time-invariant system settles at its steady-state value ($y_{ss} = 1$) before $t = 10$. However, the displacement of the time-varying system, due to the sinusoidal variation $p(t)$ of the spring coefficient, remains of harmonic type, as shown in Figure 7.4.1(b).

The MATLAB scripts for plotting Figures 7.4.1(a) and (b) are as follows:

```
% Example 7.4.1
% Time-varying system
eta0 = [0; 0];
tf = 10; npts = 4001; h = tf/(npts-1);
t = linspace(0, tf, npts);
z = zeros(2, npts);
z(:,1) = eta0;
y = zeros(1, npts);
for i = 1: npts-1
  tt = t(i); zz = z(:,i);
  f1 = force_exm741(tt, zz);
  f2 = force_exm741(tt+h/2, zz+h/2*f1);
  f3 = force_exm741(tt+h/2, zz+h/2*f2);
  f4 = force_exm741(tt+h, zz+h*f3);
  z(:,i+1) = zz +h/6*(f1+2*f2+2*f3+f4);
end
x = z(1,:);

% Time-invariant system
sgm = 0.6; wd = sqrt(16*4-1.2^2)/2;
y = 1-exp(-sgm*t).*(cos(wd*t)+sgm/wd*sin(wd*t));

% Plotting
plot(t,x,t,y,'--')
grid
xlabel('Time, t')
ylabel('Displacement')
legend('x(t)', 'y(t)')

% A function for the time-varying system
function f = force_exm741(tt,z)
m = 1; c = 1.2; k = 16;
```

```
e0 = 0.3; omg = 3.8;
f0 = 16;
f1 = z(2);
f2 = (f0-c*z(2)-(k+e0*sin(omg*tt))*z(1))/m;
f = [f1; f2];
```

Example 7.4.2

Consider the nonlinear system described by Eq. (7.4.2). Let the system parameters have the non-dimensional values: $m = 1$, $c = 1.2$, $k_0 = 16$, $k_1 = 0.5$. Assume zero initial disturbances. For a constant external force (step input) $f(t) = 16$, determine the forced response of the system by Simulink.

Solution

A Simulink model of the nonlinear system is established in Figure 7.4.2. Refer to Section 6.6 for how to build a Simulink model. Running this model yields the forced response $x(t)$ plotted in Figure 7.4.3. For comparison, the response $y(t)$ of the corresponding linear system ($k_1 = 0$) is also plotted in the figure. As observed from the figure, the linear system has a large amplitude in response. This is because the term $k_1 x^3$ in the spring force of the nonlinear system has a hardening effect.

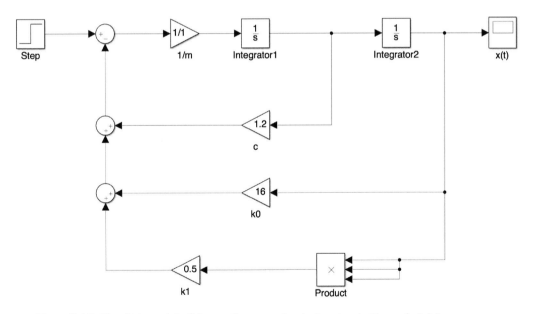

Figure 7.4.2 Simulink model of the nonlinear mechanical system in Example 7.4.2

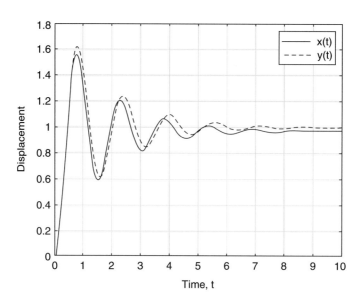

Figure 7.4.3 Step responses: solid line – nonlinear system; dashed line – linear system

MATLAB Function Blocks for Nonlinear State Equations

In Simulink, the Used-Defined Functions library has a MATLAB Function block, with the icon [MATLAB Function], which can be used to determine the solutions of time-varying and/or nonlinear state equations.

Example 7.4.3

Repeat the same problem in Example 7.4.2 by using the MATLAB Function block.

Solution
With the assigned parameter values and the constant input $f(t) = 16$, the nonlinear state equations (7.4.5) become

$$\begin{pmatrix} \dot{z}_1(t) \\ \dot{z}_2(t) \end{pmatrix} = \begin{pmatrix} z_2(t) \\ 16 - 1.2z_2(t) - 16z_1(t) - 0.5z_1^3(t) \end{pmatrix}$$

Through use of the MATLAB Function block, a Simulink model is created in Figure 7.4.4. Double-click the MATLAB Function block shows the MATLAB commands as follows:

```
function [y1, y2] = fcn(z1, z2)
y1 = z2;
y2 = 16-1.2*z2-16*z1-0.5*z1^3;
```

Running simulation with the model gives the same $x(t)$ curve as shown in Figure 7.4.3.

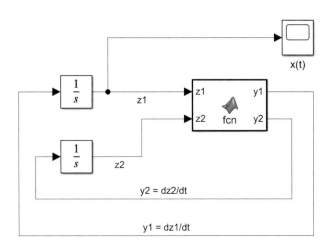

Figure 7.4.4 Simulink model using the MATLAB Function block

CHAPTER SUMMARY

This chapter addresses three important issues regarding modeling, analysis, and design of dynamic systems: system response analysis in the time domain, stability analysis, and system response analysis in the frequency domain. In time-domain analysis, system order (e.g., first order, second order, and higher orders), input type (e.g., impulse, step, ramp, and sinusoidal functions), free response and forced response, and transient response and steady-state response are classified. Analytical solutions of the impulse and step responses for first- and second-order systems are presented. In particular, four types of damping cases (no damping, underdamping, critical damping, and overdamping) of second-order systems are thoroughly examined, and the effects of damping on time response are illustrated. Also, the step response of an underdamped second-order system is evaluated and quantified through use of performance specification parameters. MATLAB and Simulink are used to compute time responses for higher-order systems, time-varying systems, and nonlinear systems.

In stability analysis, the concept of stability is first introduced; three stability definitions based on free response or impulse response are then provided; and afterwards stability conditions for a linear time-invariant system are related to the location of its transfer function poles. To detect the stability of a linear time-invariant system without having to know its transfer function poles, the Routh–Hurwitz stability criterion is introduced.

In system response analysis in the frequency domain, the steady-state behavior of a dynamic system subject to a sinusoidal input is described and the frequency response function of the system is defined. For a given dynamic system, the frequency response plots of M and ϕ against ω are derived; Bode diagrams and Nyquist plots are introduced; and the use of MATLAB in plotting frequency-response plots is demonstrated.

Upon completion of this chapter, you should be able to:

(1) Obtain the impulse and step responses of first- and second-order systems in analytical form.
(2) Determine the parameter values of an underdamped second-order system that meet the performance specifications on its step response in terms of rise time, maximum overshoot, and settling time.
(3) Compute the time response of general dynamic systems, including linear time-invariant, time-varying, and nonlinear systems, by numerical integration and via MATLAB and Simulink.
(4) Examine the stability of linear time-invariant systems by two methods: (i) examining transfer function pole locations; and (ii) applying the Routh–Hurwitz stability criterion.
(5) Determine the steady-state response of linear time-invariant systems.
(6) Plot the frequency response of linear time-invariant systems, by either analytical formulas or MATLAB.

REFERENCES

1. R. C. Dorf and R. H. Bishop, *Modern Control Systems*, 13th ed., Pearson, 2016.
2. G. Franklin, J. Powell, and A. Emami-Naeini, *Feedback Control of Dynamic Systems*, 8th ed., Pearson, 2018.
3. N. S. Nise, *Control Systems Engineering*, 8th ed., Wiley, 2020.

PROBLEMS

Section 7.1 System Response Analysis in the Time Domain

7.1 A dynamic system is governed by the differential equation

$$(D^2 + 2D + 4)y = 3(1 - e^{-5t}), \quad D = d/dt$$

and the initial conditions

$$y(0) = 0.1, \quad \dot{y}(0) = -1.$$

(a) Plot the free response of the system.
(b) Plot the forced response of the system.
Hint: You may use MATLAB or Simulink to plot the system response.

7.2 A dynamic system is governed by the differential equation

$$(D^3 + 4D^2 + 2D + 3)y = 12, \quad D = d/dt$$

and the initial conditions

$$y(0) = 0.5, \; \dot{y}(0) = 0, \; \ddot{y}(0) = 2.$$

(a) Plot the free response of the system.
(b) Plot the forced response of the system.
Hint: You may use MATLAB or Simulink to plot the system response.

7.3 For each of the following models, check if the condition of the final-value theorem is satisfied. If the condition is met, evaluate the steady-state response of the system by the final-value theorem. If the condition is not satisfied, plot the system response by MATLAB, to show that no steady-state value is reached.

(a) $(D^2 + 2D + 4)y = 3(1 - e^{-5t}), \quad D = d/dt$
$y(0) = 0, \quad \dot{y}(0) = 1$

(b) $(D^2 - 2D + 4)y = 3(1 - e^{-5t}), \quad D = d/dt$
$y(0) = 1, \quad \dot{y}(0) = 0$

(c) $(D^3 + 4D^2 + 2D + 3)y = 10, \quad D = d/dt$
$y(0) = 1 \quad \dot{y}(0) = 0, \quad \ddot{y}(0) = 0$

(d) $(D^3 + 2D^2 + 3D + 6)y = 10, \quad D = d/dt$
$y(0) = 1, \; \dot{y}(0) = 0, \; \ddot{y}(0) = 0$

7.4 For an armature-controlled DC motor in Section 6.4, its governing equations are

$$J\frac{d\omega}{dt} + c_b\omega = Ki_a - T_L \tag{i}$$

$$L_a\frac{di_a}{dt} + R_a i_a = v_{in} - K_b\omega \tag{ii}$$

where ω is the rotation speed of the motor; v_{in} and i_a are the voltage input and current of the armature circuit, respectively; K is the motor torque constant; K_b is the back electromotive force constant; and T_L denotes a load torque. Consider step inputs: $v_{in}(t) = V_0$ and $T_L = \tau_0$, where V_0 is a constant voltage and τ_0 is a constant torque. Assume that $KV_0 > R_a\tau_0$. Let the governing equations have zero initial conditions. Determine the steady-state rotation speed of the motor, $\omega_{ss} = \lim_{t \to \infty} \omega(t)$.

7.5 Figure P7.5 shows a dynamic system in a block diagram, which has two inputs and one output. Consider the step inputs: $r(t) = 5$ and $d(t) = 3$. Assume zero initial conditions.
(a) Determine the steady-state output of the system, $y_{ss} = \lim_{t \to \infty} y(t)$. Hint: You may use the final-value theorem.
(b) Plot the system output $y(t)$ by MATLAB. Show that the steady-state response in the simulation is the same as predicted in part (a).

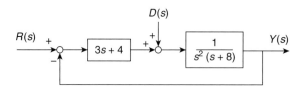

Figure P7.5

7.6 Consider the following three first-order systems:
System I: $\dot{x} + 3x = r$
System II: $4\dot{x} + 3x = r$
System III: $4\dot{x} + 3x = 2\dot{r} + r$
Every system is subject to an impulse input $r(t) = \delta(t)$, where $\delta(t)$ is the Dirac delta function. Assume zero initial conditions.
(a) Obtain the impulse responses of the system by Laplace transform.
(b) Plot the impulse responses of the systems versus time by MATLAB.

7.7 Consider the following first-order systems
System I: $\dot{x} + x = r$
System II: $2\dot{x} + x = r$
System III: $5\dot{x} + x = r$
Every system is subject to a unit step input. Assume zero initial conditions.
(a) Obtain the step response of each system by Laplace transform.
(b) Plot the step responses of the three systems versus time in one figure by MATLAB and discuss the effect of the time constant on the system response.

7.8 An RC circuit of resistance R and capacitance C is governed by

$$Ri + \frac{1}{C}\int i\, dt = v_{in}$$

$$v_{out} = \frac{1}{C}\int i\, dt$$

where v_{in} is the input voltage and v_{out} is the output voltage (capacitor voltage or system response). Let the input voltage be a constant, $v_{in}(t) = E_0$. Assume that the initial output voltage is $v_{out}(0) = V_{c0}$. Determine the free, forced, and the steady-state responses of the circuit.

7.9 According to Section 5.6, a single-tank liquid-level system, with constant cross-section area A, liquid height h, volume inflow rate q_{in}, and fluid resistance R, is governed by

$$A\frac{dh}{dt} + \frac{g}{R}h = q_{in}$$

where g is the gravitational acceleration. Let the initial liquid height be $h(0) = h_0$. Assume a step inflow rate, $q_{in} = q_0$. Determine the total response (h) of the liquid-level system. Also find the steady-state liquid head h_{ss}.

7.10 Consider a first-order system in the standard form of Eq. (7.1.17). Under a step input $r(t) = r_0$ and with the initial condition $y(0) = 2r_0$, the system response at $t = 2$ s is $y(2) = 1.5r_0$.
  (a) Determine the time constant T of the system.
  (b) With the time constant of part (a), determine the time for the system output y to reach a 5% neighborhood of its steady-state value y_{ss}.

7.11 For the following second-order systems, consider the following initial conditions: $x(0) = 1$, $\dot{x}(0) = 0$. Determine the natural frequency and damping ratio for each system. Also, plot the free responses of all the systems versus time by MATLAB.
  (a) $\ddot{x} + 4x = 0$
  (b) $\ddot{x} + \dot{x} + 4x = 0$
  (c) $\ddot{x} + 4\dot{x} + 4x = 0$
  (d) $\ddot{x} + 5\dot{x} + 4x = 0$

7.12 For the following second-order systems, consider an impulse input $r(t) = \delta(t)$, with $\delta(t)$ being the Dirac delta function. Determine the natural frequency and damping ratio for each system. Also, plot the impulse responses of all the systems versus time by MATLAB. Zero initial conditions are assumed for each system.
  (a) $\ddot{x} + 4x = r$
  (b) $\ddot{x} + \dot{x} + 4x = r$
  (c) $\ddot{x} + 4\dot{x} + 4x = r$
  (d) $\ddot{x} + 5\dot{x} + 4x = r$

7.13 For the following second-order systems, consider a step input $r(t) = 4u(t)$, with $u(t)$ being the unit step function. Determine the damping characteristics (undamped, underdamped, critically damped, or overdamped) of each system. Also, plot the step responses of the systems versus time by MATLAB. Zero initial conditions are assumed for each system.
  (a) $\ddot{x} + 4x = r$
  (b) $\ddot{x} + \dot{x} + 4x = r$
  (c) $\ddot{x} + 4\dot{x} + 4x = r$
  (d) $\ddot{x} + 5\dot{x} + 4x = r$

7.14 Consider the L-shaped rigid bar in Figure P7.14. Assume small rotation, $|\theta| \ll 1$.
  (a) Determine the natural frequency and damping ratio of the bar.
  (b) Derive the conditions on the spring and damping coefficients (k, c) and the geometric parameters of the bar for the bar to be overdamped.

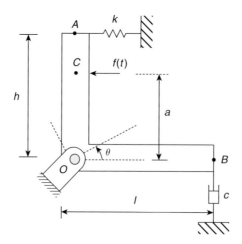

Figure P7.14

7.15 A series RLC circuit is shown in Figure P7.15, where the applied voltage v_{in} is the input and the capacitor voltage v_C is the output.
  (a) Show that the circuit is a second-order system.
  (b) Determine the natural frequency and damping ratio of the circuit in terms of parameters R, L, and C.
  (c) For a constant voltage, $v_{in} = E_0$, find the steady-state output of the circuit.

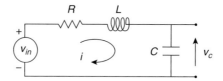

Figure P7.15

7.16 Consider the feedback control system in Figure P7.16, in which $\omega > 0$.
  (a) Determine the conditions of parameters a and b such that the system is underdamped.
  (b) With the condition derived in part (a), determine if the system has a steady-state response to a step input, $r(t) = r_0$. If so, find $y_{ss} = \lim_{t \to \infty} y(t)$.

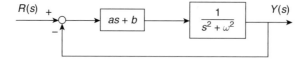

Figure P7.16

7.17 For the following second-order systems, consider a step input $r(t) = 25u(t)$, with $u(t)$ being the unit step function. Compute the 100% rise time, peak time, maximum overshoot, and settling time (with a 2% criterion) for each system. Also, plot the step responses of the systems versus time in one figure.
 (a) $\ddot{x} + 2\dot{x} + 25x = r$
 (b) $\ddot{x} + 5\dot{x} + 25x = r$
 (c) $\ddot{x} + 8\dot{x} + 25x = r$

7.18 Consider a dynamic system, with its response y governed by the differential equation
$$2\ddot{x} + (3 + a)\dot{x} + (18 + b)x = r$$
where r is an input.
 (a) Determine the values of a and b such that the natural frequency and damping ratio of the system are $\omega_n = 4$ and $\xi = 0.5$, respectively.
 (b) With a and b determined in part (a), compute the 100% rise time, maximum overshoot, and settling time (with a 5% criterion) of the step response of the system.
 (c) With the values of a and b from part (a), plot the step response of the system by MATLAB, for $r(t) = 32$.

7.19 Consider the L-shaped rigid bar in Figure P7.14, which is in small rotation, $|\theta| \ll 1$. Assume that $h = 3$ m, $l = 4$ m, $a = 2$ m, and $I_O = 100$ kg·m².
 (a) Determine the values of the spring and damping coefficients (k, c), such that the roots of the characteristic equation of the system are $-2 \pm j9$.
 (b) With the k and c from part (a), compute the 100% rise time, maximum overshoot, and settling time (with a 2% criterion) of the step response of the system.
 (c) Use MATLAB to plot the step response of the system, with $f(t) = 12$ N.

7.20 Figure P7.20 shows a dynamic system in closed-loop format.
 (a) Determine parameters a and b such that the step response of the system meets the following two performance specifications: (i) $M_p \leq 15\%$; and (ii) $t_s = 1$ s with a 1% criterion.
 (b) Taking the values of a and b from part (a), plot the system response subject to a step input $r(t) = 1$.

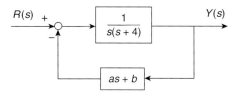

Figure P7.20

7.21 A second-order system is described by the transfer function

$$G(s) = \frac{Y(s)}{R(s)} = \frac{20(s/a + 1)}{s^2 + 4s + 20}$$

where $a > 0$. The system can be viewed as a standard second-order system with an added zero: $z = -a$. When $a \to \infty$, the transfer function becomes that for the standard second-order system described by Eq. (7.1.31). Let the input be a unit step function. In one figure, plot the step responses of the system for $a = 2, 5, 10, 20$, and ∞ and for $0 \leq t \leq 3$ s. Discuss the effect of the added zero on the system response.

7.22 A third-order system is described by the transfer function

$$G(s) = \frac{Y(s)}{R(s)} = \frac{25}{(s^2 + 6s + 25)(s/b + 1)}$$

where $b > 0$. The system can be viewed as a standard second-order system with an added pole: $p = -b$. When $b \to \infty$, the transfer function becomes that for the standard second-order system described by Eq. (7.1.31). Let the input be a unit step function. In one figure, plot the step responses of the system for $b = 3, 6, 15, 30$, and ∞ and for $0 \leq t \leq 2.5$ s. Discuss the effect of the added pole on the system response.

7.23 For a system $\dfrac{X(s)}{F(s)} = \dfrac{4(0.5s + 1)}{(s^2 + 1.2s + 4)(0.2s + 1)}$ subject to a step input $r(t) = 1$, obtain a mathematical expression of the forced response and plot the response. Hint: follow Example 7.1.10 and apply partial fraction expansion.

Section 7.2 Stability Analysis

7.24 Listed below are the characteristic equations of six dynamic systems.
System 1: $s^2 + 9 = 0$
System 2: $s^2 + 2s + 9 = 0$
System 3: $s^2 - 2s + 9 = 0$
System 4: $s^3 + 2s^2 + 5s + 9 = 0$
System 5: $s^3 + 4s^2 + 9s + 36 = 0$
System 6: $s^4 + 18s^2 + 81 = 0$
Obtain the poles for each system by MATLAB and determine the stability of the system.

7.25 Listed below are the characteristic equations of three dynamic systems.
System 1: $s^3 + 3s^2 + 5s + 7 = 0$
System 2: $s^3 + 6s^2 - 36s + 40 = 0$
System 3: $s^4 + 3s^3 + 5s^2 + 7s + 9 = 0$

By the Routh–Hurwitz stability criterion, examine the stability of each system. Also, for an unstable system, determine the number of unstable poles (poles with positive real parts) *without* solving the characteristic equation.

7.26 Consider the characteristic equations of the following two systems:

System 1: $s^3 + 6s^2 + (a - 36)s + (40 + b) = 0$

System 2: $s^4 + s^3 + 3s^2 + as + b = 0$

For each system, determine the conditions on a and b such that the system is stable.

7.27 A feedback control system is shown in Figure P7.27, where a third-order plant (a system under control) is controlled by a PID controller (Section 8.3).
(a) Determine whether or not the plant is stable.
(b) Determine the conditions on the gain constants (K_P, K_I, K_D) such that the feedback control system is stable.

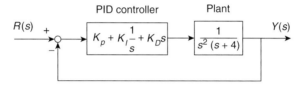

Figure P7.27

7.28 In Figure P7.28, a field-controlled DC motor is under feedback control, where $\Omega(s)$ is the motor speed; $R(s)$ is the reference input; K, J, c_b, L_f, and R_f are the motor parameters defined in Section 6.4; K_P and K_I are the gain constants of a PI controller (Section 8.3); g_a is the gain constant of the power amplifier; and g_s is the gain constant of the motor speed sensor. Determine the conditions of the controller gain constants for the control system to be stable.

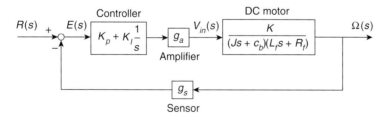

Figure P7.28

7.29 For the thermal control system in Figure 8.7.3, its characteristic equation is

$$s(RCs + 1)(\tau_a s + 1) + Rg_a g_s (K_P s + K_I + K_D s^2) = 0$$

where C and R are the thermal capacitance and resistance of the container (plant), respectively; g_s is the gain constant of the sensor; τ_a and g_a are the time constant and gain constant of the heater (actuator); and K_P, K_I, and K_D are the gain constants of the PID controller (Section 8.3). Determine the conditions of the controller gain constants such that the thermal control system is stable.

Section 7.3 System Response Analysis in the Frequency Domain

7.30 Determine the steady-state response $y_{ss}(t)$ of each of the following first-order systems to a sinusoidal input.

System 1: $\dfrac{Y(s)}{R(s)} = \dfrac{2}{s+16}$, $r(t) = 10 \sin(2t)$

System 2: $\dfrac{Y(s)}{R(s)} = \dfrac{s+16}{s+25}$, $r(t) = 2 \cos(10t + 30°)$

System 3: $\dfrac{Y(s)}{R(s)} = \dfrac{2s-3}{s+8}$, $r(t) = 4 \sin(5t) + 3 \cos(5t)$

7.31 Determine the steady-state response $y_{ss}(t)$ of each of the following second-order systems to a sinusoidal input.

System 1: $\dfrac{Y(s)}{R(s)} = \dfrac{2}{s^2 + 2s + 10}$, $r(t) = 10 \sin(3t)$

System 2: $\dfrac{Y(s)}{R(s)} = \dfrac{2(s+1)}{s^2 + 3s + 16}$, $r(t) = 5 \cos(4t + 45°)$

System 3: $\dfrac{Y(s)}{R(s)} = \dfrac{4s-1}{s^2 + 5s + 25}$, $r(t) = 4 \sin(6t) + 3 \cos(6t)$

7.32 Consider the L-shaped rigid bar Figure P7.14, which is in small rotation, $|\theta| \ll 1$. Assume the parameters of the bar as follows: $h = 3$ m, $l = 4$ m, $a = 2$ m, $I_O = 100$ kg·m², $k = 50$ N/s, and $c = 8$ N·s/m. Let the external force be $f(t) = 5 \sin(\omega t)$ N, where ω is the excitation frequency.

(a) Plot the magnitude and phase of the steady-state response $\theta_{ss}(t)$ of the bar, versus the excitation frequency ω.
(b) Following Section 7.3.4, determine the resonant frequency ω_r of the bar.

7.33 Consider the series RLC circuit in Figure P7.15, with $R = 0.1$ Ω, $L = 0.01$ H, and $C = 1$ F. Let the input voltage be $v_{in} = 12 \cos(\omega t)$ V.

(a) Plot the magnitude and phase of the steady-state response $v_{C,ss}(t)$ of the circuit.
(b) Following Section 7.3.4, determine the resonant frequency ω_r of the circuit.

7.34 Consider the control system in Figure P7.20, with $a > 0$ and $b > 0$.
  (a) For a sinusoidal input $r(t) = r_0 \sin(\omega t)$, obtain the expression of the steady-state output $y_{ss}(t)$ of the system.
  (b) For $a = 1$ and $b = 400$, plot the magnitude and phase of $y_{ss}(t)$.

7.35 Use MATLAB to plot the Bode diagrams of the following systems:

System 1: $\dfrac{Y(s)}{R(s)} = \dfrac{5(s+1)}{s+10}$

System 2: $\dfrac{Y(s)}{R(s)} = \dfrac{2s}{s^2 + 3s + 16}$

System 3: $\dfrac{Y(s)}{R(s)} = \dfrac{4(s+5)}{s^3 + 4s^2 + 9s + 36}$

7.36 Use MATLAB to plot the Nyquist plots of the following systems:

System 1: $\dfrac{Y(s)}{R(s)} = \dfrac{5(s+1)}{s+10}$

System 2: $\dfrac{Y(s)}{R(s)} = \dfrac{2s}{s^2 + 3s + 16}$

System 3: $\dfrac{Y(s)}{R(s)} = \dfrac{4(s+5)}{s^3 + 4s^2 + 10s + 36}$

Section 7.4 Time Response of Linear Time-Varying and Nonlinear Systems

7.37 Consider the cart–pendulum system in Figure P7.37, which is a nonlinear dynamic system. The system parameters are given as follows: $M = 20$ kg, $m_b = 5$ kg, $k = 500$ N/m, $c = 20$ N·s/m, and $L = 1.2$ m. A sinusoidal force is applied to the mass M: $f = 10 \sin(6t)$ N. The initial conditions of the system are as follows: $y(0) = 0$, $\dot{y}(0) = 0$; $\theta(0) = 0.02$ rad, $\dot{\theta}(0) = 0$.

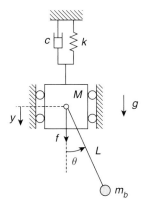

Figure P7.37

By the fixed-step Runge–Kutta method of order four, as shown in Section 7.4, compute and plot the response (y, θ) of the system for $0 \leq t \leq 10$ s. Hint: make sure to select a small enough step size h.

7.38 Repeat Problem 7.37 by building a Simulink model. Hint: you may follow Example 7.4.3.

7.39 Consider the dynamic system governed by the following differential equations

$$2\ddot{y}_1 + 3\dot{y}_1 + (18 + 2\sin 3t)y_1 - 3\dot{y}_2 - 8y_2 = f$$
$$5\ddot{y}_2 + 3\dot{y}_2 + 8y_2 + y_2^3 - 3\dot{y}_1 - 8y_1 = 0$$

which is nonlinear and time-variant. Let the input of the system be $f = 2(1 - e^{-0.5t})\sin 4t$. Assume the initial conditions as follows: $y_1(0) = 0.2$, $\dot{y}_1(0) = 0$; $y_2(0) = 0$, $\dot{y}_2(0) = -1$. By the fixed-step Runge–Kutta method of order four, as shown in Section 7.4, compute and plot the dynamic response (y_1 and y_2) for $0 \leq t \leq 8$ s.

7.40 Repeat Problem 7.39 by using the MATLAB function ode45.

8 Introduction to Feedback Control Systems

Contents

8.1	General Concepts	589
8.2	Advantages of Closed-Loop Control Systems	595
8.3	PID Control Algorithm	601
8.4	Control System Analysis	611
8.5	The Root Locus Method	619
8.6	Analysis and Design by the Root Locus Method	632
8.7	Additional Examples of Control Systems	648
	Chapter Summary	653
	References	654
	Problems	654

Controls in engineering are efforts to change, design, or modify the behaviors of dynamic systems. Automatic control is control that involves only machines and devices, and that has no human intervention. Examples of automatic control are diverse, including room-temperature control, cruise control of cars, missile guidance, trajectory control of robots, control of appliances such as washing machines and refrigerators, and control of industrial processes like papermaking and steelmaking. In this chapter, for simplicity, an automatic control system is called a control system. Because the focus of this text is on the modeling and analysis of dynamic systems, only basic concepts of feedback control are introduced. For theories and methods about feedback control systems, one may refer to standard textbooks for control courses, including those listed at the end of the chapter.

8.1 General Concepts

A control system usually consists of the following four major components or subsystems:

Plant: A plant is a dynamic system that is controlled so that its output meets certain performance requirements. The output of the plant is also the output of the control system. In engineering applications, a plant can be a machine, a device, a structure, or an industrial process.

Sensor: A sensor is a device that measures the output of the plant. Examples of sensors in engineering applications include accelerometers, tachometers, displacement transducers, strain gauges, manometers, flowmeters, magnetometers, and thermometers. As will be explained, not every control system requires a sensor.

Controller: A controller is a component that computes a control signal or control command by an algorithm. The algorithm, which is also called control logic or control law, makes use of the specified performance requirements and the measured plant output if necessary. Nowadays, a controller is usually implemented on a computer chip (logic board).

Actuator: An actuator is a device that applies a control effort (e.g., a force, a voltage, or a heat flow rate) to the plant according to the control command from the controller.

Depending on whether the plant output is measured, there are two control system configurations: open-loop control systems and closed-loop control systems, which are explained as follows.

Open-Loop Control System

An open-loop control system (simply called an open-loop system) is a control system in which the plant is controlled without measurement of the plant output; see Figure 8.1.1(a). In other words, no sensor is implemented in an open-loop system and the output of the control system has no influence on the control signal of the controller.

Closed-Loop Control System

A closed-loop control system is a control system in which the plant is controlled via measurement of the plant output and comparison of the measured output with a reference input (desired output); see Figure 8.1.1(b). A closed-loop control system is also called a closed-loop system, feedback control system, or simply feedback system. In contrast to open-loop control, the output of a closed-loop system has influence on the control signal from the controller. In such a control system, the sensor data are fed back to close a loop from the output to the reference input. Because of this, a closed-loop system is also called a feedback control system.

Control systems are seen commonly in everyday life. An electric toaster in the kitchen is an example of an open-loop control system; an air conditioner is an example of a closed-loop control system. Open-loop control, which does not require an output measurement, is easier and less expensive to implement. However, the performance of an open-loop control system may not be satisfactory. Due to the complexity of dynamic systems in many applications and because of the need for optimal performance, closed-loop control plays an essentially important role in the development and advancement of modern technology and civilization.

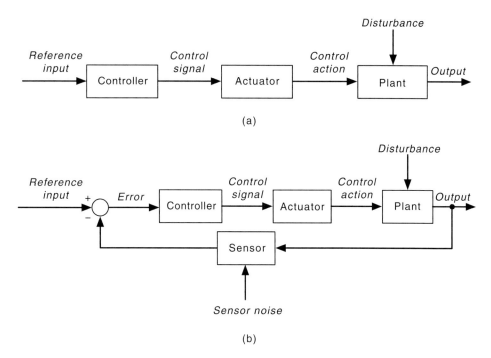

Figure 8.1.1 Control system configurations: (a) open-loop control system; and (b) closed-loop control system

Closed-loop control has three major advantages over open-loop control: (i) it can reject a disturbance and sensor noise; (ii) it reduces the influence of the variations of system parameters; and (iii) it can improve the transient and steady-state response and stability of a control system. These advantages are illustrated in Section 8.2.

Example 8.1.1

Figure 8.1.2 shows a closed-loop system for temperature control of a house in hot weather, by an air conditioner. In this control system, the temperature of a room is measured by a thermostat. A controller (logic board) compares the measured room temperature with a desired temperature (preset as a reference input) and computes a control signal. Based on the control signal (the controller output), a trio of compressor, condenser, and evaporator as the actuator removes a certain amount of heat from the house via a working fluid. This process continues until the room temperature reaches the preset temperature as desired. The thermostat, controller, compressor, condenser, and evaporator are the parts of the air conditioner.

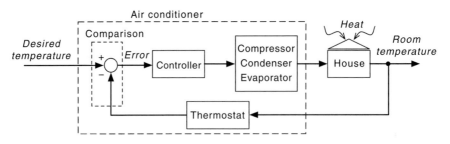

Figure 8.1.2 Block diagram of a room-temperature control system enabled by an air conditioner

Transfer Function Formulation of Control Systems

Assume that all the components of a control system can be described as linear time-invariant subsystems. The two control system configurations in Figure 8.1.1 then can be presented in the s-domain block diagrams shown in Figure 8.1.3, where $R(s)$ is the reference input, $Y(s)$ is the output, $E(s)$ is the error (for closed-loop control only), $U(s)$ is the control signal or controller output, $D(s)$ is a disturbance applied to the plant, $N(s)$ is an sensor noise, and $G_p(s)$, $G_s(s)$, $G_c(s)$, $G_a(s)$ are the transfer functions of the plant, sensor, controller, and actuator, respectively.

The output of the open-loop system, by Figure 8.1.3(a), is written as

$$Y(s) = G_p(s)G_a(s)G_c(s)R(s) + G_p(s)D(s) \qquad (8.1.1)$$

As can be seen from the previous equation, the controller $G_c(s)$ has no influence on the disturbance $D(s)$. This disadvantage is also demonstrated through an example (motor speed control) in Section 8.2.

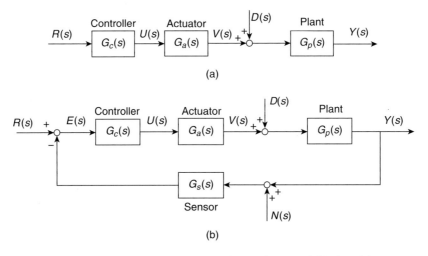

Figure 8.1.3 Block diagrams of control systems: (a) open-loop; and (b) closed-loop

The output of the closed-loop control system, Figure 8.1.3(b), is of the form

$$Y(s) = T_R(s)R(s) + T_D(s)D(s) + T_N(s)N(s) \tag{8.1.2}$$

where $T_R(s)$, $T_D(s)$, and $T_N(s)$ are the transfer functions describing the influence of the reference input, disturbance, and sensor noise on the output of the closed-loop control system, respectively. In other words, the transfer functions can be expressed as

$$T_R(s) = \left.\frac{Y(s)}{R(s)}\right|_{\substack{D(s)=0 \\ N(s)=0}}, \quad T_D = \left.\frac{Y(s)}{D(s)}\right|_{\substack{R(s)=0 \\ N(s)=0}}, \quad T_N(s) = \left.\frac{Y(s)}{N(s)}\right|_{\substack{R(s)=0 \\ D(s)=0}} \tag{8.1.3}$$

With the formulas of the block diagrams in Section 6.2.3, these transfer functions are found to be

$$T_R(s) = \frac{G_c(s)G_a(s)G_p(s)}{1+L(s)}, \quad T_D(s) = \frac{G_p(s)}{1+L(s)}, \quad T_N(s) = -\frac{L(s)}{1+L(s)} \tag{8.1.4}$$

with

$$L(s) = G_s(s)G_c(s)G_a(s)G_p(s) \tag{8.1.5}$$

Here, $L(s)$ is known as the *loop transfer function*. As can be seen from Eqs. (8.1.4) and (8.1.5), in a closed-loop control system, the controller $G_c(s)$ has influence on the disturbance $D(s)$ and noise $N(s)$. This is one main reason that closed-loop control systems are widely used in engineering applications. Accordingly, this chapter focuses on closed-loop control systems.

Example 8.1.2

The displacement of a mass–damper system is under feedback control; see Figure 8.1.4(a), where f is an external force and f_c is the control force applied to the mass by the actuator. Without loss of generality, assume that the sensor measures the displacement of the mass by $y_s(t) = g_s x(t)$ and that the actuator force is related to the control signal u by $f_c(t) = g_a u(t)$, where g_s and g_a are gain constants. A proportional plus derivative (PD) controller, which is presented in detail in Section 8.3, is implemented. The transfer function of the plant (the mass–damper system) without control is given by

$$\frac{X(s)}{F(s)} = \frac{1}{ms^2 + cs}$$

The PD controller is described by the transfer function

$$\frac{U(s)}{E(s)} = K_P + K_D s$$

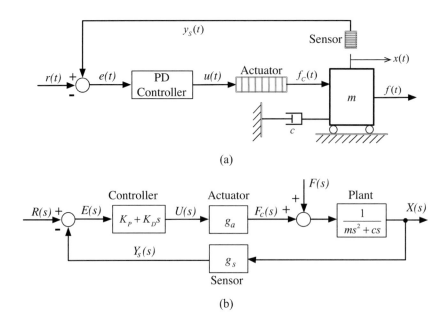

Figure 8.1.4 A mass–damper system under PD feedback control in Example 8.1.2

where K_P and K_D are the gain constants of the controller, and $E(s)$ is the Laplace transform of the error $e = r - y_s$, with r being a reference input.

With the above description, a block diagram with component transfer functions is established in Figure 8.1.4(b). By the block diagram and from Eq. (8.1.2), the s-domain output of the closed-loop system is obtained as

$$X(s) = T_R(s)R(s) + T_D(s)F(s)$$

in which the transfer functions are given by

$$T_R(s) = \frac{(K_P + K_D s)g_a \frac{1}{ms^2 + cs}}{1 + g_s(K_P + K_D s)g_a \frac{1}{ms^2 + cs}} = \frac{g_a(K_D s + K_P)}{ms^2 + (c + g_s g_a K_D)s + g_s g_a K_P}$$

$$T_D(s) = \frac{\frac{1}{ms^2 + cs}}{1 + g_s(K_P + K_D s)g_a \frac{1}{ms^2 + cs}} = \frac{1}{ms^2 + (c + g_s g_a K_D)s + g_s g_a K_P}$$

As indicated in Eq. (8.1.4), the two transfer functions have the same denominator

$$ms^2 + (c + g_s g_a K_D)s + g_s g_a K_P$$

In control system design, the control gains K_P and K_D are selected such that the system output meets certain stability and performance specifications. This will be discussed in the subsequent sections.

8.2 Advantages of Closed-Loop Control Systems

In this section, the three main advantages of closed-loop control as mentioned in the previous section are illustrated in an example of motor speed control. Figure 8.2.1 shows an open-loop control system and a closed-loop control system for the speed control of a field-controlled DC motor. For simplicity in analysis, unity feedback without sensor noise has been assumed for the closed-loop system. According to Section 6.4, a simplified model of the motor (by neglecting the inductance of the field circuit) is given by

$$\tau \frac{d\omega}{dt} + \omega = K_o(\tau_m + \tau_d) \tag{8.2.1}$$

where ω is the motor rotation speed, τ and K_o are constants, τ_m is the motor torque, and τ_d is a torque load applied to the motor shaft. Also, in the figure, g_a is the gain constant of the amplifier, and $\Omega_r(s)$ is a reference input in the s-domain, which is the desired motor speed.

Let the desired motor speed be $\omega_r = \bar{\omega}_r$, which is a prescribed constant. Let the torque load be $\tau_d = \bar{\tau}_d$, which is a constant but can be unknown. Thus, $\Omega_r(s) = \bar{\omega}_r \frac{1}{s}$ and $T_d(s) = \bar{\tau}_d \frac{1}{s}$. According to the block diagrams in Figure 8.2.1, the output of the open-loop system is given by

$$\Omega(s) = \frac{K_o}{\tau s + 1} \frac{1}{s} \{g_a C_o(s)\bar{\omega}_r + \bar{\tau}_d\} \tag{8.2.2}$$

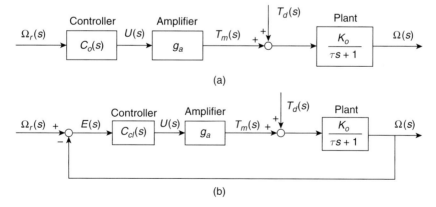

Figure 8.2.1 Speed control of a DC motor: (a) open-loop control system; and (b) closed-loop control system

and the output of the closed-loop system is given by

$$\Omega(s) = \frac{K_o}{\tau s + 1 + g_a C_{cl}(s) K_o} \frac{1}{s} \{g_a C_{cl}(s)\overline{\omega}_r + \overline{\tau}_d\} \quad (8.2.3)$$

In Eqs. (8.2.2) and (8.2.3), $C_o(s)$ and $C_{cl}(s)$ are the controller transfer functions to be designed.

Advantage 1. Reduction of the Effects of Parameter Variations

In control engineering, the ability of a closed-loop system to reduce the effects of parameter variations is also known as the robustness of the control system. For simplicity in discussion, consider a zero load, $\overline{\tau}_d = 0$. Let the open-loop controller be $C_o(s) = K_c$, where K_c is a constant gain. The steady-state speed of the motor, from Eq. (8.2.2) and by the final-value theorem (Section 2.4.3), is

$$\omega_{ss} = \lim_{s \to o} s\Omega(s) = \lim_{s \to o} s \frac{K_o}{\tau s + 1} \frac{1}{s} g_a K_c \overline{\omega}_r = K_o g_a K_c \overline{\omega}_r \quad (8.2.4)$$

Thus, if the control gain is chosen as $K_c = 1/(K_o g_a)$, the steady-state speed of the motor is the same as the desired speed; that is $\omega_{ss} = \overline{\omega}_r$.

Now we assume that the motor has parameter variations such that K_o become $K_o + \Delta K_o$, with ΔK_o being an unknown constant. Parameter variations such as ΔK_o usually come from uncertainties in system modeling or parameter drifting during operation. The steady-state speed of the motor under open-loop control with $K_c = 1/(K_o g_a)$ is

$$\omega_{ss}^{\Delta K_o} = (K_o + \Delta K_o) g_a K_c \overline{\omega}_r = \overline{\omega}_r + \frac{\Delta K_o}{K_o} \overline{\omega}_r \quad (8.2.5)$$

Define the percentage error of the motor's steady-state speed caused by the parameter variations as

$$\varepsilon_{\Delta K_o} = \frac{\Delta \omega_{ss}}{\omega_{ss}} = \frac{\omega_{ss}^{\Delta K_o} - \omega_{ss}}{\omega_{ss}} \quad (8.2.6)$$

If follows from Eq. (8.2.5) that

$$\varepsilon_{\Delta K_o} = \left(\overline{\omega}_r + \frac{\Delta K_o}{K_o} \overline{\omega}_r - \overline{\omega}_r\right) \frac{1}{\overline{\omega}_r} = \frac{\Delta K_o}{K_o} \quad (8.2.7)$$

According to Eq. (8.2.7), the percentage error of the motor speed is directly related to the percentage change of K_o, and the open-loop control has no influence on the parameter variations. For example, a 10% parameter variation ($\Delta K_o/K_o = 10\%$) renders a 10% error in the motor speed ($\Delta \omega_{ss}/\overline{\omega}_r = 10\%$), which is not acceptable in application.

Under closed-loop control, if $C_{cl}(s) = K_c$ (which describes P control as defined in Section 8.3), the steady-state speed of the motor, from Eq. (8.2.3) and using the final-value theorem, is given by

$$\omega_{ss} = \lim_{s \to 0} s \frac{K_o}{\tau s + 1 + g_a K_c K_o} \frac{1}{s} g_a K_c \overline{\omega}_r = \frac{K_o g_a K_c}{1 + K_o g_a K_c} \overline{\omega}_r \quad (8.2.8)$$

Under P control with a large enough control gain K_c, the motor speed can be close to the desired speed. With the parameter uncertainty $K_o + \Delta K_o$, the steady-state output $\omega_{ss}^{\Delta K_o}$ of the system is

$$\omega_{ss}^{\Delta K_o} = \frac{(K_o + \Delta K_o) g_a K_c}{1 + (K_o + \Delta K_o) g_a K_c} \overline{\omega}_r \quad (8.2.9)$$

The percentage error of the motor speed, defined in Eq. (8.2.6), is obtained as

$$\varepsilon_{\Delta K_o} = \frac{(K_o + \Delta K_o) g_a K_c}{1 + (K_o + \Delta K_o) g_a K_c} \frac{\overline{\omega}_r}{\omega_{ss}} - 1 = \frac{(1 + \varepsilon) K_o g_a K_c}{1 + (1 + \varepsilon) K_o g_a K_c} \frac{\overline{\omega}_r}{\omega_{ss}} - 1 \quad (8.2.10)$$

where $\varepsilon = \Delta K_o/K_o$. For relatively small parameter variations ($\varepsilon^2 \ll 1$), the first term on the right-hand side of Eq. (8.2.10) is reduced to

$$\frac{(1 + \varepsilon) K_o g_a K_c}{1 + (1 + \varepsilon) K_o g_a K_c} \frac{\overline{\omega}_r}{\omega_{ss}} = \left(\frac{K_o g_a K_c}{1 + K_o g_a K_c} \overline{\omega}_r \right) \frac{1}{\omega_{ss}} \frac{(1 + \varepsilon)}{1 + \frac{\varepsilon K_o g_a K_c}{1 + K_o g_a K_c}}$$

$$= \omega_{ss} \frac{1}{\omega_{ss}} (1 + \varepsilon) \left\{ 1 - \varepsilon \frac{K_o g_a K_c}{1 + K_o g_a K_c} + \varepsilon^2 \left(\frac{K_o g_a K_c}{1 + K_o g_a K_c} \right)^2 - \cdots \right\}$$

$$\approx 1 + \varepsilon \left\{ 1 - \frac{K_o g_a K_c}{1 + K_o g_a K_c} \right\} = 1 + \varepsilon \frac{1}{1 + K_o g_a K_c}$$

where Eq. (8.2.8) and the series $\frac{1}{1+x} = 1 - x + x^2 - x^3 + \cdots$ (for $|x| < 1$) have been used, and the ε^2-term and higher-order terms have been eliminated. It follows that

$$\varepsilon_{\Delta K_o} \approx \frac{1}{1 + K_o g_a K_c} \frac{\Delta K_o}{K_o} \quad (8.2.11)$$

This indicates that the percentage error of the motor speed can be significantly reduced by properly assigning the control gain K_c. For instance, for a 10% parameter variation, $\Delta K_o/K_o = 10\%$, the selection of K_c satisfying the condition $1 + K_o g_a K_c \geq 20$ results in $\Delta \omega_{ss}/\omega_{ss} \leq 0.5\%$, which is much smaller than that with open-loop control.

Alternatively, let us now consider PI control (Section 8.3), $C_{cl}(s) = K_P + K_I \frac{1}{s}$. It is easy to show that the steady-state motor speed without parameter variations is

$$\omega_{ss} = \lim_{s \to 0} s\Omega(s) = \lim_{s \to 0} \frac{g_a K_o (K_P s + K_I)}{\tau s^2 + s + g_a K_o (K_P s + K_I)} \overline{\omega}_r$$

$$= \frac{g_a K_o K_I}{g_a K_o K_I} \overline{\omega}_r = \overline{\omega}_r \qquad (8.2.12)$$

Here, it is easy to check that $s\Omega(s)$ satisfies the condition of the final-value theorem because all the parameters of the motor and controller are positive. Replacing K_o by $K_o + \Delta K_o$ in Eq. (8.2.12) gives the steady-state output of the control system as follows:

$$\omega_{ss}^{\Delta K_o} = \frac{g_a(K_o + \Delta K_o)K_I}{g_a(K_o + \Delta K_o)K_I} \overline{\omega}_r = \overline{\omega}_r \qquad (8.2.13)$$

Equations (8.2.12) and (8.2.13) show that, under PI control, the motor reaches the desired speed, with or without the parameter variation ΔK_o. Therefore, the percentage error of the motor speed as defined in Eq. (8.2.6) is zero for any value of ΔK_o; namely,

$$\varepsilon_{\Delta K_o} = \frac{\overline{\omega}_r - \overline{\omega}_r}{\overline{\omega}_r} = 0.$$

Advantage 2. Disturbance Rejection

We now compare open-loop control and closed-loop control in relation to a disturbance. To this end, assume the motor is subject to a constant load $\overline{\tau}_d$. For simplicity, we assume that the motor has no parameter variation. For the open-loop system with the controller $C_o(s) = K_c = 1/(K_o g_a)$, its steady-state output by Eq. (8.2.2) is

$$\omega_{ss}^{\overline{\tau}_d} = \lim_{s \to 0} \frac{K_o}{\tau s + 1} \left\{ g_a \frac{1}{K_o g_a} \overline{\omega}_r + \overline{\tau}_d \right\} = \overline{\omega}_r + K_o \overline{\tau}_d \qquad (8.2.14)$$

Without disturbance, the steady-state output of the open-loop system from Eq. (8.2.4) is $\omega_{ss} = \overline{\omega}_r$. The percentage error of the motor's steady-state speed caused by the disturbance $\overline{\tau}_d$ is defined as

$$\varepsilon_{\overline{\tau}_d} = \frac{\Delta \omega_{ss}}{\omega_{ss}} = \frac{\omega_{ss}^{\overline{\tau}_d} - \omega_{ss}}{\omega_{ss}} \qquad (8.2.15)$$

From Eq. (8.2.14),

$$\varepsilon_{\overline{\tau}_d} = \frac{\Delta \omega_{ss}}{\omega_{ss}} = \frac{\overline{\omega}_r + K_o \overline{\tau}_d - \overline{\omega}_r}{\overline{\omega}_r} = \frac{K_o}{\overline{\omega}_r} \overline{\tau}_d \qquad (8.2.16)$$

which indicates that the open-loop controller has no influence on the reduction of the output error caused by the disturbance. Indeed, an increased $\bar{\tau}_d$ yields an increased error in the motor speed.

Now, consider the closed-loop control system in Figure 8.2.1(b), with a disturbance $\bar{\tau}_d$ and $C_{cl}(s) = K_c$ (P control). The steady-state output of the control system, from Eqs. (8.2.3) and (8.2.8) and by the final-value theorem, is

$$\omega_{ss}^{\bar{\tau}_d} = \lim_{s \to 0} \frac{K_o}{\tau s + 1 + g_a K_c K_o} \{g_a K_c \bar{\omega}_r + \bar{\tau}_d\}$$

$$= \frac{g_a K_c K_o}{1 + g_a K_c K_o} \bar{\omega}_r + \frac{K_o \bar{\tau}_d}{1 + g_a K_c K_o} = \omega_{ss} + \frac{K_o \bar{\tau}_d}{1 + g_a K_c K_o} \quad (8.2.17)$$

It follows from Eq. (8.2.15) that

$$\varepsilon_{\bar{\tau}_d} = \frac{\Delta \omega_{ss}}{\omega_{ss}} = \left(\omega_{ss} + \frac{K_o \bar{\tau}_d}{1 + g_a K_c K_o} - \omega_{ss}\right) \frac{1 + K_o g_a K_c}{K_o g_a K_c \bar{\omega}_r} = \frac{\bar{\tau}_d}{\bar{\omega}_r} \frac{1}{g_a K_c} \quad (8.2.18)$$

Thus, the percentage error of the motor's steady-state speed decreases as the control gain K_c increases.

If PI control is applied, $C_{cl}(s) = K_P + K_I \frac{1}{s}$, the steady-state speed of the motor with a disturbance $\bar{\tau}_d$ is

$$\omega_{ss}^{\bar{\tau}_d} = \lim_{s \to 0} s \frac{K_o}{\tau s + 1 + g_a \left(K_P + K_I \frac{1}{s}\right) K_o} \frac{1}{s} \left\{ g_a \left(K_P + K_I \frac{1}{s}\right) \bar{\omega}_r + \bar{\tau}_d \right\} = \bar{\omega}_r$$

This means that the steady-state speed of the PI-controlled motor under the disturbance is the same as the desired speed. Also, by setting $\bar{\tau}_d = 0$ in the previous equation, it can be shown that the steady-state speed ω_{ss} of the motor without disturbance is also $\bar{\omega}_r$. It follows that the percentage error of the motor's steady-state speed is zero; namely,

$$\varepsilon_{\bar{\tau}_d} = \frac{\Delta \omega_{ss}}{\omega_{ss}} = \frac{\bar{\omega}_r - \bar{\omega}_r}{\bar{\omega}_r} = 0 \quad (8.2.19)$$

This example shows that closed-loop control can reject the effects of disturbances on the control system output.

Advantage 3. Improvement of System Response

Closed-loop (feedback) control can improve the transient response, the steady-state response, and the stability of a dynamic system. To show this, the motor parameters and amplifier gain in Figure 8.2.1 are assigned as $\tau = 0.5$, $K_o = 3$, $g_a = 16$ and the desired motor

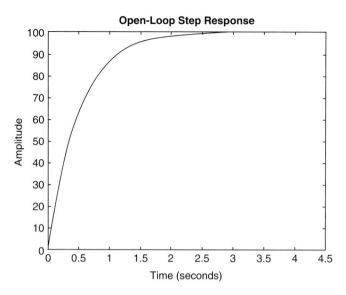

Figure 8.2.2 The output of the open-loop control system in Figure 8.2.1(a) with $C_o(s) = K_c = 1/48$ and $\bar{\tau}_d = 0$

speed as $\bar{\omega}_r = 100$. Here, nondimensional values of the parameters have been considered. Also, for simplicity, we assume that the motor does not have parameter variations, and is not subject to a disturbance.

The output of the open-loop system in Figure 8.2.1(a), with $C_o(s) = 1/(K_o g_a) = 1/48$, and $\bar{\tau}_d = 0$ in Eq. (8.2.2), is given by

$$\Omega_o(s) = \frac{1}{\tau s + 1} \frac{\bar{\omega}_r}{s} = \frac{100}{0.5s + 1} \frac{1}{s} \qquad (8.2.20)$$

which by inverse Laplace transform gives the time-domain response

$$\omega_o(t) = 100(1 - e^{-2t}) \qquad (8.2.21)$$

The steady-state output of the open-loop system is $\omega_{o,ss} = \lim_{s \to 0} \omega_o(t) = \bar{\omega}_r = 100$. The output of the open-loop system is plotted against time in Figure 8.2.2. We define a 2% settling time t_s as the time for the output to reach a 2% neighborhood of the final value $\omega_{o,ss}$. In other words, $\omega_o(t_s) = 98\% \times \omega_{o,ss}$, which from Eq. (8.2.21) gives $t_s = -0.5\ln(0.02) = 1.956$ s.

The above-defined settling time is a measure of the system performance. (Refer to Section 7.1 for system response analysis in the time domain.) A shorter settling time means that the system response to the reference input ω_r is faster. As can be seen from Eq. (8.2.20), the settling time of the open-loop system is the same as that of the plant (motor). Thus, the open-loop control gain K_c has no influence on the settling time.

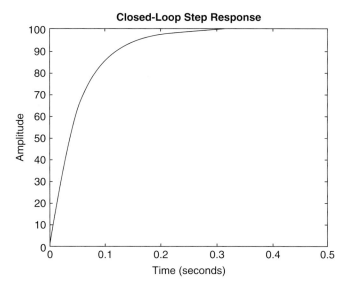

Figure 8.2.3 The output of the closed-loop control system in Figure 8.2.1(b) with $C_{cl}(s) = 0.2 + 0.4/s$ and $\bar{\tau}_d = 0$

To improve the system response with a shorter settling time, consider the closed-loop control system in Figure 8.2.1(b), with a PI controller, $C_{cl}(s) = K_P + K_I \dfrac{1}{s}$. With careful design, the controller parameters are chosen as $K_P = 0.2$, $K_I = 0.4$. The output of the closed-loop system, by Eq. (8.2.3), is obtained as

$$\Omega_{cl}(s) = \frac{g_a K_o (K_P s + K_I)}{\tau s^2 + (1 + g_a K_o K_P)s + g_a K_o K_I} \frac{\bar{\omega}_r}{s} = \frac{4800\,(0.2s + 0.4)}{0.5s^2 + 10.6s + 48 \times 0.4} \frac{1}{s} \quad (8.2.22)$$

By using the MATLAB function `step`, the output of the closed-loop system versus time is plotted in Figure 8.2.3, which shows that the 2% settling time of the system is 0.204 s. Hence, the motor speed under the closed-loop control reaches the desired value much faster than that under the open-loop control.

8.3 PID Control Algorithm

8.3.1 PID Controller

The PID controller, which is also known as the three-term controller, is probably the most commonly used control algorithm in engineering applications. The PID controller has the transfer function form

$$G_c(s) = \frac{U(s)}{E(s)} = K_P + K_I \frac{1}{s} + K_D s \qquad (8.3.1)$$

where $U(s)$ and $E(s)$ are the controller output and error, as shown in Figure 8.1.3(b), and K_P, K_I, and K_D are the control gains. It follows that the controller output in the time domain is

$$u(t) = K_P \, e(t) + K_I \int_{t_0}^{t} e(\tau) d\tau + K_D \frac{de(t)}{dt} \qquad (8.3.2)$$

The PID control algorithm given in Eq. (8.3.1) can also be written as

$$G_c(s) = K_P \left(1 + \frac{1}{T_I s} + T_D s \right) \qquad (8.3.3)$$

where T_I is the integral time and T_D is the derivative time. Comparison of Eqs. (8.3.1) and (8.3.3) shows that $K_I = K_P/T_I$ and $K_D = K_P T_D$.

As can be seen from Eq. (8.3.2), the controller output has three terms, with the first term **P**roportional to the error $e(t)$, the second term given by an **I**ntegral about the error, and the third term related to a **D**erivative of the error. Thus, the PID control represents a combination of proportional, integral, and derivative control actions; see Figure 8.3.1, where the three blocks describe the actions of P control, I control, and D control, respectively.

By setting certain control gains to zero, the PID controller is reduced to the following three commonly used controllers.

(i) proportional (P) controller:

$$G_c(s) = K_P \qquad (8.3.4)$$

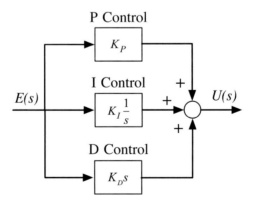

Figure 8.3.1 A block diagram of the PID controller

(ii) proportional plus integral (PI) controller:

$$G_c(s) = K_P + K_I \frac{1}{s} \quad (8.3.5)$$

(iii) proportional plus derivative (PD) controller:

$$G_c(s) = K_P + K_D s \quad (8.3.6)$$

Thus, the PID, P, PI, and PD control algorithms are proper combinations of P, I, and D control actions.

Example 8.3.1

In Figure 8.3.2, a first-order plant with parameters $T = 2$ and $b = 3$ is under PID feedback control. Consider a step reference input $r(t) = 1$. Assume a step disturbance $d(t) = 2$. Select the control gain constants K_P, K_I, and K_D such that the closed-loop system meets the following control objectives:

(i) the system is stable
(ii) the steady-state error ($e_{ss} = \lim_{t \to \infty} e(t)$) is zero
(iii) the settling time of the output (2% criterion) is less than 3 seconds
(iv) the maximum overshoot of the output is less than 10%

Solution

Write the reference input and disturbance as $r(t) = r_0$, $d(t) = d_0$. From Eqs. (8.1.2) and (8.1.4), the output of the control system is

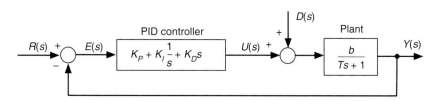

Figure 8.3.2 A first-order system under PID feedback control

$$Y(s) = \frac{\frac{1}{s}(K_D s^2 + K_P s + K_I)\frac{b}{Ts+1}R(s) + \frac{b}{Ts+1}D(s)}{1 + \frac{1}{s}(K_D s^2 + K_P s + K_I)\frac{b}{Ts+1}} \quad \text{(a)}$$

$$= \frac{b(K_D s^2 + K_P s + K_I)\frac{r_0}{s} + bd_0}{s(Ts+1) + b(K_D s^2 + K_P s + K_I)}$$

This is a second-order system with the characteristic equation

$$s^2 + \frac{1+3K_P}{2+3K_D}s + \frac{3K_I}{2+3K_D} = 0 \quad \text{(b)}$$

According to Table 7.2.1, the control system is stable if and only if all the coefficients of the characteristic equation are positive. Thus, the stability conditions are

$$(2 + 3K_D) > 0, \quad (1 + 3K_P) > 0, \quad K_I > 0 \quad \text{(c)}$$

These conditions are met if K_P, K_I, and K_D are all positive.

Under the stability conditions, the poles of the function

$$sY(s) = \frac{3(K_D s^2 + K_P s + K_I) + 6s}{(2+3K_D)s^2 + (1+3K_P)s + 3K_I} \quad \text{(d)}$$

which is obtained with $T = 2$, $b = 3$, $r_0 = 1$, and $d_0 = 2$, all have negative real parts. Thus, the steady-state response of the closed-loop system, by the final-value theorem, is

$$y_{ss} = \lim_{s \to 0} sY(s) = \lim_{s \to 0} \frac{3(K_D s^2 + K_P s + K_I) + 6s}{(2+3K_D)s^2 + (1+3K_P)s + 3K_I} = 1 \quad \text{(e)}$$

This implies that the steady-state error $e_{ss} = \lim_{t \to \infty}(r - y) = 1 - y_{ss} = 0$.

Through comparison of the characteristic equation (a) with the standard form

$$s^2 + 2\zeta\omega_n s + \omega_n^2 = 0$$

we obtain

$$2\zeta\omega_n = \frac{1+3K_P}{2+3K_D} \quad \text{(f)}$$

An estimation of the settling time, from Eq. (7.1.70), is $t_s = 4/\zeta\omega_n < 3$, which by Eq. (f) gives the condition

$$\frac{1+3K_P}{2+3K_D} > \frac{8}{3} \quad \text{(g)}$$

Note that the above estimation is only approximate because the closed-loop transfer function is not in the standard form of a second-order system and because the system has two inputs (reference and disturbance). Nevertheless, the formulas in Eq. (7.1.70) provide useful guidance in control system design.

For numerical simulation, we choose $K_I = 14$ and $K_D = 4$, for which, condition (g) gives $K_P > 12.11$. Figure 8.3.3 shows the curves of the step response of the control system for $K_P = 12$, 15, and 20, which are plotted by using MATLAB. It can be seen that

$$\text{for } K_P = 12, \ t_s = 2.16 \text{ s}, \ M_P = 11\%$$

$$\text{for } K_P = 15, \ t_s = 2.22 \text{ s}, \ M_P = 9\%$$

$$\text{for } K_P = 25, \ t_s = 3.02 \text{ s}, \ M_P = 6\%$$

where t_s and M_P are the settling time and maximum overshoot, respectively. Here, an increase in the proportional gain K_P decreases the maximum overshoot, but makes the settling time longer. It follows that for $K_P = 15$, $K_I = 14$, and $K_D = 4$, all the four control objectives are satisfied.

The MATLAB commands for computing the step response in the case of $K_P = 15$, $K_I = 14$, and $K_D = 4$ are as follows:

```
Kp = 15; Ki = 14; Kd = 4;
num = [3*Kd 3*Kp+6 3*Ki];
den = [2+3*Kd 1+3*Kp 3*Ki];
sys = tf(num, den)
step(sys)
```

where `tf` and `step` are MATLAB functions (see Appendix B).

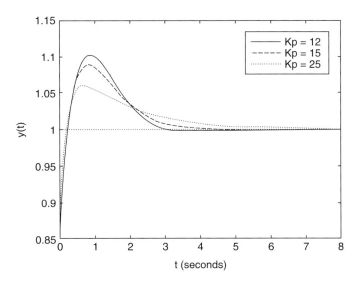

Figure 8.3.3 Step-response curves of the PID feedback control system, with $K_I = 14$, $K_D = 4$, and $K_P = 12, 15, 25$

The control system design presented in Example 8.3.1 is a trial and error approach, which is just for demonstrative purposes. The design of a closed-loop system is more efficient and systematic with the methods of analysis and design given in Sections 8.4 to 8.6. To show this, the same control system is redesigned by the root locus method in Example 8.6.6.

8.3.2 PID Gain Tuning

The PID gains have pros and cons. Consider the step response of a system under PID control. (Refer to Section 7.1.4 for the description of system step response.) The effects of proportional, integral, and derivative gains are summarized in Table 8.3.1. In general, a proportional control action is added to improve the rise time; an integral control action is added to eliminate steady-state error; and a derivative control action is added to improve the overshoot or stability margin. Note that the effects of these control gains interfere with each other. Therefore, Table 8.3.1 only provides a reference for the initial design of a PID controller. To assure the performance and stability of a closed-loop system with PID control, fine tuning of the control gains is necessary.

Given the pros and cons of P, I, and D control actions, the gains of a PID controller need to be properly tuned. The process for selection of PID gains is usually called PID tuning or simply gain tuning. There are many methods for tuning PID gains, including manual PID tuning, Ziegler–Nichols tuning, tuning by root locus, and tuning by the Nyquist stability criterion. In the remainder of this section, manual tuning and a Ziegler–Nichols tuning method are described. Additionally, PID tuning by the root locus method is introduced in Section 8.5.

Manual Tuning

Manual tuning is a trial and error approach. In gain tuning, Table 8.3.1 is used as a reference, and attention is paid to both the performance and stability of the control system. The trial and error selection of the control gains in Example 8.3.1 can be viewed as manual PID tuning.

Ziegler–Nichols Gain Tuning

In the early 1940s, John G. Ziegler and Nathaniel B. Nichols developed two heuristic methods for tuning PID controllers: one based on the ultimate gain and ultimate period of a closed-loop response, the other based on the reaction curve of an open-loop response. The

Table 8.3.1 Effects of PID gains on closed-loop responses

	Rise time	Overshoot	Settling time	Steady-state error
Increasing K_P	Decrease	Increase	Minor change	Decrease
Increasing K_I	Decrease	Increase	Increase	Eliminate
Increasing K_D	Minor change	Decrease	Decrease	No influence

purpose of these methods is to achieve a fast closed-loop response without excessive oscillations and with excellent disturbance rejection. Because gain tuning methods do not require an accurate model for the plant in consideration, they are commonly used in various control applications.

In this section, Ziegler and Nichols tuning based on the ultimate gain and ultimate period, which is also known as the Ziegler–Nichols *ultimate-cycle method*, is introduced. The method considers the step response of a closed-loop system, which can be obtained either numerically or experimentally. (For Ziegler–Nichols tuning based on reaction curves, refer to the references listed at the end of this chapter.)

The Ziegler–Nichols ultimate-cycle method takes the following three steps.

(1) Initially, set the integral and derivative gains to zero ($K_I = 0$ and $K_D = 0$) and consider the response of the closed-loop system with P control and subject to a step input. Mathematically, the system response can be expressed as

$$y(t) = \mathcal{L}^{-1}\left\{\frac{K_P G_p(s)}{1 + K_P G_p(s)} \frac{r_0}{s}\right\} \tag{8.3.7}$$

where $G_p(s)$ is the plant transfer function, r_0 is the amplitude of the step input, and $\mathcal{L}^{-1}$ is the inverse Laplace transform operator. Here, without a loss of generality, a unity-feedback control system shown in Figure 8.3.4 is considered.

(2) Starting from a small value, gradually increase the proportional gain K_P until the closed-loop response shows sustained oscillations with constant amplitude, in which case the closed-loop system becomes marginally stable (see Section 7.2 for the stability definition). Here, we assume that the feedback system becomes unstable at a higher gain. Under the circumstances, the period T_u of the oscillations is called the ultimate period; and the proportional gain is called the ultimate gain, which is given by

$$K_u = K_P|_{\text{at ultimate period}} \tag{8.3.8}$$

Analytically, the ultimate gain K_u and ultimate period T_u satisfy the closed-loop characteristic equation

$$1 + K_u G_p(j\omega_u) = 0 \tag{8.3.9}$$

where $j = \sqrt{-1}$ and $\omega_u = 2\pi/T_u$. The values of T_u and K_u can also be determined by the root locus method (see Section 8.5.3 for instance).

(3) With the ultimate gain K_u and ultimate period T_u determined in Step 2, the gains of a PID controller are given by Table 8.3.2, where P control, PI control, and PD control are also considered.

Table 8.3.2 PID tuning by the Ziegler–Nichols ultimate-cycle method

Control algorithm	K_P	K_I	K_D
P control	$0.5K_u$	–	–
PI control	$0.45K_u$	$0.54K_u/T_u$	–
PD control	$0.8K_u$	–	$0.1K_uT_u$
PID control	$0.6K_u$	$1.2K_u/T_u$	$3K_uT_u/40$

Example 8.3.2

Consider the unity-feedback system in Figure 8.3.4, where the plant transfer function is

$$G_p(s) = \frac{2}{s(s+2)(s+3)} \quad (a)$$

and a PID controller is implemented. Determine the control gains K_P, K_I, and K_D by the Ziegler–Nichols ultimate-cycle method and plot the response of the control system subject a unit step input, $r(t) = 1$.

Solution

Follow the three steps of the ultimate-cycle method. By Eq. (8.3.7), the step response of the closed-loop system with P control ($K_I = 0$ and $K_D = 0$) is computed at several values of the gain K_P. Figure 8.3.5 shows the step responses at $K_P = 12, 14, 15$, and 16, which are obtained by the MATLAB functions tf and step. As indicated by Figure 8.3.5(c), the ultimate period and the ultimate gain are $T_u = 2.565$ and $K_u = 15$.

According to Table 8.3.2, the gains of the PID controller are obtained as follows

$$K_P = 0.6K_u = 9, \quad K_I = 1.2\frac{K_u}{T_u} = 7.0173, \quad K_D = \frac{3K_uT_u}{40} = 2.8857 \quad (c)$$

The transfer function of the closed-loop system is

$$\frac{Y(s)}{R(s)} = \frac{2(K_Ds^2 + K_Ps + K_I)}{s^2(s+2)(s+3) + 2(K_Ds^2 + K_Ps + K_I)} \quad (d)$$

Using Eqs. (c) and (d), the step response of the PID-controlled system is plotted by MATLAB in Figure 8.3.6. Note that the step response has a large maximum overshoot. This is because the

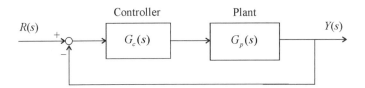

Figure 8.3.4 A unity-feedback control system

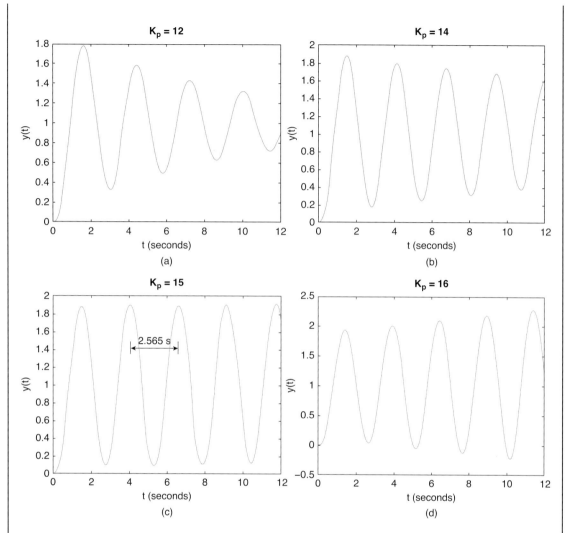

Figure 8.3.5 Step response of the closed-loop system with a P controller for K_P gains of (a) 12, (b) 14, (c) 15, and (d) 16, plotted with MATLAB

Ziegler–Nichols tuning method is developed to provide the best disturbance-rejection performance rather than the best input response performance. To reduce the overshoot, the D control gain can be increased, as indicated in Table 8.3.1. For instance, choosing $K_D = 4.0$ and keeping the other gains the same as given in Eq. (c) yields the step response in Figure 8.3.7.

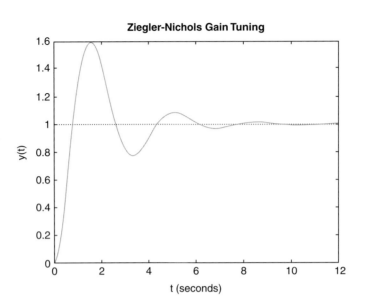

Figure 8.3.6 Closed-loop step response by the Ziegler–Nichols PID tuning: $K_P = 9$, $K_I = 7.0173$, and $K_D = 2.8857$

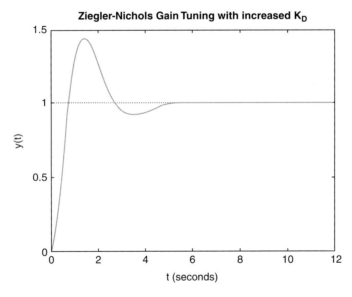

Figure 8.3.7 Closed-loop step response by the Ziegler–Nichols PID tuning with increased K_D: $K_P = 9$, $K_I = 7.0173$, and $K_D = 4$

8.3.3 System Responses by Simulink

In numerical simulation, the response of a feedback control system can be generated through use of the MATLAB functions `tf` and `step`, as shown in Examples 8.3.1 and 8.3.2. The time response of a feedback control system can also be conveniently computed by Simulink. For instance, for the feedback control system in Example 8.3.2, a Simulink model is constructed as shown in Figure 8.3.8, where the control gains in Eq. (c) of the example are assigned. Running a simulation on the model yields the step response shown in Figure 8.3.6. Also, changing the values of the control gains in the Simulink model and running another simulation again, produces the same time response plot as shown in Figure 8.3.7.

Refer to Sections 6.5 and 6.6 and Appendix B for the basics of time-domain block diagrams and Simulink models.

8.4 Control System Analysis

In the analysis and design of a closed-loop control system, the response, stability, and steady-state error of the system are examined, and the system performance is evaluated. These issues are briefly presented in this section. It is assumed that the control system in consideration is modeled by the transfer function formulation given in Section 8.1.

8.4.1 Transient Response and Steady-State Response

For a feedback control system described by Eq. (8.1.2), its response in the time domain can be obtained by inverse Laplace transform

$$y(t) = y_R(t) + y_D(t) + y_N(t) \tag{8.4.1}$$

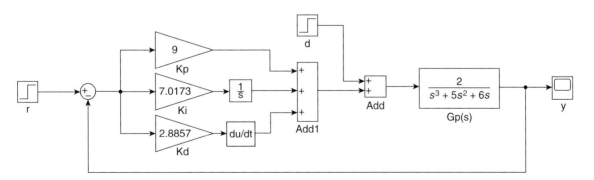

Figure 8.3.8 A Simulink model for the feedback control system in Example 8.3.2

where

$$y_R(t) = \mathcal{L}^{-1}\{T_R(s)R(s)\} = \int_0^t g_R(t-\tau)r(\tau)d\tau$$

$$y_D(t) = \mathcal{L}^{-1}\{T_D(s)D(s)\} = \int_0^t g_D(t-\tau)d(\tau)d\tau$$

$$y_N(t) = \mathcal{L}^{-1}\{T_N(s)N(s)\} = \int_0^t g_N(t-\tau)n(\tau)d\tau \qquad (8.4.2)$$

with the impulse response functions given by

$$g_R(t) = \mathcal{L}^{-1}\{T_R(s)\}, \quad g_D(t) = \mathcal{L}^{-1}\{T_D(s)\}, \quad g_N(t) = \mathcal{L}^{-1}\{T_N(s)\} \qquad (8.4.3)$$

The $y_R(t)$, $y_D(t)$, and $y_N(t)$ are the system responses caused by the reference input, disturbance, and sensor noise, respectively. In a control system design, the controller $G_c(s)$ is such that $y_D(t)$ and $y_N(t)$ are minimized and $y_R(t)$ meets specified performance requirements.

The steady-state response of the system, if there is one, can be determined by

$$y_{ss} = \lim_{s \to 0} sY(s) = \lim_{s \to 0} s\{T_R(s)R(s) + T_D(s)D(s) + T_N(s)N(s)\} \qquad (8.4.4)$$

Here, it has been assumed that all the poles of $sY(s)$ have negative real parts so that the final-value theorem can be applied (see Sections 2.4 and Section 7.1). In a control system design, the gain parameters of the controller are chosen such that y_{ss} reaches a desired value.

Example 8.4.1

Consider the feedback control system in Example 8.1.2, with the parameters having the values: $m = 1$, $c = 1$, $g_s = 1$, $g_a = 5$, $K_P = 8$, $K_D = 2$. The reference input and the external force are both step functions, $r(t) = 2$ and $F(t) = 4$. The system output in the s-domain is

$$Y(s) = \frac{g_a(K_D s + K_P)R(s) + F(s)}{ms^2 + (c + g_s g_a K_D)s + g_s g_a K_P} = \frac{10(s+4)}{s^2 + 11s + 40}\frac{2}{s} + \frac{1}{s^2 + 11s + 40}\frac{4}{s}$$

where on the right-hand side, the first term is the response caused by the reference input, and the second term by the external force. The system output is plotted in Figure 8.4.1 with MATLAB. For comparison purposes, the system output without the external force,

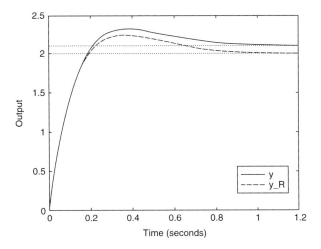

Figure 8.4.1 The time response of the feedback control system in Example 8.4.1

$$y_R(t) = \mathcal{L}^{-1}\left\{\frac{10(s+4)}{s^2+11s+40}\frac{2}{s}\right\}$$

is also plotted in the figure. It can be seen that, without the disturbance $F(t)$, the steady-state response of the control system is the same as the desired value set up by the reference input, $\lim_{t\to\infty} y_R = 2$. With the disturbance, however, the steady-state response of the closed-loop system, by Eq. (8.4.4), becomes

$$y_{ss} = \lim_{s\to 0} s\left\{\frac{10(s+4)}{s^2+11s+40}\frac{2}{s} + \frac{1}{s^2+11s+40}\frac{4}{s}\right\} = 2.1$$

8.4.2 Stability Analysis

The stability of a feedback control system is essentially important to its functionality, performance, and safe operation. Indeed, an unstable control system is not usable at all, and in the worst-case scenario it leads to catastrophic consequences. Stability concepts and stability analysis for dynamic systems are presented in Section 7.2, and the results therein can be directly applied to control systems.

For a closed-loop control system as described by Eq. (8.1.2), the transfer functions $T_R(s)$, $T_D(s)$, and $T_N(s)$ given in Eq. (8.1.4) have the same denominator. This is also seen in Example 8.4.1. Thus, the characteristic equation for a closed-loop system is given by

$$1 + L(s) = 1 + G_s(s)G_c(s)G_a(s)G_p(s) = 0 \tag{8.4.5}$$

The roots of Eq. (8.4.5) are the poles of the closed-loop system.

There are several methods for stability analysis of closed-loop control systems, including direct examination of pole locations, the Routh–Hurwitz stability criterion (Section 7.2.3), the root locus method (Sections 8.5 and 8.6), and the Nyquist stability criterion. In the following example, the Routh–Hurwitz criterion is applied.

Example 8.4.2

In Figure 8.4.2, the mass–damper system in Example 8.1.2 is controlled by a PI controller, which has the transfer function given in Eq. (8.3.5). The characteristic equation of the closed-loop system, from Eq. (8.4.5), is

$$1 + g_s\left(K_P + K_I \frac{1}{s}\right) g_a \frac{1}{ms^2 + cs} = 0$$

or

$$ms^3 + cs^2 + g_s g_a K_P s + g_s g_a K_I = 0$$

Let the parameter values be: $m = 1$, $c = 1$, $g_s = 1$, and $g_a = 5$. The characteristic equation of the closed-loop system becomes

$$s^3 + s^2 + 5K_P s + 5K_I = 0$$

According to the Routh–Hurwitz criterion (Table 7.2.1), the closed-loop control system is stable if and only if $K_P > K_I > 0$.

Choose the control gains as $K_P = 2$ and $K_I = 1$, satisfying the stability conditions. Assume the same inputs as in Example 8.4.1; namely, $r(t) = 2$ and $F(t) = 4$. The output of the closed-loop system is

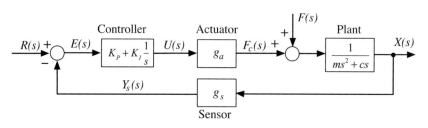

Figure 8.4.2 Mass–damper system under PI feedback control in Example 8.4.2

$$Y(s) = \frac{\frac{1}{ms^2 + cs}}{1 + g_s\left(K_P + K_I \frac{1}{s}\right)g_a \frac{1}{ms^2 + cs}} \left\{\left(K_P + K_I \frac{1}{s}\right)g_a R(s) + F(s)\right\}$$

$$= \frac{1}{s^3 + s^2 + 10s + 5}\left\{14 + 10\frac{1}{s}\right\}$$

The output of the closed-loop in the time domain is plotted in Figure 8.4.3. It can be seen that, under the influence of the external force, the output of the control system has a steady-state value that is the same as that of the system with the reference input only ($y_{ss} = 2$). However, the transient response in Figure 8.4.3 has large overshoots and a much longer settling time, compared to that shown in Figure 8.4.1.

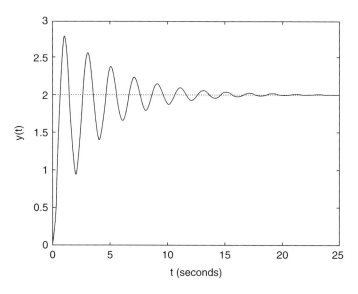

Figure 8.4.3 The time response of the feedback control system in Example 8.4.2

8.4.3 Steady-State Error

In designing a closed-loop control system, the estimation and reduction of its steady-state error is an important issue. For the closed-loop system shown in Figure 8.1.3(b), its error is

$$E(s) = R(s) - G_s(s)\Big(Y(s) + N(s)\Big)$$

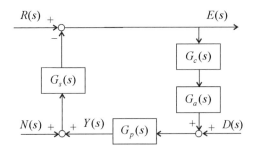

Figure 8.4.4 Error of the feedback control system shown in Figure 8.1.3

As a special case, when zero noise ($N(s) = 0$) and unity feedback ($G_s(s) = 1$) are considered, $E(s) = R(s) - Y(s)$, which means that the error is the difference between the desired output (the reference input) and the actual output.

To determine the error of the control system under the influence of the reference input, disturbance, and sensor noise, the block diagram in Figure 8.1.3(b) is re-arranged into an equivalent form in Figure 8.4.4, where $E(s)$ is selected as the output. It follows that the output (error) of the system is

$$E(s) = W_R(s)R(s) + W_D(s)D(s) + W_N(s)N(s) \tag{8.4.6}$$

where the transfer functions are given by

$$W_R(s) = \frac{1}{1+L(s)}, \quad W_D(s) = \frac{G_s(s)G_p(s)}{1+L(s)}, \quad W_N(s) = \frac{G_s(s)}{1+L(s)} \tag{8.4.7}$$

with the loop transfer function $L(s) = G_s(s)G_c(s)G_a(s)G_p(s)$.

The steady-state error of the control system is estimated by the final-value theorem as follows

$$\begin{aligned} e_{ss} &= \lim_{t\to\infty} e(t) = \lim_{s\to 0} sE(s) \\ &= \lim_{s\to 0} s\{W_R(s)R(s) + W_D(s)D(s) + W_N(s)N(s)\} \end{aligned} \tag{8.4.8}$$

Here, it has been assumed that all the poles of $sE(s)$ have negative real parts (the condition of the final-value theorem condition). If only the effect of the reference input on the error is examined, the steady-state error of the control system ($D(s) = N(s) = 0$) is

$$e_{ss} = \lim_{s\to 0} sE(s) = \lim_{s\to 0} \frac{sR(s)}{1+L(s)} \tag{8.4.9}$$

Steady-State Error Subject to a Step Input

Assume that the closed-loop system is stable, which means that all the roots of the characteristic equation $1 + L(s) = 0$ have negative real parts. For a step input $r(t) = r_0$, it is easy to show that $sE(s)$ satisfies the condition of the final-value theorem. Thus, the steady-state error of the control system can be written as

$$e_{ss} = \lim_{s \to 0} \frac{r_0}{1 + L(s)} \tag{8.4.10}$$

Here, $\lim_{s \to 0} \{1 + L(s)\} \neq 0$ because the system is stable.

If the loop transfer function does not contain a pole at the origin of the s-plane, $L(0)$ is finite and, as a result,

$$e_{ss} = \frac{r_0}{1 + L(0)} \tag{8.4.11}$$

with $1 + L(0) \neq 0$. Thus, the steady-state error is finite. If the loop transfer function has a pole at the origin, it can be written as $L(s) = \frac{1}{s} P(s)$, with $P(0) \neq 0$. By Eq. (8.4.10),

$$e_{ss} = \lim_{s \to 0} \frac{r_0 s}{s + P(s)} = \frac{\lim_{s \to 0} r_0 s}{P(0)} = 0 \tag{8.4.12}$$

Moreover, if $L(s)$ contains more than one pole at the origin, $e_{ss} = 0$ provided that the closed-loop system is stable.

Equation (8.4.11) indicates that P and PD controllers, which do not contain a pole at the origin, cannot render a zero steady-state response although they can reduce it with a large gain. On the other hand, Eq. (8.4.12) implies that PI and PID controllers, which introduce a pole at the origin, may eliminate the steady-state error. These observations are useful for control system design.

Example 8.4.3

In this example, we compare the steady-state errors of the control systems shown in Examples 8.1.2 and 8.4.2. For the system with a PD controller in Figure 8.1.4, its steady-state error is

$$E(s) = \frac{R(s) + g_s \frac{1}{ms^2 + cs} F(s)}{1 + g_s(K_P + K_D s)g_a \frac{1}{ms^2 + cs}} = \frac{(ms^2 + cs)R(s) + g_s F(s)}{ms^2 + cs + g_s(K_P + K_D s)g_a}$$

For step inputs $r(t) = r_0$ and $f(t) = f_0$,

$$sE(s) = \frac{(ms^2 + cs)r_0 + g_s f_0}{ms^2 + (c + K_D)s + g_s g_a K_P} \tag{a}$$

Assume that parameters m, c, g_s, g_a, K_D, K_P all are positive. This, by Table 7.2.1, implies that the poles of $sE(s)$ have negative real parts, satisfying the condition of the final-value theorem. It follows from Eq. (a) that the steady-state error of the control system is

$$e_{ss} = \lim_{s \to 0} sE(s) = \frac{f_0}{g_a K_P} \tag{b}$$

This result has two indications. First, under PD feedback control, the steady-state error is nonzero and it is caused by the external force (disturbance). Second, an increase in $g_a K_P$ reduces the steady-state error. Furthermore, with the data given in Example 8.4.1, $e_{ss} = \frac{4}{5 \times 8} = 0.1$, which agrees with the plots in Figure 8.4.1.

Now, consider the control system with a PI controller in Figure 8.4.2. Its error is

$$E(s) = \frac{R(s) + g_s \frac{1}{ms^2 + cs} F(s)}{1 + g_s \left(K_P + K_I \frac{1}{s}\right) g_a \frac{1}{ms^2 + cs}} = \frac{(ms^2 + cs)sR(s) + g_s s F(s)}{ms^3 + cs^2 + g_s g_a K_P s + g_s g_a K_I}$$

Under the step inputs $r(t) = r_0$ and $f(t) = f_0$,

$$sE(s) = s \frac{(ms^2 + cs)r_0 + g_s f_0}{ms^3 + cs^2 + g_s g_a K_P s + g_s g_a K_I} \tag{c}$$

To satisfy the condition of the final-value theorem, all the roots of the denominate $sE(s)$ must have negative real parts. This requires that the closed-loop system be stable. Note that parameters $m, c, g_s,$ and g_a are all positive. According to Table 7.2.1, the stability conditions for the control gain constants K_P and K_I are

$$cK_P > mK_I > 0. \tag{d}$$

Under the conditions (d), the steady-state error of the control system by Eq. (c) is zero:

$$e_{ss} = \lim_{s \to 0} sE(s) = 0. \tag{e}$$

Thus, with PI control, the steady-state error of the control system that is caused by the disturbance $f(t)$ can be eliminated. This is shown in the response plot in Figure 8.4.3.

For comparison, the errors of the control systems in Examples 8.4.1 (Example 8.1.2) and 8.4.2 are obtained with the given data, and they are

$$E_{PD}(s) = \frac{2s^2 + 2s + 4}{s(s^2 + 11s + 40)}, \quad E_{PI}(s) = \frac{2s^2 + 2s + 4}{s^3 + s^2 + 10s + 5}$$

where $E_{PD}(s)$ is the error of the PD control system in Example 8.4.1, and $E_{PI}(s)$ is the error of the PI control system in Example 8.4.2. By using MATLAB, the errors are plotted against time in Figure 8.4.5. It can be seen that $(e_{PD})_{ss} = 0.1$ and $(e_{PI})_{ss} = 0$.

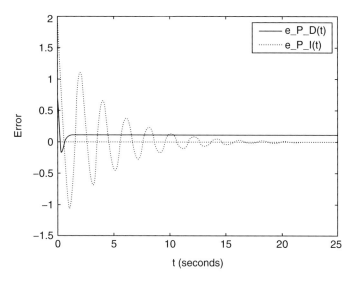

Figure 8.4.5 Comparison of the errors of the feedback control systems in Examples 8.4.1 and 8.4.2

8.5 The Root Locus Method

The root locus method is a commonly used tool for the analysis and design of linear time-invariant control systems. The root locus method, which was introduced by W. R. Evans in 1948, is a graphic technique to analyze stability and performance of closed-loop systems and to design feedback controllers. As shown in Section 7.2, the stability and performance of a feedback control system can be described in terms of the location of the roots of its characteristic equation in the complex plane (the s-plane). The root locus method is based on the correlation between root locations and system response (such as rise time, overshoot, and settling time). In this section, the basic concepts of the root locus method are presented and two methods for sketching a root locus are introduced. The utility of the method in control system design is demonstrated in Section 8.6.

8.5.1 Basic Concepts

The characteristic equation for the closed-loop system in Figure 8.1.3(b) is

$$1 + L(s) = 0 \tag{8.5.1}$$

with the loop transfer function $L(s) = G_s(s)G_c(s)G_a(s)G_p(s)$. Assume that $L(s)$ contains a parameter k, which is either a gain of the controller or a parameter of any other component (plant, sensor, or actuator). The root locus of the system is a plot of the location of the roots of the characteristic equation (8.5.1) in the complex plane as k varies from zero to infinity. In other words, the root locus is a cluster of trajectories of the closed-loop poles in the s-plane as k is varied through positive values.

As an example, consider a system with the closed-loop characteristic equation

$$s^2 + 2s + k = 0 \tag{8.5.2}$$

with k being a nonnegative parameter. The characteristic roots are

$$\begin{matrix} s_1 \\ s_2 \end{matrix} = -1 \pm \sqrt{1-k} \tag{8.5.3}$$

The roots are real and distinct if $k < 1$; they are repeated at -1 for $k = 1$; and they become complex conjugates if $k > 1$. By choosing $k = 0, 0.1, 0.2, 0.3, \ldots, 2.0$, the locations of the roots, computed by Eq. (8.5.3), are plotted in the s-plane in Figure 8.5.1(a), where the root locations for $k = 0$ are denoted by crosses ($\times$), and the root locations for $k > 0$ are denoted by dots. Of course, the root locations for other positive values of k can be determined. Connecting all the dots gives a root locus of the system as shown in Figure 8.5.1(b), where the arrows indicate the direction of root movement as k increases. The root locus plot provides a picture of the behavior of the closed-loop poles as parameter k is varied, which is useful for stability analysis, dynamic response prediction, and control system design.

In the root locus method, consider the characteristic equation of a feedback system in the following form

$$1 + k\frac{b(s)}{a(s)} = 1 + kP(s) = 0 \tag{8.5.4}$$

where k is a parameter under investigation, and $a(s)$ and $b(s)$ are polynomials of s

$$\begin{aligned} a(s) &= a_n s^n + a_{n-1} s^{n-1} + \ldots + a_1 s + a_0 \\ b(s) &= b_m s^m + b_{n-1} s^{m-1} + \ldots + b_1 s + b_0 \end{aligned} \tag{8.5.5}$$

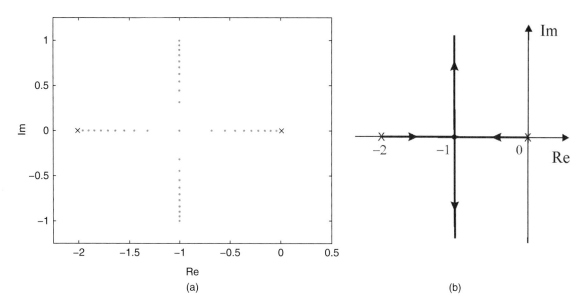

Figure 8.5.1 Plot of the root locus of $s^2 + 2s + k = 0$ versus parameter k: (a) dots – root locations at discrete values of k; (b) root locus by connecting the dots in (a)

where $a_n > 0$, $b_m > 0$, and $n \geq m$. For the purpose of generating a root locus with respect to k, $P(s) = b(s)/a(s)$ is called the open-loop transfer function. Accordingly, the roots of $a(s)$ are called the *open-loop poles* and the roots of $b(s)$ are called the *open-loop zeros*.

Depending on the problem in consideration, the open-loop transfer function $P(s)$ may or may not have the same format as the loop transfer function $L(s)$ in Eq. (8.5.1). For instance, for the closed-loop system in Figure 8.5.2(a), the characteristic equation is

$$1 + k \frac{1}{s(s+1)(s+2)} = 0 \qquad (8.5.6)$$

In this case, $L(s) = kP(s)$, without any open-loop zero and with three open-loop poles 0, −1, and −2. As another example, for the closed-loop system in Figure 8.5.2(b), the characteristic equation is

$$1 + L(s) = 0, \quad \text{with } L(s) = \frac{4(s+3)}{s(s+k)}$$

with $L(s)$ being the loop transfer function. The characteristic equation is rewritten as

$$s(s+k) + 4(s+3) = s^2 + 4s + 12 + ks = 0$$

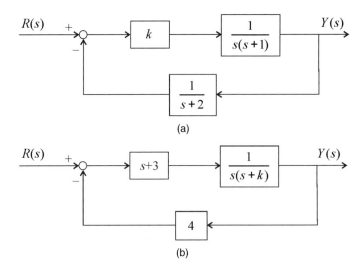

Figure 8.5.2 Two closed-loop systems with parameter k (see text for details)

It follows that an equivalent characteristic equation is

$$1 + kP(s) = 0, \text{ with } P(s) = \frac{s}{s^2 + 4s + 12} \tag{8.5.7}$$

Therefore, the open-loop transfer function $P(s)$ does not have the same format as $L(s)$. From Eq. (8.5.7), there is one open-loop zero and two open-loop poles $-2 \pm j2\sqrt{2}$.

Although Eqs. (8.5.1) and (8.5.4) may look different in format, they have the same roots, which are the poles of the closed-loop system. Also, the characteristic equation (8.5.4) can be written as an equivalent polynomial equation of degree n:

$$a(s) + kb(s) = 0 \tag{8.5.8}$$

the roots of which, namely the closed-loop poles, are dependent on the value of k. Denote the roots of Eq. (8.5.8) by $\phi_l(k)$, $l = 1, 2, \ldots, n$, where $\phi_l(k)$ are functions of k. Mathematically, the root locus of the closed-loop system with respect to k is expressed by

$$\phi_l(k), \ k : 0 \to +\infty, \ l = 1, 2, \ldots, n \tag{8.5.9}$$

8.5.2 Properties of the Root Locus

This section presents some properties of the root locus of a closed-loop system, which are conventionally used to sketch the root locus for minimum calculation. As described in Section 8.5.3, the root locus can be easily generated through use of the software MATLAB. Nevertheless, the root locus properties are still useful in analysis and design of feedback control systems, as shown in the examples in Section 8.6.

Magnitude and Angle Criteria

Let s be a point on the root locus at a value of k. The s and k meet the following criteria:

$$\text{Magnitude criterion} \quad k = \frac{|a(s)|}{|b(s)|} \quad (8.5.10a)$$

$$\text{Angle criterion} \quad \angle b(s) - \angle a(s) = (2l+1)180° \quad (8.5.10b)$$

with $l = 0, \pm 1, \pm 2, \pm 3, \ldots$, where $a(s)$ and $b(s)$ are from the characteristic equation (8.5.4).

Properties of the Root Locus

Assume that a feedback control system described by Eq. (8.5.4) has n open-loop poles $(p_1, p_2, \ldots, p_n)$ and m open-loop zeros $(z_1, z_2, \ldots, z_m)$, with $n \geq m$. The open-loop poles and zeros satisfy the polynomial equations

$$\begin{aligned} a(p_i) &= 0, \quad i = 1, 2, \ldots, n \\ b(z_j) &= 0, \quad j = 1, 2, \ldots, m \end{aligned} \quad (8.5.11)$$

In a root locus plot, the crosses ($\times$) denote open-loop pole locations and circles (°) denote open-loop zero locations. The root locus defined in Section 8.5.1 has the following five properties.

Property 1: Branches

The root locus has n branches, representing the movement of the n poles of the closed-loop system, as shown in Eq. (8.5.9). Because the roots of Eq. (8.5.4) are either real or complex conjugates, the branches are symmetric about the real axis in the s-plane.

According to Eq. (8.5.8), each branch starts at an open-loop pole. If $n = m$, all the branches end at the open-loop zeros. If $n > m$, m branches end at the open-loop zeros and the remaining $n - m$ branches go to infinity in the s-plane as $k \to \infty$. (The way in which these $n - m$ branches go to infinity is described in Property 3.) This is illustrated in Figure 8.5.3, where the arrows indicate the direction of root movement as k increases.

Property 2: Real-Axis Segments

If there exist any real open-loop poles or zeros, the root locus has segments on the real axis, which are located to the left of an odd number of the real-axis poles and zeros. Two cases of real-axis segments of root loci are shown in Figure 8.5.4.

Property: 3 Asymptotes

In the case of $n > m$, the $n - m$ branches going to infinity approach asymptotes as $k \to \infty$. These asymptotes are a cluster of radial lines with a real-axis intercept σ_a and $n - m$ angles ϕ_a, which are given by

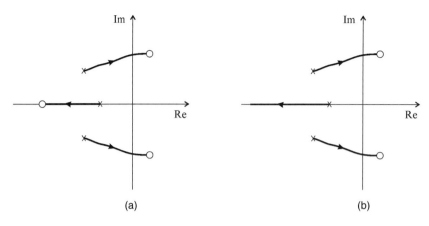

Figure 8.5.3 Branches of root loci: (a) $n = m = 3$; (a) $n = 3$ and $m = 2$, with one branch going to negative infinity on the real axis

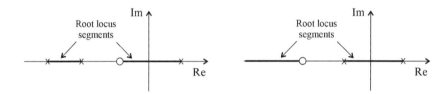

Figure 8.5.4 Two cases of real-axis segments of root loci

$$\sigma_a = \frac{1}{n-m}\left\{\sum_{i=1}^{n} p_i - \sum_{j=1}^{m} z_j\right\} \quad (8.5.12)$$

$$\phi_a = \frac{180°}{n-m}(2l-1), \ l = 1, 2, \ldots, n-m \quad (8.5.13)$$

As an example, consider a feedback control system with the characteristic equation

$$1 + k\frac{s+2}{s(s+3)(s^2+2s+5)} = 0, \ k > 0 \quad (8.5.14)$$

From Eq. (8.5.14), the open-loop poles and zero are found as follows

$$p_1 = 0, \ p_2 = -3, \ p_3 = -1+j2, \ p_4 = -1+j2, \ z_1 = -2 \quad (8.5.15)$$

Because $n - m = 4 - 1 = 3$, the root locus has three asymptotes, with the real intercept and angles given by

$$\sigma_A = \frac{1}{4-1}\{(0-3-1+j2-1-j2)-(-2)\} = -1$$

$$\phi_A = \frac{180°}{4-1}(2l-1) = 60°,\ 180°,\ 300°,\ \text{for } l = 1, 2, 3 \tag{8.5.16}$$

Figure 8.5.5 shows the asymptotes of the root locus.

Table 8.5.1 lists the asymptote angles with respect to the values of $n-m$. As observed from the table, the larger the $n-m$, the more asymptotes stretch into the right half-plane. This indicates that a feedback system with a larger $n-m$ value is more inclined to be unstable as k increases. Accordingly, adding a pole to the open-loop transfer function $P(s)$ in Eq. (8.5.4), as in I control, can make the closed-loop system less stable and adding a zero, as in PD control, can enhance the system stability.

Property 4: Breakaway and Break-in Points on the Real Axis

If a real-axis segment of the root locus is connected to two open-loop poles, there is a point on the segment at which the root locus leaves the real axis and branches out into the complex region; see Figure 8.5.6(a), where the arrows indicate the direction of root movement as k increases. This point (s_b) is called the *breakaway point*. On the other hand, if a segment of the root locus is connected to two open-loop zeros, there is a point on the segment at which the two branches of the root locus return from the complex region to the real axis; see Figure 8.5.6(b). This

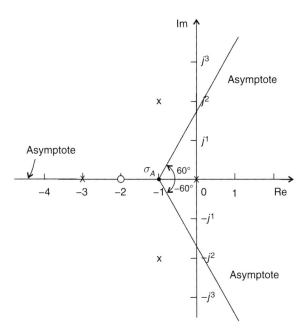

Figure 8.5.5 Asymptotes of the root locus described by Eq. (8.5.14)

Table 8.5.1 Asymptote angles versus $n - m$

$n - m$	1	2	3	4
ϕ_A	180°	90°, 270°	60°, 180°, 300°	45°, 135°, 225°, 315°
Asymptotes				

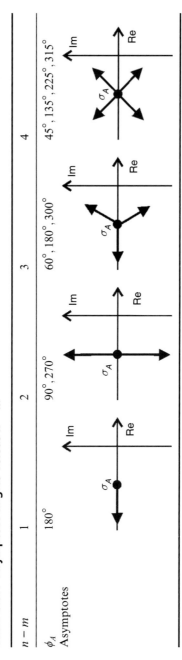

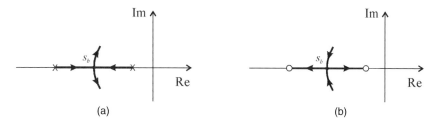

Figure 8.5.6 (a) Breakaway or (b) break-in point (s_b)

point (s_b) is called the *break-in point*. Breakaway and break-in points provide important information about the change in the pattern of system response.

The breakaway and/or break-in points are determined as follows. The characteristic equation (8.5.4) is rewritten as

$$k = F(s), \text{ with } F(s) = -\frac{a(s)}{b(s)} = -\frac{1}{P(s)} \qquad (8.5.17)$$

A breakaway or break-in point, denoted by s_b, satisfies the following two conditions

$$\frac{dk}{ds} = \frac{d}{ds} F(s_b) = 0 \qquad (8.5.18a)$$

and

$$k = F(s_b) > 0 \qquad (8.5.18b)$$

Property 5: Crossover Points

The root locus may cross the imaginary axis at certain points; see Figure 8.5.7, where the arrows indicate the direction of root movement as k increases. The points where the root locus passes through the imaginary axis are called the *crossover points* or the *imaginary axis crossings*. From Eq. (8.5.8), a crossover point ($s = j\omega$) satisfies the equation

$$a(j\omega) + kb(j\omega) = 0, \quad j = \sqrt{-1} \qquad (8.5.19)$$

where ω is called the crossover frequency. The solution of Eq. (8.5.19) for $k > 0$ gives the crossover points.

Recall that the real parts of the closed-loop poles detect the system stability (Section 7.2). Indeed, the left half of the s-plane defines a stable region. Therefore, a crossover point is a point where the stability characteristics of the closed-loop system is about to change as k increases (from being stable to unstable or vice versa). This is clearly seen in Figure 8.5.7.

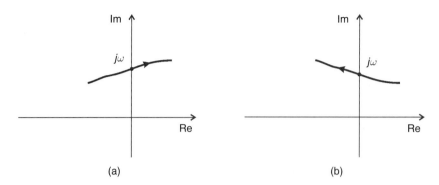

Figure 8.5.7 Crossover points: (a) from a stable region to an unstable region; (b) from an unstable region to a stable region

Also, at the crossover point, the value of ω yields the frequency of the oscillation and the value of k defines a bound of the stability region.

Besides the aforementioned properties, the angle of departure of a branch at an open-loop pole and the angle of arrival of a branch at an open-loop zero can be determined analytically. The reader may refer to the texts listed at the end of this chapter.

8.5.3 Sketching Root Locus

There are two methods for sketching a root locus: (a) using Properties 1 to 5 as guides; and (b) using the MATLAB function `rlocus`. Both the methods are often used alternatively in analysis and design of feedback control systems.

In the first method, a root locus is roughly sketched by taking the following steps.

(1) Determine the open-loop poles and zeros.
(2) Locate the segments of the root locus on the real axis.
(3) Draw the asymptotes of the root locus if $n > m$.
(4) Determine the breakaway and/or break-in points.
(5) Determine the crossover points.
(6) Complete the root locus sketch.

A root locus sketch can provide a quick preview in stability analysis and controller design, as illustrated in Section 8.6.

Example 8.5.1

Plot the root locus for the characteristic equation (8.5.14) with respect to parameter k.

Solution
Follow the above-mentioned six steps. Steps 1 and 3 (open-loop poles, zeros, and asymptotes) have been carried out previously, as shown in Eqs. (8.5.15) and (8.5.16). In Step 2, by Property 2,

two segments of the root locus on the real axis are identified: one between 0 and −2, the other between −3 and −∞. In Step 4, no breakaway or break-in point is found.

In Step 5, according to Eq. (8.5.19), the crossover points are governed by

$$j\omega(j\omega + 3)(-\omega^2 + j2\omega + 5) + k(j\omega + 2) = 0, \quad j = \sqrt{-1} \tag{a}$$

Separating the real and imaginary parts of Eq. (a) gives two coupled equations

$$\text{Real}: \quad \omega^4 - 11\omega^2 + 2k = 0 \tag{b}$$

$$\text{Imaginary}: \quad \omega(-5\omega^2 + 15 + k) = 0 \tag{c}$$

From Eq. (c), one possible solution is $\omega = 0$. However, with $\omega = 0$, Eq. (b) yields $k = 0$, which violates the condition (8.5.18b). So, $\omega \neq 0$ and Eq. (c) is reduced to

$$-5\omega^2 + 15 + k = 0 \tag{d}$$

Equations (b) and (d) are solved to obtain $\omega = \pm\sqrt{6} = \pm 2.449$ and $k = 15$. Thus, the crossover points are $\pm j2.449$. Finally, the root locus is sketched in Figure 8.5.8 in Step 6. As seen from the root locus, the feedback control system becomes unstable for $k > 15$.

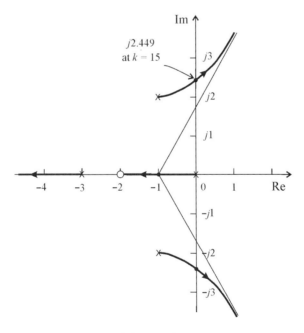

Figure 8.5.8 Root locus of the characteristic equation (8.5.14)

Example 8.5.2

Consider a closed-loop system with the characteristic equation

$$1 + \frac{2s(s+4)}{(s+7)(s^2+4s+k)} = 0, \quad k > 0 \tag{a}$$

Sketch the root locus of the system, with respect to parameter k.

Solution

Equation (a) is rewritten as

$$(s+7)(s^2+4s+k) + 2s(s+4)$$
$$= (s+7)s(s+4) + 2s(s+4) + k(s+7) = 0$$

which leads to the following equivalent characteristic equation

$$1 + k\frac{s+7}{s(s+4)(s+9)} = 0, \quad k > 0 \tag{b}$$

From Eq. (b), there are three open-loop poles at 0, −4, and −9 and one open-loop zero at −7. By Property 2, the root locus has two real segments: one between 0 and −4, and the other between −7 and −9. According to Property 3, the root locus has two asymptotes ($n - m = 2$), with the real intercept and angles given by

$$\sigma_a = \frac{1}{3-1}\{(0-4-9) - (-7)\} = -3$$

$$\phi_a = \frac{180°}{3-1}(2l-1) = 90°, 270°, \quad \text{for } l = 1, 2 \tag{c}$$

From Property 4 and Figure 8.5.6(a), there is a breakaway point between 0 and −4, which satisfies the conditions (8.5.18a) and (8.5.18b). Thus,

$$\frac{d}{ds}F(s) = \frac{d}{ds}\left[-\frac{s(s+4)(s+9)}{s+7}\right]$$

$$= -\frac{(3s^2+26s+36)(s+7) - s(s^2+13s+36)}{(s+7)^2} = 0 \tag{d}$$

Because the breakaway point is in the region $0 < s < -4$, $s + 7 \neq 0$. Thus, Eq. (d) is reduced to

$$(3s^2+26s+36)(s+7) - s(s^2+13s+36) = 0$$

or

$$s^3 + 17s^2 + 91s + 126 = 0 \tag{e}$$

Equation (e) has only one real root: −2.1187, at which $F(s) = 5.619 > 0$. Therefore, the breakaway point is $s_b = -2.1187$, at which $k = 5.619$.

To check if there is any crossover point, consider Eq. (8.5.19), which by Eq. (a) leads to two coupled equations:

$$\omega(-\omega^2 + 36 + k) = 0 \qquad \text{(f)}$$

$$13\omega^2 - 7k = 0 \qquad \text{(g)}$$

Because $\omega \neq 0$ for $k > 0$, Eq. (f) is reduced to

$$\omega^2 = 36 + k \qquad \text{(h)}$$

Substituting Eq. (h) into Eq. (g) yields $k = -78$, which contradicts the assumption that $k > 0$. Thus, the root locus does not have any crossover point.

Finally, the root locus is sketched in Figure 8.5.9.

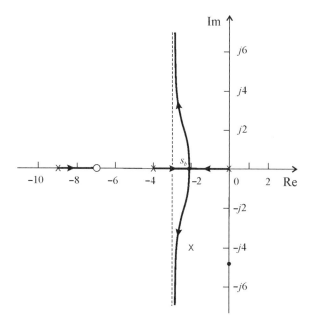

Figure 8.5.9 Root locus of the characteristic equation (a) of Example 8.5.2

In the second method, the MATLAB function `rlocus` is used to generate a root locus. For the closed-loop characteristic equation (8.5.4), the MATLAB commands

```
num = [b_m b_{m-1} ... b_1 b_0];
den = [a_n a_{n-1} ... a_1 a_0];
sys = tf(num, den)
rlocus(sys)
```

yield the root locus. Here, a_i and b_j are the coefficients of the polynomials $a(s)$ and $b(s)$ given in Eq. (8.5.5).

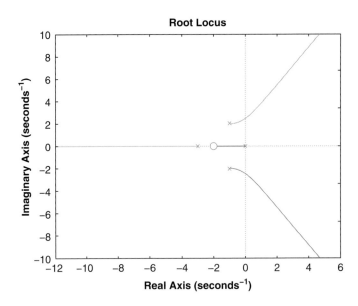

Figure 8.5.10 Root locus of the system in Example 8.5.1

For instance, for the system in Example 8.5.1, the MATLAB commands

```
num = [1 2];
den = conv([1 3 0], [1 2 5]);
sys = tf(num,den)
rlocus(sys)
```

produce a root locus in Figure 8.5.10, which is comparable with the sketch in Figure 8.5.8.

8.6 Analysis and Design by the Root Locus Method

In this section, analysis and design of feedback control systems by the root locus method is illustrated on several examples. The MATLAB function `rlocus` is used as a tool, together with the basics of stability analysis (Section 7.2) and the properties of root loci (Section 8.5).

8.6.1 Stability Analysis

As mentioned in Section 7.2.2, the stability of a feedback control system is dependent upon its pole locations. A system is stable if all its poles are in the left half-plane. A system is unstable if any of its poles are in the right half-plane. A system is marginally stable if some poles, which are not repeated, are on the imaginary axis and if all other poles are in the left half-plane. With the above-described stability concept, the stability of a feedback control system with respect to a parameter can be examined via the root locus of the system, and the corresponding stability conditions can be determined.

Example 8.6.1

Consider the closed-loop system in Figure 8.5.2(a). By plotting the root locus of the system, determine the range of k for the system to be stable.

Solution

For the closed-loop system, its characteristic equation (8.5.6) can be written as

$$1 + k\frac{1}{s^3 + 3s^2 + 2s} = 0 \tag{a}$$

By the MATLAB commands

```
num = [1];
den = [1 3 2 0];
sys = tf(num,den)
rlocus(sys)
```

the root locus is plotted in Figure 8.6.1. As seen from the figure, the root locus has three asymptotes, indicating that the system becomes unstable if k is large enough.

To determine the stability region for k, in the MATLAB figure, left-click the branch of the locus with the 60° asymptote (Table 8.5.1) at any point, which shows an information window about the point, such as location and gain. Move the point along the branch by holding the mouse button

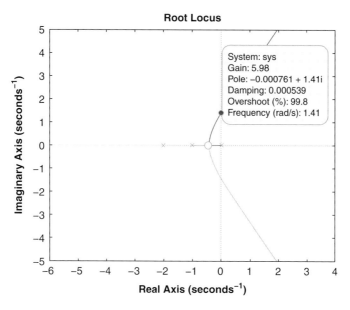

Figure 8.6.1 Root locus of the closed-loop system in Example 8.6.1

until it reaches the imaginary axis. At this point, the information window gives the approximate location of the crossover point ($j\omega = j1.41$) and the gain ($k = 5.98$). From Properties 1 and 5 in Section 8.5.2, the system is stable if $0 < k < 5.98$. A more precise analysis by Eq. (8.5.19) or the Routh–Hurwitz stability criterion (Section 7.2.3) yields the stability range as $0 < k < 6$ and the crossover point as $j6$.

Example 8.6.1 shows that the MATLAB function `rlocus` gives approximate numerical results. Nevertheless, the function is efficient and convenient in analysis and design of general feedback control systems.

Example 8.6.2

A feedback control system is shown in Figure 8.6.2, where an unstable plant is under P control. Sketch the root locus of the feedback system and find the range of the control gain k for stability.

Solution
The closed-loop characteristic equation is

$$1 + k \frac{1}{(s^2 + 8s + 25)(s-1)(s+4)} = 0 \tag{a}$$

There are four open-loop poles at $-4 \pm j3$, 1, and -4; there is no open-loop zero. With the MATLAB commands

```
g1 = tf(1, [1 8 25]);
g2 = tf(1, [1 -1]);
g3 = tf(1, [1 4]);
sys = g1*g2*g3;
rlocus(sys)
```

the root locus of the feedback system is plotted in Figure 8.6.3. Furthermore, by clicking the branches and dragging the points toward the imaginary axis, two crossover points are identified

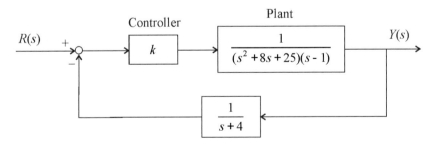

Figure 8.6.2 A feedback control system in Example 8.6.2

by the information windows as follows: $j1.97$ at $k = 260$ and 0 at $k = 100$. By the root locus properties, all the closed-loop poles lie in the left half-plane if k is larger than 100 and less than 260. Therefore, the range of k for the stability of the closed-loop system is $100 < k < 260$.

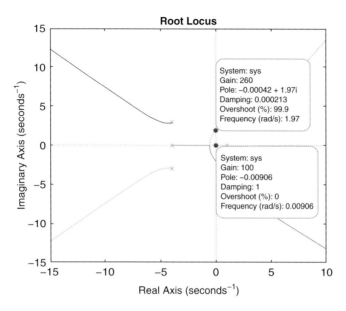

Figure 8.6.3 Root locus of Example 8.6.2

8.6.2 System Response Analysis

Consider a stable system with multiple poles. The pole of the system closest to the imaginary axis is called the *dominant pole*. The concept of dominant poles has been briefly discussed in Section 7.1.6. For a system with one dominant pole that is real ($s = -\sigma$), it may be approximated as a first-order system. For a feedback control system with two dominant poles that are a complex conjugate pair ($s = -\sigma \pm j\omega$), it may be approximated as a second-order system. With an approximate model of a dominant pole(s), the system can be designed by the root locus method to meet specific system response requirements.

Example 8.6.3

Consider the feedback system in Figure 8.6.2. Determine the value of the gain k such that the system is stable and that the system response subject to an input $r(t) = 1$ meets the following two requirements: (i) the settling time (5% criterion) is less than 8 s; and (ii) the maximum overshoot is less than 5%. With the determined k value, plot the step response of the system. Also, discuss the possibility for the control system to have a settling time less than 2 s.

Solution

The root locus of the feedback system has been plotted in Figure 8.6.3, from which it can be seen that, for $112 < k < 260$, the system has two dominant poles in a complex conjugate pair. (The other poles are also complex.) By Eq. (7.1.70), the settling time (5% criterion) for a standard second-order system is $t_s = 3/\zeta\omega_n$. Thus, we need to choose a point p on the branch with an asymptote of angle $45°$, at which $t_s < 8$ s or

$$\zeta\omega_n = -\text{Re}(p) > \frac{3}{8} = 0.375 \tag{a}$$

Following the clicking and moving approach in Example 8.6.1, a zoomed-in root locus of the system is plotted in Figure 8.6.4, where the information window shows

$$\zeta\omega_n = 0.569, \text{ at } k = 116 \tag{b}$$

By Eq. (b), the estimated settling time is $t_s = 3/0.569 = 5.27$ s, which meets condition (i). Note that the system is stable at $k = 116$. The information window also shows that the estimated overshoot is 1.06%, which satisfies condition (ii). Here, the word "estimated" is used because the k value given in Eq. (b) is based on an approximated model of the dominant poles. These conditions are yet to be validated by the step response of the feedback control system that has four poles.

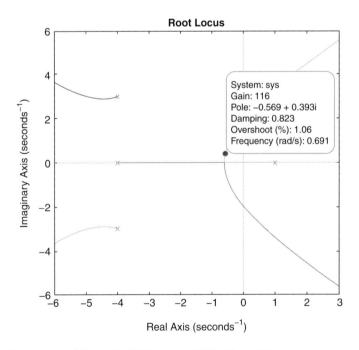

Figure 8.6.4 Root locus of Example 8.6.3: $\zeta\omega_n = 0.568$ at $k = 116$

8.6 Analysis and Design by the Root Locus Method

To plot the step response, the closed-loop transfer function is obtained as follows

$$\frac{Y(s)}{R(s)} = \frac{k(s+4)}{(s^2+8s+25)(s-1)(s+4)+k} \quad (c)$$

where Figure 8.6.2 has been used. With the MATLAB commands

```
k = 116;
num = k*[1 4];
d1 = [1 8 25]; d2 = [1 3 -4];
den = conv(d1, d2) + [0 0 0 0 k];
sys = tf(num, den)
step(sys)
```

the system step response is plotted in Figure 8.6.5.

The steady-state response of the system subject to a unit step input is

$$y_{ss} = \lim_{s \to 0} s \frac{k(s+4)}{(s^2+8s+25)(s-1)(s+4)+k} \frac{1}{s} = 29, \text{ for } k = 116$$

For a settling time of the 5% criterion, t_s is determined by $y(t_s) = 0.95 y_{ss} = 27.55$ or $y(t_s) = 1.05 y_{ss} = 30.45$. It follows from the information window in Figure 8.6.5 that the settling time is $t_s = 5.17$ s. Also seen from the figure, the maximum amplitude of the response is $y_{max} = 29.3$, which leads to a maximum overshoot of 1.03%. Thus, the two requirements for the step response are met with $k = 116$.

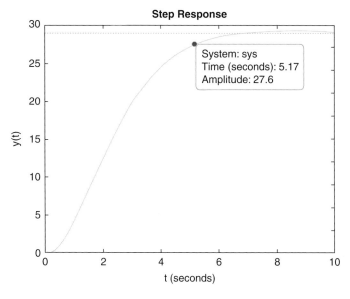

Figure 8.6.5 Step response of the feedback system in Example 8.6.3

For the feedback system to have a settling time less than 2 s, instead of condition (a), it needs to satisfy the following condition

$$\zeta\omega_n = -\text{Re}(p) > \frac{3}{2} = 1.5 \qquad (d)$$

However, as observed from the root locus in Figure 8.6.4, $\zeta\omega_n < 0.6$ for the complex dominant poles, indicating that condition (d) cannot be met for any $k > 0$. Therefore, the step response of the feedback system with P control cannot have a settling time less than 2 s. In fact, with the dominant-pole model, the shortest possible settling time is estimated as $t_s = 3/0.5 = 6$ s. This issue shall be revisited in Example 8.6.4.

8.6.3 Control System Design

The root locus method provides a useful and efficient tool for design of feedback control systems. In this section, two control system design methods are introduced: (a) an aid to PID gain tuning; and (b) control system design by zero placement.

Aid to PID Gain Tuning

In the Ziegler–Nichols ultimate-cycle method for PID gain tuning (Section 8.3.2), the ultimate period T_u and ultimate gain K_u of a closed-loop system with P control need to be determined first. According to Property 5 in Section 8.5.2, T_u and K_u are related to a crossover point, as shown in Eq. (8.3.9). By the root locus method, these parameters can be easily identified. For instance, for the unity-feedback system in Example 8.3.2, the root locus of the characteristic equation

$$1 + K_P \frac{2}{s(s+2)(s+3)} = 0 \qquad (8.6.1)$$

with respect to gain K_P is plotted in Figure 8.6.6, from which the crossover frequency is identified as $\omega = 2.45$ at $K_P = 15$. It follows that $T_u = 2\pi/\omega = 2.565$ and $K_u = K_P = 15$. This result is the same as that obtained by the step responses in Figure 8.3.5. The root locus method avoids a trial-and-error process, as demonstrated in Example 8.3.2, in which a sequence of step responses with different values of K_P has to be computed.

The Ziegler–Nichols gain tuning by the root locus method involves two steps. First, T_u and K_u are determined by a crossover point of the root locus for a corresponding feedback system with P control. Second, the PID gains are determined from Table 8.3.2.

8.6 Analysis and Design by the Root Locus Method

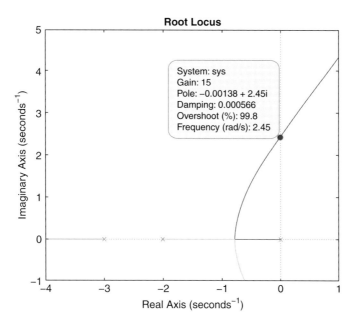

Figure 8.6.6 Root locus of Eq. (8.6.1): crossover frequency $\omega = 2.45$ at $K_P = 15$

Control System Design by Zero Placement

According to the root locus properties presented Section 8.5.2, placement of additional open-loop zeros can significantly alter the pattern of the root locus of a closed-loop system, so as to change the stability and response characteristics of the system. For instance, by Properties 1 and 3, addition of a zero reduces the number of asymptotes by increasing the number m of open-loop zeros, which in turn may improve the stability of a system. As another example, by Properties 2 and 4, addition of a real zero may remove a breakaway point from the real axis. Due to these features and others, zero placement has been widely used in design of feedback systems.

To show how zero placement works, consider PID control. From Eq. (8.3.1), the transfer function of a PID controller can be written as

$$G_c(s) = K_P + K_I \frac{1}{s} + K_D s = K_D \frac{(s-z_1)(s-z_2)}{s} \qquad (8.6.2)$$

with $z_1 + z_2 = -K_P/K_D$ and $z_1 z_2 = K_I/K_D$. As indicated by Eq. (8.6.2), the PID controller introduces one open-loop pole at the origin of the s-plane, and two open-loop zeros, which can be either real ones or a pair of complex conjugates. Assuming that the PID gains are all positive, it can be shown that the added zeros are all located in the left half-plane.

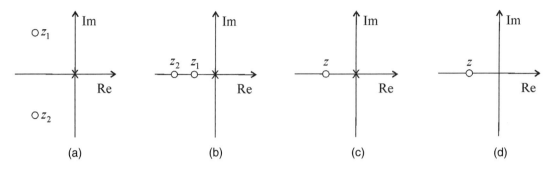

Figure 8.6.7 Zero placement: (a) PID control with complex zeros; (b) PID control with real zeros; (c) PI control; and (d) PD control

Also, a PI controller given in Eq. (8.3.5) can be expressed by

$$G_c(s) = K_P + K_I \frac{1}{s} = K_P \frac{s-z}{s} \tag{8.6.3}$$

with $z = -K_I/K_P$, and a PD controller given in Eq. (8.3.6) can be expressed by

$$G_c(s) = K_P + K_D s = K_D(s-z) \tag{8.6.4}$$

with $z = -K_P/K_D$. For positive gains, the zero introduced by a PI or PD controller is located on the negative real axis. Figure 8.6.7 illustrates zero placements by PID, PI, and PD controllers. Besides these, other controllers, including lead compensators, lag compensators, lead-lag compensators, and notch filters can also be designed by zero placement

According to the previous discussion, the design of a feedback control system by zero placement takes the following four steps.

(1) Properly locate the zero(s) of the controller in the s-plane. The zero placement here is either an initial guess of the controller zero(s) at the beginning of the control system design or in an iterative step that is requested in Step 2 or Step 4.
(2) Perform a root locus analysis via the MATLAB function `rlocus`. By moving points on the root locus branches, check if the control objectives, which are the specifications on the system stability and performance, are satisfied in terms of the dominant pole(s). It they are, go to Step 3. If not, go back to Step 1 and adjust the zero location(s).
(3) With the selected gain k in the root locus equation (8.5.4), which is obtained in Step 2, compute the controller gains, such as K_P, K_I, and K_D.

(4) With the controller gains obtained in Step 3, simulate the response of the closed-loop system to validate the control objectives. If all the specifications are met, the control system design is completed. If not, go back to Step 1 and relocate the zero(s).

Example 8.6.4

As found in Example 8.6.3, the P control for the feedback system cannot yield a step response with a settling time (5% criterion) less than 2 s. In this example, we show that PD control can resolve this issue. To this end, the P controller in Figure 8.6.2 is replaced by a PD controller as described by Eq. (8.6.3). The characteristic equation of the feedback system then is

$$1 + K_D \frac{s-z}{(s^2+8s+25)(s-1)(s+4)} = 0 \tag{a}$$

where z is the open-loop pole introduced by the controller. The control objectives are stated as follows. Under a unit step input, the response of the feedback system has: (i) a settling time (5% criterion) less than 2 s; and (ii) a maximum overshoot less than 5%.

By examining the root locus in Figure 8.6.4, one realizes that the two branches from the breakaway point (near -0.6) are too close to the imaginary axis. Due to $\xi\omega < 0.6$ in the P control, further reduction of the settling time below 2 s is impossible. Based on this observation, place the controller zero z at -3.75, in order to remove the breakaway point and to reduce the number of asymptotes. Consequently, the original branches from the breakaway point disappear and the branches starting at the complex poles $-4 \pm j3$ stretch into the right half-plane by following the $\pm 60°$ asymptotes. This way, the two dominant poles of the system now are on the new branches and they can be located further away from the imaginary axis, so as to increase the $\xi\omega_n$ value.

To show the effect of the above-mentioned zero placement ($z = -3.75$), the root locus of the feedback system is plotted in Figure 8.6.8. By moving a point on the branch starting at the pole $-4 + j3$, the complex dominant poles are selected as $-1.89 \pm j1.69$ at $K_D = 46$. Based on the approximate model of the dominant poles, the settling time is estimated as

$$t_s = \frac{3}{\xi\omega_n} = \frac{3}{1.89} = 1.59 \text{ s}$$

where Eq. (7.1.70) has been used. Also, the maximum overshoot (from the information window) is found as 2.98%. The gains of the PD controller from Eq. (8.6.4) are $K_P = 172.5$ and $K_D = 46$.

With the selected controller gains, the closed-loop transfer function is obtained as

$$\frac{Y(s)}{R(s)} = \frac{(172.5 + 46s)(s+4)}{(s^2+8s+25)(s-1)(s+4) + (172.5 + 46s)} \tag{b}$$

by which the system step response is plotted in Figure 8.6.9 and the corresponding steady-state response is found as $y_{ss} = 9.52$. From the figure, the settling time is 1.47 s, and the maximum amplitude is $y_{max} = 9.62$, which gives the maximum overshoot as 1.05%. Thus, with the PD controller $G_c(s) = 172.5 + 46s$, the control objectives are satisfied.

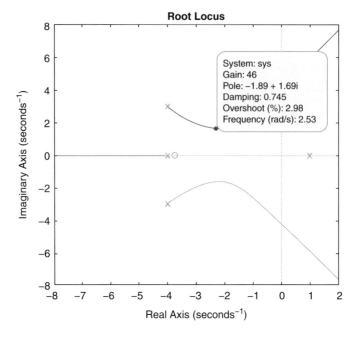

Figure 8.6.8 Root locus of the feedback system with PD control in Example 8.6.4

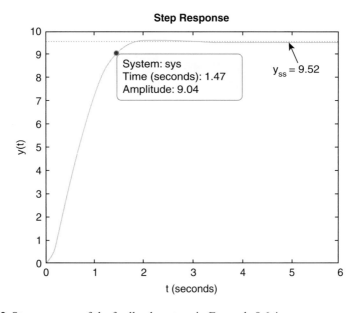

Figure 8.6.9 Step response of the feedback system in Example 8.6.4

Example 8.6.5

For the feedback control system in Figure 8.6.10, design a controller $G_c(s)$ by the root locus method such that, under a unit step input $r(t) = 1$, the system response satisfies the following control objectives:

(i) the steady-state error $e_{ss} = 0$
(ii) the settling time (5% criterion) $t_s \leq 2$ s
(iii) the maximum overshoot M_p falls in the region $5\% \leq M_p \leq 15\%$

Solution
Consider a PID controller given in Eq. (8.6.2). The closed-loop characteristic equation then is

$$1 + L(s) = 1 + K_D \frac{(s-z_1)(s-z_2)}{s(s+1)(s+2)} = 0 \quad (a)$$

where the open-loop poles are 0, −1, and −2 and the open-loop zeros are z_1 and z_2 that are to be located. Let z_1 and z_2 be a pair of complex conjugates, as shown in Figure 8.6.7(a).

With the root locus properties, a rough root locus of the closed-loop system versus K_D is sketched in Figure 8.6.11(a), where the branch starting at pole −2 lies on the negative real axis and moves to −∞; the other two branches split at a breakaway point between 0 and −1 and they eventually end at the zeros z_1 and z_2. As can be seen from the root locus, the control system is stable for any $K_D > 0$. Because the loop transfer function $L(s)$ contains a pole at the origin ($s = 0$), from Eq. (8.4.12), the steady-state error e_{ss} is zero if the closed-loop system is stable. Thus, objective (i) is automatically met for any $K_D > 0$.

Figure 8.6.11(a) also shows that, for a large enough K_D, the closed-loop system has two dominant poles p_1 and p_2 in a complex conjugate pair, and one real pole p_3 that is further away from the imaginary axis. Therefore, in the control system design, the zeros z_1 and z_2 are to be properly placed such that the dominant poles meet objectives (ii) and (iii). To this end, a design region for the dominant poles is defined as follows. Write

$$p_1 = -\xi\omega + j\sqrt{1-\xi^2}\,\omega \quad (b)$$

For objective (ii), $t_s = 3/\xi\omega \leq 2$ s, which by Eq. (7.1.70) and Eq. (b) yields

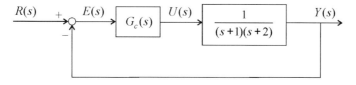

Figure 8.6.10 The feedback control system in Example 8.6.5

$$\zeta\omega \geq \frac{3}{2} = 1.5 \quad (c)$$

From Eq. (7.1.66), objective (iii) is converted to

$$0.4559 \leq \zeta \leq 0.6901 \quad (d)$$

Then, by Eq. (7.1.59), condition (d) is equivalent to $46.46° \leq \theta \leq 58.87°$, where θ is the angle made by p_1 with the negative real axis; that is, $\theta = \pi - \angle p_1$ (Figure 7.1.13). With the conditions (c) and (d), a design region for the dominant pole p_1 is drawn in Figure 8.6.11(b). A design region for p_2 is not necessary because the dominant poles are complex conjugates. Thus, in the control system design, one only needs to check if the conditions (c) and (d) are satisfied.

With the design region, the zeros of the PID controller are chosen as

$$z_1 = -2 + j2, \quad z_1 = -2 - j2, \quad j = \sqrt{-1} \quad (e)$$

by which the characteristic equation (a) becomes

$$1 + K_D \frac{s^2 + 4s + 8}{s(s+1)(s+2)} = 0 \quad (f)$$

Here, the zeros z_1 and z_2, which are the end points of the root locus, are selected to guide the branches through the design region. With Eq. (f), the root locus of the system versus K_D is plotted by MATLAB in Figure 8.6.12. The information window in the figure shows that at $K_D = 10$, $\text{Re}(p_1) = -1.77$, and $\zeta = 0.61$, satisfying conditions (c) and (d). In other words, with the selected zeros and control gain, the closed-loop pole p_1 falls in the design region. Thus, the designed PID controller is given by

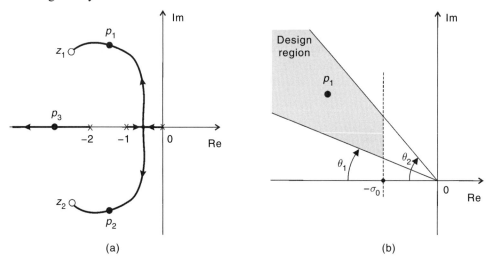

Figure 8.6.11 Design of the PID controller in Example 8.6.5: (a) a sketch of the root locus; (b) a design region (gray area) for the dominant pole p_1, where $\sigma_0 = 1.5$, $\theta_1 = 46.36°$ and $\theta_2 = 58.87°$

8.6 Analysis and Design by the Root Locus Method

$$G_c(s) = 10\frac{s^2 + 4s + 8}{s} = 40 + 80\frac{1}{s} + 10s \quad \text{(g)}$$

Finally, to validate the designed control system, the closed-loop transfer function is obtained as follows

$$\frac{Y(s)}{R(s)} = \frac{10(s^2 + 4s + 8)}{s^3 + 13s^2 + 42s + 80} \quad \text{(h)}$$

By Eq. (h), the response of the closed-loop system subject to a unit step input is plotted in Figure 8.6.13 and the corresponding steady-state response is found as $y_{ss} = 1$. From the step-response plot, it is found that the settling time is 0.91 s and the maximum overshoot is 10%. This verifies that the designed PID controller in Eq. (h) meets all the control objectives.

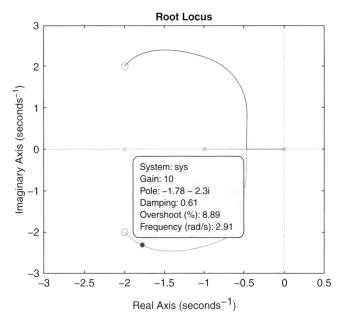

Figure 8.6.12 Root locus of the feedback system with PID control in Example 8.6.5

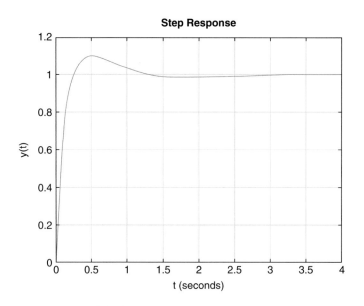

Figure 8.6.13 Step response of the feedback system in Example 8.6.5

Example 8.6.6

Redesign the control system in Example 8.3.1 by the root locus method to meet the same control objectives.

Solution
According to Figure 8.3.2, the characteristic equation of the closed-loop system is

$$1 + L(s) = 1 + K_D \frac{1.5(s - z_1)(s - z_2)}{s(s + 0.5)} = 0 \tag{a}$$

where z_1 and z_2 are the zeros of the PID controller, and K_D is the gain constant in Eq. (8.6.2). Equation (a) indicates that the root locus has two branches that start at poles 0 and -0.5 and end at zeros z_1 and z_2. Assume complex controller zeros. A rough sketch of the root locus is shown in Figure 8.6.14(a), where p_1 and p_1 are the complex closed-loop poles at a certain value of K_D to be determined. The figure shows that the closed-loop system is stable for any $K_D > 0$. Because the loop transfer function $L(s)$ contains a pole at the origin ($s = 0$), the steady-state error e_{ss} by Eq. (8.4.12) is zero if the closed-loop system is stable. Accordingly, the control objectives (i) and (ii) are met for any positive K_D.

Write $p_1 = -\xi\omega + j\sqrt{1-\xi^2}\omega$. Then, from Eq. (7.1.70), objective (iii), $t_s < 3$ s (2% criterion), is equivalent to

$$\xi\omega_n > \frac{4}{3} = 1.3333 \tag{b}$$

From Eqs. (7.1.59) and (7.1.66), objective (iv), $M_p < 10\%$, can be written as

$$\zeta > 0.5912 \quad \text{or} \quad \theta < 53.76° \tag{c}$$

where θ is the angle made by p_1 with the negative real axis, as shown Figure 8.6.14(a). By the conditions (b) and (c), select the controller zeros as $z_{1,2} = -3 \pm j1$. The root locus with the assigned zeros is plotted by MATLAB in Figure 8.6.14(b), in which the information window shows that at $K_D = 3.5$ and the closed-loop pole $p_1 = -2.56 + j1.36$. This implies that $\zeta\omega = 2.56$ and $\zeta = 0.883$, satisfying the conditions (b) and (c).

It follows that the designed PID controller is

$$\frac{U(s)}{E(s)} = K_D \frac{(s-z_1)(s-z_2)}{s} = 18.07 + 29.66\frac{1}{s} + 3.53s \tag{d}$$

showing that $K_P = 18.07$, $K_I = 29.66$, and $K_D = 3.53$. By Eq. (a) in Example 8.3.1, the step response of the PID control system is plotted in Figure 8.6.15. As shown in the figure, the system response has a maximum overshoot of 8%, a settling time of 1.38 s, and zero steady-state error ($e_{ss} = \lim_{t \to \infty}(r - y) = 0$). Thus, all the control objectives are satisfied by the designed controller. Compared with the controller design demonstrated in Example 8.3.1, the zero placement by the root locus method presented herein is systematic and avoids trial and error.

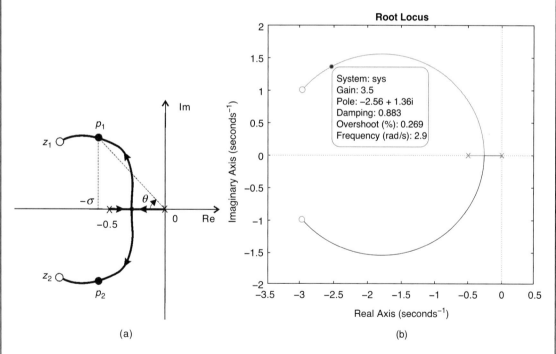

Figure 8.6.14 Root locus of the feedback system in Example 8.6.6: (a) a rough sketch, $\sigma = \zeta\omega_n$ and $\cos\theta = \zeta$; (b) a locus for the controller design

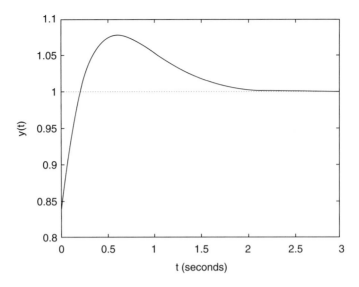

Figure 8.6.15 Step response of the PID feedback control system in Example 8.6.6, with $K_P = 18.07$, $K_I = 29.66$, and $K_D = 3.53$.

8.7 Additional Examples of Control Systems

In this section, three examples of feedback systems from control applications are presented, with the focus on transfer function formulation and block diagram representation. To analyze and design these control systems, the methods presented in Sections 7.1–7.2 and Sections 8.4–8.6 may be applied.

Example 8.7.1 Robotic System

Figure 8.7.1 shows is a single-arm robotic system, which consists of a rigid arm, a sensor, a feedback controller, an amplifier, and a field-controlled direct-current (DC) motor as an actuator. The sensor measures the rotation angle θ of the arm, y_s is the sensor output, r is the reference input, e is the error signal, u is the controller output (control command), and v is the amplifier output. Let the moment of inertia of the arm–motor-shaft assembly be I_O, and the coefficient of the motor bearing be b. The sensor equation is $y_s = k_s \theta$, where k_s is a gain constant. The amplifier equation is $v = k_a u$, where k_a is a constant. Consider PD control. Let the system input and output be r and θ, respectively.

(a) Draw a block diagram of the robotic system.
(b) With the block diagram obtained in (a), determine the transfer function $\dfrac{\Theta(s)}{R(s)}$.

Solution

The s-domain governing equations of the components of the robotic system are given as follows:

$$\text{(i) Arm:} \quad (I_O s^2 + bs)\Theta(s) = T_m(s) \tag{a}$$

where $T_m(s)$ is the motor torque

$$\text{(ii) Motor:} \quad (L_f s + R_f)I_f(s) = V(s) \tag{b}$$

$$T_m(s) = KI_f(s) \tag{c}$$

where the parameters are from Eqs. (6.4.25) and (6.4.27)

$$\text{(iii) Amplifier:} \quad V(s) = k_a U(s) \tag{d}$$

$$\text{(iv) Controller:} \quad \frac{U(s)}{E(s)} = K_P + K_D s \tag{e}$$

where K_P and K_D are constant control gains

$$\text{(v) Error:} \quad E(s) = R(s) - Y_s(s) \tag{f}$$

$$\text{(vi) Sensor:} \quad Y_s(s) = k_s \Theta(s) \tag{g}$$

Based on these equations and by following the block diagram rules in Section 6.2.3, the block diagram of the robotic system is constructed in Figure 8.7.2. From the block diagram, the transfer function of the system is obtained as follows:

$$\frac{\Theta(s)}{R(s)} = \frac{k_a K(K_P + K_D s)}{s(I_O s + b)(L_f s + R_f) + k_s k_a K(K_P + K_D s)} \tag{h}$$

The robotic system is a third-order system.

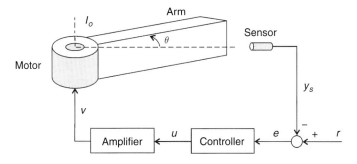

Figure 8.7.1 Schematic of the robotic control system in Example 8.7.1

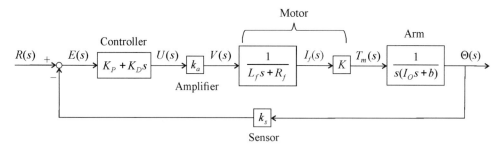

Figure 8.7.2 Block diagram of the robotic control system in Example 8.7.1

Example 8.7.2 Thermal Control System

Figure 8.7.3 shows a thermal control system regulating the temperature of a container (e.g., a thermal bottle). The operation of the control system is explained as follows. A thermometer (sensor) measures the container temperature T; a controller with a control algorithm determines its output u (control command) based on the error signal e, which is the difference between the desired temperature T_r (reference input) and the sensor output T_s; under the control command, the heater (actuator) inputs a heat flow rate q_c into the container, such that the temperature of the container, under the influence of ambient temperature T_a, is maintained as desired.

Let the thermometer and heater be described by

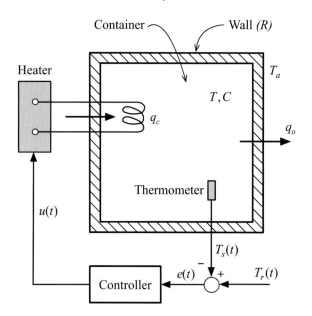

Figure 8.7.3 Thermal control system in Example 8.7.2

$$\hat{T}_s(s) = g_s\hat{T}(s),$$
$$\hat{q}_c(s) = \frac{g_a}{\tau_a s + 1}\hat{u}(s) \quad \text{(a)}$$

where $\hat{a}$ stands for the Laplace transform of a with respect to time, g_s and g_a are gain constants, and τ_a is a time constant. As shown in Eq. (a), the heater is modeled as a first-order dynamic system. Assume that PID control is adopted:

$$\hat{u}(s) = \left(K_P + K_I\frac{1}{s} + K_D s\right)\hat{e}(s) \quad \text{(b)}$$

where K_P, K_I, and K_D are control gains. Derive the transfer functions $\dfrac{\hat{T}(s)}{\hat{T}_r(s)}$ and $\dfrac{\hat{T}(s)}{\hat{T}_a(s)}$.

Solution

The dynamic equation of the container is

$$C\frac{dT}{dt} = q_c - q_o \quad \text{(c)}$$

where C is the thermal capacitance of the container, and the conductive heat transfer through the container wall with thermal resistance R is described by

$$q_o = \frac{1}{R}(T - T_a) \quad \text{(d)}$$

The error signal in the s-domain is written as

$$\hat{e}(s) = \hat{T}_r(s) - \hat{T}_s(s) \quad \text{(e)}$$

Substituting Eq. (d) into Eq. (c) yields

$$C\frac{dT}{dt} = q_c - \frac{1}{R}(T - T_a)$$

which, after Laplace transformation, becomes

$$(RCs + 1)\hat{T}(s) = R\hat{q}_c(s) + \hat{T}_a(s) \quad \text{(f)}$$

After substituting Eqs. (a), (b), and (e) into Eq. (f), we arrive at

$$(RCs + 1)\hat{T}(s) = \frac{Rg_a}{\tau_a s + 1}\hat{u}(s) + \hat{T}_a(s)$$
$$= \frac{Rg_a}{\tau_a s + 1}\left(K_P + K_I\frac{1}{s} + K_D s\right)\left(\hat{T}_r(s) - g_s\hat{T}(s)\right) + \hat{T}_a(s)$$

or

$$\{s(RCs+1)(\tau_a s+1)+Rg_a g_s(K_P s+K_I+K_D s^2)\}\hat{T}(s)$$
$$=Rg_a(K_P s+K_I+K_D s^2)\hat{T}_r(s)+s(\tau_a s+1)\hat{T}_a(s) \qquad (g)$$

It follows from Eq. (g) that the transfer functions of the control system are given by

$$\left.\frac{\hat{T}(s)}{\hat{T}_r(s)}\right|_{\hat{T}_a(s)=0}=\frac{Rg_a(K_P s+K_I+K_D s^2)}{s(RCs+1)(\tau_a s+1)+Rg_a g_s(K_P s+K_I+K_D s^2)} \qquad (h)$$

and

$$\left.\frac{\hat{T}(s)}{\hat{T}_a(s)}\right|_{\hat{T}_r(s)=0}=\frac{s(\tau_a s+1)}{s(RCs+1)(\tau_a s+1)+Rg_a g_s(K_P s+K_I+K_D s^2)} \qquad (i)$$

Note that the two transfer functions have the same denominator.

Example 8.7.3 Liquid-Level Control System

A liquid-level control system is shown in Figure 8.7.4, where the liquid height h of a tank of constant area A is measured by a sensor; a controller determines the control command u based on the error $e=r-y_s$; and with u, a valve actuator adjusts the input volume flow rate q_{in} such that the tank liquid height is maintained at a desired level specified by a reference input r. The sensor and actuator outputs are given by $y_s=\mu_s h$ and $q_{in}=\mu_a u$, respectively, with μ_s and μ_a being constants. The controller is described by $U(s)=G_c(s)E(s)$, where the transfer function $G_c(s)$

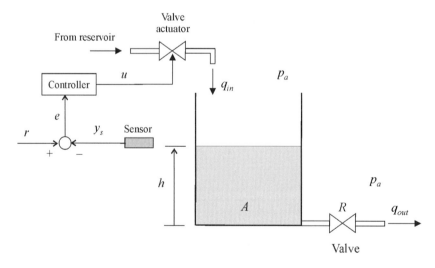

Figure 8.7.4 The liquid-level control system in Example 8.7.3

describes a control algorithm. In working principle, this control system is similar to a toilet tank, in which an assembly of a float cap, a rod, and a fill valve has the functionality of the sensor, controller, and valve actuator, respectively. The only difference is that when the toilet tank is storing water, the valve at its bottom is closed, $q_{out} = 0$.

Consider P control: $G_c(s) = K_P$. Derive the transfer function $\dfrac{H(s)}{R(s)}$ for the liquid-level control system.

Solution
The governing equations of the components of the control system are as follows.

$$\text{(i) Tank:} \qquad A\frac{dh}{dt} = q_{in} - q_{out} \tag{a}$$

$$\text{(ii) Valve:} \qquad q_{out} = \frac{g}{R}h \tag{b}$$

$$\text{(iii) Sensor:} \qquad y_s = \mu_s h \tag{c}$$

$$\text{(iv) Error:} \qquad e = r - y_s \tag{d}$$

$$\text{(v) Controller:} \qquad u = K_P e \tag{e}$$

$$\text{(vi) Actuator:} \qquad q_{in} = \mu_a u \tag{f}$$

In Eq. (b), g is the gravitational acceleration. Taking Laplace transforms of Eqs. (a) to (f) and performing substitutions, yields

$$\left(As + \frac{g}{R}\right)H(s) = Q_{in}(s) = \mu_a K_P\big(R(s) - \mu_s H(s)\big) \tag{g}$$

It follows that the transfer function of the control system is

$$\frac{H(s)}{R(s)} = \frac{\mu_a K_P}{As + \dfrac{g}{R} + \mu_s \mu_a K_P} = \frac{\beta}{Ts + 1} \tag{h}$$

with

$$T = \frac{A}{\dfrac{g}{R} + \mu_s \mu_a K_P}, \qquad \beta = \frac{\mu_a K_P}{\dfrac{g}{R} + \mu_s \mu_a K_P} \tag{i}$$

CHAPTER SUMMARY

This chapter presents the basic concepts and working principles of feedback control systems. A typical feedback control system (also known as closed-loop control system) consists of four major components: the plant, sensor, controller, and actuator. A transfer function formulation for feedback control systems with input, disturbance, and sensor noise is derived. Three advantages of closed-loop control, compared to open-loop control, are demonstrated in examples. The PID control algorithm, which is widely used in applications, and PID gain-tuning methods are introduced. In the transfer function formulation, transient and steady-state behaviors, stability, and steady-state errors of closed-loop control systems are analyzed. The root locus method, which is a commonly used tool for analysis and design of closed-loop control system, is introduced. Finally, three examples of feedback control systems in applications are presented.

Upon completion of this chapter, you should be able to:

(1) Understand the advantages of closed-loop (feedback) control, compared to open-loop control.
(2) Obtain a transfer function formulation for a feedback control system.
(3) Analyze the performance of a feedback control system in terms of transient response, stability, and steady-state error.
(4) Implement the PID control algorithm for feedback control.
(5) Use the root locus method to design feedback control systems.
(6) Compute the response of feedback control systems by MATLAB and Simulink.

REFERENCES

1. R. C. Dorf and R. H. Bishop, *Modern Control Systems*, 13th ed., Pearson, 2016.
2. G. Franklin, J. Powell, and A. Emami-Naeini, *Feedback Control of Dynamic Systems*, 8th ed., Pearson, 2018.
3. N. S. Nise, *Control Systems Engineering*, 8th ed., Wiley, 2020.

PROBLEMS

Section 8.1 General Concepts

8.1 Draw a block diagram for an open-loop system and a block diagram for a closed-loop system, which are in applications of mechanical systems. Follow the formats in Figure 8.1.1.

8.2 Draw a block diagram for an open-loop system and a block diagram for a closed-loop system, which are used in applications of electromechanical systems. Follow the formats in Figure 8.1.1.

8.3 For the control system in Figure P8.3, derive the equations of the output $Y(s)$, error $E(s)$, and actuator output $V(s)$, in terms of the inputs $R(s)$ and $D(s)$. Also, obtain the relevant transfer functions in each of the three cases, which are stated as follows:

$$Y(s) = T_R(s)R(s) + T_D(s)D(s)$$
$$E(s) = T_R^E(s)R(s) + T_D^E(s)D(s)$$
$$V(s) = T_R^V(s)R(s) + T_D^V(s)D(s)$$

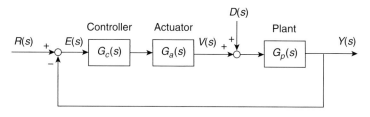

Figure P8.3

Section 8.2 Advantages of Closed-Loop Control Systems

8.4 This problem concerns reduction of the effects of parameter variations by closed-loop control. An open-loop system is shown in Figure P8.4(a), where the parameter k has a variation Δk. Let the parameter value be $k = 25$. The required steady-state response of the system to a unit step input is $y_{ss} = 1$.

(a) In the ideal case of $\Delta k = 0$ (no parameter variation), show that the open-loop system meets the requirement of $y_{ss} = 1$.

(b) For $\Delta k = -5\%$, 5% and 10% of k, plot the step responses of the open-loop system in one figure, and determine whether the requirement of $y_{ss} = 1$ is met in each case.

(c) Consider the closed-loop system in Figure P8.4(b), where P control is implemented. Show that, for a large control gain K_P, the steady-state response y_{ss} of the control system approaches 1 for $\Delta k \neq 0$. This implies that P control can reduce the effects of parameter variations.

(d) Let Δk be an unknown constant that is bounded by $|\Delta k/k| < 1$. A closed-loop system with a PI controller is shown in Figure P8.4(c). Design the controller parameters K_P and K_I to render $y_{ss} = 1$. Make sure that the closed-loop system is stable.

(e) Based on the results obtained in (d), choose appropriate values of K_P and K_I, and plot the step responses of the closed-loop system for $\Delta k = -5\%$, 5% and 10% of k in one figure. By examining the system responses, determine whether the effect of the parameter variation Δk can be eliminated by the PI controller.

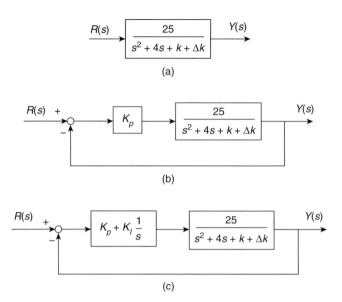

Figure P8.4

8.5 This problem concerns disturbance rejection by closed-loop control. An open-loop system is shown in Figure P8.5(a), where a second-order system is subject to a reference input $R(s)$ and a disturbance $D(s)$. Assume a constant distance, $d(t) = d_0$. The required steady-state response of the system subject to a unit step input $(r(t) = 1)$ is $y_{ss} = 1$.

(a) In the ideal case of $d_0 = 0$ (zero disturbance), show that the open-loop system meets the requirement of $y_{ss} = 1$.

(b) For $d_0 = 1$, 2 and 5, plot the step responses of the open-loop system in one figure, and determine whether the requirement of $y_{ss} = 1$ is met in each case.

(c) With $d_0 \neq 0$, consider the closed-loop system in Figure P8.5(b), where a PI controller is implemented. Show that the parameters K_P and K_I can be chosen to render the desired steady-state response, $y_{ss} = 1$. Make sure that the closed-loop system is stable.

(d) Based on the results obtained in part (c), choose the values of K_P and K_I, and plot the step responses of the closed-loop system $d_0 = 1$, 2 and 5, in one figure. By examining the response curves, show that $y_{ss} = 1$ in all the cases, and that the effect of the disturbance is rejected by the closed-loop control.

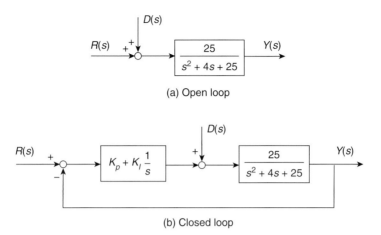

Figure P8.5

Section 8.3 PID Control Algorithm

8.6 This problem concerns improvement of system response by closed-loop control. Figure P8.6(a) shows a second-order open-loop system. It is desired to have a step response (with $r(t) = 1$) that meets the following requirements: (i) the maximum overshoot M_p is less than 30%; and (ii) the settling time t_s (with a 2% criterion) is 2 s.
(a) Compute the maximum overshoot and settling time of the open-loop system.
(b) If the open-loop system does not meet the above-mentioned requirements, consider the closed-loop system in Figure P8.6(b), where a PD controller is implemented. Show that the controller parameters K_P and K_D can be chosen to satisfy the step-response requirements.
(c) Plot the step response of the closed-loop system with the controller parameters determined in part (b).

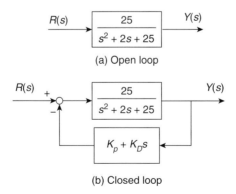

Figure P8.6

8.7 Consider the unity-feedback system in Figure P8.7, where the plant is unstable, and a PD controller is installed.
   (a) Determine the conditions of the controller gain constants (K_P and K_D) such that the closed-loop system is stable.
   (b) With the controller gain constants satisfying the stability conditions given in part (a), determine whether the steady-state error e_{ss} of the control system exists for a step input $r(t) = r_0$. If so, compute the steady-state error.

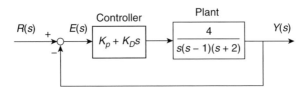

Figure P8.7

8.8 Consider the feedback control system in Figure 8.3.2, where a first-order system with parameter $T = 0.2$ and $b = 1$ is under PID feedback control. Let the reference input and disturbance be $r(t) = 1$ and $d(t) = 0$. Select the control gain constants K_P, K_I, and K_D such that the step response of the closed-loop system meets the following control objectives:
   (a) the closed-loop system is stable
   (b) the settling time of the output (5% criterion) is less than 1.5 seconds
   (c) the maximum overshoot of the output is less than 15%
   Also, with the determined control gain constants, plot the response $y(t)$ of the control system.

8.9 Consider a mass–damper system described by the transfer function

$$G(s) = \frac{Y(s)}{F(s)} = \frac{1}{s(s+4)}$$

Let the mechanical system be the plant, with force f as the input and the displacement y as the output.
   (a) Build a Simulink model of the control system shown in Figure P8.9, where the plant is controlled by a PID controller. Here, the reference input is $r(t) = 1$, and the control gain constants initially are set as $K_P = 10$, $K_I = 0$, and $K_D = 0$.
   (b) With the model obtained in part (a) and by Table 8.3.1, tune the values of the control gain constants to meet the following control objectives:

(i) the closed-loop system is stable
(ii) the displacement y reaches the desired value ($y = 1$) in less than 2 s
(iii) the maximum overshoot of the response y is less than 10%
With the control gain constants selected, plot the response of the control system.

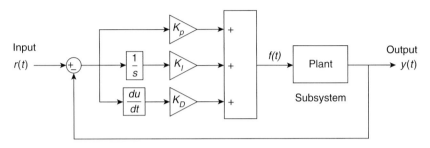

Figure P8.9

8.10 Consider the armature-controlled DC motor in Example 6.5.4, with its parameters given in Example 6.6.2. Let the motor be the plant (dynamic system to be controlled), with the applied voltage v_{in} as the input and the motor speed $\omega = \dot{\theta}$ as the output. Here, the load T_L is assumed to be zero.

(a) Build a Simulink model of the motor speed control system with a PID controller, as shown in Figure P8.9, where $f = v_{in}$, $y = \omega$, and the reference input is $r(t) = 40\pi$ rad/s (or 1200 rpm), which is a step function. For the model, the control gain constants are initially set as $K_P = 2$, $K_I = 0$, and $K_D = 0$.

(b) With the model obtained in (a) and by Table 8.3.1, tune the values of the control gain constants to meet the following control objectives:

(i) the closed-loop system is stable
(ii) the motor reaches the desired speed (40π rad/s) in less than 3 s
(iii) the maximum overshoot of the motor speed is less than 25%

With the control gain constants selected, plot the response of the control system.

8.11 Consider a unity-feedback system in Figure 8.3.4, where the plant transfer function is

$$G_p(s) = \frac{4}{s(s^2 + 8s + 25)} \tag{a}$$

and a PD controller is implemented. Determine the control gains K_P and K_D by the Ziegler–Nichols ultimate-cycle method and plot the response of the control system subject to a unit step input $r(t) = 1$.

8.12 Consider a unity-feedback system in Figure 8.3.4, where the plant transfer function is

$$G_p(s) = \frac{3(s+1)}{(s^2 + 4s + 5)(s^2 + 8s + 25)} \tag{a}$$

and a PID controller is implemented. Determine the control gains K_P, K_I, and K_D by following the steps of the Ziegler–Nichols ultimate-cycle method in Section 8.3.2. Also, with the determined control gains, plot the response of the control system subject to a unit step input $r(t) = 1$.

Section 8.4 Control System Analysis

8.13 Consider the closed-loop system in Figure 8.1.3(b), with

$$G_p(s) = \frac{1}{2s+1}, \; G_s(s) = 1, \; G_c(s) = 10 + 2s, \; G_a(s) = \frac{10}{0.2s+1}$$

(a) Derive the transfer functions $T_R(s)$, $T_D(s)$, and $T_N(s)$ in Eq. (8.1.2) and determine the poles and zeros of the transfer functions.
(b) Plot the response $y(t)$ of the system with inputs $r(t) = 2\delta(t)$, $d = 1$. and $n = 0$, where $\delta(t)$ is the Dirac delta function.

8.14 Consider the closed-loop system in Figure 8.1.3(b), with

$$G_p(s) = \frac{1}{s^2 + 4s + 16}, \; G_s(s) = 1, \; G_c(s) = 5 + \frac{2}{s}, \; G_a(s) = 8$$

(a) Derive the loop transfer function $L(s)$ as defined in Eq. (8.1.5), obtain the poles of $L(s)$, and determine the closed-loop poles.
(b) Find the steady-state response y_{ss} of the system subject to $r(t) = 4$ and $d = n = 0$.

8.15 Consider the unity-feedback control system in Figure P8.15, where $G(s)$ is the open-loop transfer function. For each of the following open-loop transfer functions, derive the stability conditions of the corresponding closed-loop system in terms of the parameters.

System 1: $G(s) = K \dfrac{s-4}{s^2 - 4s + 25}$

System 2: $G(s) = \dfrac{as+b}{s^3 + 3s^2 - 5s - 7}$

System 3: $G(s) = \left(K_P + K_I \dfrac{1}{s} + K_D s \right) \dfrac{1}{s^2 + 16}$

Figure P8.15

8.16 Consider the unity-feedback control system in Figure P8.15, where $G(s)$ is the open-loop transfer function. For each of the following open loop transfer functions, find the steady-state error e_{ss} of the corresponding closed-loop system, if any. Note that steady-state error may not always exist.

System 1: $G(s) = \dfrac{3}{s+10}$, $r(t) = 5(1 - e^{-2t})$

System 2: $G(s) = K\dfrac{s+1}{s-5}$, with $K > 10$, $r(t) = 5$

System 3: $G(s) = 3\dfrac{s+2}{s^2(s+3)}$, $r(t) = 2t$

System 4: $G(s) = \left(3 + 2\dfrac{1}{s} + s\right)\dfrac{1}{s^2+4}$, $r(t) = 5t$

8.17 In Figure P8.17, a second-order plant is under P control with gain K.
  (a) Determine the range of K for stability of the closed-loop system.
  (b) Determine the value of K such that the step response of the closed-loop system has a maximum overshoot $M_p = 10\%$.
  (c) With the K value in part (b), determine the steady-state error of the closed-loop system subject to a ramp input $r(t) = 2t$.
  (d) With the K value determined in part (b), plot the step response of the closed-loop system by MATLAB.

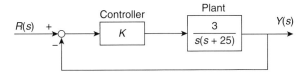

Figure P8.17

8.18 A dynamic system is governed by the coupled differential equations

$$\ddot{y} + 2\dot{y} + (\lambda - 3)y + \mu z = f$$
$$\dot{z} + z - y = 0$$

where f is the input, z is the output, and λ and μ are constant parameters to be designed.

(a) Obtain the transfer function $G(s) = Z(s)/F(s)$.
(b) With the $G(s)$ of part (a), derive the conditions on λ and μ for the system to be stable.

Section 8.5 The Root Locus Method

8.19 A unity-feedback control system, as shown in Figure 8.3.4, has a loop transfer function

$$L(s) = G_c(s)G_p(s) = \frac{K(s+3)}{(s-1)(s+2)(s+4)}$$

Sketch the root locus for $K > 0$ by following the six steps in Section 8.5.3.

8.20 A unity-feedback control system, as shown in Figure 8.3.4, has a loop transfer function

$$L(s) = G_c(s)G_p(s) = \frac{K}{s(s^2 + 8s + 25)}$$

Sketch the root locus for $K > 0$ by following the six steps in Section 8.5.3.

8.21 A feedback control system has the following characteristic equation:

$$1 + \frac{s+1}{(s+3)(s^2 + Ks + 4)} = 0$$

Sketch the root locus for $K > 0$ by following the six steps in Section 8.5.3.

8.22 Consider the unity-feedback control system in Problem 8.20.
(a) Sketch the root locus for $K > 0$ by MATLAB.
(b) Determine the range of K for the stability of the control system.
(c) Determine the gain K such that the complex closed-loop poles have a damping ratio of 0.5.
(d) With the gain of part (c), find the actual maximum overshoot and peak time by plotting the system step response.

8.23 Consider the unity-feedback control system in Problem 8.21.
(a) Sketch the root locus for $K > 0$ by MATLAB.
(b) Determine the range of K for the stability of the control system.
(c) Determine the gain K such that the complex closed-loop poles have a damping ratio of 0.707.
(d) Determine the break-in point of the root locus and the gain.

8.24 A unity-feedback control system, as shown in Figure 8.3.4, has a loop transfer function

$$L(s) = G_c(s)G_p(s) = \frac{K(s+2)}{s(s+8)(s-2)}$$

(a) Sketch the root locus for $K > 0$ by MATLAB.
(b) Determine the range of K for the stability of the control system.
(c) Determine the gain K such that the complex closed-loop poles have a damping ratio of 0.2.
(d) With the gain of part (c), find the actual maximum overshoot and settling time (with a 2% criterion) by plotting the system step response.

Section 8.6 Analysis and Design by the Root Locus Method

8.25 A closed-loop control system has the following characteristic equation:

$$1 + \frac{K(s+2)}{(s+3)(s-1)(s^2+6s+13)} = 0$$

(a) Determine the range of gain K for stability by using the Routh–Hurwitz stability criterion.
(b) Verify the stability result in part (a) by sketching the root locus by MATLAB.

8.26 Repeat Problem 8.12 using the root locus method.

8.27 Consider the feedback control system in Figure P8.27, where a third-order plant is under PD control. With the dominant complex poles, determine the gain K such that the system response subject to an input $r(t) = 1$ meets the following requirements: (i) the settling time (with a 2% criterion) is less than 2 s, and the maximum overshoot is not larger than 20%. Also, with the determined K, plot the step response of the system and identify the actual settling time and maximum overshoot.

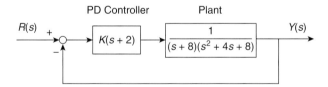

Figure P8.27

8.28 Consider the feedback control system in Figure P8.28. Design a PID controller by zero placement as shown in Section 8.6.3 such that, under a unit step input $r(t) = 1$, the response of the closed-loop system satisfies the following requirements:
 (i) the steady-state error $e_{ss} = 0$
 (ii) the settling time (5% criterion) $t_s \leq 5$ s
 (iii) the maximum overshoot $M_p \leq 10\%$

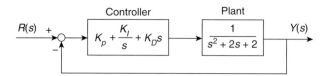

Figure P8.28

Section 8.7 Additional Examples of Control Systems

8.29 A hoist control system is shown in Figure P8.29, where a pulley with an unextendible rope carries a load (weight Mg); the pulley is connected to a shaft by a torsional spring (k); and the shaft that is supported by a bearing (b) is driven by a torque τ applied by a motor for lifting or lowering the load. The height y of the weight is measure by a sensor, with output $y_s = \mu_s y$, where μ_s is a gain constant. A controller determines a control signal u based on the error $e = r - y_s$, where r is a reference input, representing the desired height of the weight. The amplifier provides a motor voltage v by $v = \mu_a u$, where μ_a is a gain constant. The motor torque τ is governed by the differential equation $\dfrac{d\tau}{dt} + \alpha\tau = \beta v$, where α and β are constants. Let the mass moment of inertia of the motor rotor-shaft assembly be J. Consider PID control.
(a) Write down the equations for all the components of the control system.
(b) Derive the transfer function of the control system from the reference input r to the height y.
(c) Draw a block diagram of the control system, with r and Mg as the inputs and y and $\dot{\theta}_1$ as the outputs

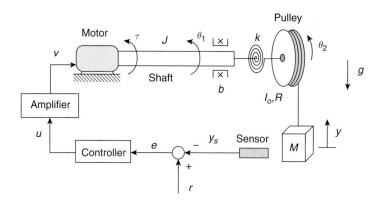

Figure P8.29

8.30 A liquid-level control system is shown in Figure P8.30, where the height h of a tank of constant area A is measured by a sensor; a controller determines a control signal u based on error $e = r - y_s$; and with u, a valve actuator adjusts the input volume flow rate q_{in} such that the tank height eventually reaches a desired level specified by the reference input r. The sensor and actuator outputs are given by $y_s = \mu_s h$ and $q_{in} = \mu_a u$, respectively, with μ_s and μ_a being constants. Consider PI control.

(a) Write down the equations for all the components of the control system.
(b) Draw a block diagram for the control system, with r as the input and h and q_{out} as the outputs.
(c) Derive the transfer function $H(s)/R(s)$ for the control system.
(d) Obtain a state-space model by using the equations obtained in part (a), with r as the input, and h and q_{out} as the outputs. Hint: with the PI controller, the control system has two state variables, and you may choose the state variables as $x_1 = h$ and $x_2 = u - K_P e$.

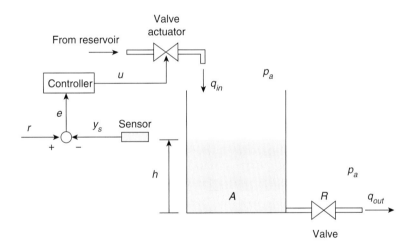

Figure P8.30

9 Application Problems

Contents

9.1	Vibration Analysis of a Car Moving on a Bumpy Road	667
9.2	Speed Control of a Coupled Engine–Propeller System	686
9.3	Modeling and Analysis of a Thermomechanical System (a Bimetallic Strip Thermometer)	697
9.4	Modeling and Analysis of an Electro-Thermo-Mechanical System (a Resistive-Heating Element)	702
9.5	Feedback Control of a Liquid-Level System for Water Purification	711
9.6	Sensors, Electroacoustic, and Piezoelectric Devices	723
	References	756

This chapter assembles six problems of combined dynamic systems from engineering applications, namely, vibration analysis of a moving car, speed control of a coupled engine–propeller system (electromechanical system), modeling and analysis of a bimetallic strip thermometer (thermomechanical system), modeling and analysis of a resistive-heating element (electro-thermo-mechanical system), feedback control of a water purification system component (liquid-level system), and certain working principles of sensors, electroacoustic, and piezoelectric devices. Each problem is presented in one section.

The purpose of this ending chapter of the book is two-fold. First, the chapter serves as a comprehensive review of the contents covered in the previous chapters, including the three-key modeling technique (fundamental principles, basic elements, and ways of analysis); the formulation of combined systems by transfer functions, block diagrams, and state-space representations; the basic concepts of dynamic response analyses (time response, stability, and frequency response); the necessity of feedback control systems; and the utility of the software packages MATLAB/Simulink and Mathematica in simulation. Second, the chapter gives the reader a taste of the model-based design of machines, devices, high-tech products, and industrial processes. Indeed, the techniques of modeling, simulation, and analysis of dynamic systems introduced in this text, with extension, are applicable to many practical problems in engineering applications.

9.1 Vibration Analysis of a Car Moving on a Bumpy Road

In Chapter 3, a one-degree-of-freedom model of an automotive suspension system, which is called a quarter-car suspension model, is derived (see Example 3.2.4). However, a quarter-car model misses the rotation of the car body, which is important in vehicle dynamics. In this section, a generalized model, which describes the rotation of the car body and the vibration of the car engine, is presented. With this model, the dynamic response of a moving car subject to two types of road condition is investigated.

9.1.1 System Description

Figure 9.1.1 shows a three-degrees-of-freedom (3-DOF) model that characterizes the motion of a car moving on a road at a constant speed v_c. This model is called a *half-car model* because two tires of the car are considered. In the figure, the two pairs of spring and damper (k and c) model the tires and suspension system; the rigid body (M_b, I_b) represents the body of the car including the chassis, with I_b being the moment of inertia of the body with respect to its center of mass (point G); the lumped mass (m_e) is the engine assembly that is mounted on the chassis at point E by the mounts that are characterized by the spring and damper pair (k_e, c_e); l_1 and l_2 are the distances of points A and B, where the tire-suspension assemblies are attached to

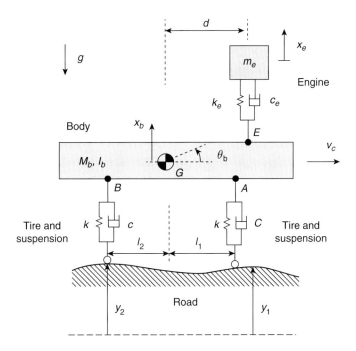

Figure 9.1.1 A half-car model

the chassis, from the mass center G, respectively; and g is the gravitational acceleration. Here, x_b and θ_b are the translation and rotation of the car body; x_e is the displacement of the engine assembly; and y_1 and y_2 are the displacement excitations to the car, which are caused by the interaction between the tires and the road surface.

For simplicity in modeling and analysis, we assume a small rotation angle θ_b of the car body: $|\theta_b| \ll 1$, which is true in normal car operating scenarios. Also, we assume that the tires are in touch with the road surface all the time. To describe a temporary separation of the tires and road surface, a complicated nonlinear dynamic model is necessary, which is beyond the scope of this text.

The condition of a road surface can be described by the displacement excitations y_1 and y_2. For a horizontal and smooth road surface (ideal case), $y_1 = y_2 = 0$. For a road surface of sinusoidal profile, the displacement excitations can be described by sinusoidal functions, such as

$$y_1(t) = y_0 \sin(\omega t)$$
$$y_2(t) = y_0 \sin(\omega(t - t_{12})) \qquad (9.1.1)$$

where y_0 is the amplitude of the surface profile, and

$$\omega = \frac{v_c}{L_r}, \quad t_{12} = \frac{l_1 + l_2}{v_c} \qquad (9.1.2)$$

with L_r being the characteristic length of the road and t_{12} the travel time for the distance between the two tires (the distance between point A and point B). The parameter L_r describes how bumpy the road surface is. For instance, for $L_r = 10$ m, the car goes up and down on the road surface for one cycle when traveling a distance of $2\pi L_r = 62.83$ m. Therefore, the smaller the L_r is, the bumpier the road is.

Potholes in a road, usually asphalt pavement, can cause severe damage to cars, including tire, wheel, and bearing damage, misalignment, damage to the car body and suspension systems, and harm to engine parts. A pothole of rectangular shape as shown in Figure 9.1.2(a) can be described by the displacement excitations as follows

Figure 9.1.2 Road surface: (a) a pothole; (b) a bump

$$y_1(t) = -h_p\left[u(t - t_0) - u\left(t - (t_0 + w_p/v_c)\right)\right]$$
$$y_2(t) = -h_p\left[u\left(t - (t_0 + t_{12})\right) - u\left(t - (t_0 + t_{12} + w_p/v_c)\right)\right]$$
(9.1.3)

where h_p and w_p are the depth and width of the pothole, respectively; t_0 is the time when the front tire of the car (the right tire in Figure 9.1.1) hits the pothole; and t_{12} is given in Eq. (9.1.2). Of course, a pothole of arbitrary shape can be similarly treated if the shape profile is known.

A road with bumps can be also be modeled via the proper assignment of the displacement excitations. For instance, a rectangular bump in road surface, as shown in Figure 9.1.2(b), can be described by

$$y_1(t) = h_b\left[u(t - t_0) - u\left(t - (t_0 + w_b/v_c)\right)\right]$$
$$y_2(t) = h_b\left[u\left(t - (t_0 + t_{12})\right) - u\left(t - (t_0 + t_{12} + w_b/v_c)\right)\right]$$
(9.1.4)

where h_b and w_b are the height and width of the bump; and t_0 is the time when the front tire of the car hits the bump.

9.1.2 Dynamic Modeling

In the development of the half-car model shown in Figure 9.1.1, three independent displacement parameters are selected to describe the motion of the 3-DOF system. Note that the selection of independent displacements for a dynamic system is not unique. Listed below are two sets of independent displacement parameters:

Set 1: x_b, θ_b, and x_e
Set 2: x_1, x_2, and x_e

where the geometric meaning of the parameters in Set 1 has been explained previously; and x_1 and x_2 are the transverse displacements of the rigid body at points A and B (see Figure 9.1.1), respectively, in the upward direction.

These two sets of parameters are equivalent in the sense that the parameters in one set can be expressed by those in the other set. For instance, with the small rotation assumption ($|\theta_b| \ll 1$), Figure 9.1.3 indicates that

$$x_b = x_1 - l_1\theta_b = x_2 + l_2\theta_b$$

It follows that the two sets of displacement parameters are related by

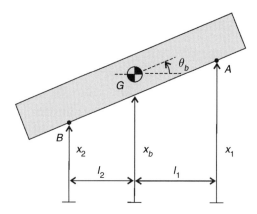

Figure 9.1.3 Rigid body of the half-car model with displacement parameters

$$x_1 = x_b + l_1\theta_b, \quad x_2 = x_b - l_2\theta_b \tag{9.1.5a}$$

or

$$x_b = \frac{l_2 x_1 + l_1 x_2}{l_1 + l_2}, \quad \theta_b = \frac{x_1 - x_2}{l_1 + l_2} \tag{9.1.5b}$$

For convenience of modeling and analysis, these two sets of parameters will be used alternatively.

As mentioned in Sections 3.2 and 3.5, the equations of motion for a mechanical system can be derived with two approaches: (i) the *Newtonian approach*, in which Newton's laws are applied to the free-body diagrams of the system; and (ii) the *Lagrangian approach*, in which Lagrange's equations are applied to the energy functions of the system. For the current problem, both approaches are used.

Modeling by the Newtonian Approach

The free-body diagrams of the car are shown in Figure 9.1.4, where f_{si} and f_{di}, for $i = 1, 2, 3$, are the internal forces of the springs and dampers, respectively. By Newton's second law and with the free-body diagrams, the governing equations of the rigid body and lumped mass are

For M_b $\quad M_b \ddot{x}_b = -M_b g - (f_{s1} + f_{d1}) - (f_{s2} + f_{d2}) + (f_{s3} + f_{d3})$ (9.1.6a)

For I_b $\quad I_b \ddot{\theta}_b = -(f_{s1} + f_{d1})l_1 + (f_{s2} + f_{d2})l_2 + (f_{s3} + f_{d3})d$ (9.1.6b)

For m_e $\quad m_e \ddot{x}_e = -m_e g - (f_{s3} + f_{d3})$ (9.1.6c)

where the moments of forces are with respect to the mass center G and in the positive rotational direction. The internal forces of the system are given by

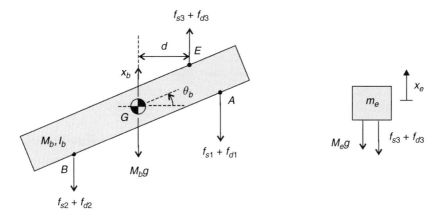

Figure 9.1.4 Free-body diagrams of the half-car model

$$\begin{aligned} f_{s1} &= k(x_1 - y_1), \quad f_{d1} = c(\dot{x}_1 - \dot{y}_1) \\ f_{s2} &= k(x_2 - y_2), \quad f_{d2} = c(\dot{x}_2 - \dot{y}_2) \\ f_{s3} &= k_e(x_e - x_E), \quad f_{d3} = c_e(\dot{x}_e - \dot{x}_E) \end{aligned} \qquad (9.1.7)$$

where x_E is the displacement of the rigid body at point E and it is given by

$$x_E = x_b + d \cdot \theta_b \qquad (9.1.8)$$

With the parameters (x_b, θ_b, x_e) of Set 1 and by using Eqs. (9.1.5), (9.1.7), and (9.1.8), Eq. (9.1.6a) is written as

$$\begin{aligned} M_b \ddot{x}_b &= -M_b g - k(x_1 - y_1) - c(\dot{x}_1 - \dot{y}_1) - k(x_2 - y_2) \\ &\quad - c(\dot{x}_2 - \dot{y}_2) + k_e(x_e - x_E) + c_e(\dot{x}_e - \dot{x}_E) \\ &= -M_b g - k(x_b + l_1 \theta_b - y_1) - c(\dot{x}_b + l_1 \dot{\theta}_b - \dot{y}_1) - k(x_b - l_2 \theta_b - y_2) \\ &\quad - c(\dot{x}_b - l_2 \dot{\theta}_b - \dot{y}_2) + k_e(x_e - x_b - d \cdot \theta_b) + c_e(\dot{x}_e - \dot{x}_b - d \cdot \dot{\theta}_b) \end{aligned}$$

which, after rearrangement, becomes

$$\begin{aligned} & M_b \ddot{x}_b + (2c + c_e)\dot{x}_b - c_e \dot{x}_e + (cl_1 - cl_2 + c_e d)\dot{\theta}_b \\ & + (2k + k_e)x_b - k_e x_e + (kl_1 - kl_2 + k_e d)\theta_b = c(\dot{y}_1 + \dot{y}_2) + k(y_1 + y_2) - M_b g \end{aligned} \qquad (9.1.9a)$$

Similarly, Eqs. (9.1.6b) and (9.1.6c) are reduced to

$$\begin{aligned} & I_b \ddot{\theta}_b + (c_e d^2 + cl_1^2 + cl_2^2)\dot{\theta}_b + (c_e d + cl_1 - cl_2)\dot{x}_b - c_e d \, \dot{x}_e \\ & + (k_e d^2 + kl_1^2 + kl_2^2)\theta_b + (k_e d + kl_1 - kl_2)x_b - k_e d x_e = cl_1 \dot{y}_1 - cl_2 \dot{y}_2 + kl_1 y_1 - kl_2 y_2 \end{aligned} \qquad (9.1.9b)$$

$$m_e \ddot{x}_e + c_e(\dot{x}_e - \dot{x}_b - d \cdot \dot{\theta}_b) + k_e(x_e - x_b - d \cdot \theta_b) = -m_e g \qquad (9.1.9c)$$

Equations (9.1.9a) to (9.1.9c) are the governing equations of motion for the half-car model.

For convenience in analysis and simulation, Eqs. (9.1.9a) to (9.1.9c) are cast in vector-matrix form as follows

$$\mathbf{M}\ddot{\mathbf{x}} + \mathbf{C}\dot{\mathbf{x}} + \mathbf{K}\mathbf{x} = \mathbf{f} \qquad (9.1.10)$$

where $\mathbf{x}$ and $\mathbf{f}$ are the displacement vector and the external force vector given by

$$\mathbf{x} = \begin{pmatrix} x_b \\ \theta_b \\ x_e \end{pmatrix}, \quad \mathbf{f} = \begin{pmatrix} c(\dot{y}_1 + \dot{y}_2) + k(y_1 + y_2) - M_b g \\ cl_1 \dot{y}_1 - cl_2 \dot{y}_2 + kl_1 y_1 - kl_2 y_2 \\ -m_e g \end{pmatrix} \qquad (9.1.11)$$

and $\mathbf{M}$, $\mathbf{C}$, and $\mathbf{K}$ are the mass, damping, and stiffness matrices, respectively, which are of the form

$$\mathbf{M} = \begin{bmatrix} M_b & 0 & 0 \\ 0 & I_b & 0 \\ 0 & 0 & m_e \end{bmatrix}, \quad \mathbf{C} = \begin{bmatrix} (2c + c_e) & c(l_1 - l_2) + c_e d & -c_e \\ c(l_1 - l_2) + c_e d & c_e d^2 + cl_1^2 + cl_2^2 & -c_e d \\ -c_e & -c_e d & c_e \dot{x}_e \end{bmatrix}$$

$$\mathbf{K} = \begin{bmatrix} (2k + k_e) & k(l_1 - l_2) + k_e d & -k_e \\ k(l_1 - l_2) + k_e d & k_e d^2 + kl_1^2 + kl_2^2 & -k_e d \\ -k_e & -k_e d & k_e \end{bmatrix} \qquad (9.1.12)$$

Note that $\mathbf{M}, \mathbf{C}$, and $\mathbf{K}$ are symmetric matrices.

Modeling by the Lagrangian Approach

The kinetic energy T, potential energy V, and Rayleigh dissipation function R of the car are obtained as follows:

$$T = \frac{1}{2} M_b \dot{x}_b^2 + \frac{1}{2} I_b \dot{\theta}_b^2 + \frac{1}{2} m_e \dot{x}_e^2$$

$$V = \frac{1}{2} k(x_b + l_1 \theta_b - y_1)^2 + \frac{1}{2} k(x_b - l_2 \theta_b - y_2)^2 + \frac{1}{2} k_e (x_b + d \cdot \theta_b - x_e)^2 \qquad (9.1.13)$$

$$+ M_b g x_b + m_e g x_e$$

$$R = \frac{1}{2} c(\dot{x}_b + l_1 \dot{\theta}_b - \dot{y}_1)^2 + \frac{1}{2} c(\dot{x}_b - l_2 \dot{\theta}_b - \dot{y}_2)^2 + \frac{1}{2} c_e (\dot{x}_b + d \cdot \dot{\theta}_b - \dot{x}_e)^2$$

where Eqs. (9.1.5a) and (9.1.8) have been used.

With the energy functions given in Eq. (9.1.13), the governing equations of the car can be obtained by Lagrange's equations as given in Section 3.5. Note that the external nonconservative forces are all zeros, namely, $Q_{x_b} = Q_{\theta_b} = Q_{x_e} = 0$. This is because the gravitational, spring, and damping forces have been included in the energy functions V and R. For x_b, the Lagrange's equation is

$$\frac{d}{dt}\left(\frac{\partial T}{\partial \dot{x}_b}\right) - \frac{\partial T}{\partial x_b} + \frac{\partial R}{\partial \dot{x}_b} + \frac{\partial V}{\partial x_b} = 0$$

or

$$\frac{d}{dt}(M_b\dot{x}_b) + c(\dot{x}_b + l_1\dot{\theta}_b - \dot{y}_1) + c(\dot{x}_b - l_2\dot{\theta}_b - \dot{y}_2) + c_e(\dot{x}_b + d\cdot\dot{\theta}_b - \dot{x}_e)$$
$$+ k(x_b + l_1\theta_b - y_1) + k(x_b - l_2\theta_b - y_2) + k_e(x_b + d\cdot\theta_b - x_e) + M_b g = 0$$

which is the same as Eq. (9.1.9a). Also, for θ_b, the Lagrange's equation is

$$\frac{d}{dt}\left(\frac{\partial T}{\partial \dot{\theta}_b}\right) - \frac{\partial T}{\partial \theta_b} + \frac{\partial R}{\partial \dot{\theta}_b} + \frac{\partial V}{\partial \theta_b} = 0$$

or

$$\frac{d}{dt}(I_b\dot{\theta}_b) + cl_1(\dot{x}_b + l_1\dot{\theta}_b - \dot{y}_1) - cl_2(\dot{x}_b - l_2\dot{\theta}_b - \dot{y}_2) + d\,c_e(\dot{x}_b + d\cdot\dot{\theta}_b - \dot{x}_e)$$
$$+ kl_1(x_b + l_1\theta_b - y_1) - kl_2(x_b - l_2\theta_b - y_2) + d\,k_e(x_b + d\cdot\theta_b - x_e) = 0$$

which is the same as Eq. (9.1.9b). Finally, for x_e, the Lagrange's equation is

$$\frac{d}{dt}\left(\frac{\partial T}{\partial \dot{x}_e}\right) - \frac{\partial T}{\partial x_e} + \frac{\partial R}{\partial \dot{x}_e} + \frac{\partial V}{\partial x_e} = 0$$

or

$$\frac{d}{dt}(m_e\dot{x}_e) - c_e(\dot{x}_b + d\cdot\dot{\theta}_b - \dot{x}_e) - k_e(x_b + d\cdot\theta_b - x_e) + m_e g\,x_e = 0$$

which is the same as Eq. (9.1.9c).

Thus, we have shown that the same equations of motion for the half-car model can be obtained by both the Newtonian and Lagrangian approaches.

9.1.3 State-Space Representations

To determine the time response of the car, a state-space representation is established. Because Eqs. (9.1.9a) to (9.1.9c) are a set of three second-order differential equations, six state variables are selected as follows:

$$\mathbf{z} = \begin{pmatrix} \mathbf{x} \\ \dot{\mathbf{x}} \end{pmatrix} = \begin{pmatrix} z_1 \\ \vdots \\ z_6 \end{pmatrix} \quad (9.1.14)$$

where $\mathbf{x}$ is the displacement vector given in Eq. (9.1.11).

According to Eq. (9.1.10),

$$\frac{d}{dt}(\dot{\mathbf{x}}) = \ddot{\mathbf{x}} = \mathbf{M}^{-1}(\mathbf{f} - \mathbf{C}\dot{\mathbf{x}} - \mathbf{K}\mathbf{x})$$

Combining the previous equation with $\frac{d}{dt}(\mathbf{x}) = \dot{\mathbf{x}}$ results in the following state equation in matrix form:

$$\dot{\mathbf{z}} = \mathbf{A}\mathbf{z} + \mathbf{B}\mathbf{f} \qquad (9.1.15)$$

with

$$\mathbf{A} = \begin{bmatrix} \mathbf{0} & \mathbf{I} \\ -\mathbf{M}^{-1}\mathbf{K} & -\mathbf{M}^{-1}\mathbf{C} \end{bmatrix}, \quad \mathbf{B} = \begin{bmatrix} \mathbf{0} \\ \mathbf{M}^{-1} \end{bmatrix} \qquad (9.1.16)$$

The assignment of the outputs in a state-space representation depends on the particular interest and requirement in the dynamic analysis. In the current problem, we are interested in the transverse vibration of the car and the resultant internal forces of the suspension system and engine mounts. Thus, six outputs are specified as follows

$$\begin{aligned}
y_1 &= x_b, \quad y_2 = \theta_b, \quad y_3 = x_e, \\
y_4 &= f_{s1} + f_{d1} = k(x_b + l_1\theta_b - y_1) + c(\dot{x}_b + l_1\dot{\theta}_b - \dot{y}_1) \\
y_5 &= f_{s2} + f_{d2} = k(x_b - l_2\theta_b - y_2) + c(\dot{x}_b - l_2\dot{\theta}_b - \dot{y}_2) \\
y_6 &= f_{s3} + f_{d3} = k_e(x_e - x_b - d\cdot\theta_b) + c_e(\dot{x}_e - \dot{x}_b - d\cdot\dot{\theta}_b)
\end{aligned} \qquad (9.1.17)$$

where Eqs. (9.1.5a), (9.1.7), and (9.1.8) have been used. It follows from Eqs. (9.1.14) and (9.1.17) that the output equations of the car are as follows:

$$\begin{aligned}
y_1 &= z_1, \quad y_2 = z_2, \quad y_3 = z_3 \\
y_4 &= k(z_1 + l_1 z_2) + c(z_4 + l_1 z_5) - c\dot{y}_1 - ky_1 \\
y_5 &= k(z_1 - l_2 z_2) + c(z_4 - l_2 z_5) - c\dot{y}_2 - ky_2 \\
y_6 &= k_e(z_3 - z_1 - d\cdot z_2) + c_e(z_6 - z_4 - d\, z_5)
\end{aligned} \qquad (9.1.18)$$

Equation (9.1.18) is cast in matrix form as follows

$$\mathbf{y} = \mathbf{D}\mathbf{z} + \mathbf{p} \qquad (9.1.19)$$

where $\mathbf{y}$ is the output vector given by

$$\mathbf{y} = (y_1 \ y_2 \ \cdots \ y_6)^T \qquad (9.1.20a)$$

and

$$\mathbf{D} = \begin{bmatrix} 1 & 0 & 0 & 0 & 0 & 0 \\ 0 & 1 & 0 & 0 & 0 & 0 \\ 0 & 0 & 1 & 0 & 0 & 0 \\ k & kl_1 & 0 & c & cl_1 & 0 \\ k & -kl_2 & 0 & c & cl_2 & 0 \\ -k_e & -k_e d & k_e & -c_e & -c_e d & c_e \end{bmatrix}, \quad \mathbf{p} = \begin{pmatrix} 0 \\ 0 \\ 0 \\ -c\dot{y}_1 - ky_1 \\ -c\dot{y}_2 - ky_2 \\ 0 \end{pmatrix} \qquad (9.1.20b)$$

Thus, a state-space representation for the half-car model in Figure 9.1.1, which consists of the state equation (9.1.15) and the output equation (9.1.19), has been established. The solution of these equations gives the dynamic response of the car traveling on a road.

9.1.4 Quasi-Static Deflections and Dynamic Response

For convenience of analysis and simulation, the effect of gravity and that of road conditions are separated. To this end, consider the following quasi-static situation: the car experiences no road excitations ($y_1 = y_2 = 0$), and it has no vibration ($\dot{\mathbf{x}} = 0$). Under these circumstances, the deflections of the car, according to Eq. (9.1.10), satisfy the following equilibrium equation:

$$\mathbf{K}\mathbf{x}_{qs} = \mathbf{f}_g \qquad (9.1.21)$$

with

$$\mathbf{x}_{qs} = \begin{pmatrix} \tilde{x}_b \\ \tilde{\theta}_b \\ \tilde{x}_e \end{pmatrix}, \quad \mathbf{f}_g = \begin{pmatrix} -M_b g \\ 0 \\ -m_e g \end{pmatrix} \qquad (9.1.22)$$

Here, $\mathbf{x}_{qs}$ contains the quasi-static deflections of the car under the gravitational forces, and it is given by

$$\mathbf{x}_{qs} = \mathbf{K}^{-1}\mathbf{f}_g \qquad (9.1.23)$$

Note that the quasi-static deflections are independent of the car speed.
Write

$$\mathbf{x} = \mathbf{x}_{qs} + \mathbf{u}, \quad \mathbf{f} = \mathbf{f}_g + \mathbf{q} \qquad (9.1.24)$$

where $\mathbf{u}$ contains the displacements measured from the quasi-static equilibrium configuration, and $\mathbf{q}$ only contains the displacement excitations. Equation (9.1.24) means that the total response of the car is the sum of its quasi-static response by gravity and dynamic response by road excitation. Substituting Eq. (9.1.24) into Eq. (9.1.10), and using Eq. (9.1.21), yields the governing equation of the car dynamic response as follows:

$$\mathbf{M}\ddot{\mathbf{u}} + \mathbf{C}\dot{\mathbf{u}} + \mathbf{K}\mathbf{u} = \mathbf{q} \qquad (9.1.25)$$

The resultant internal forces of the car suspension systems can be obtained by Eq. (9.1.17), with the solution of Eq. (9.1.25) and the expression in Eq. (9.1.24).

In the subsequent discussion, the symbol **u** for the dynamic response is replaced by **x** for simplicity.

9.1.5 Steady-State Vibration of a Car Moving on Road of Sinusoidal Profile

In this section, the dynamic response of the car moving on a road with the sinusoidal surface profile described by Eq. (9.1.1) is determined. In particular, the steady-state vibration of the car excited by such displacement excitations is studied. As mentioned in Section 9.1.4, the effect of gravity is neglected for dynamic analysis. The quasi-static deflections can always be added by using Eq. (9.1.23).

In formulating the vibration problem, substitute Eq. (9.1.1) into Eq. (9.1.25), to obtain

$$\mathbf{M}\ddot{\mathbf{x}} + \mathbf{C}\dot{\mathbf{x}} + \mathbf{K}\mathbf{x} = y_0 \Big\{ k\mathbf{q}_1 \sin(\omega t) + k\mathbf{q}_2 \sin\big(\omega(t - t_{12})\big) \\ + c\omega\mathbf{q}_1 \cos(\omega t) + c\omega\mathbf{q}_2 \cos\big(\omega(t - t_{12})\big) \Big\} \quad (9.1.26)$$

$$\mathbf{q}_1 = \begin{pmatrix} 1 \\ l_1 \\ 0 \end{pmatrix}, \quad \mathbf{q}_2 = \begin{pmatrix} 1 \\ -l_2 \\ 0 \end{pmatrix} \quad (9.1.27)$$

where **u** has been replaced by **x**. Equation (9.1.26) can be written as

$$\mathbf{M}\ddot{\mathbf{x}} + \mathbf{C}\dot{\mathbf{x}} + \mathbf{K}\mathbf{x} = \mathrm{Im}\{y_0 \mathbf{Q}_0(\omega) e^{j\omega t}\} \quad (9.1.28)$$

with

$$\mathbf{Q}_0(\omega) = (k + jc\omega)\big(\mathbf{q}_1 + e^{-j\omega t_{12}} \mathbf{q}_2\big) \quad (9.1.29)$$

where $\cos\theta = \mathrm{Im}(je^{j\theta})$ has been used. Note that the frequency-dependent vector $\mathbf{Q}_0(\omega)$ is not a function of time. Thus, from the theory of complex numbers, the steady-state solution of Eq. (9.1.26) is expressed by

$$\mathbf{x}_{ss}(t) = \mathrm{Im}\big(\mathbf{y}_{ss}(t)\big) \quad (9.1.30)$$

where $\mathbf{y}_{ss}(t)$ is the steady-state solution of the differential equation

$$\mathbf{M}\ddot{\mathbf{y}} + \mathbf{C}\dot{\mathbf{y}} + \mathbf{K}\mathbf{y} = y_0 \, \mathbf{Q}_0(\omega) e^{j\omega t} \quad (9.1.31)$$

The steady-state solution of Eq. (9.1.31), in the theory of differential equations (Section 2.6), is of the form

9.1 Vibration Analysis of a Car Moving on a Bumpy Road

$$\mathbf{y}_{ss}(t) = \mathbf{Y} e^{j\omega t} \tag{9.1.32}$$

Plugging Eq. (9.1.32) into Eq. (9.1.31) gives

$$\mathbf{Y} = y_0 \mathbf{G}(j\omega) \mathbf{Q}_0(\omega) \tag{9.1.33}$$

where the matrix $\mathbf{G}(j\omega)$ is a transfer function of the car, and it is given by

$$\mathbf{G}(j\omega) = \left(-\omega^2 \mathbf{M} + j\omega \mathbf{C} + \mathbf{K}\right)^{-1} \tag{9.1.34}$$

It follows that the steady-state response of the car is given by

$$\mathbf{x}_{ss}(t) = y_0 \operatorname{Im}\left(\mathbf{G}(j\omega) \mathbf{Q}_0(\omega) e^{j\omega t}\right) \tag{9.1.35}$$

We write

$$\mathbf{x}_{ss}(t) = \begin{pmatrix} x_{b,ss}(t) \\ \theta_{b,ss}(t) \\ x_{e,ss}(t) \end{pmatrix}, \quad \mathbf{G}(j\omega) \mathbf{Q}_0(\omega) = \begin{pmatrix} X_b(\omega) \\ \Theta_b(\omega) \\ X_e(\omega) \end{pmatrix} \tag{9.1.36}$$

where $x_{b,ss}(t)$, $\theta_{b,ss}(t)$, and $x_{e,ss}(t)$ are the steady-state displacements of the car; and $X_b(\omega)$, $\Theta_b(\omega)$, and $X_e(\omega)$ are complex functions of the excitation frequency ω. Thus, the steady-state displacements of the car can be written as

$$\begin{aligned} x_{b,ss}(t) &= y_0 |X_b(\omega)| \sin\left(\omega t + \angle X_b(\omega)\right) \\ \theta_{b,ss}(t) &= y_0 |\Theta_b(\omega)| \sin\left(\omega t + \angle \Theta_b(\omega)\right) \\ x_{e,ss}(t) &= y_0 |X_e(\omega)| \sin\left(\omega t + \angle X_e(\omega)\right) \end{aligned} \tag{9.1.37}$$

where $|z|$ and $\angle z$ are the magnitude and phase angle of the complex function z, respectively. According to Section 7.3, $X_b(\omega)$, $\Theta_b(\omega)$, and $X_e(\omega)$ are the *frequency response functions* of the car.

For numerical simulation, the values of the car parameters are chosen as follows:

Tire-suspension assemblies:

$$k = 2.54 \times 10^5 \text{ N/m}, \quad c = 2.73 \times 10^3 \text{ kg/s}, \quad l_1 = 1.46 \text{ m}, \quad l_2 = 1.34 \text{ m}$$

Car body: $M_b = 1{,}750$ kg, $I_b = 2{,}530$ kg-m^2

Engine and engine mounts:

$$m_e = 250 \text{ kg}, \quad k_e = 1.32 \times 10^5 \text{ N/m}, \quad c_e = 1.03 \times 10^3 \text{ kg/s}, \quad d = 1.65 \text{ m}$$

Also, for the road condition, assume that $y_0 = 0.05$ m and $L_r = 3$ m, which describe a bumpy road.

With Eq. (9.1.37), the steady-state displacements of the car are plotted versus the car speed in Figures 9.1.5 to 9.1.7, for $0 \leq v_c \leq 50$ m/s. In Figure 9.1.5, the amplitude and phase of $x_{b,ss}(t)$ are given by $y_0|X_b(\omega)|$ and $\angle X_b(\omega)$. The amplitudes and phases in Figure 9.1.6 and 9.1.7 are similarly obtained from Eq. (9.1.37). As observed from the figures, an increase in the car speed leads to increased steady-state vibration amplitudes of the car.

Let the acceptable amplitudes of the car displacements, for operation safety and driver comfort, be specified as follows:

$$|x_{e,ss}(t)| \leq 0.15 \text{ m}, \quad |\theta_{b,ss}(t)| \leq 6 \text{ deg}, \quad |x_{e,ss}(t)| \leq 0.15 \text{ m}$$

As indicated by the dashed lines in Figures 9.1.5 to 9.1.7, the car speed must be under 22.25 m/s (80.1 km/h), in order to meet the requirement of car displacement amplitudes.

For a less bumpy road, say, $y_0 = 0.05$ m and $L_r = 5$ m., the amplitudes of the steady-state displacements of the car versus the car speed are plotted in Figure 9.1.8. It can be seen that the upper limit of the car speed is 33.85 m/s (121.9 km/h) in order to meet the amplitude requirement.

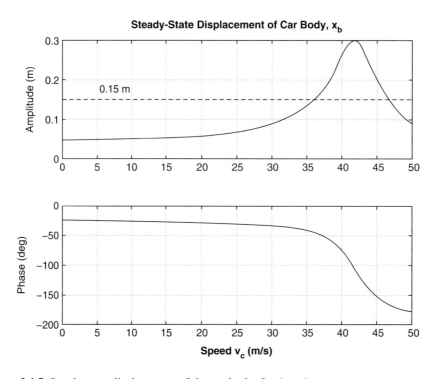

Figure 9.1.5 Steady-state displacement of the car body, for $L_r = 3$ m

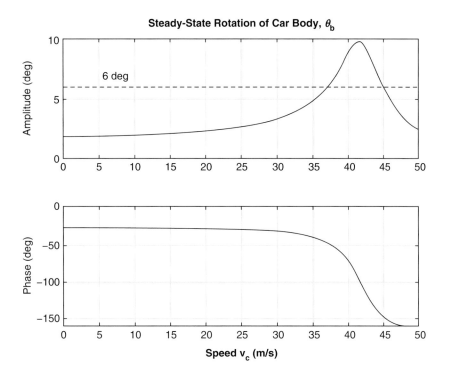

Figure 9.1.6 Steady-state rotation of the car body, for $L_r = 3$ m

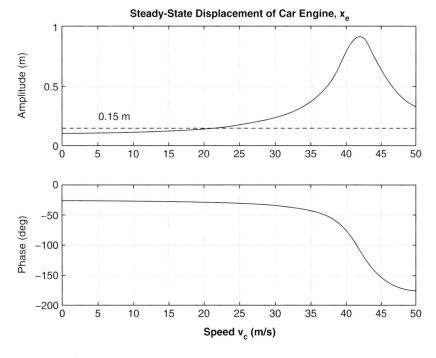

Figure 9.1.7 Steady-state displacement of the car engine, for $L_r = 3$ m

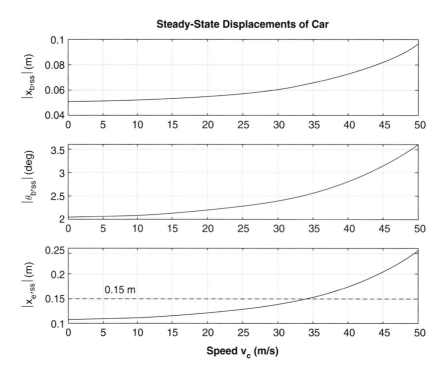

Figure 9.1.8 Steady-state displacement amplitudes of the car, for $L_r = 5$ m

In generating the steady-state displacement plots shown in Figures 9.1.5 to 9.1.8, the following MATLAB commands were used:

```
% Car parameters
me = 250; ke = 1.32e5; ce = 1.03e3; d = 1.65;
Mb = 1750; Ib = 2530;
k = 2.54e5; c = 2.73e3; l1 = 1.34; l2 = 1.46;

% Road condition
y0 = 0.05; Lr = 3;

% Parameter matrices
MM = diag([Mb Ib me]);
CC = [2*c+ce,            ce*d+c*(l1-l2),          -ce;
      ce*d+c*(l1-l2),    ce*d^2+c*(l1^2+l2^2),    -ce*d;
      -ce,               -ce*d,                    ce];
KK = [2*k+ke,            ke*d+k*(l1-l2),          -ke;
      ke*d+k*(l1-l2),    ke*d^2+k*(l1^2+l2^2),    -ke*d;
      -ke,               -ke*d,                    ke];
```

```
% Initialization
npts = 1001;
vc = linspace(0,50,npts);  % Car speed range
omega = vc/Lr;  % Excitation frequency
t12 = (l1+l2) ./ vc;
Amplitude_xe = zeros(npts,1);
Phase_Angle_xe = zeros(npts,1);
Amplitude_xb = zeros(npts,1);
Phase_Angle_xb = zeros(npts,1);
Amplitude_Thetab = zeros(npts,1);
Phase_Angle_Thetab = zeros(npts,1);
% Computation of dynamic response
q1 = [1; l1; 0];
q2 = [1; l2;0];
for index = 1:1:npts
    omg = omega(index);
    t0 = t12(index);
    Q0 = (k+c*1 j*omg)*(q1+exp(-1 j*omg*t0)*q2);
    xss = y0*inv(-omg^2*MM + 1 j*omg*CC + KK) * Q0;
    Amplitude_xe(index,1) = abs(xss(3,1));
    Phase_Angle_xe(index,1) = angle(xss(3,1));
    Amplitude_xb(index,1) = abs(xss(1,1));
    Phase_Angle_xb(index,1) = angle(xss(1,1));
    Amplitude_Thetab(index,1) = abs(xss(2,1));
    Phase_Angle_Thetab(index,1) = angle(xss(2,1));
end
```

9.1.6 Time Response of a Car Moving over a Pothole

In this section, the dynamic response of a car moving on a road surface containing a pothole is investigated. As described by Eq. (9.1.3), for a pothole, vectors **f** and **p** in the state equation (9.1.15) and output equation (9.1.19) become

$$\mathbf{f} = \begin{pmatrix} c(v_1 + v_2) + k(y_1 + y_2) \\ cl_1 v_1 - cl_2 v_2 + kl_1 y_1 - kl_2 y_2 \\ 0 \end{pmatrix}, \quad \mathbf{p} = \begin{pmatrix} 0 \\ 0 \\ 0 \\ -cv_1 - ky_1 \\ -cv_2 - ky_2 \\ 0 \end{pmatrix} \quad (9.1.38)$$

where
$$v_1(t) = -h_p[\delta(t - t_0) - \delta(t - (t_0 + w_p/v_c))]$$
$$v_2(t) = -h_p[\delta(t - (t_0 + t_{12})) - \delta(t - (t_0 + t_{12} + w_p/v_c))] \quad (9.1.39)$$

with $\delta(t)$ being the Dirac delta function. As discussed in Section 9.1.4, gravity is not included in dynamic analysis because the quasi-static deflections of a car can be added by Eq. (9.1.23). In addition, the initial condition

$$\mathbf{z}(0) = \mathbf{z}_0 = \left(x_b(0) \quad \theta_b(0) \quad x_e(0) \quad \dot{x}_b(0) \quad \dot{\theta}_b(0) \quad \dot{x}_e(0) \right)^T \quad (9.1.40)$$

needs to be specified in determination of the time response of the car.

Several analytical and numerical methods are available for solution of the state equation (9.1.15). In this undergraduate text, the following two solution methods are introduced.

Method 1: Numerical Solution via Runge–Kutta Method

A fixed Runge–Kutta method of order four is presented in Section 2.8. Solution by this method takes the following three steps.

(1) Assign a solution region in the time domain: $0 \leq t \leq t_f$, where t_f is the final time of simulation.
(2) Divide the solution region into N identical time intervals: $t_k = kh, k = 0, 1, \ldots, N$, with the time-step size h given by $h = t_f/N$. The step size h must small enough to assure the convergence of the numerical solution.
(3) The state vector at t_{k+1}, which is $\mathbf{z}_{k+1} = \mathbf{z}(t_{k+1})$, is given by the iterative formula

$$\mathbf{z}_{k+1} = \mathbf{z}_k + \frac{h}{6}(\mathbf{f}_1 + 2\mathbf{f}_2 + 2\mathbf{f}_3 + \mathbf{f}_4), \quad k = 0, 1, 2, \ldots N - 1 \quad (9.1.41)$$

where
$$\begin{aligned}
\mathbf{f}_1 &= \mathbf{\Phi}(t_k, \mathbf{z}_k) \\
\mathbf{f}_2 &= \mathbf{\Phi}\left(t_k + \frac{h}{2}, \mathbf{z}_k + \frac{h}{2}\mathbf{f}_1 \right) \\
\mathbf{f}_3 &= \mathbf{\Phi}\left(t_k + \frac{h}{2}, \mathbf{z}_k + \frac{h}{2}\mathbf{f}_2 \right) \\
\mathbf{f}_4 &= \mathbf{\Phi}(t_k + h, \mathbf{z}_k + h\mathbf{f}_3)
\end{aligned} \quad (9.1.42)$$

with
$$\mathbf{\Phi}(t, \mathbf{z}(t)) = \mathbf{A}\mathbf{z}(t) + \mathbf{B}\mathbf{f}(t) \quad (9.1.43)$$

The MATLAB commands for implementing the above formula are given in Section 2.8.

When using the Dirac delta function, as shown in Eq. (9.1.39), the following approximation can be made

$$\delta(t_k - t_*) = \begin{cases} \dfrac{1}{h}, & \text{if } t_k \leq t_* < t_{k+1} \\ 0, & \text{otherwise} \end{cases} \tag{9.1.44}$$

where t_* is a fixed nonnegative number.

Method 2: Solution by MATLAB and Mathematica

The state equation (9.1.15) can also be solved by using the Runge–Kutta solvers in software packages, such as the MATLAB function `ode45` and the Mathematica function `NDSolve`. Refer to Section 2.8 for some demonstrative examples.

In numerical simulation, the same car parameters given in Section 9.1.5 are used. Assume that the parameters of the pothole in Eq. (9.1.3) have the following values: $h_p = 0.1$ m and $w_p = 1$ m. The car travels at speed $v_c = 12$ m/s (43.2 km/h or 26.8 mph), and hits the pothole at time $t = 4$ s. Select the final time of simulation as $t_f = 10$ s. Let the initial conditions of the car be:

$$x_b(0) = -0.01 \text{ m}, \quad \theta_b(0) = 0.005 \text{ rad}, \quad x_e(0) = -0.02 \text{ m}$$
$$\dot{x}_b(0) = 0, \quad \dot{\theta}_b(0) = 0, \quad \dot{x}_e(0) = 0$$

The time response of the car is computed by Method 1 and the results on the car displacements are plotted in Figure 9.1.9. As shown in the figure, the vibration of the car caused by the initial disturbances is quickly dampened out. When the car hits the pothole at $t = 4$ s, it experiences relatively large jumps in its displacements and rotation, especially the displacement x_e of the engine. After going over the pothole, it takes several seconds for the vibration to settle down.

The resultant suspension forces are denoted by $f_{Rj} = f_{sj} + f_{dj}$, $j = 1, 2, 3$, where f_{sj} and f_{dj} are the spring and damping forces given in Eq. (9.1.17). With the output equation (9.1.19), the suspension forces are computed, and plotted against time in Figure 9.1.10. The upper subplot of the figure shows a big jump in f_{R1} of the first suspension at $t = 4$ s when the front tire hits the pothole, and shortly after, there is another jump in the opposite direction as the tire is about to leave the pothole. A similar pattern of double jumps is found for f_{R2} of the rear tire, as shown in the middle subplot. These two pairs of double jumps are separated by a time interval $t_{12} = (l_1 + l_2)/v_c = 0.233$ s, which is clearly seen on comparing the upper and middle subplots. For the internal force f_{R3} of the engine mounts, shown in the lower subplot, the appearance of jumps is not as swift as for f_{R1} and f_{R2}. Note that the maximum amplitude of f_{R3} is 5.5 times as large as the weight of the engine (250 kgf = 2452.5 N), and that the impulses in f_{R1} and f_{R2} are at least 25 times as large as the total weight of the car (2000 kgf = 19620 N). The

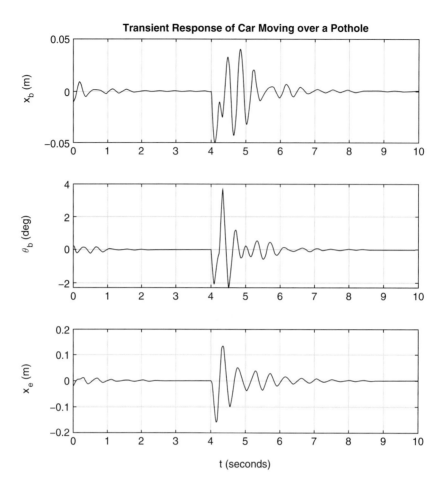

Figure 9.1.9 Time response of the car moving over a pothole, at speed $v_c = 12$ m/s

high-amplitude displacements and impulsive internal forces shown in Figures 9.1.9 and 9.1.10 provide an explanation for car damage caused by potholes.

To generate the plots in Figures 9.1.9 and 9.1.10 by Method 1 (numerical solution via the Runge–Kutta Method) a MATLAB function RKfun is created to obtain $\mathbf{\Phi}(t, \mathbf{z})$ in Eq. (9.1.43), which is given as follows:

```
function z = RKfun(t, y, h_step, AA, BB, Par_Matrix)
% Purpose: to calculate A*y + B*f in dy/dt = A*y + B*f
% Here, t = time, y = state vector
%        AA and BB = matrices A and B in Eq. (9.1.15)
%        h_step = time step size in Eq. (9.1.37)
%        Par_Matrix = [vc c k l1 l2 hb wb tb];
% Input the car parameters
```

9.1 Vibration Analysis of a Car Moving on a Bumpy Road

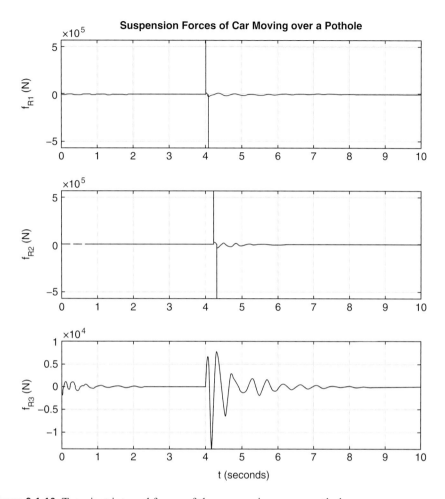

Figure 9.1.10 Transient internal forces of the car moving over a pothole

```
vc = Par_Matrix(1); c = Par_Matrix(2); k = Par_Matrix(3);
l1 = Par_Matrix(4); l2 = Par_Matrix(5);
% Input the pothole parameters
hb = Par_Matrix(6); wb = Par_Matrix(7);
tb = Par_Matrix(8); % the time when the car hits the pothole
t12 = (l1+l2)/vc;   % travel time between two suspensions

% Compute Heaviside and Dirac Delta functions for the pothole
tt1 = t - tb; tt1d = tt1 - wb/vc;
tt2 = tt1 - t12; tt2d = tt2 - wb/vc;
% To set heaviside(0) = 0. By default, MATLAB assigns heaviside(0) = 0.5.
if tt1 == 0, tt1 = tt1 - eps; end
```

```
if tt2 == 0, tt2 = tt2 - eps; end
if tt1d == 0, tt1d = tt1d - eps; end
if tt2d == 0, tt2d = tt2d - eps; end
% Heaviside functions
u1 = heaviside(tt1); u1d = heaviside(tt1d);
u2 = heaviside(tt2); u2d = heaviside(tt2d);
% Dirac delta functions
delta1 = 0; delta1d = 0; delta2 = 0; delta2d = 0;
if tt1 > 0 & tt1 <= h_step, delta1 = 1/h_step; end
if tt1d > 0 & tt1d <= h_step, delta1d = 1/h_step; end
if tt2 > 0 & tt2 <= h_step, delta2 = 1/h_step; end
if tt2d > 0 & tt2d <= h_step, delta2d = 1/h_step; end

% Compute f(t) given in Eq. (9.1.11)
f = zeros(3,1);
y1 = -hb*(u1 - u1d);
y2 = -hb*(u2 - u2d);
v1 = -hb*(delta1 - delta1d);
v2 = -hb*(delta2 - delta2d);
f(1) = c*(v1+v2) + k*(y1+y2);
f(2) = c*(l1*v1-l2*v2) +k*(l1*y1-l2*y2);

% Assign the function output
z = AA*y + BB*f;s
```

In the simulation, the function RKfun is used with MATLAB commands given in Section 2.8.

9.2 Speed Control of a Coupled Engine–Propeller System

In Example 3.3.1, the equations of motion of a rotor (motor–propeller assembly) are derived. This type of dynamic system finds applications in helicopters and unmanned aerial vehicles. In this section, feedback control of the rotation speed of the rotor is examined. To understand the subsequent presentation, some knowledge on electromechanical systems (Chapter 6) and feedback control systems (Chapter 8) is required.

The rotor system in consideration consists of a motor, a propeller, and an elastic shaft connecting the motor and propeller. Figure 9.2.1(a) shows an open-loop system and Figure 9.2.1(b) shows a closed-loop system for the rotor, where θ_p is the rotation angle of the propeller, τ_m is the torque applied to the motor shaft by the electric motor, v_m is a voltage applied to the motor by the motor driver (amplifier), and u is the command input to the driver. For the open-loop system, the rotation speed $\dot{\theta}_p$ of the propeller is controlled by

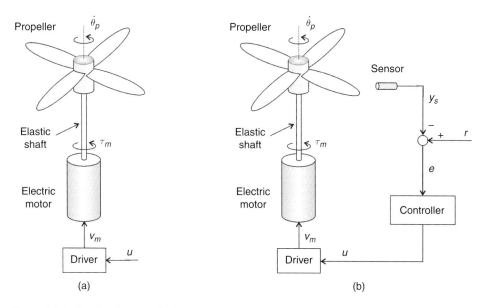

Figure 9.2.1 Feedback control of a rotor system (motor-propeller assembly): (a) open-loop system; and (b) closed-loop system

a user-specified command input u. For the closed-loop system, the propeller rotation speed is manipulated by u that is determined by a controller according to the error e between the reference input r (desired rotation speed) and the sensor output y_s (measured rotation speed). In this study, a field-controlled direct-current (DC) motor is used for the rotor system.

The main purpose of implementing a feedback controller in the rotor system is to maintain a desired rotation speed of the propeller under the influence of system parameter variations and air damping. Below, open-loop and closed-loop systems are formulated, and the necessity of feedback control is demonstrated through numerical simulations by using MATLAB and Simulink.

9.2.1 Open-Loop System

The open-loop system in Figure 9.2.1(a) has four main parts: (a) a motor driver; (b) an electrical subsystem; (c) a mechanical subsystem; and (d) a coupling mechanism between the electrical and mechanical subsystems. A block diagram of the open-loop system with these parts is shown in Figure 9.2.2.

The governing equations of the four parts of the open-loop system are as follows:

(a) Motor driver: $v_m = K_a u$ (9.2.1a)

where K_a is an amplifier gain.

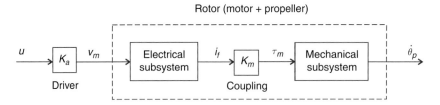

Figure 9.2.2 Block diagram of the open-loop system

(b) Electrical subsystem: $\quad v_m = R_f\, i_f + L_f \dfrac{di_f}{dt}$ (9.2.1b)

where i_f is the current of the field circuit of the field controlled motor (see Section 6.4.2).

(c) Mechanical subsystem: $\quad J_m \ddot{\theta}_m + b\dot{\theta}_m + k_T \theta_m - k_T \theta_p = \tau_m$ (9.2.1c)

$$J_p \ddot{\theta}_p + b_D \dot{\theta}_p^2 \operatorname{sgn}(\dot{\theta}_p) + k_T \theta_p - k_T \theta_m = 0 \tag{9.2.1d}$$

where θ_m is the rotation angle of the motor armature shaft; k_T is the coefficient of a torsional spring, modeling the elastic shaft connecting the motor to the propeller; and the term $b_D \dot{\theta}_p^2 \operatorname{sgn}(\dot{\theta}_p)$ is a nonlinear drag torque due to air damping. Refer to Example 3.3.1 for more details on the derivation of Eqs. (9.2.1c) and (9.2.1d).

Finally, there is

(d) Electromechanical coupling: $\quad \tau_m = K_m\, i_f$ (9.2.1e)

where K_m is the motor torque constant.

Ideal Case

In this case, it is assumed that the nonlinear damping torque $b_D \dot{\theta}_p^2 \operatorname{sgn}(\dot{\theta}_p)$ in Eq. (9.2.1d) is negligible. This assumption, of course, is not applicable to those applications in which aerodynamic forces play an essential role. However, solutions in the ideal case can serve as a reference for dynamic analysis and control of the rotor system in practical situations.

With this assumption, the open-loop system becomes linear with the transfer function

$$G_o(s) = \frac{\Omega_p(s)}{U(s)} = \frac{k_T K_a K_m}{(R_f + L_f s)[J_m J_p s^3 + b J_p s^2 + k_T(J_m + J_p)s + b k_T]} \tag{9.2.2}$$

where $\Omega_p(s) = \mathcal{L}[\dot\theta_p] = s\Theta_p(s)$.

Denote the desired rotation speed of the propeller by Ω_0 and consider a constant command input $u(t) = U_0$. From Eq. (9.2.2), it can be shown that all the poles of $sG_o(s)U(s)$ have negative real parts, satisfying the condition of the final-value theorem, Eq. (7.1.15). It follows that the steady-state rotation speed is

$$\omega_{p,ss} = \lim_{t\to\infty} \dot\theta_p = \lim_{s\to 0} sG_o(s)U(s) = \frac{K_a K_m U_0}{R_f b} \quad (9.2.3)$$

By setting $(\omega_p)_{ss} = \Omega_0$ in Eq. (9.2.3), the command input for the system to reach the desired speed is obtained as

$$U_0 = \frac{R_f b}{K_a K_m} \Omega_0 \quad (9.2.4)$$

Equation (9.2.4) shall also serve as a formula for unit conversion between the rotation speed Ω_0 (in rad/s) and the command input U_0 (in V).

To simulate the time response of the open-loop system, the parameters of the rotor system are given in Table 9.2.1. We let the desired rotation speed of the propeller be $\Omega_0 = 12000$ rpm $= 400\pi$ rad/s. By the following MATLAB commands:

```
% Rotor parameters
Ka = 10; Rf = 1; Lf = 0.002; Km = 60;
Jm = 1.2; Jp = 1.1; bm = 5; kT = 4500;

% Desired propeller speed and reference input
Omega0 = 12000;        % Desired speed in RPM
W0 = 1200/60*(2*pi);   % Conversion from RPM to rad/s
U0 = W0*Rf*bm/Ka/Km;   % Reference input U0, by Eq. (9.2.4)
```

Table 9.2.1 Parameters of the rotor system

Parameter	Symbol	Value
Polar moment of inertia of DC motor	J_m	1.2 kg·m²
Damping coefficient of the motor bearing	b	5 N·m·s
Field resistance	R_f	1 Ω
Field inductance	L_f	0.002 H
Motor torque constant	K_m	60 N·m/A
Amplifier gain	K_a	10
Torsion spring coefficient of the shaft	k_T	4500 N·m
Polar moment of inertia of the propeller	J_p	1.1 kg·m²

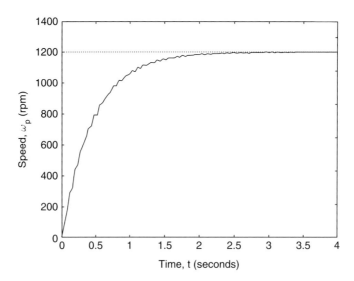

Figure 9.2.3 Rotation speed of the propeller versus time

```
% Time response
P = [Lf Rf];
Q = [Jm*Jp bm*Jp kT*(Jm+Jp) bm*kT];
den = conv(P,Q);
num = kT*Ka*Km*U0/(2*pi)*60;     % Conversion from rad/s to RPM
rotor_tf = tf(num,den);
step(rotor_tf);
```

the time response (propeller speed) of the rotor system under the input given by Eq. (9.2.4) is plotted against time in Figure 9.2.3, where $\omega_p = \dot{\theta}_p$. Note that the simulation results on the propeller speed are presented in rpm, for which the conversion between rpm and rad/s has been implemented in a MATLAB command.

The curve in Figure 9.2.3 shows wiggles, which are due to the elasticity (k_T) of the shaft connecting the motor and propeller. The elastic effect of the shaft on the system response is further illustrated in Figure 9.2.4, with $k_T = 700, 2000,$ and 8000 N·m. As observed from the figure, a larger spring coefficient k_T renders a response curve with smaller wiggles.

Parameter Variations

In operation, the motor of the rotor system may experience parameter variations, which can be caused by thermal effects, external loads, manufacturing errors, and uncertainties in the system modeling. Mathematically, a parameter p with variation Δp can be expressed by $p + \Delta p$, where Δp usually varies in a range but is unknown. Thus, if the parameters K_m, R_f,

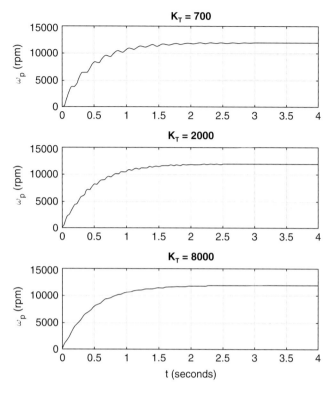

Figure 9.2.4 Rotation speed of the propeller versus time, with k_T = 700 N·m, 2000 N·m, and 8000 N·m

and b have constant variations, the actual steady-state rotation speed of the propeller, under the command input U_0 predicted by the ideal case, will be

$$(\omega_{p,ss})_{actual} = \frac{K_a(K_m + \Delta K_m)}{(R_f + \Delta R_f)(b + \Delta b)} U_0 = \frac{K_a(K_m + \Delta K_m)}{(R_f + \Delta R_f)(b + \Delta b)} \frac{R_f b}{K_a K_m} \Omega_0$$
$$= \frac{(1 + \Delta K_m/K_m)}{(1 + \Delta R_f/R_f)(1 + \Delta b/b)} \Omega_0 = \frac{(1 + \varepsilon_{K_m})}{(1 + \varepsilon_{R_f})(1 + \varepsilon_b)} \Omega_0 \quad (9.2.5)$$

where $\varepsilon_p = \Delta p/p$ is the percentage variation of parameter p. Obviously, the actual steady-state speed is not the same as Ω_0 in most cases unless $(1 + \varepsilon_{K_m}) = (1 + \varepsilon_{R_f})(1 + \varepsilon_b)$. For instance, for $\varepsilon_{K_m} = -1\%$, $\varepsilon_{R_f} = 1\%$, and $\varepsilon_b = 2\%$, Eq. (9.2.5) gives

$$(\omega_{p,ss})_{actual} = \frac{(1 - 0.01)}{(1 + 0.01)(1 + 0.02)} \Omega_0 = 96.1\% \times \Omega_0$$

For $\Omega_0 = 12\,000$ rpm, $(\omega_{p,ss})_{actual} = 11532$ rpm; see Figure 9.2.5 (with $k_T = 4500$ N·m). Thus, the open-loop system in general cannot eliminate or reduce the effects of parameter variations. This disadvantage of open-loop control is also discussed in Section 8.2.

9 Application Problems

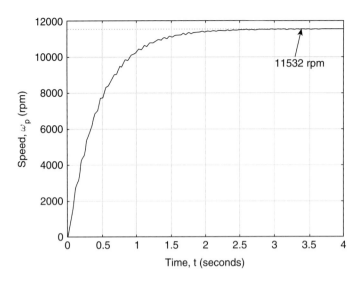

Figure 9.2.5 Rotation speed of the propeller versus time, for $\Omega_0 = 12000$ rpm and with parameter variations: $\varepsilon_{K_m} = -1\%$, $\varepsilon_{R_f} = 1\%$, and $\varepsilon_b = 2\%$

Effect of Nonlinear Air Damping

If the drag torque $b_D \dot{\theta}_p^2 \mathrm{sgn}(\dot{\theta}_p)$ in Eq. (9.2.1d) cannot be ignored, as in the applications of helicopters and unmanned aerial vehicles, the rotor system becomes nonlinear. In this case, transfer function formulations and block diagrams in the s-domain are invalid for modeling and simulation. However, state-space representations and block diagrams in the time domain are applicable to this system.

In this subsection, we shall use a time-domain block diagram as a tool for modeling and simulation of the nonlinear rotor system. To this end, we rewrite Eqs. (9.2.1b)–(9.2.1d) as follows

$$Di_f = \frac{1}{L_f}\left(v_m - R_f\, i_f\right)$$

$$D^2 \theta_m = \frac{1}{J_m}\left(\tau_m - bD\theta_m - \tau_s\right) \quad (9.2.6)$$

$$D^2 \theta_p = \frac{1}{J_p}\left(\tau_s - \tau_D\right)$$

where $D = \dfrac{d}{dt}$, $v_m = K_a u$, and τ_s and τ_D are the spring torque and drag torque given by $\tau_s = k_T(\theta_m - \theta_p)$ and $\tau_D = b_D \dot{\theta}_p^2 \mathrm{sgn}(\dot{\theta}_p)$.

As in the linear case, the desired propeller speed is $\Omega_0 = 12000$ rpm, and the command input is $u(t) = U_0$, with Ω_0 and U_0 related by Eq. (9.2.4). Following the steps presented in Section 6.5, a time-domain block diagram of the rotor system is obtained in Figure 9.2.6.

9.2 Speed Control of a Coupled Engine–Propeller System

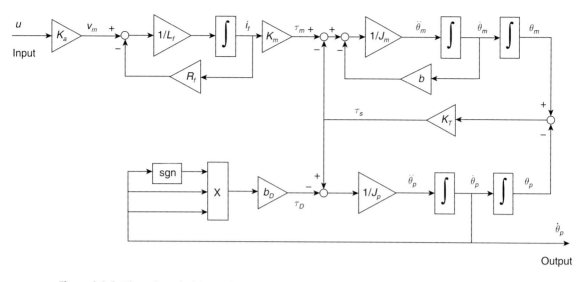

Figure 9.2.6 Time-domain block diagram of the rotor system with a nonlinear air-damping torque

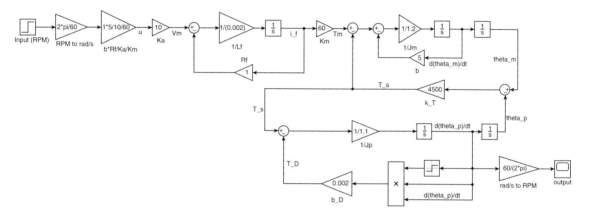

Figure 9.2.7 A Simulink model of the rotor system with a nonlinear air-damping torque

With the time-domain block diagram, a Simulink model of the open-loop system is easily built; see Figure 9.2.7, where the input is the desired rotation speed (1200 rpm), the output shown in the scope block is the propeller speed in rpm, and the gain $R_f b/(K_a K_m)$ is the conversion factor given in Eq. (9.2.4). To present the propeller speed in rpm, two gains for conversion between rpm and rad/s, $2\pi/60$ and $60/(2\pi)$, are installed in the Simulink model.

For simplicity in the subsequent analysis, the blocks from u to $\dot{\theta}_p$ in Figure 9.2.7 can be grouped into a subsystem; see Figure 9.2.8, where the subsystem is named *Rotor*. To see or to edit the subsystem, simply double-click the box of the subsystem. The block diagrams in Figures 9.2.7 and 9.2.8 are equivalent.

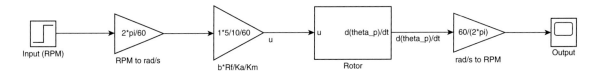

Figure 9.2.8 An equivalent Simulink subsystem model of the rotor system

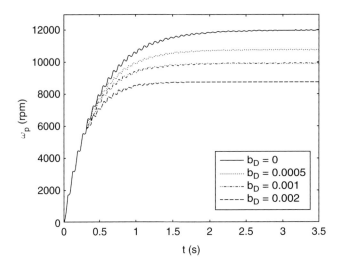

Figure 9.2.9 Rotation speed of the propeller versus time, without feedback control: $b_D = 0, 0.0005, 0.001$, and 0.002 N·m·s^2

In a numerical simulation, the following four cases of air-damping coefficient are considered: $b_D = 0, 0.0005, 0.001$, and 0.002 N·m·s^2. Here, the inclusion of the case of $b_D = 0$ is two-fold: to validate the Simulink model with the linear model given by Eq. (9.2.2), and to investigate the effect of air damping on the system response. In all four cases, the system parameters in Table 9.2.1 are used.

Figure 9.2.9 shows the time histories of the propeller rotation speed ($\omega_p = \dot{\theta}_p$) in the four air-damping cases, which are plotted by running the Simulink model in Figure 9.2.7. It can be seen that the final (steady-state) value of the propeller speed decreases as the air-damping coefficient increases. Therefore, under the influence of the drag torque, the open-loop system, which is designed based on the ideal case ($b_D = 0$), cannot reach the desired rotation speed (12000 rpm). Note that the curve with $b_D = 0$ (no air damping) is the same as that obtained by the transfer function model in Figure 9.2.3.

Because the air-damping coefficient b_D is difficult to predict beforehand, it is not efficient to adjust the command input u in real time for regulating the propeller speed in this open-loop configuration. This disadvantage of open-loop control, along with the previously mentioned issue of parameter variations, necessitates feedback control of the rotor system.

9.2.2 Closed-Loop System

To deal with the effects of parameter variations and nonlinear air damping, feedback control of the rotor system is installed in Simulink. Figure 9.2.10 shows a Simulink model of a closed-loop system that consists of the subsystem *Rotor* in Figure 9.2.8 and a PID controller block. It is assumed that a noncontact speed sensor is installed to measure the rotation speed of the propeller shaft (not shown in the figure). The PID controller block implements the following PID control algorithm with a low-pass filter on D (derivative) action:

$$K_P + K_I \frac{1}{D} + K_D \frac{N}{1 + N/D}, \quad D = \frac{d}{dt} \quad (9.2.7)$$

where N is a filter coefficient. Theoretically, as N goes to infinity, the third term in the expression (9.2.7) becomes classic D action, $K_D D$. Refer to Section 8.3 for the basic concepts of PID control. Again, to show the simulation results in rpm, two gain blocks for conversion between rpm and rad/s are installed in the Simulink model.

In the simulation using the model in Figure 9.2.10, the parameters of the rotor listed in Table 9.2.1 are used and the controller parameters are chosen as follows: $K_P = 0.7$, $K_I = 4$, $K_D = 0.0007$, and $N = 200$. The propeller speed of the controlled rotor is plotted against time in Figure 9.2.11, for four different cases of air damping: $b_D = 0.0005, 0.001, 0.002$, and 0.005 N·m·s². Comparison of Figures 9.2.9 and 9.2.11 shows that the PID controller can effectively regulate the propeller speed to the reference input (desired speed) $\Omega_0 = 12000$ rpm. In the case of $b_D = 0.005$ N·m·s², which describes a relatively large drag torque, running the Simulink model in Figure 9.2.7 gives a steady-state propeller speed of the open-loop system around 6947 rpm, which is much lower than the desired 12 000 rpm. However, under the feedback controller, the propeller speed eventually reaches the desired speed, as shown in Figure 9.2.11(d).

In the above-mentioned four cases of air damping, the PID control gains are the same. This indicates that even if the air damping is unpredictable, the PID feedback controller is able to maintain the speed of the propeller as desired.

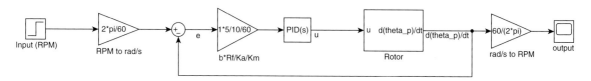

Figure 9.2.10 Simulink model of a closed-loop system

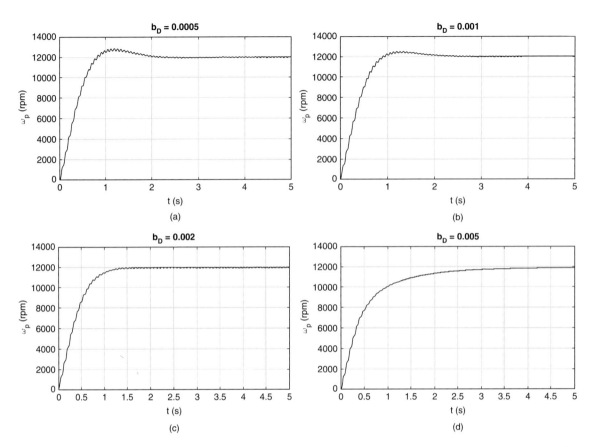

Figure 9.2.11 Time histories of the propeller speed, with PID control: (a) $b_D = 0.0005$ N·m·s²; (b) $b_D = 0.001$ N·m·s²; (c) $b_D = 0.002$ N·m·s², (d) $b_D = 0.005$ N·m·s²

Finally, a rotor system with the combined effects of air damping and parameter variations is simulated. The damping coefficient $b_D = 0.002$ N·m·s² and the percentage parameter variations defined in Eq. (9.2.5) are $\varepsilon_{K_m} = 0$, $\varepsilon_{R_f} = 5\%$, and $\varepsilon_b = 5\%$. In the closed-loop system, the same PID control gains ($K_P = 0.7$, $K_I = 4$, $K_D = 0.0007$, and $N = 200$) are assigned. Figure 9.2.12 shows the time histories of the propeller rotation speed, without and with feedback control. It can be seen that the propeller speed without control is significantly reduced to 8201 rpm. Under the PID controller, the propeller speed reaches the desired value quickly (in about 2.5 seconds).

In summary of this section, the rotation-speed regulation of a coupled motor–propeller system via PID feedback control has been presented, and the utility of MATLAB/Simulink in the relevant modeling and simulation has been demonstrated. Because the rotor system in consideration has nonlinearity due to air damping, the gains of a feedback controller depend on the reference input (the desired rotation speed). One practical way to deal with different

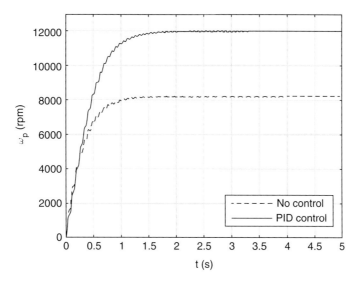

Figure 9.2.12 Time histories of the propeller speed of the rotor with air damping ($b_D = 0.002$ N·m·s^2) and parameter variations ($\varepsilon_{K_m} = 0$, $\varepsilon_{R_f} = 5\%$ and $\varepsilon_b = 5\%$): dashed line – no control; solid line – PID control

rotation speeds in operation is to build up a look-up table for different scenarios, which can also be conveniently done with MATLAB/Simulink.

9.3 Modeling and Analysis of a Thermomechanical System (a Bimetallic Strip Thermometer)

A bimetallic strip thermometer is a cantilever beam created by bonding together two layers of metals with different thermal expansion coefficients. A typical pair of metals is brass and steel, with thermal expansion coefficients of approximately 18.7 and 12 ppm/°C (parts per million per degree Celsius). The bonding is done along the length of the metal strip, and is achieved by either fusing them together or joining them mechanically with rivets. When heated, the two metals expand differently, thus making the cantilever beam bend in the direction of the lesser coefficient of thermal expansion. The deflection of the free end of a bimetallic strip is proportional to the temperature change, making this device one of the most durable thermometers. To improve its performance, the cantilever bimetallic beam is made very long and formed into a spiral as shown in Figure 9.3.1.

Some applications of this technology include the popular meat thermometers, analog thermostats, electrical breakers, and many others.

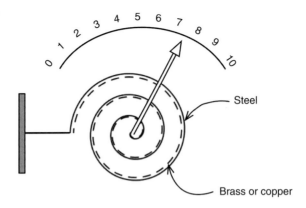

Figure 9.3.1 Bimetallic spiral strip thermometer

9.3.1 Dynamic Modeling of a Simple Bimetallic Strip

Consider a cantilevered bimetallic strip as shown in Figure 9.3.2(a).

With increasing temperature, the metal layers of this cantilever beam expand. Since the thermal expansion coefficient of the top copper layer is larger than that of the bottom steel layer, the beam bends downward. The radius of curvature of the deformed beam is related to the temperature increase as follows:

$$R = \frac{h}{(\alpha_1 - \alpha_2)\Delta T} \tag{9.3.1}$$

where h is the thickness of the bimetallic strip, α_1 and α_2 are the thermal expansion coefficients, and ΔT is the temperature increase.

As stated in the literature on mechanics of materials, the inverse of the radius of curvature equals the second derivative of the free-end deflection of the cantilever beam (x is measured from the clamped end of the cantilever beam):

$$\frac{1}{R} = \frac{d^2 z(x)}{dx^2} \tag{9.3.2}$$

Substituting the expression for R (Eq. (9.3.1)) into Eq. (9.3.2) and integrating twice, the expressions for the slope of the bending beam axis are obtained:

$$\frac{dz(x)}{dx} = \frac{(\alpha_1 - \alpha_2)\Delta T}{h} x + C_1 \tag{9.3.3}$$

and for the deflection of the beam free end:

$$z(x) = \frac{(\alpha_1 - \alpha_2)\Delta T}{2h} x^2 + C_1 x + C_2 \tag{9.3.4}$$

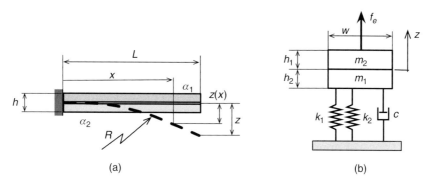

Figure 9.3.2 Bimetallic strip thermometer: (a) schematic; (b) lumped-parameter model

At the clamped end ($x = 0$) both the slope and deflection are zero; thus, both integration constants $C_1 = C_2 = 0$ are derived.

Maximum deflection is at the free end of the cantilever beam ($x = L$), and it is obtained from Eq. (9.3.4) as

$$z_{max} = z(L) = \frac{(\alpha_1 - \alpha_2)L^2}{2h} \Delta T \tag{9.3.5}$$

Equation (9.3.5) constitutes a static mathematical model of a bimetallic strip thermometer. It shows that the measured maximum deflection is proportional to the temperature change, where the constant coefficient $\frac{(\alpha_1 - \alpha_2)L^2}{2h}$ depends on the geometry and materials of the thermometer.

A bimetallic strip thermometer is a dynamic system. When its cantilever beam is subject to a temperature change, it starts bending and continues to do so until settling into a steady state. The steady-state value of the free-end deflection is the parameter z_{max}, derived above. To analyze the dynamic system and ascertain its convergence to a steady state for a constant value of ΔT, when a reading is taken, we create a lumped-parameter model as shown in Figure 9.3.2(b). Two metal strips are represented by two collocated masses m_1 and m_2, supported by springs in a parallel configuration (these springs represent elastic properties of the strips). The effect of temperature change is represented by the action of a virtual force f_e; applied to the masses, it causes the same displacement as $z(L)$. Energy losses in this lumped-parameter model are represented by a viscous damper with the damping coefficient c.

Considering that each strip is a thin parallelepiped with length L, width w, height h_1 and h_2, respectively, and density ρ_1 and ρ_2, respectively, the equivalent mass and stiffness are computed according to the tables in Chapter 3 and the literature on the strength of materials. The moments of inertia of each strip compiling the cantilever beam are $I_1 = \frac{wh_1^3}{12}$ and $I_2 = \frac{wh_2^3}{12}$, respectively.

$$m = m_1 + m_2 = \frac{33}{140} wL(\rho_1 h_1 + \rho_2 h_2)$$

$$k = k_1 + k_2 = \frac{3}{L^3}(E_1 I_1 + E_2 I_2) = \frac{w}{4L^3}\left(E_1 h_1^3 + E_2 h_2^3\right) \quad (9.3.6)$$

where E_1 and E_2 indicate the elasticity modulus (Young's modulus) for each strip.

At steady state (equilibrium), the assumed virtual force f_e equals the spring force, which is computed using Eqs. (9.3.5) and (9.3.6) as follows:

$$f_e = kz(L) = \frac{w}{4L^3}\left(E_1 h_1^3 + E_2 h_2^3\right) \frac{(\alpha_1 - \alpha_2)L^2}{2h} \Delta T = \frac{w(\alpha_1 - \alpha_2)\left(E_1 h_1^3 + E_2 h_2^3\right)}{8L(h_1 + h_2)} \Delta T \quad (9.3.7)$$

The thermometer coefficient,

$$K_T = \frac{w(\alpha_1 - \alpha_2)\left(E_1 h_1^3 + E_2 h_2^3\right)}{8L(h_1 + h_2)}$$

is a constant that depends on the design of the given thermometer (its geometry and materials).

When the temperature change is a function of time, the displacement of the free end of the cantilever bimetallic beam is also a function of time, and, consequently, the virtual force becomes $f_e = kz(t) = K_T \Delta T(t)$. Using Newton's second law to derive the governing equation for this dynamic system (see Chapter 3 for detailed discussion on such derivations) we obtain

$$m\ddot{z} = f_e - c\dot{z} - kz \Rightarrow m\ddot{z} + c\dot{z} + kz = K_T \Delta T(t) \quad (9.3.8)$$

Since this is a linear time-invariant second-order system, a transfer function model can be used for generating a response. Taking the Laplace transform of Eq. (9.3.8) the following transfer function is derived, considering that the temperature change $\Delta T(t)$ is the system input, and the deflection of the free end of the cantilever bimetallic beam $z(t)$ is the output:

$$\frac{Z(s)}{\Delta T(s)} = \frac{K_T}{ms^2 + cs + k} \quad (9.3.9)$$

The poles of this system are

$$p_{1,2} = -\frac{c}{2m} \pm \frac{1}{2m}\sqrt{c^2 - 4km}$$

They may be real or complex conjugates, depending on the values of k, c, and m, but since the system is built in such a way that $c^2 - 4km \neq 0$, in all cases the real parts of these poles are strictly negative. That means that this system is stable (see Chapter 7 for a discussion on stability). Therefore, the final-value theorem can be used to find the expected steady-state value, assuming that this system is subject to a step input $\Delta T(s) = T/s$:

$$z_{ss} = \lim_{s \to 0}\left(sZ(s)\right) = \lim_{s \to 0}\left(s\frac{K_T}{ms^2 + cs + k} \cdot \frac{T}{s}\right) = \frac{TK_T}{k} \qquad (9.3.10)$$

9.3.2 Numerical Simulations: Comparison of Two Bimetallic Strip Thermometers

Given the numerical values of the system parameters, let us compare two thermometers with bimetallic strips of the same size but different material pairs – copper–steel and aluminum–tungsten. Both thermometers are subject to a temperature increase $\Delta T = 20\ °C$.

Copper–steel thermometer:

Strip 1 (copper): $\alpha_1 = 17.8 \times 10^{-6}\ \mathrm{deg}^{-1}$, $E_1 = 117 \times 10^9\ \mathrm{N/m^2}$, $\rho_1 = 8300\ \mathrm{kg/m^3}$,
Strip 2 (stainless steel): $\alpha_2 = 10.8 \times 10^{-6}\ \mathrm{deg}^{-1}$, $E_2 = 180 \times 10^9\ \mathrm{N/m^2}$, $\rho_2 = 7480\ \mathrm{kg/m^3}$

Aluminum–tungsten thermometer:

Strip 1 (aluminum): $\alpha_1 = 21.3 \times 10^{-6}\ \mathrm{deg}^{-1}$, $E_1 = 69 \times 10^9\ \mathrm{N/m^2}$, $\rho_1 = 2700\ \mathrm{kg/m^3}$,
Strip 2 (tungsten): $\alpha_2 = 4.5 \times 10^{-6}\ \mathrm{deg}^{-1}$, $E_2 = 400 \times 10^9\ \mathrm{N/m^2}$, $\rho_2 = 19300\ \mathrm{kg/m^3}$

The common geometry and viscous damping for both thermometers is as follows: $w = 50 \times 10^{-6}\ \mathrm{m}$, $L = 350 \times 10^{-6}\ \mathrm{m}$, $c = 2.3 \times 10^{-7}\ \mathrm{N \cdot sec/m}$, $h_1 = 100 \times 10^{-9}\ \mathrm{m}$, $h_2 = 1 \times 10^{-6}\ \mathrm{m}$

The derived mathematical model is implemented in Mathematica using the functions `TransferFunctionModel` and `OutputResponse` (see Figure 9.3.4 for the code); the generated dynamic response of both thermometers is shown in Figure 9.3.3.

As seen in Figure 9.3.3, the copper–steel thermometer produces a reading (settles into a steady state) much faster than the aluminum–tungsten one. Additionally, these two thermometers need to have different scales due to the magnitude of their respective steady-state readings.

Computing the model parameters (damping ratio and natural frequency), as detailed in Chapter 7, allows us to find the settling time (Eq. (7.1.70)) and the percentage overshoot (Eq. (7.1.65)) for both thermometers analytically. These results are presented in Table 9.3.1.

Table 9.3.1 Numerical results for the comparison of the two thermometers

	ζ	ω_n, rad/s	t_s, s	M_P, %	$z(L)$, m
Copper–steel	0.086	39139	0.0012	76.3	7.8×10^{-6}
Aluminum–tungsten	0.037	38011	0.0028	88.9	18.7×10^{-6}

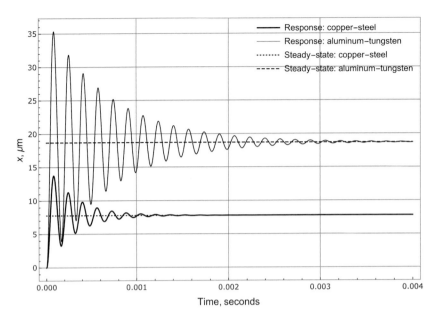

Figure 9.3.3 Dynamic response of two bimetallic strip thermometers: copper-steel and aluminum-tungsten

While different scales and a shorter settling time do not necessarily make one bimetallic strip thermometer better than the other (the choice of device depends on the intended application), the greater overshoot may render the aluminum–tungsten thermometer less desirable. A greater overshoot coupled with the larger steady-state value means that, while the bimetallic strip is bending, the pointer will experience significant rotation (see Figure 9.3.1), which will strain the device and may cause damage.

9.4 Modeling and Analysis of an Electro-Thermo-Mechanical System (a Resistive-Heating Element)

Resistive-heating elements are used in many devices and industrial processes. An electric fuse, which melts due to Joule effect and breaks the circuit when the flowing current exceeds the designated threshold value, is a commonly used safety device. Electronic cigarettes vaporize propylene glycol and vegetable glycerin by utilizing the heating filaments. Common appliances and tools such as electric stoves, electric space heaters, cartridge heaters, hair dryers, and soldering irons, just to name a few, contain resistive-heating elements to fulfill their intended functions. The simulation and analysis of the problem presented in this section gives the reader an insight into the operation of such devices.

9.4 Modeling and Analysis of an Electro-Thermo-mechanical System

```
(* set up the expressions *)
kt = (w*(e1*h1^3 + e2*h2^3)*(a1 - a2))/(8*l*(h1 + h2));
m = 33/140 *w*l*(rho1*h1 + rho2*h2);
k = w/(4*l^3) * (e1*h1^3 + e2*h2^3);
eqn = kt/(m*s^2 + c*s + k);
(* numerical values for copper-steel thermometer *)
rule1 = {w → 50*10^-6, l → 350*10^-6, h1 → 100*10^-9, h2 → 1*10^-6, e1 → 117*10^9,
    e2 → 180*10^9, a1 → 17.8*10^-6, a2 → 10.8*10^-6, rho1 → 8300, rho2 → 7480};
(* numerical values for aluminum-tungsten thermometer *)
rule2 = {w → 50*10^-6, l → 350*10^-6, h1 → 100*10^-9, h2 → 1*10^-6, e1 → 69*10^9, e2 → 400*10^9,
    a1 → 21.3*10^-6, a2 → 4.5*10^-6, rho1 → 2700, rho2 → 19 300};
(* input *)
input = 20*UnitStep[t];
(* damping *)
c = 2.3*10^-7;
(* derive transfer functions *)
tfm1 = TransferFunctionModel[eqn /. rule1, s];

tfm2 = TransferFunctionModel[eqn /. rule2, s];

(* responses *)
resp1 = OutputResponse[tfm1, input, {t, 0, 0.004}];
resp2 = OutputResponse[tfm2, input, {t, 0, 0.004}];
Plot[{10^6*resp1, 10^6*resp2, (10^6*20*kt)/k /. rule1, (10^6*20*kt)/k /. rule2}, {t, 0, 0.004},
 PlotRange → All,
 PlotStyle → {{Black, Thick}, {Black, Thickness[0.002]}, {Black, Dotted}, {Black, Dashed}},
 GridLines → Automatic, GridLinesStyle → Gray, Frame → True, FrameStyle → Gray,
 FrameTicks → {{Automatic, None}, {Automatic, None}},
 FrameTicksStyle → Directive["Label", Black, 12], ImageSize → 400, AspectRatio → 1/1.5,
 FrameLabel → {Style["Time, seconds", Black, 16], Style["x, μm", Black, 16]},
 PlotLegends →
  Placed[{Style["Response: copper-steel", 14], Style["Response: aluminum-tungsten", 14],
    Style["Steady-state: copper-steel", 14], Style["Steady-state: aluminum-tungsten", 14]},
   {Right, Top}]]
```

Figure 9.3.4 Generating dynamic system response: Mathematica code

9.4.1 System Description

A stainless-steel heating element is represented by a deformable solid cylinder of radius r and length L_0; it is treated as a resistor with initial resistance R_0. The element is heated by means of an external electrical circuit as shown in Figure 9.4.1. The applied voltage v generates heat due to the Joule effect and, subsequently, elongation of the heating element.

The heating element is immersed in a fluid, which is at constant temperature T_∞; heat loss occurs through convection with the convective coefficient h. The surface temperature of the heating element is T_s.

The known properties of the heating element are: mass m, mass density ρ, specific heat c, electrical resistivity ρ_{el}, thermal coefficient of resistance α_R, and linear coefficient of thermal expansion α.

The following assumptions can be made to simplify the derivations without losing the expressiveness of the model:

(a) The change of the cross-sectional area and dimensions with the temperature is negligible.
(b) The change of the mass density with the temperature is negligible.
(c) The electrical resistivity is constant, and the electrical resistance is linearly increasing with the change of temperature.
(d) The electrical-to-mechanical energy conversion is instantaneous.
(e) One end of the heating element is fixed; elongation is axial and occurs only at the free end.

9.4.2 Dynamic Modeling

As described in Chapter 6, for every combined system the initial step in modeling involves the derivation of the governing equations for every subsystem.

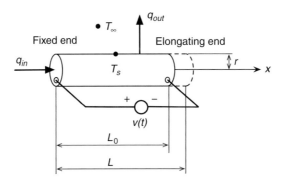

Figure 9.4.1 Heating element model

9.4 Modeling and Analysis of an Electro-Thermo-mechanical System

Consider the thermal subsystem first. The governing equation for such a system is derived in Chapter 5 as follows (Eq. (5.2.3)):

$$C\frac{dT}{dt} = q_{in} - q_{out} \qquad (9.4.1)$$

where C is the thermal capacitance of the heating element, T describes the temperature of the heating element, and $q_{in} - q_{out}$ is the net heat flow. Since the specific heat of the given heating element is not a function of temperature, and the mass is assumed to be constant, the thermal capacitance is defined using Eq. (5.2.2) as

$$C = mc = \rho V c = \rho(\pi r^2 L_0)\, c \qquad (9.4.2)$$

where m is mass of the heating element, c is the specific heat, ρ is the mass density, V is the volume, r is the radius, and L_0 is the initial length.

The output heat flow represents heat loss through convection, as derived in Eq. (5.1.3). The expression for q_{out} must also consider two important factors: (a) the surface area involved in this heat transfer is the lateral area of the heating element; (b) the elongation of the heating element is one-dimensional along the x-axis:

$$q_{out} = h(2\pi r L)(T_s - T_\infty) = 2\pi r h L_0\left(1 + \alpha(T_s - T_\infty)\right)(T_s - T_\infty) \qquad (9.4.3)$$

where h is the convective coefficient, α is the linear coefficient of thermal expansion, T_s is the surface temperature of the heating element, and T_∞ is the temperature of the fluid at some distance away from the surface.

The input heat flow occurs because of the Joule effect – the electrical energy dissipated by the resistor that models the given heating element is transformed into heat. Assuming lossless wires, the coupling equation describing the relationship between the thermal and electrical subsystems is derived as follows:

$$q_{in} = P \qquad (9.4.4)$$

where q_{in} is the input heat flow rate, and P is the electrical power absorbed by the heating element and converted into heat.

Since the given heating element is modeled as a linear resistor, the electrical power dissipated by it as heat is described with Eq. (4.1.11). Then the input heat flow rate becomes

$$q_{in} = P = \frac{v^2}{R} \qquad (9.4.5)$$

where v is the applied voltage, and R is the resistance of the heating element at the temperature T_s.

Considering that the electrical resistance of the heating element linearly increases with the change of temperature, this resistance at the temperature T_s can be expressed in terms of the initial resistance as follows:

$$R = R_0 + \alpha_R R_0 (T_s - T_\infty) = R_0 \left(1 + \alpha_R (T_s - T_\infty)\right) \tag{9.4.6}$$

The initial resistance is computed with Eq. (4.1.10) as follows:

$$R_0 = \frac{\rho_{el} L_0}{\pi r^2} \tag{9.4.7}$$

Substituting Eqs. (9.4.6) and (9.4.7) into Eq. (9.4.5) the expression for the input heat flow is obtained:

$$q_{in} = v^2 \frac{\pi r^2}{\rho_{el} L_0 \left(1 + \alpha_R (T_s - T_\infty)\right)} \tag{9.4.8}$$

Considering that the fluid temperature T_∞ is constant, the variable temperature T that describes surface temperature of the heating element can be written as $T = T_s - T_\infty$.

Then, substituting Eqs. (9.4.2), (9.4.3), and (9.4.8) into Eq. (9.4.1), we obtain the first governing equation for this system:

$$\pi r^2 L_0 \rho c \dot{T} = v^2 \frac{\pi r^2}{\rho_{el} L_0 (1 + \alpha_R T)} - 2\pi r h L_0 (1 + \alpha T) T$$

which, upon simplifying, becomes

$$r L_0^2 \rho \rho_{el} c \, \dot{T} + r L_0^2 \rho \rho_{el} c \, \alpha_R \dot{T} \, T + 2 h L_0^2 \rho_{el} \alpha_R \alpha T^3 +$$
$$+ 2 h L_0^2 \rho_{el} (\alpha_R + \alpha) T^2 + 2 h L_0^2 \rho_{el} T = r v^2 \tag{9.4.9}$$

The second governing equation describes the axial elongation of the heating element due to the temperature change:

$$x(t) = \alpha L_0 T(t) \tag{9.4.10}$$

These two equations – Eq. (9.4.9) and Eq. (9.4.10) – constitute a mathematical model of the given dynamic system. They are clearly coupled: when the nonlinear Eq. (9.4.9), which describes the electro-thermal component, is solved for $T(t)$, the mechanical component – the deformation of the heating element – can be computed.

9.4.3 Time Response of the Heating Element: Temperature Change and Elongation

The analytical solution of a nonlinear governing equation (Eq. (9.4.9)) does not exist. Therefore, numerical simulation needs to be performed. The following system parameters are known:

Geometry and density:

$$L_0 = 0.1 \text{ m}, \; r = 2 \text{ mm} = 2 \times 10^{-3} \text{ m}, \; \rho = 7930 \text{ kg/m}^3$$

Thermal properties:

$$c = 490 \text{ J/kg} \cdot \text{deg}, \; h = 32 \text{ J/m}^2 \cdot \text{sec} \cdot \text{deg},$$
$$\alpha = 12 \times 10^{-6} \text{ deg}^{-1}, \; \alpha_R = 0.001 \text{ deg}^{-1}$$

Electrical properties:

$$\rho_{el} = 6.9 \times 10^{-7} \; \Omega \cdot \text{m}, \; v = 0.015 \text{ V}$$

To simplify the computations, divide the first governing equation (Eq. (9.4.9)) by the coefficient $rL_0^2\rho\rho_{el}c$ and rewrite it in the following form:

$$\dot{T} + \alpha_R \dot{T}T + \frac{2h\alpha_R\alpha}{r\rho c}T^3 + \frac{2h(\alpha_R + \alpha)}{r\rho c}T^2 + \frac{2h}{r\rho c}T = \frac{1}{L_0^2\rho\rho_{el}c}v^2$$

Note that since $T = T_s - T_\infty$, when voltage is applied and heating starts, $T(0) = 0$ (the temperature of the heating element is the same as the temperature of the surrounding fluid).

The numerical solution is found using either the Simulink or Mathematica function `NDSolve` or its variant `NDSolveValue`. The code for deriving the solution and generating the graphs of temperature difference $T = T_s - T_\infty$ and elongation of the heating element is presented in Figure 9.4.2.

The generated graphs are shown in Figure 9.4.3.

Figure 9.4.3(a) shows that within approximately 800 seconds the temperature difference between the heating element and the surrounding fluid reaches a steady state: a constant value of approximately 1.02 degrees. The deformation of a stainless-steel heating element also exhibits a steady-state response of approximately 1.22 μm. While a small deformation is desirable, the achieved steady-state temperature difference is clearly insufficient for meaningful heating. In order to improve the performance of the heating element and increase the generated temperature difference, the net heat flow needs to be increased. Decreasing the convective heat loss is not possible since it depends on the parameters of the fluid, into which the heating element is immersed: this fluid is specified in the technical requirements and cannot be changed. Then, designing a more efficient heating element involves increasing the input heat flow rate q_{in} by either requiring a higher input voltage or changing the material of the heating element to one with smaller electrical resistivity ρ_{el}.

Changing the geometry of the heating element does not provide the desired temperature difference. Increasing the length decreases the magnitude of the input, thus decreasing the steady-state response. While decreasing the length to a half of the original value (setting

```
In[*]:= (* setting the numerical values *)
       lo = 0.1; r = 2 * 10^-3; rho = 7930;
       c = 490; h = 32; a = 12 * 10^-6; ar = 0.001;
       rhoel = 6.9 * 10^-7; v = 0.015;

       (* computing the a_i coefficients *)
       a1 = (2 * h * ar * a) / (r * rho * c);
       a2 = (2 * h * (ar + a)) / (r * rho * c);
       a3 = (2 * h) / (r * rho * c);
       a4 = 1 / (lo^2 * rho * rhoel * c);

       (* solving the nonlinear governing equation *)
       sol =
       NDSolveValue[
         {T'[t] + ar * T'[t] * T[t] + a1 * T[t]^3 + a2 * T[t]^2 + a3 * T[t] == a4 * v^2,
          T[0] == 0}, T, {t, 0, 1000}]

In[*]:= (* plotting temperature vs. time *)
       Plot[sol[t], {t, 0, 1000}, AxesOrigin → {0, 0},
         PlotStyle → {Black, Thick}, GridLines → Automatic,
         GridLinesStyle → Gray, Frame → True, FrameStyle → Gray,
         PlotRange → All, FrameTicks → {{Automatic, None}, {Automatic, None}},
         FrameTicksStyle → Directive["Label", Black, 12], ImageSize → 400,
         AspectRatio → 1 / 1.5,
         FrameLabel → {Style["Time (seconds)", Black, 16],
           Style["Temperature (degrees)", Black, 16]}]

In[*]:= (* plotting deformation vs. time *)
       Plot[10^5 * a * lo * sol[t], {t, 0, 1000}, AxesOrigin → {0, 0},
         PlotStyle → {Black, Thick}, GridLines → Automatic,
         GridLinesStyle → Gray, Frame → True, FrameStyle → Gray,
         PlotRange → All, FrameTicks → {{Automatic, None}, {Automatic, None}},
         FrameTicksStyle → Directive["Label", Black, 12], ImageSize → 400,
         AspectRatio → 1 / 1.5,
         FrameLabel → {Style["Time (seconds)", Black, 16],
           Style["Elongation (meters*10^-5)", Black, 16]}]
```

Figure 9.4.2 Generating the graphs of the temperature and elongation of the heating element: Mathematica code

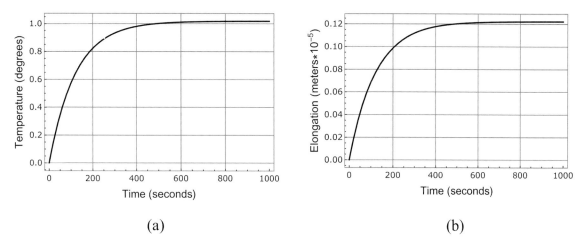

Figure 9.4.3 Dynamic response of the heating element: (a) temperature; (b) elongation

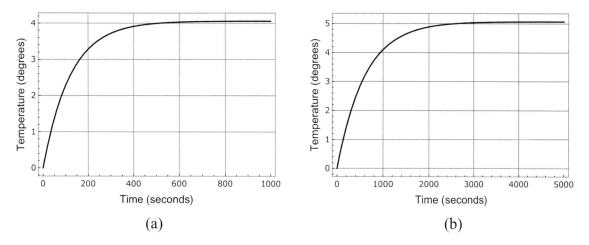

Figure 9.4.4 Dynamic response of the heating element with modified geometry: (a) $L_0 = 0.05$ m; (b) $r = 10$ mm

$L_0 = 0.05$ m) helps to increase the input heat flow as seen in Figure 9.4.4(a), this increase is insufficient, and making the heating element even shorter is impractical.

Increasing the radius makes the settling into a steady state significantly slower, while providing a marginal increase in the temperature difference; see Figure 9.4.4(b).

Increasing the input voltage to 0.075 V generates a temperature difference of approximately 25.8 degrees and a deformation of 29.8 μm, as shown in Figure 9.4.5.

In many practical applications a specific temperature difference is required to be produced by the heating element. For example, assume that the required temperature difference for the given heating element is at least 80 °C. Numerical simulations show that this is achieved by

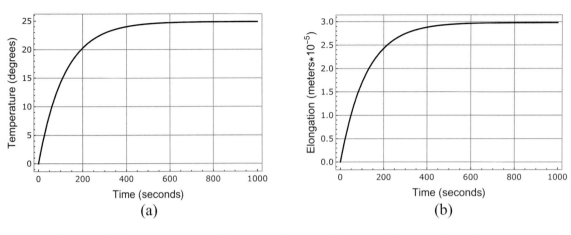

Figure 9.4.5 Dynamic response of the heating element with increased input voltage: (a) temperature, (b) elongation

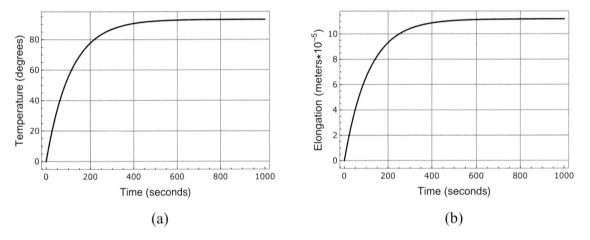

Figure 9.4.6 Dynamic response of the heating element to an input of 0.15 V: (a) temperature, (b) elongation

applying an input voltage of approximately 0.15 V (see Figure 9.4.6(a)), which is ten times larger than the originally intended input. The elongation of the heating element in response to this input is still small – only 0.1 mm (see Figure 9.4.6(b)).

If a higher input voltage is unavailable, a voltage-isolation amplifier could be added to the electrical subsystem to increase the given voltage as required (see Section 4.7.1 for detailed information). Nonetheless, a higher voltage in the electrical circuit of the heating element system may be undesirable for safety reasons. In this case, changing the material of a heating element becomes necessary. If the heating element is made of copper instead of stainless steel,

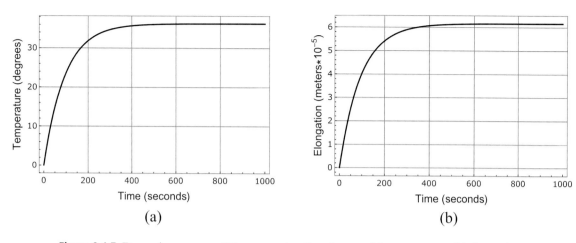

Figure 9.4.7 Dynamic response of the copper heating element: (a) temperature; (b) elongation

a performance improvement is achieved with lower input voltage, as shown in the following simulations.

The copper parameters are:

$$\rho = 8900 \text{ kg/m}^3, \; c = 385 \text{ J/kg} \cdot \text{deg},$$
$$\alpha = 17 \times 10^{-6} \text{deg}^{-1}, \; \alpha_R = 0.0039 \text{ deg}^{-1}, \; \rho_{el} = 17 \times 10^{-9} \; \Omega \cdot \text{m}$$

Note that copper has an approximately 100 times smaller electrical resistivity than stainless steel, with the other parameters comparable in order of magnitude.

By using copper, the temperature difference for the same size of heating element and the same input voltage (0.015 V) increases by more than 30 times (from 1.02 to 36.2 degrees), as evidenced in Figure 9.4.7(a). Deformation also increases significantly, but it is still in microns (approximately 61.6 μm) as seen in Figure 9.4.7(b).

Alas, the input voltage still needs to be increased to reach the objective of at least an 80 °C temperature difference, but that increase does not have to be as drastic as it was in the case of a stainless-steel heating element. Doubling the input voltage (to 0.03 V) yields a temperature difference of 114.23 degrees and elongation of 0.19 mm. as seen in Figure 9.4.8.

9.5 Feedback Control of a Liquid-Level System for Water Purification

Liquid-level systems, which are assemblies of tanks, pipes, valves, pumps, and other components, have extensive applications in various industrial processes. A detailed discussion on these systems is presented in Chapter 5. An example of a two-tank liquid-level system,

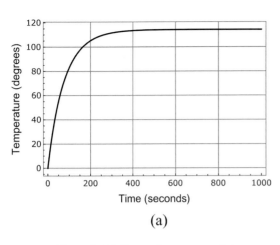
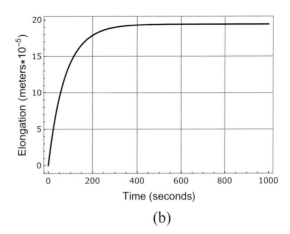

Figure 9.4.8 Dynamic response of the copper heating element to an input of 0.03 V: (a) temperature; (b) elongation

commonly used for a small-scale water purification (two-stage biochemical process), is presented in this section.

9.5.1 System Description

The system, shown in Figure 9.5.1(a), consists of two cylindrical tanks, the circular bottom areas of which are A_1 and A_2, respectively. Raw water is pumped into the first tank for pretreatment at a flow rate of $q_{in}(t)$. A mix of aerobic microorganisms that can digest organic matter in water resides in both tanks, producing an effluent that is relatively free from suspended micro-organics. Treated water moves to the second tank through a large orifice with the cross-sectional area a_1. In the second tank, the water undergoes additional treatment with a different mix of aerobic bacteria, leaving through an orifice with the cross-sectional area a_2 and flowing into a clarifying vessel where the agglomerated bacterial particles are removed. The orifice in each tank is a slit in the cylinder wall, as shown in Figure 9.5.1(b), defined by the height d, and the angle φ of the circular sector generating the upper and lower boundaries of the slit.

During the operation of the system, the liquid level in tank 1 is $h_1(t)$, and in tank 2 it is $h_2(t)$. The discharge coefficient C_d is the same for both tanks.

The liquid level in tank 2 is crucial to the process efficiency and quality of the effluent due to the nature of the biochemical reactions occurring in this tank. A linear controller needs to be designed, with a piecewise-continuous reference value r defined in accordance to the specifics of the process.

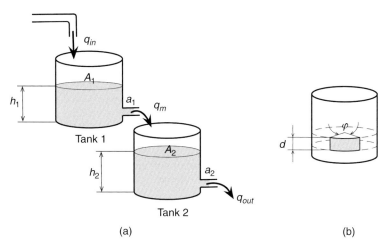

Figure 9.5.1 Liquid-level system: (a) description, (b) orifice

9.5.2 Dynamic Modeling

A mathematical model of the liquid-level system needs to be derived and analyzed prior to the controller design. As derived in Chapter 5 (Eq. (5.5.12)), the governing equations for each tank are as follows:

$$\begin{cases} A_1 \dfrac{dh_1}{dt} = q_{in}(t) - q_m \\ A_2 \dfrac{dh_2}{dt} = q_m - q_{out} \end{cases} \quad (9.5.1)$$

where $q_{in}(t)$ is the inflow of raw water into tank 1, q_m is the outflow from tank 1 through the orifice a_1, h_1 and h_2 are the liquid levels in tank 1 and tank 2, respectively, and q_{out} is the outflow from tank 2 through the orifice a_2.

According to Torricelli's principle. the outflow q_{max} through the orifice of area a cannot exceed $a\rho v_{max}$, where ρ is the density of the fluid, and v_{max} is its maximum speed. This maximum speed can be derived from the law of the conservation of energy as $v_{max} = \sqrt{2gh}$, where h is the height of the fluid above the orifice, and g is the gravitational acceleration. Hence, the outflow through the orifice cannot exceed the following value: $q_{max} = a\rho\sqrt{2gh}$. The actual outflow will be less than this because of frictional effects and is derived as

$$q_{actual} = C_d a \sqrt{2gh} \quad (9.5.2)$$

where C_d is the discharge coefficient.

Applying the derived expression for the outflow through an orifice (Eq. (9.5.2)) to the governing equations (Eq. (9.5.1)) we obtain

$$\begin{cases} A_1 \dfrac{dh_1}{dt} = q_{in}(t) - C_d a_1 \sqrt{2gh_1} \\ A_2 \dfrac{dh_2}{dt} = C_d a_1 \sqrt{2gh_1} - C_d a_2 \sqrt{2gh_2} \end{cases} \quad (9.5.3)$$

The form of these first-order ordinary differential equations strongly resembles the form of state equations. Defining h_1 and h_2 as state variables, we obtain the state-space model of this system, considering that the output is the liquid level in tank 2:

$$\begin{cases} \dfrac{dh_1}{dt} = \dfrac{1}{A_1} q_{in}(t) - \dfrac{C_d a_1}{A_1} \sqrt{2gh_1} \\ \dfrac{dh_2}{dt} = \dfrac{C_d a_1}{A_2} \sqrt{2gh_1} - \dfrac{C_d a_2}{A_2} \sqrt{2gh_2} \end{cases}$$

$$y = h_2 \quad (9.5.4)$$

For the stability of the biochemical reactions, the volumes of liquid in both tanks must stay constant at all times, which means that the liquid levels must stay constant. Thus, a steady-state response of the system is desired.

The tanks need to be filled up to a predefined liquid level for the bacterial mix to be added and reactions to start. This moment in time is considered the starting point $t = 0$, and liquid levels $h_1(0)$ and $h_2(0)$ are the initial conditions which need to be incorporated into the derived state-space model.

This nonlinear state-space model does not have an analytical solution; hence, it needs to be solved via numerical integration using Runge–Kutta methods as described earlier in Section 2.8. These methods are implemented in computational software packages such as MATLAB (function `ode45`) and Mathematica (function `NDSolve` and its variants). It is also possible to use Mathematica functions unique to state-space system representations such as `NonlinearSystemModel`, `StateResponse`, and `OutputResponse` (note: the function `NonlinearSystemModel` is only available for Mathematica version 12.0 or later).

A numerical simulation is run to determine whether this system reaches a steady state in response to a constant (step) input, i.e., constant inflow of the raw water. These steady-state values can be estimated analytically prior to the simulation in the following manner. Assuming a step input $q_{in} = qu(t)$, consider that, at steady state, $\dfrac{dh_1}{dt} = 0$ and $\dfrac{dh_2}{dt} = 0$. Then, the system of two algebraic equations:

$$\begin{cases} 0 = \dfrac{1}{A_1} q - \dfrac{C_d a_1}{A_1} \sqrt{2gh_1} \\ 0 = \dfrac{C_d a_1}{A_2} \sqrt{2gh_1} - \dfrac{C_d a_2}{A_2} \sqrt{2gh_2} \end{cases} \quad (9.5.5)$$

can be solved for the expected steady-state liquid levels:

$$\begin{cases} h_1 = \dfrac{q^2}{2C_d^2 a_1^2 g} \\ h_2 = \dfrac{q^2}{2C_d^2 a_2^2 g} \end{cases} \quad (9.5.6)$$

The following system parameters are given:

$q_{in} = 0.1u(t)$ m^3/sec, $C_d = 0.62$, $g = 9.8$ m/s^2,
Tank 1 : $\quad r_1 = 0.4$ m, $\quad d = 3$ cm $= 0.03$ m, $\quad \varphi = 60°$, $h_1(0) = 1.0$ m,
Tank 2 : $\quad r_2 = 0.42$ m, $\quad d = 3.5$ cm $= 0.035$ m, $\quad \varphi = 50°$, $h_2(0) = 0.5$ m.

The bottom area and orifice area for each tank are computed as follows:

$$A = \pi r^2 \text{ and } a_{orifice} = \frac{2\pi r}{360/\varphi} d \quad (9.5.7)$$

The Mathematica code for deriving the solution of state equations by numerical integration and generating the graphs of liquid levels vs. time for both tanks is shown in Figure 9.5.2. Note that the state-space model is set using the function NonlinearSystemModel, and the solutions are generated with StateResponse, since both $h_1(t)$ and $h_2(t)$ variables are of interest.

The output of the generated nonlinear state-space model (see Figure 9.5.3) can be verified vs. the derived by-hand result – note the incorporation of the initial conditions.

The generated liquid-level graphs are shown in Figure 9.5.4.

Finding the numerical values for the steady-state response is straightforward with the Mathematica algebraic and transcendental equation solver Solve. It is also possible to approximate the settling time with the FindRoot function, considering a 0.5% allowed error and using the graphs to estimate this settling time at approximately 400 seconds for the first tank, and 600 seconds for the second tank. Code for these operations is shown in Figure 9.5.5.

The results are as follows:

Steady-state liquid level: 8.41 m (tank 1) and 8.1 m (tank 2)
Settling time: 415.3 s $\cong$ 6.92 min (tank 1) and 602.8 s $\cong$ 10.05 min (tank 2)

As the simulation demonstrates, this system reaches a steady state in response to a step input. The magnitude of the steady-state liquid levels depends on the input, as evidenced by Eq. (9.5.6), and shown further in Figure 9.5.6.

The magnitude of the steady-state response to a specific step input may or may not be optimal for the biochemical reactions occurring in the liquid. Additionally, environmental

716   9 Application Problems

```
In[ ]:= (* setting up the equations *)
eqsSSM = {h1'[t] == 1/A1 * q[t] - (c*a1)/A1 * √(2 g * h1[t]),
          h2'[t] == (c*a1)/A2 * √(2 g * h1[t]) - (c*a2)/A2 * √(2 g * h2[t])};
(* deriving the nonlinear state-space model *)
ssm = NonlinearStateSpaceModel[eqsSSM, {{h1[t], h1o}, {h2[t], h2o}}, {{q[t]}}, h2[t], t]

In[ ]:= (* setting up the numerics *)
pars = {c → 0.62, g → 9.8, A1 → π * 0.4^2, a1 → (2*Pi*0.4*0.03)/6, A2 → π * 0.42^2,
        a2 → (2*Pi*0.42*0.035)/7.2, h1o → 1, h2o → 0.5};
(* computing the state variables h₁ and h₂ *)
sr = StateResponse[ssm /. pars, 0.1*UnitStep[t], {t, 0, 1000}]

In[ ]:= (* generating the graph of the liquid levels for both tanks *)
Plot[sr, {t, 0, 1000}, AxesOrigin → {0, 0},
  PlotStyle → {{Black, Thick, Dashed}, {Black, Thick}}, GridLines → Automatic,
  GridLinesStyle → Gray, Frame → True, FrameStyle → Gray, PlotRange → All,
  FrameTicks → {{Automatic, None}, {Automatic, None}},
  FrameTicksStyle → Directive["Label", Black, 12], ImageSize → 400, AspectRatio → 1/1.5,
  FrameLabel → {Style["Time (seconds)", Black, 16],
    Style["Liquid Level (meters)", Black, 16]},
  Epilog → {Inset[Style["h₁(t)", 14], {60, 7}], Inset[Style["h₂(t)", 14], {300, 6.5}]}]
```

Figure 9.5.2 Generating the graphs of liquid levels for both tanks: Mathematica code

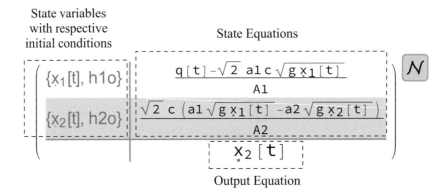

Figure 9.5.3 Nonlinear state-space model in general form

9.5 Feedback Control of a Liquid-Level System for Water Purification

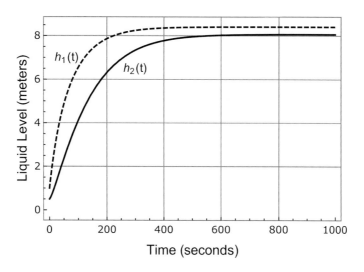

Figure 9.5.4 Dynamic response of the system: liquid levels in both tanks

```
(* estimating the steady-state values for h1 and h2 in general form *)
ssvals = Solve[eqsSSM /. {h1'[t] → 0, h2'[t] → 0}, {h1[t], h2[t]}] /. q[t]² → Q²

(* finding the numerical values for the steady-state values for h1 and h2 *)
nums = ssvals /. pars /. Q → 0.1

(* finding settling time considering 0.5% allowed error;
   from the graph estimate that settling time should be around 400 seconds
   for Tank 1, and around 600 seconds for Tank 2 *)
st1 = FindRoot[sr[[1]] == 0.995 * nums[[1, 1, 2]], {t, 400}]

st2 = FindRoot[sr[[2]] == 0.995 * nums[[1, 2, 2]], {t, 600}]

(* settling time in minutes *)
{st1[[1, 2]], st2[[1, 2]]} / 60
```

Figure 9.5.5 Estimating steady-state values and settling times for liquid levels in both tanks: Mathematica code

factors and the variability of the system parameters due to manufacturing imperfections, uncertainties in the system modeling, accumulation of microscopic bacterial slime around the orifices, and other factors make maintaining the optimal operating conditions impossible in the open-loop system. Hence, a feedback control system needs to be designed for

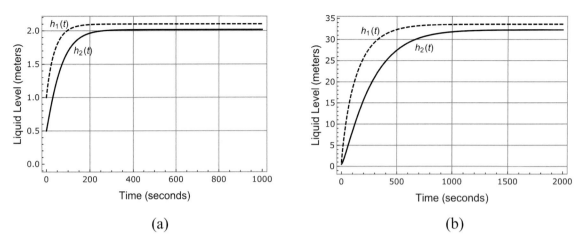

Figure 9.5.6 Dynamic response of the system to different inputs: (a) $q_{in} = 0.05u(t)$; (b) $q_{in} = 0.2u(t)$

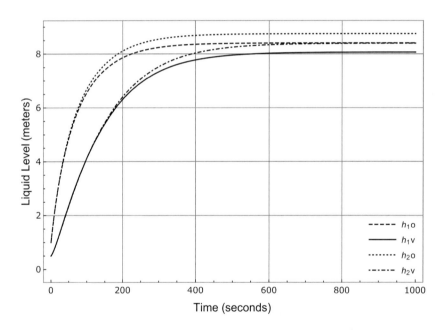

Figure 9.5.7 Influence of the system parameters variability on the dynamic response

providing a consistent quality of effluent regardless of the fluctuations of the raw water inflow, the variability of the operating requirements, and other disturbances.

The influence of the variability of the system parameters is clearly seen in Figure 9.5.7. A mere $\pm 1\%$ change in the values of system parameters causes a visible (approximately 4.3%) difference in the steady-state liquid levels: 0.35 m for tank 1, and 0.34 m for tank 2.

9.5 Feedback Control of a Liquid-Level System for Water Purification

The given system is nonlinear, represented by the nonlinear state-space model (Eq. (9.5.4)), the state equations of which can be written in a general form as follows:

$$\dot{h}(t) = \Phi(h(t), u(t)) \quad (9.5.8)$$

where $h(t) = \begin{bmatrix} h_1(t) \\ h_2(t) \end{bmatrix}$ is a state vector, Φ is a nonlinear function, and $u(t)$ is an input vector. The behavior of nonlinear equations is often complicated. The analysis of the time response of a nonlinear system and the design of an efficient control law for it require specific approaches, the presentation of which is typically reserved for graduate-level classes; hence, this is beyond the scope of this book.

If a nonlinear system has an *equilibrium point* – a point at which the system is in a steady state, then it can be reliably approximated with a linear system, and the analysis of the system behavior around such equilibrium point can be performed. This process is called *linearization*. We assume that for some input $u_s(t)$ and some initial state there exists a solution of the nonlinear system $h_s(t)$ and for input u_0 there exists an equilibrium point h_0. We let this input to be slightly perturbed to become $u_s(t) + \Delta u(t)$, and let the initial state to be also slightly perturbed. The solution of the system with this perturbation is expressed as $h_s(t) + \Delta h(t)$, where $\Delta h(t)$ is small in the vicinity of the equilibrium point for all values of t. Plugging this solution into the nonlinear state equation (Eq. (9.5.8)) and using a Taylor series expansion of the function Φ we derive the linearized state equation at the equilibrium point (h_0, u_0):

$$\frac{d}{dt}h_s(t) + \frac{d}{dt}\Delta h(t) = \Phi(h_s(t) + \Delta h(t), u_s(t) + \Delta u(t))\Big|_{(h_0, u_0)} = \Phi(h_0 + \Delta h, u_0 + \Delta u)$$

$$= \Phi(h_0, u_0) + \frac{\partial \Phi}{\partial h}\Big|_{(h_0, u_0)} \Delta h + \frac{\partial \Phi}{\partial u}\Big|_{(h_0, u_0)} \Delta u + \left(\frac{\partial^2 \Phi}{\partial h^2}\Big|_{(h_0, u_0)} \frac{\Delta h^2}{2!} + \frac{\partial^2 \Phi}{\partial u^2}\Big|_{(h_0, u_0)} \frac{\Delta u^2}{2!} + \cdots \right)$$

(9.5.9)

Considering that $\Phi(h_0, u_0)$ is the solution of the original unperturbed state equation, i.e. $\frac{dh_s(t)}{dt} = \Phi(h_0, u_0)$, these components are canceled in Eq. (9.5.9). The higher-order terms (starting from $\frac{\Delta h^2}{2!}$ and $\frac{\Delta u^2}{2!}$) can be neglected since these perturbations are assumed to be small.

The partial derivative terms $\frac{\partial \Phi}{\partial h}\Big|_{(h_0, u_0)}$ and $\frac{\partial \Phi}{\partial u}\Big|_{(h_0, u_0)}$ represent constant matrices, denoted A and B, respectively (they are called Jacobian matrices). Redefining $\Delta h \triangleq h$ and $\Delta u \triangleq u$, the linearized system is obtained:

$$\dot{h} = A \cdot h + B \cdot u \quad (9.5.10)$$

where

$$A = \frac{\partial \Phi}{\partial h}\Big|_{(h_0, u_0)}$$

and

$$B = \frac{\partial \Phi}{\partial u}\bigg|_{(h_0, u_0)}$$

For the given system these Jacobian matrices are:

$$A = \begin{bmatrix} \frac{\partial}{\partial h_1}\left(\frac{1}{A_1}q_{in}(t) - \frac{C_d a_1}{A_1}\sqrt{2gh_1}\right) & \frac{\partial}{\partial h_2}\left(\frac{1}{A_1}q_{in}(t) - \frac{C_d a_1}{A_1}\sqrt{2gh_1}\right) \\ \frac{\partial}{\partial h_1}\left(\frac{C_d a_1}{A_2}\sqrt{2gh_1} - \frac{C_d a_2}{A_2}\sqrt{2gh_2}\right) & \frac{\partial}{\partial h_2}\left(\frac{C_d a_1}{A_2}\sqrt{2gh_1} - \frac{C_d a_2}{A_2}\sqrt{2gh_2}\right) \end{bmatrix}\bigg|_{(h_{1e},\,h_{2e})}$$

$$= \begin{bmatrix} -\dfrac{C_d a_1 g}{A_1\sqrt{2gh_{1e}}} & 0 \\ \dfrac{C_d a_1 g}{A_2\sqrt{2gh_{1e}}} & -\dfrac{C_d a_2 g}{A_2\sqrt{2gh_{2e}}} \end{bmatrix} \tag{9.5.11a}$$

$$B = \begin{bmatrix} \dfrac{\partial}{\partial q_{in}}\left(\dfrac{1}{A_1}q_{in}(t) - \dfrac{C_d a_1}{A_1}\sqrt{2gh_1}\right) \\ \dfrac{\partial}{\partial q_{in}}\left(\dfrac{C_d a_1}{A_2}\sqrt{2gh_1} - \dfrac{C_d a_2}{A_2}\sqrt{2gh_2}\right) \end{bmatrix}\bigg|_{(h_{1e},\,h_{2e})} = \begin{bmatrix} \dfrac{1}{A_1} \\ 0 \end{bmatrix} \tag{9.5.11b}$$

Then the state-space model of the linearized system in expanded matrix form becomes

$$\begin{bmatrix} \dfrac{dh_1}{dt} \\ \dfrac{dh_2}{dt} \end{bmatrix} = \begin{bmatrix} -\dfrac{C_d a_1 g}{A_1\sqrt{2gh_{1e}}} & 0 \\ \dfrac{C_d a_1 g}{A_2\sqrt{2gh_{1e}}} & -\dfrac{C_d a_2 g}{A_2\sqrt{2gh_{2e}}} \end{bmatrix} \cdot \begin{bmatrix} h_1 \\ h_2 \end{bmatrix} + \begin{bmatrix} \dfrac{1}{A_1} \\ 0 \end{bmatrix} q_{in}(t) \tag{9.5.12}$$

where the equilibrium values h_{1e} and h_{2e} for the input q are given by Eq. (9.5.6).

Taking the Laplace transform of the linear equation (9.5.12) and recalling that the required output is $h_2(t)$ the transfer function of this system is derived. To simplify the derivations, denote

$$b_{11} = \frac{C_d a_1 g}{A_1\sqrt{2gh_{1e}}}, \quad b_{21} = \frac{C_d a_1 g}{A_2\sqrt{2gh_{1e}}}, \quad b_{22} = \frac{C_d a_2 g}{A_2\sqrt{2gh_{2e}}}$$

Then, the Laplace transform of the linearized state equations becomes

$$\begin{cases} sH_1(s) = \dfrac{1}{A_1}Q_{in}(s) - b_{11}H_1(s) \\ sH_2(s) = b_{21}H_1(s) - b_{22}H_2(s) \end{cases} \Rightarrow H_1(s) = \frac{s + b_{22}}{b_{21}}H_2(s) \Rightarrow (s + b_{11})\frac{s + b_{22}}{b_{21}}H_2(s) = \frac{1}{A_1}Q_{in}(s)$$

$$G(s) = \frac{H_2(s)}{Q_{in}(s)} = \frac{b_{21}}{A_1(s + b_{11})(s + b_{22})}$$

9.5 Feedback Control of a Liquid-Level System for Water Purification

Substituting the above values for the coefficients b_{11}, b_{21}, b_{22} and steady-state values h_{1e} and h_{2e}, the following transfer function is obtained:

$$G(s) = \frac{H_2(s)}{Q_{in}(s)} = \frac{C_d^2 a_1^2 g Q}{(C_d^2 a_1^2 g + A_1 Q s)(C_d^2 a_2^2 g + A_2 Q s)} \quad (9.5.13)$$

where Q is the value of the input (0.1 m³/s)) at which the steady-state values were computed.

This transfer function is used to design the proportional–integral (PI) control – $G_{PI}(s) = K_P + K_I \frac{1}{s}$ – using the Ziegler–Nichols method. Applying the proportional control K_P helps to alleviate the effects of parameter variations of the given system, as well as to reduce the output error caused by the disturbances arising from the environmental influence on the system. While improving the system robustness the proportional control is not sufficient to make the system response converge to the required value. Adding the integral control $K_I \frac{1}{s}$ improves the response by driving the steady-state error to zero. A detailed discussion of the advantages and design of feedback control systems is presented in Sections 8.2 and 8.3.

In Mathematica, the generation of the system transfer function and design of PI control are done using the functions `TransferFunctionModel`, which automatically linearizes the derived nonlinear state-space model, and `PIDTune`, which generates the K_P and K_I gains (see Section 8.3.2 for an explanation of Ziegler–Nichols gain tuning). This code is shown in Figure 9.5.8.

The derived control gains are as follows: $K_I = 7.95 \times 10^{-5}$, $K_P = 0.013$.

To visualize the performance of the controlled system, we use the operational parameter (reference input) defined by a simple linear function $h_{2\_reference} = 6$ m $\forall\, t > 0$. The closed-loop system model is generated with the Mathematica functions `SystemsModelSeriesConnect`

```
(* re-write the nonlinear state-space model incorporating the steady-state values as
   initial conditions *)
ssm4L = NonlinearStateSpaceModel[eqsSSM,
  {{h1[t], ssvals[[1, 1, 2]]}, {h2[t], ssvals[[1, 2, 2]]}}, {{q[t], Q}}, h2[t], t]

(* derive the transfer function of the linearized model in the general form
   to be used in controller design *)
tfm = TransferFunctionModel[ssm4L] // FullSimplify

(* numerical value for the transfer function of the linearized model *)
tfmV = tfm /. Append[Most[Most[pars]], Q → 0.1]

(* generate a PI controller using Ziegler-Nichols tuning rule *)
pid = PIDTune[tfmV]
```

Figure 9.5.8 Designing PI control using transfer functions of linearized system: Mathematica code

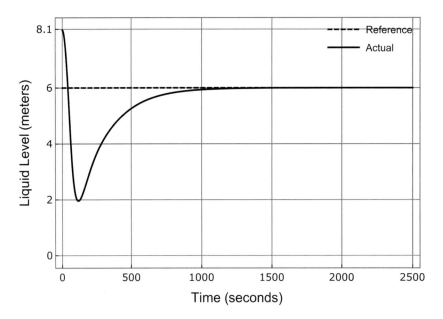

Figure 9.5.9 Dynamic response of the nonlinear system under PI control to a reference input of 6 m

and `SystemsModelFeedbackConnect`. The system response is then generated with the function `OutputResponse` and plotted together with the reference input (see Figure 9.5.9).

As this simulation demonstrates, PI control successfully makes the system follow the required operational parameter. If the settling time is unacceptably long, a differential component can be added to make the controlled system faster.

The designed PI control is capable of following the operational parameter defined by a piecewise-continuous function, such as

$$h_{2\_reference} = \begin{cases} 7.2 \text{ m} & 0 \leq t \leq 800 \text{ s} \\ 9.3 \text{ m} & 800 < t \leq 1200 \text{ s} \\ 8.1 \text{ m} & 1200 < t \leq 1800 \text{ s} \\ 5.9 \text{ m} & 1800 < t \leq 2500 \text{ s} \\ 3.5 \text{ m} & t > 2500 \text{ s} \end{cases}$$

In this case, convergence will be more difficult to achieve, but the controlled system will still try to follow the reference trajectory as closely as possible (see the dynamic response graph in Figure 9.5.10). The Mathematica code for this simulation is shown in Figure 9.5.11.

The validity of the original assumption that this system can be reliably approximated with a linearized model can be illustrated with a comparison of the performance of two systems, both controlled with the same PI control: nonlinear and linearized. As seen in Figure 9.5.12, for all but one segment of the reference trajectory both systems exhibit nearly identical behavior.

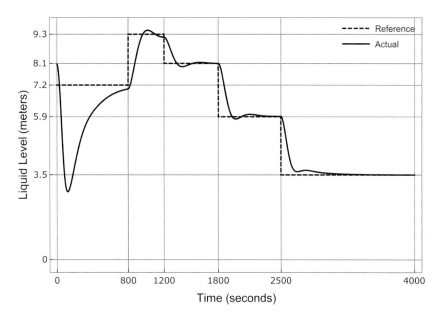

Figure 9.5.10 Dynamic response of the nonlinear system under PI control to a piecewise-continuous reference input

The code for this simulation is shown in Figure 9.5.13 (note the different options used in `PIDTune`).

9.6 Sensors, Electroacoustic, and Piezoelectric Devices

9.6.1 Sensors

Sensors are pervasive in nature and in all fields of human activity. The generic definition of a *sensor* as given by the *Merriam-Webster Dictionary* is "a device that responds to a physical stimulus (as heat, light, sound, pressure, magnetism, or a particular motion) and transmits a resulting impulse (as for measurement or operating a control)." Sensors that are most commonly encountered in engineering systems detect a change in a physical parameter such as, for example, acceleration, heat flux, or pressure, and transform it into an electrical signal. Sensors typically include a sensing element and a *transducer* – circuitry that performs energy conversion from one form to another, hence producing the signal suitable for processing.

Obviously, since sensor operation involves energy conversion, such a device is a compound dynamic system. An example of a sensor that is an electromechanical system is a galvanometer (see Section 6.3.2), where the change in electrical current is converted to mechanical energy (rotation of a pointer) that is subsequently measured.

```
(* set up the reference value for liquid levels in Tank 2 *)
       ⎡ 7.2  0 ≤ t ≤ 800
       ⎢ 9.3  800 < t ≤ 1200
ref = ⎨ 8.1  1200 < t ≤ 1800;
       ⎢ 5.9  1800 < t ≤ 2500
       ⎣ 3.5  True

(* re-write the derived nonlinear system model in the numeric form,
   setting the value of Q = 0.1 -- this is the value that generated the analyzed
   previously steady-state response *)
dtank = ssm4L /. Append[Most[Most[pars]], Q → 0.1]

(* create a transfer function of the controlled linearized system;
   linearization of the nonlinear state-space model and application of the controller
   is done automatically *)
controlledSystem = SystemsModelFeedbackConnect[SystemsModelSeriesConnect[pid, dtank]]

(* generating the response of the controller system to the piecewise-continuous
   reference input *)
response = OutputResponse[controlledSystem, ref, {t, 0, 4000}];

(* plotting controlled system response and the reference input *)
Plot[{ref, response}, {t, 0, 4000}, PlotRange → All,
 PlotLegends → Placed[{"Reference", "Actual"}, {Right, Top}], Exclusions → None,
 AxesOrigin → {0, 0}, PlotStyle → {{Black, Thick, Dashed}, {Black, Thick}},
 GridLines → {{800, 1200, 1800, 2500, 4000}, {3.5, 5.9, 7.2, 8.1, 9.3}},
 GridLinesStyle → Gray, Frame → True, FrameStyle → Gray, PlotRange → All,
 FrameTicks → {{{3.5, 5.9, 7.2, 8.1, 9.3}, None}, {{800, 1200, 1800, 2500, 4000}, None}},
 FrameTicksStyle → Directive["Label", Black, 12], ImageSize → 400, AspectRatio → 1/1.5,
 FrameLabel → {Style["Time (seconds)", Black, 16],
   Style["Liquid Level (meters)", Black, 16]}]
```

Figure 9.5.11 Applying PI control and plotting the system response to a piecewise-continuous reference input: Mathematica code

Other electromechanical sensing devices of interest are strain gauges and accelerometers. A *strain gauge*, shown in Figure 9.6.1, is used to measure deflection, stress, and pressure. A metallic strain gauge consists of a metallic foil or a fine wire, arranged in a grid pattern that is bonded to a thin backing (*carrier*). The carrier is attached to an element, the strain of which needs to be measured. The gauge bends together with the tested element, and the length of the gauge grid changes. This elongation translates into a linear change in the gauge electrical resistance. The device's electrical subsystem consists of a specific circuit called the *Wheatstone bridge*, which is used to measure the electrical resistance of the gauge terminals.

9.6 Sensors, Electroacoustic, and Piezoelectric Devices

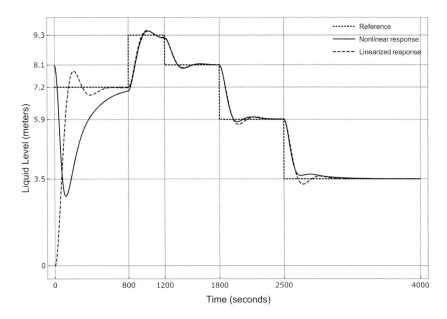

Figure 9.5.12 Dynamic response of the nonlinear and linear systems under PI control to a piecewise-continuous reference input

```
(* comparing performance of the controlled nonlinear and linearized systems *)
(* controlled nonlinear system *)
nlContr = PIDTune[dtank, Automatic, "ReferenceOutput"]

(* response of the controlled nonlinear system to the reference input *)
nlor = OutputResponse[nlContr, ref, {t, 0, 4000}];

(* controlled linearized system *)
lContr = PIDTune[TransferFunctionModel[dtank], Automatic, "ReferenceOutput"]

(* response of controlled linearized system to the reference input *)
lor = OutputResponse[lContr, ref, {t, 0, 4000}];

(* plotting responses and reference input *)
Plot[{ref, nlor, lor}, {t, 0, 4000}, PlotRange → All,
 PlotLegends → Placed[{"Reference", "Nonlinear response", "Linearized response"}, {Right, Top}],
 Exclusions → None, AxesOrigin → {0, 0}, PlotStyle → {{Black, Thick, Dotted}, {Black}, {Black, Dashed}},
 GridLines → {{0, 800, 1200, 1800, 2500, 4000}, {0, 3.5, 5.9, 7.2, 8.1, 9.3}}, GridLinesStyle → Gray,
 Frame → True, FrameStyle → Gray, PlotRange → All,
 FrameTicks → {{{0, 3.5, 5.9, 7.2, 8.1, 9.3}, None}, {{0, 800, 1200, 1800, 2500, 4000}, None}},
 FrameTicksStyle → Directive["Label", Black, 12], ImageSize → 400, AspectRatio → 1/1.5,
 FrameLabel → {Style["Time (seconds)", Black, 16], Style["Liquid Level (meters)", Black, 16]}]
```

Figure 9.5.13 Applying PI control and plotting the nonlinear and linearized system response: Mathematica code

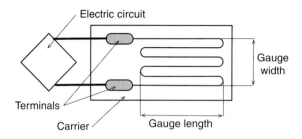

Figure 9.6.1 Strain gauge

Example 9.6.1 Displacement sensor to detect a sinusoidal motion

Consider a displacement sensor for the detection of a sinusoidal motion. The sensor uses a microcantilever beam, and four strain gauges connected in a Wheatstone bridge (see Figure 9.6.2). The sinusoidal motion of the supporting structure (the motion that needs to be detected) generates oscillatory bending of the microcantilever beam in the shown direction of u (see Figure 9.6.2(a)). Two strain gauges are installed on each face of the microcantilever beam (face 1 is shown in Figure 9.6.2(a), face 2 is the opposite one) at its root.

Modeling of this electromechanical system starts with the formulation of the governing equations for each subsystem separately, proceeds with the derivation of the relationship between the axial strain and the input/output voltages of the Wheatstone bridge, and finally synthesizes all the information into a coherent system model that relates the sensor sensitivity K to the sinusoidal input frequency ω. The last step encompasses a numerical simulation, the generation of the graph of K vs. ω, and analysis of the results.

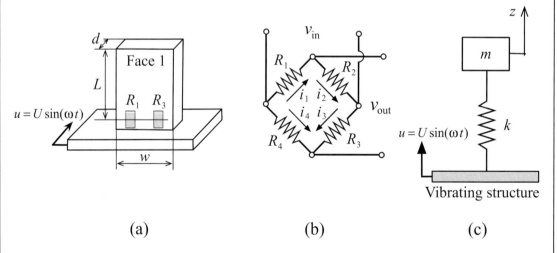

Figure 9.6.2 Displacement sensor: (a) strain gauges and microcantilever-beam assembly; (b) Wheatstone bridge; (c) lumped-parameter model

9.6 Sensors, Electroacoustic, and Piezoelectric Devices

The sensitivity of a strain-gauge displacement sensor is related to the variation in the Wheatstone-bridge output voltage as follows:

$$\Delta v_0 = Kx \qquad (9.6.1)$$

where K is sensitivity and x is the variation of the mechanical parameter that is being monitored. For this problem, x is the relative displacement of the free end of the microcantilever beam with respect to the oscillating supporting structure.

A lumped-parameter model of the microcantilever beam and strain gauges assembly is shown in Figure 9.6.2(c). In this model, m and k are the effective mass and effective stiffness (refer to Chapter 3 for a detailed discussion on finding the effective mass and stiffness for a continuous medium such as a beam). The governing equation for this mechanical subsystem is derived using Newton's second law:

$$m\ddot{z} = -k(z - u) \qquad (9.6.2)$$

where u is the oscillatory displacement of the supporting structure, and z is the displacement of the free end of the microcantilever beam.

Denoting the relative displacement of this free end as $x = z - u$, the governing equation is rewritten as follows:

$$m\frac{d^2}{dt^2}(x+u) = -kx \quad \Rightarrow \quad m\ddot{x} + kx = -m\ddot{u} \qquad (9.6.3)$$

Since the input u is a periodic function, the frequency response needs to be derived (refer to Chapter 7 for details). Taking the Laplace transform of Eq (9.6.3):

$$G(s) = \frac{X(s)}{U(s)} = -\frac{ms^2}{ms^2 + k} \qquad (9.6.4)$$

and finding the corresponding $G(j\omega) = \frac{m\omega^2}{-m\omega^2 + k}$, which is real, its modulus therefore equals the function itself. Then, from Eq. (9.6.4) and Eq. (7.3.5) the magnitude of the steady-state response of the given system under a sinusoidal input is

$$X = U|G(j\omega)| = UG(j\omega) = \frac{m\omega^2}{-m\omega^2 + k}U \qquad (9.6.5)$$

In order to derive the expression for X, we assume that the supporting structure is stationary, but there is a virtual force f, applied at the free end of the microcantilever beam, that causes the same out-of-plane bending as did the sinusoidal motion of the supporting structure. As derived from literature on the strength of materials, a force applied to the free end of a cantilever beam produces a deflection:

$$x = \frac{fl^3}{3EI} \tag{9.6.6}$$

where f is the applied force, l is the length of the beam, E is elasticity modulus (Young's modulus), and I is the moment of inertia. EI is often called the *bending stiffness*.

For the given microcantilever beam, $I = \frac{wd^3}{12}$, where w and d are the width and thickness of the cantilever beam (see Figure 9.6.2(a)); hence, Eq. (9.6.6) becomes

$$x = \frac{fL^3}{3E\left(\dfrac{wd^3}{12}\right)} = \frac{4fL^3}{Ewd^3} \tag{9.6.7}$$

Using Eq. (9.6.7) to express f in the frequency domain and substituting the value of X as derived in Eq. (9.6.5), we obtain

$$F = \frac{Ewd^3}{4L^3} X = \frac{Ewd^3 m\omega^2}{4L^3(-m\omega^2 + k)} U \tag{9.6.8}$$

The microcantilever beam is experiencing a deformation and, consequently, strain. This strain exists on both faces of the beam. Assuming that face 1 extends under bending while face 2 compresses, their strains will have opposite signs – positive for face 1, and negative for face 2. Working under the previous assumption of a virtual force f applied at the free end of the microcantilever beam, we find the bending moment caused by this force $M_b = fL$.

According to Hooke's law, the stress due to bending, perpendicular to any cross-section wd of the microcantilever beam, is related to the axial strain in the following manner:

$$\sigma = E\varepsilon \tag{9.6.9}$$

where σ is the normal stress, E is Young's modulus, and ε is the axial (aligned with the generating force) strain.

The stress is computed using Navier's equation:

$$\sigma = \frac{M_b d}{2I} \tag{9.6.10}$$

where M_b is the bending moment due to the assumed virtual force f, d is the thickness of the beam, and I is the cross-sectional moment of inertia about the bending axis.

For a rectangular cross-section beam of width w and thickness d, this moment of inertia is

$$I = \frac{wd^3}{12} \tag{9.6.11}$$

Then, the magnitude of strain from Eqs. (9.6.8)–(9.6.10) is derived as

$$\varepsilon = \frac{\sigma}{E} = \frac{M_b d/2I}{E} = \frac{f\,Ld}{2IE} = \frac{f\,Ld}{2(wd^3/12)E} = \frac{6f\,L}{Ewd^2} \quad (9.6.12)$$

Converting Eq. (9.6.12) into the frequency domain and substituting the expression for the virtual force from Eq. (9.6.8) into it, we obtain

$$\varepsilon = \frac{6L}{\varepsilon wd^2}\left(\frac{\varepsilon wd^3 m\omega^2}{4L^3(-m\omega^2 + k)}U\right) = \frac{3dm\omega^2}{2L^2(-m\omega^2 + k)}U \quad (9.6.13)$$

Equation (9.6.13) relates the strain to be sensed to the frequency and magnitude of the sinusoidal input. Now, we need to derive the expression relating this strain to the variation of the voltage output v_{out} of the Wheatstone bridge.

For a strain gauge connected to an electrical resistance, the following resistance–strain relationship exists:

$$\frac{\Delta R}{R} = K_g \varepsilon \quad (9.6.14)$$

where ε is the axial strain, K_g is the strain-gauge sensitivity, R is the electrical resistance, and ΔR is the change in electrical resistance due to the strain.

Assuming that the four strain gauges attached to the microcantilever beam are identical, the following relationship between the input voltage v_{in} and the variation of the voltage output v_{out} exists (see Chapter 4 for derivation):

$$\Delta v_{out} = \frac{1}{4}\left(\frac{\Delta R_1}{R_1} - \frac{\Delta R_2}{R_2} + \frac{\Delta R_3}{R_3} - \frac{\Delta R_4}{R_4}\right)v_{in} \quad (9.6.15)$$

From Eq. (9.6.14):

$$\frac{\Delta R_1}{R_1} = -\frac{\Delta R_2}{R_2} = \frac{\Delta R_3}{R_3} = -\frac{\Delta R_4}{R_4} = K_g \varepsilon \quad (9.6.16)$$

Therefore, Eq. (9.6.15) becomes

$$\Delta v_{out} = K_g \varepsilon v_{in} \quad (9.6.17)$$

Converting to the frequency domain and using the expression for the strain described by Eq. (9.6.13), the relationship between input parameters – the amplitude and frequency – and the system parameters is derived:

$$\Delta V_{out} = K_g \left(\frac{3dm\omega^2}{2L^2(-m\omega^2 + k)}\right)UV_{in} \quad (9.6.18)$$

Equation (9.6.18) constitutes the frequency domain model of the displacement sensor, consisting of a microcantilever beam and four strain gauges connected in a Wheatstone bridge.

Recalling that the variation of the voltage output of the Wheatstone bridge is proportional to the mechanical displacement being measured (Eq. (9.6.1)), we obtain the expression of sensitivity as a function of the input frequency:

$$K = \frac{3dmK_g\omega^2 V_{in}}{2L^2(-m\omega^2 + k)} \qquad (9.6.19)$$

For a numerical simulation consider the following values of system parameters:

Microcantilever-beam geometry: $L = 200 \times 10^{-6}$m, $d = 100 \times 10^{-9}$m, $w = 40 \times 10^{-6}$m
$\rho = 5200$ kg/m^3

Electrical and mechanical parameters: $K_g = 2.2$, $E = 1.6 \times 10^{11}$ N/m^2, $v_{in} = 10$ V

From the tables in Chapter 3 and the literature on the strength of materials, the equivalent mass and equivalent stiffness of the microcantilever beam are found to be as follows:

$$m_{eq} = \frac{33}{140}m = \frac{33}{140}Ldw\rho = 9.81 \times 10^{-13}\text{kg}$$

$$k_{eq} = \frac{3EI}{L^3} = \frac{3E}{L^3}\left(\frac{wd^3}{12}\right) = 2 \times 10^{-4}\text{N/m}$$

Only below-resonance input frequencies need to be considered. As derived in Chapter 7, the natural frequency of this second-order system is $\omega_n = \sqrt{\frac{k_e}{m_e}} = 14281.6$ rad/s

The numerical simulation results are presented in Figure 9.6.3, and the Mathematica code for generating these results is straightforward, as shown in Figure 9.6.4.

As seen in Figure 9.6.3, the sensitivity of this displacement sensor increases with the increase of input frequency and goes to infinity when the system approaches resonance.

Using this sensor for detecting sinusoidal motion with a close-to-resonant frequency is not recommended since the very high sensitivity will cause a significant increase in the output voltage of the Wheatstone bridge, thus damaging the strain gauges.

An *accelerometer*, as its name suggests, is used to measure the acceleration. This measurement serves as an input for many types of control systems. There are different types of accelerometers, depending on the physical effect employed. An electromechanical accelerometer is shown in Figure 9.6.5, where A indicates an accelerometer device, and B denotes the element that experiences the acceleration that needs to be measured. The accelerometer is attached to the tested element.

The mechanical subsystem of this device is represented by a mass m (*seismic* or *proof mass*) attached to the accelerometer's body with a linear spring of stiffness k and a linear damper

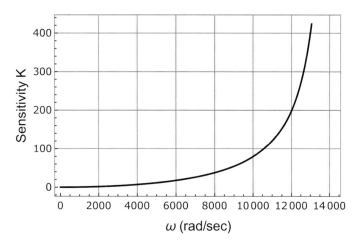

Figure 9.6.3 Sensitivity of a displacement sensor vs. the sinusoidal input frequency

```
(* compute mass and equivalent mass *)
l = 200 * 10^-6; w = 40 * 10^-6; d = 100 * 10^-9; rho = 5200;
me = 33./140 * l * w * d * rho

(* compute equivalent stiffness *)
eE = 1.6 * 10^11; ii = w * d^3 / 12;
ke = 3 * eE * ii / l^3

(* compute natural frequency *)
omegan = √(ke / me)

(* plot the relationship between sensitivity and input
   frequency that is always sub-resonant) *)
vin = 10; kg = 2.2;
k = 3 * d * me * kg * ω^2 * vin / (2 * l^2 * (ke - me * ω^2));

Plot[k, {ω, 0, omegan}, PlotStyle → {Black, Thick}, Frame → True,
  FrameStyle → Gray, GridLines → Automatic, GridLinesStyle → Gray,
  FrameTicks → {{Automatic, None}, {Automatic, None}},
  FrameTicksStyle → Directive["Label", Black, 12], ImageSize → 400,
  FrameLabel → {Style["ω (rad/sec)", Black, 16],
    Style["Sensitivity K", Black, 16]}]
```

Figure 9.6.4 Sensitivity of a displacement sensor vs. the sinusoidal input frequency: Mathematica code

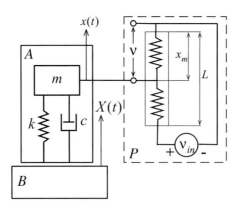

Figure 9.6.5 Electromechanical accelerometer

element with a damping coefficient c. The damping element may be represented by a viscous fluid rather than a dashpot. The electrical system that is responsible for the actual measurements is represented by a potentiometer P. In some implementations it may also include an amplifier.

When the tested element B starts moving with acceleration $\ddot{X}(t)$, it pulls the attached accelerometer A and the proof mass along with it. If the displacement of the proof mass is $x(t)$, and the displacement of the tested element is $X(t)$, then applying the Newton's second law to the proof mass, we obtain

$$m\ddot{x}(t) + c\left(\dot{x}(t) - \dot{X}(t)\right) + k\left(x(t) - X(t)\right) = 0 \tag{9.6.19}$$

Let the relative displacement of the proof mass be $x_m(t) = x(t) - X(t)$. Then, Eq. (9.6.19) becomes

$$m\ddot{X}(t) + m\ddot{x}_m(t) + c\dot{x}_m(t) + kx_m(t) = 0 \tag{9.6.20}$$

Separating the components that contain the relative displacement at the left-hand side, we obtain

$$m\ddot{x}_m(t) + c\dot{x}_m(t) + kx_m(t) = -m\ddot{X}(t) \tag{9.6.21}$$

If the potentiometer is calibrated to read this relative displacement, its output voltage is linearly dependent on it as follows (see Section 4.2.3, Example 4.2.4 for detailed derivations):

$$v = v_{in}\frac{x_m}{L} = K_L x_m \tag{9.6.22}$$

where L is the potentiometer length, v_{in} is the input voltage, and $K_L = \dfrac{v_{in}}{L}$ is the potentiometer gain.

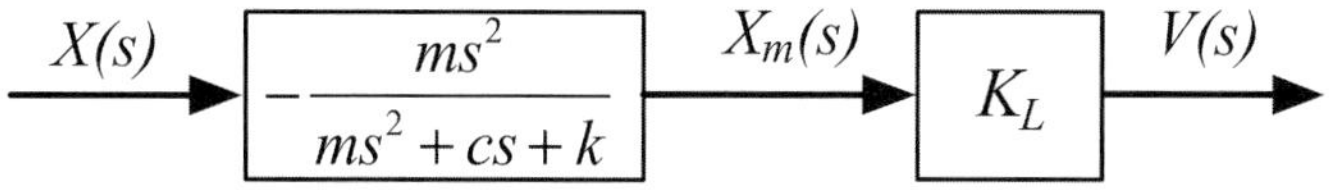

$$X(s) \longrightarrow \boxed{-\frac{ms^2}{ms^2+cs+k}} \xrightarrow{X_m(s)} \boxed{K_L} \xrightarrow{V(s)}$$

Figure 9.6.6 Electromechanical accelerometer: block diagram

Thus, the accelerometer dynamic system is modeled by two governing equations: Eqs. (9.6.21) and (9.6.22). Its block diagram, considering that the system input is the acceleration of the tested element $\ddot{X}(t)$ and the output is the voltage readout, is shown in Figure 9.6.6.

Depending on the selection of the system parameters m, c, and k, the device can be used as an accelerometer as discussed above, or as a *vibrometer* that measures the amplitude of sinusoidal displacement. The most commonly known application of a vibrometer is a *seismograph* – the device to measure ground motion resulting from an earthquake.

Example 9.6.2 Modeling of a vibrometer and accelerometer

For the accelerometer discussed above, in order to measure the amplitude of sinusoidal displacement the amplitude of the device's response must be close to the amplitude of the displacement input.

As derived in Chapter 7 (Eqs. (7.3.5) and (7.3.6)), the amplitude of a steady-state frequency response of a second-order system equals $r_0|G(j\omega)|$, where r_0 is the input amplitude and $|G(j\omega)| = \sqrt{[\text{Re}(G(j\omega))]^2 + [\text{Im}(G(j\omega))]^2}$ is the magnitude of the system transfer function in the frequency domain.

For the electromechanical accelerometer discussed above, the transfer function can be written as

$$G(s) = \frac{-K_L s^2}{s^2 + \frac{c}{m}s + \frac{k}{m}} \qquad (9.6.23)$$

Recalling that $\frac{c}{m} = 2\xi\omega_n$ and $\frac{k}{m} = \omega_n^2$, this transfer function is converted to the frequency domain as follows:

$$G(j\omega) = \frac{K_L \omega^2}{(\omega_n^2 - \omega^2) + 2\xi\omega_n j} \qquad (9.6.24)$$

Multiplying the numerator and denominator by the complex conjugate of the denominator, and separating the real and imaginary parts, we obtain

$$G(j\omega) = K_L\omega^2 \frac{\omega_n^2 - \omega^2}{(\omega_n^2 - \omega^2)^2 + 4\xi^2\omega^2\omega_n^2} - K_L\omega^2 \frac{2\xi\omega\omega_n}{(\omega_n^2 - \omega^2)^2 + 4\xi^2\omega^2\omega_n^2}j \qquad (9.6.25)$$

Then, using the expressions for the natural frequency and damping ratio, the magnitude of this transfer function in terms of the system physical parameters m, k, and c is

$$|G(j\omega)| = \frac{K_L \omega^2}{\sqrt{(k/m - \omega^2)^2 + (c/m)^2 \omega^2}} \tag{9.6.26}$$

In order for the device to work as a vibrometer $r_0|G(j\omega)| \cong r_0$, hence $|G(j\omega)| \cong 1$ for the specific frequency(ies) of interest. Otherwise, this sensor will be functioning as an accelerometer.

This can be illustrated by the following numerical simulation.

Assume the following values of the system parameters: $\frac{c}{m} = 115$, $\frac{k}{m} = 70$, $K_L = 2.15$, $\omega = 60$. Then, $|G(j\omega)| = 0.9986 \cong 1$, which means that the magnitude of the system steady-state frequency response $r_0|G(j\omega)|$ is almost equal to the magnitude r_0 of the sinusoidal displacement input with the frequency of 60 rad/s – the sensor works as a vibrometer.

Changing the first two parameters as $\frac{c}{m} = 345$, $\frac{k}{m} = 210$, the magnitude $|G(j\omega)|$ decreases to 0.369, making the sensor work as an accelerometer for the sinusoidal input of the same frequency. Assuming that $r_0 = 1$, for simplicity, compute the acceleration of this sinusoidal input as 3600. Then, the amplitude of response $r_0|G(j\omega)| = 0.369$ is 1.02×10^{-4} times the acceleration of the input. Hence, this device works as an accelerometer with the gain 1.02×10^{-4}.

In addition to the described spring–damper construction, accelerometers can employ other technologies that use the change of some physical property due to acceleration of the studied element as a measure of the acceleration that is experienced by that element. For example, in a capacitive device, the degree of acceleration is indicated by the change of capacitance, while in a piezoelectric one it is measured by the change in the voltage output of the piezoelectric crystal, and in a piezoresistive device the degree of acceleration is measured by the change in resistance.

Capacitive accelerometers are typically constructed as microdevices or MEMS (*microelectromechanical systems*). The accuracy and high sensitivity of MEMS accelerometers, as well as their very small size (micrometers in size in silicon) and low cost make such devices very attractive to the industry. Currently, the biggest market demand for such devices is exerted by the automotive industry, the manufacturers of personal electronic devices, and the aerospace industry. MEMS accelerometers are used in numerous applications such as image stabilization in camcorders and built-in cameras, some user-interface control in smartphones and tablets, navigation units and anti-skidding systems in cars, contactless game controllers, and real-time controls for smart weapon systems.

A capacitive accelerometer is illustrated in Figure 9.6.7. The assembly is used for a *balanced force micro-accelerometer* (MEMS device), widely used in the air-bag deployment sensor in automobiles. Since there is no room in a microdevice for a coil spring, damper, or

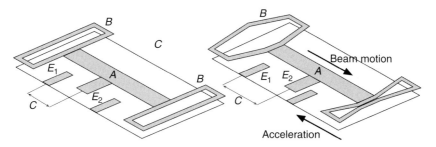

Figure 9.6.7 Balanced force micro-accelerometer: (a) the device configuration at rest; (b) the configuration in response to an applied acceleration

even a seismic mass, miniaturizing an accelerometer to the microscale requires a different approach. In the device shown in Figure 9.6.7 a plate beam *A* serves as a substitute for the seismic mass. It is supported by the two folded tethers (denoted by *B*), which are firmly fixed to the supporting plate. These tethers play the role of the spring, and the surrounding air provides damping.

The capacitance *C* is measured between the pair of electrodes E_1 and E_2, where E_1 is the stationary electrode, and E_2 is the moving one. When an acceleration is applied to the device, the plate beam moves in the direction opposite to the acceleration, and the distance between the electrodes E_1 and E_2 changes, thus changing the capacitance *C*. The relation between the change in the measured capacitance and the experienced acceleration depends on the geometry and materials of the accelerometer, and the device scale is calibrated accordingly.

Piezoelectric and piezoresistive accelerometers are presented in the next section.

9.6.2 Piezoelectric Devices

Piezoelectric crystals that are at the core of piezoelectric devices represent a subset of a wider class of materials that possess an oriented internal structure – *electrets*. Electrets are usually solid dielectrics that have a permanent, fixed electrical polarization, just like permanent magnets have a magnetic polarization. The molecules of an electret are dipoles, aligned in such a way that the material possesses opposing electrical poles of positive and negative charges of the same magnitude. The existence of these poles results in a discernible fixed electrostatic surface charge, or static potential. The *real-charge electrets* exhibit a relatively strong external electric field, while *dipolar-charge electrets* are electrically neutral even though they still possess aligned dipolar charges. The charge and dipole orientation of an electret depend on many factors, such as, for example, the chemical constitution of the material, macromolecular arrangements, the presence of impurities, etc.

While both electret types are widely used in different applications such as, for example, electret microphones, headphones, speakers, transducers, air filters, and other engineering devices, the real-charge ones have a disadvantage. Exposed to the air, this type of electret attracts the charged particles floating in the air as ions and dust. The accumulation of these particles neutralizes the surface charge of this electret, rendering it unusable.

A piezoelectric crystal is an electret in which the unit cells of its lattice contain electric dipoles; these unit cells are oriented such that the dipoles are aligned. There is no net charge on the faces of a piezoelectric crystal since the electric dipole moments exactly cancel out.

Upon the application of a sufficient force, a piezoelectric crystal deforms, causing a disruption in the alignment of its unit cells, resulting in a re-orientation of the electric dipoles. This misalignment forces the dipole moments out of balance, subsequently generating an electrical charge on the surface of the crystal. Therefore, a voltage is developed across the piezoelectric crystal: mechanical energy due to the applied force is converted into electrical energy.

The described *piezoelectric effect* is employed in sensors that measure strains, pressures, and forces.

The *inverse piezoelectric effect* performs the opposite energy conversion: a voltage, applied to a piezoelectric crystal, causes its deformation, thus producing a force. It is used primarily in actuators. Among the everyday devices that use the inverse piezoelectric effect are quartz watches. In these watches, the electrical energy, which is supplied by a battery, makes the piezoelectric crystal oscillate with a very high frequency. Then, the electronic circuitry of the watch modifies the frequency of these oscillations to one hertz (Hz, one cycle per second) and uses them as an input to the gear system that drives the watch hands around the clock face.

Piezoelectric materials can be *monocrystalline*, such as, for example, quartz and zinc oxide, or *ferroelectric*. Ferroelectric materials are represented by various ceramics (for example, PZT – lead zirconate titanate – and barium titanate), and polymers. The ceramics, in particularly PZT, are widely used in medical technology as ultrasonic transducers, which convert electrical energy into high-frequency vibrations. Some of the best-known naturally occurring piezoelectric crystals are quartz, Rochelle salt, and sucrose (cane sugar).

The piezoelectric effect of monocrystalline materials depends on their crystalline structures, and it is relatively small, while the effect for ferroelectrics is considerably more pronounced, which makes them a better fit for the majority of applications. Additionally, forcing an inclusion of metal ions (for example, nickel, bismuth, or niobium) into the crystalline lattice of piezoelectric ceramics such as PZT allows optimization of the piezoelectric parameters of the ceramic according to the desired specifications. Nonetheless, ferroelectrics have some specific disadvantages. Unlike the monocrystalline materials, the fabricated ferroelectrics need to be polarized with a strong electrical field to gain the piezoelectric capacity. Consequently, if exposed to an electric field opposite to their polarization, the ferroelectrics can become depolarized, losing their piezoelectric qualities. Exposure to high temperatures (greater than a specific threshold that depends on the material) also causes depolarization.

The large modulus of elasticity (Young's modulus) of piezoelectric materials is responsible for the linear relationship between the force and the voltage for a wide range of inputs. It is also responsible for the small deformation of such materials, which makes them particularly well suited for micro- and nano-devices, but unfit for measuring static forces.

The moderate-to-low cost of piezoelectric devices, their small size, sufficient accuracy, and the lack of need for an external power source make them very attractive for a wide spectrum of applications, from simple applications such as spark lighters and inkjet printers to sophisticated medical technology and scanning tunneling microscopes.

Piezoelectric accelerometers are typically used in civil structures that operate in dynamic settings. The construction of such an accelerometer is shown in Figure 9.6.8(a), where A denotes an accelerometer device and B indicates the element that experiences the acceleration to be measured.

The accelerometer body is secured to the tested element. The piezoelectric crystal that constitutes a core of this accelerometer is mounted to the accelerometer body and sandwiched between two metal plates (electrodes). These electrodes are connected to a voltage-measuring device such as a voltmeter or potentiometer. The acceleration applied to B results in the motion of the seismic mass m, which causes a deformation of the piezoelectric crystal. The piezoelectric effect is responsible for the voltage v being generated across the crystal. Within the operating envelope of the specific accelerometer the generated voltage is linearly dependent on the experienced acceleration.

A piezoresistive accelerometer is illustrated in Figure 9.6.8(b), where A denotes an accelerometer device, B indicates the element that experiences the acceleration to be measured, C is a silicon cantilever beam, and P denotes a piezoresistor. Like the balanced force microaccelerometer discussed earlier, the *cantilever-beam accelerometer*, shown in Figure 9.6.8(b), is a MEMS device. Its operation is based on the piezoresistive effect – the resistivity of a material changes in response to the applied mechanical stress.

In the design shown, the accelerometer body is secured to the tested element. When an acceleration is applied to the tested element, the proof mass m starts moving, causing

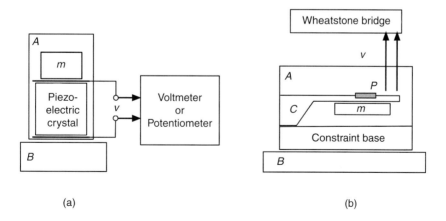

Figure 9.6.8 (a) Piezoelectric accelerometer; (b) piezoresistive accelerometer

a mechanical strain in the cantilever beam C. The deformation of the beam results in a strained piezoresistor P and a consequent change in its resistivity. This change is detected and measured by the electrical subsystem, typically represented by a Wheatstone-bridge circuit.

The cantilever-beam accelerometer's volume is often filled with a viscous fluid for damping the motion of the proof mass.

The dependency between the strain, experienced by P, and the change of resistivity is linear.

Piezoresistive accelerometers are widely used for low-frequency vibrations since they can measure very small accelerations. They also show satisfactory performance in shock environments. These devices exhibit high sensitivity and typically do not require pre-amplification. The major disadvantage of piezoresistive accelerometers is temperature sensitivity and limited high-frequency response.

Example 9.6.3 A piezoelectric accelerometer

A piezoelectric accelerometer to measure an amplitude of sinusoidal displacement input is shown in Figure 9.6.9(a). This representation assumes that the piezoelectric plate is modeled as a mass–spring–damper system and that the total system losses are lumped into a known single viscous damping coefficient c_e. The other known system parameters are the seismic mass m, spring stiffness k, density of the piezoelectric crystal ρ_{pz}, Young's modulus of the piezoelectric crystal E_{pz}, piezoelectric voltage constant g_{pz}, and the amplitude v of the voltage read by the piezoelectric sensor. The piezoelectric crystal is a thin square plate, a side of which measures w and its height measures h.

The model of this system is an expression that relates the magnitude of the vibrating-platform acceleration and magnitude of the sensed voltage v.

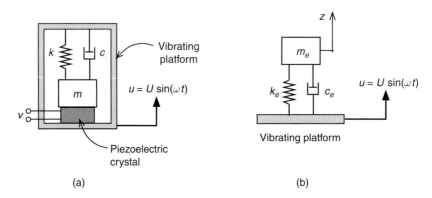

Figure 9.6.9 Piezoelectric accelerometer: (a) schematic; (b) lumped-parameter model

Since the seismic mass is in contact with the piezoelectric crystal at all times, and they move together, the lumped-parameter model is simplified to a single equivalent mass. The influence of the piezoelectric crystal on the motion of this equivalent mass is represented by a spring with a stiffness coefficient k_{pz}. Therefore, the lumped-parameter model can be further simplified to contain a single equivalent spring exerting the same effort as a combination of the given spring k and the spring k_{pz}. Note that the piezoelectric crystal is undergoing axial longitudinal deformation.

As derived in the literature on the strength of materials and shown in Chapter 3, the equivalent mass and equivalent stiffness, indicated in the lumped-parameter model shown in Figure 9.6.9(b), are computed as follows:

$$\begin{cases} m_e = m + m_{pz} = m + \frac{1}{3}\rho_{pz}hw^2 \\ k_e = k + k_{pz} = k + \frac{E_{pz}w^2}{h} \end{cases} \qquad (9.6.27)$$

where m_{pz} and k_{pz} are the equivalent mass and stiffness of the piezoelectric crystal respectively.

The governing equation for the mechanical subsystem of this dynamic system is derived using the Newton's second law:

$$m_e \ddot{z} = -c_e(\dot{z} - \dot{u}) - k_e(z - u) \qquad (9.6.28)$$

where u is the oscillatory displacement of the vibrating platform, and z is the displacement of the lumped mass m_e.

Denoting the relative motion of the lumped mass with respect to the vibrating platform as $x = z - u$, the governing equation is rewritten as follows:

$$m_e \frac{d^2}{dt^2}(x + u) = -c_e \dot{x} - k_e x \quad \Rightarrow \quad m_e \ddot{x} + c_e \dot{x} + k_e x = -m_e \ddot{u} \qquad (9.6.29)$$

As stated in references on the dynamics of MEMS, the sensed voltage v is proportional to the relative displacement x

$$v = g_{pz} E_{pz} x \qquad (9.6.30)$$

Deriving x from Eq. (9.6.30) and substituting it into Eq. (9.6.29), we obtain

$$\frac{m_e}{g_{pz}E_{pz}} \ddot{v} + \frac{c_e}{g_{pz}E_{pz}} \dot{v} + \frac{k_e}{g_{pz}E_{pz}} v = -m_e \ddot{u} \qquad (9.6.31)$$

which generates the following transfer function:

$$G(s) = \frac{V(s)}{U(s)} = -\frac{g_{pz}E_{pz}m_e s^2}{m_e s^2 + c_e s + k_e} \qquad (9.6.32)$$

One of the most important characteristics of an accelerometer is the *transmissibility* – the ratio of the amplitude of the sensed voltage to the magnitude of displacement input. The transmissibility is a modulus of the transfer function $G(j\omega)$ as expressed in the frequency domain. Recalling the relationship between the system and model parameters: $\omega_n = \sqrt{\dfrac{k_e}{m_e}}$ and $\zeta = \dfrac{c_e}{2m_e\omega_n} = \dfrac{c_e}{2\sqrt{k_e m_e}}$ (see Chapter 7 for details), the expression for transmissibility is derived as follows:

$$T = |G(j\omega)| = \frac{g_{pz} E_{pz} \omega^2}{\sqrt{(k/m - \omega^2)^2 + (c/m)^2 \omega^2}} = \frac{g_{pz} E_{pz} \omega^2}{\sqrt{(\omega_n^2 - \omega^2)^2 + 4\zeta^2 \omega_n^2 \omega^2}} \quad (9.6.33)$$

We divide numerator and denominator of $|G(j\omega)|$ by ω_n^2, and denote ratio $r = \dfrac{\omega}{\omega_n}$. Then, the transmissibility becomes

$$T = \frac{g_{pz} E_{pz} r^2}{\sqrt{(1 - r^2)^2 + (2\zeta r)^2}} \quad (9.6.34)$$

The transmissibility, being the modulus of the transfer function $G(j\omega)$, can be viewed as the *gain* of the Bode plot – a graph of the $|G(j\omega)|$ vs. frequency ω, where the frequency is in radians per second, and $|G(j\omega)|$ is measured in decibels, $|G(j\omega)|$, dB $= 20\log_{10}(|G(j\omega)|)$. The Bode plot consists of two graphs: *gain* ($|G(j\omega)|$ vs. ω, where $|G(j\omega)|$ is related to the magnitude of the frequency response) and *phase* (the phase shift of the frequency response). It is of utmost importance for the system frequency response analysis and control design (see Chapter 7 for more information).

Given the following numerical values of the system parameters:

Piezoelectric crystal parameters:

$$w = 0.015 \text{ m}, \ h = 1.5 \times 10^{-3} \text{ m}, \ \rho_{pz} = 7500 \text{ kg/m}^3$$
$$E_{pz} = 5 \times 10^{10} \text{ N/m}^2, \ g_{pz} = 0.074 \text{ m}^3/\text{C}, \ v = 0.1 \text{ V'}$$

Seismic mass and supporting elements:

$$m = 0.075 \text{ kg}, \ k = 80 \text{ N/m}, \ c_e = 135 \text{ N·s/m},$$

the Bode plot is generated (gain only, see Figure 9.6.10).

The peak gain (G_{max}) occurs at the resonant frequency ω_r. The numerical values for the resonant frequency and peak gain can be read from the graph or can be derived analytically using the procedure presented in Section 7.3.4. The resonant frequency is computed as

9.6 Sensors, Electroacoustic, and Piezoelectric Devices

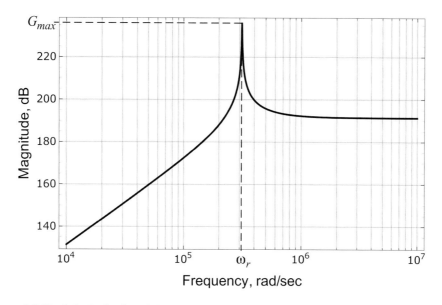

Figure 9.6.10 Gain Bode plot of the piezoelectric accelerometer system

$\omega_r = \dfrac{\omega_n}{\sqrt{1-2\xi^2}} \cong 3.14 \times 10^5$ rad/s; at the resonant frequency peak gain is $G_{max} \cong 236.3$ dB. The Mathematica code for constructing the Bode plot and analytically deriving the resonant frequency and peak gain is shown in Figure 9.6.11.

An accelerometer is typically constructed to have a high natural frequency. Assuming that for a given piezoelectric device $\omega_n \gg \omega$, the denominator of the expression for transmissibility (Eq. (9.6.34)) becomes approximately equal to 1.

Since $T = |G(j\omega)|$, from Eq. (9.6.32) we can conclude that

$$|G(j\omega)| = \frac{|V(j\omega)|}{|U(j\omega)|} = T \qquad (9.6.35)$$

Then, the relationship between the voltage reading and periodic input amplitude is

$$V = TU \cong g_{pz} E_{pz} \frac{\omega^2}{\omega_n^2} U \qquad (9.6.36)$$

where $\omega^2 U = A$ is the magnitude of acceleration of the sinusoidal displacement input.

Hence, the relationship between the sensed acceleration and the voltage reading becomes

$$A = \frac{\omega_n^2}{g_{pz} E_{pz}} V \qquad (9.6.37)$$

where the constant $\dfrac{\omega_n^2}{g_{pz} E_{pz}}$ is defined by the accelerometer construction.

```
(* set up the numerics *)
w = 0.015; h = 1.5*10^-3; rho = 7500; epz = 5*10^10;
gpz = 0.074; m = 0.075; k = 80; ce = 135;
v = 0.1;

(* compute mₑ and kₑ *)
me = m + 1/3 * rho*h*w^2;

ke = k + epz*w^2/h;

tfm = TransferFunctionModel[{{-(gpz*epz*me*s^2)/(me*s^2 + ce*s + ke)}}, s]

bp = BodePlot[tfm, {10^4, 10^7}, PlotLayout → "Magnitude", PlotStyle → {Black, Thick},
   GridLines → Automatic, FrameTicksStyle → Directive["Label", Black, 12],
   FrameLabel → {Style["Frequency, rad/sec", Black, 16],
     Style["Magnitude, dB", Black, 16]}]

(* analytically derive resonant frequency *)
tf = (gp*ep*x^2)/Sqrt[(wnn^2 - x^2)^2 + 4*z^2*wnn^2*x^2];
dtf = ∂_x tf // FullSimplify;
rfreq = Solve[dtf == 0, x]
num = rfreq[[3, 1, 2]] /. {wnn → wn, z → zeta};

(* analytically derive max gain *)
FullSimplify[20*Log10[tf /. {x → num, ep → epz, gp → gpz, wnn → wn, z → zeta}]]
```

Figure 9.6.11 Generating a Bode plot and computing the resonant frequency and peak gain: Mathematica code

Another important characteristic of an accelerometer is its very low damping. Computing the damping ratio of the given piezoelectric accelerometer, we find $\xi = 0.00283$. With such a low damping ratio the resonant frequency is approximately equal to the natural frequency of the system (which can be easily shown numerically), and is used to derive the numeric expression relating A to V.

For a given device, Eq. (9.6.37) reads $A = 26.73\ V$. If, as specified, the voltage reading equals 0.1 V, the acceleration of the measured sinusoidal input equals 2.673 m/s^2.

9.6.3 Electroacoustic Devices

Electroacoustic devices are transducers that convert sound pressure into electrical signals (microphones), or electrical signals into sounds waves (speakers).

Sound information may be visualized as a pattern of air pressure. Microphones convert this information into a corresponding voltage pattern, while loudspeakers perform the opposite operation. A concept of the *fidelity* of audio equipment describes the accuracy of such transformations. High-fidelity audio systems generate minimal amounts of noise and sound distortion, providing a high quality recording and reproduction of sound.

There are two types of microphones – *passive* and *active*. Passive microphones derive the output solely from the absorbed acoustic power, while active microphones use an additional external source of power to generate an output. Any passive microphone can be inverted to make a speaker, although design modifications to adjust its acoustical and electrical characteristics are often necessary. An active microphone typically has no speaker analogy. Similarly, there are certain speaker designs that cannot be converted into a microphone.

A detailed discussion of technological approaches used for microphone and speaker design belongs to the field of acoustic engineering and is beyond the scope of this book. The subjects of interest in the current section are dynamic microphones and their analogs – loudspeakers – and two types of condenser microphones.

A *dynamic microphone*, illustrated in Figure 9.6.12, is an example of a passive device and one of the most commonly used acoustic-to-electric transducers. Its operation is based on the phenomenon of electromagnetic induction, described earlier in Chapter 6.

As shown in Figure 9.6.12, a dynamic microphone has three major parts – coupled magnets, a thin metallic diaphragm in the form of a cone, and a coil of wire (in some literature sources it is called a voice coil) fastened to the diaphragm. The sound waves represented by the changing air pressure cause movement of the cone, which in turn provides

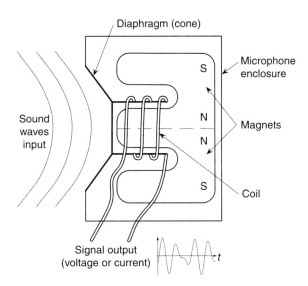

Figure 9.6.12 Dynamic microphone: schematic

for the motion of the coil in the magnetic field, generated by the two coupled permanent magnets. According to Faraday's law of induction, this change in the magnetic environment of the coil (conductor) causes a voltage to be induced in the coil. The induced voltage is proportional to the negative rate of change of the magnetic flux (see Section 6.3.1).

Modeling of this dynamic system starts with considering its mechanical and electrical subsystems separately. The mechanical subsystem is represented by the diaphragm and the coil. It can be approximated by a typical mass supported with a spring–damper assembly, as shown in Figure 9.6.13(a), where m represents the combined mass of the diaphragm and the coil, and k and c denote the stiffness and damping coefficients of the diaphragm, respectively. These constants depend on the material and geometry of the diaphragm. The parameter $F_S(t)$ is the force exerted on the diaphragm by the incoming sound waves, and $f_M(t)$ is the force exerted on the coil by the magnetic field.

The electrical subsystem is represented by the same diaphragm and coil assembly. It is modeled as a simple circuit, shown in Figure 9.6.13(b), where R is the combined resistance of the diaphragm and the coil, L denotes the inductance of the coil solenoid, v_b is the induced voltage, and $i(t)$ is the coil current, which is also the current in the circuit.

We let the displacement of the diaphragm-coil assembly be $x(t)$. Then, applying the Newton's second law to the mechanical subsystem, we obtain

$$m\ddot{x}(t) + c\dot{x}(t) + kx(t) = F_S(t) + f_M(t) \tag{9.6.38}$$

Recalling the discussion on magnetic forces in Section 6.3.1, we derive the expression for $f_M(t)$ using Eq. (6.3.9):

$$f_M(t) = nlBi(t) \tag{9.6.39}$$

where B is the magnetic flux density, $i(t)$ is the induced current, l is the length of a single coil segment, and n denotes the number of loops in the coil.

For the depicted microphone it can be safely assumed that the magnetic flux density is uniform, and perpendicular to the current vector.

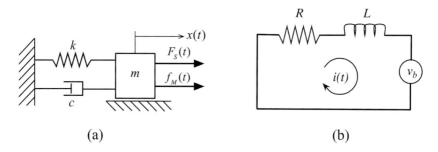

(a) (b)

Figure 9.6.13 Dynamic microphone: (a) mechanical and (b) electrical subsystems

9.6 Sensors, Electroacoustic, and Piezoelectric Devices

Since B, l, and n are constants, $K_M = nlB$. Then, the governing equation for the mechanical subsystem becomes

$$m\ddot{x}(t) + c\dot{x}(t) + kx(t) = F_S(t) + K_M i(t) \tag{9.6.40}$$

As derived earlier (Eq. (6.3.13)), the induced voltage is proportional to the velocity of the conductor, moving in the magnetic field:

$$v_b = K_b \dot{x}(t) \tag{9.6.41}$$

where K_b is a constant coefficient that depends on the coil geometry and magnetic flux density.

Since there is no voltage input in the system, the governing equation for the electrical subsystem is

$$Ri + L\frac{di}{dt} + v_b = 0 \tag{9.6.42}$$

The microphone dynamics is modeled by Eqs. (9.6.40)–(9.6.42).

If the input into the system is the force exerted by the sound waves, and the output is the induced voltage, then the system block diagram and transfer function representations are derived as follows.

Taking the Laplace transform of the governing equations, we obtain

$$\begin{cases} (ms^2 + cs + k)X(s) = F_S(s) + K_M I(s) \\ V_b = K_b\, sX(s) \\ (R + Ls)I(s) + V_b = 0 \end{cases} \tag{9.6.43}$$

The expression for V_b is substituted into the last line of Eq. (9.6.43) and we solve for $I(s)$, expressing it in terms of $X(s)$. Then, we substitute the result into the first line of Eq. (9.6.43) and derive the transfer function $\dfrac{X(s)}{F_S(s)}$. The resulting block diagram is shown in Figure 9.6.14.

The derivation of the system transfer function from this block diagram is straightforward:

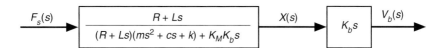

Figure 9.6.14 Dynamic microphone with voltage output: block diagram

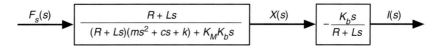

Figure 9.6.15 Dynamic microphone with current output: block diagram

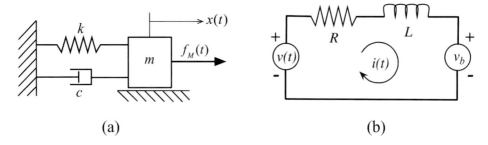

Figure 9.6.16 Loudspeaker: (a) mechanical and (b) electrical subsystems

$$\frac{V_b(s)}{F_S(s)} = \frac{(R+Ls)K_b s}{(R+Ls)(ms^2+cs+k)+K_M K_b s} \qquad (9.6.44)$$

If the system output is the induced current in the coil, the block diagram is as shown in Figure 9.6.15, and the corresponding system transfer function is

$$\frac{I(s)}{F_S(s)} = -\frac{K_b s}{(R+Ls)(ms^2+cs+k)+K_M K_b s} \qquad (9.6.45)$$

Since the induced voltage is proportional to the speed of motion of the diaphragm–coil assembly, dynamic microphones are also known as *velocity-sensitive microphones* or simply as *velocity microphones*. They are low-noise microphones, require no external power inputs, have a relatively rugged construction, and can be miniaturized easily. Combined with a reasonably low cost, these advantages make dynamic microphones very attractive for broadcasting and sound recording. The disadvantage of dynamic microphones is their low sensitivity, which calls for an obligatory signal amplification.

The inverse of a dynamic microphone is a loudspeaker. The speaker construction is the same as depicted in Figure 9.6.12, only the input is an electric signal, and the output is the sound waves. The dynamic system of a loudspeaker is shown in Figure 9.6.16.

The only force acting on the diaphragm–coil assembly is the magnetic force $f_M(t)$. Similarly to dynamic microphones, the expression for $f_M(t)$ is derived using Eq. (6.3.9):

$$f_M(t) = nlBi(t) \qquad (9.6.46)$$

9.6 Sensors, Electroacoustic, and Piezoelectric Devices

where B is the magnetic flux density, $i(t)$ is the circuit current, l is the length of a single coil segment, and n denotes the number of loops in the coil.

As before, letting $K_M = nlB$, we obtain the governing equation for the mechanical subsystem:

$$m\ddot{x}(t) + c\dot{x}(t) + kx(t) = f_M(t) = K_M i(t) \quad (9.6.47)$$

The governing equation for the electrical subsystem is

$$Ri + L\frac{di}{dt} + v_b = v(t) \quad (9.6.48)$$

Considering the expression derived earlier for the induced voltage (Eq. (9.6.41)), the loudspeaker model is a set of three equations: Eqs. (9.6.47), (9.6.48), and (9.6.41). In the Laplace domain it becomes

$$\begin{cases} (ms^2 + cs + k)X(s) = K_M I(s) \\ V_b = K_b s\, X(s) \\ (R + Ls)I(s) + V_b = V(s) \end{cases} \quad (9.6.49)$$

Using the same procedure as described earlier for the dynamic microphone, the block diagram representation is constructed and the system transfer function derived. For the speaker system there is a single option of the input–output pair – the applied voltage is the input, and the generated sound signal, represented by the displacement of the diaphragm, is the output.

The block diagram is shown in Figure 9.6.17, and the corresponding system transfer function is

$$G(s) = \frac{X(s)}{V(s)} = \frac{K_M}{(R + Ls)(ms^2 + cs + k) + K_M K_b s} \quad (9.6.50)$$

Even though this speaker is an inverted dynamic microphone, its proportions and physical construction are dissimilar, since this device is optimized for the different set of electrical and acoustical characteristics.

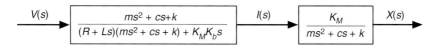

Figure 9.6.17 Loudspeaker: block diagram

Example 9.6.4 Dynamic response of a loudspeaker

The derived loudspeaker transfer function presented in Eq. (9.6.50) is of third order. Given some numerical values for the system parameters and voltage input:

```
(* setting up the numerics *)
km = 16.0; r = 12.1; l = 10^-3; m = 0.002; c = 0.02; k = 5.9*10^5; kb = 12.5;
(* setting up the expression *)
expr = km / ((r + l*s) * (m*s^2 + c*s + k) + km*kb*s);
(* derive the transfer function *)
tfm = TransferFunctionModel[expr, s] // Simplify
(* derive and plot system response x *)
resp = OutputResponse[tfm, 0.5*UnitStep[t], {t, 0, 0.01}];
pl1 = Plot[10^6 * resp, {t, 0, 0.01}, PlotRange → All, PlotStyle → {Black, Thick},
   GridLines → Automatic, GridLinesStyle → Gray, Frame → True, FrameStyle → Gray,
   FrameTicks → {{Automatic, None}, {Automatic, None}},
   FrameTicksStyle → Directive["Label", Black, 12], ImageSize → 400,
   AspectRatio → 1/1.5,
   FrameLabel → {Style["Time, seconds", Black, 16], Style["x, μm", Black, 16]}]
```

(a)

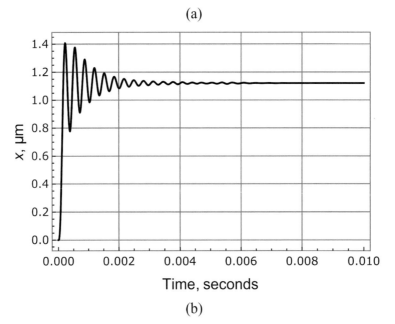

(b)

Figure 9.6.18 Dynamic response of the loudspeaker: (a) Mathematica code; (b) response graph

$$K_M = 16.0 \text{ N/A}, \quad R = 12.1 \text{ }\Omega, \quad L = 10^{-3}\text{H}, \quad K_b = 12.5 \text{ V·s/m}$$
$$m = 0.002 \text{ kg}, \quad k = 5.9 \times 10^5 \text{ N/m}, \quad c = 0.02 \text{ N·s/m}, \quad v = 0.5 \text{ V}$$

We can derive the dynamic response using the Mathematica functions `TransferFunctionModel` and `OutputResponse`.

The code for generating the dynamic response is shown in Figure 9.6.18(a) and the response itself is shown in Figure 9.6.18(b).

As seen from the graph in Figure 9.6.18(b), this system is fast – it reaches the steady state in approximately 6 milliseconds.

The input voltage to a loudspeaker can be viewed as an electrical "image" of the sound, emitted by an audio signal source, typically either a recording or a microphone. This electrical image is a signal that has the same frequency, amplitude, and waveform (harmonic content) as the original sound. The constant voltage input used for this simulation is just a single component of that image. When a voltage is applied to the voice coil of the loudspeaker, the electrical current starts flowing through this coil in the magnetic field. A magnetic force is generated and the voice coil starts moving, driving the cone of the loudspeaker and producing sound in the air. One of the conditions of reproducing the sound pressure variations of the original signal is a fast convergence of the system response. To better visualize the importance of fast convergence, a simulation demonstrates the system response to a piecewise-continuous input that is a square-wave approximation of a small segment of a sound wave (see Figure 9.6.19).

As the simulation results show, while a response convergence to the input is, in general, achieved, significant imperfections are observed. They are particularly pronounced at the points of input discontinuity. Also, when the duration of some constant input is small (for example, inputs applied from 0 to 0.01 s. and from 0.03 s to 0.04 s), the response still exhibits visible oscillations instead of converging to a constant steady state.

Returning to the transfer function of this loudspeaker system, we compute its poles using the Mathematica function `TransferFunctionPoles` obtaining the following results: $p_1 = -9617.4, \quad p_{2,3} = -1246.28 \pm 19225j$. This dynamic system has one real pole and two complex ones that are complex conjugates. Since the real parts of all three poles are strictly negative, the system is stable (see Section 7.2.2 for details). The ratio of the real parts of these poles $\dfrac{Re(p_1)}{Re(p_{2,3})} = \dfrac{-9617.4}{-1246.28} \simeq 7.7$ shows that the complex poles are dominant, and this system can be approximated by the second-order system. The transfer function of such approximation is derived in the following way (see Section 7.1.6 for details).

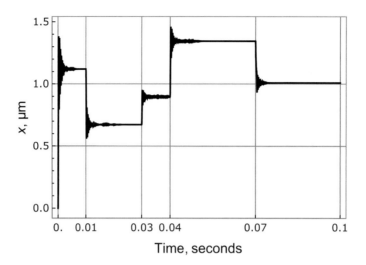

Figure 9.6.19 Dynamic response of the loudspeaker to a piecewise-continuous input

For a third-order transfer function

$$G(s) = \frac{a\omega_n^2}{(s-a)(s^2 + 2\zeta\omega_n s + \omega_n^2)}$$

where the complex conjugate poles $p_{cc} = -\zeta\omega_n \pm j\omega_n\sqrt{1-\zeta^2}$ are dominant, the second-order system approximation transfer function is

$$G_{appr}(s) = \frac{\omega_n^2}{s^2 + 2\zeta\omega_n s + \omega_n^2}$$

(the contribution of the nondominant real pole is neglected, and the numerator of the transfer function is divided by the magnitude of the nondominant pole).

Therefore, for the original transfer function of the given loudspeaker dynamic system:

$$G(s) = \frac{8 \times 10^6}{s^3 + 12110s^2 + 3.95 \times 10^8 s + 3.57 \times 10^{12}} = \frac{8 \times 10^6}{(s + 9617.43)(s^2 + 2492.57s + 3.71 \times 10^8)}$$

the approximation is

$$G_{appr}(s) = \frac{8 \times 10^6 / 9617.43}{s^2 + 2492.57s + 3.71 \times 10^8} = \frac{831.823}{s^2 + 2492.57s + 3.71 \times 10^8}$$

This approximation is possible, but since the ratio of the real parts of the dominant and nondominant poles is less than 10, the quality of such an approximation is not very good, which is evidenced by the simulation results in Figure 9.6.20.

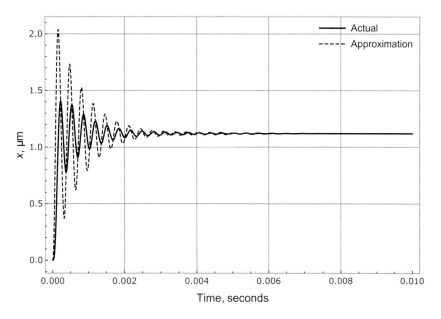

Figure 9.6.20 Dynamic response of the loudspeaker: actual system vs. a second-order approximation

If the inductance of the given loudspeaker system was so small that it could be neglected, the system would become second order instead of the third order, with the transfer function

$$G_{2nd}(s) = \frac{K_M}{R(ms^2 + cs + k) + K_M K_b\, s} = \frac{661.16}{s^2 + 8274.5s + 2.95 \times 10^8}$$

Its settling time decreases (see Figure 9.6.21(a)), and it will follow a piecewise-continuous input much more closely (see Figure 9.6.21(b)), even though there are still significant imperfections at the points of input discontinuity. These "spikes" are higher than they were for the original third-order system because of the higher overshoot of the second-order system (see Figure 9.6.21(a)).

An amplifier is an integral part of any high-fidelity speaker, since the electrical signal input needs to be large enough to provide for the coil vibrations of sufficient amplitude to reproduce the sound pressure pattern of the recorded signal.

Some important considerations of a good speaker design include an even frequency response so that no frequency is unduly weakened or emphasized, and a minimal sound distortion, i.e., minimal creation of new frequencies in the output signal. To achieve the typical frequency range of the high-fidelity reproduction of sound (50 Hz to 15 kHz) in a single speaker is difficult since a flawless generation of high and low frequencies requires different designs of the diaphragm. To produce high frequencies, a loudspeaker diaphragm needs to be very light to be able to rapidly respond to the input signal. For low frequencies, a large and relatively heavy diaphragm performs best. This calls for more power to be

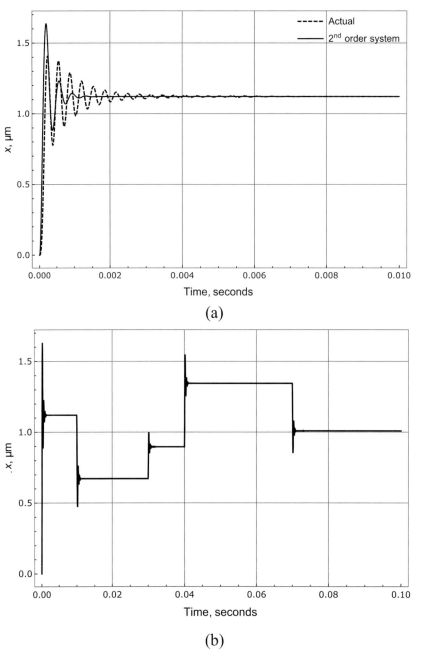

Figure 9.6.21 Dynamic response of the loudspeaker: (a) original system vs. the second-order system; (b) response of the second-order system to a piecewise-continuous input

9.6 Sensors, Electroacoustic, and Piezoelectric Devices

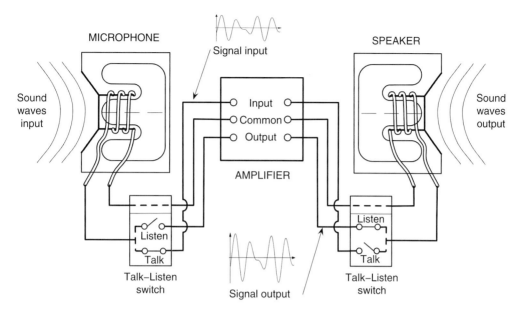

Figure 9.6.22 Simple intercom assembly

supplied to make this massive structure vibrate. Additionally, since human sound perception discriminates against lower frequencies, they require more acoustic power to be sufficiently audible. Hence, the customary design approach involves subdividing the full range of sound into two or three overlapping parts, and using multiple speakers, each optimized for its own frequency range. An assembly, which is relatively inexpensive but of sufficient sound quality, has two speakers, one covering the 50–1000 Hz range, the other the 500–15 kHz range. The higher-end speakers employ at least three-speaker assembly – a low-frequency (*woofer*), high-frequency (*tweeter*), and mid-range speaker to provide for a smoother frequency response.

In a simple intercom assembly, a small dynamic speaker, described above, can be used at both ends – as a microphone and as a loudspeaker, as illustrated in Figure 9.6.22.

Depending on the position of the "*talk–listen*" switch at the end device of the depicted intercom, this device acts either as a microphone or a speaker. A signal from the microphone end serves as an input for the amplifier (typically an op-amp, as described in Section 4.7), and the resulting amplified electrical signal becomes an input into the speaker end of the intercom. The low cost, relatively rugged construction, and simplicity of this intercom assembly make it very attractive for a variety of uses, despite its at-best average sound quality.

A *condenser microphone*, illustrated in Figure 9.6.23, is an example of an active device. Its superior frequency response makes it the microphone of choice for many sound-recording applications despite its high cost and the need for an external power supply.

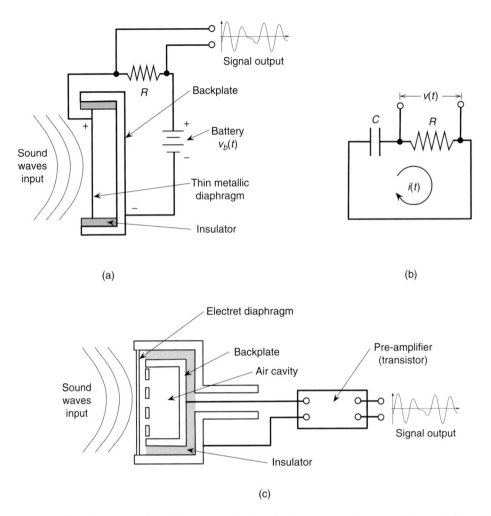

Figure 9.6.23 Condenser microphone: (a) traditional design – schematic; (b) traditional design – electrical subsystem; (c) electret-based design

In a condenser microphone the sound pressure changes a distance between two charged plates, subsequently changing the capacitance. This approach to microphone design employs two major approaches: (a) traditional, where the capacitor is represented by a diaphragm and a metal backplate (see Figure 9.6.23(a)), and (b) electret-based (see Figure 9.6.23(c)).

In a traditional condenser microphone, the diaphragm and the backplate form a capacitor, which is charged by a battery as shown in Figure 9.6.23(a). As discussed in Section 4.1.2, the capacitance of a parallel-plate capacitor depends on its geometry and the material, and is represented mathematically with Eq. (4.1.13). The sound pressure, acting on the microphone, moves the diaphragm, modifying the distance between it and the backplate.

Considering that, in a typical construction of this microphone, air is the dielectric substance between the capacitor plates, the capacitance is expressed as

$$C = \varepsilon A \frac{1}{d} = \frac{Q}{v_b} \qquad (9.6.51)$$

where Q is the stored charge, v_b denotes the voltage provided by the battery, ε is the permittivity, A is the area of the plates, and d is the distance between plates.

Since A, ε, and v_b are constant, it can be stated that the charge on the capacitor plates is inversely proportional to the distance between them:

$$Q = \varepsilon A \, v_b \frac{1}{d} \qquad (9.6.52)$$

When the distance between the diaphragm and the backplate changes, the stored charge also changes, causing the electric current in the microphone circuit, shown in Figure 9.6.19(b). From the definition of current (Eq. (4.1.1)):

$$i = \frac{\Delta Q}{\Delta t} = \varepsilon A v_b \frac{\Delta(1/d)}{\Delta t} \qquad (9.6.53)$$

Then, the voltage (output signal) measured across the resistor R (Eq. (4.1.9)) is

$$v(t) = iR = \varepsilon A v_b R \frac{\Delta(1/d)}{\Delta t} \qquad (9.6.54)$$

and is proportional to the inverse of the distance between the diaphragm and the backplate of the microphone. Hence, $v(t)$ can be viewed as an image of the pattern of the sound pressure that moves the diaphragm.

If we assign the following numerical values to the microphone parameters:

$\varepsilon = 8.854 \times 10^{-12}$ F/m, $A = 1 \times 10^{-4}$ m^2, $R = 100$ MΩ, $v_b = 0.2$ V and assume that the inverse of the distance between the diaphragm and the backplate changes in a periodic fashion is $\sin(t) + \sin(0.5t) + \sin(2.5t)$, the resulting image is shown in Figure 9.6.24.

Electret-based condenser microphones, as shown in Figure 9.6.23(c), use an electret material for their diaphragms. Being very thin (with a thickness of the order of a thousandth of an inch), an electret diaphragm is typically backed by a thin metallic membrane.

Since electrets have a permanent fixed electrical polarization (see Section 9.6.2), there is no need for an internal battery to charge the capacitor plates as in the traditional condenser microphones – the backplate is charged by the electrical field of the electret diaphragm.

An electret-based microphone operation is the same as discussed above: the sound pressure moves the diaphragm, thus changing the capacitance and causing an electric current in the microphone circuit. The recorded output signal is an image of the applied sound. Since

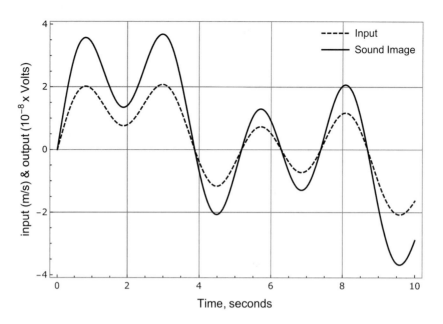

Figure 9.6.24 Dynamic response of a condenser microphone

the capacitance of an electret-based microphone is small, the output signal requires amplification. Hence, a pre-amplifier is an integral part of this microphone's electrical subsystem.

While the recorded sound quality of electret-based microphones is average, these devices are attractive due to their low cost, robust structure, and no need for an external power source. Since electret materials can be dependably manufactured according to exact specifications, electret-based microphones can be custom-made for a specific application or project. They also can be easily miniaturized, which prompts their main use for portable tape-recorders, hearing aids, and other small-scale devices.

The major disadvantage of electret-based microphones lies in the fact that electrets may lose their charge with time, thus making the device nonoperational.

REFERENCES

1. F. P. Beer, E. R. Johnston, Jr., P. Cornwell, and B. Self, *Vector Mechanics for Engineers: Dynamics*, 11th ed., McGraw-Hill Education, 2015.
2. F. P. Beer, E. R. Johnston Jr., and D. F. Mazurek, *Vector Mechanics for Engineers: Statics*, 11th ed., McGraw-Hill Publications, 2015.
3. J. L. Meriam, L. G. Kraige, and J. N. Bolton, *Engineering Mechanics: Dynamics*, 8th ed., Wiley, 2015.
4. Y. Cengel, M. Boles, and M. Kanoglu, *Thermodynamics: An Engineering Approach*, 9th ed., McGraw-Hill Education, 2018.
5. R. C. Hibbeler, *Fluid Mechanics*, 2nd ed., Pearson, 2017.

6. N. Lobontiu and E. Garcia, *Mechanics of Microelectromechanical Systems*, Kluwer, 2004.
7. N. Lobontiu, *System Dynamics for Engineering Students: Concepts and Applications*, 2nd ed., Academic Press, 2017.
8. N. Lobontiu, *Dynamics of Microelectromechanical Systems*, Springer, 2007.
9. R. S. Figliola and D. E. Beasley, *Theory and Design for Mechanical Measurements*, 5th ed., John Wiley & Sons, 2007.
10. W. J. Palm III, *System Dynamics*, 3rd ed., McGraw-Hill, 2014.
11. C. M. Close, D. K. Frederick, and J. C. Newell, *Modeling and Analysis of Dynamic Systems*, Wiley, 3rd ed., 2002.
12. D. G. Luenberger, *Introduction to Dynamic Systems: Theory, Models, and Applications*, Wiley, 1979.
13. R. S. Esfandiari and B. Lu, *Modeling and Analysis of Dynamic Systems*, CRC Press, 2010.
14. R. L. Woods and K. L. Lawrence, *Modeling and Simulation of Dynamic Systems*, Prentice Hall, 1997.

APPENDIX A
Units and Conversion Table

This appendix consists of five tables. Table A1 is about six base units of the International System (SI) that are used in this book. Tables A2 to A4 list some derived SI units for mechanical, electrical, magnetic, and thermal and fluid systems. Table A5 gives factors for unit conversion between the SI and the US customary systems.

Table A1 SI base units*

Quantity	Unit name	Symbol
Length	Meter	m
Mass	Kilogram	kg
Amount of substance	Mole	mol
Time	Second	s
Temperature	Kelvin	K
Electric current	Ampere	A

* The International System of Units comprises a total of seven base units. Besides the six units listed in Table A1, the base unit that is not used in this text is the candela (cd) for luminous intensity.

Table A2 Derived SI units for mechanical systems

Quantity	Unit name	Symbol	Derivation
Velocity (v)	Meter per second	m/s	1 km/h = 0.2778 m/s
	Kilometer per hour	km/h	
Acceleration (a)		m/s^2	
Angle (θ)	Radian	rad	1 rad = 57.2958 deg
	Degree	deg	1 deg = 0.017453 rad
Angular velocity (ω)	Radian per second	rad/s	1 rad/s = 9.5493 RPM
	Revolutions per minute	RPM	1 RPM = 0.1047 rad/s
Agular acceleration (α)		rad/s^2	
Mass moment of inertia (I)		kg·m^2	
Mass density (ρ)		kg/m^3	
Force (f)	Newton	N	1 N = 1 kg·m/s^2
Moment of force (M)	Newton-meter	N·m	
Torque (τ)			

Table A2 (cont.)

Quantity	Unit name	Symbol	Derivation
Impulse (I)	Newton-second	N·s	$I = \int F dt$
Pressure (p, P)	Pascal	Pa	1 Pa = 1 N/m^2
Mechanical stress (σ)			
Energy	Joule	J	1 J = 1 N·m
Work			
Power	Watt	W	1 W = 1 J/s
Coefficient of a translational spring (k)		N/m	
Coefficient of a viscous damper (c)		N·s/m	
Frequency (f)	Hertz	Hz	1 Hz = 1/s
Circular frequency (ω)		rad/s	

Table A3 Derived SI units for electrical and magnetic systems

Quantity	Unit name	Symbol	Derivation
Electrical charge (q)	Coulomb	C	1 C = 1 A·s
Voltage (v)	Volt	V	1 V = 1 W/A
Electric potential (e)			
Electrical resistance (R)	Ohm	Ω	1 Ω = 1 V/A
Electrical capacitance (C)	Farad	F	
Electrical inductance (L)	Henry	H	1 H = 1 V·s/A
Electrical power	Watt	W	1 W = J/s
Magnetic flux	Weber	Wb	1 Wb = 1 V·s
Magnetic flux density	Tesla	T	1 T = 1 Wb/m^2

Table A4 Derived SI units for thermal and fluid systems

Quantity	Unit name	Symbol	Derivation
Temperature	Celsius	°C	°C = K − 273.15
Heat (E)	Joule	J	1 J = 1 N·m
Heat flow rate (q_h)	Watt	W	1 W = 1 J/s
Thermal conductivity (k)		W/(m·K) W/(m·°C)	
Convective coefficient (h)		W/(m^2·K) W/(m^2·°C)	
Specific heat (c, c_P, c_V)		J/(kg·K) J/(kg·°C)	

Table A4 (cont.)

Quantity	Unit name	Symbol	Derivation
Thermal capacitance (C)		J/K	
Thermal resistance (R)		K·s/J	
Volume (V)		m^3	
Mass density (ρ)		kg/m^3	
Volume flow rate (q)		m^3/s	
Mass flow rate (q_m)		kg/s	$q_m = \rho q$
Pressure (p, P)		Pa	1 Pa = 1 N/m^2
Liquid head (h)		m	
Specific gas constant (R_g)		J/(kg·K)	
Fluid capacitance (C)		kg·m^2/N	
Fluid resistance (R)		Pa·s/kg	
Fluid inductance (I)		kg/m^4	
Pneumatic capacitance (C)		kg·m^2/N	
Pneumatic resistance (R)		Pa·s/kg	

Table A5 Unit conversion factors

Quantity	SI unit to US unit	US unit to SI unit
Length	1 m = 3.281 ft	1 ft = 0.3048 m
	1 km = 1000 m = 0.62137 mile	1 mile = 5280 ft = 1609.34 m
Volume	1 liter (L) = 1 dm^3 = 1.0564 quart	1 gal = 4 quarts = 3.875 L
Mass	1 kg = 1000 g = 0.06852 slug	1 slug = 14.594 kg
	1 kg = 2.20462 lb$_m$	1 lb$_m$ = 16 oz = 0.45359 kg
Force	1 N = 0.2248 lb$_f$	1 lb$_f$ = 4.4484 N
Pressure	1 Pa = N/m^2 = 0.9869 × 10^{-5} atm	1 atm = 101325 Pa
Energy	1 J = 1 N·m = 0.239 cal	1 cal = 4.184 J
	1 J = 0.7376 lb$_f$·ft	1 lb$_f$·ft = 1.3558 J
Power	1 W = 1 J/s = 3.413 Btu/h	1 Btu/h = 0.293071 W
	1 kW = 1000 W = 1.34102 hp	1 hp = 745.7 W
Temperature	K = °C + 273.16	°F = 1.8 °C + 32
	°C = (°F − 32)/1.8	°F = 1.8 K − 459.69
Heat	1 kJ = 1000 J = 0.947817 Btu	1 Btu = 1055.056 J

APPENDIX B
A Brief Introduction to MATLAB and Simulink

This appendix gives a brief introduction to MATLAB and Simulink. A detailed tutorial about these software tools is available on the publisher's website.

MATLAB is a programming platform designed specifically for engineers and scientists. It provides an interactive environment for technical computation, data analysis, graphics, and visualization. Simulink, which is built on top of MATLAB, is a block diagram environment for modeling, simulation, analysis, and model-based design of multi-component multi-field dynamic systems. It provides a graphical editor, customizable block libraries, and various solvers for algebraic and differential equations.

B.1 MATLAB Operators and Functions

Listed in Tables B1–B9 are some commonly used MATLAB functions. In the Command Window, typing `help Func` will display what the function named `Func` is, and how to use it.

B.2 Simulink Models

Because it is based on and integrated with MATLAB, Simulink is used with MATLAB. In other words, to use Simulink, MATLAB must be started first. Once the MATLAB Command Window is on, Simulink can be launched in one of the following two ways:

- Type `simulink` at the prompt; or
- Click on the Simulink icon on the MATLAB toolbar.

This will open a separate window – Simulink Start Page – from which one can create a new Simulink model, work on an existing model, see some demo examples on modeling and simulation with Simulink, and take other actions.

Model Window and Library Browser

To create a new Simulink model, click the "Blank Model" tab on the Simulink Start Page. This opens a model window. On the toolbar of the model window, click the Library Browser

Table B1 Operators and special characters

+	Addition	=	Assignment operator
−	Subtraction	>	Less than
*	Multiplication	<=	Less than or equal to
/	Division	>	Greater than
\	Left division	>=	Greater than or equal to
^	Power	==	Equal to
.*	Array multiplication operator	~=	Not equal to
./	Array division operator	&	Logical operator: AND
.^	Array exponential operator	|	Logical operator: OR
%	Percentage; comment	~	Logical operator: NOT

Table B2 Commonly used mathematical functions of MATLAB

Function	Description	Function	Description
`sqrt(x)`	Square root	`exp(x)`	Exponential
`sin(x)`	Sine	`log(x)`	Natural logarithm
		`log10(x)`	Common logarithm
`cos(x)`	Cosine	`abs(x)`	Absolute value
`tan(x)`	Tangent	`sign(x)`	Signum function
`cot(x)`	Cotangent	`max(x)`	Maximum value
`asin(x)`	Arcsine	`min(x)`	Minimum value
`acos(x)`	Arcosine	`real(x)`	Complex real part
`atan(x)`	Arctangent	`imag(x)`	Complex imaginary part
`sinh(x)`	Hyperbolic sine	`angle(x)`	Phase angle
`cosh(x)`	Hyperbolic cosine	`conj(x)`	Complex conjugate
`tang(x)`	Hyperbolic tangent		

Table B3 Matrix functions

Function	Utility
`det`	Determinant
`diag`	Diagonal matrices and diagonals of a matrix
`eig`	Eigenvalues and eigenvectors
`expm`	Matrix exponentials
`eye`	Identity matrix
`inv`	Matrix inverse
`norm`	Matrix and vector norms
`ones`	Matrix containing all elements ones
`zeros`	Matrix containing all elements zeros

Table B4 Functions for graphics

Function	Utility
plot	Generate xy plot
subplot	Divide a graphic window into multiple ones
grid	Add grid lines
legend	Graph legend
title	Place graph title
xlabel	x-axis label
ylabel	y-axis label

Table B5 Equation solvers

Function	Description
linsolve	Solve linear algebraic equations in matrix form
roots	Find polynomial roots
fzero	Find roots of nonlinear function of one variable
fsolve	Solve system of nonlinear algebraic equations of multiple variables
ode45	Solve linear and nonlinear differential equations

Table B6 Transfer function formulation

Function	Description
tf	Create a transfer function model
zpk	Create a zero-pole-gain model
series	Series interconnection of the two blocks (transfer functions)
parallel	Parallel interconnection of the two blocks (transfer functions)
feedback	Obtain the transfer function of a feedback control system

Table B7 State-space formulation

Function	Description
ss	Create a state-space model
ss2tf	State-space to transfer function conversion
tf2ss	Transfer function to state-space conversion

Table B8 Time response

Function	Description
`impulse`	Impulse response
`step`	Step response
`lsim`	Response to arbitrary inputs

Table B9 Frequency response

Function	Description
`freqresp`	Frequency response
`bode`	Bode plot of the frequency response
`nyquist`	Nyquist plot
`margin`	Gain and phase margins

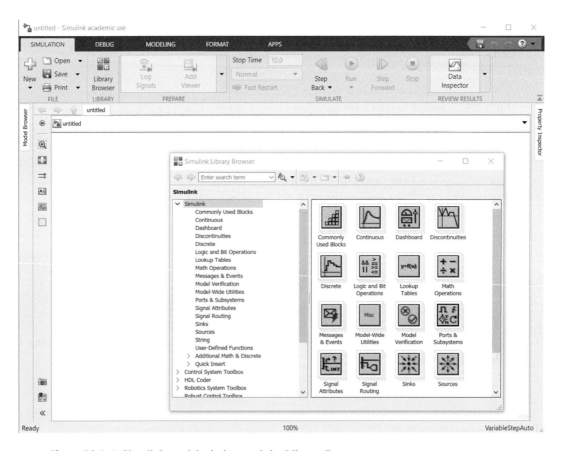

Figure B2.1 A Simulink model window and the Library Browser

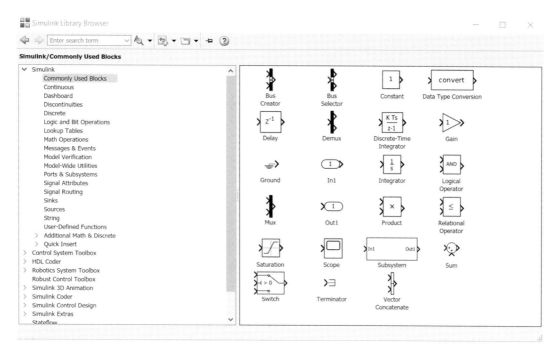

Figure B2.2 The Commonly Used Blocks Library

icon to open another window showing a collection of Simulink block libraries; see Figure B2.1. If a Simulink model has been created, the user can double-click the Simulink model file that is stored in the computer. This will open an existing model window.

In Simulink, a model is to be constructed in the model window by using these Simulink libraries. For instance, clicking on Commonly Used Blocks on the left panel opens the window with a set of blocks as shown in Figure B2.2. As can be seen from the figure, the blocks Gain, Sum, Integrator, and Product are similar to some components listed in Table 6.5.1, but the library has many other blocks with different functionalities. Clicking another entry on the panel, say Continuous, opens another library containing blocks like Differentiator, Transfer Fcn, State-Space, and PID Controller. There are many other block libraries on the panel.

Creation of a Model

As mentioned previously, a model is built in a model window. This is done by taking the following actions:

1. Select a wanted block from a block library and place it in the model window. This is done by first clicking and holding the block, and then dragging and dropping it anywhere in the window.

2. Connect two blocks by selecting the output port (with symbol >) and moving it to the input port (also with symbol >). Once the connection is made, a connecting line shows up, and the input and output ports disappear.
3. Double-click a block to assign numerical values or specifications if needed.
4. Add the Scope block from the Sinks library for outputting the simulation results on a selected variable.

These actions don't have to follow the order listed above, and they can be taken alternatively at the user's convenience.

There are two points regarding this process of model building. First, because Simulink is MATLAB-based, some blocks, such as Gain and Integrator, have default numerical values to begin with. These values often need to be changed according to the problem in consideration. To do this, double-click the block and assign the desired value(s). Second, when constructing a model, the user is recommended to follow the five-step procedure presented in Section 6.5. This is simply because the formation of a time-domain block diagram is quite similar to that for a Simulink model.

Simulation with a Model

A simulation with a Simulink model can be run in one of the following two ways:

- Click on the Run icon on the toolbar of the model window
- In the MATLAB Command Window, call the model at the prompt by typing sim ('model_name'), where model_name is the file name of the model.

On the toolbar of the model window, the user can specify a desired stop time for simulation. The default stop time is 10.0. After the simulation is executed, double-clicking the Scope blocks will show the simulation results.

As a demonstrative example, consider a spring–mass–damper system given by

$$m\ddot{x} + c\dot{x} + kx = q$$

with $m = 2$, $c = 3$, and $k = 32$. Assume a constant external force (input), $q(t) = 16\,u(t)$, where $u(t)$ is a unit step function. The initial conditions of the system are given as follows $x(0) = 0.2$, $\dot{x}(0) = -1$. Let the system outputs be the displacement x and the damping force $f_d = c\dot{x}$. A Simulink model for the system is built to compute the system outputs for $0 \leq t \leq 8.0$.

According to the time-domain block diagram of the system shown in Figure 6.5.2, we obtain the corresponding block components by the Library Browser and place them in a model window (saved with name SMD_Model); see Figure B2.3, where the far-left block is the Step block from the Sources library.

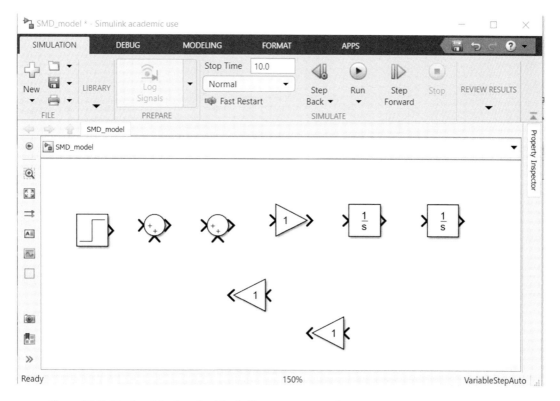

Figure B2.3 Blocks picked up for block diagram construction

Connecting all the blocks in the model window, naming the blocks, double-clicking each block for updates, and assigning the parameter values of the Gain blocks gives an incomplete block diagram shown in Figure B2.4.

Finally, adding a Gain block and two Scope blocks for the system outputs completes the construction of the block diagram for the spring–mass–damper system; see Figure B2.5.

Before running a simulation, three things need to be done. First, set up the stop time as 8.0 on the toolbar. Second, assign the initial conditions of the system, which is done by double-clicking each of the integrators and assigning an initial value (Figure B2.6). Third, for a small enough step size in the simulation, click on the Configuration Parameters icon ⚙ on the toolbar of the MODELING tab and, in the open window, choose the Fixed-step for Solver selection and set the Fixed-step size to a small number, say 0.005. Afterwards, save all changes made on the model.

Now, run the Simulink model. Once the job is done, double-click the scopes to see the system outputs in Figure B2.7. Here, in Figure B2.7(a), the x plot is obtained by the command Copy to Clipboard from the File menu of the Scope, and the curve in Figure B2.7(a) is an original plot of f_d from the Scope block.

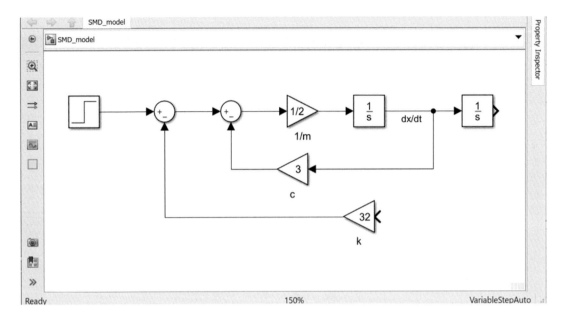

Figure B2.4 Connected blocks of an incomplete block diagram

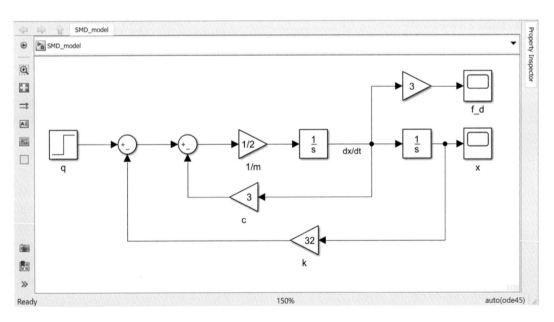

Figure B2.5 Completed block diagram of the spring-mass-damper system

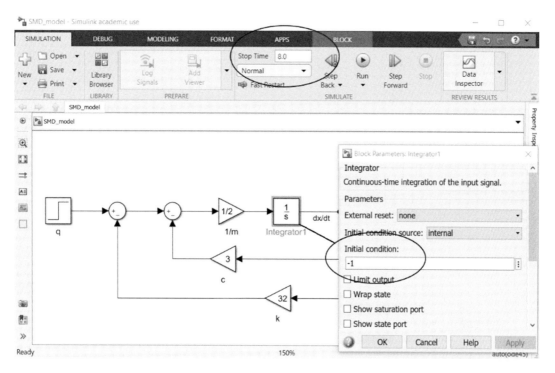

Figure B2.6 Completed block diagram for the spring–mass–damper system

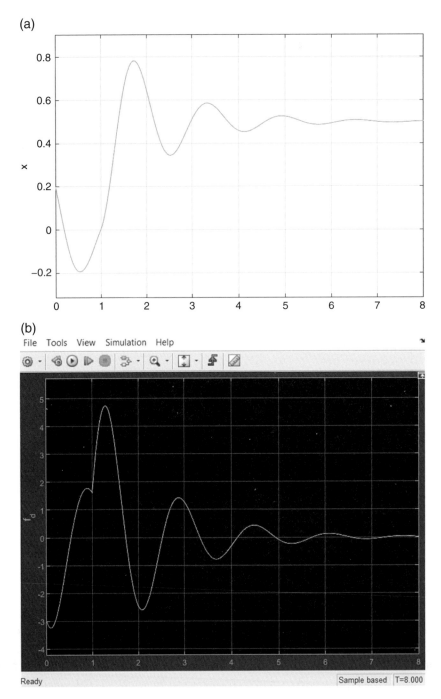

Figure B2.7 The outputs of the spring–mass–damper system: (a) displacement x; and (b) damping force f_d

APPENDIX C
A Brief Introduction to Wolfram Mathematica

This appendix gives a brief introduction to Wolfram Mathematica. A detailed tutorial about these software tools is available on the publisher's website and through Wolfram Research Institute documentation.

Wolfram Mathematica is a state-of-the art software tool for the widest variety of symbolic and numerical computations, as well as data visualization and analysis. Among scientists, Mathematica has been long renowned as the ultimate computation platform due to its capacity for superior symbolic manipulations, numeric computations, data analysis, graphics, and visualization.

Wolfram Language – the programming language of Mathematica – is a functional language, which allows for massive parallelization of computations. It is a knowledge-based interpreted language, with the scripts (notebooks) evaluating from top to bottom. Well-designed intuitive interfaces provide for a high degree of interactivity and programming ease. A vast library of built-in functions (version 12 of Mathematica has 6981 functions), a boundless capacity for extending the built-in capabilities with custom functions, and the high level of abstraction of the code allow for the fast development of relatively large projects by small teams.

C.1 Mathematica Operators and Functions

Mathematica uses a "notebook" interface to receive input and produce symbolic, numerical, graphical, or interactive output. When the software is launched, an empty notebook with a blinking cursor is displayed. Useful data-entering and notebook-formatting assistants – palettes – allow for typing equations just like they are written on paper. These palettes are available from the drop-down menu as shown in Figure C1 below (one of the most popular palettes – *Basic Math Input* – is shown at the left of the notebook).

Listed in Tables C1 to C9 are some commonly use Mathematica functions. Typing ? FunctionName in the notebook will display what the function named FunctionName is, and how to use it. If the name of the sought function is not known, Mathematica help allows to type a partial name, in response to which it will yield a list of possible matches.

772   Appendix C

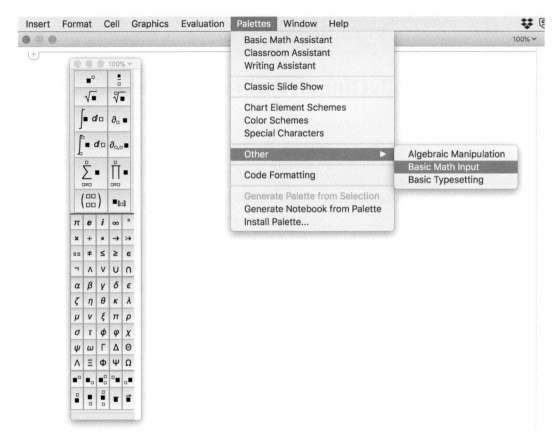

Figure C1 Mathematica palette: drop-down menu and palette

Table C1 Operators and special characters

+	Addition	=	Assignment operator
−	Subtraction	>	Less than
*	Multiplication	≤	Less than or equal to
/	Division	>	Greater than
^	Power	≥	Greater than or equal to
.	Matrix multiplication operator	==	Equal to
(* *)	Code comment	!=	Not equal to
{ }	indicates a list (non-scalar variable)	&&	Logical operator: AND
[ ]	argument of a function	||	Logical operator: OR
[[ ]]	accessing list element by address	!	Logical operator: NOT

Table C2 Commonly used mathematical functions of Mathematica (*x* indicates scalar argument, *xList* indicates list argument)

Function	Description
Sqrt [x]	Square root
Surd [x, n]	Real-valued *n*-th root of x; if x is symbolic, it is assumed real-valued
Exp [x]	Exponential
Log [x]	Natural logarithm (base-*e*)
Log10 [x]	Base-10 logarithm
Sign [x]	Signum function
Abs [x]	Absolute value
Round [x]	Yields the integer closest to the value of x
Floor [x]	Rounds x to the closest integer $n \leq x$
Ceiling [x]	Rounds x to the closest integer $n \geq x$
Mod [x, y]	Yields the remainder of division of x by y
Min [xList]	Minimum value of *xList*
Max [xList]	Maximum value of *xList*
Mean [xList]	Statistical mean of *xList*
Sum [xList]	Sum of elements of *xList*
Re [x]	Complex real part
Im [x]	Complex imaginary part
Arg [x]	Phase angle in radians
Conjugate [x]	Complex conjugate
ComplexExpand [x]	Expands symbolic expression x into real and imaginary parts
ExpToTrig [x]	Converts complex exponential x into trigonometric functions using Euler formula
TrigToExp [x]	Converts a trigonometric expression x into a sum of complex exponentials using Euler formula
Apart [x]	Decomposes a rational expression into partial fractions
Together [x]	Takes a sum of fractions to a common denominator, canceling factors in the result
Collect [expr, x]	Collects together terms in expression *expr* involving the same powers of objects matching x
Sin [x]	Sine
Cos [x]	Cosine
Tan [x]	Tangent
ArcSin [x]	Arcsine
ArcCos [x]	Arccosine
ArcTan [x]	Arctangent
Sinh [x]	Hyperbolic sine
Cosh [x]	Hyperbolic cosine
Tanh [x]	Hyperbolic tangent

Table C3 Matrix functions (*xMat* argument is a two-dimensional list that is a full array)

Function	Utility
`Det [xMat]`	Determinant
`DiagonalMatrix [xList]`	Creates a diagonal matrix with the elements of *xList* on the main diagonal
`DiagonalMatrix [xList,k]`	Creates a diagonal matrix with the elements of *xList* on the *k*-th diagonal
`IdentityMatrix [n]`	Creates an identity matrix of the size $n \times n$
`Dot`	Scalar product of matrices, automatically handling row and column vectors
`Cross [u, v]`	Vector (cross) product of two vectors **u** and **v**
`VectorAngle [u, v]`	Angle between vectors **u** and **v**
`Transpose [xMat]`	Matrix transpose
`Inverse [xMat]`	Matrix inverse
`MatrixPower [xMat]`	Power of numeric or symbolic matrix
`MatrixExp [xMat]`	Matrix exponential
`MatrixLog [xMat]`	Matrix natural logarithm
`Eigensystem [xMat]`	Generates a list of eigenvalues and eigenvectors
`Eigenvalues [xMat]`	Generates a list of eigenvalues
`Eigenvectors [xMat]`	Generates a list of eigenvectors
`Norm [xMat]`	Matrix and vector norm
`Normalize [xMat]`	Normalize vector *xMat* to unit length

Table C4 Graphics generation and styling functions

Function	Utility
`Plot [f, {x, x`$_{min}$`, x`$_{max}$`}]`	Generates a plot of a single function $f(x)$ on the interval $x \in [x_{min}, x_{max}]$
`Plot [{f`$_1$`, f`$_2$`, ..., f`$_n$`}, {x, x`$_{min}$`, x`$_{max}$`}]`	Generates a plot of n functions f_i of an independent variable x on the interval $x \in [x_{min}, x_{max}]$
`ParametricPlot [{f`$_x$`, f`$_y$`}, {t, t`$_{min}$`, t`$_{max}$`}]`	Generates a plot of a single function f defined parametrically, where its x- and y-components $f_x(t)$ and $f_y(t)$ are functions of an independent variable t, on the interval $t \in [t_{min}, t_{max}]$
`ParametricPlot [{{f`$_{1x}$`, f`$_{1y}$`}, {f`$_{2x}$`, f`$_{2y}$`}, ..., {f`$_{nx}$`, f`$_{ny}$`}}, {t, t`$_{min}$`, t`$_{max}$`}]`	Generates a plot of n parametric functions f_i, where the interval for the independent variable is $t \in [t_{min}, t_{max}]$
`PolarPlot [r, {`θ`, `θ_{min}`, `θ_{max}`}]`	Generates a polar plot of a single function $r(\theta)$, where r is the polar radius and θ is the polar angle, on the interval $\theta \in [\theta_{min}, \theta_{max}]$
`LogPlot [f, {x, x`$_{min}$`, x`$_{max}$`}]`	Generates a linear-logarithmic plot of a function $f(x)$, where x-axis is linearly scaled and y-axis is logarithmically scaled, on the interval $x \in [x_{min}, x_{max}]$

Table C4 (cont.)

Function	Utility
`LogLinearPlot[f,{x,x`$_{min}$`,x`$_{max}$`}]`	Generates a log-linear plot of a function $f(x)$, where x-axis is logarithmically scaled and y-axis is linearly scaled, on the interval $x \in [x_{min}, x_{max}]$
`LogLogPlot[f,{x,x`$_{min}$`,x`$_{max}$`}]`	Generates a log-log plot of a function $f(x)$, where both axes are logarithmically scaled, on the interval $x \in [x_{min}, x_{max}]$
`Graphics[g_primitives, styling_opts]`	Generates styled graphic primitives such as circles, polygons, lines, etc.
`Plot3D[f,{x,x`$_{min}$`,x`$_{max}$`}, {y,y`$_{min}$`,y`$_{max}$`}]`	Generates a plot of a function $z = f(x,y)$ on the intervals for its independent variables $x \in [x_{min}, x_{max}]$ and $y \in [y_{min}, y_{max}]$
`Plot3D[{f`$_1$`,f`$_2$`,...,f`$_n$`}, {x,x`$_{min}$`,x`$_{max}$`},{y,y`$_{min}$`,y`$_{max}$`}]`	Generates a plot of n functions f_i of two independent variables x and y on the interval $x \in [x_{min}, x_{max}]$ and $y \in [y_{min}, y_{max}]$
`ParametricPlot3D[{f`$_x$`,f`$_y$`,f`$_z$`}, {t,t`$_{min}$`,t`$_{max}$`}]`	Generates a plot of function f defined parametrically, where its x-, y-, and z- components $f_x(t), f_y(t)$, and $f_z(t)$ are functions of an independent variable t, on the interval $t \in [t_{min}, t_{max}]$
`Graphics3D[g_primitives, styling_opts]`	Generates styled three-dimensional graphic primitives such as spheres, cylinders, polyhedrons, etc.
	Plot styling options
`PlotRange`	Specifies the range of coordinates to be included in the generated plot
`PlotStyle`	Specifies styles in which plot elements need to be generated; common directives apply to line color, type (continuous, dashed, etc.), thickness, surface color and opacity, and point size
`PlotLegends`	For multiple graphs on the same plot specifies the labels for each graph and legend location on the plot figure
`PlotLabel`	Specifies plot title and its appearance
`Axes`	Specifies whether axes need to be drawn
`AxesOrigin`	Specifies where the drawn axes need to cross
`AxesLabel`	Specifies labels for the drawn axes
`AxesEdge`	For three-dimensional graphics specifies on which edges of the bounding box to draw the axes
`Frame`	Specifies whether to draw a frame around the two-dimensional plot object
`FrameLabel`	Specifies whether labels need to be placed on the edges of a frame, and on which edges they need to be
`GridLines`	For two-dimensional graphics specifies whether to draw the grid lines, and grid line spacing in x- and y- directions
`FaceGrids`	For three-dimensional graphics specifies on which faces of the bounding box to draw the grid lines, and grid line spacing in x-, y-, and z- directions

Table C5 Equation solvers

Function	Description
Solve [eqs,vars]	Solves a system *eqs* of algebraic equations or inequalities for the variables *vars*; yields a closed-form solution
NSolve [eqs,vars]	Finds numerical approximations to the solutions of the system *eqs* of algebraic equations or inequalities for the variables *vars*
FindRoot [eqs,x0]	Finds numerically local solutions of the system *eqs* of algebraic equations, starting from the point $x = x_0$
FindInstance [eqs, vars]	Finds numerically an instance of the variable *vars* values that satisfy the system *eqs* of algebraic equations
LinearSolve	Solves a system of linear algebraic equations in matrix form
Roots	Finds roots of a single polynomial
DSolve	Finds exact solutions to differential, delay, and hybrid equations
DSolveValue	
NDSolve	Finds numerical approximations to the solutions to linear and nonlinear ordinary differential equations (ODEs), partial differential equations (PDEs), differential algebraic equations (DAEs), delay differential equations (DDEs), integral equations, integro-differential equations, and hybrid differential equations
NDSolveValue	
ParametricNDSolve	Finds numerical approximation to the solution to ordinary and partial differential equations with parameters

Table C6 Transfer function formulation

Function	Description
TransferFunctionModel	Creates a transfer function model; can take a linear state-space model as an argument to generate the transfer function representation
TransferFunctionPoles	Gives a matrix of poles of the transfer function
TransferFunctionZeros	Gives a matrix of zeros of the transfer function
SystemsModelSeriesConnect	Series interconnection of the two blocks (transfer functions)
SystemModelParallelConnect	Parallel interconnection of the two blocks (transfer functions)
SystemModelFeedbackConnect	Obtains the transfer function of a feedback control system

Table C7 State-space formulation

Function	Description
`StateSpaceModel`	Creates a linear state-space model; can take a transfer function model as an argument to generate the state-space representation
`NonlinearStateSpaceModel`	Creates a nonlinear state-space model

Table C8 Response

Function	Description
`StateResponse`	Gives the numeric or symbolic state response of the state-space model to a given input
`OutputResponse`	Gives the numeric or symbolic output response of the state-space model to a given input

Note: to view a symbolic response in its simplest form, applying functions `Simplify` and `FullSimplify` to the generated response expression is recommended.

Table C9 Classical analysis and design

Function	Description
`RootLocusPlot`	Plot root locations as a parameter varies
`BodePlot`	Bode plot (magnitude and phase) of the frequency response
`NyquistPlot`	Nyquist plot (polar plot of the frequency response)
`GainPhaseMargins`	Gain and phase margins
`PIDTune`	Automatic PID tuning for any linear systems

Index

accelerometers 730
  capacitive 734–5
  electromechanical 730, 732–3
  piezoelectric 735–42
  vibrometer and 733–4
active-circuit analysis 277
  isolation amplifiers 277–81
  operational amplifiers 281–97
active elements 223, 230
active microphones 743
actuator 353, 590, 592
adder circuit 291, 292
adjoint matrix 30–1
algebraic manipulations 18–19, 36
alternating current (AC) 433, 558
aluminum–tungsten thermometer 701–2
analog devices 433
analogous relations
  element laws 12–13
  flow and effort 11–12
  variables and equivalent coefficients 13, 14
analysis, physical systems 4
anti-commutative property, cross product 22–3
armature 431–4
armature-controlled DC motors 436, 438
  angular velocity 440
  block diagram 441
  electrical subsystem 439
  expanded block diagram 441–2
  Kirchhoff's voltage law 439
  with load 448, 452–3
  mechanical subsystem 439
  state-space representation 442–3
  transfer function 441
asymptotes 623
  angles *vs.* $n - m$ 626
  root locus 624–5
asymptotically stable system 545–6
automatic control system 589

back emf, DC motor 439–40, 444
balanced force micro-accelerometer 734–5
band-pass passive filter 274–7
bandwidth-limited differentiator circuit 289
bandwidth-limited integrator circuit 287, 288
Bernoulli equation 370–4
bimetallic strip thermometer 697
  copper–steel and aluminum–tungsten 701–3
  dynamic modeling 698–701
  schematic and lumped-parameter model 699
  spiral 697, 698
Biot number 345–6
block diagrams 396, 398–9, 401
  armature-controlled DC-motor 441–2
  branch/takeoff point 414
  closed-loop 592
  DC motor with load 455
  dynamic system with 417
  electrical circuit 258–65
  equivalent transformation 415–16, 418–19, 423–4
  expanded 441–2, 445–6
  field-controlled DC-motor 446
  linear time-invariant dynamic system 414–15
  mechanical systems 181–5
  open-loop 592, 688
  parts of 462–3
  PID controller 602
  in *s*-domain 9–10, 182, 457, 461
  second-order system 420–1
  signal distribution 417–18
  summation point 414
  system transfer function 418–19
  in time domain 10–11, 457–69
  time-invariant systems 97–8

Bode diagram/plot 562–3
  first-order system 564–6
  second-order system 568–71
bounded input–bounded output (BIBO) 545–6
branches 623–4
branch point 414–15
breakaway point 625, 627
break frequency 565, 569
break-in point 625, 627

cantilever-beam accelerometer 737–8
capacitive accelerometers 734–5
capacitors 227–8
Cauchy form 401
cause-and-effect approach 414
closed-loop control system 590–2
  advantages 591
  block diagrams 592
  configuration 591
  coupled engine–propeller system 695–7
  disturbance rejection 598–9
  output of 600
  parameter variations 596–8
  PID gains effect 606
  root locus of 621–2
  speed control, DC motor 595
  step response 609
  system response improvement 599–601
commutator 435–6
complementary homogeneous equation 80–2
complete solution, linear differential equation 91–2
complex numbers
  algebraic manipulations 36
  complex conjugates 35–6
  definition 34
  magnitude and phase angle 34–5
  phase angles, quadrants 35

Index 779

trigonometric manipulations 36–7
computation software 13, 15
condenser microphone 753–6
conduction 336, 337
conservative force 119
constant-pressure process 355
constant-temperature process 356
constant-volume process 356
continuous-time models *vs.* discrete-time models 5
controlled mechanical system 399–401
control logic 590
control systems 589
   actuator 590
   closed-loop 590–2
   controller 590
   liquid-level 652–3
   open-loop 590
   PID gain tuning 638–9
   plant 589–90
   robotic system 648–50
   robustness of 596
   room-temperature 591–2
   sensor 590
   stability analysis 613–15
   steady-state error 615–19
   steady-state response 611–13
   thermal 650–2
   transfer function formulation 592–5
   transient response 611–13
   by zero placement 639–48
convection 337, 341
convolution theorem 52–3
copper–steel thermometer 701–2
corner frequency 565, 569
coupled engine–propeller system 686–7
   closed-loop system 695–7
   open-loop system 687–95
coupling relations 397
cover-up method 54
Cramer's rule 28–30, 144
critically damped system 517, 520, 523, 525
crossover points 627–8
cross product 21–3
current divider rule 237–9
current element 426–7
current-isolation amplifiers 279
current-to-voltage converter 296
cutoff frequency 271, 272

damping elements, rotational systems 149–51
damping force 123, 125, 770
damping ratio 518, 521, 532, 567–8, 571
damping states, second-order system 517–18
D'Arsonval meter/movement 430–3
DC electrical circuit 225
DC motors. *See* direct current (DC) motors
degrees of freedom (DOFs) 160, 179, 667, 669
diagonal matrix 24
differential equations
   applications 63
   linear and nonlinear 63–4
   linear ordinary 70–95
   partial 66–7
   spring–mass–damper dynamic system 64–5
   state equations 67–9
   two-mass dynamic system 65–6
   See also specific differential equations
differentiator circuits 288–90
dipolar-charge electrets 735
direct current (DC) motors 396
   armature-controlled 438–43
   bearings 434
   electrical subsystems 437–9
   elements 434
   field-controlled 443–7
   with load 447–56
   mechanical subsystems 437–9
   modeling 436–8
   operations 433–6
   rotor 434
   shaft 434
   simplification 456–7
   stator 434
   torque–current relations 437–8
direct integration 71–2
   first-order differential equation 72
   second-order differential equation 73
discrete-time models 5
displacement transmissibility 570, 571
distributed models *vs.* lumped models 4
DOFs. *See* degrees of freedom (DOFs)
dominant poles 543–4, 635
dot product 21–3

dry friction 125–7
dynamic modeling, half-car model 669–70
   Lagrangian approach 672–3
   Newtonian approach 670–2
dynamic responses
   Mathematica 198–203, 297–306
   MATLAB 191–8, 378–83
dynamic systems
   analogous relations 11–13
   application problems 666–756
   classification 4–5
   combined systems and system modeling techniques 396–485
   components 1–2
   computation software 13, 15
   electrical systems 222–306
   feedback control systems 589–653
   input and output 2–3
   mathematics fundamentals 16–102
   mechanical systems 114–203
   system model representations 6–11
   system response analysis 502–77
   thermal and fluid systems 334–83
   units 15

electret(s)
   -based microphones 755–6
   dipolar-charge 735
   piezoelectric crystal 736
   real-charge 735
electrical circuit 223. *See also* electrical systems
electrical subsystem 397–8, 425, 442
   condenser microphone 754
   DC motor 437–9
   dynamic microphone 744
   field-controlled DC motor 443–5
   galvanometer 432
   loudspeaker 746
   open-loop system 687, 688
electrical systems 222–3
   active-circuit analysis 277–97
   block diagrams 258–65
   current and voltage 224–5
   definition 223
   fundamentals 223–30
   impedance 230–41
   Kirchhoff's laws 242–4

electrical systems (cont.)
   Mathematica, dynamic responses 297–306
   passive-circuit analysis 244–57
   passive filters 271–7
   state-space representation 265–70
electroacoustic devices 742–56
electromagnetism 425–30
electromechanical systems 396, 425
   D'Arsonval meter 430–3
   electromagnetism 425–30
electromotive force (emf) 429
electronic cigarettes 702
energy conservation 338–9
energy functions 179
   kinetic energy 177–8
   potential energy 119, 178
   Rayleigh dissipation function 178
equations of motion derivation
   rotational systems 145–53
   translational systems 130–8
error signal 416
Euler's formula 34, 37, 44, 520
Evans, W. R. 619
exponential function 43–4, 47

Faraday's law of induction 429, 744
feedback control systems 590
   root locus 619–48
   rotor system 686–7
   Simulink model 611
   step response 637, 642, 646
   time response 613, 615
   by zero placement 640–1
   See also closed-loop control system; control systems
ferroelectrics 736
field-controlled DC motor 436
   block diagram 446
   with load 450
   mechanical and electrical subsystems 443–5
   state-space representation 446–7
final-value theorem 52, 507–8
finite-dimensional system 4
first-order differential equations
   direct integration 72
   particular solutions 82–4
   piecewise-continuous forcing function 78–9

separation of variables 73–4
first-order high-pass filter 289
first-order systems 457
   applications 508
   Bode diagram 564–6
   forced response 509–10
   free response 509
   frequency response 564–6
   general inputs, response to 512–13
   impulse response 510–11
   PID controller 603
   ramp response 511–12
   step response 511
   time constant estimation 513–14
fluid capacitance 356, 358–9
fluid inertance 353, 356–7, 368–70
fluid resistance 356, 360–2
fluid systems
   duct/pipe 354–5
   gases, equation of state 355–6
   hydraulic systems 353
   mass conservation 354
   pneumatic systems 353
   properties 334
   SI units 759
   See also liquid-level systems
forced response 70–1
   boundedness 550
   first-order systems 509–10
   second-order system 522
   stability by 545
   system response analysis 503–5
forces and moments 116–17
force transmissibility 570–1
free-body diagrams and auxiliary plots
   equations of motion derivation 130–6
   motor–propeller assembly 151–3
   rigid-body system 172
   rotational systems 151–3
   translational systems 127–9
free response 70–1
   damping cases 520–2
   first-order systems 509
   stability by 545–6
   system response analysis 503–5
frequency domain, system response analysis 558–71
frequency response 558
   advantages 558

   Bode diagram/plot 562–3
   first-order systems 564–6
   general concepts 559–61
   Nyquist plot 563–4
   plots 561–4
   second-order systems 567–71
   utilities 560
friction forces 125–7, 150–1

galvanometer 430–3
geared systems 162, 164–8
   kinematic relation 163
   spur gears 162–3
   torque ratio 163–4
   work done by torques 164
general differential equations 90

half-car model 667, 669–71
heat energy 336, 338
heating elements 704
   governing equations 704–6
   properties of 704
   stainless-steel 704
   temperature change and elongation 706–12
heat transfer
   conduction 336, 337
   convection 337
   radiation 337–8
   thermal resistance 340–1
higher-order systems, time responses 539–44
   dominant poles 543–4
   general 541–2
   model reduction 543–4
   third-order systems 540–1
high-fidelity audio systems 743
high-pass filters 272–4, 275
Hooke's law 121, 728
hydraulic systems 353, 370

ideal current source 223
ideal gas law 355
ideal voltage source 223
identity matrix 24
imaginary axis crossings 627
impedance
   circuit analysis 231
   current divider rule 237–9
   definition 230–1

Index   781

differentiator op-amp circuit 288, 289
equivalent form 231, 233–5
integrator op-amp circuit 285, 287
inverter op-amp circuit 285
lead-lag circuit 290–1
low-pass filter 271
op-amp 281–3
parallel connection 232–3
series connection 231–2
voltage divider rule 235–7, 239–41
voltage-isolation amplifier 278–9
impulse response
    damping cases 522–4
    first-order systems 510–11
    mechanical systems 549
inductors 228–9
inertia effect 368
infinite-dimensional system 4
initial-value theorem 51–2
input–output differential-equations 399
input–output relations 2–3, 258
integrator circuits 285–8
International System of Units (SI) 15
    base units 758
    conversion factors 760
    electrical and magnetic systems 759
    fluid capacitance 358
    mechanical systems 758–9
    thermal and fluid systems 759–60
    thermal capacitance 339
inverse Laplace transform 38
    imaginary and complex poles 58–9
    partial fraction expansion 53–8
    time-domain response 76
inverse piezoelectric effect 736
inverter (sign-changer) circuits 285
invertible matrix 30
isentropic process 356
isobaric process 355
isochoric process 356
isolation amplifiers 277–81
    non-inverting op-amps 294–5
isothermal process 356

kinetic energy 177–8
Kirchhoff's current law (KCL) 242, 247, 283, 284, 291, 293
Kirchhoff's voltage law (KVL) 242, 245, 258, 271, 276, 278, 279, 432, 439, 444

lag circuit 287
lag/firstorder low-pass filter 287
lag-lead op-amp circuit 290
Lagrange's equations 127, 179–81
Lagrangian approach 177, 670, 672–3
laminar flow 361
Laplace transform 37, 398, 440
    convolution theorem 52–3
    definition 38–9
    differentiation (derivative property) 48–9
    evaluation 42
    exponential function 43–4
    factoring 42
    final-value theorem 52
    functions, derivation 46–8
    governing equations 409, 412, 421
    initial-value theorem 51–2
    integration (integral property) 49–50
    linearity property 42
    multiplication by exponential and t 46
    pairs 39, 40
    piecewise-continuous functions 60–2
    poles and zeros, function $F(s)$ 39, 41
    ramp function 43
    sinusoidal function 44
    step function 43
    time-shifted function 44–5
lead circuit 289, 290
lead-lag op-amp circuit 290
Lenz's law 429
lever–spring system 155–6
linear differential equation 63–4
linear integro-differential expression 50–1
Linear models vs. nonlinear models 4–5
linear ordinary differential equations
    direct integration 71–3
    free response and forced response 70–1
    Laplace transform 74–9
    method of undetermined coefficients 80–92
    piecewise-continuous forcing functions 93–5
    separation of variables 73–4
linear resistor 227
linear state-space representations 402–3, 408

linear time-invariant systems 398
    block diagram 97, 414–15
    Laplace transform for 398
    spring–mass–damper system in 460–1
    stability states of 545–6
    steady-state response 414, 558
    time response 571–7
liquid-level systems
    dynamic modeling 363–8
    energy sources 363
    engineering applications 334
    flow source 363
    fluid capacitance 358–9
    fluid resistance 360–2
    hydraulic systems 356
    mass conservation 359–60
    orifice resistance formulas 362–3
    pressure source 363
    storage tanks 357
    water purification 711–23
loading effect 281
loop currents 245
loop method 245–7, 253
Lorentz's force law 426, 429
loudspeaker 743
    actual system vs. second-order approximation 750–1
    block diagram 747
    dynamic response 748–9
    input voltage 749
    mechanical and electrical subsystems 746
    original system vs. second-order system 751, 752
    piecewise-continuous input 749, 750
    simple intercom assembly 753
    transfer function 749–50
low-pass filters 271–3
lumped masses 120–1, 175

magnetic field strength 425
magnetic force 426–8
manual PID tuning 606
marginally stable system 546, 548
mass elements 147, 172
MATLAB 3, 99–100, 397, 573–6, 761
    bode diagrams 562
    dynamic responses 191–8, 378–83
    equation solvers 763

MATLAB (cont.)
  frequency response 764
  function blocks, nonlinear state equations 576–7
  functions for graphics 763
  linear/nonlinear systems, time response 573–7
  mathematical functions 762
  matrix functions 762
  Nyquist plots 563
  operators and special characters 762
  root locus 628, 631–2
  state-space formulation 763
  time response of car 683–6
  transfer function formulation 763
  See also Simulink
matrix algebra
  definition 24
  determinants 25–30
  inverse of 30–1
  operations 32–3
  single-matrix operations 24–5
matrix determinants 25–30
matrix inverse 30–1
maximum overshoot 530–2
mechanical subsystem 397–401, 425, 442
  DC motor 437–9
  DC motor with load 447
  field-controlled DC motor 443–5
  open-loop system 687, 688
mechanical systems 114
  block diagrams 181–5
  energy approach 177–81
  fundamental principles 115–20
  Mathematica 198–203
  MATLAB 191–8
  rigid-body systems, plane motion 169–77
  rotational systems 144–68
  state-space representations 185–91
  translational systems 120–44
mechanical workhorse 434
metric system 15
micro-electromechanical systems (MEMS) accelerometers 734, 737
microphones, dynamic 743
  condenser 753–6
  current output 746
  fidelity 743

loudspeaker 746–53
mechanical and electrical subsystems 744–6
parts 743
passive and active 743
sound waves 743–4
velocity-sensitive 746
voltage output 745
Modelica System Model 481–2
modeling and simulation 3
  by MATLAB/Simulink 464–75
  by Wolfram Mathematica 475–85
model parameters 515, 535
  graphic interpretation 529
moment of force 116–17
monocrystalline 736
motor–propeller assembly 152–3, 686–7
multiple inputs and multiple outputs (MIMO systems) 2, 411, 414, 421
multiplier (amplifier) circuits 283–5

NDSolve functionality 99, 100
net currents 245
Newtonian approach 177, 670–2
Newton's laws of motion 115–16
Nichols, N. B. 606
node method 247–9
node voltages 247
non-conservative force 119
non-inverting op-amp circuits 293–5
nonlinear air damping effect 692–5
nonlinear differential equation 63
nonlinear state equations, MATLAB 576–7
nonlinear systems, time response 571–7
nonsingular matrix 30
Norton's theorem 251, 255–6
null matrix 24
numerical integration method 572
Nyquist plot 563–4

Ohm's law 226, 227, 245, 247, 251, 278, 291, 293
open-loop control system 590
  block diagrams 592, 688
  configuration 591
  coupled engine–propeller system 687–95
  ideal case 688–90
  nonlinear air damping effect 692–5

output of 600
parameter variations 690–2
parts of 687–8
speed control, DC motor 595
open-loop poles/zeros 621, 623
operational amplifiers (op-amps) circuits
  adder, subtractor and comparator 291–2
  amplification factor 282
  converters 295–6
  differentiator 288–90
  ideal inverting 281
  integrator 285–8
  integrator–differentiator combination 290–1
  inverter (sign-changer) 285
  with load 296–7
  multiplier (amplifier) 283–5
  non-inverting amplifiers 293–5
  non-inverting input 281, 283
  typical inverting 282
ordinary differential equations (ODEs) 4
orifice resistance 362–3
overdamped system 517, 520–1, 523, 525

parallel-axis theorem 146, 147
parallel connection, impedance 232–3
partial differential equations (PDEs) 4, 66–7
partial fraction expansion, inverse Laplace transform 53–8
particle kinematics 117–18
particular solution 80, 82
  first-order differential equations 82–4
  second-order differential equations 85–91
passive-circuit analysis 244
  loop method 245–7, 253
  node method 247–9
  Norton's theorem 251, 255–6
  superposition theorem 249–50, 256–7
  Thevenin's theorem 250–1, 253–5
passive elements 223
passive filters 271
  band-pass filter 274–7
  high-pass filter 272–4, 275
  low-pass filter 271–2, 273

passive microphones 743
peak time 530, 531
performance specification parameters 529–36
permanent-magnet moving-coil movement 431
phase angles 34, 35, 529, 531, 569, 677
PID controller. *See* proportional–integral–derivative (PID) controller
PID feedback control law 50
piecewise-continuous forcing functions 93–5
piecewise-continuous functions 60–2
piezoelectric accelerometers 738–42
  construction of 737
  crystals 735–7
  electrets 735–6
  ferroelectrics 736
  inverse piezoelectric effect 736
  moderate-to-low cost 737
  monocrystalline materials 736
  Young's modulus 736–7
piezoelectric effect 736, 737
piezoresistive accelerometers 737–8
plant 397, 589–90
pneumatic systems 353
  basic elements 374–6
  fundamental principle 374
  modeling 376–8
polytropic process 356
potential energy 119, 178, 357
potentiometers 239–40
principle of impulse and momentum 120, 147
principle of work and energy 118–19, 146–7
proportional–integral–derivative (PID) controller 601–6
  algorithm 601–11
  block diagram 602
  first-order system 603
  gain tuning 606–10, 638–9
  manual tuning 606
  proportional (P) 602
  proportional plus derivative (PD) 593–4, 603
  proportional plus integral (PI) 603
  Simulink, system responses 611
  step-response curves 605
  tuning 606

Ziegler–Nichols gain tuning 606–10
pulleys 153–4

quarter-car suspension model 133–5, 142, 179, 667
quasi-static deflections of car 675–6

radiation 337–8, 341
ramp function 43
ramp response, first-order systems 511–12
Rayleigh dissipation function 178
real-charge electrets 735
rectangular pulse, piecewise-continuous function 61–2
resistive-heating elements 702. *See also* heating elements
resistors 226–7
resonant frequency 567–8
resonant response 538
resonant vibration 85–8
revisable adiabatic process 356
right-hand rule 21–3, 145, 426, 427, 430
rigid-body systems, plane motion
  free-body diagrams and auxiliary plots 172
  fundamental principle 171–2
  lumped masses and 175–7
  mass center and relative motion 169–71
  mass elements 172
  pulleys/disks with movable pins 172–3
  spring and damping elements 172
  wheel in rolling and sliding motion 173–5
rise time 530, 531
robotic system 648–50
root locus method 619
  analysis and design 632–48
  angle criteria 623
  asymptotes 623–6
  branches 623–4
  breakaway point 625, 627
  break-in point 625, 627
  closed-loop system 621–2
  concepts 620–2
  control system design 638–48
  crossover points 627–8
  feedback system 634–5

magnitude criteria 623
  vs. parameter k 620–1
  plot 620–1
  properties 622–8
  real-axis segments 623–4
  sketching 628–32
  stability analysis 632–5
  system response analysis 635–8
rotating shafts 150, 151, 157
rotational systems 144
  damping elements 149–51
  double-headed arrows 145
  free-body diagrams and auxiliary plots 151–3
  geared systems 162–8
  lever 155–6
  mass elements 147
  parallel-axis theorem 146, 147
  principle of impulse and momentum 147
  principle of work and energy 146–7
  pulleys 153–4
  rigid bodies 145–6
  rotating shafts 157
  simple pendulum 154–5
  slewing rigid bar 156–7
  spinning rigid disk 157
  spring elements 147–9
  transfer function formulation 168
  translational elements and 158–62
rotor system 686–90, 695–6
  nonlinear air damping 692–3
  parameters of 689
  simulink model 694
Routh array 551–4, 556
Routh–Hurwitz stability criterion 550–1, 614
  application 552, 554–5
  features 551
  necessary condition 551–2
  Routh array 552–4
  rules 553
  system orders two to four 555–8
Runge–Kutta method 99, 101, 573, 682–3

scalar product 21
s-domain block diagrams 9–10, 182, 457, 461
second-order band-pass filter 290–1

784   Index

second-order differential equations 49, 51, 291, 445
   direct integration 73
   Laplace transform 76–7
   particular solutions 85–91
second-order systems
   damping cases 517–19
   free response 520–2
   frequency response 567–71
   general form 526–8
   impulse response 522–4
   sinusoidal input to 538–9
   step response 525–6
   time responses 515–28, 537–9
   underdamped 528–37
seismograph 733
sensors 590
   accelerometers 730, 732–5
   definition 723
   displacement, sinusoidal motion 726–31
   strain gauge 724, 726
   Wheatstone bridge 724
series connection, impedance 231–2
settling time 530–4
simple pendulum 154–5
Simulink 3, 378–83, 464
   closed-loop system 695
   DC motor 471–3
   feedback control system 611
   MATLAB function blocks, nonlinear state equations 576–7
   model building 765–6
   model window and Library Browser 761, 764–5
   nonlinear system 575–6
   open-loop system 693–4
   simulation 766–70
   solution by 573–6
   system responses by 611
   on time-domain block diagrams 465–9
   transfer functions and state-space blocks 469–75
single-input single-output (SISO) system 2
single-matrix operations 24–5
sinusoidal function 44, 668
sinusoidal input, second-order systems 537
   damped system 538–9

undamped systems 537–8
SI units. *See* International System of Units (SI)
slewing rigid bar 156–7
specific heat capacity 338
spinning rigid disk 157
spring elements 147–9, 172
spring–mass–damper systems 4, 6, 95, 571
   block diagram of 9, 97, 767–9
   capacitance, resistance, and inertance 12–13
   free-body diagrams 128, 129
   free response and forced response 71
   mechanical systems and fluid systems 13, 14
   nonlinear differential equation 11
   second-order linear differential equation 64–5
   second-order systems 515
   state variables 8
   transfer function 7, 95
spur gears 162–3
stability analysis 544–5
   conditions in pole locations 546–50
   definitions 545–6
   feedback control system 613–15
   root locus method 632–5
   Routh–Hurwitz stability criterion 550–8
   states of 545–6
   systems of order 555–8
stable system 546–8
state equations 67–9, 186
   linear and nonlinear 401
   numerical integration 98–102, 572
state-space representations 7–9, 399, 414
   differential equation 405–7
   dynamic system 402
   electrical systems 265–70
   mechanical systems 185–91
   state/output equations 401–2
   vibration analysis of car 673–5
stator 434
steady-state error 615–19
steady-state response
   final-value theorem 507–8
   frequency response 560
   linear time-invariant system 558
   transient response and 505–7

steady-state vibration of car 676–81
step function 43
step response
   closed-loop system 609
   curves PID controller 605
   damping cases 525–6
   feedback system 637
   first-order systems 511
   third-order systems 544
   underdamped second-order systems 528–37
storage tanks 357
subtractor circuit 292
summation point 414
superposition theorem 249–50, 256–7
system analysis 398
system-level modeling 396–8
system modeling techniques 398–401
   block diagrams 414–24
   governing equations 399
   input–output differential-equations 399
   state-space representations 401–8
   transfer function formulations 408–13
system model representations
   s-domain block diagram 9–10
   state-space representation 7–9
   time domain, block diagram in 10–11
   transfer function formulation 6–7
system order and input type 508

takeoff point 414
thermal control system 650–2
thermal systems
   Biot number 345–6
   energy conservation 338–9
   engineering applications 334
   free-body diagrams and auxiliary plots 346–53
   heat energy 338
   heat transfer 336–8
   linear lumped-parameter models 335
   MATLAB and Simulink, dynamic responses 378–83
   properties 334
   SI units 759
   thermal capacitance 339–40
   thermal energy 335–6
   thermal resistance 340–5
Thevenin's theorem 250–1, 253–5

Index   785

third-order differential equation 77–8
third-order systems 540–1
three-degrees-of-freedom (3-DOF) model 667, 669
three-term controller 601
time constant, first-order systems 513–14
time-delay parameter 44
time-domain block diagram 457–64
    nonlinear system 11
    Simulink models on 465–9
time domain, system response analysis 502–44
    concepts 503–8
    first-order systems 508–14
    higher-order systems 539–44
    second-order systems 515–28
    sinusoidal input, second-order systems to 537–9
    underdamped second-order systems 528–37
time-invariant linear system 408–9
time-invariant models *vs.* time-variant models 5
time-invariant systems 95
    block diagrams 97–8
    transfer functions 95–7
time responses
    car moving over pothole 681–6
    feedback control system 613, 615
    first-order systems 508–14
    higher-order systems 539–44
    second-order systems 515–28, 537–9
time-shifted function 44–5
time-variant system 8, 136, 457, 471, 482
torsional springs 147–9
traditional condenser microphone 754–5
trans-conductance amplifier 295–6
transfer functions 6–7, 168, 399
    control systems, formulation 592–5
    derivation of 139
    formulations 401, 408–13
    negative feedback configuration 416–17
    quarter-car suspension model 142
    spring–mass–damper system 140–4
    state-space representations and 186, 469–75
    time-invariant linear system 408

time-invariant systems 95–7, 139
    for two-input–two-output system 411–13
transfer matrix 411
transient response 505–7
translational springs 121–2
translational systems
    derivation of equations of motion 130–8
    effective spring coefficients, elastic bodies 122–4
    free-body diagrams and auxiliary plots 127–9
    friction forces 125–7
    lumped masses 120–1
    transfer function formulation 138–44
    translational springs 121–2
    viscous dampers 123, 125
transmissibility parameters 570–1
trans-resistance amplifier 296
triangular pulse, piecewise-continuous function 61, 62
trigonometric manipulations, complex numbers 36–7
Tsiolkovsky rocket equation 137
turbulent flow 361
two-mass dynamic system 65–6
two-segment commutator 435–7
typical inverting op-amp circuit 282

undamped system 517, 520, 522, 525, 537–8
underdamped second-order systems 528
    model parameters 529
    performance specification parameters 529–35
    physical parameters 535–6
    sets of parameters 535–7
underdamped system 517, 520, 523, 525
unit conversion factors 760
unit matrix 24
unit ramp function 43
unit step function 43
unit vector
    decomposition 19–21
    definition 17
unity-feedback control system 608

unity-gain buffer 294
unstable system 546–8

vector algebra 16–17
    addition and subtraction 18
    definition 17
    multiplication operations 21–3
    scalar multiplication 19
    unit-vector decomposition 19–21
vector multiplication operations 21–3
velocity-sensitive microphones 746
vibration analysis of car moving 667
    dynamic modeling 669–73
    potholes, time response 681–6
    quasi-static deflections and dynamic response 675–6
    sinusoidal profile, steady-state vibration 676–81
    state-space representations 673–5
    system description 667–9
viscous dampers 123, 125
viscous friction 125–7
voltage divider rule 235–7
    in potentiometer 239–40
    voltage drop 238–41
voltage follower 294
voltage-isolation amplifier 277–9
voltage-to-current converter circuit 295–6

water hammer 368
water purification (liquid-level system) 711–12
    actual outflow 713
    description 712–13
    equilibrium point 719
    generated liquid-level graphs 715, 717
    governing equations 713
    linearization 719
    Mathematica code 715, 716
    nonlinear and linearized system, PI control 722, 725
    nonlinear state-space model 714–16
    piecewise-continuous function, PI control 722–3, 724
    state-space model 714
    steady-state liquid levels 715, 717, 718

water purification (cont.)
  system parameters variability 718
  transfer function and PI control 721
Wheatstone bridge 724, 726, 727, 730
Wolfram Mathematica 3, 99, 100, 198–203, 297–306
  classical analysis and design 777
  drop-down menu 771, 772

dynamic system, modeling/simulation 475–85
equation solvers 776
graphics generation and styling functions 774–5
mathematical functions 773
matrix functions 774
operators and special characters 772
response 777
state-space formulation 777

time response of car 683–6
transfer function formulation 776

zero matrix 24
zero-pole-gain form 41, 409
zero vector 17
Ziegler, J. G. 606
Ziegler–Nichols ultimate-cycle method 607–8, 638